Control Systems

K Alice Mary
Professor
Department of Electrical and Electronics Engineering
Gudlavalleru Engineering College
Gudlavalleru
Andhra Pradesh

P Ramana
Professor
Department of Electrical and Electronics Engineering
GMR Institute of Technology
Rajam
Andhra Pradesh

Universities Press

We dedicate this book to our family members

Control Systems

UNIVERSITIES PRESS (INDIA) PRIVATE LIMITED

Registered office
3-6-747/1/A & 3-6-754/1, Himayatnagar, Hyderabad 500 029, Telangana, India
info@universitiespress.com; www.universitiespress.com

Distributed by
Orient Blackswan Private limited

Registered office
3-6-752 Himayatnagar, Hyderabad 500 029, Telangana, India

Other offices
Bengaluru / Bhopal / Chennai / Guwahati / Hyderabad / Jaipur / Kolkata
Lucknow / Mumbai / New Delhi / Noida / Patna / Visakhapatnam

First published 2016
Reprinted 2019

ISBN: 978-81-7371-985-1

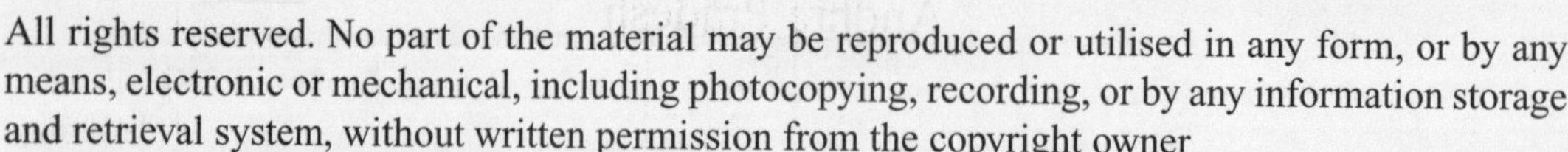

Typeset in Times New Roman PS 10.5/12.5 *by*
Cameo Corporate Services, Coimbatore, Tamil Nadu, India

Printed in India at
Navaneetham Enterprises, Chennai

Published by
Universities Press (India) Private Limited
3-6-747/1/A & 3-6-754/1, Himayatnagar, Hyderabad 500 029, Telangana, India

About the Authors

K Alice Mary is Professor at Gudlavalleru Engineering College, Gudlavalleru, Andhra Pradesh. She received her B.E. degree from Mysore University and her master's degree from University of Roorkee (now IIT Roorkee), Uttar Pradesh, before she progressed to earn her Ph.D. degree from IIT Kharagpur, West Bengal. She has been teaching for the past 35 years and has to her credit, more than 80 research papers published in reputed journals within India and abroad. A keen enthusiast of active research, her areas of special interest include control system application to power electronics and electrical machine drives.

Dr Alice Mary was conferred the Mahila Jyothi Award, the Mother Teresa Excellence Award, the Shastra, Pratibha and the Vijeta awards and the Best Teacher Award for overall academic excellence. She has undertaken and completed seven AICTE/MHRD projects funded by the government. She is a fellow member of Institute of Engineers, Institute of Electronics and Telecommunication Engineers and member of several academic bodies. An able administrator, she works towards receiving NBA, NAAC, ISO and other permanent affiliation accreditations for technical institutions.

P Ramana is Professor at GMR Institute of Technology, Rajam, Andhra Pradesh, where he has been teaching for the past 18 years. He received his B. Tech. degree in EEE and his M. Tech. degree in electrical power engineering from JNTU Hyderabad, before earning his Ph.D. degree in EEE, also from the same university. He has published/presented 45 papers in international and national journals/conferences of repute. His areas of interest are control systems, power systems and electrical machine drives. A fellow member of the Institute of Engineers, India, and member of several academic bodies, he has co-authored several textbooks with reputed publishers.

Contents

Preface

This book aims to discuss the essential elements of control system engineering in simple language that could be understood by students with urban as well as rural background. It is designed as a textbook for undergraduate students who study control system engineering as a core subject in their curriculum and presents excellent pedagogical features that will help to drive home the concepts with ease.

With several solved examples, multiple-choice questions, short questions and unsolved problems for practice, the book is a repository that delves into the subject's governing principles for the student's benefit. In addition, it includes MATLAB fundamentals and some basic programs to enable first-time learners gain a fair exposure to the practical utility of this high-performance programming language as a computational tool in nearly all fields of science and engineering. The contents of the book have been selected to suit the syllabus of all major technical universities across India and organised as follows.

Chapter 1 deals with the fundamentals and basic concepts of control systems. Chapter 2 focuses on the mathematical preliminaries such as Laplace Transforms and Matrices. Chapter 3 explains mathematical modelling, transfer function and analogous systems. Chapter 4 elucidates transfer function using block diagram reduction. Chapter 5 focuses on the use of Mason's gain formula for transfer functions. Chapter 6 describes various control system components such as servo motors, synchros, magnetic amplifiers, potentiometers and tacho-generators. Chapter 7 reveals the effect of feedback on control systems and various controllers. Chapter 8 is devoted to the time domain analysis of various systems, particularly the second-order system. Chapter 9 details the stability of control systems and discusses Routh–Hurwitz stability criterion in detail to get absolute stability of a system. Chapter 10 illustrates the root locus technique which provides knowledge of movement of poles by graphical method in s-plane. Chapter 11 expounds on frequency domain analysis and Bode plot method and calculation of transfer functions from Bode plot. Chapter 12 delineates the applications of the polar plot and the Nyquist stability criterion. Chapter 13 spells out the use of state–space analysis for modelling any process or a physical system. This chapter presents various state–space canonical forms, transformation of system models, solution of linear time invariant state equations, Eigen values and Eigen vectors, state transition matrix and controllability and observability. Chapter 14 sheds light on the need of compensators in an unstable system and elaborates on compensator design using Bode plot. Chapter 15 outlines the advent of digital control systems, the z-transform and the inverse z-transform.

We welcome constructive suggestions and feedback that could be used for the book's improvement.

ACKNOWLEDGEMENTS

We express our wholehearted thanks to all those who helped us, directly or indirectly, in turning this book into reality. We are grateful to our beloved Chairman Dr G M Rao and Chief Executive Officer Col. G Rajendra Prasad, GMRVF; Principal Dr C L V R S V Prasad and Vice-principal Dr J Raja Murugadoss, GMR Institute of Technology (GMRIT), as also the management members of

GMRIT. We are indebted to Dr Nageswara Rao Vallurupalli, Chairman, and Dr Potluri Ravindra Babu, Principal, Gudlavalleru Engineering College, Gudlavalleru, Andhra Pradesh, for providing us a conductive work atmosphere and concurrent support in writing this book.

We are obliged to Dr K B Madhu Sahu, Principal, AITAM–Tekkali andDr M Suryakalavathi, Professor, Department of EEE, JNTUCE, Hyderabad, for their suggestions and invaluable help.

We thank our colleagues and students at the Department of EEE, GMRIT Rajam, and GEC Gudlavalleru, who have helped us, directly or indirectly, in bringing out this book in its present form.

We thank Mr M R Ramesh, Editor, Orient BlackSwan Pvt Ltd, for providing us excellent editorial suggestions for publishing this book.

Last but not the least, we are deeply indebted to our respective family members for their encouragement, support and co-operation at every step of bringing this book into being.

K Alice Mary
P Ramana

Fundamentals of Control Systems

Contents

1.1 INTRODUCTION

Control system plays an important role in our daily life. In this age of automation and product development, it is interesting to note the constant reference to the control system that is required or being used. Automatic control systems play a vital role in advancement and improvement of engineering skills and in the working of space vehicles, satellites, guiding missiles, industrial processes, home appliances, etc. The study of automatic controls is important since it provides a basic understanding of the behaviour of all dynamic systems, which leads to the better appreciation and utilisation of fundamental laws of nature.

1.2 BASIC DEFINITIONS

System: A system is a coordinated unit of individual elements performing a specific function.

***Example*:** A class room, a lamp

Control: It means to regulate, direct or command a system so that a desired objective is obtained.

Control system: A control system is a combination of elements arranged in a planned manner where in each element causes an effect to produce a desired output.

Plant or process: The portion of the system that is to be controlled or regulated is called plant or process.

Input: The applied signal or excitation signal that is applied to the control system to get required output is called an input.

Output: The actual response that is obtained from a control system with the application of input is called as an output.

Controller: The element of the system itself or external to the system that controls the plant or the process is called controller.

Disturbance: The signal that has some adverse effect on the value of the output of a system is called disturbance. If a disturbance is produced within the system, it is termed as an internal disturbance; otherwise, it is known as an external disturbance.

Actuating or **Error Signal:** It is the difference between the reference input and the feedback signal.

1.3 CLASSIFICATION OF CONTROL SYSTEMS

1.3.1 Natural and Manmade Control Systems

The various control systems that are designed and developed by man are known as manmade control systems.

Example: An automobile system with gears, accelerator and breaking system.

The system inside a human being or a biological system is known as natural control system.

Example: The perspiration system inside the human being.

The combination of a natural control system and manmade control system is an example of combinational control system.

Example: The driver driving a car.

1.3.2 Time-varying and Time in varying Control Systems

If the parameters of the control system are varying with respect to time, then it is called time-varying control system.

Example: A space vehicle leaving the earth.

If the parameters of the control system are not varying with respect to time, then it is called time in varying control system.

Example: The elements of an electrical network such as resistance, inductance and capacitance are not time varying.

1.3.3 Linear and Non-linear Control Systems

A control system is known as linear if it satisfies the principle of superposition (additive property as well as homogeneous property).

Example: $f(x) = x$

Those systems that do not obey superposition principle are known as non-linear control system.

Example: $f(x) = x^2$

The principle of superposition is a combination of the following two properties:

1. **Additive property:** If x and y belong to the domain of the function f, we can write

$$f(x+y) = f(x) + f(y)$$

2. **Homogeneous property:** For any x belonging to the domain of the function f and for any scalar constant k, we can write $f(kx) = kf(x)$

All physical systems are non-linear in nature. If the presence of a certain non-linearity does not affect the performance of the system much, the presence of non-linearity can be neglected and the system can be treated as a linear system.

1.3.4 Continuous-time and Discrete-time Control Systems

If all the system variables of a control system are functions of time, it is termed as a continuous-time control system.

***Example*:** The speed control of a DC motor with tachogenerator feedback.

If one or more system variables of control system are known at a certain discrete time, it is termed as discrete-time control system.

***Example*:** The microprocessor- or computer-based system.

1.3.5 Deterministic and Stochastic Control Systems

A control system is deterministic if the response to input and to external disturbances is predictable and repeatable.

Any control system is called stochastic if the response is unpredictable.

1.3.6 Static and Dynamic Control Systems

A system is called dynamic if its present output depends on the past input, while a static system is the one whose current output depends only on the current input.

1.3.7 Lumped and Distributed Parameter Control Systems

If a control system can be represented by ordinary differential equations, such a control system is called lumped parameter control system.

***Example*:** In the case of electrical networks, parameters such as resistance, inductance and capacitance are lumped parameter systems.

If a control system can be represented by partial differential equations, such a control system is known as distributed parameter control system.

***Example*:** In a transmission line, its parameters such as resistance, inductance and capacitance are totally distributed along its length.

1.3.8 Single Input–Single Output and Multiple Input–Multiple Output Systems

If a control system has one input and one output, it is termed as single input–single output (SISO) system.

***Example*:** A position control system has only one input (desired position) and one output (actual position).

If a control system has multiple inputs and multiple outputs, it is known as multiple input–multiple output (MIMO) system.

1.3.9 Open-loop and Closed-loop Control Systems

This is another classification of control systems. Both types are given in detail in the following section.

1.4 OPEN-LOOP AND CLOSED-LOOP CONTROL SYSTEMS

1.4.1 Open-loop Control System

A system in which control action does not depend on output is known as open-loop control system. In other words, in an open-loop control system, the output is neither measured nor feedback given for comparison with the input. A practical example is the washing machine. Soaking, washing and rinsing in the washer operate on a time base. The machine does not measure the output signal, that is, the cleanliness of the clothes.

The block diagram representation of an open-loop control system is shown in Figure 1.1. In the presence of disturbances, an open-loop control system will not perform the desired task. Open-loop control systems are non-feedback control systems. In any open-loop control system, the output is not compared with the reference input. Thus, to each reference input, there corresponds a fixed operating condition; as a result, the accuracy of the system depends on calibration.

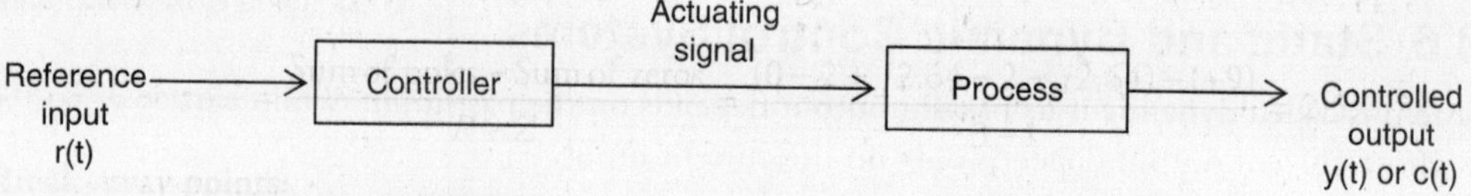

Figure 1.1 Open-loop control system

The drawback of an open-loop control system is that it is incapable of making automatic adjustments. Even when the magnitude of the desired output is too big or too small, the system will not make the appropriate adjustments. For this reason, an open-loop control system is not suitable for use as a complex control system. Sometimes, it may even require monitoring and response from the user. For example, when a washing machine finishes cleaning the clothes, the user will need to check whether the clothes are clean or not; if they are not, the clothes will have to be put back into the machine to be washed again.

Advantages

1. These systems are simple in construction and design.
2. These systems are more economic.
3. These systems are generally stable.
4. Such systems are easy for maintenance point of view.
5. These systems are convenient to use when the output is difficult to measure.

Disadvantages

1. These systems are inaccurate and unreliable.
2. In these systems, the changes in the output due to external disturbances are not corrected automatically.
3. To maintain the quality and accuracy, recalibration of the controller is necessary time to time.

1.4.2 Closed-loop Control System

A system in which the control action depends on output through feedback is known as closed-loop control system.

***Example*:** An automatic electric iron, a voltage stabiliser, a person moving on a desired road with open eyes, a missile launcher and a room temperature control system.

In a room temperature control system, by measuring the actual room temperature and comparing it with reference temperature (desired temperature), the thermostat turns the heating equipment or cooling equipment on or off in such a way as to ensure that the room temperature remains at comfortable level regardless of outside conditions.

The block diagram representation of closed-loop control system is shown in the following Figure 1.2. The block diagram shown in Figure 1.2 consists of a forward path containing output controller and a plant or a process, driven by the error sensor, which compares the actual output with the reference input. The function of the output is prescribed by the feedback path element.

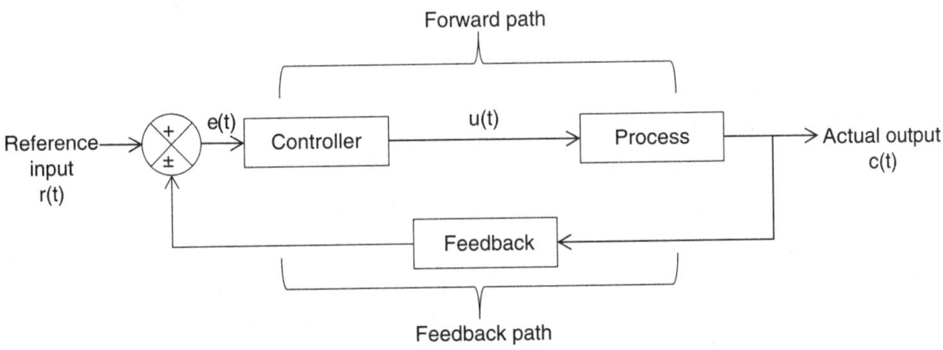

Figure 1.2 Closed-loop control system

The control action is actuated by an error signal 'e', which is the difference between the input signal 'r' and the output signal 'c'. This process of comparison between the output and the input maintains the output at a desired level through control action process. Feedback control system is often referred to as closed-loop control system. The term closed-loop control always implies the use of feedback control action in order to reduce the system error.

A system that maintains the prescribed relationship between the output and the reference input by comparing them and using the difference as a means of control is called a feedback control system.

The closed-loop control system adjusts its output automatically by feeding the output signal back to the input. When the load changes, the error signals generated by the system will adjust the output. However, closed-loop control systems are generally more complicated, and thus, more expensive to make.

Advantages

1. They are more accurate due to correction of any arising error.
2. The effect of external disturbance signal can be made very small.
3. These systems have high bandwidth, that is, high operating frequency zone.
4. Speed of the response can be greatly increased.

5. In such system, there is reduced effect of non-linearities and distortions.
6. The sensitivity of the system may be made small to make the system more stable.

Disadvantages

1. They are more complex and expensive.
2. The maintenance cost is high.
3. These systems may become unstable.
4. The feedback in closed loop system may lead to oscillatory response.
5. The feedback reduces the overall gain of the system.

1.4.3 Comparison between Open-loop and Closed-loop Control Systems

S.No.	Open-loop control system	Closed-loop control system
1	In these systems, the output has no effect on control action, that is, feedback does not exist.	In these systems, the output has an effect on control action through feedback
2	Output measurement is not required for operation of system.	Output measurement is necessary.
3	Simple to construct and cheap.	Complicated to construct and costlier.
4	Inaccurate and unreliable.	Accurate and reliable.
5	These systems are stable.	Generally, great efforts are needed to design a stable system.
6	Error detector is absent.	Error detector is presented.
7	Bandwidth is small.	Bandwidth is large.
8	Highly affected by non-linearities.	Reduced effect of non-linearities.

1.5 EXAMPLES OF CONTROL SYSTEM

1.5.1 Traffic Control System

Open-loop Control System

Traffic control by means of traffic signals operated on a time basis constitutes an open-loop control system. The sequence of control signals are based on a traffic study. The system will not measure the density of the traffic before giving the signals. Since the time slots do not change according to traffic density, the system is open loop system.

Closed-loop Control System

Traffic control system can be made as a closed loop system if the time slots of the signals are decided based on the density of the traffic is measured on all the sides and the information is fed to a computer. The timings of the control signals are decided by the computer based on the density of traffic. Since the closed-loop system dynamically changes the timings, the flow of vehicles will be better than that in an open-loop system.

1.5.2 A Man Walking on a Road

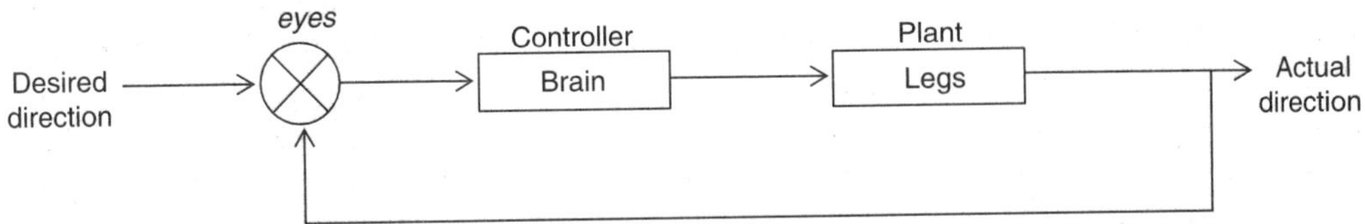

Figure 1.3 Control system for moving on a desired road

A man walking on a road with his eyes opened is an example of a closed-loop control system as shown in Figure. 1.3. The major components involved in walking are the brain, the eyes and the feet. The input may be chosen as the desired walk direction and the output is the actual walk direction. The control action is determined by the eyes, which detects the difference between the input and the output and sends this information to the brain. The brain commands the legs and feet to walk in the prescribed direction. Walking is a closed-loop operation because the control action is a function of the output. If the eyes are closed, the feedback loop is broken and the system becomes open-loop system.

1.5.3 Target Aiming System of a Gun

The aiming of a gun to the target is represented by the system is shown in Figure 1.4. In this process, the eyes compare the position of the target and the direction of the gun. The difference is detected by brain, which then moves the arms, and hence, the gun through body muscles in such a way as to reduce the error between the target position and the gun direction.

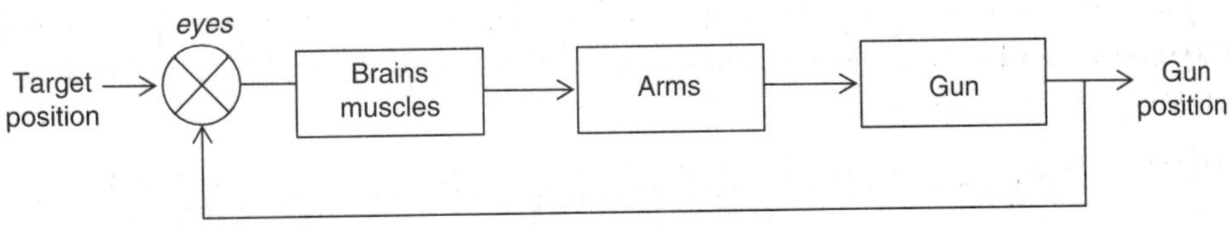

Figure 1.4 Target aiming system of a gun

The input of this system is the target position, while output is the gun direction. The system components are the eyes, brain and muscles, arms and gun. This is an example of negative feedback control system as the system output is subtracted from the initial input through the feedback path.

The system becomes an open-loop system when the gun is aimed to the target with eyes closed as this eliminates the feedback path from the system.

1.5.4 Economic Inflation System

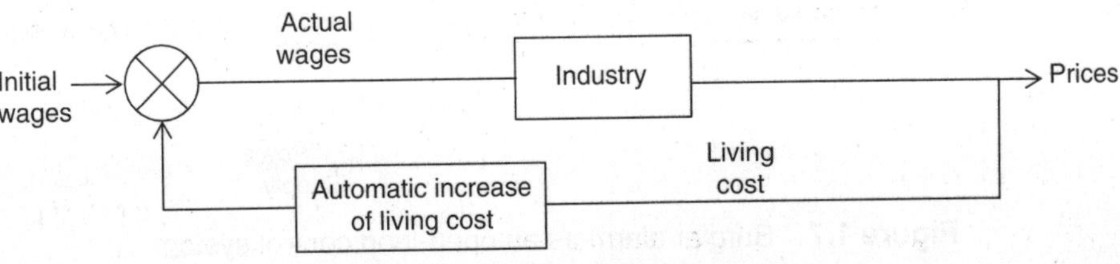

Figure 1.5 Economic inflation system

Feedback control systems need not always be negative feedback control systems. Figure 1.5 shows an economic inflation system in nature, which is a positive feedback system. A positive feedback control system adds the feedback signal to the input signal and the resultant is used as the input to the controlling process. In this process, the price of the commodity is fixed by the wages of the worker in the industry to produce it. This price affects the cost of living of each worker. Thus, an increase in price increases the cost of living automatically. This increase in the cost of living adds to initial wages of the worker. The actual wages for the industry worker increase their initial wages and this increase depends upon the increase in commodity prices. Hence, in a positive feedback control system, the input to controlling process with feedback will always be larger than the input without feedback. Here, the system input and output, respectively, are the initial wages and commodity prices. The system components are industry process, workers in industry and management. The system parameters are wages, prices and the cost of living.

1.5.5 Washing Machine

Nowadays, many families use fully automatic washing machines. There are numerous preset washing procedures available for the users. When we have chosen the suitable washing procedures, the machine automatically starts to pour water, add washing powder, spin and wash clothes, discharge waste water, etc. After the completion of all the procedures, the washing machine will stop the operation.

Fully automatic washing machine only requires the user to input a suitable procedure to complete the whole washing process, and thus, it saves much of the user's time. However, this kind of machine only operates according to the preset time to complete the whole washing process. It ignores the cleanliness of the clothes and does not generate feedback. Therefore, this kind of washing machine uses an open-loop control system. The block diagram of control system of the washing machine is shown in Figure 1.6.

Figure 1.6 Washing machine as open-loop control system

1.5.6 Burglar Alarm

A more sophisticated example of open-loop control system is the burglar alarm system, as shown in Figure 1.7. The function of the sensor is to collect data regarding the concerned house. When the electronic sensor is triggered off (for example, by the entry of an unauthorised person), it will send a signal to the receiver. The receiver will then activate the alarm, which will in turn generate an alarm signal. The alarm signal will not cease until the alarm is stopped manually.

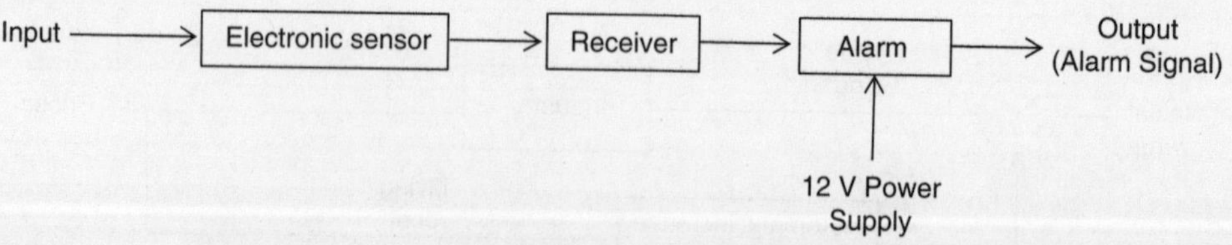

Figure 1.7 Burglar alarm as an open-loop control system

1.5.7 Air Conditioner

Nowadays, there are many families using automatic control system for the temperature of the air conditioner. Figure 1.8 shows the interior structure of an air conditioner. The coolant circulated in the machine will absorb the heat indoor; then it will transport the heat from the vapourisation device to the cooling device. The hot air is then blown outdoor by a fan. The air conditioner is equipped with an adjustable temperature device for the users to adjust the extent of cooling. When the temperature of the cool air is lower than the preset one, the controller of the air conditioner will stop the operation of the compressor to cease the circulation of the coolant. The temperature sensor installed near the vaporisation device will continuously measure the indoor temperature and send the results to the controller for further processing.

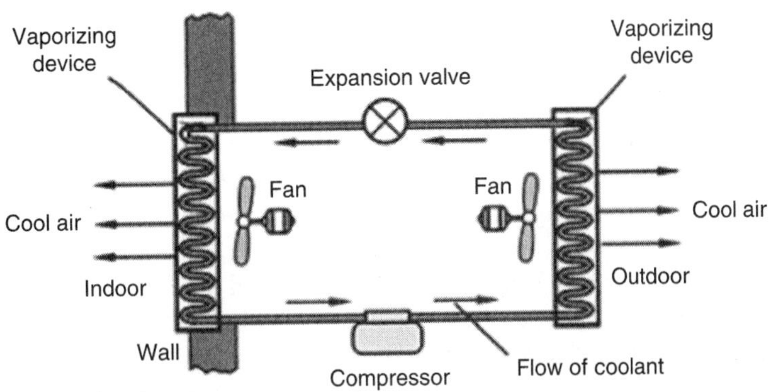

Figure 1.8 Internal structure of an air conditioner

Since the output of cool air by the air conditioner will affect its working, the control system of the machine is said to be in a closed loop. The block diagram of the control system for air conditioners is as shown in Figure 1.9.

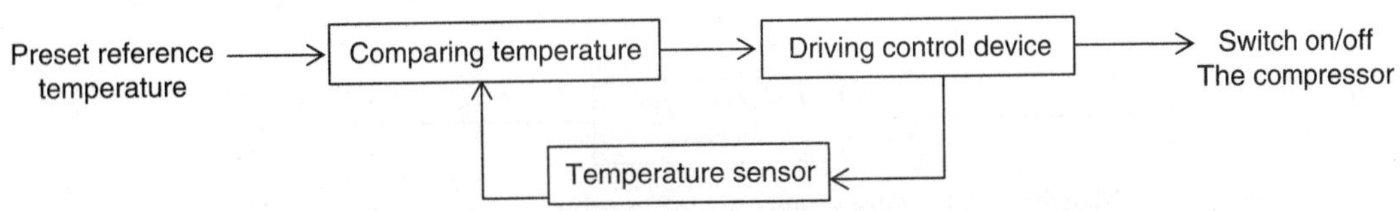

Figure 1.9 An air conditioner as a closed-loop control system

1.5.8 Temperature Control System

Figure 1.10 shows a diagram of temperature control of an electric furnace. The temperature in the electric furnace is measured by a thermometer, which is an analogue device. The analogue temperature is converted to digital temperature by an A/D converter. The digital temperature is fed to a controller through an interface. This digital temperature is compared with the programmed input temperature, and if there is any error, the controller sends out a signal to the heater through an interface, amplifier and relay to bring the furnace temperature to a desired value.

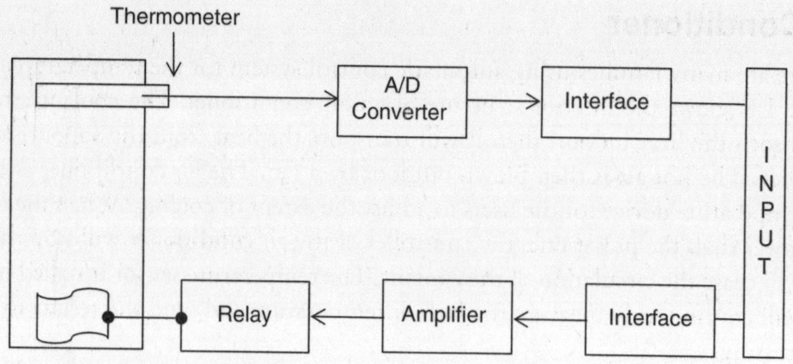

Figure 1.10 Temperature control as a closed-loop system

1.6 FEEDBACK CHARACTERISTICS OF CONTROL SYSTEM

A system that maintains a prescribed relationship between the output and the reference input by comparing them and using the difference as a means of control is called feedback control system. Feedback exists whenever there is a closed sequence of cause-and-effect relationship. Feedback control systems are often referred to as closed-loop control systems.

Negative feedback results in better stability in steady state and rejects any disturbance signals. It also has low sensitivity to parameter variations. Hence, negative feedback is preferred in closed-loop systems. On the other hand, positive feedback increases the error signal and drives the output to instability. However, sometimes positive feedback is used in minor loops in control systems to amplify certain internal signals or parameters.

The general form of the block diagram of a feedback control system is shown in Figure 1.11. The components of feedback control system are plant, feedback path element, error detector and controller.

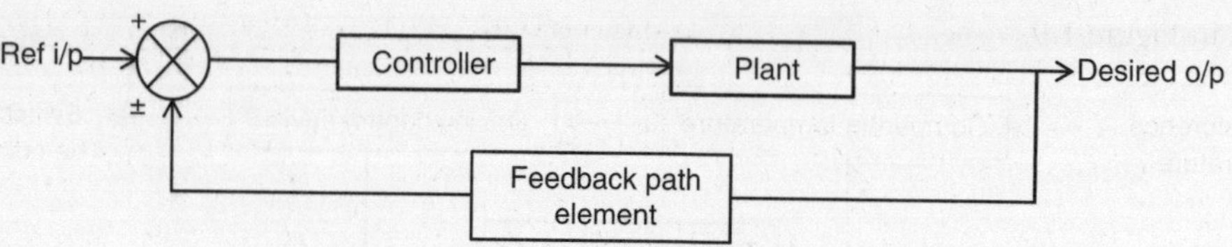

Figure 1.11 Block diagram of feedback control system

The block diagram consists of a forward path containing a controller and a plant or a process driven by the output of the error sensor that compares a function of the output with reference input. The function of the output is prescribed by the feedback path element.

Advantages

The following are the advantages of feedback control system:

1. Reduction in sensitivity to variations of parameters in the forward path.
2. Control over the transient response as well as the steady state accuracy by adjusting the loop gain.
3. Reduction in the effect of noise or disturbance of the output level.

Disadvantages

The following are the disadvantages of feedback control system:

1. There is a need for additional hardware that consequently increases the cost.
2. The introduction of feedback may lead to instability of the closed-loop system even though the open-loop system may be stable.

REVIEW QUESTIONS

1. Explain the basic components of control systems.
2. Explain the classification of control systems.
3. Write the important differences between open loop and closed loop systems with suitable examples.
4. Explain the limitations of closed loop system over open loop system.
5. By means of relevant diagrams, explain the working principles of a practical closed-loop control systems.
6. Explain feedback characteristics of closed-loop systems.
7. List out the limitations of open loop systems over closed-loop systems.
8. Explain the traffic control system concepts using open loop as well as closed-loop system.
9. Why negative feedback is invariably preferred in closed-loop systems?
10. What is control system? Explain various types of control systems with examples and their advantages.
11. Explain the positive and negative feedback in a closed-loop control system. What are the differences between them and how do they affect the control system?

OBJECTIVE TYPE QUESTIONS

1. In an open-loop control system,
   (a) control action is independent of the output
   (b) control action depends on the human adjustment
   (c) Both (a) and (b)
   (d) internal system changes are automatically taken care
2. In a closed-loop control system,
   (a) control action depends on the output and also on the input command
   (b) output signal feedback to be compared with the reference signal
   (c) Both (a) and (b)
   (d) None
3. The main drawback of the feedback system is
   (a) inaccuracy
   (b) inefficient
   (c) insensitivity
   (d) instability
4. Which of the following system is a closed loop system?
   (a) Electric switch
   (b) Car starter
   (c) AC generator
   (d) Autopilot for an aircraft

5. Which one of the following is not the example of closed-loop control system?
   (a) Man walking on a road with his eyes open
   (b) Room temperature control system
   (c) Automatic iron
   (d) Traffic signal system based on the time slots
6. As compared to closed-loop system, an open-loop system is
   (a) more stable and more accurate
   (b) more stable and less accurate
   (c) less stable and more accurate
   (d) less stable and less accurate
7. The principle of homogeneity and superposition is applied to
   (a) linear time-variant system
   (b) non-linear time-variant system
   (c) linear time-invariant system
   (d) non-linear time-invariant system
8. Consider the following statements regarding a linear system $y = f(x)$
   1. $f(x_1 + x_2) = f(x_1) + f(x_2)$
   2. $f[x(t + T)] = f[x(t)] + f[x(T)]$
   3. $f(Kx) = Kf(x)$ of these statements
   (a) 1, 2 and 3 are correct
   (b) 1 and 2 are correct
   (c) 3 alone is correct
   (d) 1 and 3 are correct
9. A driver maintains the speed of a car at 100 km/h. Here, the feedback element is
   (a) eyes
   (b) steering wheel
   (c) needle of the speedometer
   (d) all
10. An automatic washing machine is
   (a) a single feedback control system
   (b) a multi-feedback control system
   (c) an open-loop control system
   (d) none
11. The difference between the reference input and the actual output is called
   (a) error signal
   (b) controlling signal
   (c) actuating signal
   (d) transfer function
12. Which one of the following statements regarding open-loop control system is incorrect?
   (a) less expensive
   (b) simple construction and ease of maintenance
   (c) more accurate
   (d) stability can be assured
13. In a closed loop system, the source power is modulated with
   (a) feedback signal
   (b) error signal

(c) actuating signal
(d) reference signal

14. When a human being tries to approach an object, then his brain acts as a
(a) controller
(b) actuator
(c) amplifier
(d) error measuring device

15. Human eye can be considered as
(a) an open loop system
(b) a closed loop system with single feedback
(c) a closed loop system with multivariable feedback
(d) none

16. Which system has tendency to oscillate?
(a) open loop system
(b) closed loop system
(c) both a and b
(c) none

17. The error detect element in a control system gives
(a) the sum of the reference signal and the feedback signal
(b) the sum of the reference signal and the error signal
(c) the difference of the reference signal and the feedback signal
(d) none

ANSWERS

1. c	**2.** c	**3.** d	**4.** d	**5.** d	**6.** b	**7.** c	**8.** d	**9.** c	**10.** c
11. a	**12.** c	**13.** b	**14.** a	**15.** c	**16.** b	**17.** c			

CHAPTER

2 Mathematical Preliminaries

Contents

2.1 INTRODUCTION

For a proper understanding of the behaviour of a system, one has to represent the system in the form of mathematical relations. The voltage equations in algebraic form or the differential equation of an electrical circuit are a few such mathematical relations. The Laplace transformation offers a very simple and elegant method of solving linear or linearised differential equations, which result from mathematical modelling of a process of a system. The method of transforming a function from time domain into s-domain is known as Laplace transform, where s is a complex operator denoted by $s = \sigma + j\omega$. In a control system, setting up of system equations and their solutions are required. Laplace transform is a very handy tool to solve differential equations because it automatically includes initial conditions when compared to classical methods. Matrix algebra has its wide application in modern control engineering. The aim of this chapter is to explore Laplace transform and matrix algebra.

2.2 LAPLACE TRANSFORMATION

If $f(t)$ is a function of t defined for all positive values of t, then the Laplace transform of $f(t)$ is denoted by $L[f(t)]$ and is defined as

$$L[f(t)] = F(s) = \int_0^\infty e^{-st} f(t)dt \tag{2.1}$$

where s is a parameter that may be real or complex number.

Examples:

1. Laplace transform of e^{at}

$$L[e^{at}] = \int_0^\infty e^{-st} e^{at}\, dt$$
$$= \int_0^\infty e^{(a-s)t}\, dt$$
$$= \int_0^\infty e^{-(s-a)t}\, dt$$
$$= \left[\frac{e^{-(s-a)t}}{-(s-a)}\right]_0^\infty$$
$$= \frac{1}{a-s}[e^{-\infty} - e^{0}] = \frac{1}{a-s} = \frac{1}{s-a}$$

2. Laplace transform of cos at

$$L[\cos at] = \int_0^\infty e^{-st} \cos atdt$$
$$= \frac{e^{-st}}{s^2+a^2}[-s\cos at + a\sin at]_0^\infty$$
$$= -\frac{1}{s^2+a^2}(-s) = \frac{s}{s^2+a^2}$$

3. Laplace transform of sinh at

$$L[\sinh at] = L\left(\frac{e^{at}-e^{-at}}{2}\right)$$
$$= \frac{1}{2}L[e^{at}] - \frac{1}{2}L[e^{-at}]$$
$$= \frac{1}{2}\frac{1}{s-a} - \frac{1}{2}\frac{1}{s+a}$$
$$= \frac{1}{2}\left(\frac{1}{s-a} - \frac{1}{s+a}\right)$$
$$= \frac{1}{2}\left(\frac{s+a-s+a}{s^2-a^2}\right) = \frac{a}{s^2-a^2}$$

4. Laplace transform of k

$$L(k) = \int_0^\infty e^{-st} \times kdt$$
$$= k\left[\frac{e^{-st}}{-s}\right]_0^\infty$$

$$= \frac{-k}{s}(e^{-\infty} - e^{-0})$$

$$= \frac{k}{s}$$

Laplace transforms of some standard functions are given in the following table:

$f(t)$	$F(s)$
(i) k	$\frac{k}{s}$
(ii) $e^{\pm at}$	$\frac{1}{s \mp a}$
(iii) $\sin at$	$\frac{a}{s^2 + a^2}$
(iv) $\cos at$	$\frac{s}{s^2 + a^2}$
(v) t	$\frac{1}{s^2}$
(vi) t^n	$\frac{n!}{s^{n+1}}$
(vii) $\sinh at$	$\frac{a}{s^2 - a^2}$
(viii) $\cosh at$	$\frac{s}{s^2 - a^2}$
(ix) $e^{-at} \cos bt$	$\frac{s + a}{(s + a)^2 + b^2}$
(x) $e^{-at} \sin bt$	$\frac{b}{(s + a)^2 + b^2}$
(xi) $e^{-at} t^n$	$\frac{n!}{(s + a)^{n+1}}$
(xii) $\delta(t)$	1

2.3 PROPERTIES OF LAPLACE TRANSFORMS

1. **Linear property:** If a, b and c are any constants and f, g and h are any functions of t, then

$$\text{L}[af(t) + bg(t) + ch(t)] = a\text{L}[f(t)] + b\text{L}[g(t)] + c\text{L}[h(t)] \tag{2.2}$$

2. **First shifting property:** If $\text{L}[f(t)] = \bar{f}(s) = F(s)$, then

$$\text{L}[e^{at} f(t)] = \bar{f}(s - a) = F(s - a) \tag{2.3}$$

If we know the Laplace transform $F(s)$ of $f(t)$, we can write the transform of $e^{at} f(t)$ by simply replacing 's' with $(s-a)$ to get $F(s-a)$.

3. **Change of scale property:** If $L[f(t) = F(s)$, then

$$L[f(at)] = \frac{1}{a} F(s/a) \tag{2.4}$$

4. **Scaling theorem:** If k is a constant, then Laplace transform of $k\,f(t)$ is given as k times the Laplace transform of $f(t)$.

i.e.,
$$L[k\,f(t)] = k\,L[f(t)] = k\,F(s) \tag{2.5}$$

5. **Real differentiation:** If $F(s)$ be the Laplace transform of $f(t)$, then

$$L\left[\frac{df(t)}{dt}\right] = sf(s) - f(\bar{0}) \tag{2.6}$$

where $f(\bar{0})$ = the value of $f(t)$ at $t = \bar{0}$, that is, just before the instant $t = 0$.

6. **Real integration:** If $F(s)$ is the Laplace transform of $f(t)$, then

$$L\left[\int_0^t f(t)dt\right] = \frac{F(s)}{s} \tag{2.7}$$

7. **Differentiation by *s*:** If $F(s)$ is the Laplace transform of $f(t)$, then the differentiation by s in the complex frequency domain corresponds to the multiplication by t in the time domain:

$$L[tf(t)] = -\frac{dF(s)}{ds} \tag{2.8}$$

8. **Real translation (shifting theorem):** This theorem is useful to obtain the Laplace transform of the shifted or delayed function of time.

If $F(s)$ is the Laplace transform of $f(t)$, then the Laplace transform of the function delayed by time T is

$$L[f(t-T)] = e^{-Ts} F(s) \tag{2.9}$$

2.4 INITIAL AND FINAL VALUE THEOREMS

Initial value theorem: The Laplace transform is very useful to find the initial value of the time function $f(t)$. Thus, if $F(s)$ is the Laplace transform of $f(t)$, then

$$f(0) = \underset{t\to 0}{Lt}\, f(t) = \underset{s\to\infty}{Lt}\, sF(s) \tag{2.10}$$

Final value theorem: The Laplace transform is also useful to find the final value of the time function $f(t)$. Thus, if $F(s)$ is the Laplace transform of $f(t)$, then

$$f(\infty) = \underset{t\to\infty}{Lt}\, f(t) = \underset{s\to 0}{Lt}\, sF(s) \tag{2.11}$$

2.5 LAPLACE TRANSFORM OF PERIODIC FUNCTION

If $f(t)$ is a periodic function with period T, that is, $f(t+nT) = f(t)$, then

$$L[f(t)] = \frac{\int_0^T e^{-st} f(t)dt}{1-e^{-sT}} \tag{2.12}$$

2.6 INVERSE LAPLACE TRANSFORMS

Inverse Laplace transform is calculated by partial fraction method rather than complex integration evaluation. Let $F(s)$ be the Laplace transform of $f(t)$. Then, the inverse Laplace transform is denoted as

$$f(t) = L^{-1}[F(s)]$$

2.7 APPLICATION TO SOLUTION OF DIFFERENTIAL EQUATION

Let us consider the nth order differential equation as

$$\frac{d^n y}{dt^n} + k_1 \frac{d^{n-1} y}{dt^{n-2}} + k_2 \frac{dn^{n-2} y}{dt^{n-2}} + \ldots + k_{n-1}\frac{dy}{dt} + k_n y = x(t) \tag{2.13}$$

By taking Laplace transform on both sides

$$s^n Y(s) - s^{n-1} y[0] - s^{n-2} y'(0) \ldots - y^{n-1}(0) + k_1[s^{n-1}Y(s) - s^{n-2}y(0) - s^{n-3}y'(0)]$$

$$\ldots y^{n-1(0)} + \ldots k_n Y(s) = X(s)$$

$$\therefore \quad Y(s)[s^n + k_1 s^{n-1} + k_2 s^{n-2} + \ldots + k_n] = X(s) + s^{n-1}y(0) + \ldots + y^{n-1}(0)$$

$$Y(s) = \frac{X(s) + s^{n-1}y(0) + s^{n-2}y'(0) + \ldots + y^{n-1}(0)}{s^n + k_1 s^{n-1} + k_2 s^{n-2} + \ldots + k_n} \tag{2.14}$$

In Eq. (2.14), $y[0]$, $y'[0]$, $y^2[0]$, y^{n-1} are the initial conditions for the given differential equation. Thus, we see that the advantages listed in Section 2.1, viz, the differential equation is converted into algebraic equation, initial conditions are used in the beginning of the problem, the total solution is obtained at one stroke, are well sustained.

The problem is not still completely solved. The time function $y(t)$ is required to be found by inverse transformation. First, we resolve the partial fractions to the above equation $Y(s)$. After that, by taking inverse Laplace transforms on both sides, we can get the time function $y(t)$.

2.8 HEAVISIDE'S PARTIAL FRACTION EXPANSION TECHNIQUE

Let $F(s) = \dfrac{P(s)}{Q(s)} = \dfrac{\text{Numerator polynomial in } s}{\text{Denominator polynomial in } s}$

Case (i): Roots of $Q(s)$ are distinct. Then, $\dfrac{P(s)}{Q(s)}$ can be written as

$$\frac{P(s)}{Q(s)} = \frac{k_1}{s-s_1} + \frac{k_2}{s-s_2} + \ldots + \frac{k_n}{s-s_n}$$

where

$$k_1 = \underset{s \to s_1}{Lt}\,(s-s_1)\frac{P(s)}{Q(s)}$$

$$k_2 = \underset{s \to s_2}{Lt}\,(s-s_2)\frac{P(s)}{Q(s)}$$

$$\vdots \qquad \vdots$$

In general,

$$k_i = \underset{s \to s_i}{Lt}\,(s-s_i)\frac{P(s)}{Q(s)}$$

Case (ii): Repeated roots

$$\frac{P(s)}{Q(s)} = \frac{P(s)}{(s-s_0)^n} = \frac{k_1}{s-s_0} + \ldots + \frac{k_n}{(s-s_0)^n}$$

Suppose, if $n = 3$

$$k_3 = \underset{s \to s_0}{Lt}\left[(s-s_0)^3\frac{P(s)}{Q(s)}\right]$$

$$k_2 = \underset{s \to s_0}{Lt}\left[\frac{d}{ds}(s-s_o)^3\frac{p(s)}{Q(s)}\right]$$

$$k_1 = \underset{s \to s_0}{Lt}\left[\frac{1}{2}\frac{d^2}{ds^2}(s-s_0)^3\frac{P(s)}{Q(s)}\right]$$

In general

$$k_m = \underset{s \to s_0}{Lt}\left[\frac{1}{(n-m)!}\frac{d^{n-m}}{ds^{n-m}}(s-s_0)^n\frac{P(s)}{Q(s)}\right]$$

where n = order of the multiplicity

Note: The order of the numerator polynomial $P(s)$ should be less than the order of the denominator polynomial, $Q(s)$.

2.9 MATRIX

It is a rectangular array of variables or constants in horizontal rows and vertical columns enclosed in brackets.

Example:

$$A = \begin{bmatrix} a_{11} & a_{12} & \ldots & a_{1n} \\ a_{21} & a_{22} & \ldots & a_{2n} \\ \ldots & \ldots & \ldots & \\ a_{m1} & a_{m2} & \ldots & a_{mn} \end{bmatrix} = \{A_{ij}\}$$

1. **Element:** Each value in a matrix is either a number or a constant.
2. **Dimension:** Number of rows by number of columns of a matrix. A matrix is named by its dimensions.

2.9.1 Different Types of Matrices

1. **Column matrix:** A matrix with only one column.

   Example:

$$X = \begin{bmatrix} a_1 \\ a_2 \\ \dots \\ a_n \end{bmatrix} = \{a_i\}$$

2. **Row matrix:** A matrix with only one row.

   Example:

$$Y = \begin{bmatrix} a_1, & a_{2,} & \dots & a_n \end{bmatrix} = \{a_j\}$$

3. **Square matrix:** A matrix that has the same number of rows and columns.

   Example:

$$B = \begin{bmatrix} 5 & 4 & 7 \\ 3 & 6 & 1 \\ 2 & 1 & 3 \end{bmatrix}$$

4. **Equal matrices:** Two matrices that have the same dimensions; the element of one matrix is equal to the corresponding element of the other matrix.
5. **Identity matrix:** Square matrix with ones on the diagonal and zeros elsewhere.

   Example:

$$I = \begin{bmatrix} 1 & 0 & 0 & 0 \\ 0 & 1 & 0 & 0 \\ 0 & 0 & 1 & 0 \\ 0 & 0 & 0 & 1 \end{bmatrix}$$

6. **Transpose matrix:** Rows become columns and columns become rows.

   Example:

$$A' = \begin{bmatrix} a_{11} & a_{21} & \dots & a_{m1} \\ a_{12} & a_{22} & \dots & a_{m2} \\ \dots & \dots & \dots & \\ a_{1n} & a_{2n} & \dots & a_{mn} \end{bmatrix}$$

2.9.2 Matrix Addition and Subtraction

Matrix addition

If A and B are $m \times n$ matrices, then the sum of A and B, denoted as $C = A + B$, is a matrix obtained by adding corresponding elements of A and B.

Here, $C = A + B$ is defined by

$$\{C_{ij}\} = \{A_{ij}\} + \{B_{ij}\}$$

Example:

If $A = \begin{bmatrix} a_{11} & a_{12} \\ a_{21} & a_{22} \end{bmatrix}$ and $B = \begin{bmatrix} b_{11} & b_{12} \\ b_{21} & b_{22} \end{bmatrix}$, then $C = \begin{bmatrix} a_{11}+b_{11} & a_{12}+b_{12} \\ a_{21}+b_{21} & a_{22}+b_{22} \end{bmatrix}$

Properties of matrix addition

1. Associative property of addition: $(A + B) + C = A + (B + C)$
2. Commutative property of addition: $A + B = B + A$
3. Distributive property of addition and subtraction: $S(A + B) = SA + SB$

$$S(A - B) = SA - SB$$

Matrix subtraction

If A and B are $m \times n$ matrices, then the sum of A and $-B$, denoted as $C = A - B$, is a matrix obtained by adding corresponding elements of A and $-B$.

Here, $C = A - B$ is defined by

$$\{C_{ij}\} = \{A_{ij}\} - \{B_{ij}\}$$

Example:

If $A = \begin{bmatrix} a_{11} & a_{12} \\ a_{21} & a_{22} \end{bmatrix}$ and $B = \begin{bmatrix} b_{11} & b_{12} \\ b_{21} & b_{22} \end{bmatrix}$, then $C = \begin{bmatrix} a_{11}-b_{11} & a_{12}-b_{12} \\ a_{21}-b_{21} & a_{22}-b_{22} \end{bmatrix}$

Note: All three matrices are of the same dimension.

2.9.3 Scalar Multiplication of Matrices

If A is an $m \times n$ matrix and k is a scalar, then let kA be the matrix obtained by multiplying every element of A by k. This procedure is called scalar multiplication.

Example:

If $A = \begin{bmatrix} 1 & -2 & 2 \\ 0 & -1 & 3 \end{bmatrix}$, then $3A = \begin{bmatrix} 3(1) & 3(-2) & 3(2) \\ 3(0) & 3(-1) & 3(3) \end{bmatrix} = \begin{bmatrix} 3 & -6 & 6 \\ 0 & -3 & 9 \end{bmatrix}$

Properties of scalar multiplication

1. $k(hA) = (kh)A$
2. $(k+h)A = kA + hA$
3. $k(A+B) = kA + kB$

2.9.4 Multiplication of Matrices

If $A = \left[a_{ij}\right]_{m\times n}$ and $B = \left[b_{ij}\right]_{n\times p}$, then $C = AB = \sum_{l=1}^{k} a_{il} b_{lj}$

where $i = 1, 2 \ldots m$ and $j = 1, 2, \ldots n$

Example:

If $A = \begin{bmatrix} a_{11} & a_{12} \\ a_{21} & a_{22} \end{bmatrix}$ and $B = \begin{bmatrix} b_{11} & b_{12} & b_{13} \\ b_{21} & b_{22} & b_{23} \end{bmatrix}$, then

$$C = A \times B = \begin{bmatrix} a_{11}b_{11} + a_{12}b_{21} & a_{11}b_{12} + a_{12}b_{22} & a_{11}b_{13} + a_{12}b_{23} \\ a_{21}b_{11} + a_{22}b_{21} & a_{21}b_{12} + a_{22}b_{22} & a_{21}b_{13} + a_{22}b_{23} \end{bmatrix}$$

Note: If the number of columns of ***A*** does not equal the number of rows of ***B***, then the product ***AB*** is undefined.

Properties of matrix multiplication

1. The associative law: $A(BC) = (AB)C$
2. The distributive laws: $A(B + C) = AB + AC$ and $(A + B)C = AC + BC$

2.9.5 Rank of a Matrix

A matrix ***A*** has a rank ***m***, if there exists an $m \times m$ sub-matrix ***B*** of ***A*** such that its determinant is non-zero and the determinant of every $k \times k$ sub-matrix, where $k \geq m + 1$ of ***A*** is zero.

Example:

If $A = \begin{bmatrix} 1 & 2 & -1 \\ 3 & 1 & -2 \\ 1 & 4 & 3 \end{bmatrix}$, then $|A| = 1(1 \times 3 + 2 \times 4) - 2(3 \times 3 + 1 \times 2) - (3 \times 4 - 1 \times 1)$

$$= 11 - 22 - 11 = -22$$

which is not equal to zero, and hence, the rank is 3.

2.9.6 Co-factor

The co-factor A_{ij} of the element a_{ij} of a matrix is defined by

$$A_{ij} = (-1)^i + jM_{ij}$$

2.9.7 Determinant of a Matrix

The association of a real number with square matrices of any dimension (order) is called the determinant of the matrix. Determinants can be distinguished from a matrix because they are always enclosed within a single set of vertical lines '||'. The entries are called elements of the determinant and the number of entries in any row or column is called the order of the determinant. The value of the real number defined by the array is calculated by following definite rules referred to as 'expanding the determinant'.

Example:

If $A = \begin{bmatrix} a & b \\ c & d \end{bmatrix}$, then $\Delta A = |A| = \begin{vmatrix} a & b \\ c & d \end{vmatrix} = ad - bc$

For a 3×3 matrix, the determinant can be calculated as follows:

If $A = \begin{bmatrix} 1 & 2 & -1 \\ 3 & 1 & -2 \\ 1 & 4 & 3 \end{bmatrix}$, then $|A| = 1(1\times 3 + 2\times 4) - 2(3\times 3 + 1\times 2) - 1(3\times 4 - 1\times 1)$

$$= 11 - 22 - 11 = -22$$

Note: There exists a determinant for every matrix if the matrix is square.

2.9.8 Adjoint of a Matrix

Adjoint of matrix A is defined as a matrix where the element of the ith row and j^{th} column equals A_{ji}.

Example:

If $A = \begin{bmatrix} 1 & 2 & 1 \\ -1 & -4 & -3 \\ -1 & 1 & 3 \end{bmatrix}$, then the adjoint of matrix A is given by

$$\text{Adj}(A) = \begin{bmatrix} -12+3 & (3+3) & (-1-4) \\ -(6-1) & (3+1) & -(1+2) \\ -6+4 & -(-3+1) & -4+2 \end{bmatrix}^T = \begin{bmatrix} -9 & 6 & -5 \\ -5 & 4 & -3 \\ -2 & 2 & -2 \end{bmatrix}^T = \begin{bmatrix} -9 & -5 & -2 \\ 6 & 4 & 2 \\ -5 & -3 & -2 \end{bmatrix}$$

2.9.9 Inverse of a Matrix

Let A be an $n \times n$ matrix. If there exists a matrix B such that $AB = BA = I$, then we call this matrix the inverse of A and denote it A^{-1}.

If A has an inverse, we say that A is non-singular.

If A^{-1} does not exist, we say A is singular.

By definition, the inverse of matrix $[A]_{m\times n}$, where $m = n$ is given by

$$A^{-1} = \frac{Adj(A)}{|A|} \text{ where } |A| \neq 0$$

Example:

If $A = \begin{bmatrix} 1 & 2 & 1 \\ -1 & -4 & -3 \\ -1 & 1 & 3 \end{bmatrix}$, then the inverse of matrix A is given by

$$|A| = \begin{vmatrix} 1 & 2 & 1 \\ -1 & -4 & -3 \\ -1 & 1 & 3 \end{vmatrix} = 1(-12+3) - 2(-3-3) + 1(-1-4) = -9 + 12 - 5 = -2$$

$$\text{Adj}(A) = \begin{bmatrix} -12+3 & -(-3-3) & (-1-4) \\ -(6-1) & (3+1) & -(1+2) \\ -(6+4) & -(-3+1) & (-4+2) \end{bmatrix}^T = \begin{bmatrix} -9 & 6 & -5 \\ -5 & 4 & -3 \\ -2 & 2 & -2 \end{bmatrix}^T = \begin{bmatrix} -9 & -5 & -2 \\ 6 & 4 & 2 \\ -5 & -3 & -2 \end{bmatrix}$$

$$A^{-1} = \frac{Adj(A)}{|A|} = \frac{1}{-2}\begin{bmatrix} -9 & -5 & -2 \\ 6 & 4 & 2 \\ -5 & -3 & -2 \end{bmatrix} = \begin{bmatrix} 4.5 & 2.5 & 1 \\ -3 & -2 & -1 \\ 2.5 & 1.5 & 1 \end{bmatrix}$$

Problem-2.1: Find the Laplace transform of the function $f(t) = \sin^2 t$.

Solution:

$$f(t) = \sin^2(t) = \frac{1-\cos 2t}{2}$$

$$F(x) = \text{L}[f(t)] = \frac{1}{2}\text{L}(1-\cos 2t)$$

$$= \frac{1}{2}[\text{L}(1) - \text{L}[\cos 2t]]$$

$$= \frac{1}{2}\left(\frac{1}{s} - \frac{s}{s^2+2^2}\right)$$

$$= \frac{1}{2}\left[\frac{s^2+4-s^2}{s(s^2+4)}\right] = \frac{2}{s(s^2+4)}$$

Problem-2.2: Find the Laplace transform of the function $f(t) = 1 - t + 4t^2 + 2t^3$.

Solution:

$$F(s) = \text{L}[f(\text{t})] = \text{L}[1 - t + 4t^2 + 2t^3]$$

$$= \text{L}[1] - \text{L}[t] + 4\,\text{L}[t^2] + 2\text{L}[t^3]$$

$$= \frac{1}{s} - \frac{1}{s^2} + 4\frac{2}{s^3} + 2\times\frac{6}{s^4}$$

$$= \frac{s^3 - s^2 + 8s + 12}{s^4}$$

Problem-2.3: Find the inverse Laplace transform for the following functions.

(a) $F(s) = \dfrac{2s+3}{s^2+3s+2}$

Solution:

$$F(s) = \frac{2s+3}{s^2+3s+2} = \frac{2s+3}{(s+1)(s+2)} = \frac{1}{s+1} + \frac{1}{s+2}$$

$$f(t) = \text{L}^{-1}[f(s)] = \text{L}^{-1}\left[\frac{1}{s+1} + \frac{1}{s+2}\right] = e^{-t} + e^{-2t}$$

(b) $F(s) = \dfrac{2s^2+3s+2}{(s+1)^3}$

Solution:

$$F(s) = \frac{2s^2+3s+2}{(s+1)^3} = \frac{k_1}{s+1} + \frac{k_2}{(s+1)^2} + \frac{k_3}{(s+1)^3}$$

$$k_3 = \underset{S\to -1}{Lt}\, \cancel{(s+1)^3}\, \frac{(2s^2+3s+2)}{\cancel{(s+1)^3}} = 4 - 3 = 1$$

$$k_2 = \underset{S\to -1}{Lt}\left\{\frac{d}{ds}\left[(s+1)^3\frac{(2s^2+3s+2)}{(s+1)^3}\right]\right\}$$

$$= \underset{S\to -1}{Lt}\ 4s+3 = -4+3 = -1$$

$$k_1 = \underset{S\to -1}{Lt}\left\{\frac{1}{2}\frac{d^2}{ds^2}\left[(s+1)^3\frac{2s^2+3s+2}{(s+1)^3}\right]\right\}$$

$$= \frac{1}{2}\underset{S\to -1}{Lt}\left[\frac{d}{ds}(4s+3)\right] = 2$$

$$F(s) = \frac{2s^2+3s+2}{(s+1)^3} = \frac{2}{s+1} - \frac{1}{(s+1)^2} + \frac{1}{(s+1)^3}$$

$$f(t) = \mathrm{L}^{-1}[F(s)] = \mathrm{L}^{-1}\left[\frac{2}{s+1} - \frac{1}{(s+1)^2} + \frac{1}{(s+1)^3}\right]$$

$$= 2e^{-t} - te^{-t} + \frac{1}{2}t^2e^{-t}$$

(c) $F(s) = \dfrac{s+1}{s(s^2+4s+4)}$

Solution:

$$F(s) = \frac{s+1}{s(s^2+4s+4)} = \frac{s+1}{s(s+2)^2} = \frac{k_0}{s} + \frac{k_1}{s+2} + \frac{k_2}{(s+2)^2}$$

$$k_0 = \underset{S\to 0}{Lt}\ sF(s) = \underset{s\to 0}{Lt}\frac{s+1}{s^2+4s+4} = 1/4$$

$$k_2 = \underset{S\to -2}{Lt}\left[(s+2)^2\frac{(s+1)}{s(s+2)^2}\right] = \frac{-2+1}{-2} = 1/2$$

$$k_1 = \underset{S\to -2}{Lt}\left\{\frac{d}{ds}\left[(s+2)^2\frac{(s+1)}{s(s+2)^2}\right]\right\}$$

$$= \underset{S\to -2}{Lt}\left\{\frac{d}{ds}\left[\frac{s+1}{s}\right]\right\}$$

$$= \underset{S\to -2}{Lt}\left[\frac{s-(s+1)}{s^2}\right] = -1/4$$

$$F(s) = \frac{s+1}{s(s+2)^2} = \frac{1}{4}\frac{1}{s} - \frac{1}{4}\frac{1}{s+2} + \frac{1}{2(s+2)^2}$$

$$f(t) = L^{-1}[F(s)] = L^{-1}\left[\frac{1}{4s} - \frac{1}{4(s+2)} + \frac{1}{2(s+2)^2}\right]$$

$$= \frac{1}{4} - \frac{1}{4}e^{-2t} + \frac{1}{2}te^{-2t}$$

(d) $F(s) = \dfrac{s-3}{s^2+4s+13}$

Solution:

$$F(s) = \frac{s-3}{s^2+4s+13} = \frac{s-3}{(s+2)^2+3^2} = \frac{(s+2)-5}{(s+2)^2+3^2}$$

$$= \frac{s+2}{(s+2)^2+3^2} - \frac{5}{(s+2)^2+3^2}$$

$$f(t) = L^{-1}[F(s)] = L^{-1}\left[\frac{s+2}{(s+2)^2+3^2} - \frac{5}{(s+2)^2+3^2}\right]$$

$$= e^{-2t}\cos 3t - \frac{5}{3}e^{-2t}\sin 3t$$

$$= e^{-2t}\left[\cos 3t - \frac{5}{3}\sin 3t\right]$$

Problem-2.4: Verify the initial value and final value theorems for the function $f(t) = e^{-t}(\sin 3t + \cos 5t)$.

Solution:

$$f(t) = e^{-t}(\sin 3t + \cos 5t)$$

$$= e^{-t}\sin 3t + e^{-t}\cos 5t$$

$$L[e^{-t}\sin 3t] = \frac{3}{(s+1)^2+3^2}$$

$$L[e^{-t}\cos 5t] = \frac{s+1}{(s+1)^2+5^2}$$

$$F(s) = L[f(t)] = L\left[e^{-t}\sin 3t + e^{-t}\cos 5t\right]$$

$$= \frac{3}{(s+1)^2+3^2} + \frac{s+1}{(s+1)^2+5^2}$$

$$\underset{t\to 0}{Lt}\, f(t) = \underset{t\to 0}{Lt}\, e^{-t}(\sin 3t + \cos 5t)$$

$$= e^{-0}(0+1) = 1$$

$$\underset{s\to\infty}{Lt}\, sF(s) = \underset{s\to\infty}{Lt}\left[\frac{3s}{s^2+2s+10} + \frac{s(s+1)}{s^2+2s+26}\right]$$

$$= 0 + 1 = 1$$

$\therefore \underset{t\to 0}{Lt}\, f(t) = \underset{s\to\infty}{Lt}\, sF(s) = 1$, initial value theorem verified.

$$\underset{t\to\infty}{Lt}\, f(t) = \underset{t\to\infty}{Lt}\, e^{-t}(\sin 3t + \cos 5t) = 0$$

$$\underset{s\to 0}{Lt}\, sF(s) = \underset{s\to 0}{Lt}\left[\frac{3s}{s^2+2s+10} + \frac{s^2+s}{s^2+2s+26}\right] = 0$$

$\therefore \underset{t\to\infty}{Lt}\, f(t) = \underset{s\to 0}{Lt}\, sF(s) = 0$, final value theorem is verified.

Problem-2.5: Find the value of $i(0)$ using the initial value theorem for the following Laplace transform:

$$I(s) = \frac{2s+3}{(s+1)(s+3)}.$$

Verify the result by solving it for $i(t)$.

Solution:

$$i(0) = \underset{t\to 0}{Lt}\, i(t) = \underset{s\to\infty}{Lt}\, sI(s)$$

$$= \underset{s\to\infty}{Lt}\, \frac{s(2s+3)}{(s+1)(s+3)} = \frac{2}{1\times 1} = 2$$

$$I(s) = \frac{2s+3}{(s+1)(s+3)} = \frac{1/2}{s+1} + \frac{3/2}{s+3}$$

$$i(t) = L^{-1}[I(s)] = L^{-1}\left[\frac{1}{2}\frac{1}{s+1} + \frac{3}{2}\frac{1}{s+3}\right]$$

$$= \frac{1}{2}e^{-t} + \frac{3}{2}e^{-3t}$$

$\Rightarrow$
$$i(0) = \frac{1}{2} + \frac{3}{2} = \frac{4}{2} = 2$$

ADDITIONAL SOLVED PROBLEMS

ASP-1: Find the inverse Laplace transform of the function $F(s) = \dfrac{s+1}{(s+2)(s+3)}$.

Solution:

Given that

$$F(s) = \frac{s+1}{(s+2)(s+3)} = \frac{A}{s+2} + \frac{B}{s+3}$$

$$\Rightarrow \frac{s+1}{(s+2)(s+3)} = \frac{A(s+3)+B(s+2)}{(s+2)(s+3)}$$

From the above equation,

Equating s-quotients both sides $A + B = 1$ (1)

Equating constant terms both sides $3A + 2B = 1$ (2)

After solving eqs (1) and (2),

$$A = -1,\ B = 2$$

$$F(s) = \frac{-1}{s+2} + \frac{2}{s+3}$$

On taking inverse Laplace transform on both sides

$$f(t) = -e^{-2t} + 2e^{-3t}$$

ASP-2: Solve the following differential equation $\frac{d^2y}{dt^2} + 6\frac{dy}{dt} + 8y = 2;\ y(0) = y'0) = 0.$

Solution:

Given that

$$\frac{d^2y}{dt^2} + 6\frac{dy}{dt} + 8y = 2;\ y(0) = y'(0) = 0$$

By applying Laplace transform to each term

$$s^2Y(s) + 6sY(s) + 8Y(s) = 2/s$$

$$\Rightarrow Y(s) = \frac{2}{s(s+2)(s+4)}$$

By applying partial fraction expansion,

$$Y(s) = \frac{1}{4s} + \frac{-1}{2(s+2)} + \frac{1}{4(s+4)}$$

On taking inverse Laplace transform on both sides,

$$y(t) = \frac{1}{4} - \frac{e^{-2t}}{2} + \frac{e^{-4t}}{4}$$

ASP-3: Find the initial and final values for the function $F(s) = \frac{2}{s(s+2)(s+4)}$.

Solution:

Given that

$$F(s) = \frac{2}{s(s+2)(s+4)}$$

According to initial value theorem,

$$f(0) = \lim_{s\to\infty} sF(s) = \lim_{s\to\infty} \frac{2}{(s+2)(s+4)} = \frac{2}{(\infty+2)(\infty+4)} = 0$$

According to final value theorem,

$$f(\infty) = \lim_{t\to\infty} f(t) = \lim_{s\to 0} sF(s) = \lim_{s\to 0} \frac{2}{(s+2)(s+4)} = \frac{2}{(0+2)(0+4)} = \frac{1}{4}$$

ASP-4: Find $x(t)$, where $X(s) = \frac{s+8}{s^2+6s+13}$

Solution:

Given that

$$X(s) = \frac{s+8}{s^2+6s+13} = \frac{(s+3)+5}{s^2+6s+9+4}$$

$$= \frac{s+3}{(s+3)^2+2^2} + \frac{(5/2)\times 2}{(s+3)^2+2^2}$$

$$x(t) = \mathrm{L}^{-1}[X(s)] = e^{-3t}\cos 2t + \frac{5}{2}e^{-3t}\sin 2t \qquad (t > 0)$$

EXERCISE PROBLEMS

1. Find the Laplace transforms of the following functions:

(i) $f(t) = t\cos at$

(ii) $f(t) = t\sin at$

(iii) $f(t) = e^{at}\cos bt$

(iv) $f(t) = e^{at}\sin bt$

2. Find the inverse Laplace transforms of the following functions:

(i) $F(s) = \dfrac{s+2}{s^2+2s+2}$

(ii) $F(s) = \dfrac{2s+2}{s(s-1)(s+1)}$

(iii) $F(s) = \dfrac{1}{(s+1)^2(s+2)}$

3. Solve by Laplace's method the initial value problem

(i) $x'' + x = 0.\ x(0) = 0,\ x'(0) = 1$

(ii) $x'' + 2x' + 2x = 0.\ x(0) = 1,\ x'(0) = -1$

4. Find the minor of elements a_{11}, a_{23}, and also find the cofactors of all the elements and hence obtain the adjoint of matrix A, given

$$A = \begin{bmatrix} 1 & 0 & 3 \\ 2 & 4 & 1 \\ 3 & 2 & 2 \end{bmatrix}$$

5. Find the inverse of matrix $A = \begin{bmatrix} 2 & 3 \\ 4 & 1 \end{bmatrix}$.

ANSWERS

1. (i) $\dfrac{s^2-a^2}{(s^2+a^2)^2}$ (ii) $\dfrac{2sa}{(s^2+a^2)^2}$ (iii) $\dfrac{s-a}{(s-a)^2+b^2}$ (iv) $\dfrac{b}{(s-a)^2+b^2}$

2. (i) $e^{-t}(\sin t + \cos t)$, (ii) $-2 + 2e^t$, (iii) $e^{-t}(t-1) + e^{-2t}$

3. (i) $x(t) = \sin t$ (ii) $x(t) = e^{-t}(\cos t - 4\sin t)$

4. $\text{adj}(A) = \begin{bmatrix} 6 & 6 & -12 \\ -1 & -7 & 5 \\ -8 & -2 & 4 \end{bmatrix}$

5. $A^{-1} = \dfrac{1}{10}\begin{bmatrix} -1 & 3 \\ 4 & -2 \end{bmatrix}$

CHAPTER

3

Mathematical Modelling of Physical Systems

Contents

3.1 INTRODUCTION

For the analysis and design of control systems, we need to formulate a mathematical description of the system. The process of obtaining the desired mathematical description of the system is known as modelling. The basic models of dynamic physical systems are differential equations obtained by application of the appropriate laws of nature. These equations may be linear or non-linear depending on the system being modelled. Moreover, they are inconvenient for analysis and design manipulations, and therefore, the Laplace transformation is used to convert the differential equations into algebraic equations.

To study and examine a physical system, it is always necessary to have some types of equivalent representations, which will describe the various components and their relations in the system. Any physical system that is mathematically described by a set of equations is known as system equations. These equations are simplified by using Laplace transformations under the assumption that all initial conditions of the system are set to zero. Further, these simplified system equations are manipulated to determine the ratio of the Laplace transform of output to the Laplace transform of input and to investigate the system properties. This ratio is known as the transfer function of the system. The complete transfer function of the system is obtained by determining the transfer function of the various components in the system, and then combining them according to their relationships.

3.2 TRANSFER FUNCTION

In any system, first, the system parameters are designed and their values are selected as per requirement. Next, the input is selected to see the performance of the designed system, as shown in Figure 3.1.

Figure 3.1 An open-loop system

The performance of the system can be expressed in terms of its output as

$$\text{output} = \text{effect of system parameters on the selected input}$$

$$\therefore \text{ output} = \text{input} \times \text{effect of system parameters}$$

Mathematically, a function explaining the effect of system parameters on input to produce an output is called 'transfer function'. Due to the characteristics of the system parameters, the input is transferred into the output, once it is applied to the system. This is the concept of transfer function.

Mathematically, transfer function is defined as the ratio of Laplace transform of output of the system (i.e., response) to the Laplace transform of input (i.e., excitation or driving function) under the assumption that all initial conditions are zero.

Symbolically, transfer function can be represented as shown in Figure 3.2.

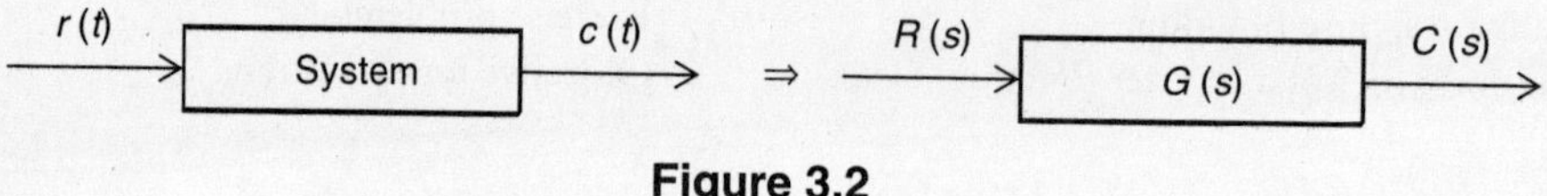

Figure 3.2

If $G(s)$ is the transfer function of the system, then

$$G(s) = \frac{\text{Laplace transform of output}}{\text{Laplace transform of input}} = \frac{C(s)}{R(s)}$$

3.2.1 Advantages of Transfer Function

1. It gives mathematical models of all system components, and hence, of the overall system.
2. It converts integro-differential time domain equations into simple algebraic equations.
3. Transfer function is the property and characteristics of the system itself. Its value is dependent on the parameters of the system and independent of the values of inputs.
4. Once transfer function is known, the output response for any type of the reference input can be calculated.
5. It helps in determining important information about the system, that is, poles, zeros, characteristic equation, etc.
6. It helps in the stability analysis of the system.

3.2.2 Limitations of Transfer Function

1. It is only applicable to linear time-invariant systems.
2. It does not provide any information concerning the physical structure of the system.

3. Effects arising due to initial conditions are totally neglected. Hence, initial conditions lose their importance.

3.2.3 Procedure to Obtain Transfer Function

1. Write down the time domain equations for the system by introducing different variables in the system.
2. Take the Laplace transform of the system equations assuming all initial conditions are set to zero.
3. Identify system input and output variables.
4. Eliminate the introduced variables to obtain the resultant equation in terms of input and output variables.
5. Take the ratio of Laplace transform of output variable to Laplace transform of input variable to obtain the transfer function of the system.

3.3 POLES AND ZEROS OF A SYSTEM

In general, the transfer function of any linear time-invariant system can be expressed as

$$G(s) = \frac{C(s)}{R(s)} = \frac{a_0 s^n + a_1 s^{n-1} + a_2 s^{n-2} + \cdots + a_n}{b_0 s^m + b_1 s^{m-1} + b_2 s^{m-2} + \cdots + b_m}, \; m \geq n \qquad (3.1)$$

The numerator and denominator can be factorised to get

$$G(s) = K \frac{(s - z_1)(s - z_2) \cdots (s - z_n)}{(s - p_1)(s - p_2) \cdots (s - p_m)} \qquad (3.2)$$

where K is called system gain factor or scalar factor.

Poles: The values of s that makes the transfer function infinite are called poles of that transfer function.

∴ Here, values $p_1, p_2, \ldots p_m$ are called poles of the transfer function.

These poles are nothing but the roots of the equation obtained by equating denominator of the transfer function to zero.

Characteristic Equation: The equation obtained by equating the denominator of a transfer function to zero, whose roots are the poles of that transfer function, is called characteristic equation of that system.

i.e., $F(s) = b_0 s^m + b_1 s^{m-1} + b_2 s^{m-2} + \cdots + b_m = 0$ is called the characteristic equation.

Zeros: The values of s that make the transfer function zero are called zeros of that transfer function.

∴ Here, the values $z_1, z_2 \cdots z_n$ are called zeros of the transfer function.

Poles and zeros may be real or complex conjugates or combination of both types. Further, poles and zeros may be located at the origin in s-plane.

3.4 IMPULSE RESPONSE FUNCTION

Consider the output or response of a system to a unit-impulse input when the initial conditions are zero. Since the Laplace transform of the unit-impulse input is unity, Laplace transform of output of the system is

$$C(s) = G(s) \tag{3.3}$$

The inverse Laplace transform of the output, as given in (3.3), gives the impulse response of the system and is given as

$$L^{-1}[G(s)] = g(t) \tag{3.4}$$

Thus, the impulse response function $g(t)$ is the response of a linear system to a unit-impulse input when all the initial conditions are zero.

***Problem-3.1*:** The dynamic behaviour of the system is described by the equation $\frac{dc}{dt} - 10c = 40e$, where e is the input and c is the output. Determine the transfer function of the system.

***Solution*:**

Given that

the dynamic behaviour of the system is $\frac{dc}{dt} - 10c = 40e$, where e is the input and c is the output.

By taking the Laplace transform on both sides and assuming that all initial conditions are zero, we get

$$sC(s) + 10C(s) = 40E(s)$$

$$\Rightarrow (s+10)C(s) = 40E(s)$$

$$\Rightarrow T(s) = \frac{C(s)}{E(s)} = \frac{40}{s+10}$$

***Problem-3.2*:** Find the transfer function $\frac{C(s)}{R(s)}$ of a system having differential equation $9\frac{d^2c(t)}{dt^2} + 12\frac{dc(t)}{dt} + c(t) = r(t) + 2r(t-1)$.

***Solution*:**

Given that

$$9\frac{d^2c(t)}{dt^2} + 12\frac{dc(t)}{dt} + c(t) = r(t) + 2r(t-1)$$

By taking the Laplace transform on both sides and assuming that all initial conditions are zero, we get

$$(9s^2 + 12s + 1)C(s) = R(s) + 2e^{-s}R(s)$$

$$\Rightarrow \quad (9s^2 + 12s + 1)C(s) = R(s)[1 + 2e^{-s}]$$

$$\Rightarrow \quad T(s) = \frac{C(s)}{R(s)} = \frac{1+2e^{-s}}{9s^2 + 12s + 1}.$$

Hence, the desired transfer function is given as

$$\frac{C(s)}{R(s)} = \frac{1+2e^{-s}}{9s^2 + 12^s + 1}.$$

3.5 TYPE AND ORDER OF A TRANSFER FUNCTION

Type: It is defined as the number of poles located at origin in a given transfer function.
Order: It is defined as the highest power in the denominator polynomial of a given transfer function.

***Example*:** For the transfer function, $T(s)=\dfrac{4s+1}{s^3(s^2+4s+10)}$; type = 3 and order = 5.

Type is the characteristic property of open-loop transfer function, whereas order is the characteristic property of closed-loop transfer function.

***Problem-3.3*:** Determine the type and order of the following systems.

(i) $G(s)H(s)=\dfrac{100}{s(s^2+4s+200)}$ (ii) $G(s)H(s)=\dfrac{200}{s^2(s^2+10s+200)}$

(iii) $G(s)H(s)=\dfrac{4(s^2+10s+100)}{s(s+3)(s^2+2s+10)}$ (iv) $G(s)H(s)=\dfrac{200}{(1+0.1s)(1+0.5s)}$

***Solution*:**

(i) $G(s)H(s)=\dfrac{100}{s(s^2+4s+200)}$

Type = 1, Order = 3

(ii) $G(s)H(s)=\dfrac{200}{s^2(s^2+10s+200)}$

Type = 2, Order = 4

(iii) $G(s)H(s)=\dfrac{4(s^2+10s+100)}{s(s+3)(s^2+2s+10)}$

Type = 1, Order = 4

(iv) $G(s)H(s)=\dfrac{200}{(1+0.1s)(1+0.5s)}$

Type = 0, Order = 2

3.6 MATHEMATICAL MODELS

3.6.1 Electrical Systems

Electrical systems have either lumped or distributed parameters depending upon the system properties. In general, electrical systems with lumped parameters can be represented by linear equation of first order. The elements of electrical systems are divided into two categories: active elements and passive elements. The elements that are sources of electrical energy are known as active elements, while those elements that consume energy are known as passive elements. The sources of energy may be voltage or current source.

The energy source that maintains a particular magnitude of voltage and a defined waveform across its terminals and is independent of the current flowing through it is known as the ideal voltage source. This is represented by the symbol $v(t)$ along with reference polarity, as shown in Figure 3.3(a) and (b).

Similarly, the ideal current source is defined as that source which maintains a current of particular magnitude and waveform at its terminals and is independent of the potential difference across the terminals. This is represented by $i(t)$ associated with a reference direction, as shown in Figure 3.3(c). Ideal passive elements with their properties are shown in Figure 3.4. Resistors, inductors and capacitors are bilateral linear passive elements.

Various electrical elements with describing equations are shown in Table 3.1.

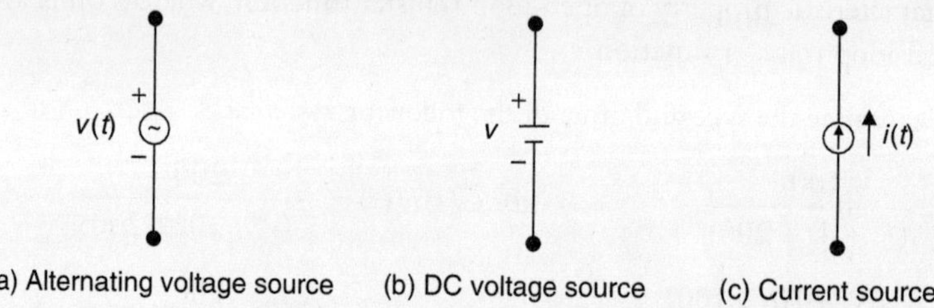

(a) Alternating voltage source (b) DC voltage source (c) Current source

Figure 3.3 Active elements

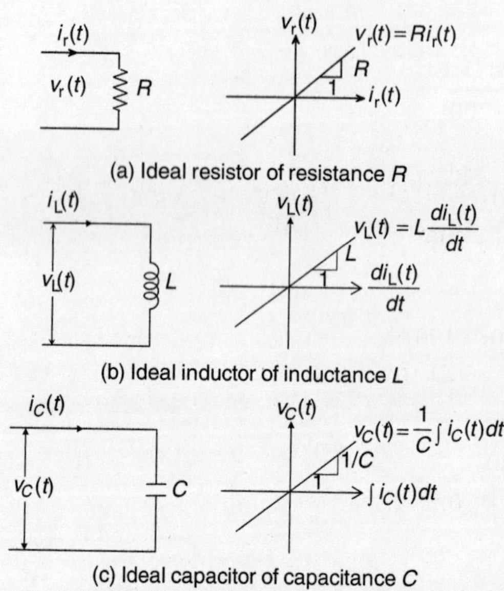

Figure 3.4 Passive elements

Table 3.1 Basic electrical elements with describing equations

Passive elements	Describing equations	Associated power or energy
Resistor $i_r(t)$ $v_2(t)$ — R — $v_1(t)$	$v_r(t) = v_2(t) - v_1(t)$ $i_r(t) = \frac{v_r(t)}{R}$ $v_r(t) = Ri_r(t)$	Power $(P) = \frac{v_r^2(t)}{R}$

(Continued)

Table 3.1 *(Continued)*

Passive elements	Describing equations	Associated power or energy
Inductor $i_L(t)$, $v_2(t)$ — L — $v_1(t)$	$v_L(t) = v_2(t) - v_1(t)$ $i_L(t) = \frac{1}{L}\int_0^t v_L(t)dt + i_L(0)$ $v_L(t) = L\frac{di_L(t)}{dt}$	Energy $(E) = \frac{1}{2}Li_L^2(t)$
Capacitor $i_C(t)$, $v_2(t)$ — C — $v_1(t)$	$v_C(t) = v_2(t) - v_1(t)$ $i_c(t) = C\frac{dv_C(t)}{dt}$ $v_C(t) = \frac{1}{C}\int_0^t i_c(t)dt + v_C(0)$	Energy $(E) = \frac{1}{2}Cv_C^2(t)$

The following section describes the procedure for finding system equations for electrical system.

Procedure

1. Define variables in the system, that is, voltages and currents in the system or network. Impose the constraints on these variables.
2. Obtain loop equations by applying Kirchhoff's voltage and current laws to the loops formed by various elements in the network.
3. Apply physical voltage–current relationships to individual elements.
4. From loop equations in Step 2 and voltages–current relationship in Step 3, obtain the system equations by algebraic manipulations.

Problem-3.4: Find the transfer function of the network, as shown in the figure.

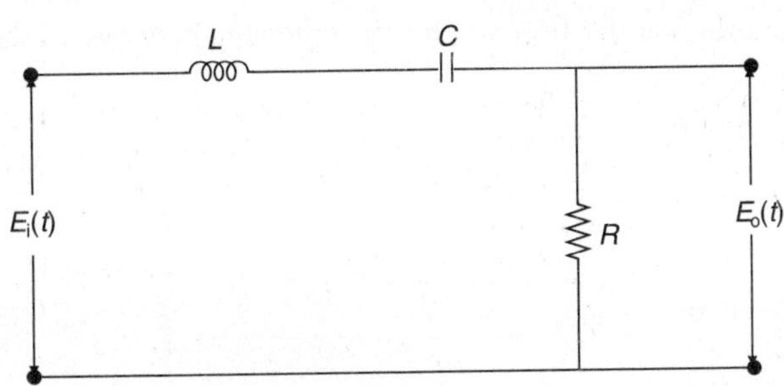

Solution:
The given circuit is

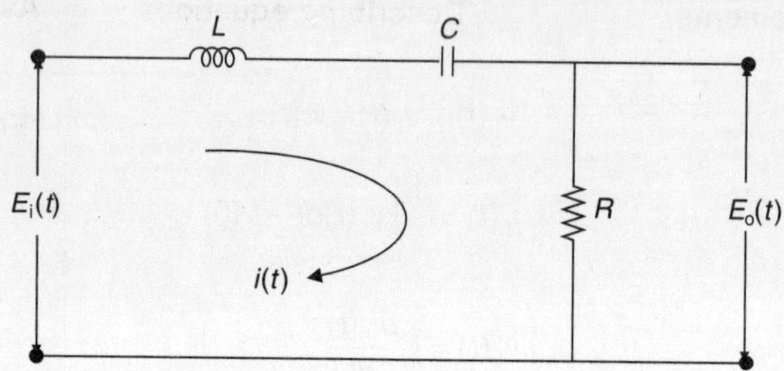

By applying KVL to the above circuit, we get

$$E_i(t) = L\frac{di(t)}{dt} + \frac{1}{C}\int i(t)dt + i(t)R.$$

While $$E_0(t) = i(t)R$$

By taking the Laplace transform on both sides of the above equation under the assumption of zero initial conditions, we get

$$E_i(s) = sLI(s) + \frac{1}{C}\frac{I(s)}{s} + RI(s) \tag{1}$$

$$E_0(s) = I(s)R \tag{2}$$

$$\frac{(2)}{(1)} \Rightarrow \frac{E_0(s)}{E_i(s)} = \frac{R}{\left[sL + \frac{1}{sC} + R\right]} = \frac{sRC}{s^2LC + sRC + 1}$$

$$\Rightarrow \qquad \frac{E_0(s)}{E_i(s)} = \frac{sRc}{s^2LC + sRC + 1}.$$

Hence, the final transfer function is given as

$$\frac{E_0(s)}{E_i(s)} = \frac{sRC}{s^2LC + sRC + 1}.$$

Problem-3.5: Obtain the transfer function for the following network, as shown in the figure.

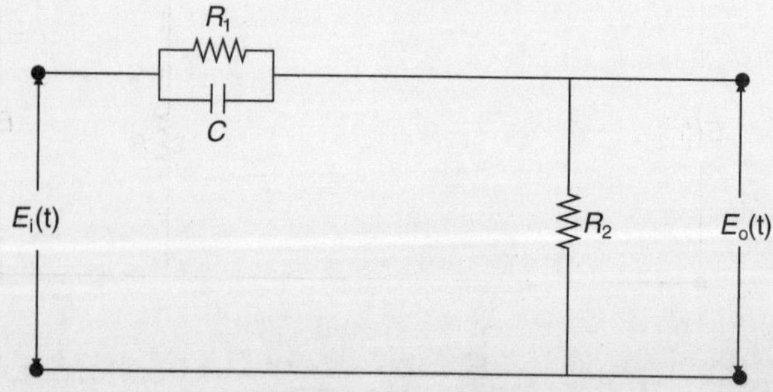

Solution:
The s-domain equivalent network of the given circuit is

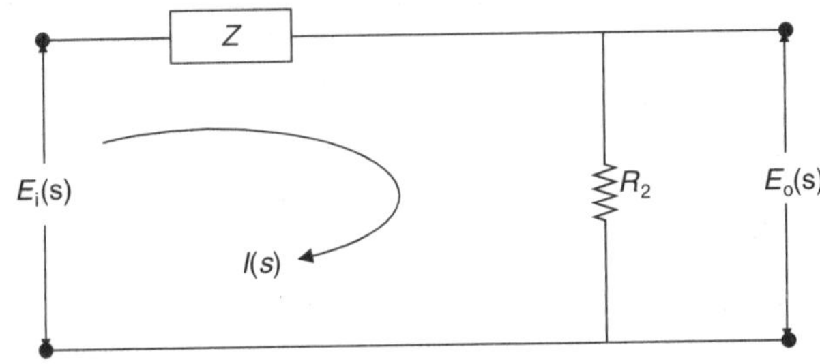

The parallel combination of R_1 and $1/sC$ gives impedance of

$$Z = \frac{Z_1 Z_2}{Z_1 + Z_2} = \frac{R_1 \times \dfrac{1}{sC}}{R_1 + \dfrac{1}{sC}} = \frac{R_1}{1 + sR_1C}.$$

By applying KVL to the circuit, we get

$$E_i(s) = ZI(s) + I(s)R_2 \quad (1)$$

$$E_0(s) = I(s)R_2 \quad (2)$$

$$\frac{(2)}{(1)} \Rightarrow \frac{E_0(s)}{E_i(s)} = \frac{R_2}{[Z + R_2]}.$$

By substituting Z in the above equation, we get

$$\frac{E_0(s)}{E_i(s)} = \frac{R_2}{\dfrac{R_1}{1 + sR_1C} + R_2} = \frac{sR_1R_2C + R_2}{R_1 + sR_1R_2C + R_2}.$$

Problem-3.6: Find the transfer function for the following electrical network, as shown in the figure.

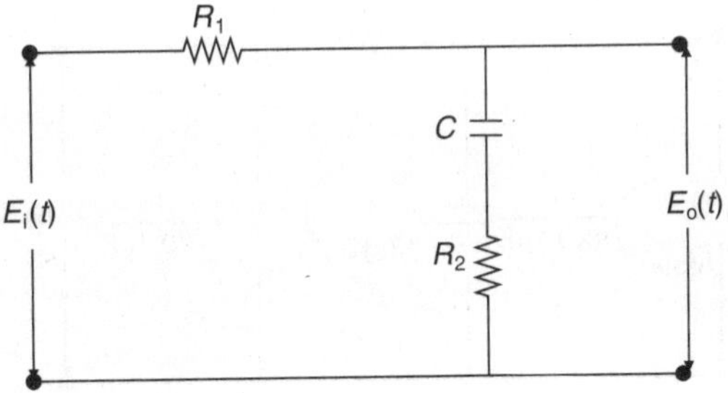

Solution:
The s-domain equivalent circuit for the given network is

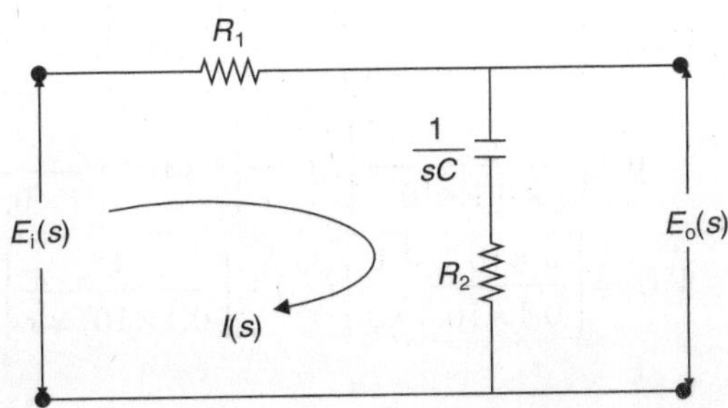

By applying KVL to the network, we get

$$E_i(s) = R_1 I(s) + \frac{1}{sC} I(s) + I(s) R_2 \tag{1}$$

$$E_0(s) = \frac{1}{sC} I(s) + R_2 I(s)$$

$$\Rightarrow \qquad E_0(s) = \left[\frac{1}{sC} + R_2\right] I(s) \tag{2}$$

$$\frac{(2)}{(1)} \Rightarrow \frac{E_0(s)}{E_i(s)} = \frac{\left[\frac{1}{sC} + R_2\right]}{\left[R_1 + \frac{1}{sC} + R_2\right]} = \frac{1 + sCR_2}{1 + sC[R_1 + R_2]}.$$

Hence, the final transfer function is

$$\frac{E_0(s)}{E_1(s)} = \frac{1 + sCR_2}{1 + sC[R_1 + R_2]}.$$

Problem-3.7: Find the transfer function of the following system as shown in the figure.

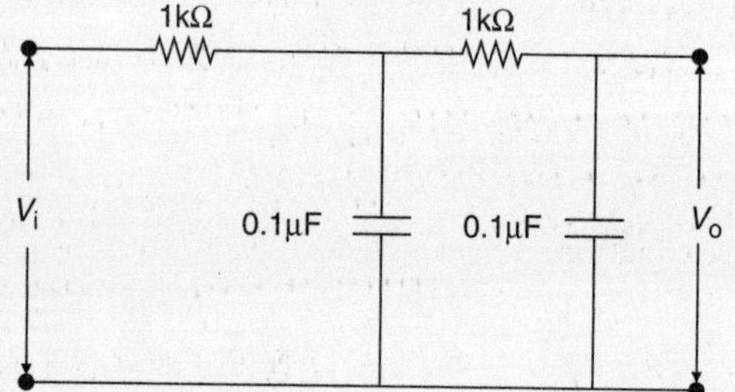

Solution:
The *s*-domain equivalent circuit for the given network is

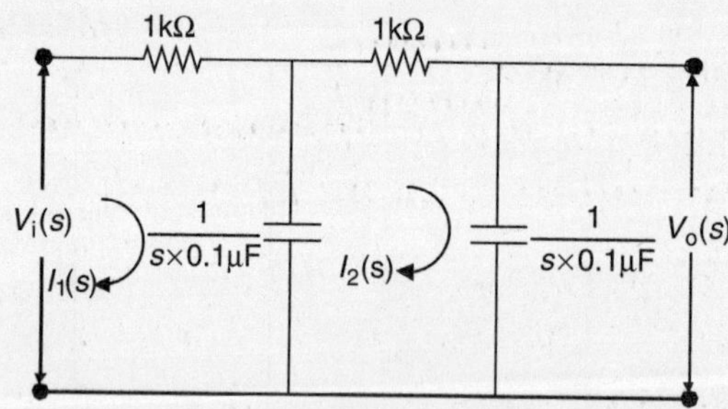

By applying KVL to the above circuit, we will get the following equation:

$$V_i(s) = \left[1\times10^3 + \frac{1}{s\times0.1\times10^{-6}}\right] I_1(s) - \left[\frac{1}{s\times0.1\times10^{-6}}\right] I_2(s)$$

and

$$0 = \left[\frac{1}{s\times0.1\times10^{-6}}\right] I_1(s) - \left[1\times10^3 + \frac{2}{s\times0.1\times10^{-6}}\right] I_2(s)$$

i.e.,

$$V_i(s) = \left[\frac{1\times10^{-4} s + 1}{0.1\times10^{-6}\times s}\right] I_1(s) - \left[\frac{1}{0.1\times10^{6}\times s}\right] I_2(s)$$

and $$0=\left[\frac{1}{0.1\times10^{-6}\times s}\right]I_1(s)-\left[\frac{1\times10^{-4}s+2}{0.1\times10^{-6}\times s}\right]I_2(s).$$

We can solve the above-mentioned two equations by using Cramer's rule to obtain the solution.

$$\Delta=\begin{vmatrix}\dfrac{1\times10^{-4}s+1}{0.1\times10^{-6}s} & -\dfrac{1}{0.1\times10^{-6}\times s}\\ \dfrac{1}{0.1\times10^{-6}\times s} & -\dfrac{1\times10^{-4}s+2}{0.1\times10^{-6}s}\end{vmatrix}=\frac{-[10^{-8}s^2+3\times10^{-4}s+1]}{1\times10^{-14}s^2}$$

$$\Delta_2=\begin{vmatrix}\dfrac{1\times10^{-4}s+1}{0.1\times10^{-6}s} & V_i(s)\\ \dfrac{1}{0.1\times10^{-6}\times s} & 0\end{vmatrix}=\frac{-V_i(s)}{0.1\times10^{-6}s^2}$$

$$I_2(s)=\frac{\Delta_2}{\Delta}=\frac{-V_i(s)}{0.1\times10^{-6}s^2}\times\frac{1\times10^{-14}s^2}{-[10^{-8}s^2+3\times10^{-4}s+1]}$$

$$=\frac{V_i(s)\times1\times10^{-7}s}{1\times10^{-8}s^2+3\times10^{-4}s+1}$$

Now, the output equation is

$$V_o(s)=\frac{1}{0.1\times10^{-6}s}I_2(s)=\frac{1}{0.1\times10^{-6}s}\times\frac{V_i(s)\times1\times10^{-7}s}{1\times10^{-8}s^2+3\times10^{-4}s+1}$$

$$\therefore\quad\frac{V_o(s)}{V_i(s)}=\frac{1}{1\times10^{-8}s^2+3\times10^{-4}s+1}=\frac{1\times10^8}{s^2+3\times10^4s+1\times10^8}$$

Hence, the final transfer function is $\dfrac{V_o(s)}{V_i(s)}=\dfrac{1\times10^8}{s^2+3\times10^4s+1\times10^8}$

3.6.2 Mechanical Systems

Translational mechanical systems

A system containing mechanical elements is known as mechanical system. These systems are classified as translational and rotational mechanical systems depending upon the type of operation involved. In practice, the properties of mechanical elements are not linear and are investigated after making various assumptions that make them linear. The motion along a straight line is defined as translational motion and is described by displacement, velocity and acceleration. Systems having translational mechanical motion are known as translational mechanical systems. Their active elements are force and velocity, while the passive elements are mass, stiffness and viscous friction. However, passive elements consume mechanical power by opposing the motion in the system. Further, the motion in this type of mechanical systems is governed by Newton's laws of motion and D'Alemberts principle.

Newton's laws of motion

The Newton's laws of motion are stated as follows:

1. Every body continues in its state of rest or of uniform motion in a straight line unless it is coupled to change the state by a force impressed on it.

2. The change in motion is proportional to the motive force impressed and is made in the direction of the line in which the motive force is applied. In other words, force of mass M is equal to the product of mass and its acceleration.
3. To every action, there is an equal and opposite reaction.

D'Alemberts principle

This principle states that for any rigid body, the algebraic sum of externally applied forces and the forces resisting the motion in a given direction is zero. From Newton's Second Law of motion, the inertia force of a body, F_m, in the system is

$$F_m = M\frac{d^2x}{dt^2} = M\frac{dv}{dt} \tag{3.5}$$

where M, x and v are mass, displacement and velocity of the body under consideration.

The coefficient M in (3.5), which is the mass of moving body has unit of force per unit acceleration. This is illustrated in Figure 3.5(a)

The second force that resists the body motion is the restoring force of a linear translational spring that is stretched within its elastic limits. This restoring force created inside the spring is proportional to displacement

i.e.,

$$F_{spr} = Kx \tag{3.6}$$

where K is the spring stiffness and x is displacement, as illustrated in Figure 3.5(b)

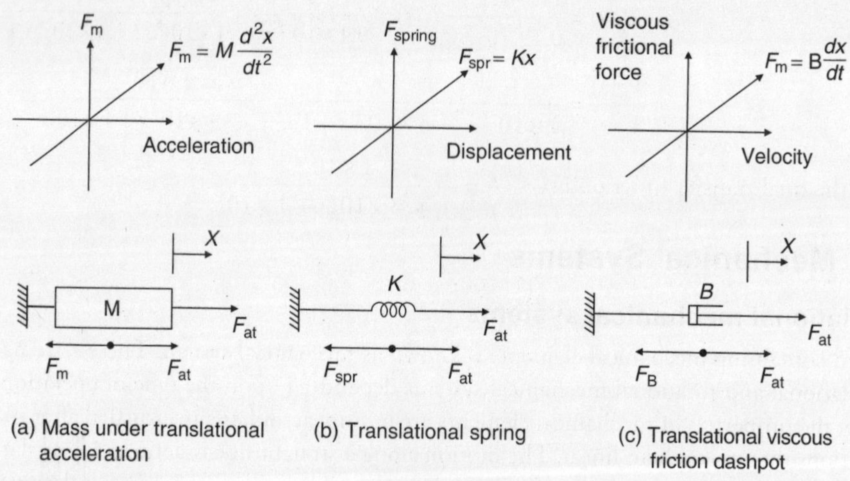

Figure 3.5

When a body moves on the surface, an opposing force is created between the contact surfaces that contribute the third passive element to translational mechanical systems. This opposing force is known as frictional force. As frictional force always opposes the motion of the body, its direction always changes with the direction of the moving body. This force which opposes the motion of the body has a magnitude F_v and is directly proportional to the body velocity. It is known as viscous friction and is defined as follows:

$$F_v = B\frac{dx}{dt} \tag{3.7}$$

where B is the coefficient of viscous friction, as illustrated in Figure 3.5(c).

The various elements of translational mechanical systems are listed in Table 3.2.

Table 3.2 Basic transitional mechanical elements with describing equations

Type of elements	Elements and its symbol	Describing equation	Associated energy or power
Active elements	1) Force →F 2) Velocity→V		
Passive elements	1) Mass (M) F, v_2, M, v_1	$F = M\frac{dv_{21}}{dt}$ $v_{21} = \frac{1}{M}\int_0^t F dt + v_{21}(0)$	Kinetic energy $E_k = \frac{1}{2}Mv_{21}^2$
	2) Spring stiffness (K) F, v_2, K, v_1, F	$F = K\int_0^t v_{21} dt + v_{21}(0)$ $v_{21} = \frac{1}{K}\frac{dF}{dt}$	Potential energy $E_p = \frac{1}{2}\frac{F^2}{K}$
	3) Viscous friction F, v_2, B, v_1	$F = Bv_{21}$ $v_{21} = \frac{1}{B}F$	Power dissipated $P = Bv_{21}^2$

The equations for translational mechanical systems are obtained by using the following procedure.

Procedure

1. Draw the physical system into an equivalent configuration with respect to a reference and define various coordinates with their positive direction.
2. Draw free body diagram with all forces present on the mass in the system. Draw separate diagrams for each mass in the system.
3. Write equation of equilibrium for all forces acting on elements using D'Alemberts principle for translational system.
4. Write force geometry relations for individual elements.
5. Combine relations of Steps 3 and 4 to obtain the system equations.

Rotational mechanical system

When a body moves around a fixed axis with a constant distance from the axis, the motion of the body is known as rotation. Systems having this type of motion are called rotational mechanical systems and are described by angular velocity and angular acceleration. The active elements of this type of system are torque and angular acceleration. Rotational motion of body generates three types of torques in the system that contribute passive elements in the system. These are known as inertia torque, spring torque and damping torque.

Inertia torque of a body is defined as the product of the moment of inertia and the angular acceleration of the body and is due to the body inertia. The inertia torque is

$$T = J\frac{d^2\theta}{dt^2} \tag{3.8}$$

where J is the moment of inertia of the body and θ is the angular displacement.

The spring or restoring torque is defined as the product of angular displacement and the spring stiffness in the system. The spring torque is

$$T_s = K\theta \tag{3.9}$$

where K is the spring stiffness.

The damping or frictional torque is defined as the product of angular velocity and the coefficient of rotational friction. This is due to friction when the body rotates. The frictional torque is

$$T_B = B\frac{d\theta}{dt} \tag{3.10}$$

where B is the coefficient of angular friction.

The passive elements of systems from (3.10) are J, K and B.

The procedure for determining the system equations for rotational mechanical system is described in the following section.

Procedure

1. Draw the physical system into a configuration with respect to reference and define coordinates with positive directions.
2. Draw free body diagram.
3. Write equations of equilibrium for all torques acting on various elements using D'Alemberts principle for rotational systems.
4. Write torque geometry equations for individual elements in the system.
5. Combine equation in Steps 3 and 4 to obtain the system equations.

Table 3.3 describes the active and passive elements of rotational mechanical systems.

Table 3.3 Basic rotational mechanical elements with describing equations

Types of elements	Elements and its symbol	Describing equation	Associated energy or power
Active elements	Torque T Angular velocity, ω		
Passive elements	1) Moment of inertia (J) T, ω_2, θ_2, J, ω_1, θ_1	$T = J\frac{d^2\theta_{21}}{dt^2}$ $\omega_{21} = \frac{1}{J}\int_0^t T dt + \omega_{21}(0)$	Kinetic energy $E_k = \frac{1}{2}J\omega_{21}^2$

(Continued)

Table 3.3 (*Continued*)

Types of elements	Elements and its symbol	Describing equation	Associated energy or power
	2) Stiffness K ω_2 K ω_1 T T	$T = K\int_0^t \omega_{21}dt + T_{12}(0)$ $\omega_{21} = \frac{1}{K}\frac{dT}{dt}$	Potential energy $E_p = \frac{1}{2}\frac{T^2}{K}$
	3) Rotational friction f ω_2 B ω_1 T T	$T = B\omega_{21}$ $\omega_{21} = \frac{1}{B}T$	Power dissipated $P = B\omega_{21}^2$

3.7 ANALOGOUS SYSTEM

There exists a fixed analogy and similarity between the equilibrium equations of electrical and mechanical systems. Due to this reason, it is possible to draw an electrical system that will behave exactly similar to the given mechanical system and vice versa. It is always advantageous to obtain 'electrical analogues' of the given mechanical system as we are more familiar with the methods of anlysing electrical network than mechanical network.

There are two methods of obtaining electrical analogous networks, namely

1. Force–voltage analogy (i.e., direct analogy)
2. Force–current analogy (i.e., inverse analogy)

Let us consider a simple mechanical system as shown in Figure 3.6.

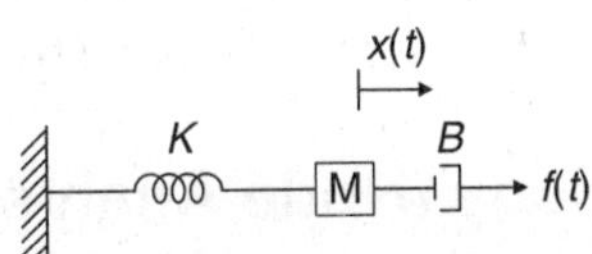

Figure 3.6 Basic mechanical system

Due to the applied force, mass M will displace by an amount of $x(t)$ in the direction of the force $f(t)$, as shown in Figure 3.6. According to Newton's law of motion, applied force will cause displacement $x(t)$ in the spring, and acceleration to mass M against frictional force having constant B.

$$\therefore \qquad f(t) = M\frac{d^2x}{dt^2} + B\frac{dx}{dt} + Kx. \tag{3.11}$$

3.7.1 Force–Voltage Analogy (Loop Analysis)

The analogous term for force in mechanical system is the voltage in electrical system. Let us consider the electrical network shown in Figure 3.7.

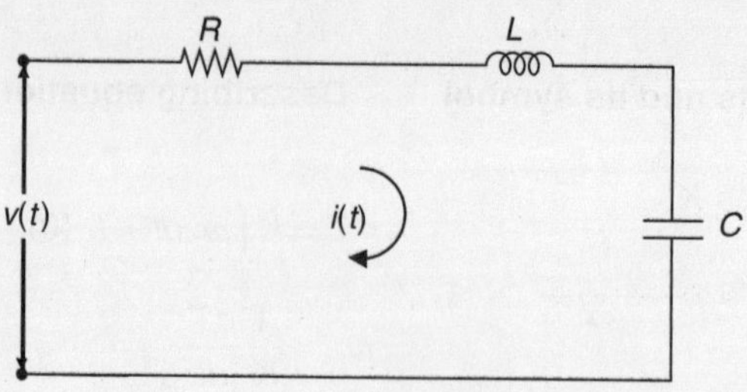

Figure 3.7 RLC series circuit

By applying KVL to the circuit,

$$v = Ri + L\frac{di}{dt} + \frac{1}{C}\int i dt$$

However, $i = \frac{dq}{dt}$

$$v = R\frac{dq}{dt} + L\frac{d^2q}{dt^2} + \frac{q}{C} \tag{3.12}$$

By comparing (3.11) and (3.12), it is clear that

1. voltage is analogous to force.
2. inductance L is analogous to M (mass).
3. resistance R is analogous to f or B (friction).
4. reciprocal of capacitor, that is $1/C$ is analogous to K (spring constant).

3.7.2 Force–Current Analogy (Node Analysis)

In this method, current is treated as analogous quantity to force in the mechanical system. Therefore, force is replaced by the current source in the system, as shown in Figure 3.8.

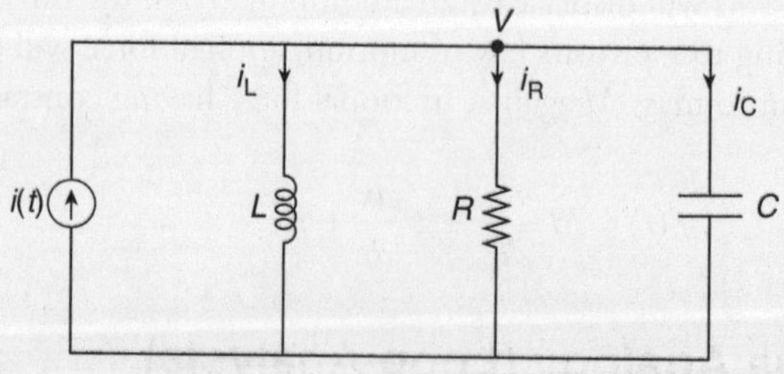

Figure 3.8 RLC parallel circuit

By applying KCL to the circuit,

$$i = i_R + i_L + i_C$$

$$= \frac{v}{R} + \frac{1}{L}\int v dt + C\frac{dv}{dt}.$$

However,

$$v = \frac{d\psi}{dt}$$

$$\therefore \qquad i = \frac{1}{R}\frac{d\psi}{dt} + C\frac{d^2\psi}{dt^2} + \frac{\psi}{L} \tag{3.13}$$

By comparing (3.11) and (3.13), it is clear that

1. current is analogous to force.
2. capacitor C is analogous to mass M.
3. reciprocal of resistance $1/R$ is analogous to frictional constant B.
4. reciprocal of inductance I/L is analogous to spring constant K.

Important Note: The elements that are in series in *F–V* analogy get connected in parallel in *F–I* analogous network, while elements that are in parallel in *F–V* analogy get connected in series in *F–I* analogous network.

Table 3.4 gives the analogous elements between the electrical and the mechanical systems.

Table 3.4 Analogous elements

Translational	Rotational	Electrical	
		F–V	***F–I***
1. Force F	Torque T	Voltage V	Current I
2. Mass M	Inertia J	Inductance L	Capacitance C
3. Viscous friction f (or) B	Torsional friction constant f (or) B	Resistance R	$1/R$
4. Spring constant K N/m	Torsional spring constant K (Nm/rad)	$1/C$	$1/L$
5. Displacement x	Angular displacement θ	Charge q	Flux ϕ
6. Velocity $V = \dot{x} = \frac{dx}{dt}$	Angular velocity $\dot{\theta} = \frac{d\theta}{dt} = \omega$	Current $i = \frac{dq}{dt}$	Voltage $e = \frac{d\Psi}{dt}$

3.7.3 Steps to Solve Problems on Analogous Systems

The following procedure is adapted to obtain electrical analogous systems for given mechanical systems

Step 1: Identify all the displacements due to the applied forces. The elements like spring and friction between the two moving surfaces cause change in displacement.

Step 2: Draw the equivalent mechanical system based on node basics. To draw equivalent mechanical system, first a reference line is drawn to represent the ground line, and then, the number of displacements (i.e., number of nodes) are identified. The elements under the same displacement will get connected in parallel under that node and reference line. An element causing change in displacement (either friction or spring) is always connected between the two nodes.

Step 3: Write the equilibrium equations, that is, the algebraic sum of all the forces acting at each node must be equal to zero.

Step 4: In *F–V* analogy, use the following replacements and rewrite the equations.

$$f \to V, M \to L, B \to R, K \to 1/C, x \to q$$

Step 5: Simulate the equations using loop method. Here, the number of displacements is equal to number of loop currents.

Step-6: In *F–I* analogy, use the following replacements and rewrite the equations.

$$f \to I, M \to C, B \to 1/R, K \to 1/L, x \to \psi$$

Step 7: Simulate the equations using node basis. Here, the number of displacements is equal to number of node voltages. In fact, the system will be exactly same as equivalent mechanical system obtained in Step 2 with appropriate replacements.

Note: The same method can also be applicable to rotational mechanical system.

Problem-3.8: Find the transfer function for the following system.

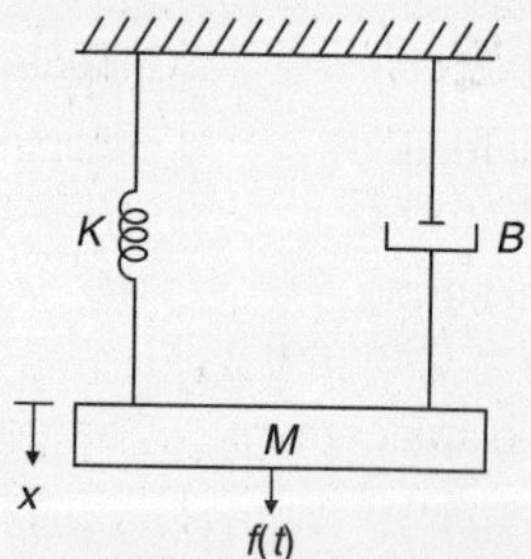

where M is the mass of the system; K is the spring deflection; and B is the coefficient of viscous damping.

Solution:

By applying Newton's second law, the equation of motion is given as

$$M\frac{d^2x}{dt^2} + B\frac{dx}{dt} + Kx = f(t)$$

By taking Laplace transform on both sides under the assumption of zero initial conditions, we get

$$\Rightarrow \quad Ms^2X(s)+BsX(s)+KX(s)=F(s)$$

$$\Rightarrow \quad \left[Ms^2+Bs+K\right]X(s)=F(s)$$

$$\Rightarrow \quad \frac{X(s)}{F(s)}=\frac{1}{Ms^2+Bs+K}.$$

Hence, the transfer function is $\dfrac{X(s)}{F(s)}=\dfrac{1}{Ms^2+Bs+K}$.

Problem-3.9: Obtain the transfer function relating the displacement y and x for the following system, as shown in the figure.

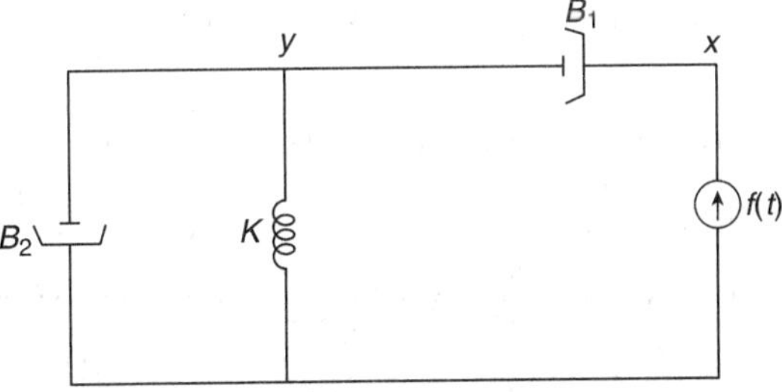

Solution:

By applying Newton's second law, the equations for the system can be written as

$$B_2\frac{dy}{dt}+Ky+B_1\frac{d}{dt}(y-x)=0$$

$$\Rightarrow \quad B_i\frac{d}{dt}(y-x)=f(t).$$

By taking Laplace transform on both sides under the assumption of zero initial conditions, we get

$$B_2sY(s)+KY(s)+B_1s[Y(s)-X(s)]=0$$

$$\Rightarrow \quad Y(s)[B_2s+K+B_1s]=B_1sX(s)$$

$$\Rightarrow \quad \frac{Y(s)}{X(s)}=\frac{B_1s}{(B_2+B_1)s+k}.$$

Hence, the transfer function relating the displacement y and x is given as $\dfrac{Y(s)}{X(s)}=\dfrac{B_1s}{(B_2+B_1)s+k}$.

Problem-3.10: Obtain the transfer function $\dfrac{X_0(s)}{X_1(s)}$ of the following system.

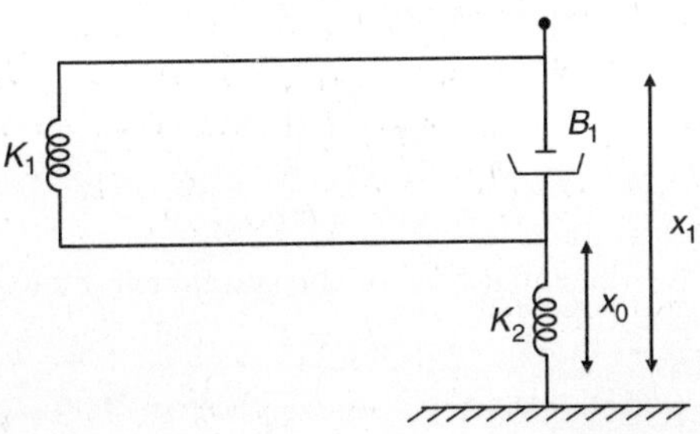

Solution:

By applying Newton's second law, the equation for the system can be written as

$$B_1\left(\frac{dx_1}{dt}-\frac{dx_0}{dt}\right)+K_1(x_1-x_0)=K_2x_0$$

By taking Laplace transform on both sides under the assumption that all initial conditions are zero, we get

$$\Rightarrow \quad (sB_1+K_1)X_1(s)-(sB_1+K_1)X_0(s)=K_2X_0(s)$$

$$\Rightarrow \quad X_0(s)[sB_1+K_1+K_2]=[sB_1+K_1]X_1(s)$$

$$\Rightarrow \quad \frac{X_0(s)}{X_1(s)}=\frac{sB_1+K_1}{sB_1+K_1+K_2}.$$

Hence, the transfer function relating the displacement X_0 and X_1 is given as

$$\frac{X_0(s)}{X_1(s)}=\frac{sB_1+K_1}{sB_1+K_1+K_2}$$

Problem-3.11: Find the transfer function $\frac{\theta(s)}{T(s)}$ for the following system.

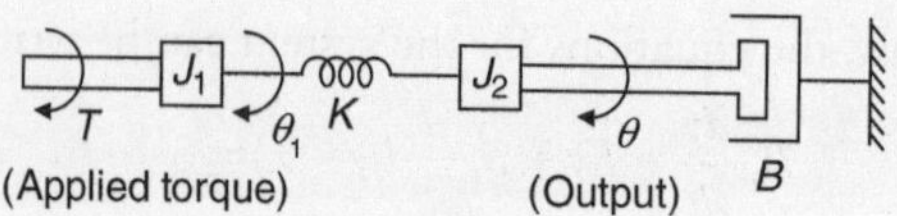

Solution:

The circuit diagram for the abovementioned network is given as

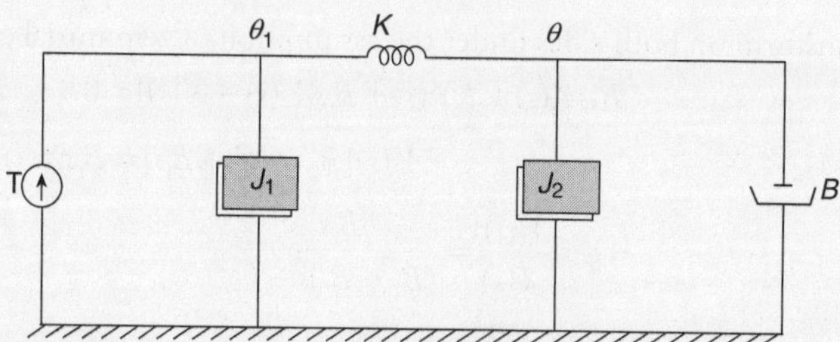

Here, applied torque T is the input and angular displacement θ is the output. The system has two nodes and they are masses with moment of inertia J_1 and J_2. At these nodes, the differential equations governing the system are given by torque balance equations.

Let the opposing torques acting on J_1 be marked as T_{j1} and T_k.

$$T_{j1}=J_1\frac{d^2\theta_1}{dt^2}$$

$$T_k=K(\theta_1-\theta)$$

By applying Newton's second law, the equations for the system can be written as

$$T_{j1}+T_k=T$$

$$J_1 \frac{d^2\theta_1}{dt^2} + K(\theta_1 - \theta) = T \qquad (1)$$

By applying Laplace transform on both the sides assuming that all initial conditions are zero, we get

$$J_1 s^2 \theta_1(s) + K\theta_1(s) - K\theta(s) = T(s)$$

$$(J_1 s^2 + K)\theta_1(s) - K\theta(s) = T(s). \qquad (2)$$

Let the opposing torques acting on J_2 be T_{j2}, T_b and T_k.

$$T_{j2} = J_2 \frac{d^2\theta}{dt^2}$$

$$T_b = B\frac{d\theta}{dt}$$

$$T_k = K(\theta - \theta_1) \cdot$$

Using Newton's second law,

$$T_{j2} + T_b + T_k = 0$$

$$J_2 \frac{d^2\theta}{dt^2} + B\frac{d\theta}{dt} + K(\theta - \theta_1) = 0. \qquad (3)$$

By applying Laplace transform on both sides assuming that all initial conditions are zero, we get

$$J_2 s^2 \theta(s) + Bs\theta(s) + K\theta(s) - K\theta_1(s) = 0$$

$$(J_2 s^2 + Bs + K)\theta(s) - K\theta_1(s) = 0$$

$$\theta_1(s) = \frac{J_2 s^2 + Bs + K}{K}\theta(s). \qquad (4)$$

From (2) and (4), we get

$$(J_1 s^2 + K)\frac{J_2 s^2 + Bs + K}{K}\theta(s) - K\theta(s) = T(s)$$

$$\left[\frac{(J_1 s^2 + K)(J_2 s^2 + Bs + K) - K^2}{K}\right]\theta(s) = T(s)$$

$$\therefore \quad \frac{\theta(s)}{T(s)} = \frac{K}{(J_1 s^2 + K)(J_2 s^2 + Bs + K) - K^2}.$$

Hence, the final transfer function is

$$\frac{\theta(s)}{T(s)} = \frac{K}{(J_1 s^2 + K)(J_2 s^2 + Bs + K) - K^2}.$$

Problem-3.12: Derive the transfer function for the following rotational mechanical systems:

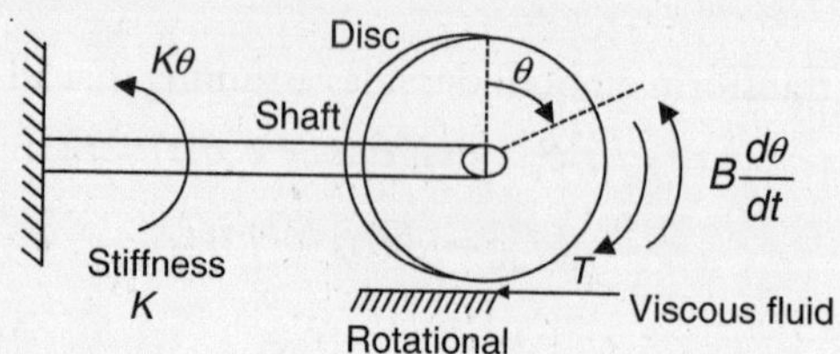

Solution:

Let T be the applied torque that tends to rotate the disc.

The system equation can be written using the relation:

Masses of inertia × Acceleration = Forces acting on mass.

The equivalent circuit diagram of the mechanical system is shown in the following figure:

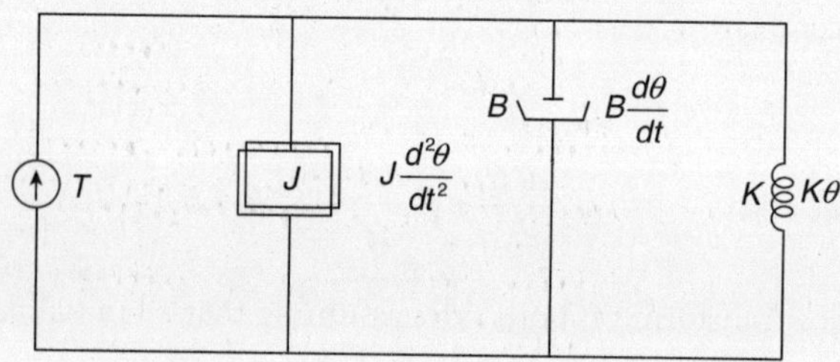

By applying Newton's second law, the equation for the system can be written as

$$J\frac{d^2\theta}{dt^2} = -B\frac{d\theta}{dt} - K\theta + T$$

or

$$T = J\frac{d^2\theta}{dt^2} + B\frac{d\theta}{dt} + K\theta.$$

By applying Laplace transform on both sides assuming zero initial conditions, we get

$$T(s) = Js^2\theta(s) + Bs\theta(s) + K\theta(s)$$

or

$$T(s) = (Js^2 + Bs + K)\theta(s)$$

or

$$\frac{\theta(s)}{T(s)} = \frac{1}{(Js^2 + Bs + K)}.$$

Hence, the final transfer function is

$$\frac{\theta(s)}{T(s)} = \frac{1}{(Js^2 + Bs + K)}$$

Problem-3.13: Determine the system equations for the given system as shown in the figure. Draw *F–V* and *F–I* analogous circuits.

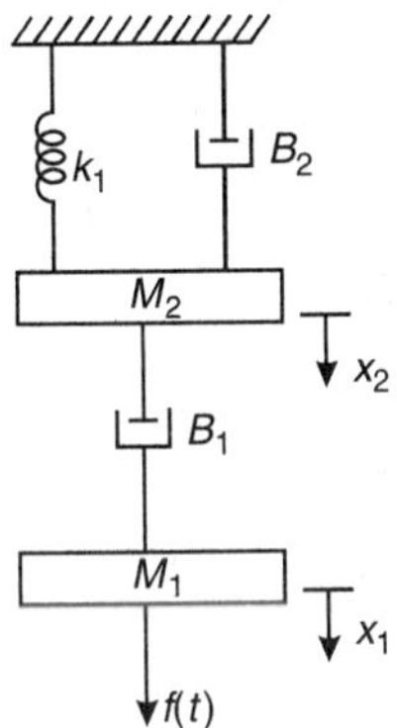

Solution:

The equivalent network diagram of the given mechanical system is

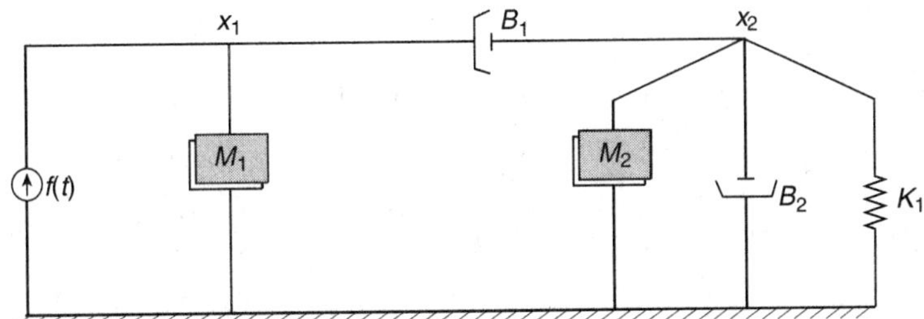

By applying force balance equation at node 1, we get

$$f(t) = M_1 \frac{d^2x_1}{dt^2} + B_1 \frac{d}{dt}(x_1 - x_2) \tag{1}$$

By applying force balance equation at node 2, we get

$$0 = M_2 \frac{d^2x_2}{dt^2} + B_2 \frac{dx_2}{dt} + K_1x_2 + B_1 \frac{d}{dt}(x_2 - x_1) \tag{2}$$

***F–V* analogous circuit:**

$$f \to v, M \to L, B \to R, K \to 1/C, x \to q$$

From (1), we have

$$v = L_1 \frac{d^2q_1}{dt^2} + R_1 \frac{d}{dt}(q_1 - q_2)$$

$$v = L_1 \frac{di_1}{dt} + R_1(i_1 - i_2) \tag{3}$$

From (2), we have

$$0 = L_2 \frac{d^2q_2}{dt^2} + R_2 \frac{dq_2}{dt} + \frac{1}{C_1} q_2 + R_1 \frac{d}{dt}(q_2 - q_1)$$

$$0 = L_2 \frac{di_2}{dt} + R_2 i_2 + \frac{1}{C_1} \int i_2 dt + R_1(i_2 - i_1) \tag{4}$$

The $F–V$ analogous circuit by using (3) and (4) can be drawn as shown in the following figure.

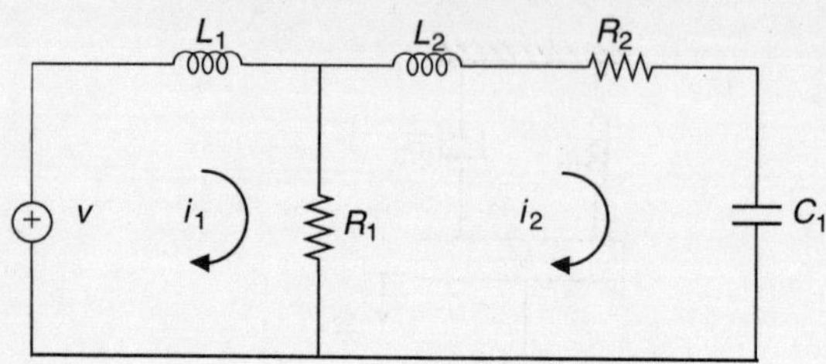

***F–I* analogous circuit:**

$$f \to i, M \to C, B \to 1/R, K \to 1/L, x \to \psi$$

From (1), we have

$$i = C_1 \frac{d^2\psi_1}{dt^2} + \frac{1}{R_1}\frac{d}{dt}(\psi_1 - \psi_2)$$

$$i = C_1 \frac{dv_1}{dt} + \frac{1}{R_1}(v_1 - v_2) \tag{5}$$

From (2), we have

$$0 = C_2 \frac{d^2\psi_2}{dt^2} + \frac{1}{R_2}\frac{d\psi_2}{dt} + \frac{1}{L_1}q_2 + \frac{1}{R_1}\frac{d}{dt}(\psi_2 - \psi_1)$$

$$0 = C_2 \frac{dv_2}{dt} + \frac{v_2}{R_2} + \frac{1}{L_1}\int v_2 dt + R_1(v_2 - v_1). \tag{6}$$

The $F–I$ analogous circuit by using (5) and (6) can be drawn as shown in the following figure.

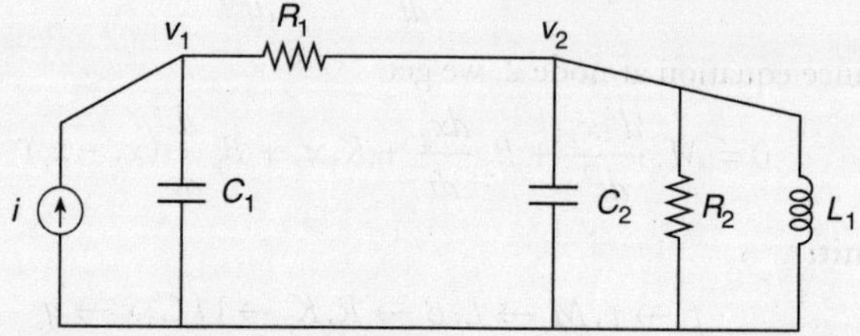

ADDITIONAL SOLVED PROBLEMS

ASP-1: Determine the transfer function $\frac{X_1(s)}{F(s)}$ and $\frac{X_2(s)}{F(s)}$ for the following system:

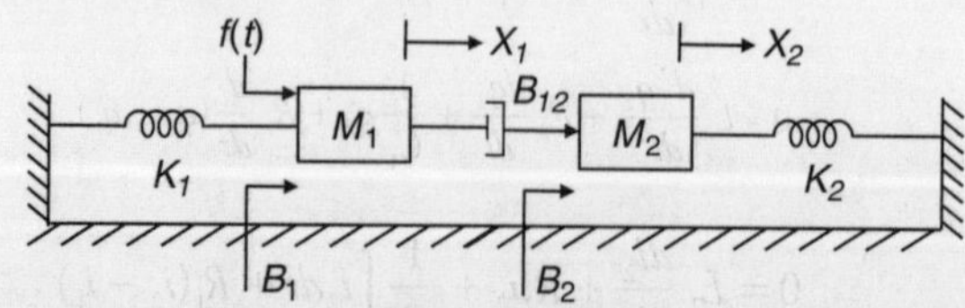

Solution:

The system has two nodes; these nodes are masses M_1 and M_2. The differential equations governing the system are the force balance equations at these nodes.

$$f_{m1} = M_1 \frac{d^2 x_1}{dt^2}; f_{b1} = B_1 \frac{dx_1}{dt}; f_{b12} = B_{12} \frac{d(x_1 - x_2)}{dt}; f_{k1} = K_1 x_1; f_{m2} = M_2 \frac{d^2 x_2}{dt^2}$$

$$f_{b2} = B_2 \frac{dx_2}{dt}; f_{k2} = K_2 x_2$$

By applying force balance equation at mass M_1, we get

$$f_{m1} + f_{b1} + f_{b12} + f_{k1} = f(t)$$

$$M_1 \frac{d^2 x_1}{dt^2} + B_1 \frac{dx_1}{dt} + B_{12} \frac{d(x_1 - x_2)}{dt} + K_1 x_1 = f(t)$$

On taking Laplace transform on both sides under the assumption of zero initial conditions, we get

$$M_1 s^2 X_1(s) + B_1 s X_1(s) + B_{12} s\,[X_1(s) - X_2(s)] + K_1 X_1(s) = F(s)$$

$$\Rightarrow \quad X_1(s)[M_1 s^2 + (B_1 + B_{12})s + K_1] - sB_{12} X_2(s) = F(s). \quad (1)$$

By applying force balance equation at mass M_2, we get

$$f_{m2} + f_{b2} + f_{b12} + f_{k2} = 0$$

$$\Rightarrow \quad M_2 \frac{d^2 X_2}{dt^2} + B_2 \frac{dx_2}{dt} \quad B_{12} \frac{d(x_2 - x_1)}{dt} + K_2 x_2 = 0.$$

On taking Laplace transform on both sides under the assumption of zero initial conditions, we get

$$M_2 s^2 X_2(s) + sB_2 X_2(s) + B_{12} s[X_2(s) - X_1(s)] + K_2 X_2(s) = 0$$

$$\Rightarrow \quad X_2(s)[M_2 s^2 + (B_2 + B_{12})s + K_2] = B_{12} s X_1(s) \quad (2)$$

$$\Rightarrow \quad X_2(s) = \frac{B_{12} s X_1(s)}{[M_2 s^2 + (B_2 + B_{12})s + K_2]}. \quad (3)$$

From (1) and (3), we will get

$$X_1(s)[M_1 s^2 + (B_1 + B_{12})s + K_1] - \frac{(B_{12} s)^2 X_1(s)}{[M_2 s^2 + B_2 + B_{12})s + K_2]} = F(s)$$

$$\Rightarrow \quad \frac{X_1(s)\{[M_1 s^2 + (B_1 + B_{12})s + K_1][M_2 s^2 + B_2 + B_{12})s + K_2] - (B_{12} s)^2\}}{M_2 s^2 + (B_2 + B_{12})s + K_2} = F(s)$$

$$\Rightarrow \quad \frac{X_1(s)}{F(s)} = \frac{M_2 s^2 + (B_2 + B_{12})s + K_2}{[M_1 s^2 + (B_2 + B_{12})s + K_1][M_2 s^2 + (B_2 + B_{12})s + K_2] - (B_{12} s)^2)}.$$

From (2), we will get

$$X_1(s) = \frac{[M_2 s^2 + (B_2 + B_{12})s + K_2] X_2(s)}{B_{12} s}. \quad (4)$$

From (1) and (4), we will get

$$X_2(s)\times\left[\frac{[M_2s^2+(B_2+B_{12})s+K_2][M_1s^2+(B_2+B_{12})s+K_1]-(B_{12}s)^2}{B_{12}s}\right]=F(s)$$

$$\Rightarrow \qquad \frac{X_2(s)}{F(s)}=\frac{B_{12}s}{[M_2s^2+(B_2+B_{12})s+K_2][M_1s^2+(B_1+B_{12})s+K_1]-(B_{12}s)^2}.$$

Hence, finally, the transfer functions $\frac{X_1(s)}{F(s)}$ and $\frac{X_2(s)}{F(s)}$ are as follows:

$$\frac{X_1(s)}{F(s)}=\frac{M_2s^2+(B_2+B_{12})s+K_2}{[M_1s^2+(B_1+B_{12})s+K_1][M_2s^2+(B_2+B_{12})s+K_2]-(B_{12}s)^2}$$

$$\frac{X_2(s)}{F(s)}=\frac{B_{12}s}{[M_2s^2+(B_2+B_{12})s+K_2][M_1s^2+(B_1+B_{12})s+K_1]-(B_{12}s)^2}.$$

ASP-2: Find the transfer function of the following system.

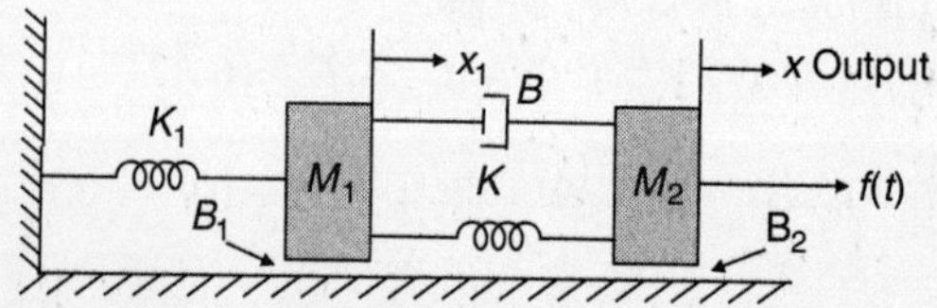

Solution:

The equivalent network diagram of the given mechanical system may be shown as follows:

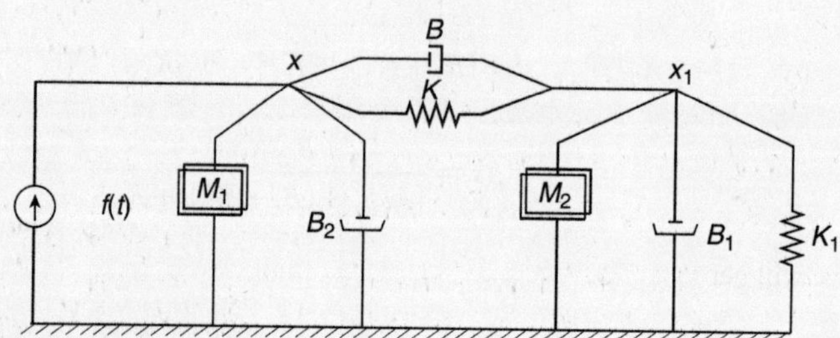

Writing the nodal equation (i.e., for balance equation) at each node, we get the following equations.
At node x,

$$M_2\ddot{x}+B_2\dot{x}+B\ \dot{x}-\dot{x}_1\ +K(x-x_1)=f(t).$$

At node x_1,

$$M_1\ddot{x}_1+B_1\dot{x}_1+K_1x_1=B(\dot{x}-\dot{x}_1)+K(x-x_1).$$

By taking the Laplace transform of abovementioned two equations under the assumption of zero initial conditions, we get

$$(M_2s^2+B_2s+Bs+K)X(s)-(Bs+K)X_1(s)=F(s)$$

and

$$(M_1s^2+B_1s+Bs+K+K_1)X_1(s)=(Bs+K)X(s).$$

By solving above-mentioned two equations, we will get

$$(M_2s^2 + B_2s + Bs + K)X(s) - \frac{(Bs+K)(Bs+K)X(s)}{(M_1s^2 + B_1s + Bs + K + K_1)} = F(s)$$

$$\Rightarrow \quad [(M_2s^2 + B_2s + Bs + K)(M_1s^2 + B_1s + Bs + K + K_1) - (Bs + K)^2]X(s)$$
$$= (M_1s^2 + B_1s + Bs + K + K_1)F(s)\cdot$$

Hence, the desired transfer function $X(s)/F(s)$ is given as

$$\frac{X(s)}{F(s)} = \frac{(M_1s^2 + B_1s + Bs + K + K_1)}{(M_2s^2 + B_2s + Bs + K)(M_1s^2 + B_1s + Bs + K + K_1) - (Bs + K)^2}.$$

ASP-3: Derive the transfer function for the following network.

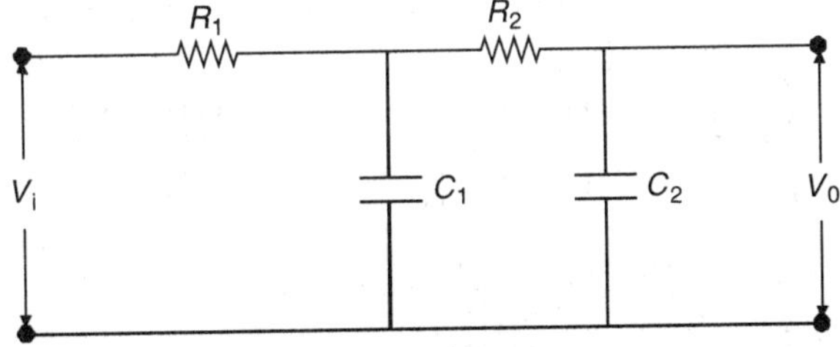

Solution:
The given network is

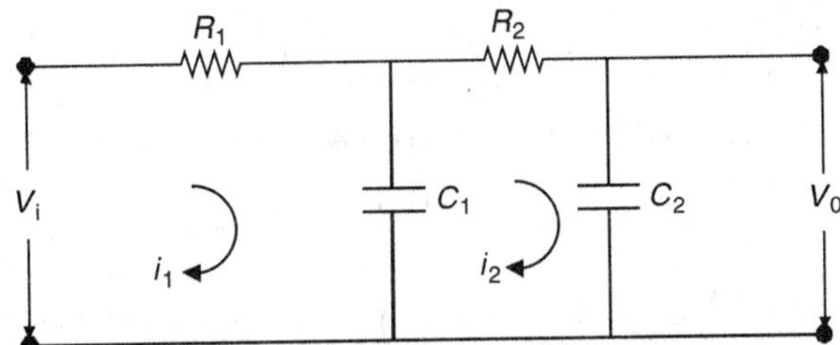

By applying KVL to Loop 1, we get

$$V_i(t) = R_1i_1(t) + \frac{1}{C_1}\int[i_1(t) - i_2(t)]dt.$$

By taking Laplace transform on both sides under the assumption of zero initial conditions, we get

$$V_i(s) = \left(R_1 + \frac{1}{sC_1}\right)I_1(s) - \frac{1}{sC_1}I_2(s). \quad (1)$$

By applying *KVL* to Loop 2, we obtain

$$R_2i_2(t) + \frac{1}{C_1}\int i_2(t)dt + \frac{1}{C_2}\int i_2(t)dt - \frac{1}{C_1}\int i_1(t)dt = 0.$$

By taking Laplace transform on both sides under the assumption of zero initial conditions, we get

$$\left(R_2 + \frac{1}{sC_1} + \frac{1}{sC_2}\right)I_2(s) - \frac{1}{sC_1}I_1(s) = 0$$

$$\Rightarrow \qquad \frac{1}{sC_1} I_1(s) = \left(R_2 + \frac{1}{sC_1} + \frac{1}{sC_2} \right) I_2(s)$$

$$\Rightarrow \qquad I_1(s) = sC_1 \left(R_2 + \frac{1}{sC_1} + \frac{1}{sC_2} \right) I_2(s). \tag{2}$$

Using (1) and (2), we obtain

$$V_i(s) = \left(R_1 + \frac{1}{sC_1} \right) sC_1 \left(R_2 + \frac{1}{sC_1} + \frac{1}{sC_2} \right) I_2(s) - \frac{1}{sC_1} I_2(s)$$

$$\Rightarrow \qquad V_i(s) = \left[\frac{(sC_1R_1+1)(sR_2C_1C_2+C_1+C_2)-C_2}{sC_1C_2} \right] I_2(s). \tag{3}$$

Again, $\quad V_0(s) = \dfrac{I_2(s)}{sC_2}$. $\hfill (4)$

Substituting (4) in (3), we get

$$V_i(s) = \left[\frac{(sC_1R_1+1)(sR_2C_1C_2+C_1+C_2)-C_2}{C_1} \right] V_0(s)$$

$$\frac{V_o(s)}{V_i(s)} = \left[\frac{C_1}{(sC_1R_1+1)(sR_2C_1C_2+C_1+C_2)-C_2} \right]$$

$$= \frac{C_1}{s^2C_1^{\,2}R_1R_2C_2 + sC_1^{\,2}R_1 + sR_1C_1C_2 + sR_2C_1C_2 + C_1 + C_2 - C_2}.$$

$$= \frac{1}{s^2C_1R_1R_2C_2 + sC_1\,R_1 + sR_1C_2 + \,sR_2C_2 + 1}$$

$$= \frac{1}{s^2C_1R_1R_2C_2 + s(R_1C_1 + R_1C_2 + R_2C_2) + sR_2C_2 + 1}$$

Hence, the transfer function for the given network is given as

$$\frac{V_o(s)}{V_i(s)} = \frac{1}{s^2C_1R_1R_2C_2 + s(R_1C_1 + R_1C_2 + R_2C_2) + 1}.$$

ASP-4: Obtain the transfer function for the following system and draw its analogous electrical circuit.

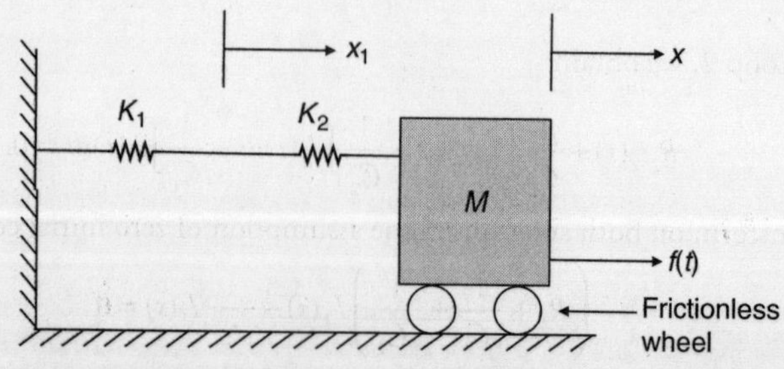

Solution:
The mechanical network for the given system is shown in the following figure.

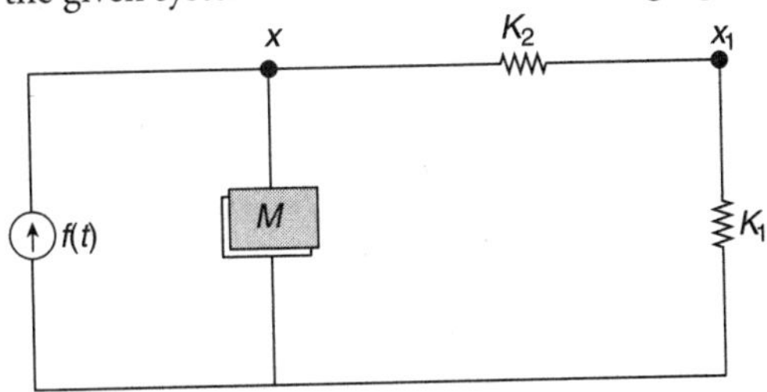

The force balance equation at node x is

$$M\ddot{x} + K_2(x - x_1) = f(t). \tag{1}$$

The force balance equation at node x_1 is

$$K_1x_1 = K_2(x - x_1) \tag{2}$$

Substituting the value of x_1 from (2) in (1) gives

$$M\ddot{x} + K_2x - \frac{K_2^2}{K_1 + K_2}x = f(t)$$

$$\Rightarrow \qquad M\ddot{x} + \frac{K_2K_1}{K_1 + K_2}x = f(t).$$

By taking the Laplace transform of abovementioned two equations under the assumption of zero initial conditions, we get

$$\left[Ms^2 + \frac{K_2K_1}{K_1 + K_2}\right]X(s) = F(s)$$

$$\Rightarrow \qquad \frac{X(s)}{F(s)} = \left[\frac{1}{Ms^2 + \frac{K_2K_1}{K_1 + K_2}}\right].$$

Hence, the desired transfer function $X(s)/F(s)$ is given as

$$\frac{X(s)}{F(s)} = \left[\frac{1}{Ms^2 + \frac{K_2K_1}{K_1 + K_2}}\right].$$

The analogous electrical circuit of the given circuit is

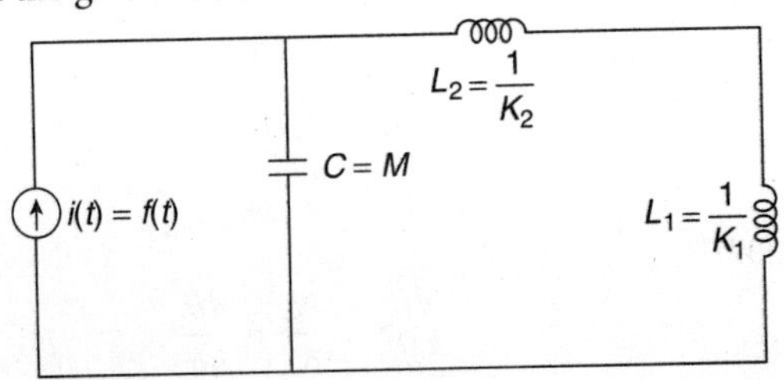

ASP-5: A torque T_1 Nm is applied to a load through a gear having a gear ratio N_2/N_1, as shown in the figure. If the moment of inertia and coefficient of viscous friction of the load is J_2 kgm^2 and f_2 Nm/(rad/s), respectively, determine the transfer function relating the applied torque T_1 and the angular shift 1 at the primary shaft side.

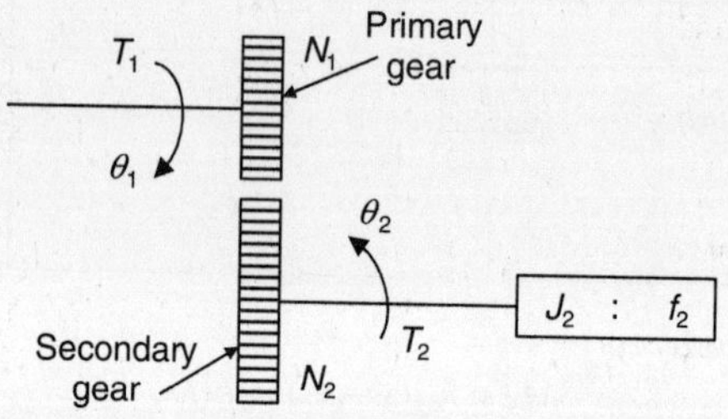

Solution:
Given that

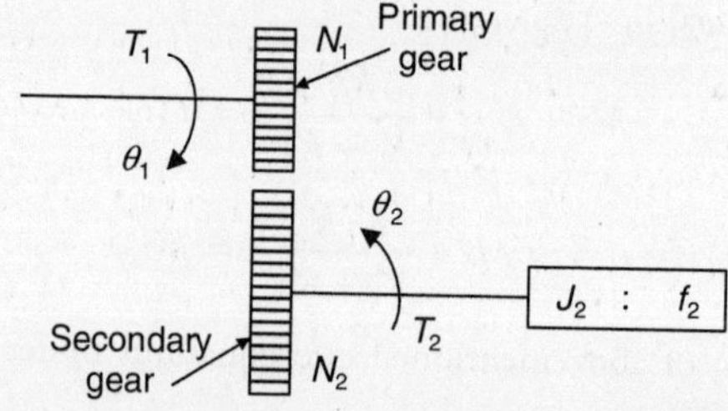

The effect of applied torque T_1 through the gear results in application of torque T_2 to the load. The torques T_1 and T_2 are related as follows:

$$\frac{T_2}{T_1} = \frac{N_1}{N_2}$$

$$T_2 = \frac{N_1}{N_2} T_1. \tag{1}$$

The equation relating the torque T_2 and the angular shift θ_2 is formulated as

$$T_2 = J_2 \frac{d^2\theta_2}{dt^2} + f_2 \frac{d\theta_2}{dt}. \tag{2}$$

From (1) and (2)

$$\frac{N_2}{N_1} T_1 = J_2 \frac{d^2\theta_2}{dt^2} + f_2 \frac{d\theta_2}{dt}. \tag{3}$$

Again, since $N_1\theta_1 = N_2\theta_2$

$\Rightarrow$
$$\theta_2 = \frac{N_1}{N_2}\theta_1. \tag{4}$$

Differentiating (4) wrt t gives

$$\frac{d\theta_2}{dt} = \frac{N_1}{N_2} \times \frac{d\theta_1}{dt}. \tag{5}$$

Further, differentiating (5)

$$\frac{d^2\theta_2}{dt^2} = \frac{N_1}{N_2}\frac{d^2\theta_1}{dt^2}. \tag{6}$$

From (3), (4), (5) and (6), we will get

$$\frac{N_2}{N_1}T_1 = J_2\frac{N_1}{N_2}\frac{d^2\theta_1}{dt^2} + f_2\frac{N_1}{N_2}\frac{d\theta_1}{dt}. \tag{7}$$

Simplifying (7), we will get

$$T_1 = J_2\left[\frac{N_1}{N_2}\right]^2\frac{d^2\theta_1}{dt^2} + f_2\left[\frac{N_1}{N_2}\right]^2\frac{d\theta_1}{dt}. \tag{8}$$

By taking Laplace transform on both sides and zero initial conditions, we get

$$T_1(s) = J_2\left[\frac{N_1}{N_2}\right]^2 s^2\,\theta_1(s) + f_2\left[\frac{N_1}{N_2}\right]^2 s\theta_1(s). \tag{9}$$

Finally, the required transfer function is

$$\frac{\theta_1(s)}{T_1(s)} = \frac{1}{s\left[J_2\left(\frac{N_1}{N_2}\right)^2 s + f_2\left(\frac{N_1}{N_2}\right)^2\right]}.$$

ASP-6: The transfer function of the system is $G(s) = \frac{4(s+1)(s+2)}{s(s+3)(s^2+2s+5)}$. Determine the poles and zeros and show the pole–zero configuration in the s-plane.

Solution:

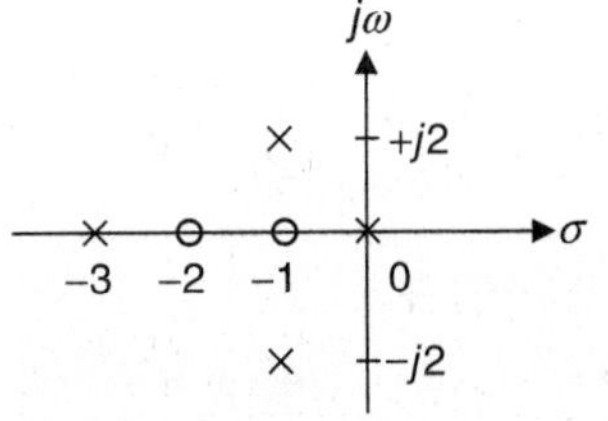

jω
+j2
σ
−3 −2 −1 0
−j2

The given transfer function is:

$$G(s) = \frac{4(s+1)(s+2)}{s(s+3)(s^2+2s+5)}.$$

The poles are given as

$$s(s+3)(s^2+2s+5) = 0$$

$$s = 0, s = -3, s^2 + 2s + 5 = 0$$

$$s = \frac{-2 \pm \sqrt{4-20}}{2}$$

$$= -1 \pm j2$$

∴ Poles are $P_1 = 0, P_2 = -3, P_3 = -1 + j2, P_4 = -1 - j2.$

The zeros are given as

$$(s+1)(s+2)=0$$

$$s=-1, s=-2$$

Zeros are $z_1 = -1, Z_2 = -2$

The pole–zero configuration is as shown in the figure.

ASP-7: The pole–zero configuration of a transfer function is shown in the following figure. The value of the transfer function at $s = 1$ is found to be 6.4. Determine the transfer function.

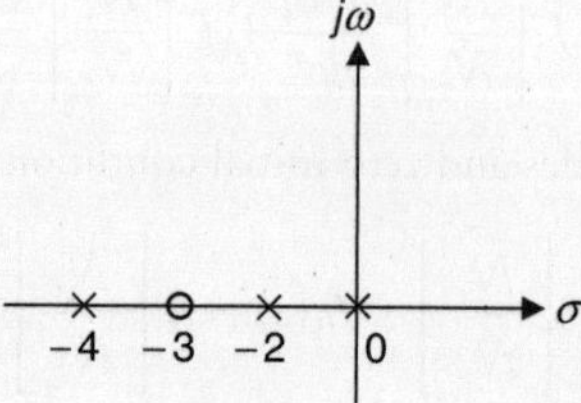

Solution:

The poles of the system are

$$p_1 = 0, p_2 = -2, p_3 = -4$$

The zeros of the system are $z_1 = -3$.

The transfer function of the system can be written as

$$G(s) = \frac{k(s+3)}{s(s+2)(s+4)}.$$

However, as per the given data at $s = 1$, $G(s) = 6.4$

$$\Rightarrow \qquad G(1) = 6.4 = \frac{k(1+3)}{1(1+2)(1+4)} = \frac{k \times 4}{15}$$

$$\Rightarrow \qquad k = 6.4 \times 15/4 = 24.$$

∴ The transfer function of the system is given as

$$G(s) = \frac{24(s+3)}{s(s+2)(s+4)}.$$

ASP-8: Find the transfer function of the network, as shown in the figure. For $R = C = 1$ and draw its pole–zero plot on the s-plane

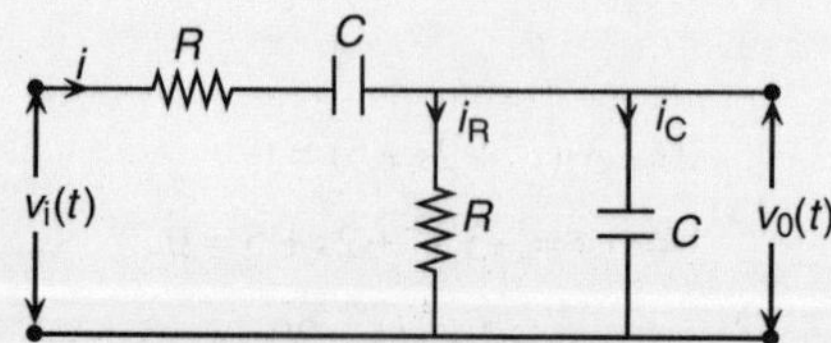

Solution:

The s-domain equivalent circuit of the above figure is

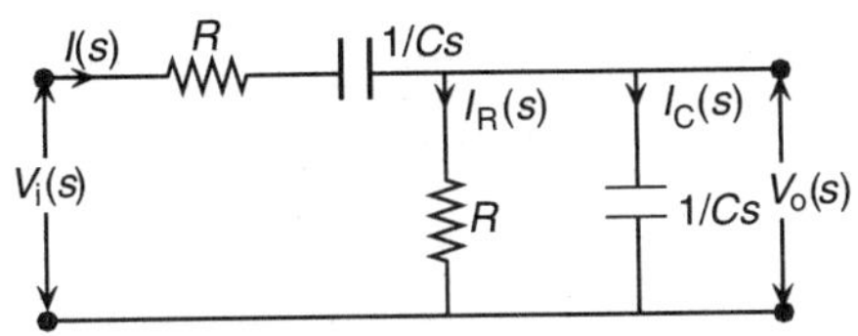

From the equivalent circuit,

$$I(s) = I_R(s) + I_C(s). \tag{1}$$

By applying KVL to Loop (1),

$$V_i(s) = \left(R + \frac{1}{Cs}\right)I(s) + V_0(s). \tag{2}$$

We know that
$$V_0(s) = I_R(s)R = \frac{1}{Cs}I_C(s).$$

⇒
$$I_R(s) = \frac{V_0(s)}{R}, I_C(s) = CsV_0(s) \tag{3}$$

From (1) and (3)

$$I(s) = \frac{V_0(s)}{R} + CsV_0(s)$$
$$= \left(\frac{1}{R} + Cs\right)V_0(s)$$
$$= \left(\frac{1 + RCs}{R}\right)V_0(s). \tag{4}$$

From (2) and (4),

$$V_i(s) = \left(\frac{RCs+1}{Cs}\right)\left(\frac{RCs+1}{R}\right)V_0(s) + V_0(s)$$
$$= \left[\frac{(1+RCs)^2 + RCs}{RCs}\right]V_0(s)$$

⇒
$$\frac{V_0(s)}{V_i(s)} = \frac{RCs}{(1+RCs)^2 + RCs}.$$

Substitute $R = C = 1$

$$G(S) = \frac{V_0(s)}{V_i(s)} = \frac{s}{(1+s)^2 + s} = \frac{s}{s^2 + 3s + 1}.$$

The poles are given as

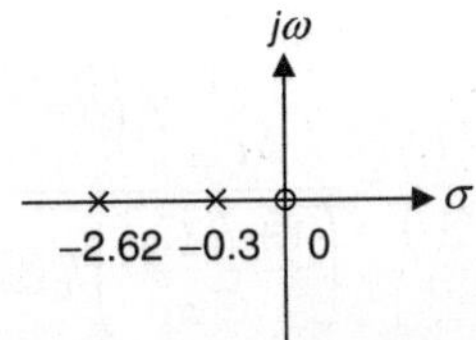

$$s^2+3s+1=0$$

$$s=\frac{-3\pm\sqrt{9-4}}{2}=\frac{-3}{2}\pm\frac{\sqrt{5}}{2}=-0.3,-2.62$$

$\Rightarrow$ $$p_1=-0.3,\ p_2=-2.62\ .$$

The zeros are given as

$$s=0$$

$\Rightarrow$ $$z_1=0.$$

ASP-9: Derive the transfer function of the system, as shown in the figure, with K being the amplifier gain.

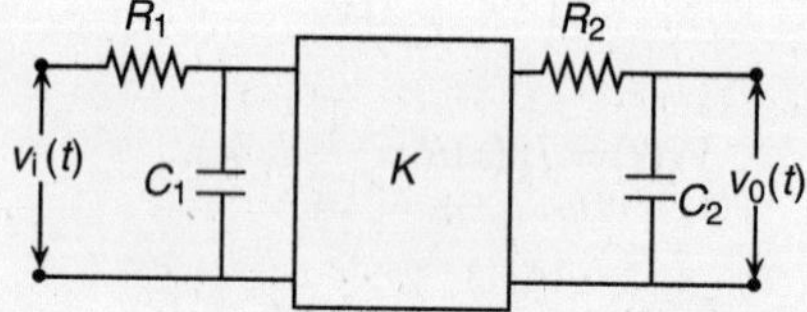

Solution:

The s-domain equivalent circuit for the abovementioned figure is shown in the following figure.

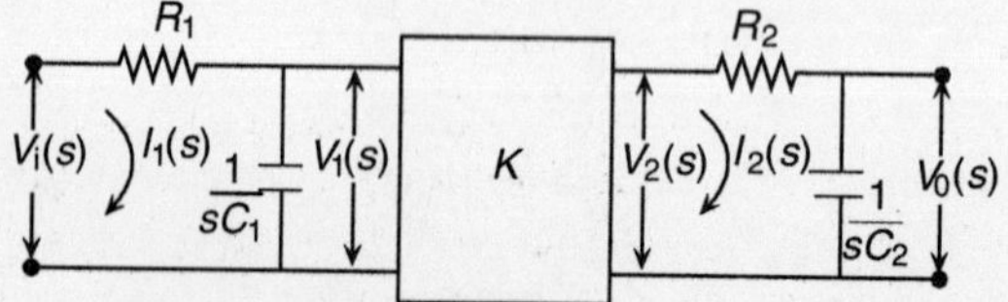

From the above figure,

$$V_i(s)=\left(R_1+\frac{1}{C_1s}\right)I_1(s)$$

$$=\left(\frac{1+R_1C_1s}{C_1s}\right)I_1(s)$$

$$V_1(s)=\frac{1}{C_1s}I_1(s)$$

$$\frac{V_2(s)}{V_1(s)}=K\Rightarrow V_2(s)=KV_1(s)$$

$$V_2(s)=\left(R_2+\frac{1}{C_2s}\right)I_2(s)=\left(\frac{1+R_2C_2s}{C_2s}\right)I_2(s)$$

$$V_0(s)=\frac{1}{C_2s}I_2(s)$$

$\therefore$ $$\frac{V_0(s)}{V_i(s)}=\frac{\frac{1}{C_2s}I_2(s)}{\left(\frac{1+R_2C_2s}{C_2s}\right)I_2(s)}$$

$$\frac{V_0(s)}{V_i(s)} = \frac{C_1}{C_2} \times \frac{1}{1+R_1C_1s} \frac{I_2(s)}{I_1(s)} \tag{1}$$

$$\frac{V_2(s)}{V_1(s)} = \frac{\left(\dfrac{1+R_2C_2s}{C_2s}\right)I_2(s)}{\dfrac{1}{C_1s}I_1(s)}$$

$$\Rightarrow \qquad K = \frac{C_1}{C_2}(1+R_2C_2s)\frac{I_2(s)}{I_1(s)}$$

$$\frac{I_2(s)}{I_1(s)} = K \times \frac{C_2}{C_1} \times \frac{1}{1+R_2C_2s}. \tag{2}$$

From (1) and (2),

$$\frac{V_0(s)}{V_i(s)} = \frac{C_1}{C_2} \times \frac{1}{(1+R_1C_1s)} \times K \times \frac{C_2}{C_1} \times \frac{1}{(1+R_2C_2s)}.$$

$$= \frac{K}{(1+R_1C_1s)(1+R_2C_2s)}$$

∴ The required transfer function

$$G(s) = \frac{V_0(s)}{V_i(s)} = \frac{K}{(1+R_1C_1s)(1+R_2C_2s)}.$$

EXERCISE PROBLEMS

1. Obtain the transfer function for the electrical systems, as shown in the figures.

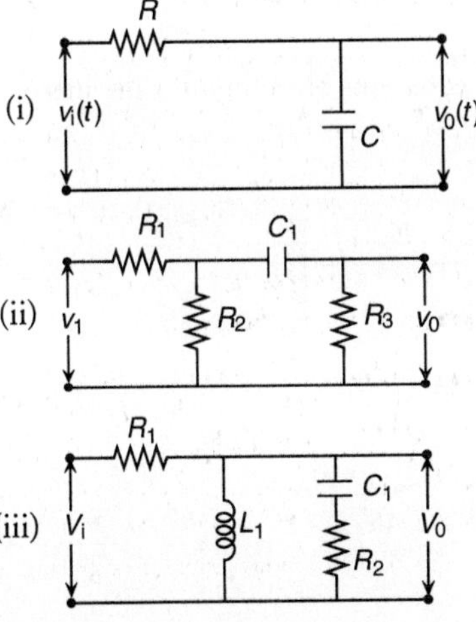

2. Determine the transfer function relating $V_0(s)$ to $V_i(s)$ for the network, as shown in the figure. Calculate the output voltage for a unit step input at $t = 0$.

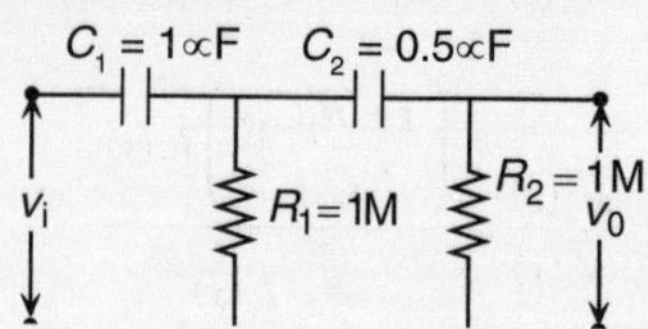

3. A system having input x and output y is represented by $\frac{d^3y}{dt^2}+4\frac{d^2y}{dt^2}+2\frac{dy}{dt}+y=\frac{d^2x}{dt^2}+\frac{dx}{dt}+2x$ .Find the transfer function of the system.

4. Draw the pole–zero plot for the transfer function $F(s)=\frac{s^2+3s+2}{s^2+7s+12}$ and find $f(t)$.

5. The transfer function of a system is given as $F(s)=\frac{10(s+8)}{s(s+4)(s^2+6s+25)}$. Obtain (i) poles, (ii) zeros, (iii) type and order (iv) pole–zero plot.

6. Obtain the transfer function for the system shown in the figure.

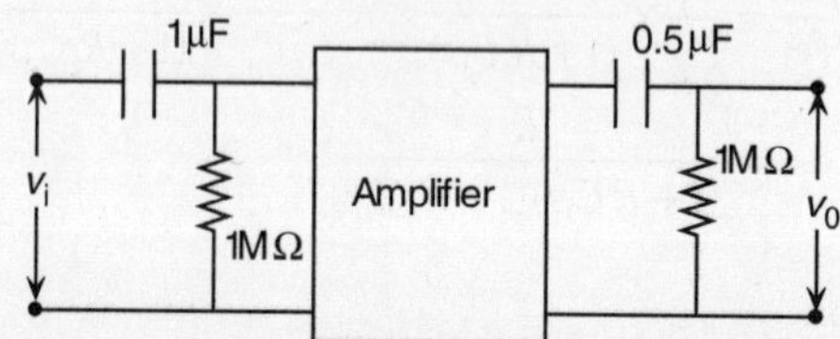

7. If the transfer function of the system is $\frac{Y(s)}{X(s)}=\frac{s+3}{s\ +2s+12}$, then obtain the differential equation representing the system.

8. For the mechanical system, as shown in the figure, find the transfer function 1 $\frac{X_1(s)}{F(s)}$.

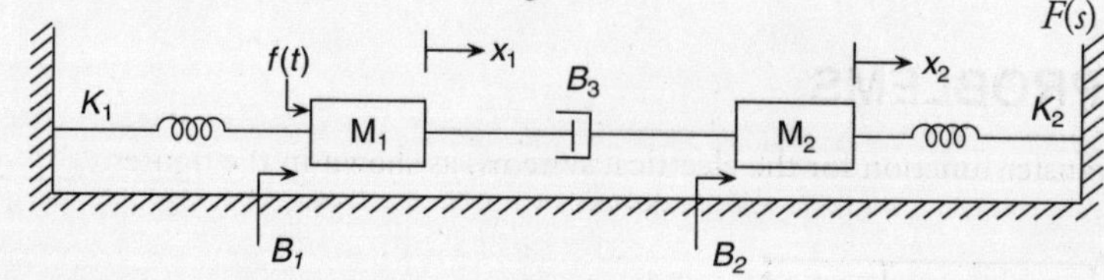

9. For the rotational system as shown in the figure, obtain the equations of motion for the inertial J_1 and J_2 find the transfer function $\frac{\theta_1(s)}{T(s)}$.

ANSWERS

1. (i) $T(s)=\frac{V_0(s)}{V_i(s)}=\frac{1}{1+RCs}$

(ii) $T(s)=\frac{V_0(s)}{V_i(s)}=\frac{sR_2R_3C}{(R_1+R_2)+Cs(R_1R_2+R_2R_3+R_3R_1)}$

(iii) $\frac{V_0(s)}{V_i(s)}=\frac{sL_1(1+sR_2C)}{s^2L_1C_1\ (R_1+R_2)sL_1+sR_1R_2C_1+R_1}$

2. $T(s) = \dfrac{V_0(s)}{V_i(s)} = \dfrac{s^2}{s^2+4s+2}$, output voltage = $1.2e^{-3.4t} - 0.21e^{-0.6t}$

3. $T(s) = \dfrac{Y(s)}{X(s)} = \dfrac{s^2+s+2}{s^3+4s^2+2s+1}$

4.

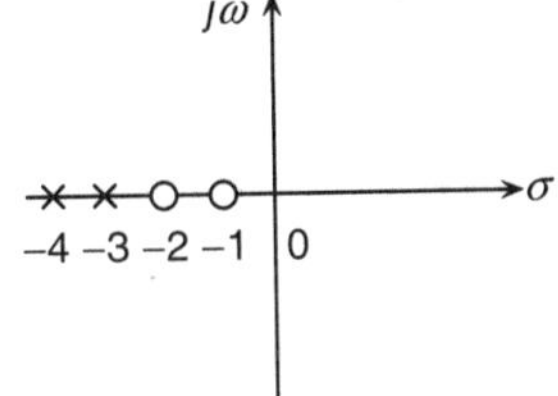

$f(t) = 2e^{-3t} - 6e^{-4t}$

5. (i) poles: 0, −4, $-3 \pm j4$ (ii) zero: −8 (iii) type = 1, order = 4

6. $T(s) = \dfrac{V_0(s)}{V_i(s)} = \dfrac{s^2}{(s+1)(s+2)}$

7. $\dfrac{d^2y}{dt^2} + \dfrac{2dy}{dt} + 12y = \dfrac{dx}{dt} + 3x$

8. $T(s) = \dfrac{X_1(s)}{F(s)} = \dfrac{M_2s^2 + B_2s + B_3s + K_2}{(M_1s^2 + B_1s + B_3s + K_1)(M_2s^2 + B_2s + B_3s + K_2) - B_3^2s^2}$

9. At node $\theta_1 = \dfrac{J_1d^2\theta_1}{dt} + K_1(\theta_1 - \theta_2) = T(t)$.

At node $\theta_2 = \dfrac{J_2d^2\theta_2}{dt^2} + K_2\theta_2 + K_1(\theta_1 - \theta_2) = 0$.

Transfer function = $\dfrac{\theta_1(s)}{T(s)} = \dfrac{J_1s^2 + K_1 + K_2}{(J_1s^2 + K_1)(J_2s^2 + K_1 + K_2) - K_1^2}$

OBJECTIVE TYPE QUESTIONS

1. The transfer function is defined as
   (a) the ratio of Laplace transform of output to Laplace transform of input considering initial conditions to be zero
   (b) the ratio of Laplace transform of input to Laplace transform of output considering initial condition as zero
   (c) the ratio of input to output
   (d) the ratio of output to input
2. The transfer function of a control system is defined as
   (a) Fourier transform of impulse response
   (b) Fourier transform of unit step response
   (c) Laplace transform of impulse response
   (d) Laplace transform of unit step response

3. In case of force–voltage analogy, the analogous quantity of mass is
(a) R (b) C (c) L (d) None

4. In the case of force–current analogy, the analogous quantity of friction dashpot is
(a) $1/R$ (b) $1/C$ (c) L (d) None

5. A system is represented by the differential equation $M.\frac{d^2x}{dt^2}+F.\frac{dx}{dt}+Kx=u(t)$. The transfer function relating output X(s) and input U(s) is ________

(a) $\frac{M}{Ms^2+Fs+K}$ (b) $\frac{K}{Ms^2+Fs+K}$

(c) $\frac{1}{Ms^2+Fs+K}$ (d) None

6. The transfer function relating output I(s) and input V(s) in a series RLC circuit is given as

(a) $\frac{Cs}{LCs^2+RCs+LC}$ (b) $\frac{Cs}{LCs^2+RCs+1}$

(c) $\frac{Cs}{s^2+LCs+R}$ (d) None

7. The Laplace transform of df/dt is given as
(a) $sF(s)-f(0)$ (b) $sF(s)+f(0)$ (c) $F(s)-f(0)$ (d) None

8. If a system has a transfer function $(1-s)/(1+s)$, then its gain at *1* rad/s is ________
(a) 0. (b) 1 (c) –1 (d) ½

9. The pole–zero configuration is given in the following figure.

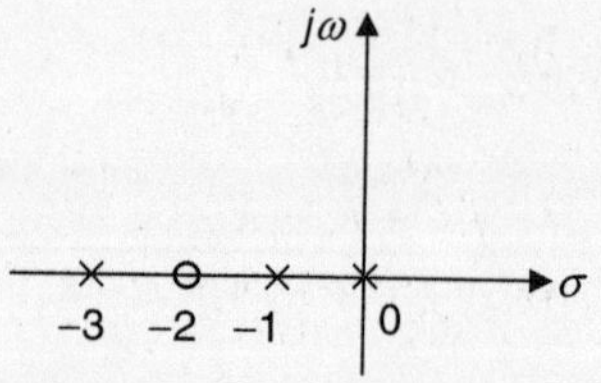

Gain factor, if the value of transfer function at $s = 1$ is 2.
(a) 2. (b) 4 (c) 6 (d) 5.33

10. The transfer function of the abovementioned system is ________

(a) $\frac{2(s+2)}{s(s+1)(s+3)}$ (b) $\frac{4(s+2)}{s(s+1)(s+3)}$ (c) $\frac{5.33(s+2)}{s(s+1)(s+3)}$ (d) $\frac{7.33(s+2)}{s(s+1)(s+3)}$

11. If the open-loop transfer function of a unity feedback control system is $G(s)=\frac{1}{(s+1)^2}$, then the poles of the closed-loop transfer function are
(a) –2, –2 (b) –2, 2 (c) 2,2 (d) $-1+j, -1-j$

12. Poles are the complex frequencies of a transfer function where the response becomes
(a) infinite (b) zero (c) oscillatory (d) decaying

13. Zeros are the complex frequencies of a transfer function where the response becomes
(a) infinite (b) zero (c) oscillatory (d) decaying

14. The transfer function of a system can be used to study the
(a) transient behaviour (b) steady state behaviour
(c) both (a) and (b) (d) none

15. The transfer function of the following network is

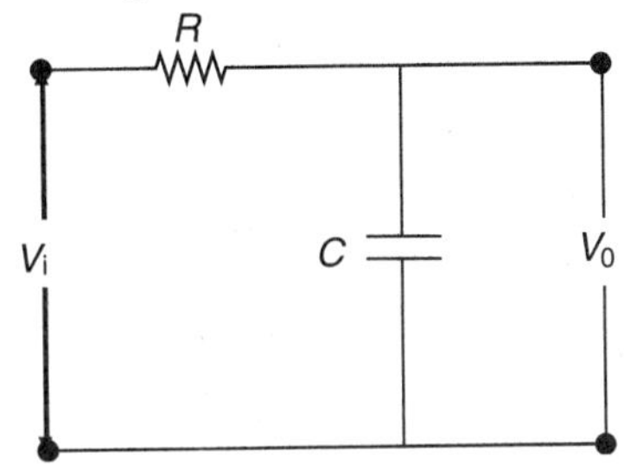

(a) $\frac{1}{1+sRC}$ (b) $\frac{sRC}{1+sRC}$ (c) $\frac{RC}{1+sRC}$ (d) none

16. The system transfer function is $G(s)=\frac{2(s+2)}{s(s+1)(s+3)}$. The type and order of the system are

(a) 2, 2 (b) 1, 2 (c) 1, 3 (d) 3, 3

17. If the unit step response of a system is $y(t) = 1 - e^{-4t}$, then the transfer function of the system is

(a) $\frac{s}{s+4}$ (b) $\frac{4s}{s+4}$ (c) $\frac{4}{s+4}$ (d) $\frac{s}{1+4s}$

18. The impulse response of a system having transfer function $G(s)=\frac{4}{s(s+1)(s+2)}$ is given as

(a) $2-4\bar{e}^{t}+2\,\bar{e}^{2t}$ (b) $2-2e^{-t}+e^{-2t}$ (c) $2-2e^{-t}+e^{-2t}$ (d) none

19. If the unit impulse response of a unity feedback system is given as $c(t)=1+e^{-2t}$, then the open-loop transfer function is

(a) $\frac{s}{s+2}$ (b) $\frac{2}{s+2}$ (c) $\frac{2(s+1)}{s(s+2)}$ (d) $\frac{s}{s(s+2)}$

20. If a system having input x and output y is represented by $\frac{d^3y}{dt^3}+2\frac{d^2y}{dt^2}+3\frac{dy}{dt}+y=\frac{dx}{dt}+2x$, then the transfer function of the system is

(a) $\frac{s+2}{s^3+2s^2+3s+1}$ (b) $\frac{s}{s^3+2s^2+3s+1}$ (c) $\frac{2}{s^3+2s^2+3s+1}$ (d) none

ANSWERS

1. a	**2.** c	**3.** c	**4.** a	**5.** c	**6.** b	**7.** a	**8.** b	**9.** d	**10.** c
11. d	**12.** a	**13.** b	**14.** c	**15.** a	**16.** c	**17.** c	**18.** a	**19.** c	**20.** a

CHAPTER

4 Block Diagrams

Contents

4.1 INTRODUCTION

When a given system is complicated, it is difficult to analyse it as a whole. Using the transfer function approach, we can find the transfer function of each and every element of the complicated system; further, by showing the connection between the elements, the complete system can be split into different blocks and can be analysed conveniently.

Basically, block diagram is a pictorial representation of the given system. It is a very simple way of representing the given complicated practical system. It explains the cause-and-effect relationship existing between the input and the output of the system through the blocks. To draw the block diagram of a practical system, each element of the practical system is represented by a block called functional block. In other words, block explains the mathematical operation on the input by the element to produce the corresponding output. The actual mathematical function is indicated by inserting corresponding transfer function of the element inside the block.

Block diagram does not include any information about the physical construction of the system. In general, source of energy is not shown in the block diagram. Therefore, different block diagrams can be drawn depending upon the point of view of analysis. Further, the block diagram for the given system is not unique.

4.2 BASIC ELEMENTS OF A BLOCK DIAGRAM

A block diagram has the following six basic elements associated with it.

1. **Blocks:** The shorthand pictorial representation of the cause-and-effect relationship between the input and the output of a physical system is known as functional block. Figure 4.1 shows a functional block.

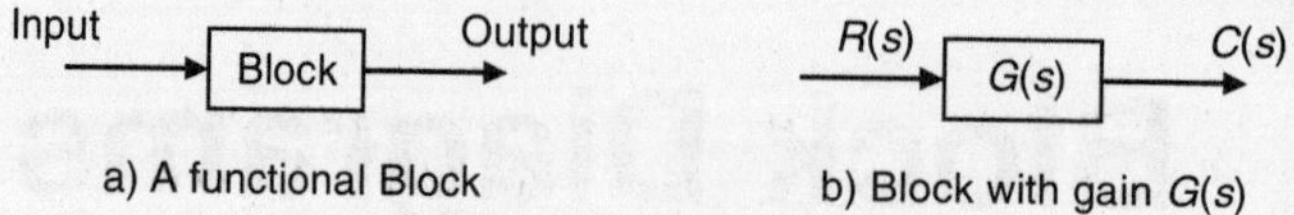

a) A functional Block   b) Block with gain $G(s)$

Figure 4.1 A simple block

2. **Summing point:** At summing point, two or more signals can be added or subtracted. Figure 4.2 represents a summing point. From the figure, we have

$$C(s) = X(s) + Y(s) - Z(s)$$

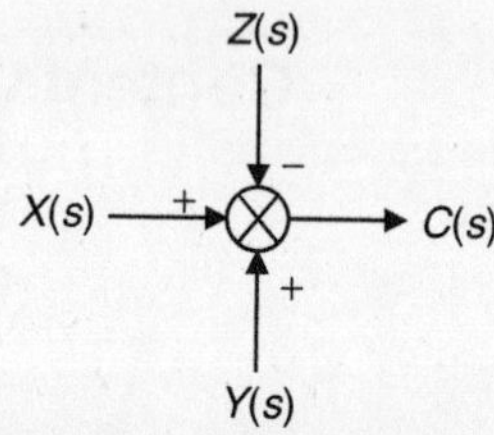

Figure 4.2 A summing point

3. **Take-off point or branch point:** Take-off point or branch point is the point from which the signal from a block goes concurrently to other blocks or summing points. Figure 4.3 represents a take-off point.

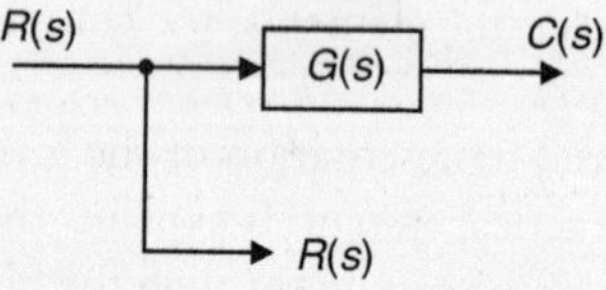

Figure 4.3 A take-off point

4. **Signal arrows:** The block diagram contains arrows indicating the unidirectional flow of signal in these diagrams.
5. **Forward path:** The direction of flow of signal from the input to the output is known as forward path. Figure 4.4 shows a forward path.

Input → $G_1(s)$ → $G_2(s)$ → $G_3(s)$ → Output

Figure 4.4 Forward path

6. **Feedback path:** The direction of flow of signal from the output to the input is known as feedback path. Figure 4.5 shows a feedback path.

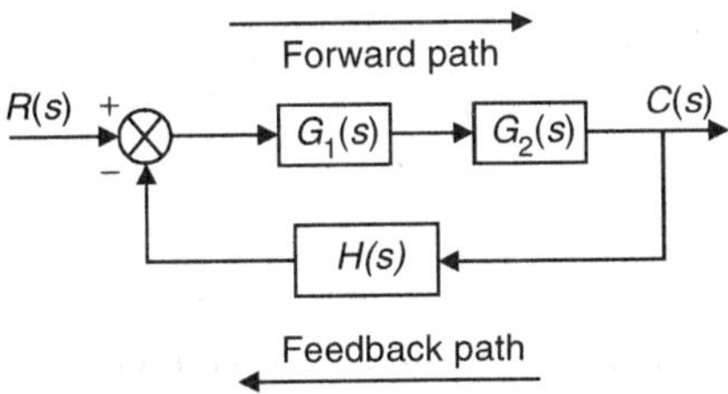

Figure 4.5 Feedback path

4.3 BLOCK DIAGRAM REDUCTION RULES

4.3.1 Rule 1: Blocks Connected in Cascade or Series

In cascade connections, blocks are connected in series as shown in Figure 4.6. The number of blocks connected are *n*, in which the input and the output of the system are $R(s)$ and $C(s)$, respectively.

$R(s) = R_1(s)$ $R_2(s)$ $R_3(s)$ $C(s)$
$G_1(s)$ $G_2(s)$ $G_3(s)$ ··· $G_n(s)$

Figure 4.6 Blocks connected in series

The overall transfer function of the system is given as

$$\frac{C(s)}{R(s)} = \frac{R_2(s)}{R_1(s)} \cdot \frac{R_3(s)}{R_2(s)} \cdots \frac{C(s)}{R_n(s)} = G_1(s).G_2(s).G_3(s) \cdots G_n(s)$$

4.3.2 Rule 2: Blocks Connected in Parallel

Figure 4.7 shows the blocks connected in parallel. Each of these blocks forms a parallel path in the system. The overall transfer function of the system is obtained by adding transfer functions of each block. Thus, from Figure 4.7, the overall transfer function is given as

$$C(s) = G_1(s)R(s) + G_2(s)R(s) + G_3(s)R(s) \cdots + G_n(s)R(s)$$

$$C(s) = [G_1(s) + G_2(s) + G_3(s) \cdots + G_n(s)]R(s)$$

$$\frac{C(s)}{R(s)} = G_1(s) + G_2(s) + G_3(s) \cdots + G_n(s)$$

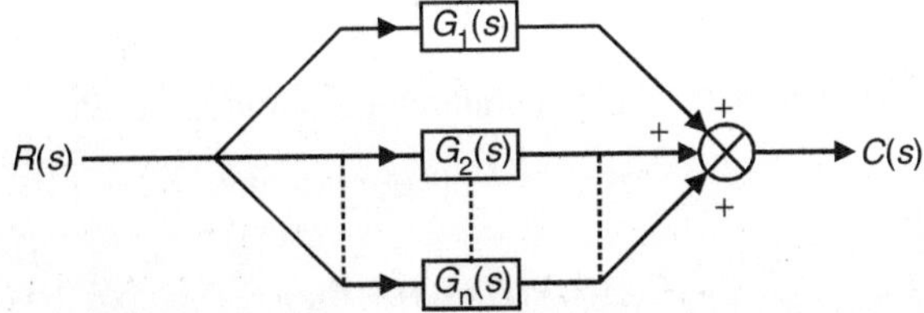

Figure 4.7 Blocks connected in parallel

4.3.3 Rule 3: Eliminating a Feedback Loop

When two blocks are connected to form a closed loop, the block diagram is known as canonical form of block diagram, as shown in Figure 4.8. In this diagram, positive and negative signs represent the respective feedback in the system. The following defines the various quantities in the block diagram.

$R(s)$ = Laplace of reference input $r(t)$
$C(s)$ = Laplace of controlled output $c(t)$
$E(s)$ = Laplace of error signal $e(t)$
$B(s)$ = Laplace of feedback signal $b(t)$
$G(s)$ = Equivalent forward path transfer function
$H(s)$ = Equivalent feedback path transfer function

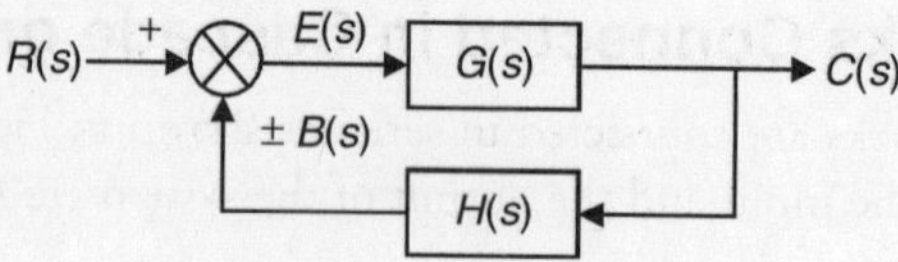

Figure 4.8 Feedback loop

From Figure 4.8, we get

$$E(s) = R(s) \pm B(s) = R(s) \pm C(s)H(s) \tag{4.1}$$

$$C(s) = E(s)G(s) \tag{4.2}$$

From (4.1) and (4.2)

$$C(s) = [R(s) \pm C(s)H(s)]G(s)$$
$$= R(s)G(s) \pm C(s)G(s)H(s)$$
$$C(s)[1 \mp G(s)H(s)] = G(s)R(s)$$

Therefore, the transfer function of the abovementioned system is given as

$$\frac{C(s)}{R(s)} = \frac{G(s)}{1 \mp G(s)H(s)}$$

4.3.4 Rule 4: Shifting a Summing Point Behind the Block

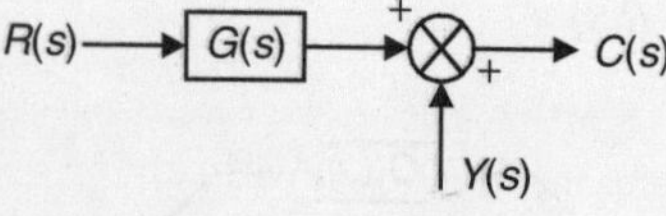

Figure 4.9 Before shifting the summing point

From Figure 4.9, we get

$$C(s) = R(s)\ G(s) + Y(s)$$

To shift the summing point ahead of the block, the value of the output signal should remain same.

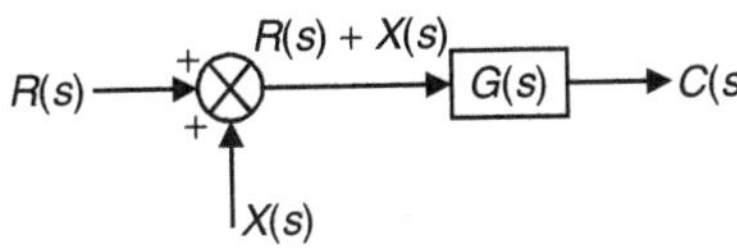

Figure 4.10 After shifting the summing point

From Figure 4.10, we have

$$[R(s)+X(s)]G(s)=C(s)$$

To find $X(s)$,

$$\begin{aligned}[R(s)+X(s)]G(s) &= R(s)G(s)+Y(s)\\ R(s)G(s)+X(s)G(s) &= R(s)G(s)+Y(s)\\ X(s)G(s) &= Y(s) \text{ or } X(s)=[1/G(s)]Y(s)\end{aligned}$$

Thus, the equivalent block diagram for moving a summing point ahead of the block is as shown in Figure 4.11

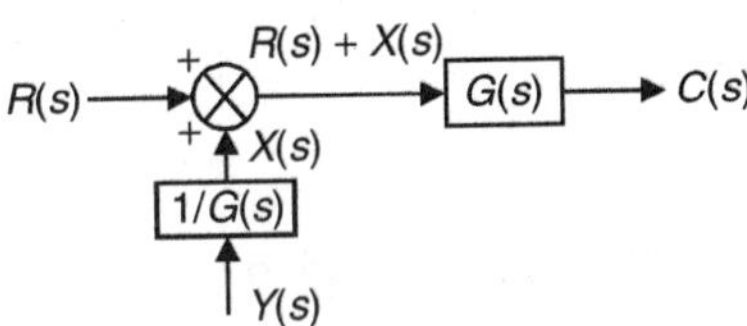

Figure 4.11 Equivalent block diagram after shifting the summing point

4.3.5 Rule 5: Shifting a Summing Point beyond (after) the Block

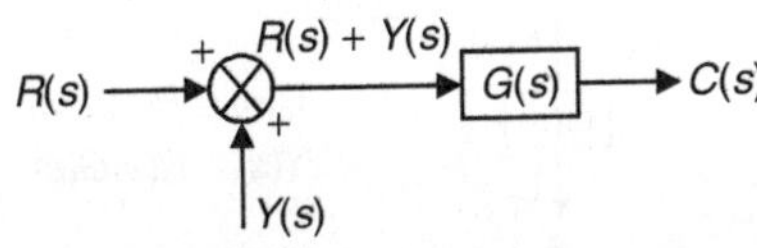

Figure 4.12 Before shifting the summing point

From Figure 4.12, we get

$$C(s)=R(s)G(s)+Y(s)G(s)$$

Now, to shift the summing point after the block, the value of output should remain same; consider the shifted summing point without any change as shown in Figure 4.13

Figure 4.13 After shifting the summing point

To find $X(s)$,

$$R(s)G(s)+X(s)=C(s)=R(s)G(s)+Y(s)G(s)$$

or

$$X(s)=Y(s)G(s)\cdot$$

Thus, the equivalent block diagram for shifting the summing point beyond the block is given as shown in Figure 4.14

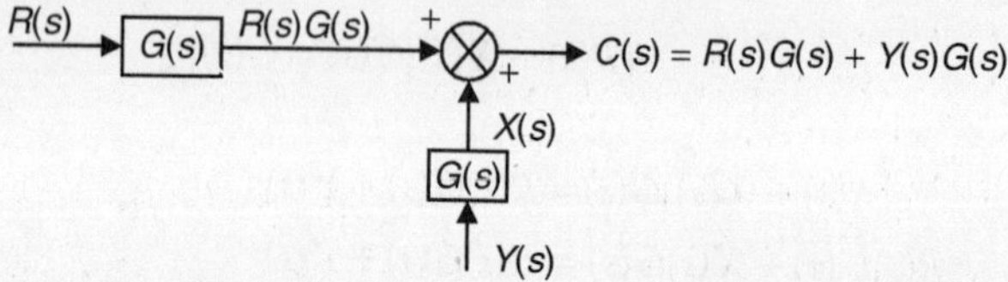

Figure 4.14 Equivalent block diagram after shifting the summing point

4.3.6 Rule 6: Shifting a Take-off Point behind the Block

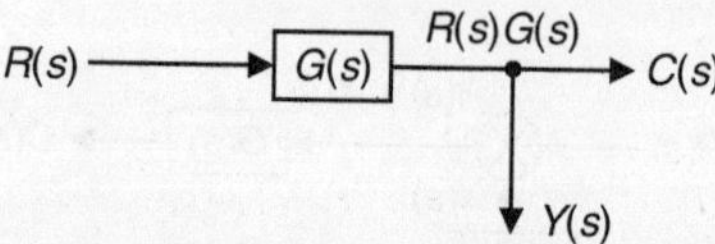

Figure 4.15 Before shifting the take-off point

From Figure 4.15, we get

$$C(s) = Y(s) = R(s)G(s).$$

To move the take-off point ahead of the block, the value of taking-off signal must remain same, as shown in Figure 4.16

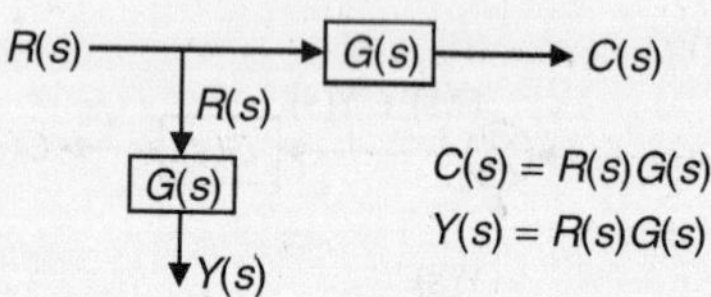

Figure 4.16 After shifting the take-off point

4.3.7 Rule 7: Shifting a Take-off Point beyond the Block

R(s) G(s) C(s) = R(s)G(s)

R(s)

Y(s) = R(s)

Figure 4.17 Before shifting the take-off point

From Figure 4.17, we get

$$C(s)=R(s)G(s)$$

To shift take-off point beyond the block, the value of $Y(s)$ must remain same as shown in Figure 4.18

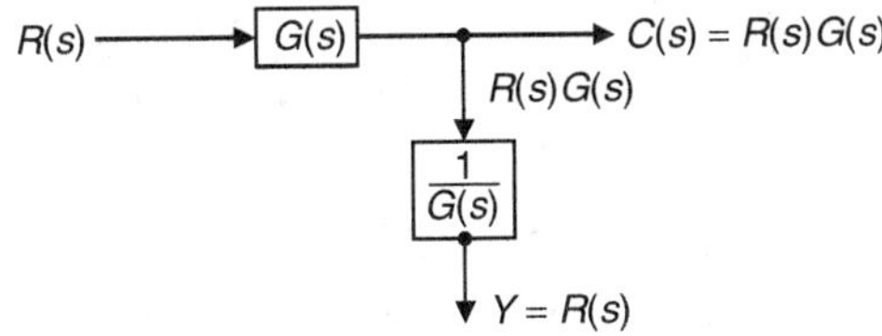

Figure 4.18 After moving the take-off point

4.3.8 Rule 8: For Multiple-input–Multiple-output System

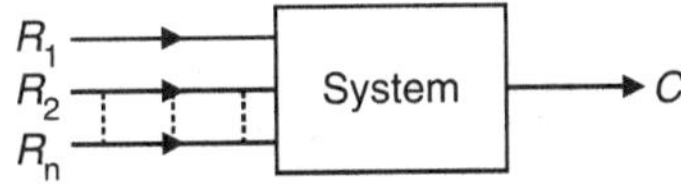

Figure 4.19 Multiple-input, multiple-output system

Superposition theorem is used for multiple-input, multiple-output system as shown in Figure 4.19. Here, only one input is considered at a time and all other inputs are set to zero. First, let us assume $R_1, R_2 = R_3 = \cdots = R_n = 0$, and find output C_1. Then, we assume $R_2, R_1 = R_3 = \cdots R_n = 0$ and find output C_2. Finally, when all inputs are covered, take algebraic sum of all the outputs.

$\therefore$ Total output $C = C_1 + C_2 + \cdots + C_n$.

Similar logic can be extended to find the outputs if it is a multiple-input multiple-output system. Further, separate ratio of each output with each input is to be calculated, assuming that all other inputs and outputs are zero. Therefore, such components of outputs can be added to get resultant output of the system.

4.3.9 Rule 9: Interchanging Summing Point

Consecutive summing points can be interchanged, as this interchange does not alter the output signal, as shown in Figures 4.20 and 4.21.

Figure 4.20 Before interchanging

Figure 4.21 After interchanging

4.3.10 Rule 10: Shifting Take-off Point after a Summing Point

Shifting a take-off point after a summing point is illustrated in Figure 4.22.

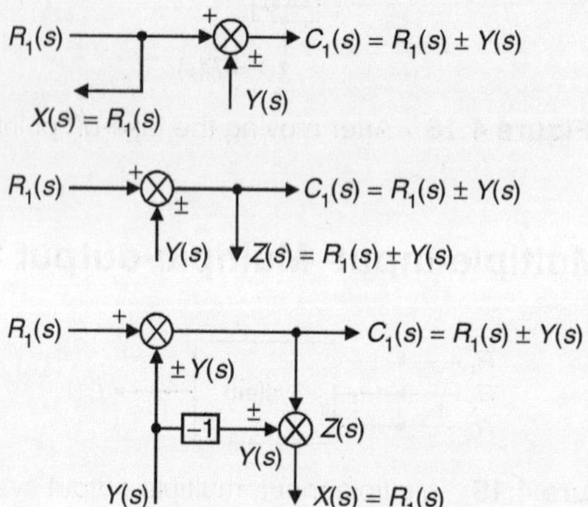

Figure 4.22 Shifting a take-off point after a summing point

4.3.11 Rule 11: Shifting Take-off Point before a Summing Point

Shifting a take-off point before a summing point is illustrated in Figure 4.23.

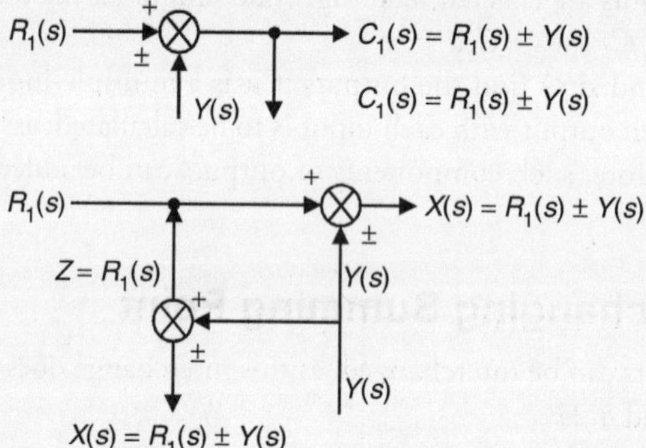

Figure 4.23 Shifting a take-off point before summing point

Thus, it can be noticed that by keeping the value of output signal and the take-off point signal same, and shifting a take-off point before or after a summing point add an additional summing point in the block diagram, and therefore, this complicates the block diagram.

4.4 BLOCK DIAGRAM REDUCTION PROCEDURE

The following procedure is adopted to obtain the simplest form of block diagram:

1. Reduce the blocks connected in series.
2. Reduce the blocks connected in parallel.
3. Reduce the minor internal feedback loops.
4. As far as possible, try to shift take-off points towards the right and summing points to the left.

5. Repeat all the above-mentioned steps until the simple form is obtained.
6. Find the transfer function of the overall system using the formula $C(s)/R(s)$.

4.4.1 Procedure for Multiple Inputs

1. Here, make all but one input to zero.
2. Repeat the above step until all the inputs are covered.
3. Find the resultant output using Superposition theorem.

Problem-4.1: Determine the transfer function $\frac{C(s)}{R(s)}$ for the following block diagram.

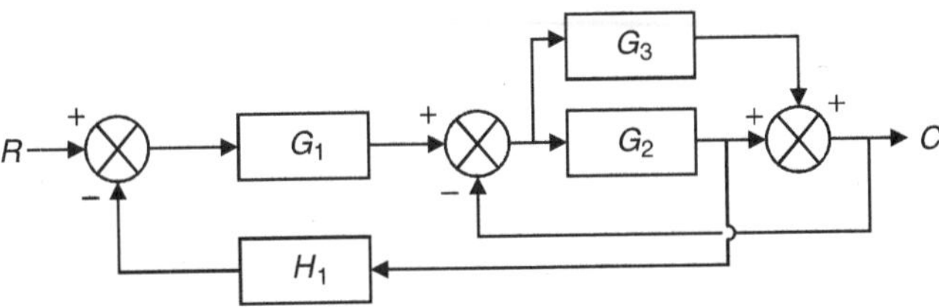

Solution:

The first step is to shift the take-off point after block G_2 to a position before of block G_2; this gives the following block diagram.

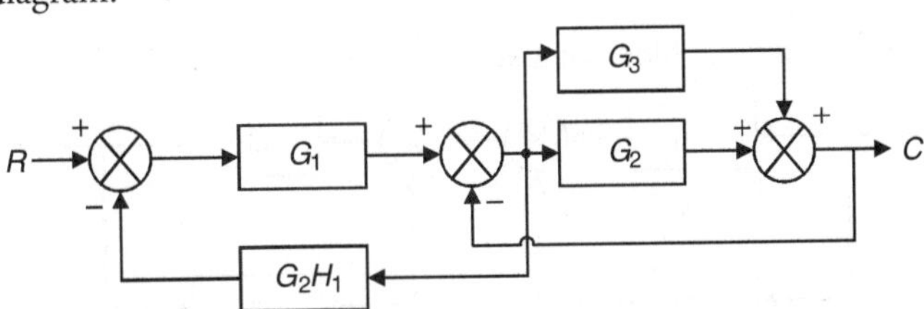

The next step is to eliminate the summing point after block G_2 and shift the take-off point before block $(G_2 + G_3)$ to a position behind block $(G_2 + G_3)$. Thus, we get the following block diagram.

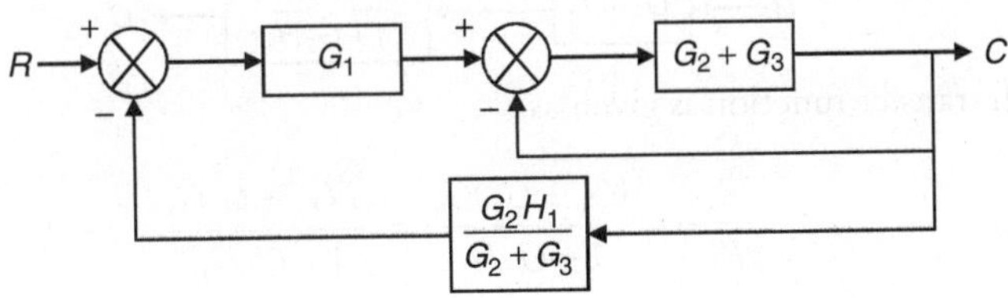

By eliminating the summing point ahead of block $(G_2 + G_3)$, we get the following block diagram.

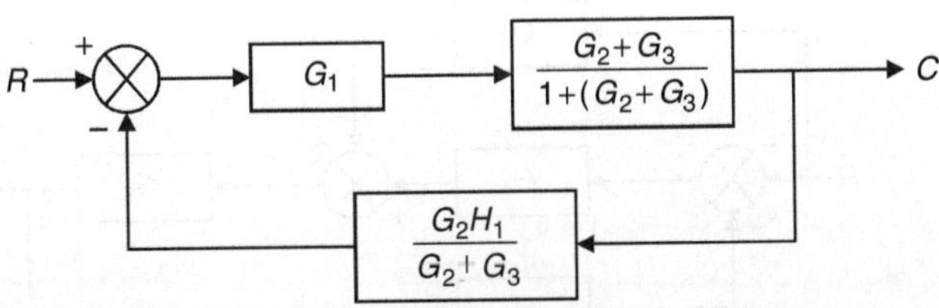

Finally, after reducing the above block diagram, we will get

$$\frac{C(s)}{R(s)} = \frac{\dfrac{G_1(G_2+G_3)}{[1+(G_2+G_3)]}}{1+\dfrac{G_1(G_2+G_3)}{[1+(G_2+G_3)]}\times\dfrac{G_2H_1}{(G_2+G_3)}} = \frac{G_1G_2+G_1G_3}{1+G_2+G_3+G_1G_2H_1}$$

Hence, the final transfer function is

$$\frac{C(s)}{R(s)} = \frac{G_1G_2+G_1G_3}{1+G_2+G_3+G_1G_2H_1}$$

Problem-4.2: Determine the transfer function $\dfrac{C(s)}{R(s)}$ for the following block diagram.

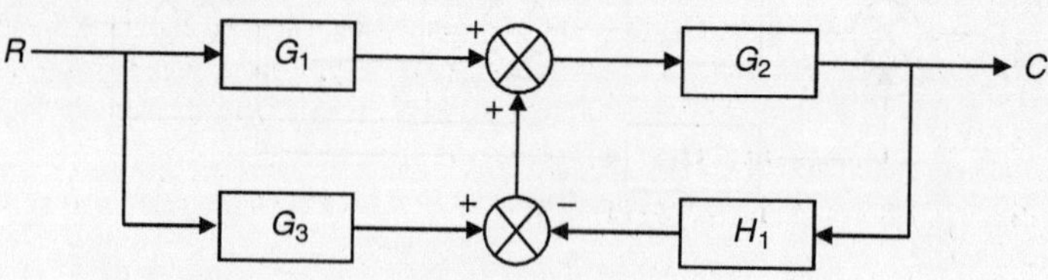

Solution:

Here, the first step is to redraw the block diagram as follows:

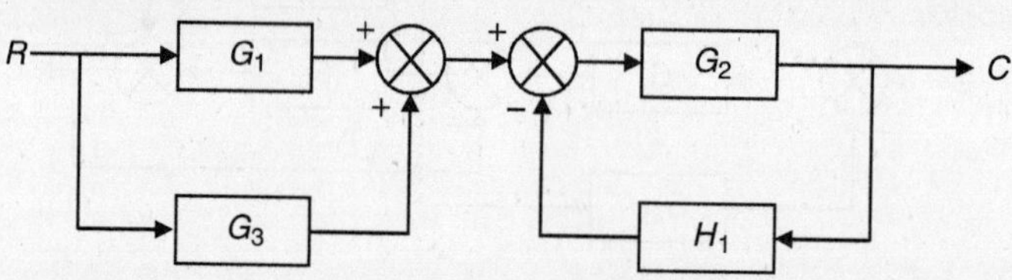

By eliminating two summing points, we get the following reduced diagram:

$$R \rightarrow \boxed{G_1+G_3} \rightarrow \boxed{\frac{G_2}{1+G_2H_1}} \rightarrow C$$

Hence, the overall transfer function is given as

$$\frac{C(s)}{R(s)} = \frac{(G_1+G_3)G_2}{1+G_2H_1} = \frac{G_1G_2+G_2G_3}{1+G_2H_1}$$

Problem-4.3: Determine the transfer function $\dfrac{C(s)}{R(s)}$ for the following block diagram.

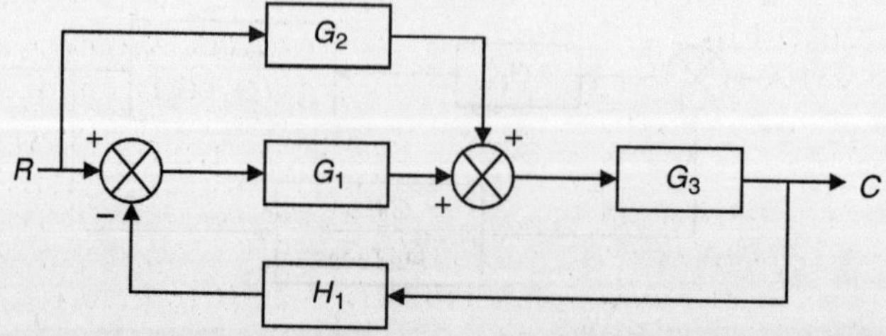

Solution:

Here, the first step is to move the summing point before block G_1 to a position after block G_1.

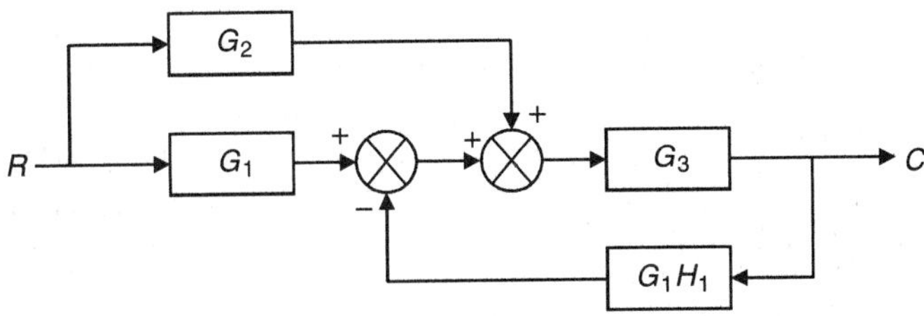

The next step is to interchange consecutive summing points after block G_1.

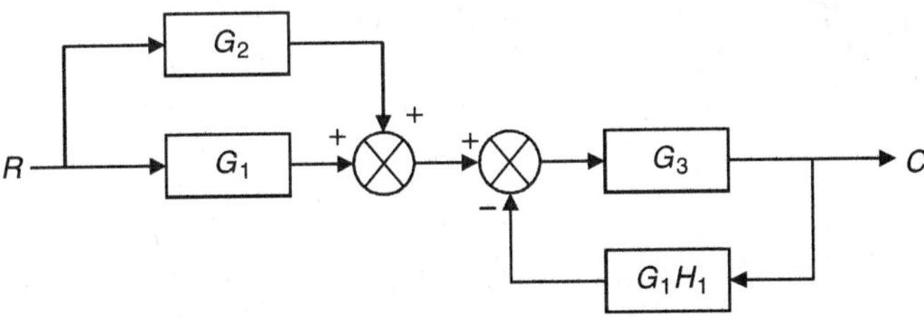

Next, we eliminate the summing point after block G_1.

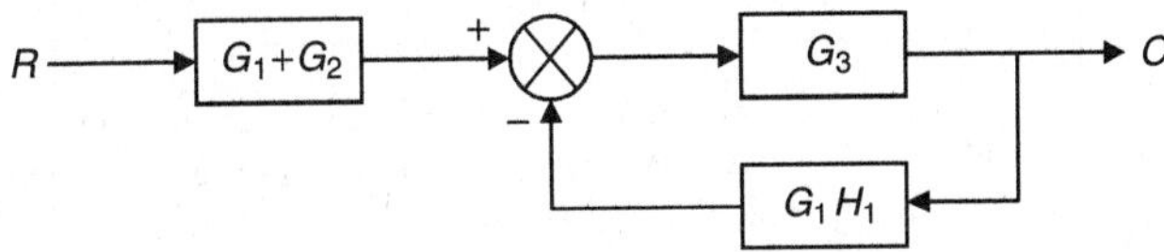

Now, shift the summing point after block $(G_1 + G_2)$ to a position before block $(G_1 + G_2)$ and combine blocks $(G_1 + G_2)$ and G_3.

Hence, finally the transfer function is given as

$$\frac{C(s)}{R(s)} = \frac{(G_1 + G_2)G_3}{1 + (G_1 + G_2)G_3 \cdot \dfrac{G_1 H_1}{(G_1 + G_2)}} = \frac{G_1 G_3 + G_2 G_3}{1 + G_1 G_3 H_1}.$$

Problem-4.4: Determine the overall transfer function relating C and R for the system whose block diagram is given in the following figure.

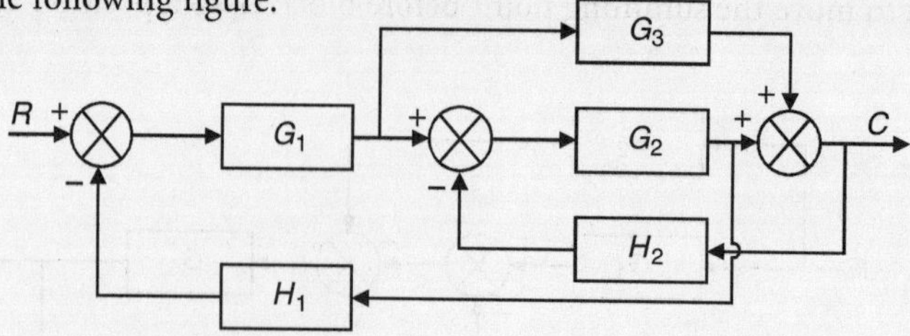

Solution:

Here, the first step is to shift the take-off point after block G_2 to a position before block G_2.

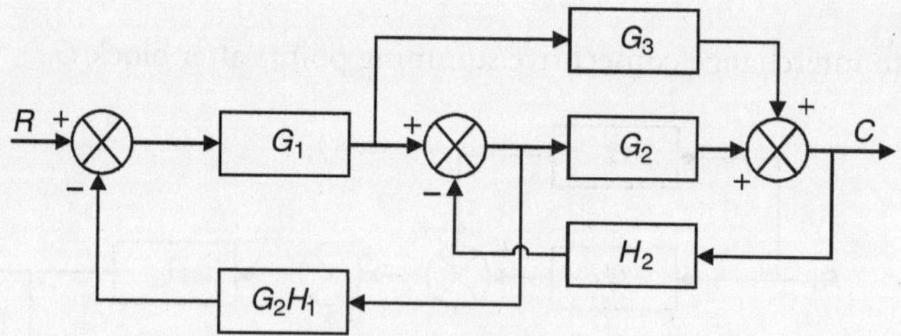

Next, shift the take-off point before block G_2 to a position after block G_1.

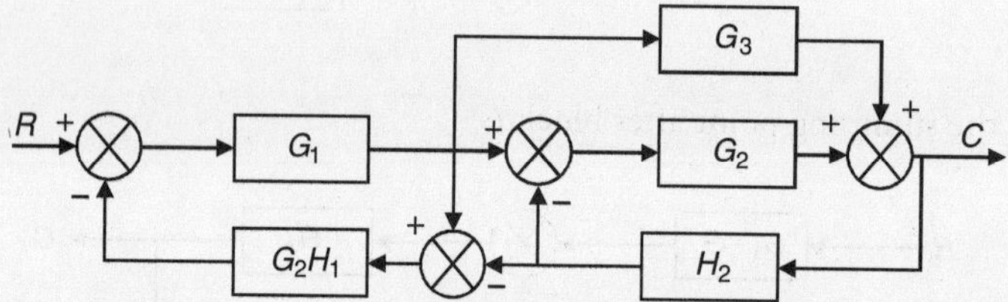

Now, shift the summing point before block G_2 to a position after block G_2.

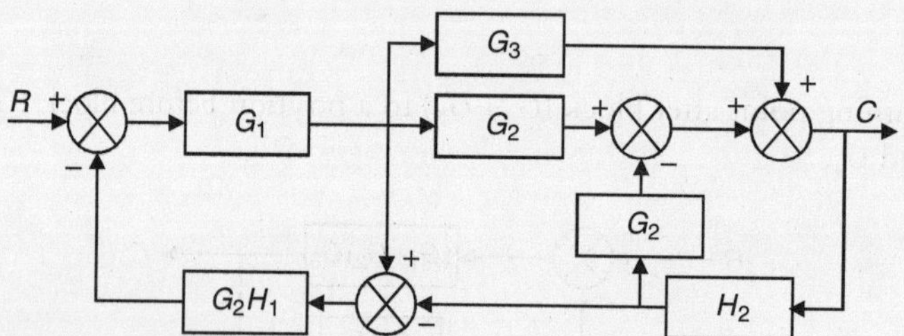

Here, interchange the consecutive summing points after block G_2.

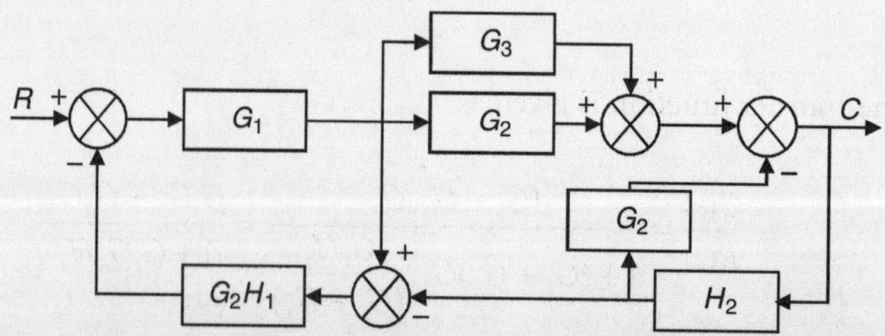

By shifting the take-off point after block H_2 to a position before block H_2, we get the following figure.

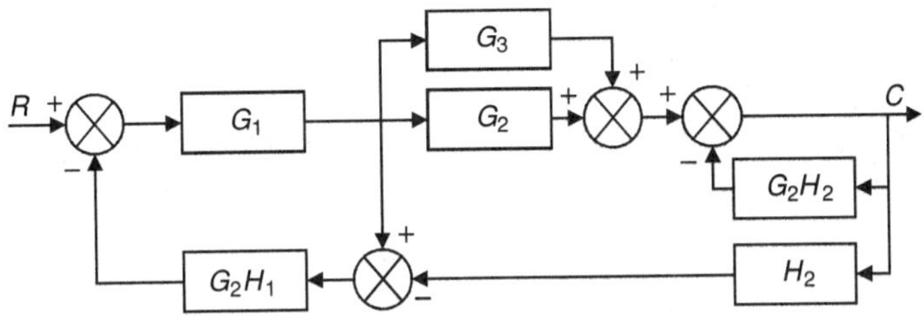

The above block diagram can be rearranged as

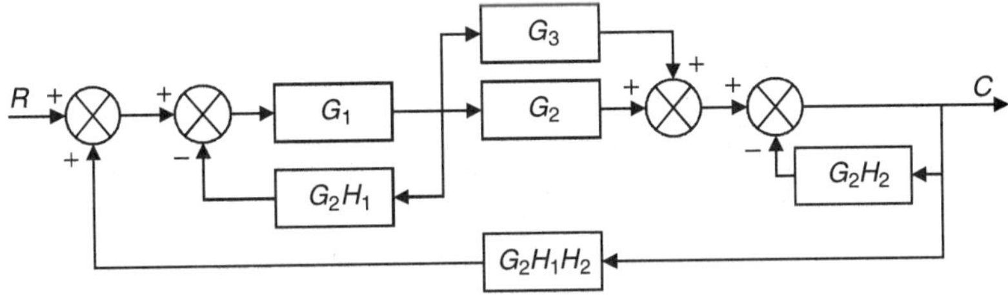

Now, eliminate two consecutive summing points after block G_2.

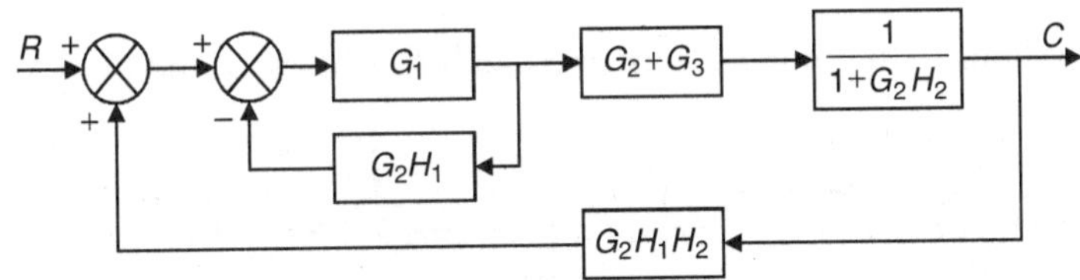

When we eliminate the summing points before block G_1, we get

R $\frac{G_1}{1+G_1G_2H_1}$ G_2+G_3 $\frac{1}{1+G_2H_2}$ C

$G_2H_1H_2$

Therefore, the overall transfer function is given as

$$\frac{C(s)}{R(s)} = \frac{\dfrac{G_1}{1+G_1G_2H_1} \times (G_2+G_3) \times \dfrac{1}{(1+G_2H_2)}}{1 - \dfrac{G_1}{(1+G_1G_2H_1)} \times (G_2+G_3) \times \dfrac{1}{(1+G_2H_2)} \times G_2H_1H_2}$$

$$= \frac{G_1G_2 + G_1G_3}{1 + G_2H_2 + G_1G_2H_1 - G_1G_2G_3H_1H_2}$$

Problem-4.5: Determine the ratio $\frac{C(s)}{R(s)}$ and $\frac{C(s)}{D(s)}$ for the following diagram.

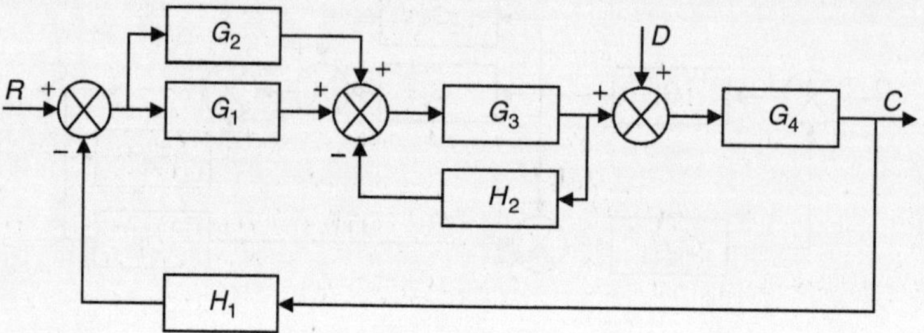

Solution:

Here, the first step is to redraw the block diagram as per the question:

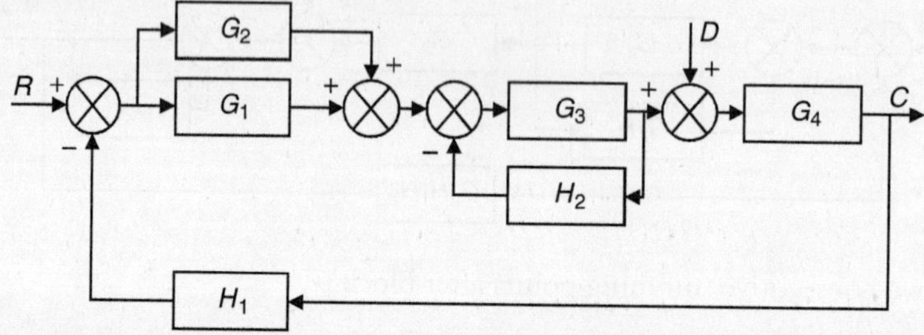

The next step is to eliminate summing points after and before block G_1 and block G_3, respectively.

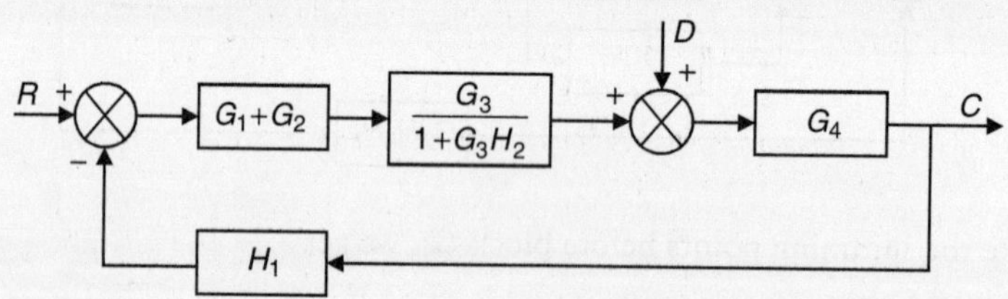

Then, combine blocks (G_1+G_2) and $\frac{G_3}{1+G_3H_2}$

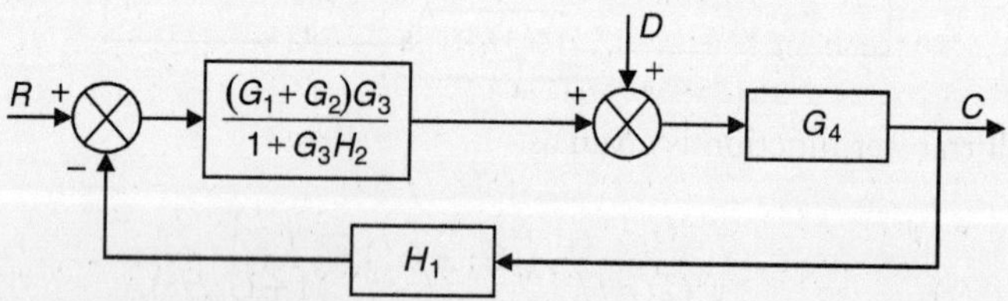

(i) In order to determine $C(s)/R(s)$, let us assume $D = 0$. Therefore, the above block diagram takes the following form:

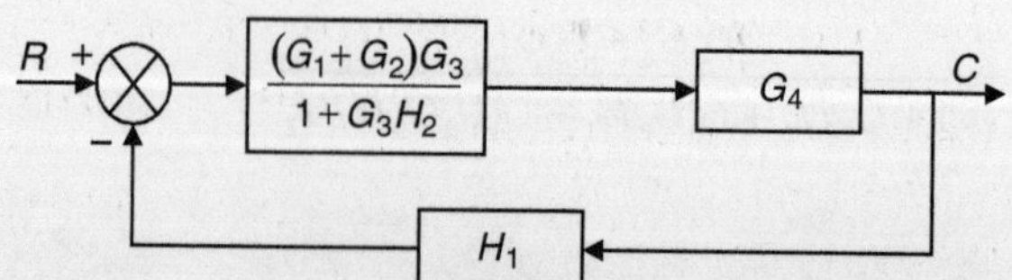

According to the above block diagram shown, the ratio C/R is given as

$$\frac{C(s)}{R(s)}=\frac{\frac{(G_1+G_2)G_3}{(1+G_3H_2)}\times G_4}{1+\frac{(G_1+G_2)G_3}{(1+G_3H_2)}\times G_4\times H_1}=\frac{G_1G_3G_4+G_2G_3G_4}{1+G_3H_2+G_1G_3G_4H_1+G_2G_3G_4H_1}$$

(ii) In order to determine $C(s)/D(s)$, let us assume $R=0$. Therefore, the above block diagram takes the following form:

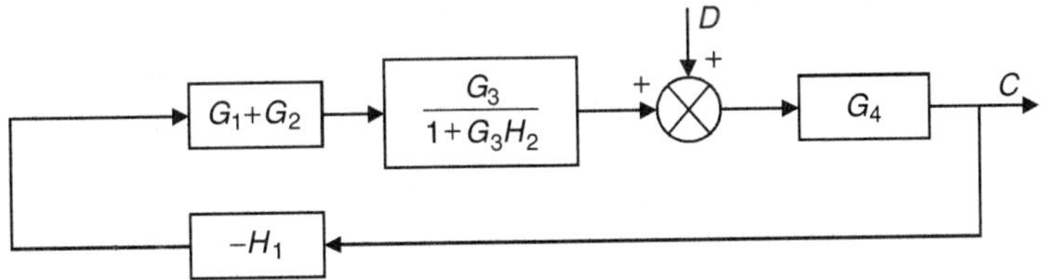

The above block diagram can be redrawn as

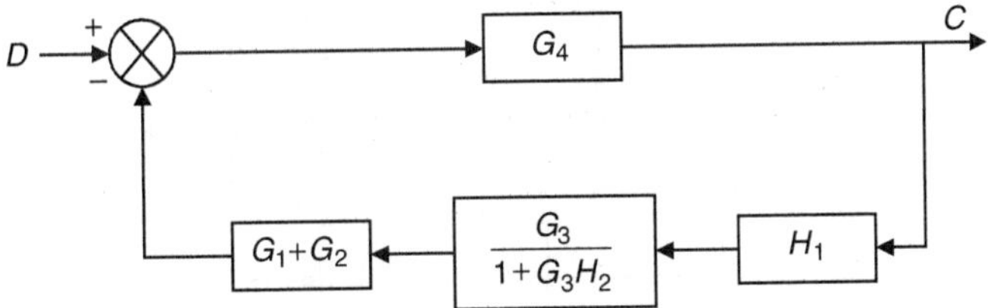

According to the above block diagram, we get

$$\frac{C(s)}{D(s)}=\frac{G_4}{1+G_4\frac{(G_1+G_2)G_3}{(1+G_3H_2)}\times H_1}=\frac{G_4(1+G_3H_2)}{1+G_3H_2+G_1G_3G_4H_1+G_2G_3G_4H_1}.$$

Problem-4.6: Reduce the given block diagram and obtain the transfer function $\frac{C(s)}{R(s)}$.

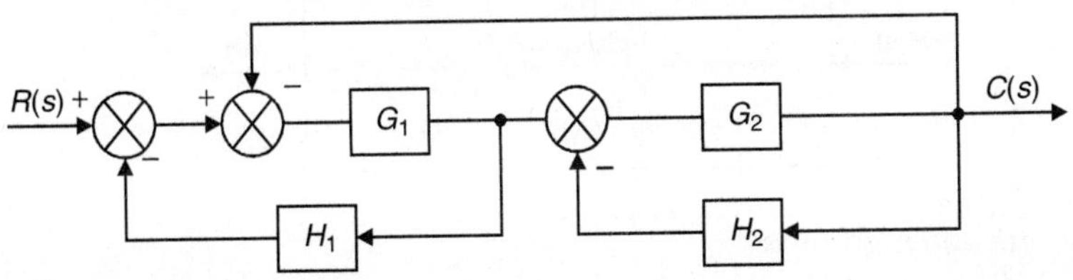

Solution:

Here, the first step is to eliminate the minor feedback loop. Therefore, we get

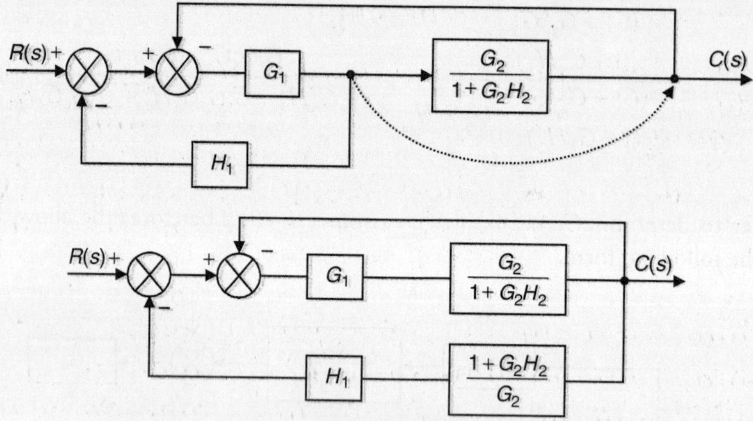

Block $\dfrac{G_1G_2}{1+G_2H_2}$ with unity-feedback transfer function forms a minor feedback loop and therefore, eliminate it.

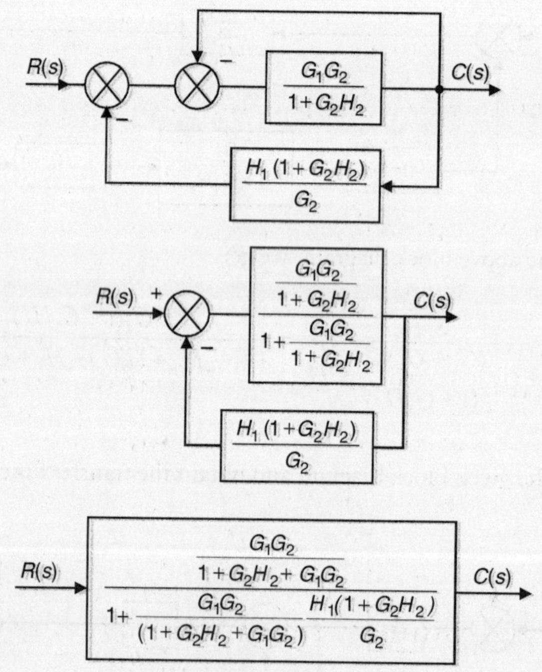

Finally, the transfer function is given as

$$\frac{C(s)}{R(s)} = \frac{G_1G_2}{1+G_1G_2+G_2H_2+G_1H_1+G_1G_2H_1H_2}$$

ADDITIONAL SOLVED PROBLEMS

ASP-1: Reduce the given block diagram and obtain the transfer function $\dfrac{C(s)}{R(s)}$.

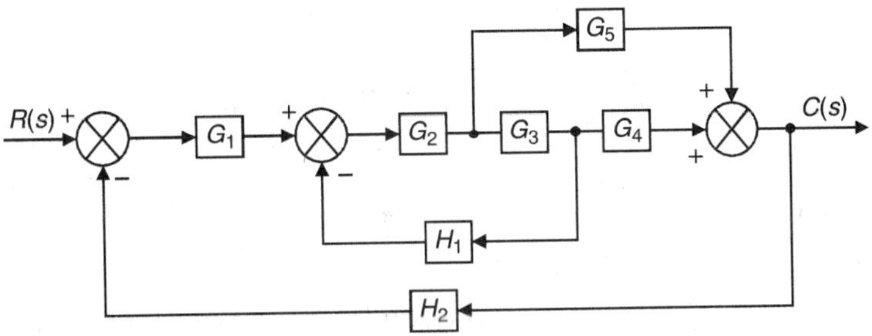

Solution:

Here, the first step is to shift the take-off point towards the right, as shown in the following figure.

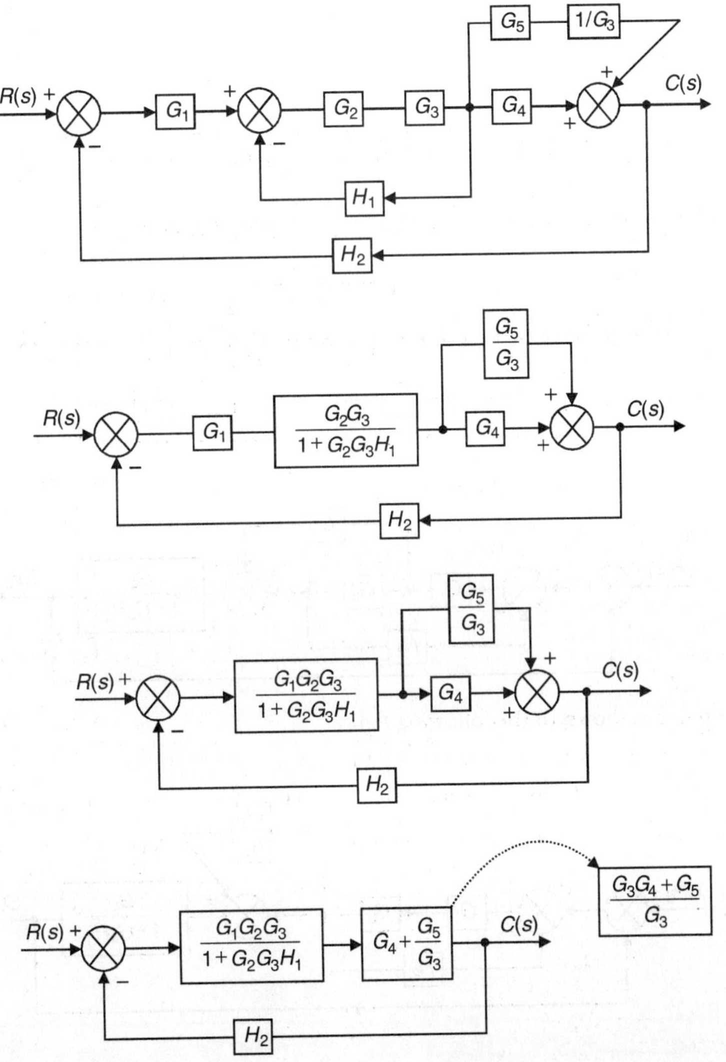

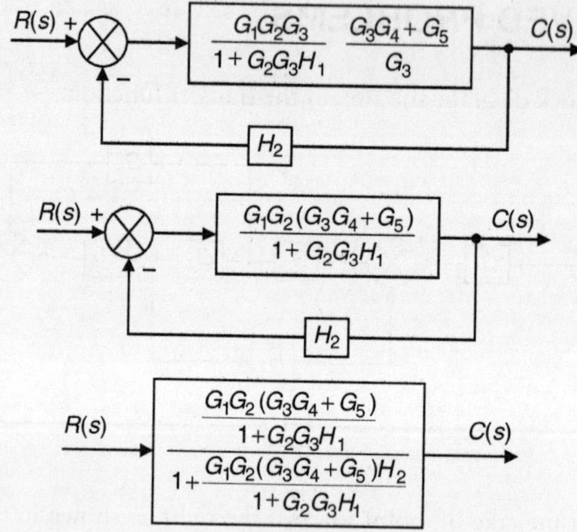

Hence, finally, the required transfer function given as

$$\frac{C(s)}{R(s)} = \frac{G_1G_2(G_3G_4+G_5)}{1+G_2G_3H_1+G_1G_2(G_3G_4+G_5)H_2}.$$

ASP-2: Reduce the given block diagram and obtain the transfer function $\frac{C(s)}{R(s)}$.

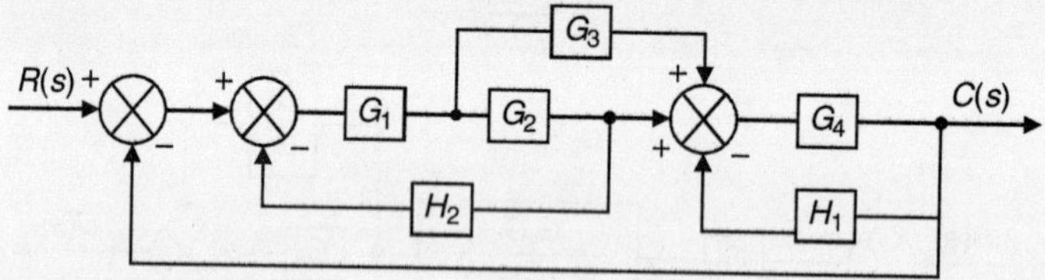

Solution:

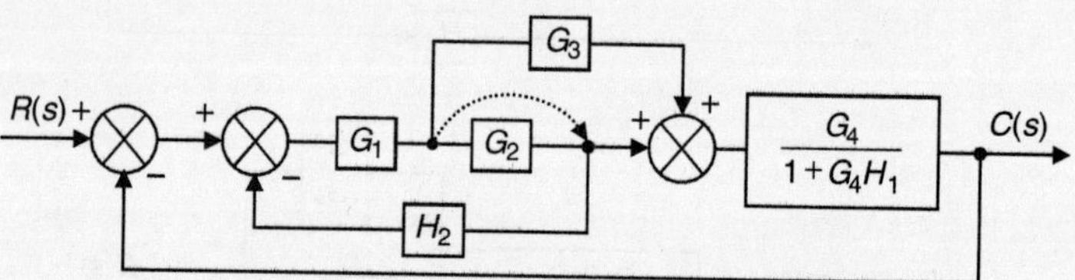

Shift the take-off point as shown in the following figure.

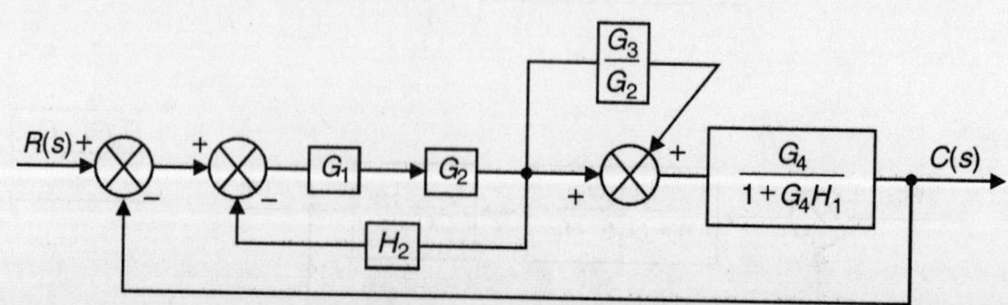

The above block diagram can be reduced as

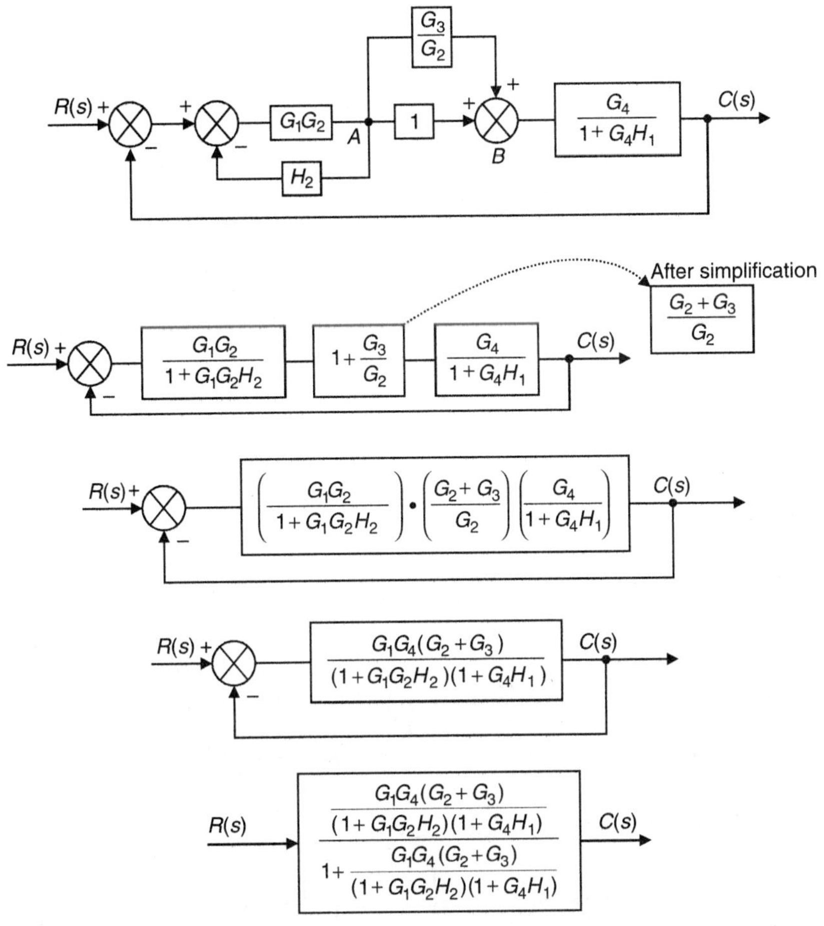

Finally, the required transfer function is given as

$$\frac{C(s)}{R(s)}=\frac{G_1G_4(G_2+G_3)}{1+G_1G_2H_2+G_4H_1+G_1G_2G_4H_1H_2+G_1G_4(G_2+G_3)}.$$

ASP-3: Reduce the given block diagram and obtain the transfer function $\frac{C(s)}{R(s)}$.

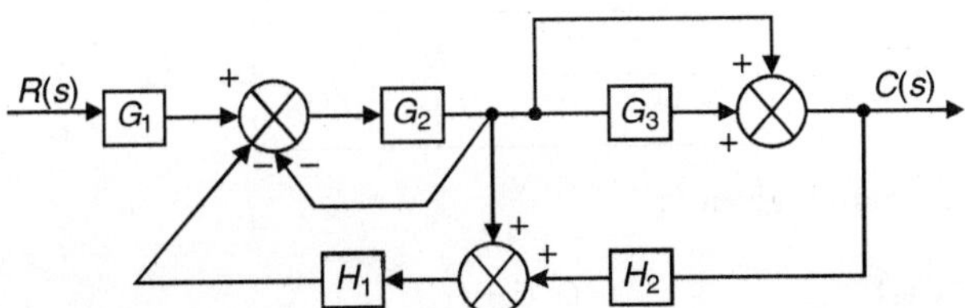

***Solution*:**

The first step is to combine the block with $G_3$and unity gain block that are in parallel; therefore, by combining them, we get

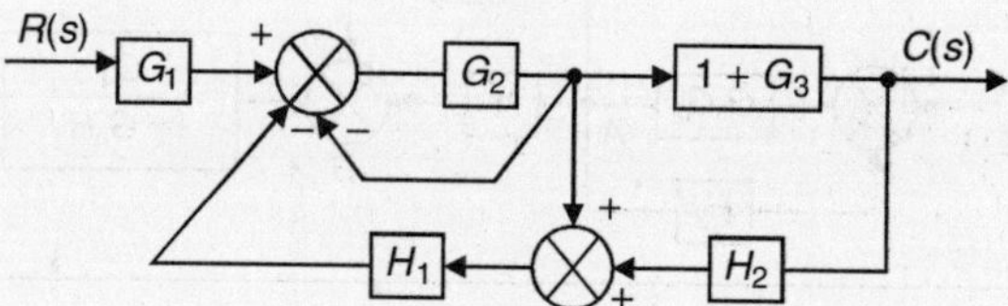

By reducing minor feedback loop of block G_2 and unity feedback, the reduced block diagram is given as

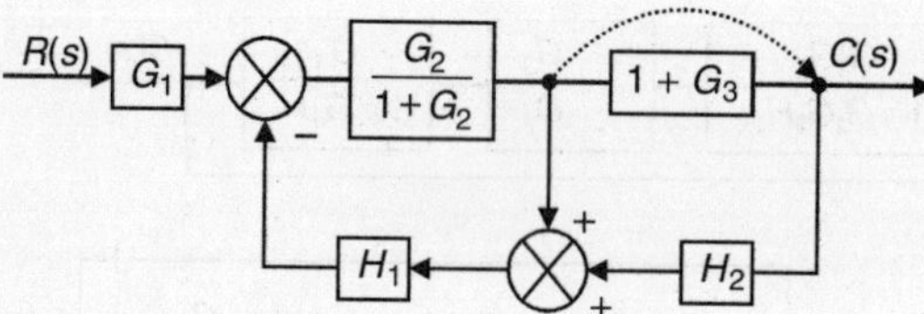

By shifting the take-off point, we get

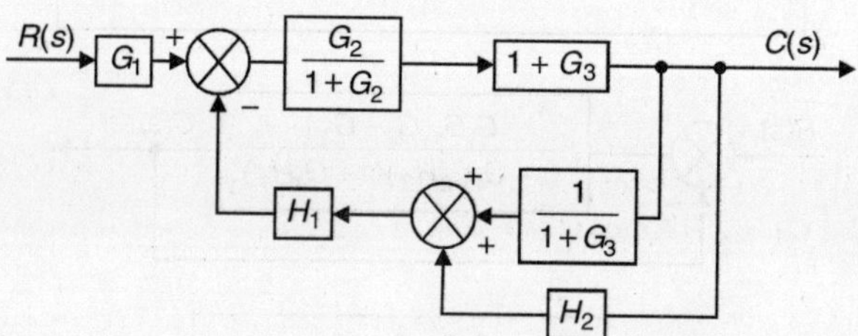

By adding the parallel blocks H_2 and $1/(1+G_3)$, we get

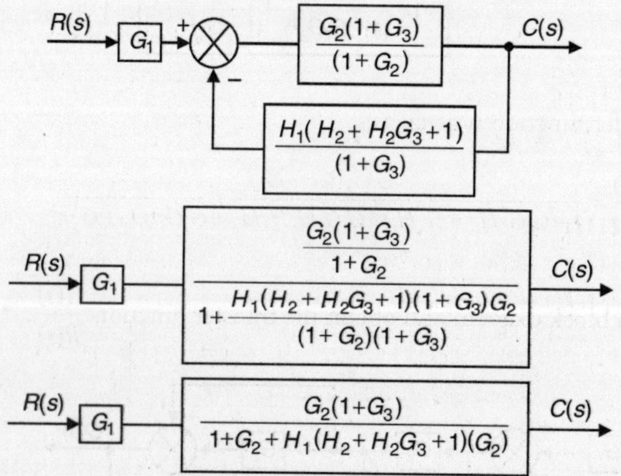

Finally, the required transfer function is given as

$$\frac{C(s)}{R(s)} = \frac{G_1G_2(1+G_3)}{1+G_2+H_1G_2+H_1H_2G_2+H_1H_2G_2G_3}$$

ASP-4: Reduce the given block diagram and obtain the transfer function.

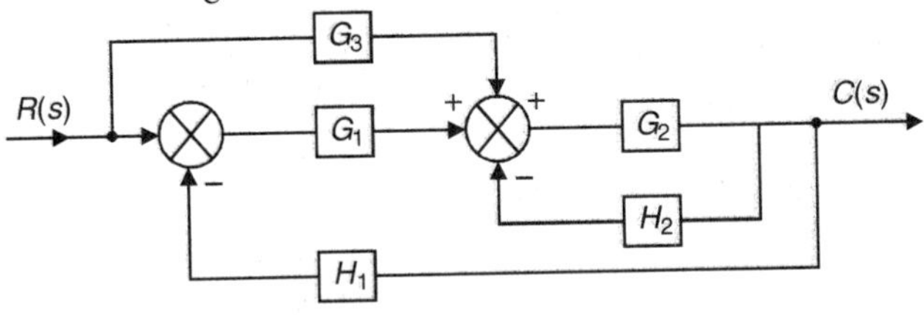

Solution:

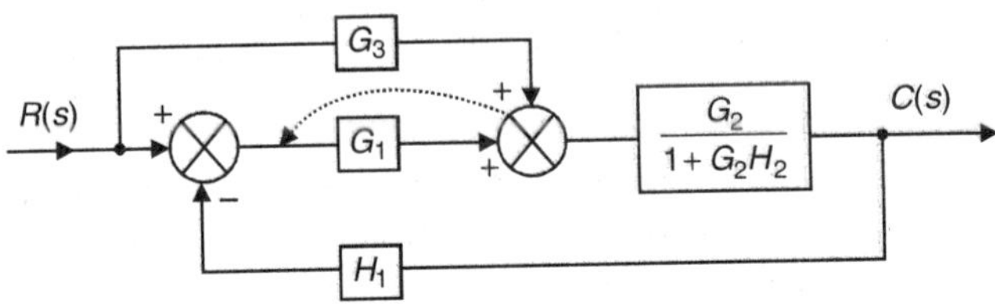

By shifting the summing point to the left gives the following reduced block diagram:

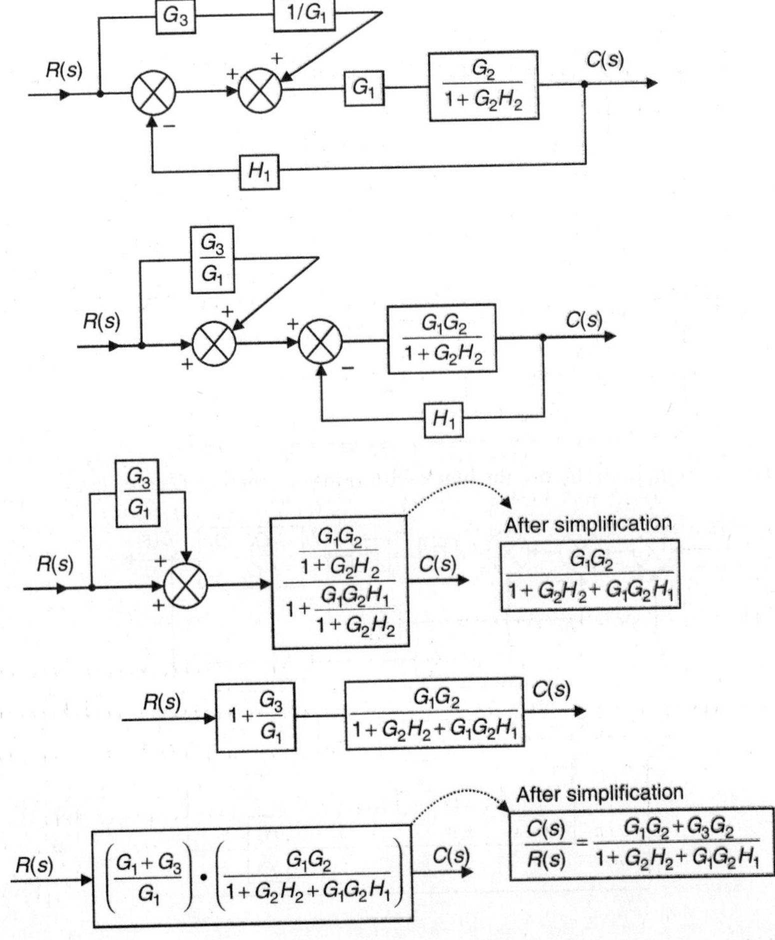

Finally, the required transfer function is given as

$$\frac{C(s)}{R(s)} = \frac{G_1G_2 + G_3G_2}{1 + G_2H_2 + G_1G_2H_1}.$$

ASP-5: Reduce the given block diagram and obtain the transfer function.

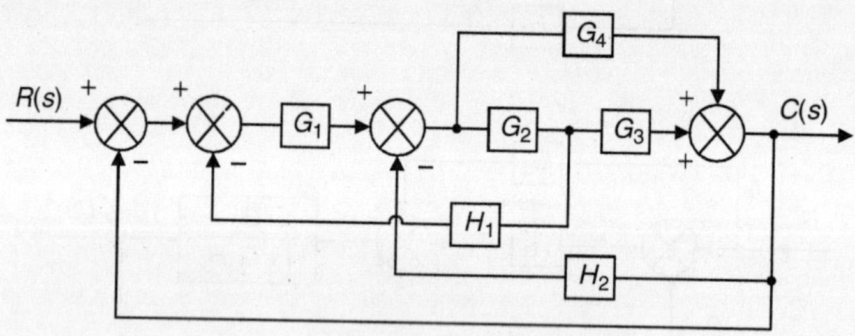

Solution:

By shifting the take-off point after the block having transfer function G_2, we get

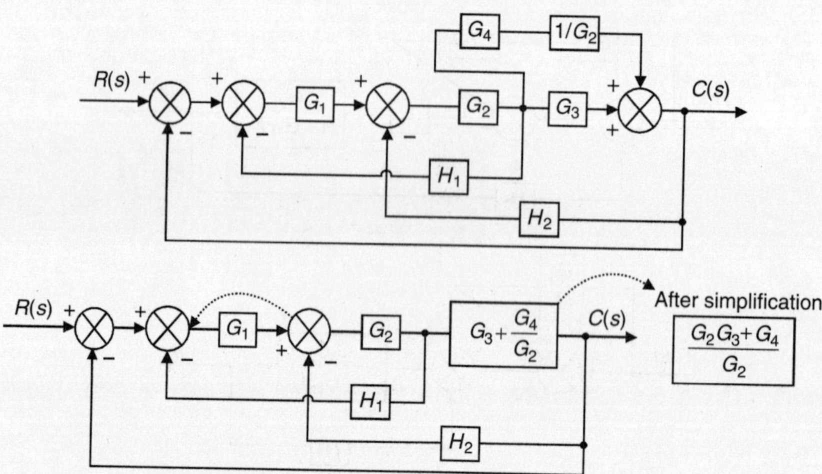

Shifting the summing point before the block with transfer function G_1, we get

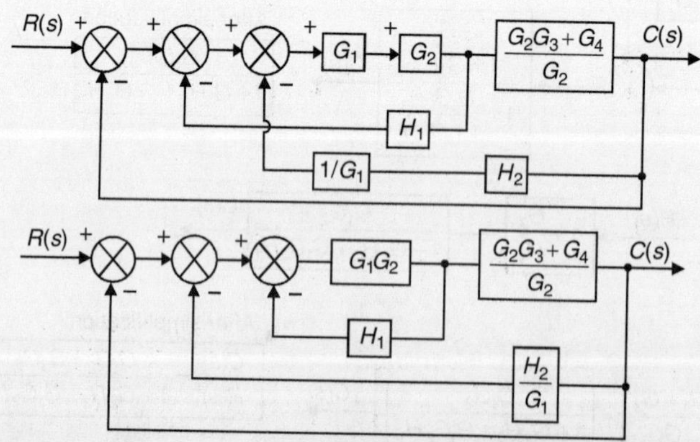

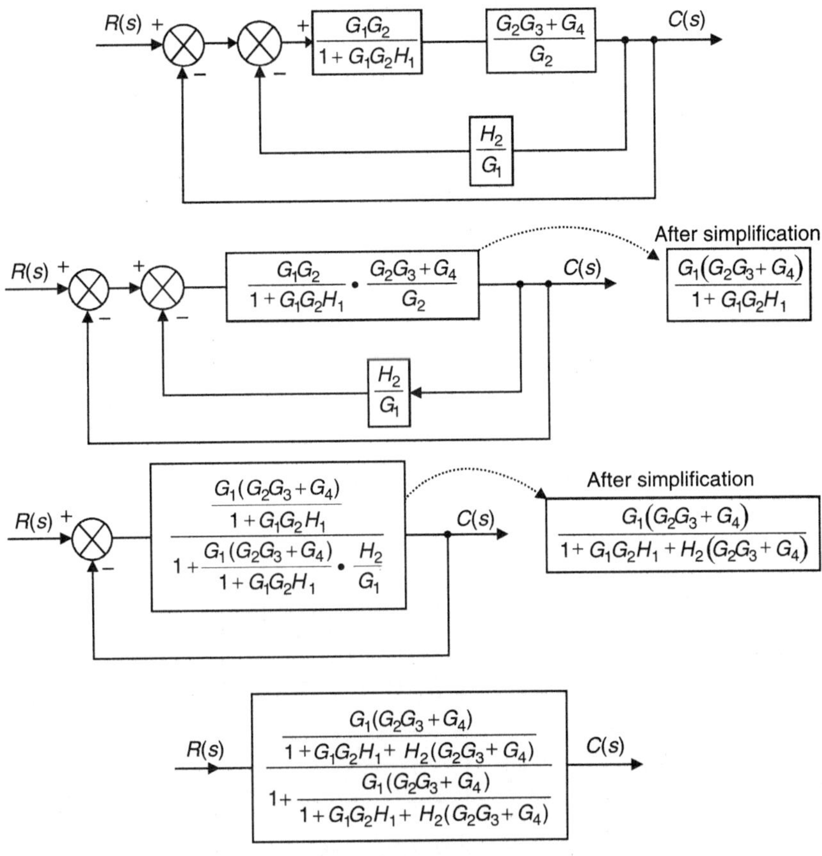

Finally, the required transfer function is given as

$$\frac{C(s)}{R(s)} = \frac{G_1G_2G_3 + G_1G_4}{1 + G_1G_2H_1 + H_2G_2G_3 + H_2G_4 + G_1G_2G_3 + G_1G_4}$$

ASP-6: Reduce the given block diagram into its simplest form and obtain the transfer function $\frac{C(s)}{R(s)}$.

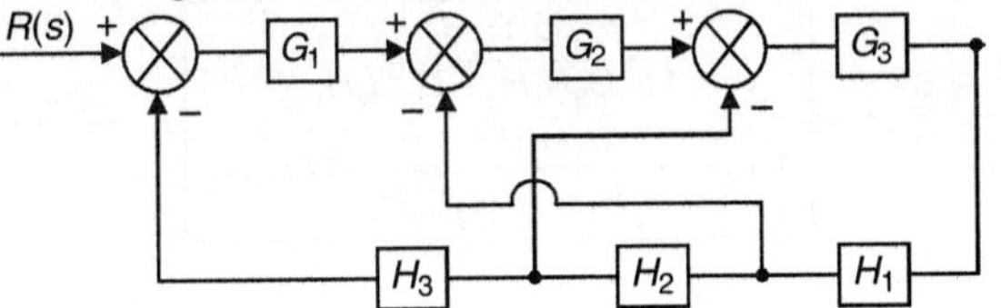

Solution:

By separating the feedback paths at different summing points, we can rearrange the above block diagram as

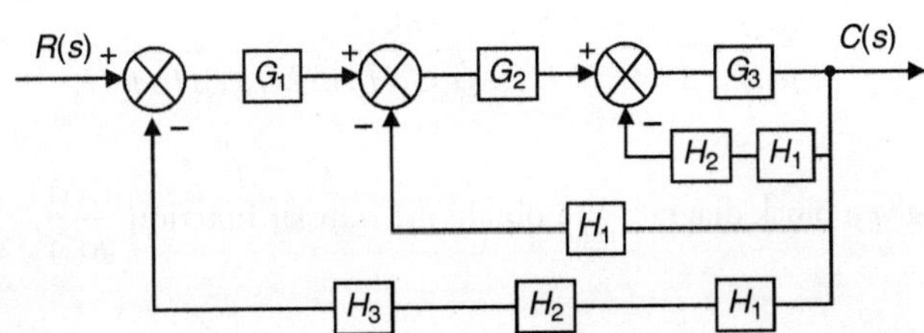

The above block diagram can be drawn as in the following figure. After simplifying, we get

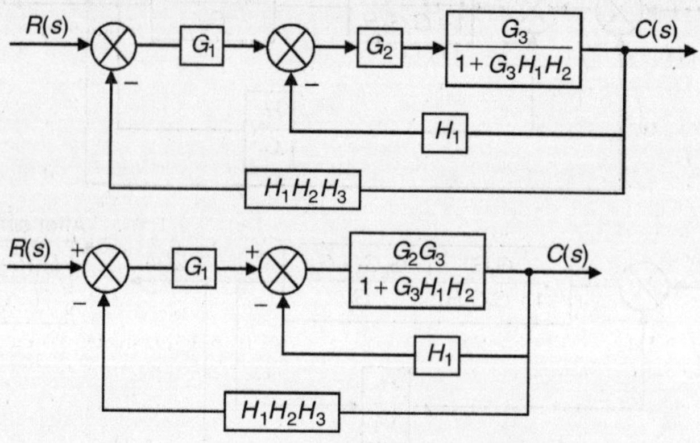

By eliminating the feedback element H_1, we get the following block diagram.

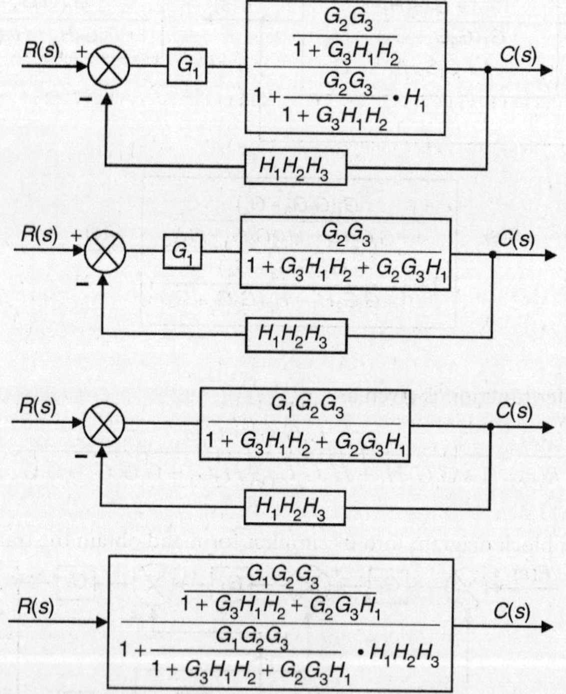

The required transfer function is as given as follows:

$$\frac{C(s)}{R(s)} = \frac{G_1G_2G_3}{1+G_3H_1H_2+G_2G_3H_1+G_1G_2G_3H_1H_2H_3}.$$

ASP-7: Reduce the given block diagram and obtain the transfer function $\dfrac{C(s)}{R(s)}$.

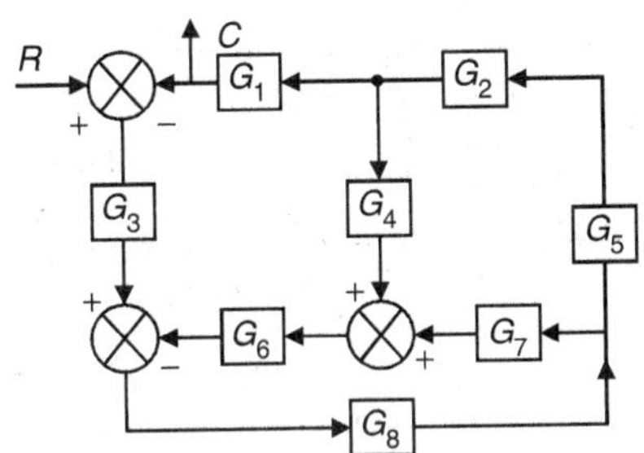

Solution:

By rearranging the given block diagram, we can get

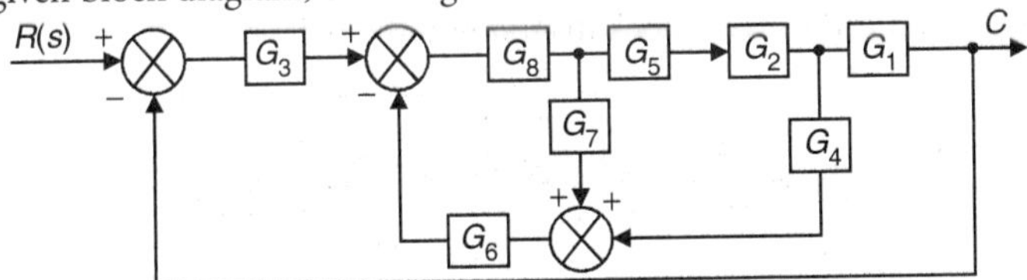

By separating the feedback paths through G_4 and G_6 and G_7 and G_6, we get the following reduced block diagram:

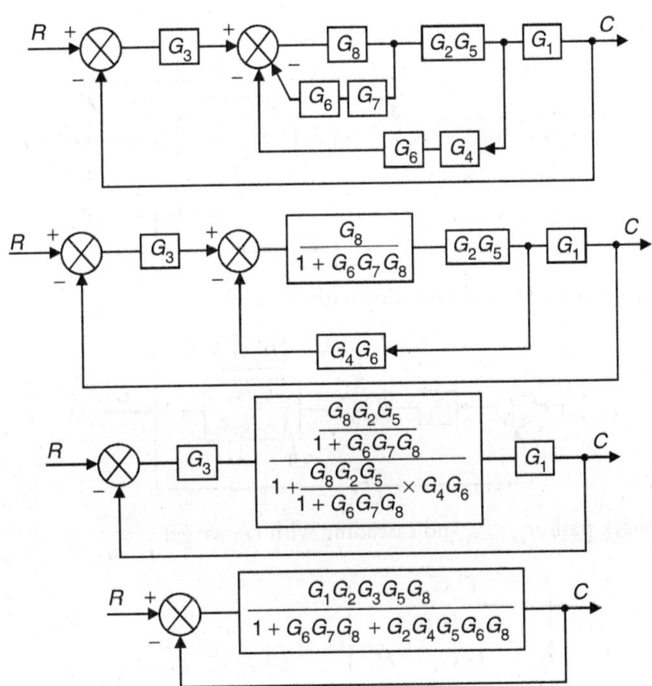

The required transfer function $C(s)/R(s)$ is as given as

$$\frac{C(s)}{R(s)}=\frac{\dfrac{G_1G_2G_3G_5G_8}{1+G_6G_7G_8+G_2G_4G_5G_6G_8}}{1+\dfrac{G_1G_2G_3G_5G_8}{1+G_6G_7G_8+G_2G_4G_5G_6G_8}}=\frac{G_1G_2G_3G_5G_8}{1+G_6G_7G_8+G_2G_4G_5G_6G_8+G_1G_2G_3G_5G_8}.$$

ASP-8: For the system represented by the block diagram shown in the figure, evaluate the closed-loop transfer function, when input R is (1) at Station I and (2) at Station II.

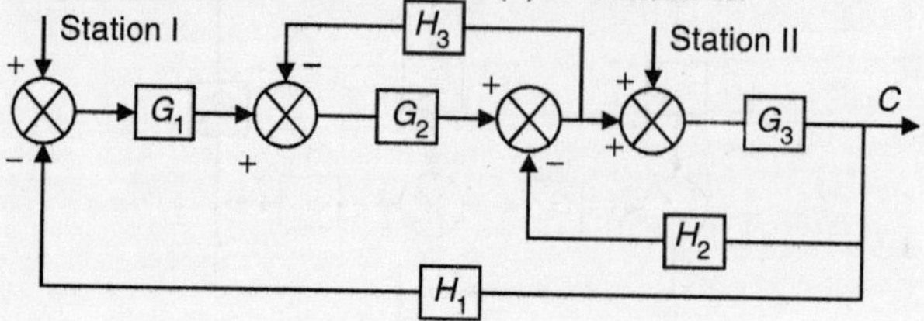

Solution:

1. **Let R be at Station I:**

When input R is at Station I, the given block diagram can be redrawn as

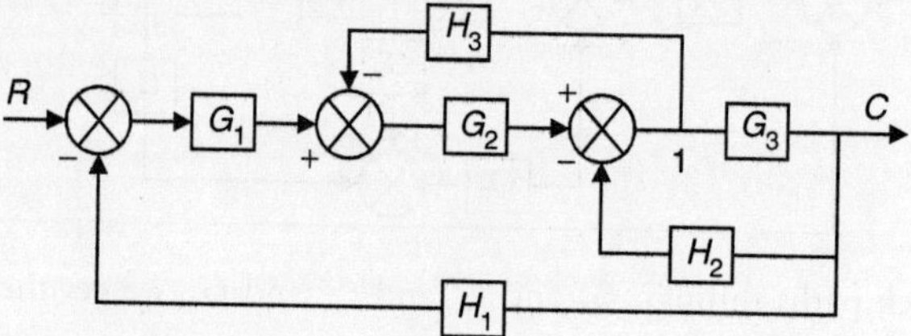

By moving the take-off point 1 after block G_3, we get

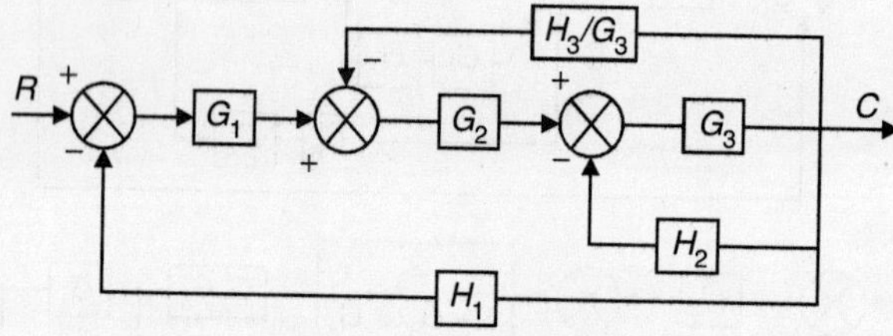

By eliminating feedback path H_2 and cascading with G_2, we have

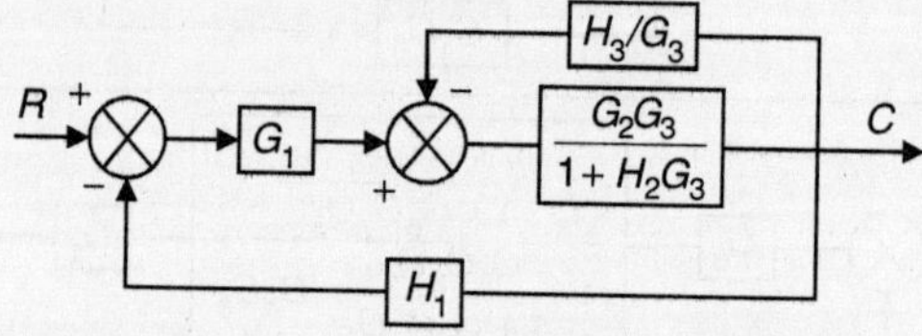

By eliminating feedback path H_3/G_3 and cascading with G_1, we get

$$\left(\frac{\dfrac{G_2G_3}{1+H_2G_3}}{1+\dfrac{G_2G_3}{1+H_2G_3}\cdot\dfrac{H_3}{G_3}}\right)G_1 = \frac{G_1G_2G_3}{1+H_2G_3+G_2H_3}$$

The required closed transfer function is

$$T(s) = \frac{C}{R} = \frac{\dfrac{G_1G_2G_3}{1 + H_2G_3 + G_2H_3}}{1 + \dfrac{G_1G_2G_3}{1 + H_2G_3 + G_2H_3} \times H_1} = \frac{G_1G_2G_3}{1 + H_2G_3 + G_2H_3 + G_1G_2G_3H_1}.$$

2. **Let 'R' be at Station II:**
When input R is at Station II, the given block diagram can be redrawn as

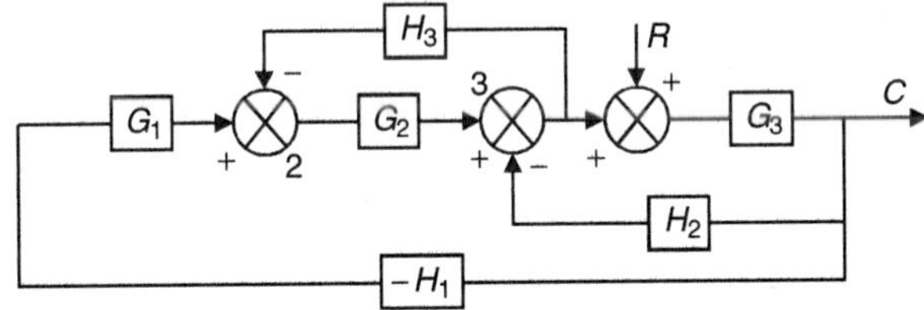

By moving the summing point 2 behind block G_2, we get the following reduced block diagram:

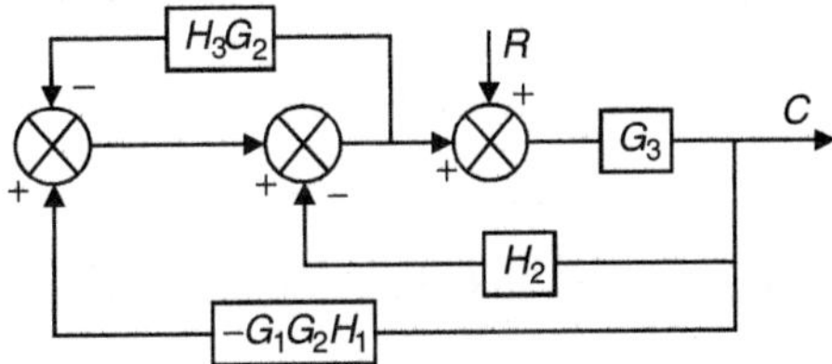

By rearranging the summing points, we have

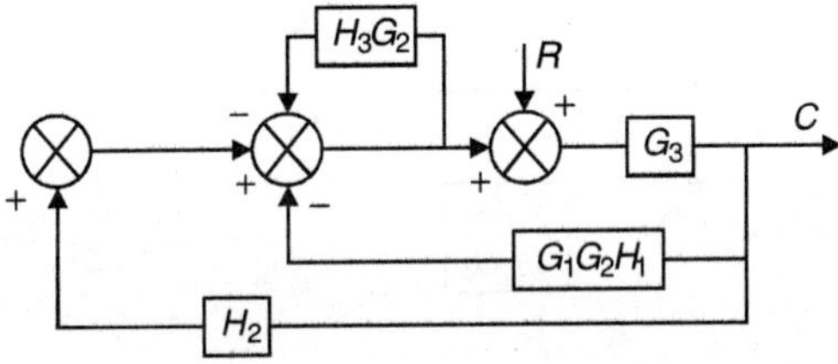

By simplifying the above figure, we get

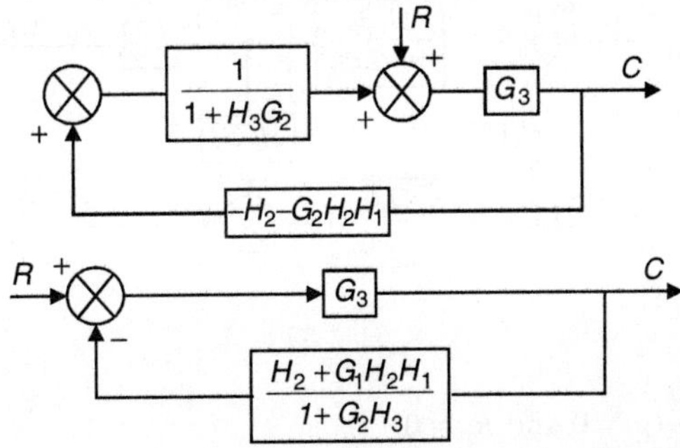

The required closed-loop transfer function is

$$T = \frac{C}{R} = \frac{G_3}{1+G_3\left(\dfrac{H_2+G_1H_2H_1}{1+G_2H_3}\right)}$$

$$= \frac{G_2(1+G_2H_3)}{1+G_2H_3+G_3H_2+G_1G_3H_1H_2}$$

ASP-9: Evaluate $\frac{C_1}{R_1}$ and $\frac{C_2}{R_2}$ for a system whose block diagram, is as shown in the figure.

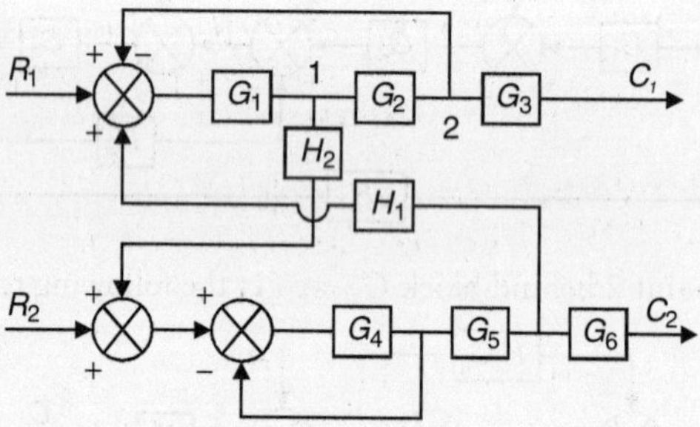

Solution:

By shifting the take-off point 1 after block G_2 and cascading with G_1, we have

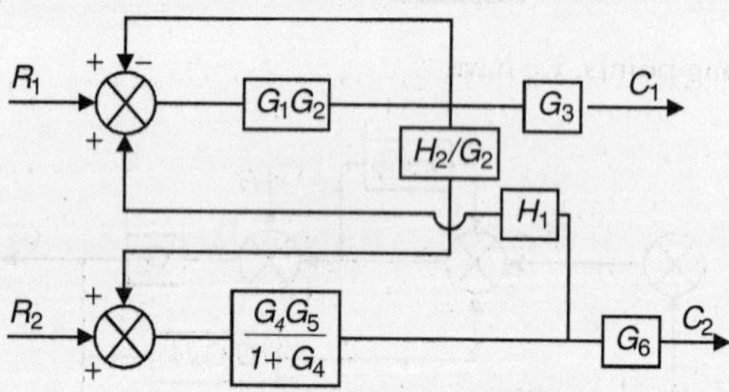

By eliminating the unity feedback and rearranging the figure, we get

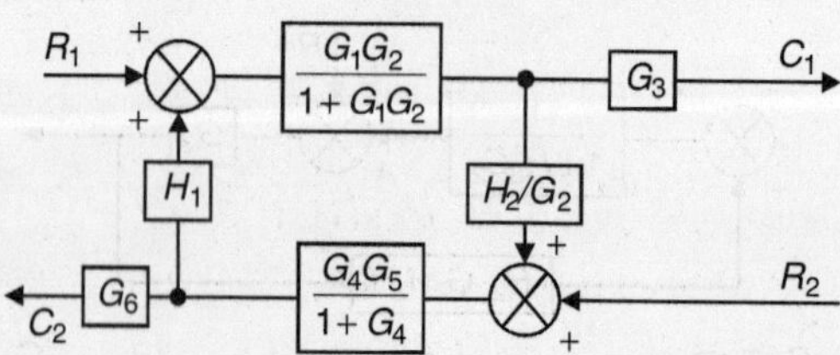

Evaluation of C_1 / R_1:

To find C_1 / R_1, let us assume $C_2 = 0$ and $R_2 = 0$

The above block diagram can be reduced to

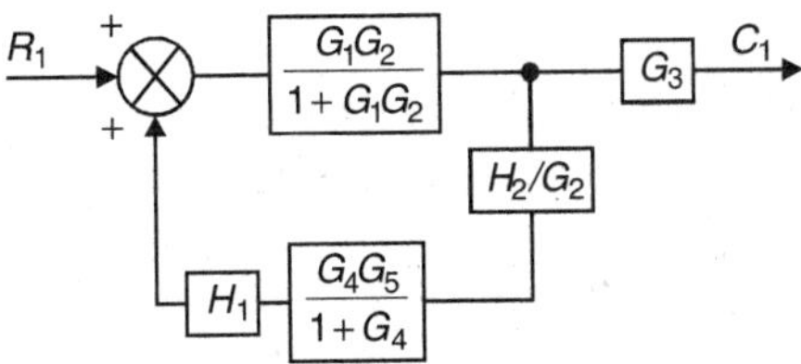

The required transfer function is

$$T_1 = \frac{C_1}{R_1} = \left(\frac{\dfrac{G_1G_2}{1+G_1G_2}}{1 - \dfrac{G_1G_2}{1+G_1G_2} \cdot \dfrac{H_1G_4G_5}{1+G_4} \cdot \dfrac{H_2}{G_2}} \right) G_3 .$$

$$= \frac{G_1G_2G_3(1+G_4)}{(1+G_1G_2)(1+G_4) - G_1G_4G_5H_1H_2}$$

Evaluation of C_2/R_2 :

To find C_2/R_2, let us assume $C_1 = 0$ and $R_1 = 0$.

The above block diagram can be reduced to

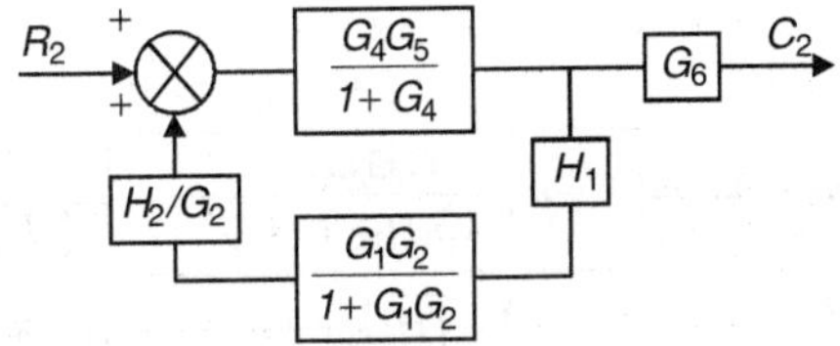

The required transfer function is

$$T_2 = \frac{C_2}{R_2} = \left(\frac{\dfrac{G_4G_5}{1+G_4}}{1 - \dfrac{G_4G_5}{1+G_4} \cdot \dfrac{H_2}{G_2} \cdot \dfrac{G_1G_2}{1+G_1G_2} . H_1} \right) \times G_6$$

$$= \frac{G_4G_5(1+G_1G_2)G_6}{(1+G_4)(1+G_1G_2) - G_1G_4G_5H_1H_2}$$

ASP-10: Reduce the given block diagram into its simplest form and obtain the transfer function $\dfrac{C(s)}{R(s)}$.

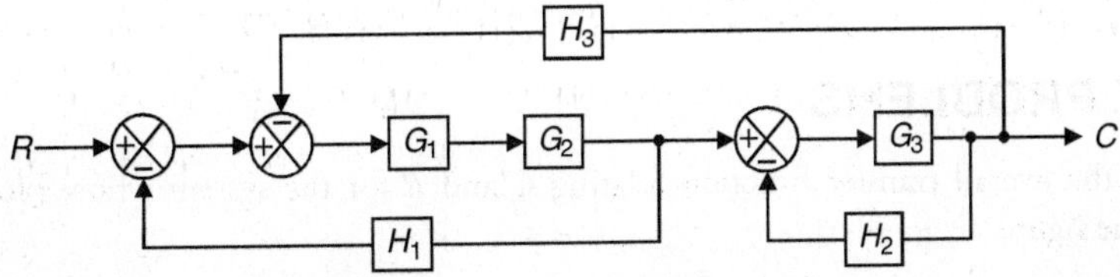

Solution:

By removing the feedback loop having feedback path transfer function H_2, we get

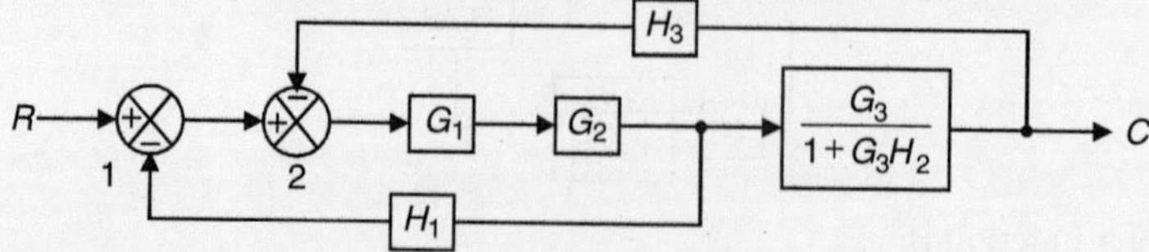

Interchanging summing points 1 and 2 as well as replacing the cascade block by its equivalent block, we get

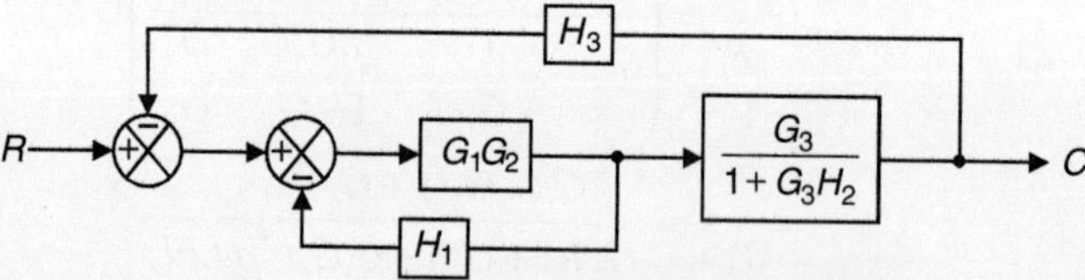

After reducing the above block diagram, we have

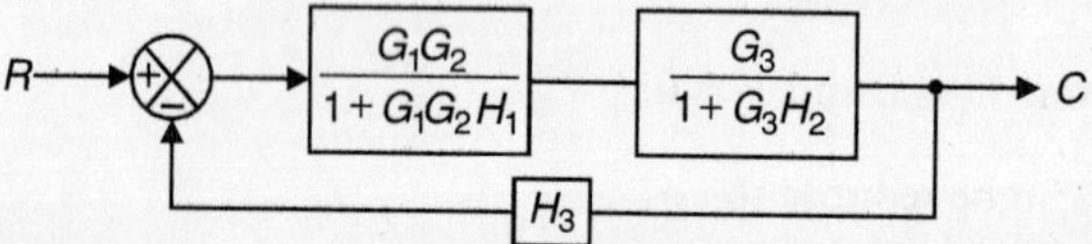

By cascading the two blocks, we get

R → $\dfrac{G_1G_2G_3}{(1+G_1G_2H_1)(1+G_3H_2)}$ → C, with feedback H_3

After reducing the above block diagram, we have the following block diagram.

R → $\dfrac{\dfrac{G_1G_2G_3}{(1+G_1G_2H_1)(1+G_3H_2)}}{1+\dfrac{G_1G_2G_3H_3}{(1+G_1G_2H_1)(1+G_3H_2)}}$ → C

Hence, the required transfer function is

$$\frac{C(s)}{R(s)}=\frac{G_1G_2G_3}{(1+G_1G_2H_1)(1+G_3H_2)+G_1G_2G_3H_3}.$$

EXERCISE PROBLEMS

1. Determine the overall transfer function relating C and R for the system whose block diagram is as shown in the figure.

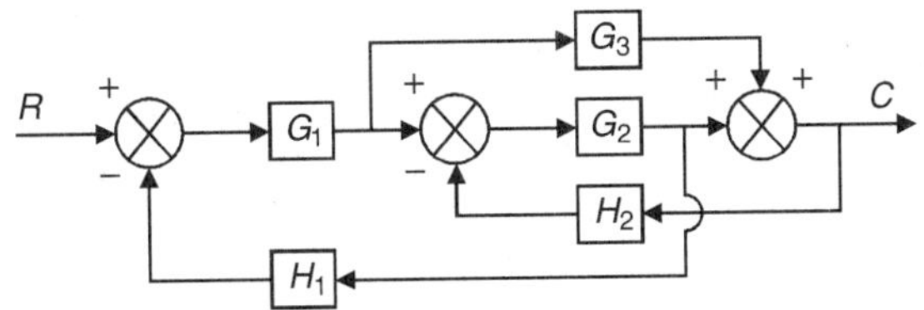

2. Reduce the given block diagram and obtain the transfer function $\frac{C(s)}{R(s)}$. Verify the same by using signal flow graph.

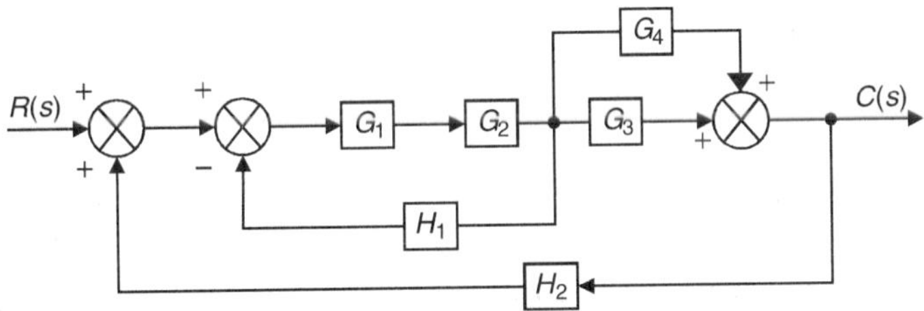

3. Reduce the given diagram and obtain the transfer function $\frac{C(s)}{R(s)}$.

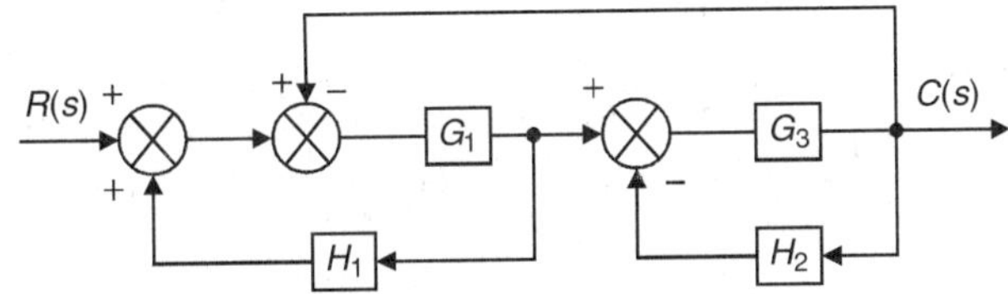

4. Reduce the given block diagram and obtain the transfer function $\frac{C(s)}{R(s)}$. Verify the same by using signal-flow graph.

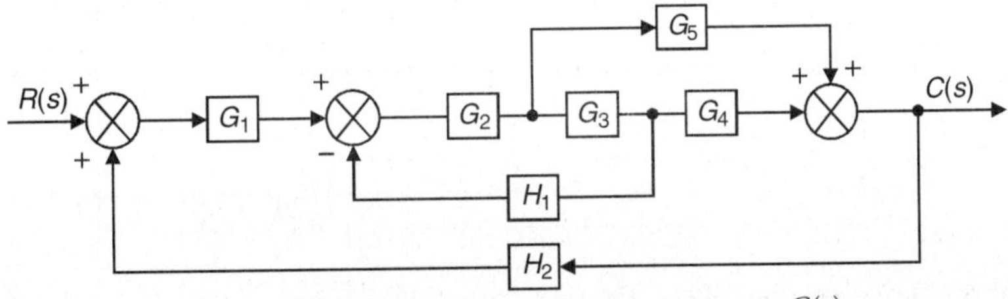

5. Reduce the given block diagram and obtain the transfer function $\frac{C(s)}{R(s)}$. Verify the same by using signal-flow graph.

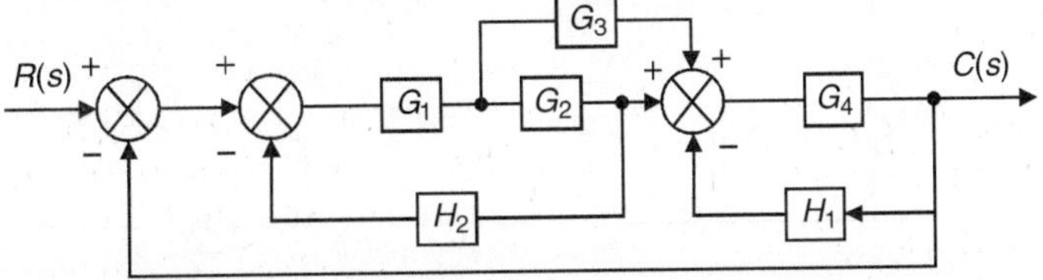

6. Reduce the given block diagram and obtain the transfer function $\frac{C(s)}{R(s)}$. Verify the same by using signal-flow graph.

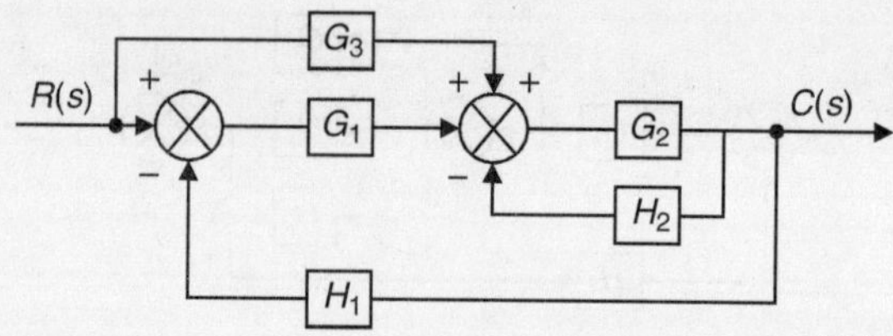

7. Reduce the given block diagram and obtain the transfer function $\frac{C(s)}{R(s)}$. Verify the same by using signal-flow graph.

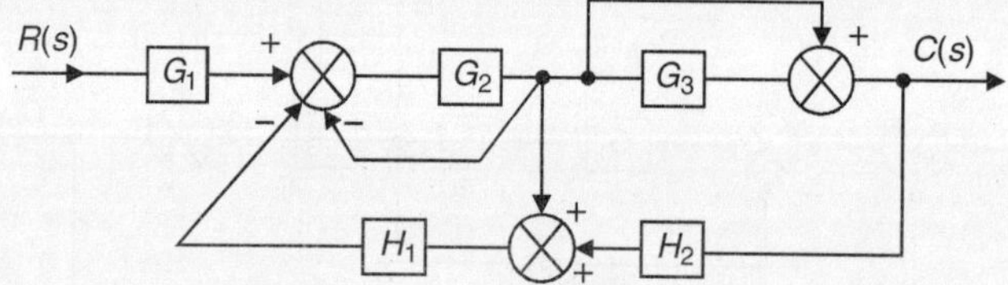

8. Determine the ratio $\frac{C(s)}{R(s)}$ and $\frac{C(s)}{D(s)}$ for the following block diagram.

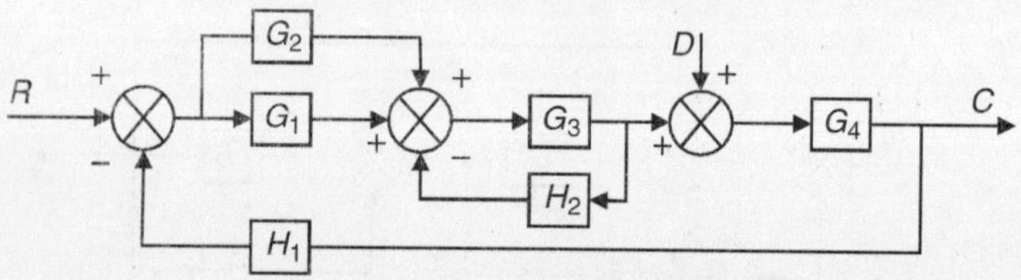

9. Reduce the given diagram and obtain the transfer function $\frac{C(s)}{R(s)}$.

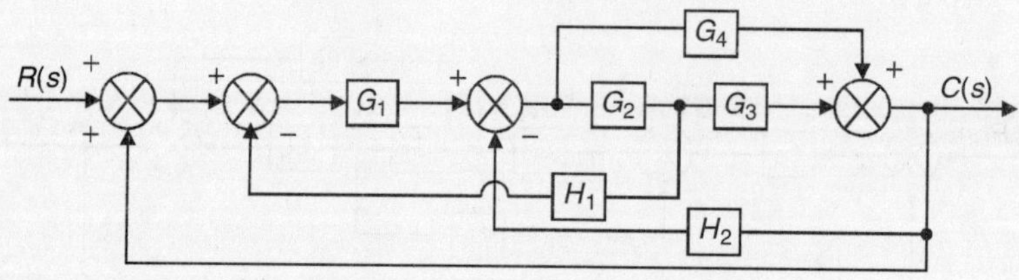

10. Reduce the given diagram and obtain the transfer function $\frac{C(s)}{R(s)}$.

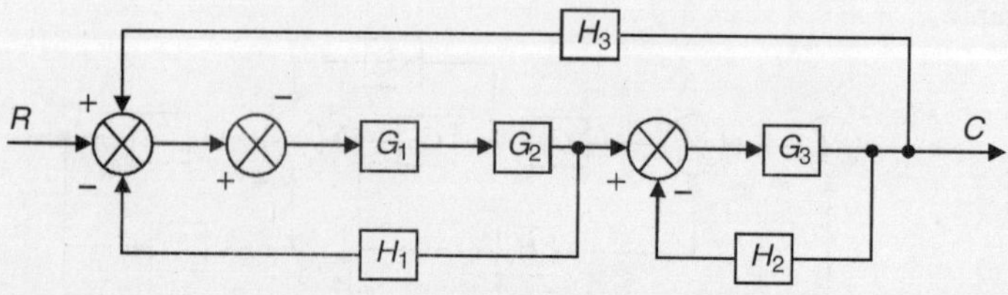

11. Reduce the given diagram and obtain the transfer function $\frac{C(s)}{R(s)}$.

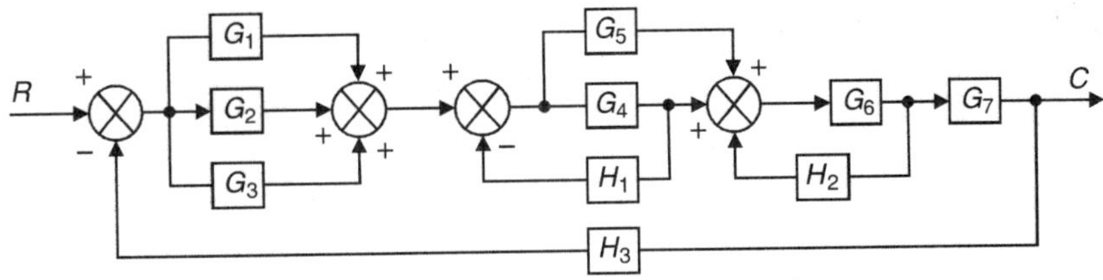

ANSWERS

1. $T(s) = \dfrac{C(s)}{R(s)} = \dfrac{G_1G_2 + G_1G_3}{1 + G_2H_2 + G_1G_2H_1 + G_1G_2G_3H_1H_2}$

2. $T(s) = \dfrac{C(s)}{R(s)} = \dfrac{G_1G_2(G_3 + G_4)}{1 + G_1G_2H_1 - G_1G_2(G_3 + G_4)H_2}$

3. $T(s) = \dfrac{C(s)}{R(s)} = \dfrac{G_1G_2}{1 + G_1G_2 + G_2G_2 + G_1H_1 + G_1G_2H_1H_2}$

4. $T(s) = \dfrac{C(s)}{R(s)} = \dfrac{G_1G_2(G_3G_4 + G_5)}{1 + G_1G_3H_1 + G_1G_2(G_3G_4 + G_5)H_2}$

5. $\dfrac{C(s)}{R(s)} = \dfrac{G_1G_4(G_2 + G_3)}{1 + G_2G_2H_2 + G_4H_1 + G_1G_2G_4H_1H_2 + G_1G_4(G_4 + G_3)}$

6. $T(s) = \dfrac{C(s)}{R(s)} = \dfrac{G_1G_2 + G_3G_2}{1 + G_2H_2 + G_1G_2H_1}$

7. $T(s) = \dfrac{C(s)}{R(s)} = \dfrac{G_1G_2(1 + G_3)}{1 + G_2H_1G_2 + H_1H_2G_2 + H_1H_2G_2G_3}$

8. $\dfrac{C(s)}{R(s)} = \dfrac{G_3G_4(G_1 + G_2)}{1 + G_3H_2 + G_1G_3G_4H_1 + G_2G_3G_4H_1}$

$\dfrac{C(s)}{D(s)} = \dfrac{G_4(1 + G_3H_2)}{1 + G_3H_2 + G_1G_3G_4H_1 + G_2G_3G_4H_1}$

9. $\dfrac{C(s)}{R(s)} = \dfrac{G_1G_2G_3 + G_1G_4}{1 + G_1G_2H_1 + H_2G_2G_3 + H_2G_4 + G_2G_2G_3 + G_2G_4}$

10. $\dfrac{C(s)}{R(s)} = \dfrac{G_1G_2G_3}{(1 + G_1G_2H_1)(1 + G_3H_2) + G_1G_2G_3H_3}$

11. $\dfrac{C(s)}{R(s)} = \dfrac{(G_1G_2G_3)(G_4 + G_5)G_6G_7}{(1 + G_4H_1)(1 + G_6H_2) + (G_1 + G_2 + G_3)(G_4 + G_5)G_6G_7H_3}$

OBJECTIVE TYPE QUESTIONS

1. The transfer function of $C(s)/R(s)$ of the system, as shown in the figure, is

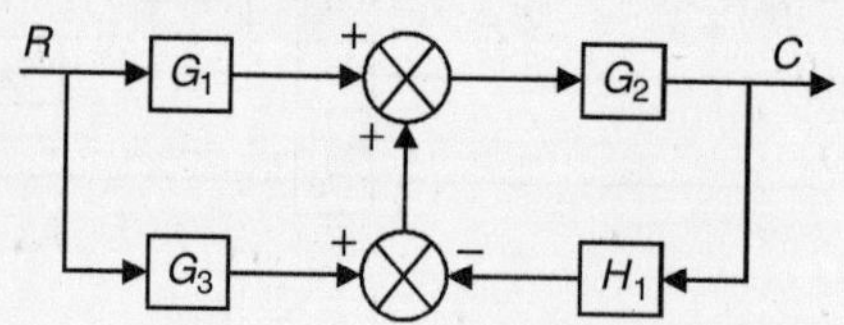

(a) $\dfrac{G_1G_2 + G_2G_3}{1+G_2H_1}$ (b) $\dfrac{G_1G_2 - G_2G_3}{1+G_2H_1}$

(c) $\dfrac{G_1 + G_2 + G_3}{1+G_2H_1}$ (d) None

2. The transfer function $C(s)/R(s)$ of the system, as shown in the figure, is

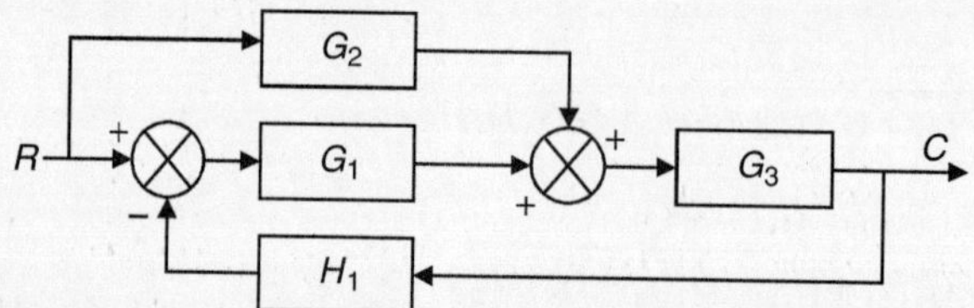

(a) $\dfrac{G_1G_3 - G_2G_3}{1-G_1G_3H_1}$ (b) $\dfrac{G_1G_3 + G_2G_3}{1+G_1G_3H_1}$

(c) $\dfrac{G_1 + G_2 + G_3}{1+G_1G_3H_1}$ (d) None

3. The transfer function of the block diagram, as shown in the figure, is

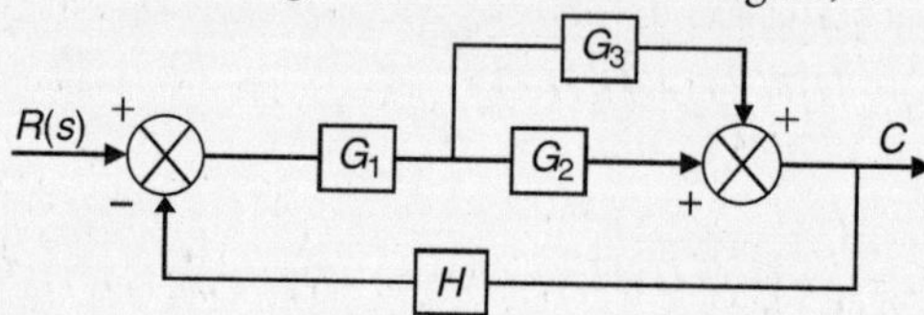

(a) $\dfrac{G_2(G_1 + G_3)}{1+G_1G_2H + G_1G_3H}$ (b) $\dfrac{G_1(G_2 + G_3)}{1+G_1G_2H + G_1G_3H}$

(c) $\dfrac{G_1 + (G_2 - G_3)}{1+G_1H + G_2H}$ (d) None

4. The transfer function of a system in regenerative feedback is given as

(a) $\dfrac{C(s)}{R(s)} = \dfrac{G(s)}{1+G(s)H(s)}$ (b) $\dfrac{C(s)}{R(s)} = \dfrac{G(s)H(s)}{1+G(s)H(s)}$

(c) $\dfrac{C(s)}{R(s)} = \dfrac{G(s)}{1-G(s)H(s)}$ (d) None

5. Regenerative feedback means the output feedback with
   (a) a positive sign (b) a negative sign
   (c) step (d) none
6. The transfer of the system, as shown in the figure, is

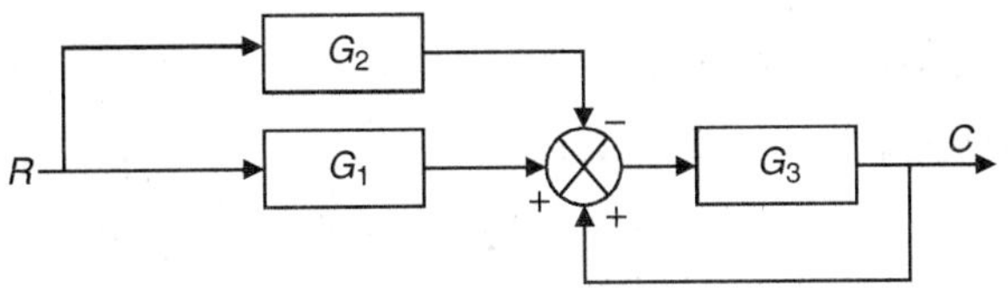

(a) $\frac{(G_1 + G_2)G_3}{1 + G_3}$ (b) $\frac{G_1G_2 + G_3}{1 + G_3}$

(c) $\frac{(G_1 + G_3)G_2}{1 + G_3}$ (d) None

7. The transfer function $C(s)/R(s)$ of the system, as shown in the figure, is

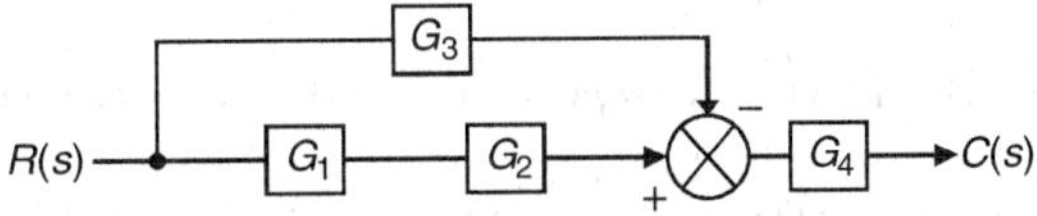

(a) $G_1G_2G_3 + G_4$ (b) $G_1G_2 + G_3 + G_4$

(c) $(G_1G_2 - G_3)\ G_4$ (d) none

8. For the following figure, output Y is

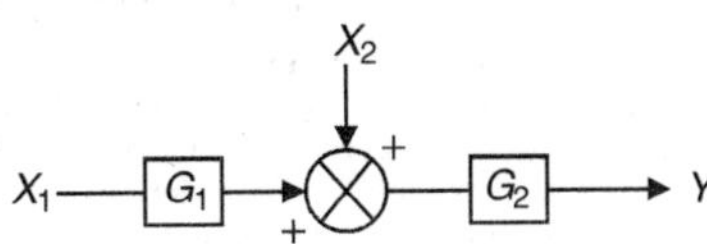

(a) $(X_1 + X_2)G_3$ (b) $(X_1 - X_2)\ G_2$

(c) $(G_1X_1 + X_2)G_2$ (d) $(G_1X_1 - X_2)\ G_2$

ANSWERS

1. a **2.** b **3.** b **4.** c **5.** a **6.** a **7.** c **8.** c

CHAPTER 5

Signal-Flow Graphs

Contents

5.1 INTRODUCTION

If the set of equations describing a system is known, it is possible to represent the system in another pictorial form, which is known as signal-flow graph. An alternative to block diagram is the signal-flow graph. S. J. Mason introduced the signal-flow graph for representing the cause and effect of linear systems that are modelled by algebraic equations. A signal-flow graph is a diagram that represents a set of simultaneous linear algebraic equations. Each signal-flow graph consists of a network in which nodes are connected by directed branches. Each node represents a system variable, and each branch acts as a signal multiplier. The signal flows in the direction indicated by the arrow. The aim of this chapter is to introduce signal-flow graph and discuss its application to find the transfer function of complicated systems.

5.2 TERMINOLOGY USED IN SIGNAL-FLOW GRAPH

The following gives the definition of various terms used in a signal-flow graph, which is illustrated in Figure 5.1.

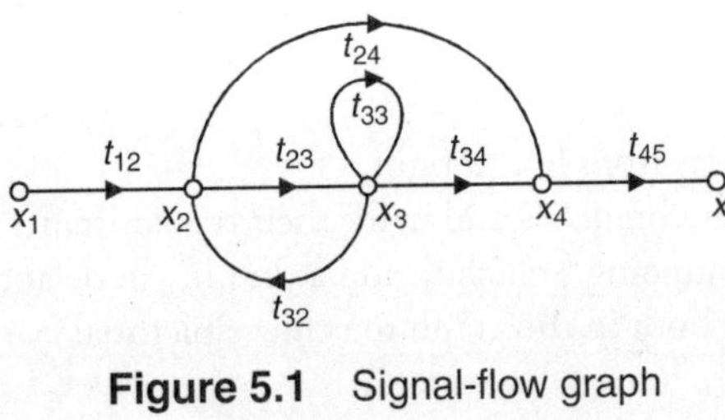

Figure 5.1 Signal-flow graph

1. **Node:** A node is a point representing a variable or signal.
2. **Branch:** A branch is a directed line segment joining two nodes.
3. **Transmittance:** It is the gain between two nodes.
4. **Path:** A path is a traversal of connected branches in the direction of the branch arrow.
5. **Input node or source node:** A node that has only outgoing branch(es). It is also called as source and corresponds to independent variable.
   Example: In the above figure, x_1 is the input node.
6. **Output node or sink node:** A node that has only incoming branches. This is also called as sink and corresponds to dependent variable.
   Example: In the above figure, x_5 is the output node.
7. **Mixed node or chain node:** A node that has incoming and outgoing branches.
   Example: In the above figure, x_2, x_3 and x_4 are mixed nodes.
8. **Forward path:** A path from source to sink without traversing any node more than once.
   Example: In the above figure, x_1 to x_2 to x_3 to x_4 to x_5 is the forward path.
9. **Feedback path:** A path that originates and terminates at the same node.
   Example: In the above figure, x_2 to x_3 to x_2 is a feedback path.
10. **Path gain:** Product of branch gains in a forward path.
    Example: In the above figure, path gain for path $x_1x_2x_3x_4x_5$ is $t_{12}\ t_{23}\ t_{34}t_{45}$.
11. **Loop:** A loop is a closed path.
    Example: In the above figure, x_2 to x_3 to x_2 is called a loop.
12. **Self-loop:** It is a feedback loop consisting of single branch.
    Example: In the above figure, t_{33} is the self-loop.
13. **Loop gain:** The loop gain is the product of branch transmittances of the loop.
    Example: In the above figure, the loop gain of loop x_2 to x_3 to x_2 is $t_{23}.t_{32}$.
14. **Non-touching loops:** Loops that do not possess a common node.
    Using above definitions, the signal-flow graphs for systems are constructed.

5.3 GUIDELINES TO CONSTRUCT SIGNAL-FLOW GRAPH

A signal-flow graph of a system is constructed from its describing equations, or by direct reference to the block diagram of the system. Each variable of the block diagram becomes a node and each block becomes a branch. The procedure used for the construction of a signal-flow graph of a linear system is described below.

1. Determine the system equations and write them into the following forms after finding the variables in the system.

$$x_1 = t_{11}x_1 + t_{12}x_2 + t_{13}x_3 + \cdots + t_{1n}x_n$$
$$x_2 = t_{21}x_1 + t_{22}x_2 + t_{23}x_3 + \cdots + t_{2n}x_n$$
$$\vdots$$
$$x_m = t_{m1}x_1 + t_{m2}x_2 + t_{m3}x_3 + \cdots + t_{mn}x_n$$

2. Arrange the input to output nodes from left to right.
3. Connect the nodes by appropriate branches and mark their transmittances $t_{11}, t_{12}, \ldots\ldots$, etc
4. If the desired output node has outgoing branches, add a dummy node and a unity gain branch.
5. Rearrange the nodes or loops or both in the graph to achieve pictorial clarity.

5.4 SIGNAL-FLOW GRAPH ALGEBRA

1. **Addition rule:** The value of the variable designated by a node is equal to the sum of all signals entering the node, as shown in Figure 5.2

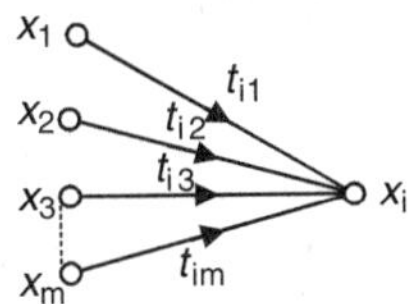

Figure 5.2 Addition rule

2. **Transmission rule:** The value of the variable designated by a node is transmitted on every branch leaving the node, as shown in Figure 5.3

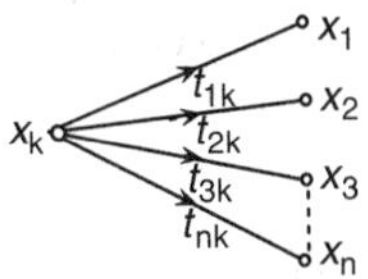

Figure 5.3 Transmission rule

3. **Multiplication rule:** A cascaded connection of $n - 1$ branches with transmission functions can be replaced by a single branch with new transmission function equal to the product of the old branches, as shown in Figure 5.4.

t_{21} t_{32} — x_1 x_2 x_3 x_m $\equiv$ $t_{21}\, t_{32} \cdots\cdots t_{m(m-1)}$ — x_1 x_m

Figure 5.4 Multiplication rules in signal-flow graphs

***Problem-5.1*:** Draw the signal-flow graph of the system of equations.

$$X_1 = a_{11}X_1 + a_{12}X_2 + a_{13}X_3 + b_1u_1$$
$$X_2 = a_{21}X_1 + a_{22}X_2 + a_{23}X_3 + b_2u_2$$
$$X_3 = a_{31}X_1 + a_{32}X_2 + a_{33}X_3$$

***Solution*:**
Variables X_1, X_2, X_3, u_1 and u_2 choose five nodes representing the variables. Now, connect the various nodes choosing appropriate branch gain in accordance with the equations. The signal-flow graph is shown in the figure.

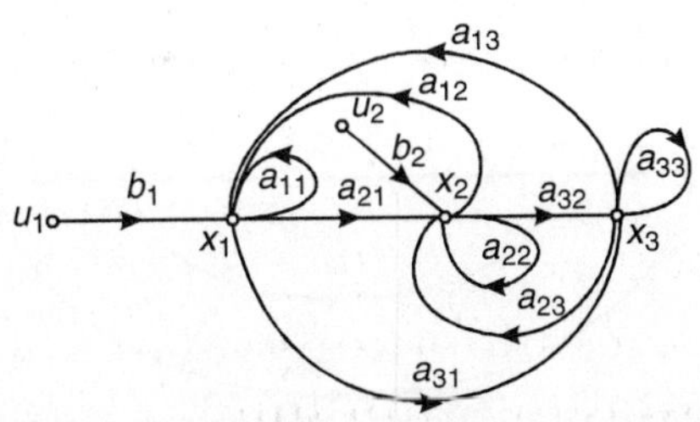

5.5 CONSTRUCTION OF SIGNAL-FLOW GRAPH FROM BLOCK DIAGRAM

A block diagram of complex linear system can easily be analysed by converting them into equivalent signal-flow graphs, and then applying direct gain formulas. Hence, generally, the signal-flow graphs are obtained from the block diagrams of the system before analysing them.

While drawing a signal-flow graph from the given block diagram, the adjacent summing point and take-off point (but not a take-off point preceding a summing point in the direction of signal flow) are represented by a node and the block transfer function is represented by a line joining the respective nodes. The direction of signal flow is indicated by an arrow on the line. However, in the direction of signal flow, if a take-off point precedes a summing point, then such points are represented by two separate nodes with a transmittance of unity between them.

Problem-5.2: Draw the signal-flow graph of the block diagram shown in the figure.

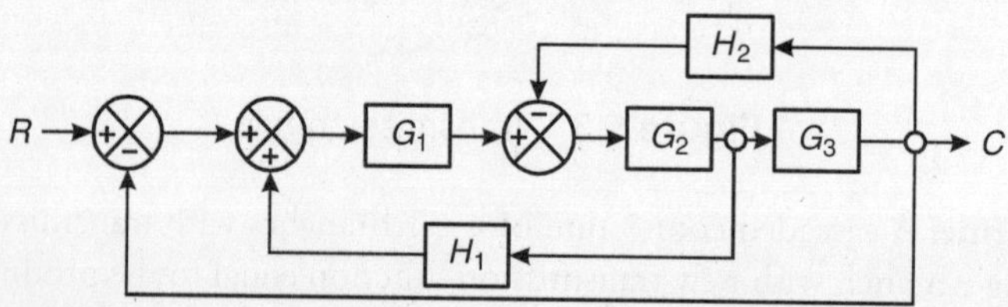

Solution:

Choose the nodes to represent the variables, say X_1 to X_5, as shown in the block diagram. Connect the nodes with appropriate gain along the branch. The signal-flow graph is shown in the figure.

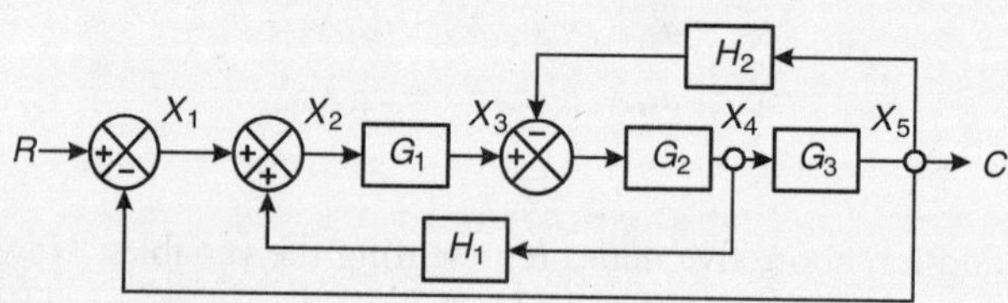

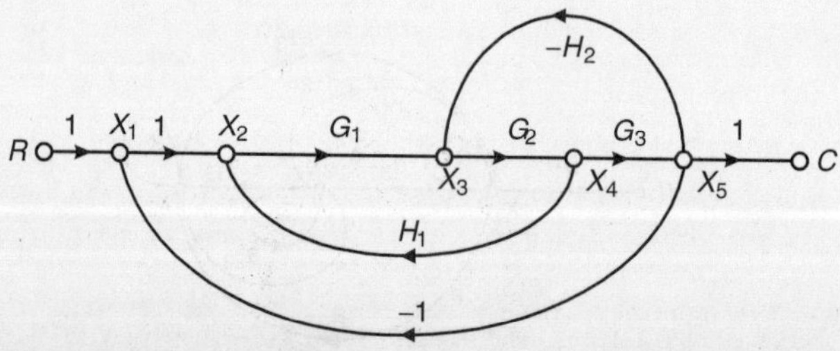

***Problem-5.3*:** Draw the signal-flow graph of the block diagram shown in the figure.

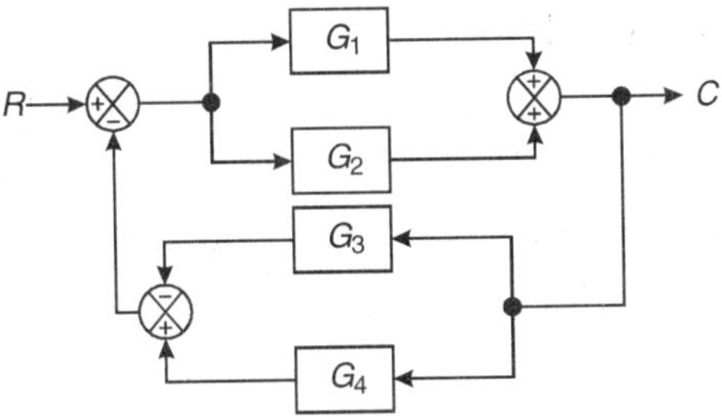

***Solution*:**
The nodal variables are X_1, X_2 and X_3. The signal-flow graph is shown in the following figure.

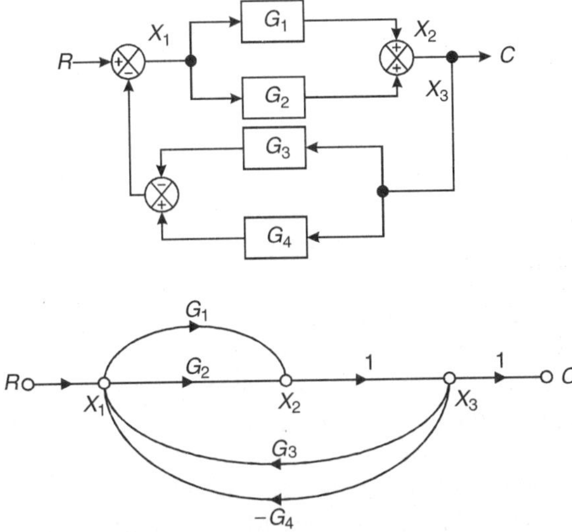

5.6 PROPERTIES OF SIGNAL-FLOW GRAPHS

The various properties of signal-flow graph are listed as follows:

1. It is applicable to linear systems.
2. The equations based on which a signal-flow graph is drawn must be algebraic equations in the form of effects as a function of causes. Nodes are used to represent variables. Normally, the nodes are arranged left to right, following a succession of causes and effects through the system.
3. Signals travel along the branches only in the direction described by the arrows of the branches.
4. A branch indicates functional dependence of one signal on the other.
5. The signal-flow graph of the system is not unique. By rearranging the system equations, different types of signal-flow graphs can be drawn for a given system.
6. A signal-flow graph can represent a block diagram, but the reverse is not true.

5.7 MASON'S GAIN FORMULA FOR SIGNAL-FLOW GRAPH

The overall gain of the system is obtained from the block diagram by reducing the diagram into a single block having the same input and output as the system. The overall gain can also be obtained by converting the block diagram into the corresponding signal-flow graphs, and then, by applying the gain calculation methods available for signal-flow graphs. The direct gain formula for calculating the gain of signal-flow graph is preferred as compared to reduction methods because it is less time consuming.

The relationship between an input variable and an output variable of a signal-flow graph is given by the net gain between the input and the output nodes and is known as the overall gain of the system. Mason's gain formula is used to obtain the overall gain (transfer function) of signal-flow graphs.

The Mason's gain formula states that

$$T(s) = \frac{\sum M_k \Delta_k}{\Delta} \tag{5.1}$$

where M_k is the gain of the kth forward path, $\Delta = 1 -$ (Sum of the gain products of all individual loops) + (Sum of the gain products two non-touching loops) – (Sum of the gain products three non-touching loops) + ... and Δ_k= same as Δ formed by loops but not touching k[th] forward path.

The procedure for determining the gain (transfer function) of a signal-flow graph by using Mason's gain formula is described as follows:

1. Determine the number of forward paths present in the signal-flow graph and calculate the forward gain of each path.
2. Determine the number of feedback loops present in the signal-flow graph and calculate the loop gain of each loop. The sum of all loop gains will be the second term of equation Δ.
3. Determine the possible combinations of having two non-touching loops simultaneously and obtain the product of loop gains. The sum of all loop gains for all possible combinations will be the third term of the equation Δ. Similarly, obtain the possible combinations of having three non-touching loops simultaneously and calculate the product of all loop gains. The sum of all possible combinations will be the fourth term of equation Δ and so on.
4. Examine the forward paths obtained in Step 1 and see which of the paths are not touching the loops. Then, for each forward path determine the value of Δ_k.
5. Substitute various values of M_k, Δ_k and Δ in the Mason's gain formula (5.1) to obtain the overall gain (transfer function) of the signal-flow graph.

Problem-5.4: For the signal-flow graph shown in the figure, determine the ratio $\frac{x_5}{x_1}$. Use Mason's gain formula for signal-flow graphs.

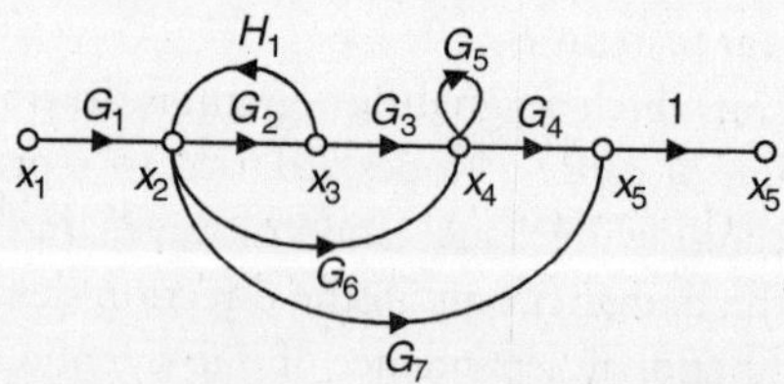

Solution:

1. There are three forward paths as shown in Figure 1 (a), (b) and (c) for which the path gains are

$$M_1 = G_1.G_2.G_3.G_4$$
$$M_2 = G_1.G_6.G_4$$
$$M_3 = G_1.G_7$$

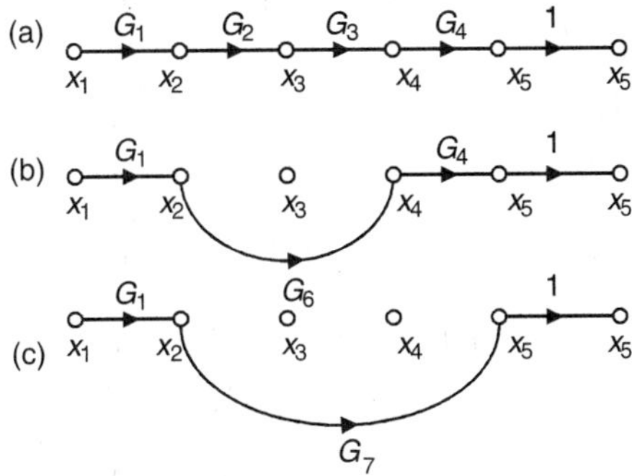

Figure 1 (a), (b) and (c) Forward paths

2. There are two non-touching loops, as shown in Figure 1(d) and (e). The loop gains are

$$L_{11} = G_2.H_1$$
$$L_{21} = G_5$$

Figure 1 (d) and (e) Non-touching feedback loops

3. There is only one set of two non-touching loops, as shown in Figure 1(d) and (e). The product of loop gains is

$$G_2G_5H_1.$$

4. As there are no sets of three, four, five or six non-touching loop combinations present in the signal-flow graph terms, we have

$$M_{j3} = M_{j4} = M_{j5} = M_{j6} = M_{jk} = 0.$$

5. From equation, Δ can be calculated as follows:

$$\Delta = 1-(G_2H_1 + G_5) + G_2H_1G_5.$$

6. To determine the value of Δ_k, the forward path of signal-flow graph is examined. It is found that only forward path M_3 in Figure 1(c) does not touch the loops, while forward paths M_1 and M_2 touch the loops. Hence, value of Δ_k for paths M_1 and M_2 will have only unity term, while other terms will be zero, that is,

$$\Delta_1 = 1$$
$$\Delta_2 = 1$$

For forward path M_3 from Figure 1(f) and (g), the value of Δ_3 is

$$\Delta_3 = 1 - G_5.$$

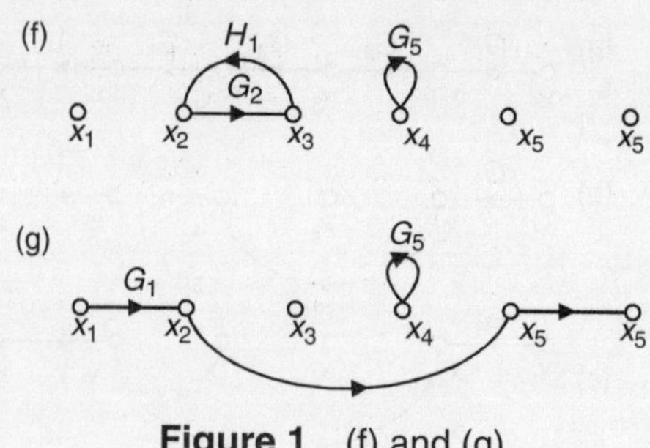

Figure 1 (f) and (g)

7. From Mason's gain formula, we have

$$\frac{x_5}{x_1} = \frac{M_1\Delta_1 + M_2\Delta_2 + M_3\Delta_3}{\Delta}.$$

8. Substituting all the values from previous steps, we get

$$\frac{x_5}{x_1} = \frac{G_1G_2G_3G_4 + G_1G_4G_6 + G_1G_7(1 - G_5)}{1 - G_2H_1 + G_2G_5H_1 - G_5}.$$

Problem-5.5: Using Mason's direct gain formula, determine the gain $\frac{x_o}{x_i}$ in the signal-flow graph, as shown in the figure.

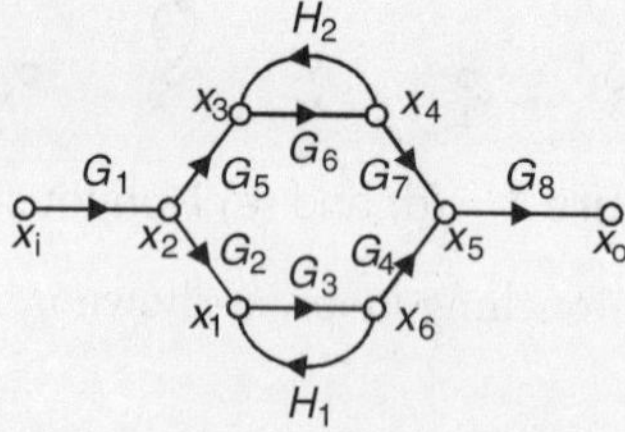

Solution: There are two forward paths in the given signal-flow graph, as shown in Figure 2(a) and (b).

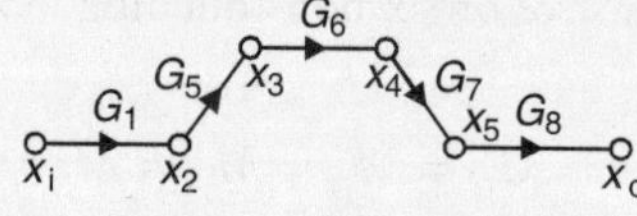

(a) Forward path for M_1

(b) Forward path for M_2

(c) Feedback path for L_1(loop)

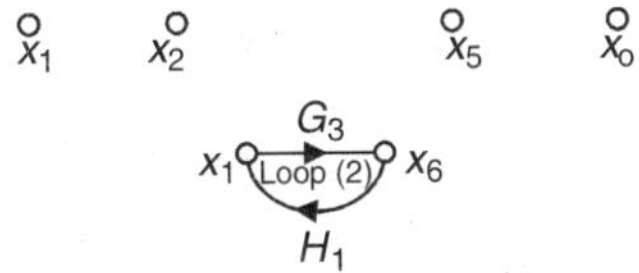

(d) Feedback path for L_2(loop)

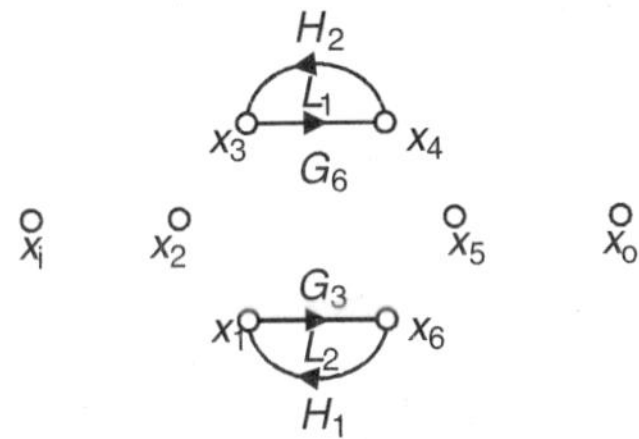

(e) Two non-touching loops

Figure 2

1. The path gains are

$$M_1 = G_1G_5G_6G_7G_8$$
$$M_2 = G_1G_2G_3G_4G_8.$$

2. There are two non-touching loops in the signal-flow graph that are shown in Figure 2(c) and (d). The loop gains for these two loops are

$$L_1 = G_6H_2$$
$$L_2 = G_3H_1.$$

3. There is only one possible combination of two non-touching loops shown in Figure 2(e). The product of these loop gains are

$$L_{12} = G_3G_6H_1H_2.$$

4. Substituting values from the above-mentioned steps in Δ equation, we get

$$\Delta = 1 - (G_3H_1 + G_6H_2) + G_3G_6H_1H_2\,.$$

5. To determine the value of Δ_K, the forward paths for M_1 and M_2 are examined whether they touch loops (1) and (2) or not. It is observed that M_1 path does not touch loop (2), whereas M_2 path does not touch loop (1). Hence,

$$\Delta_1 = 1 - G_3H_1$$
$$\Delta_2 = 1 - G_6H_2$$

6. Therefore, Mason's gain formula for this graph is

$$\frac{x_0}{x_i} = \frac{M_1\Delta_1 + M_2\Delta_2}{\Delta}$$

$$= \frac{G_1G_5G_6G_7G_8(1 - G_3H_1) + G_1G_2G_3G_4G_8(1 - G_6H_2)}{1 - (G_3H_1 + G_6H_2) + G_3G_6H_1H_2}$$

Problem-5.6: For the signal-flow graph of a multiple loop system shown in the figure, determine $\frac{C}{R}$.

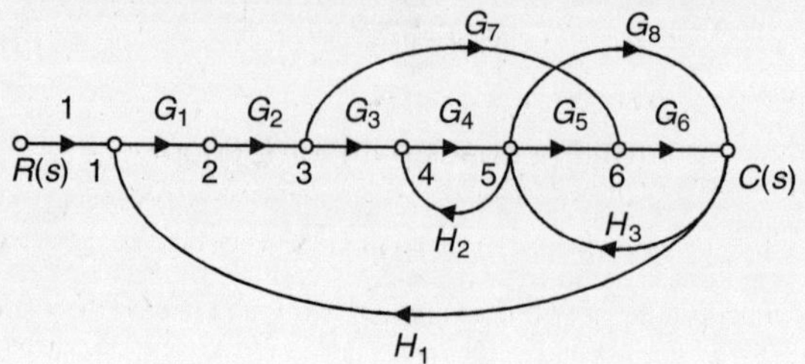

Solution:

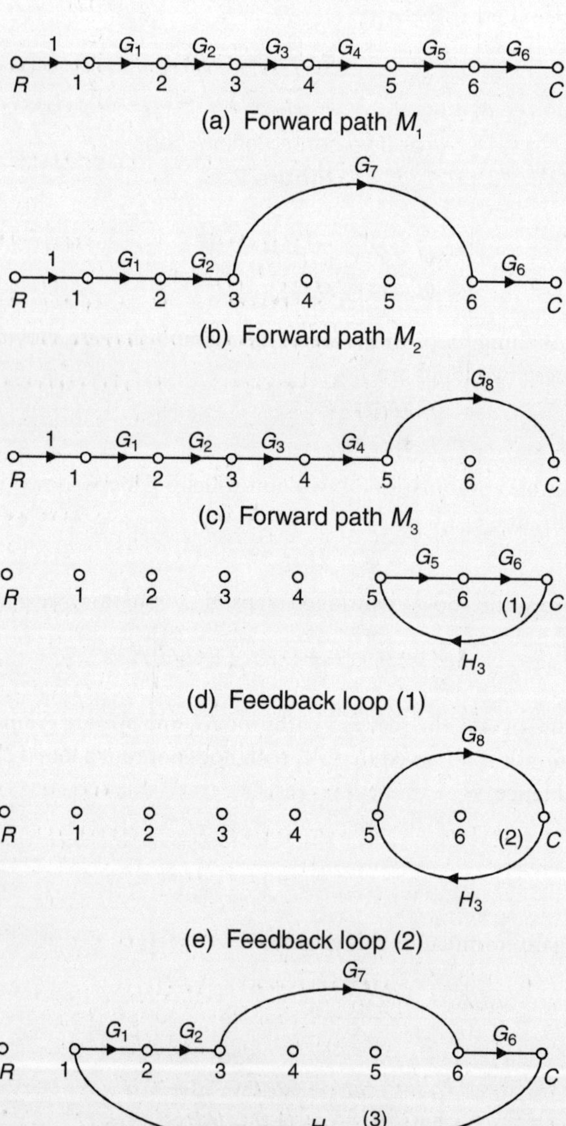

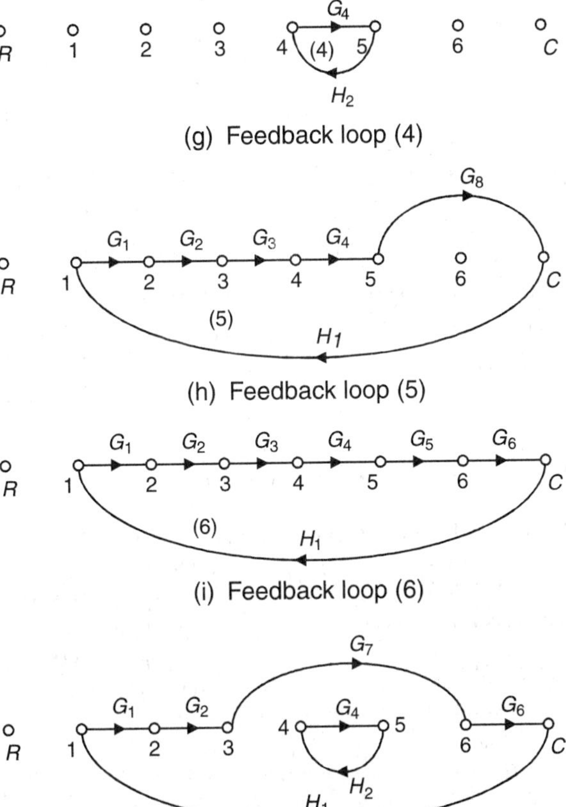

(g) Feedback loop (4)

(h) Feedback loop (5)

(i) Feedback loop (6)

(j) A combination of two non-touching loops

Figure 3

1. The forward paths are

$$M_1 = G_1G_2G_3G_4G_5G_6$$
$$M_2 = G_1G_2G_7G_6$$
$$M_3 = G_1G_2G_3G_4G_8$$

2. There are six non-touching feedback loops in the signal-flow graph. The loop transfer functions are as follows:

$$L_1 = G_5G_6H_3$$
$$L_2 = G_8H_3$$
$$L_3 = G_1G_2G_7G_6H_1$$
$$L_4 = G_4H_2$$
$$L_5 = G_1G_2G_3G_4G_8H_1$$
$$L_6 = G_1G_2G_3G_4G_5G_6H_1$$

3. The only possible combinations of two non-touching loops of the signal-flow graphs is shown in Figure 3 (j). The product of loop gain is

$$L_{12} = G_1G_2G_6G_7H_1G_4H_2.$$

4. Δ is determined as

$$\Delta = 1-[L_1 + L_2 + L_3 + L_4 + L_5 + L_6] + L_{12}$$
$$= 1-[G_5G_6H_3 + G_8H_3 + G_1G_2G_6G_7H_1 + G_4H_2$$
$$+ G_1G_2G_3G_4G_8H_1 + G_1G_2G_3G_4G_5G_6H_1]$$
$$+ G_1G_2G_4G_6G_7H_1H_2$$

5. All forward paths except one shown in Figure 3(b) does not touch loop L_4. Hence, values of Δ_K for forward paths M_1, M_2 and M_3 are

$$\Delta_1 = 1$$
$$\Delta_2 = 1 - G_4H_2.$$
$$\Delta_3 = 1$$

6. The Mason's gain formula for this signal-flow graph is

$$\frac{C}{R} = \frac{M_1\Delta_1 + M_2\Delta_2 + M_3\Delta_3}{\Delta}$$
$$\frac{C}{R} = \frac{G_1G_2G_3G_4G_5G_6 + G_1G_2G_7G_6(1 - H_2G_4) + G_1G_2G_3G_4G_8}{\Delta}$$

Problem-5.7: From the block diagram shown below, draw the corresponding signal-flow graph and evaluate closed-loop transfer function relating the output and the input.

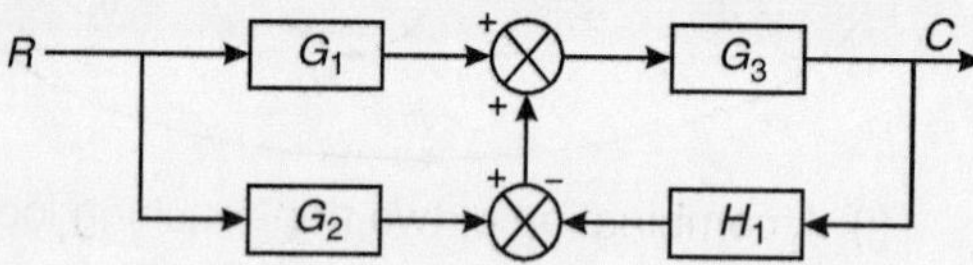

Solution:

The signal-flow graph for the above block diagram can be drawn as follows:

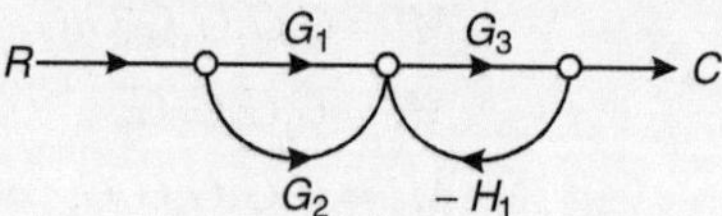

The forward paths are

$$M_1 = G_1G_3,\ \Delta_1 = 1$$
$$M_2 = G_2G_3,\ \Delta_2 = 1$$

The individual loop is

$$L_1 = -G_3H_1.$$

The graph determinant is

$$\Delta = 1-(L_1) = 1-(-G_3H_1) = 1 + G_3H_1.$$

According to Mason's gain formula, the closed-loop transfer is

$$T(s) = \frac{C(s)}{R(s)} = \frac{M_1\Delta_1 + M_2\Delta_2}{\Delta} = \frac{G_1G_3 + G_2G_3}{1 + G_3H_1}$$

Problem-5.8: Represent the following set of equations by a signal-flow graph and determine the overall gain relating x_5 and x_1.

$$x_2 = ax_1 + fx_2$$
$$x_3 = bx_2 + ex_4$$
$$x_4 = cx_3 + hx_3$$
$$x_5 = dx_4 + gx_2$$

Solution:

Variables x_1, x_2, x_3, x_4 and x_5 are represented by nodes and respective transmittances are joined by lines to draw the following signal-flow graph.

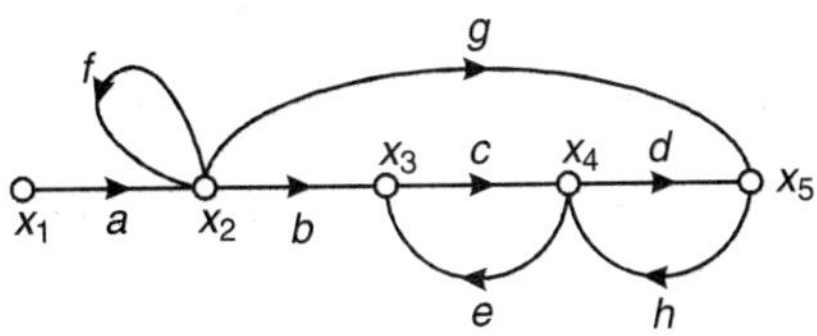

The forward paths are

$$M_1 = abcd, \Delta_1 = 1$$
$$M_2 = ag, \Delta_2 = 1 - L_1 = 1 - ce$$

The individual loops are

$$L_1 = ce, L_2 = dh, L_3 = f$$

Two pairs of non-touching loops are

$$L_1L_3 = fce, L_2L_3 = fdh$$

The graph determinant Δ is given as

$$\begin{aligned}\Delta &= 1 - (L_1 + L_2 + L_3) + (L_1L_2 + L_1L_3) \\ &= 1 - (f + ce + dh) + (fce + fdh) \\ &= 1 - f - ce - dh + fce + fdh\end{aligned}$$

By applying Mason's gain formula, the overall gain relating x_5 and x_1 is

$$\begin{aligned}\frac{x_5}{x_1} &= \frac{M_1\Delta_1 + M_2\Delta_2}{\Delta} \\ &= \frac{abcd + ag(1 - ce)}{1 - f - ce - dh + fce + fdh} \\ &= \frac{abcd + ag - agce}{1 - f - ce - dh + fce + fdh}\end{aligned}$$

5.8 ADVANTAGES OF SIGNAL-FLOW GRAPH OVER BLOCK DIAGRAM REDUCTION TECHNIQUE

The following are the advantages of signal-flow graph technique over block diagram reduction technique:

1. In block diagram reduction technique, the resultant transfer function is obtained by block diagram reduction rules, in which number of steps are to be performed. On the other hand, in signal-flow graph technique, the resultant transfer function is directly obtained by using Mason's gain formula.
2. The block diagram reduction technique is slightly complicated and time consuming because block diagram is to be drawn after each step. In the signal-flow graph, there is no need for such repetition of drawings for calculating the resultant transfer function.
3. In block diagram reduction technique, formula $\frac{G(s)}{1 \pm G(s)H(s)}$ is used for minor feedback loop. In contrast, in signal-flow graph, no such formula is necessary, as gain of various forward paths and feedback loops are just the product of associated branch gains.

ADDITIONAL SOLVED PROBLEMS

ASP-1: The signal-flow graph shown below has one forward path and single isolated loop. Determine the overall transmittance relating x_6 and x_1.

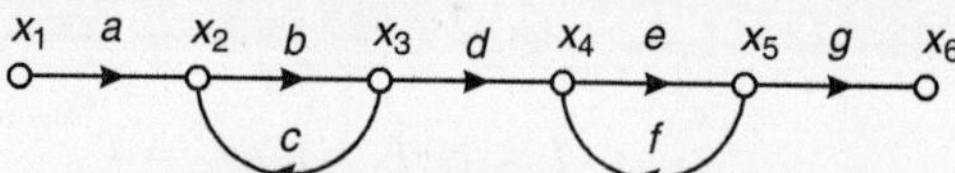

Solution:
Given that

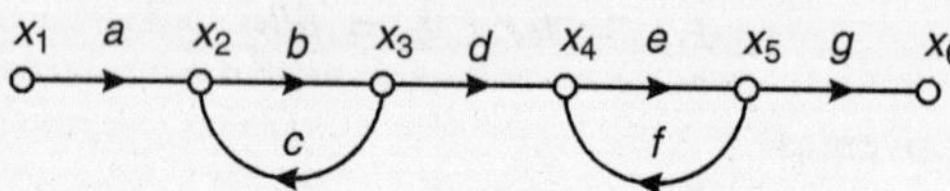

The equations for the above signal-flow graph are given as

$$x_2 = ax_1 + cx_3$$
$$x_3 = bx_2$$
$$x_4 = dx_3 + fx_5$$
$$x_5 = ex_4$$
$$x_6 = gx_5$$

By eliminating the intermediate variables, that is, x_2, x_3, x_4 and x_5 from the above-mentioned equations, the overall transmittance is determined as

$$\frac{x_6}{x_1} = \frac{abdeg}{1-(bc+ef)+bcef}$$

ASP-2: The signal-flow graph shown below has one forward path and two non-touching loops. Determine the overall transmittance relating x_6 and x_1.

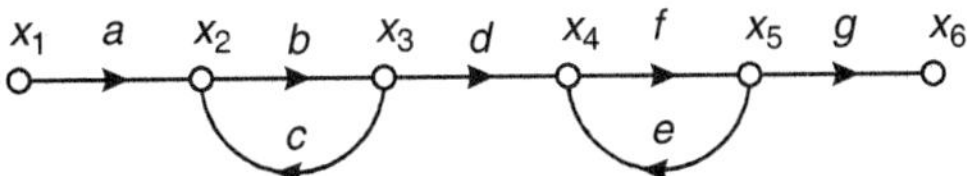

Solution:

Given that

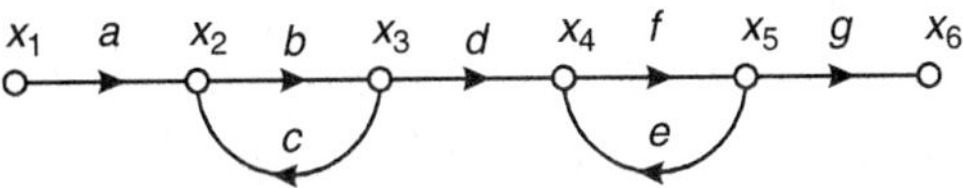

The equations for the above signal-flow graph are given as

$$x_2 = ax_1 + cx_3$$
$$x_3 = bx_2$$
$$x_4 = dx_3 + ex_5$$
$$x_5 = fx_4$$
$$x_6 = gx_5$$

By eliminating the intermediate variables, that is, x_2, x_3, x_4 and x_5, the overall transmittance is determined as

$$\frac{x_6}{x_1} = \frac{abdfg}{1-(bc+de)}$$

ASP-3: From the block diagram given below, draw the corresponding signal-flow graph and evaluate closed-loop transfer function relating the output and the input.

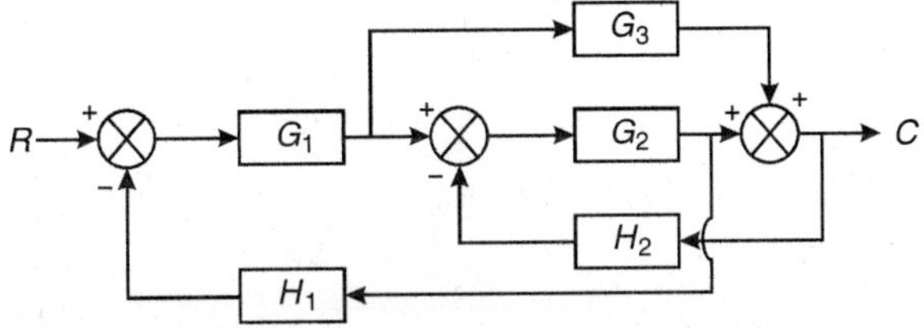

Solution:

The take-off points after blocks G_1 and G_2, respectively, are preceding a summing point, and hence, they are represented by separate nodes and a transmittance of unity is incorporated between them. The signal-flow graph for the block diagram can be drawn as follows.

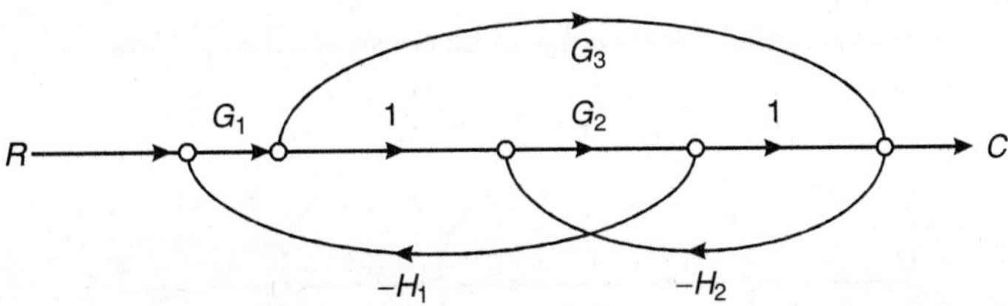

The forward paths are

$$M_1 = G_1G_2, \Delta_1 = 1$$
$$M_2 = G_1G_3, \Delta_2 = 1$$

The individual loops are

$$L_1 = -G_1G_2H_1$$
$$L_2 = -G_2H_2$$
$$L_3 = G_1G_3(-H_2)G_2(-H_1) = G_1G_2G_3H_1H_2$$

The graph determinant is

$$\Delta = 1-(L_1+L_2+L_3)$$
$$= 1-(-G_2H_2 - G_1G_2H_1 + G_1G_2G_3H_1H_2).$$
$$= 1+G_2H_2 + G_1G_2H_1 - G_1G_2G_3H_1H_2$$

By applying Mason's gain formula, the relation $C(s)/R(s)$ can be determined as

$$T(s) = \frac{C(s)}{R(s)} = \frac{M_1\Delta_1 + M_2\Delta_2}{\Delta} = \frac{G_1G_2 + G_2G_3}{1+G_2H_2 + G_1G_2H_1 - G_1G_2G_3H_1H_2}$$

ASP-4: A system is represented by the following set of equations. Construct the signal-flow graph and find the closed-loop transfer function.

$$x = x_1 + t_3u$$
$$\overset{o}{x_1} = -g_1x_1 + x_2 + t_2u$$
$$\overset{o}{x_2} = -g_2x_1 + t_1u$$

Solution:

Let us assume x as the output and u as the input.

By writing the above mentioned equations in the s-domain, we get

$$X = X_1 + t_3U \tag{1}$$
$$sX_1 = -g_1X_1 + X_2 + t_2U$$
$$(s+g)X_1 = X_2 + t_2U$$
$$X_1 = \frac{X_2}{s+g_1} + \frac{t_2}{s+g_1}U \tag{2}$$
$$sX_2 = -g_2X_1 + t_1U$$
$$X_2 = \frac{-g_2}{s}X_1 + \frac{t_1}{s}U \tag{3}$$

By using the three equations, the signal-flow graph can be constructed as follows:

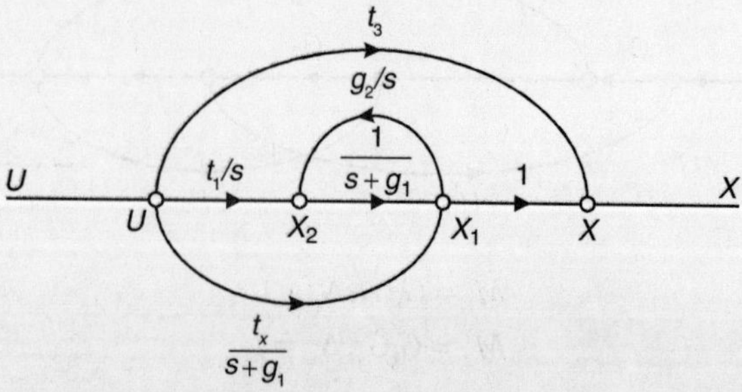

The forward paths are

$$M_1 = \frac{t_1}{s}\frac{1}{s+g_1} = \frac{t_1}{s(s+g_1)}, \quad \Delta_1 = 1$$

$$M_2 = t_3, \quad \Delta_2 = 1 + \frac{g_2}{s(s+g_1)}$$

$$M_3 = t_2 / s + g_1, \quad \Delta_3 = 1$$

The graph determinant is

$$\Delta = 1 + \frac{g_2}{s(s+g_1)}$$

The loops are

$$L_1 = \frac{-g_2}{s(s+g_1)}$$

According to Mason's gain formula, the required transfer function is

$$T = \frac{M_1\Delta_1 + M_2\Delta_2 + M_3\Delta_3}{\Delta}$$

$$= \frac{\frac{t_1}{s(s+g_1)}1 + t_3\left[1 + \frac{g_2}{s(s+g_1)}\right] + \frac{t_2}{s+g_1}}{1 + \frac{g_2}{s(s+g_1)}}$$

$$= \frac{t_1 + t_3[s(s+g_1) + g_2] + st_2}{s(s+g_1) + g_2}$$

$$= \frac{t_1 + t_3(s^2 + g_1 s + g_2) + st_2}{s^2 + g_1 s + g_2}$$

ASP-5: Find C/R for the system as shown in the figure.

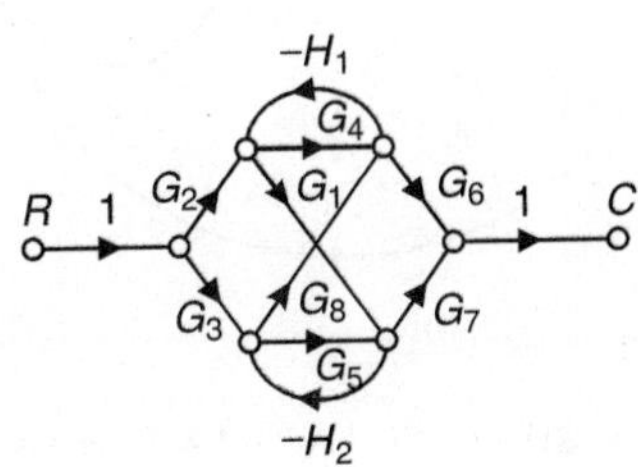

Solution:

Forward paths are

$$M_1 = G_2G_4G_6, \quad \Delta_1 = 1 + G_5H_2$$

$$M_2 = G_1G_2G_7, \quad \Delta_2 = 1$$

$$M_3 = G_3G_5G_7, \quad \Delta_3 = 1 + G_4H_1$$

$$M_4 = G_3G_6G_8, \quad \Delta_4 = 1$$

$$M_5 = -G_2G_1H_2G_8G_6, \quad \Delta_5 = 1$$

$$M_6 = -G_3G_8H_1G_1G_7, \quad \Delta_6 = 1$$

Individual loops are

$$L_1 = -G_4H_1$$
$$L_2 = -G_5H_2$$
$$L_3 = G_1H_2G_8H_1$$

Non-touching loops are

$$L_1L_2 = G_4H_1G_5H_2$$

The graph determinant is

$$\begin{aligned}\Delta &= 1-(-G_4H_1 - G_5H_2 + G_1H_2G_8H_1)+(G_4H_1G_5H_2)\\ &= 1+G_4H_1 + G_5H_2 - G_1H_2G_8H_1 + G_4H_1G_5H_2\end{aligned}$$

According to Mason's gain formula, the required transfer function is

$$T = \frac{C}{R} = \frac{M_1\Delta_1 + M_2\Delta_2 + M_3\Delta_3 + M_4\Delta_4 + M_5\Delta_5 + M_6\Delta_6}{\Delta}$$

$$= \frac{\begin{matrix}G_2G_4G_6(1+G_5H_2)+G_1G_2G_7 + G_3G_5G_7(1+G_4H_1)\\ +G_3G_6G_8 - G_2G_1G_6G_8H_2 - G_1G_3G_5G_8H_1\end{matrix}}{1+G_4H_1 + G_5H_2 - G_1G_8H_1H_2 + G_4G_5H_1H_2}$$

ASP-6: Find the transfer function for the signal-flow graph, as shown in the following figure.

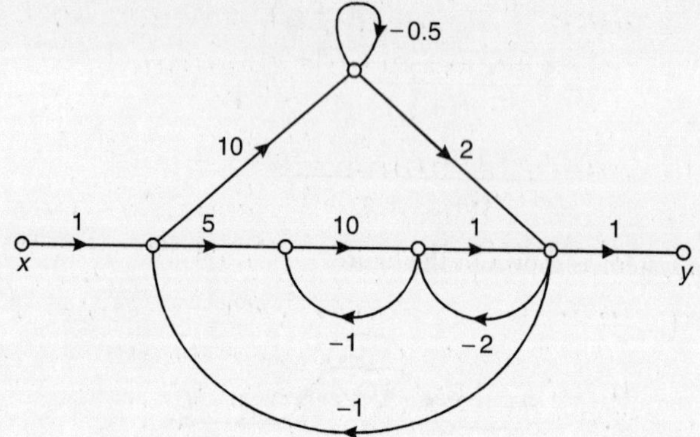

Solution:
Forward paths are

$$M_1 = 1\times10\times2 = 20, \qquad \Delta_1 = 1 - L_3 = 1+10 = 11$$
$$M_2 = 1\times5\times10\times1\times1 = 50, \quad \Delta_2 = 1 - L_2 = 1+0.5 = 1.5$$

Individual loops are

$$L_1 = 10\times2\times-1 = -20$$
$$L_2 = -0.5$$
$$L_3 = 10\times-1 = -10$$
$$L_4 = 1\times-2 = -2$$
$$L_5 = 5\times10\times1\times-1 = -50$$

Non-touching loop pairs are

$$L_1L_3 = -20 \times -10 = 200$$
$$L_2L_3 = -0.5 \times -10 = 5$$
$$L_2L_4 = -0.5 \times 2 = 1$$
$$L_2L_5 = -0.5 \times -50 = 25$$

The graph determinant is

$$\begin{aligned}\Delta &= 1 + (20 + 0.5 + 10 + 2 + 50) + (200 + 5 + 1 + 25) \\ &= 1 + 82.5 + 231 \\ &= 314.5\end{aligned}$$

According to Mason's gain formula, the required transfer function is

$$\begin{aligned}T &= \frac{Y}{X} = \frac{M_1\Delta_1 + M_2\Delta_2}{\Delta} \\ &= \frac{20 \times 11 + 50 \times 1.5}{314.5} \\ &= 0.9379\end{aligned}$$

EXERCISE PROBLEMS

1. Determine the transfer function $\frac{C(s)}{R(s)}$ for the following signal-flow graph by using Mason's gain formula.

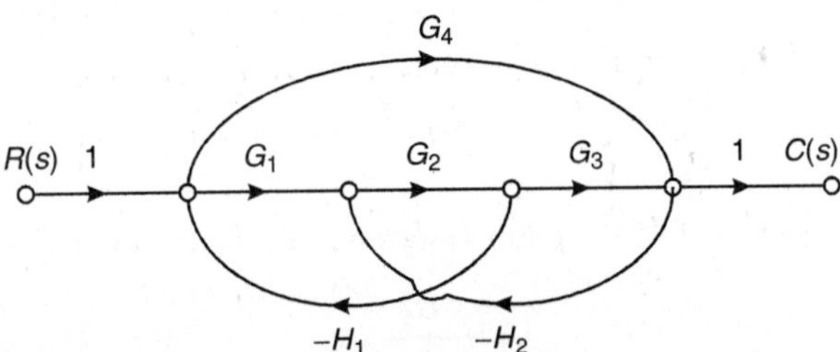

2. Find the transfer function for the following signal-flow graph by using Mason's gain formula.

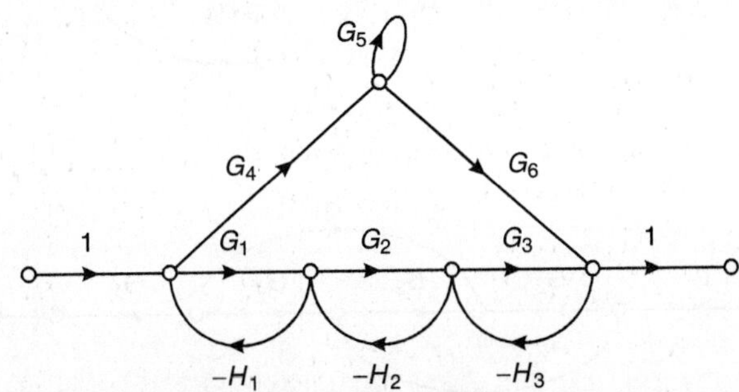

3. Determine the transfer function of the system as shown in the figure by using Mason's gain formula.

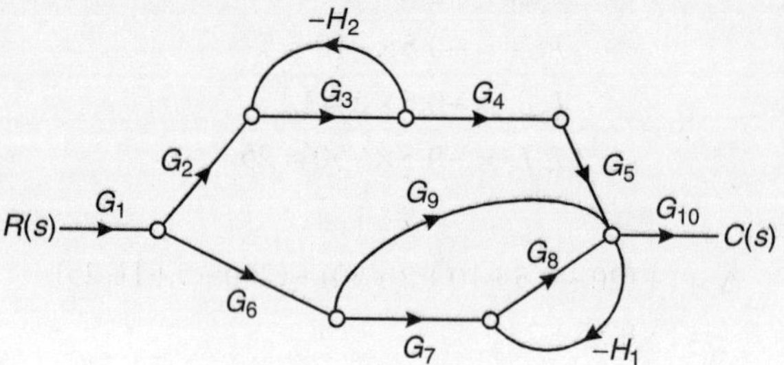

4. Identify the forward paths and individual loops. Determine the relation $\dfrac{C(s)}{R(s)}$.

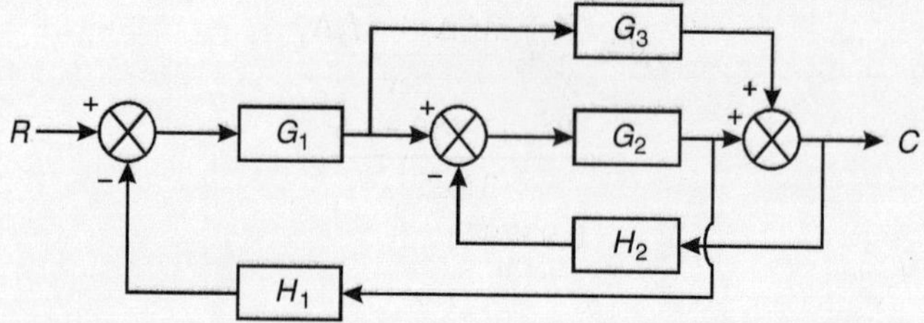

5. Find the transfer function for the system as shown in the figure using Mason's gain formula.

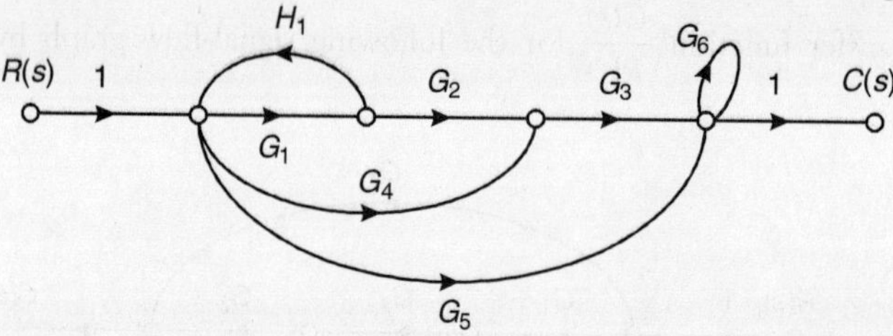

6. Find the transfer function for the system as shown in the figure using Mason's gain formula.

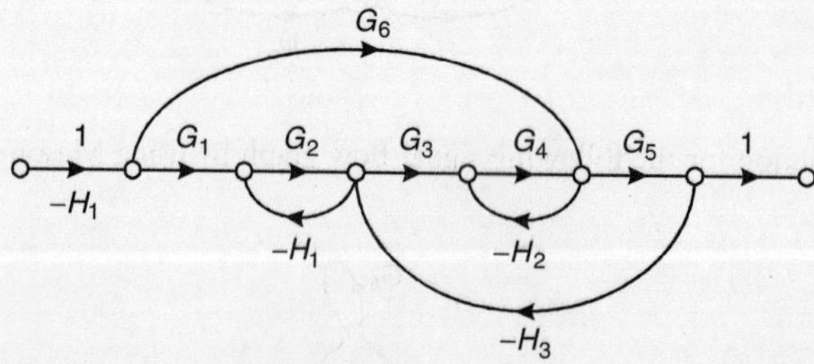

7. Find the transfer function for the system as shown in the figure using Mason's gain formula.

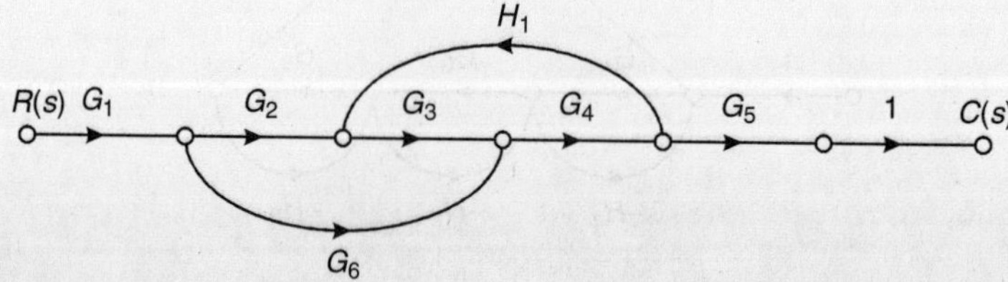

8. Find the transfer function for the system as shown in the figure using Mason's gain formula.

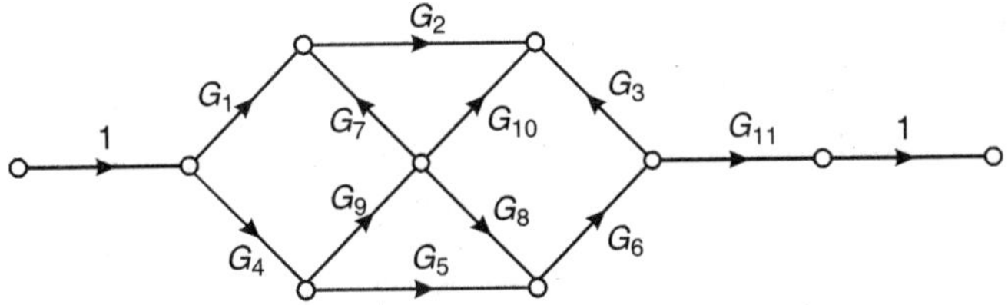

ANSWERS

1. $\dfrac{C(s)}{R(s)} = \dfrac{G_1G_2G_3 + G_4}{1 + G_1G_2H_1 + G_2G_3H_2}$

2. $\dfrac{G_1G_2G_3(1-G_5) + G_4G_6(1+G_2H_2)}{1 + G_1H_1 + G_2H_2 + G_3H_3 - G_5 + G_1G_2H_1H_2 - G_1G_5H_1 - G_2G_5H_2 - G_4G_6G_1H_2H_3 - G_3G_5H_3}$

3. $\dfrac{C(s)}{R(s)} = \dfrac{G_1G_2G_3G_4G_5G_{10} + G_1G_6G_9G_{10}(1-G_3H_2) + G_1G_6G_7G_8G_{10}(1+G_2H_2)}{1 + G_3H_2 + G_8H_1 + G_3G_8H_1H_2}$

4. $\dfrac{C(s)}{R(s)} = \dfrac{G_1G_2 + G_2G_3}{1 + G_2H_2 + G_1G_2H_1 - G_1G_2G_3H_1H_2}$

5. $\dfrac{C(s)}{R(s)} = \dfrac{G_1G_2G_3 + G_3G_4 + G_5}{1 - G_1H_1 - G_6 + G_1G_6H_1}$

6. $\dfrac{G_1G_2G_3G_5 + G_5G_6(1+G_2H_1)}{1 + G_2H_1 + G_4H_2 + G_3G_4G_5H_3 + G_2G_4H_1H_2}$

7. $\dfrac{G_1G_2G_3G_4G_5 + G_1G_4G_5G_6}{1 - G_3H_4H_1}$

8. $$\frac{\begin{array}{l} G_1G_2G_3G_{11}(1 + G_5H_2 + G_8G_9H_2) + G_1G_7G_{10}G_3G_{11}(1 + G_5H_2) + G_1G_7G_{10}G_3G_{11} + \\ G_4G_6G_8G_9G_{11}(1 + G_2H_1 + G_7G_{10}H_1) + G_3G_4G_9G_{10}G_{11} \end{array}}{\begin{array}{l} 1 + G_2H_1G_5H_2 + G_7G_{10}H_1 + G_8G_9H_2 + G_{11}H_3 + G_2G_5H_1H_2 + G_2G_8G_9H_1H_2 + \\ G_2G_{11}H_1H_3 + G_2G_7G_{10}H_1H_2 + G_5G_{11}H_2H_3 + G_7G_{10}G_{11}H_1H_3 + G_8G_9G_{11}H_1H_3 + \\ G_2G_8G_9G_{11}H_1H_2H_3 + G_5G_7G_{10}G_{11}H_1H_2H_3 \end{array}}$$

OBJECTIVE TYPE QUESTIONS

1. A sink in a signal-flow graph is
   (a) input node (b) output node (c) both (d) none
2. The transfer function for the system, as shown in the figure,

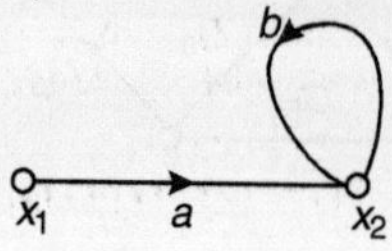

   (a) $\frac{a}{1-b}$ (b) $\frac{a}{1+b}$ (c) $\frac{b}{1-a}$ (d) $\frac{b}{1+a}$

3. The signal-flow graph is used to find the
   (a) stability of the system (b) transfer function of the system
   (c) controllability of the system (d) observability of the system
4. The transfer function of a signal-flow graph, as shown in the figure, is

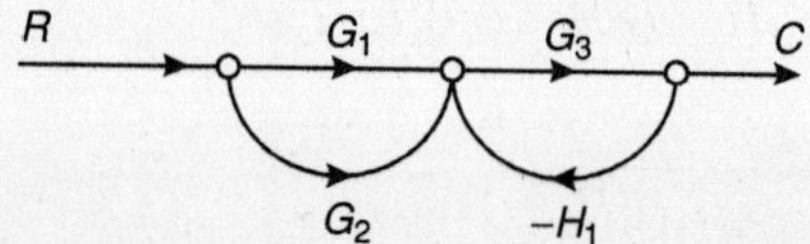

   (a) $\frac{G_1G_2G_3}{1+G_3H_1}$ (b) $\frac{G_1G_3+G_2G_3}{1+G_3H_1}$
   (c) $\frac{G_1+G_2+G_3}{1+G_3H_1}$ (d) None

Q. For the following signal-flow graph

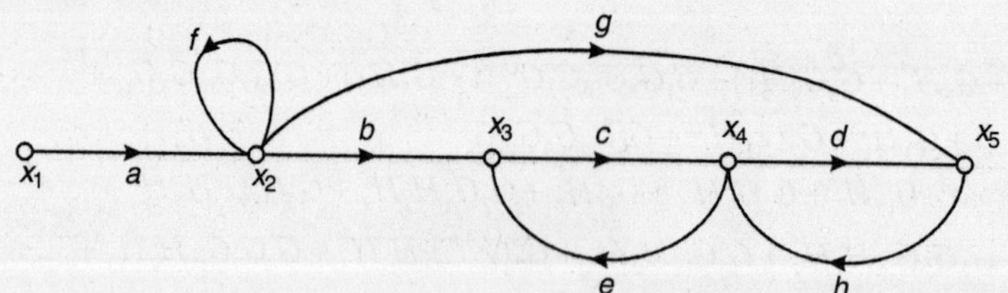

5. The number of forward paths are
   (a) 2 (b) 4 (c) 3 (d) none
6. The number of closed loops are
   (a) 1 (b) 2 (c) 3 (d) 6
7. The number of non-touching loops (pairs) are
   (a) 1 (b) 2 (c) 3 (d) none

8. The transfer function C/R of signal-flow graph, as shown in the figure, is

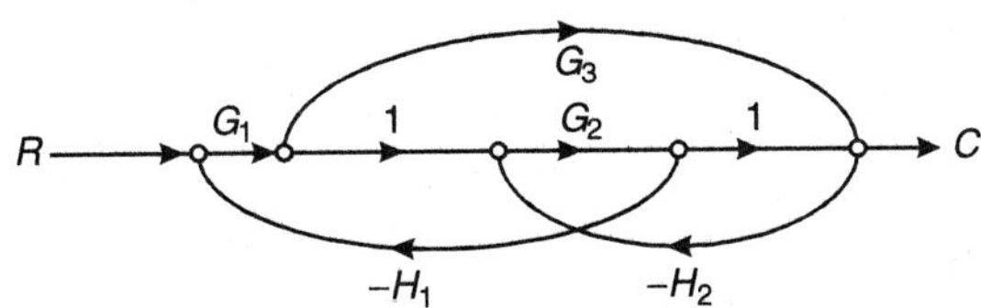

(a) $\dfrac{G_1G_2 - G_2G_3}{1+G_2H_2+G_1G_2H_1-G_1G_2G_3H_1H_2}$

(b) $\dfrac{G_1G_2 + G_1G_3}{1+G_2H_2+G_1G_2H_1-G_1G_2G_3H_1H_2}$

(c) $\dfrac{G_1G_2 + G_2G_3}{1-G_2H_2-G_1G_2H_1-G_1G_2G_3H_1H_2}$

(d) none

9. A source in a signal-flow graph is
(a) input node (b) output node (c) both (d) none

10. In the signal-flow graph as shown in the figure, the value of transfer function C/R is

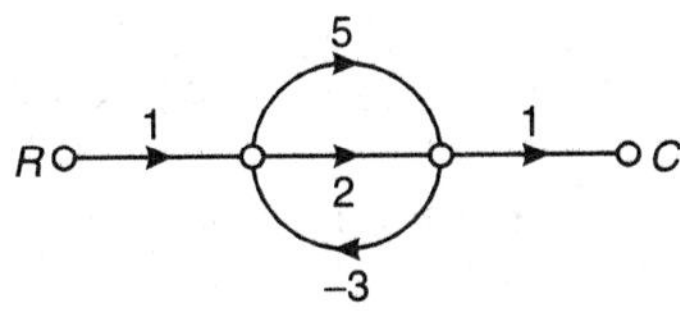

(a) 9/22 (b) 7/22 (c) 3/16 (d) 1/16

ANSWERS

1. b **2.** a **3.** b **4.** b **5.** a **6.** c **7.** b **8.** b **9.** a **10.** b

CHAPTER

6

Control System and Its Components

Contents

6.1 INTRODUCTION

Every activity in our day-to-day life is influenced by some sort of control systems. Control systems are now integral parts of industrial processes and home appliances. Hence, control systems and their components play an important role in our daily life.

A generalised unity feedback control system comprises three functional components: error detector, amplifier and controller. The error detector in a control system produces an error signal when there is a difference between the actual value of the output and the desired value. In some systems, the error signal is directly proportional to the difference between the actual value of the output and the desired value. In other systems, error signal is proportional to the rate at which the output is deviating the desired value. Furthermore, in still other systems, an error signal is produced only when the output has deviated by a predetermined value. However, the basic function of error detector is to provide an error signal or actuating signal to initiate a corrective action in the system. Some of the popular sensing elements are potentiometer, synchros and transducers. In this chapter, we will study the various components of the control system.

6.2 SERVOMOTORS

The motors used in automatic control system are called 'servomotors'. When the objective of the system is to control the position of an object, then the system is called servomechanism. In a control system, a servomotor is used to convert an electrical signal (control voltage) applied to it into mechanical displacement, velocity, torque, etc., as the desired output. There are varieties of servomotors available for control system applications. The suitability of a motor for a particular application depends on the characteristic of the system, the purpose of the system and its operating conditions.

In general, a servomotor should have the following features:

1. Linear relationship between the speed and the electric control signal.
2. Linearity of mechanical characteristics throughout the entire speed range.
3. Fast response and wide range of speed control.
4. Low mechanical and electrical inertia.

Servomotor can be either DC or AC motors depending on the supply required to run the motor. The DC motors are expensive than AC motor, but the DC servomotors have linear characteristics, and therefore, it is easier to control the speed. They are generally used for large power applications such as in machine tools and robotics.

The advantages of DC servomotors are as follows:

1. Higher output when compared to that from an AC motor of the same size.
2. Easy speed control: from zero speed to full load speed in both the directions.
3. High torque-to-inertia ratio that enables them to respond quickly to control signal.
4. Easy achievement of linear characteristics.
5. Low electrical and mechanical time constants.
6. The DC servomotors have light weight, low inertia and low inductance armature that can respond quickly to commands for a change in speed or position.

The advantages of AC servomotors are low cost, high efficiency and less maintenance due to the absence of brushes and commutators. The disadvantages of AC motors are their characteristics are quite non-linear and these motors are more difficult to control especially for positioning applications.

6.2.1 DC Servomotor

DC servomotors are specially designed DC motors with high starting torque and low inertia. They are used in systems where high starting torque and wide speed range may be required to operate in either direction of rotation. DC servomotors are generally used for large power applications such as in mechanical tools and robotics.

DC motor with separate excitations are used to control the power because the power can be easily controlled by varying the field or armature current of the motor. Depending upon the current that controls the motor output, the motors are classified as 'field-controlled DC motor' and 'armature-controlled DC motor'.

1. **Field-controlled DC Motor:** It is a DC shunt motor designed to satisfy the requirement of a servomotor. In this motor, the armature is supplied with a constant current or voltage. When armature current is constant, the torque is directly proportional to field flux ($T \propto \phi I_a$). Since the field current is proportional to flux, the torque of the motor is controlled by controlling the field current. Therefore, it is called field-controlled DC motor. Further, reversible operation is possible by reversing the field current.

Derivation of Transfer Function: Figure 6.1 illustrates a field-controlled DC motor that converts electrical energy into mechanical energy. The electrical energy supplied at the armature terminals is converted into controlled mechanical energy. The controlled energy is supplied at the field terminals of the motor. The armature current $i_a(t)$ is maintained to a constant value I_a, while the field voltage $e_f(t)$ is varied to control the speed or torque of the motor. Thus, the input of the motor is field voltage $e_f(t)$, while the output is the motor shaft and load displacement $\theta_m(t)$.

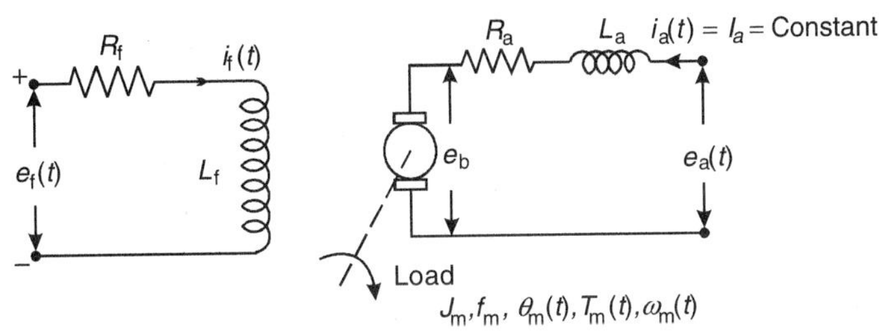

Figure 6.1 Schematic diagram of a field-controlled DC motor

The speed control system is an electromechanical control system. The electrical system consists of armature and field circuit; however, for the purpose of analysis, only field circuit is considered because the armature is excited by the constant DC voltage. The mechanical system consists of rotating part of the motor and the load connected to the shaft of the motor.

Let us consider the following variables:

R_f = Field resistance in Ω

L_f = Field inductance in H

i_f = Field current in A

e_f = Field voltage in V

T_m = Torque developed by motor in Nm

f_m = Coefficient of viscous friction in Nm/rad/s

J_m = Moment of inertia of motor and load in kgm²/rad.

The analysis of this motor is made linear by assuming that

1. The armature is supplied with a constant armature current

   i.e., $$i_a(t) = I_a \text{ constant} \tag{6.1}$$

2. The air gap flux $\phi(t)$ is directly proportional to field current $i_f(t)$

   i.e., $$\phi(t) = k_f i_f(t) \tag{6.2}$$

   where k_f is the constant of motor field.

3. Torque $T_m(t)$ transmitted by the motor shaft is directly proportional to the air gap flux $\phi(t)$ and the armature current I_a

   i.e., $$T_m(t) = k_{m1} I_a \phi(t) \tag{6.3}$$

   where k_{m1} is a constant of proportionality.

   From (6.2) and (6.3), we get

$$T_m(t) = k_{m1} I_a k_f i_f(t)$$

$$= k_m k_f i_f(t) \tag{6.4}$$

where $$k_m = k_{m1} \times I_a.$$

By applying Kirchhoff's voltage law to the field circuit, we get

$$L_f \frac{di_f(t)}{dt} + R_f i_f(t) = e_f(t). \tag{6.5}$$

Torque $T_m(t)$ developed by the motor is used for displacing the load. Hence, the torque equation for mechanical element in the system is

$$J_m \frac{d^2\theta_m(t)}{dt^2} + f_m \frac{d\theta_m(t)}{dt} = T_m(t). \tag{6.6}$$

The Laplace transforms of (6.4), (6.5) and (6.6) with zero initial conditions are

$$T_m(s) = k_m k_f I_f(s) \tag{6.7}$$

$$E_f(s) = (sL_f + R_f)I_f(s) \tag{6.8}$$

$$T_m(s) = (s^2 J_m + sf_m)\theta_m(s) \tag{6.9}$$

By eliminating I_f from (6.7) and (6.8), we get

$$T_m(s) = \frac{k_m k_f E_f(s)}{(sL_f + R_f)} \tag{6.10}$$

By eliminating $T_m(s)$ from (6.9) and (6.10), we have

$$(s^2 J_m + sf_m)\theta_m(s) = \frac{k_m k_f E_f(s)}{(sL_f + R_f)}. \tag{6.11}$$

Therefore, the required transfer function is

$$\frac{\theta_m(s)}{E_f(s)} = \frac{k_m k_f}{(R_f + sL_f)(s^2 J_m + sf_m)}$$

$$= \frac{k_m k_f}{R_f\left(1 + s\frac{L_f}{R_f}\right) sf_m\left(1 + s\frac{J_m}{f_m}\right)}$$

$$= \frac{k_m k_f}{R_f . s . f_m (1 + s\tau_f)(1 + s\tau_m)}$$

$$\Rightarrow \quad \frac{\theta_m(s)}{E_f(s)} = \frac{k_f}{R_f(s\tau_f + 1)} \cdot \frac{k_m}{f_m(s\tau_m + 1)} \cdot \frac{1}{s} \tag{6.12}$$

where $\tau_m = \dfrac{J_m}{f_m}$ is the mechanical time constant and $\tau_f = \dfrac{L_f}{R_f}$ is the field time constant.

The block diagram of the motor from (6.12) is drawn, as shown in Figure 6.2.

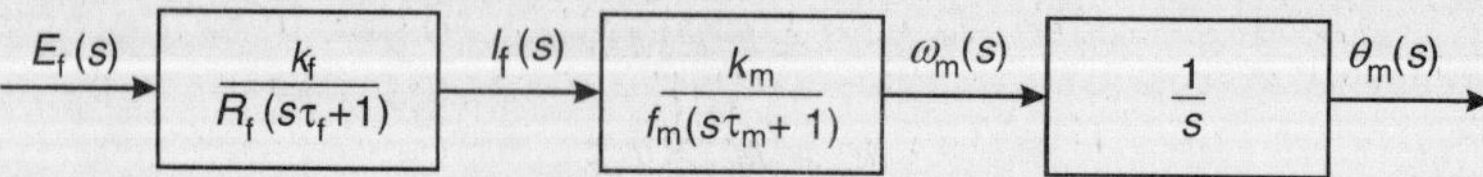

Figure 6.2 Block diagram representation of a field-controlled DC motor

2. **Armature-controlled DC motor:** It is a DC shunt motor designed to satisfy the requirement of a servomotor. The field is excited by a constant DC supply. If the field current is constant, then the speed is directly proportional to armature voltage and torque is directly proportional to armature current. Hence, the torque and speed can be controlled by armature voltage. The reversible operation is possible by reversing the armature voltage.

Derivation of transfer function Figure 6.3 illustrates an armature-controlled DC motor that converts electrical energy into mechanical energy. The field current $i_f(t)$ is maintained constant by keeping the voltage $e_f(t)$ to a constant value E_f and the armature current $i_a(t)$ is varied to control the speed or torque of the motor. Thus, the input of the motor is armature voltage $e_a(t)$ and the output is the motor shaft and load displacement $\theta_m(t)$.

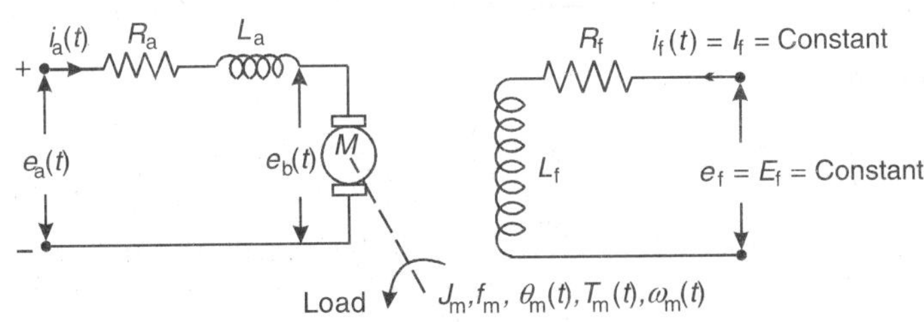

Figure 6.3 Schematic diagram of an armature-controlled DC motor

The speed control system is an electromechanical control system. The electrical system consists of the armature and the field circuit. However, for the purpose of analysis, only the armature is considered because the field is excited by a constant voltage. The mechanical system consists of the rotating part of the motor and load connected to the shaft of the motor.

Let R_f = Field resistance in Ω
R_a = Armature resistance in Ω
L_a = Armature inductance in H
i_a = Armature current in A
T_m = Torque development by the motor in Nm
θ_m = Angular displacement of the shaft
J_m = Moment of inertia of the motor and load
k_b = Back e.m.f. constant.

The analysis of this motor is made linear by assuming

1. The air gap flux ϕ is directly proportional to the field current

   i.e.,
   $$\phi = k_f I_f \quad (6.13)$$

   where k_f is the field constant of the motor.

2. The torque developed by the motor is directly proportional to the flux ϕ and the armature current $i_a(t)$.

   i.e.,
   $$T_m(t) = k_{m1}\phi i_a(t) \quad (6.14)$$

3. The back e.m.f., $e_b(t)$ is directly proportional to the motor speed $\omega_m(t)$.

   i.e.,
   $$e_b(t) = k_b \omega_m(t) = k_b \frac{d\theta_m(t)}{dt} \quad (6.15)$$

From (6.13) and (6.14), we have

$$T_{\mathrm{m}}(t)=k_{\mathrm{m1}}k_{\mathrm{f}}I_{\mathrm{f}}i_{\mathrm{a}}(t)=k_{\mathrm{m}}i_{\mathrm{a}}(t) \tag{6.16}$$

where $k_{\mathrm{m}}=k_{\mathrm{m1}}k_{\mathrm{f}}I_{\mathrm{f}}=\text{constant}$.

By applying Kirchhoff's voltage law to the armature circuit, we get

$$e_{\mathrm{a}}(t)=L_{\mathrm{a}}\frac{di_{\mathrm{a}}(t)}{dt}+R_{\mathrm{a}}i_{\mathrm{a}}(t)+e_{\mathrm{b}}(t). \tag{6.17}$$

The torque developed in the motor is used to displace the load. Hence, the torque equation for mechanical element in the system is

$$T_{\mathrm{m}}(t)=J_{\mathrm{m}}\frac{d^2\theta_{\mathrm{m}}}{dt^2}+f_{\mathrm{m}}\frac{d\theta_{\mathrm{m}}}{dt}. \tag{6.18}$$

By eliminating $e_{\mathrm{b}}(t)$ from (6.15) and (6.17), we have

$$e_{\mathrm{a}}(t)=L_{\mathrm{a}}\frac{di_{\mathrm{a}}(t)}{dt}+R_{\mathrm{a}}i_{\mathrm{a}}(t)+k_{\mathrm{b}}\frac{d\theta_{\mathrm{m}}(t)}{dt}. \tag{6.19}$$

The Laplace transform of (6.16), (6.18) and (6.19) with zero initial conditions are

$$T_{\mathrm{m}}(s)=k_{\mathrm{m}}I_{\mathrm{a}}(s) \tag{6.20}$$

$$T_{\mathrm{m}}(s)=(s^2J_{\mathrm{m}}+sf_{\mathrm{m}})\theta_{\mathrm{m}}(s) \tag{6.21}$$

$$E_{\mathrm{a}}(s)=(sL_{\mathrm{a}}+R_{\mathrm{a}})I_{\mathrm{a}}(s)+sk_{\mathrm{b}}\theta_{\mathrm{m}}(s) \tag{6.22}$$

By eliminating $T(s)$ from (6.20) and (6.21), we get

$$k_{\mathrm{m}}I_{\mathrm{a}}(s)=(s^2J_{\mathrm{m}}+sf_{\mathrm{m}})\theta_{\mathrm{m}}(s) \tag{6.23}$$

$$I_{\mathrm{a}}(s)=\left(\frac{s^2J_{\mathrm{m}}+sf_{\mathrm{m}}}{k_{\mathrm{m}}}\right)\theta_{\mathrm{m}}(s). \tag{6.24}$$

By eliminating $I_{\mathrm{a}}(s)$ from (6.22) and (6.24), we have

$$E_{\mathrm{a}}(s)=(sL_{\mathrm{a}}+R_{\mathrm{a}})\left(\frac{s^2J_{\mathrm{m}}+sf_{\mathrm{m}}}{k_{\mathrm{m}}}\right)\theta_{\mathrm{m}}(s)+sk_{\mathrm{b}}\theta_{\mathrm{m}}(s)$$

$$E_{\mathrm{a}}(s)=\left[\frac{(sL_{\mathrm{a}}+R_{\mathrm{a}})(s^2J_{\mathrm{m}}+sf_{\mathrm{m}})+sk_{\mathrm{b}}k_{\mathrm{m}}}{k_{\mathrm{m}}}\right]\theta_{\mathrm{m}}(s)$$

$$\Rightarrow \quad \frac{\theta_{\mathrm{m}}(s)}{E_{\mathrm{a}}(s)}=\frac{k_{\mathrm{m}}}{(s^2J_{\mathrm{m}}+sf_{\mathrm{m}})(sL_{\mathrm{a}}+R_{\mathrm{a}})+sk_{\mathrm{b}}k_{\mathrm{m}}}$$

$$=\frac{k_{\mathrm{m}}}{sf_{\mathrm{m}}\left(1+s\dfrac{J_{\mathrm{m}}}{f_{\mathrm{m}}}\right)R_{\mathrm{a}}\left(1+s\dfrac{L_{\mathrm{a}}}{R_{\mathrm{a}}}\right)+sk_{\mathrm{b}}k_{\mathrm{m}}}$$

$$=\frac{k_{\mathrm{m}}}{sR_{\mathrm{a}}f_{\mathrm{m}}(1+s\tau_{\mathrm{m}})(1+s\tau_{\mathrm{a}})+sk_{\mathrm{b}}k_{\mathrm{m}}}$$

Therefore, the required transfer function is

$$\frac{\theta_m(s)}{E_a(s)} = \frac{k_m}{sR_a f_m(1+s\tau_m)(1+s\tau_a)+sk_b k_m} \tag{6.25}$$

where $\tau_m = \frac{J_m}{f_m}$ is the mechanical time constant and $\tau_a = \frac{L_a}{R_a}$ is the armature time constant.

Rearranging (6.25), we get

$$\frac{\theta_m(s)}{E_a(s)} = \frac{\dfrac{k_m}{sR_a f_m(1+s\tau_m)(1+s\tau_a)}}{1+sk_b\dfrac{k_m}{sR_a f_m(1+s\tau_m)(1+s\tau_a)}}. \tag{6.26}$$

The block diagram of the motor from (6.26) is drawn as shown in Figure 6.4.

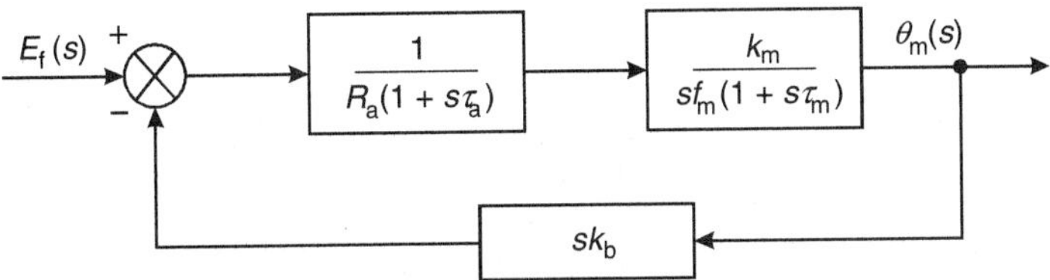

Figure 6.4 Block diagram representation of an armature-controlled DC motor

6.2.2 Comparison Between Armature-controlled and Field-controlled DC Servomotors

S. no	Armature-controlled DC motor	Field-controlled DC motor
1	Control voltage is applied to the armature.	Control voltage is applied to the field.
2	Time constant is small.	Time constant is large.
3	It is a closed-loop system.	It is an open-loop system.
4	Field current is kept constant.	Armature current is kept constant.
5	Better efficiency.	Poor efficiency.
6	Suitable for high rated motors.	Suitable for low rated motors.
7	High power amplifiers are required to design.	Due to low power applications, amplifiers are simple to design.

6.2.3 AC Servomotor

For low-power applications, AC servomotors are preferred because they are light in weight, rugged and there are no brush contacts. An AC servomotor is basically a two-phase induction motor except for certain special design features. A two-phase induction motor consists of two stator windings oriented 90° apart in space and excited by AC voltages that differ in time phase by 90°.

A two-phase servomotor differs in two ways from normal induction motor:

1. The rotor of the servomotor is built with high resistance so that its X/R (inductive reactance-to-resistance ratio) is small. This results in linear speed–torque characteristic, as shown by curve B in Figure 6.5. However, the conventional induction motors have large X/R ratio that results in high efficiency but in highly nonlinear torque–speed characteristics, as shown by curve 'A' in Figure 6.5. Further, because of the positive slope for part of the characteristic, the system using such a motor becomes unstable.
2. In servomotor, the voltages applied to the two-stator winding should have a phase difference of 90°. One of the phases is known as the reference phase and it is excited by constant voltage; while the other phase is known as the control phase, and it is excited by a variable magnitude and polarity source with respect to the reference phase.

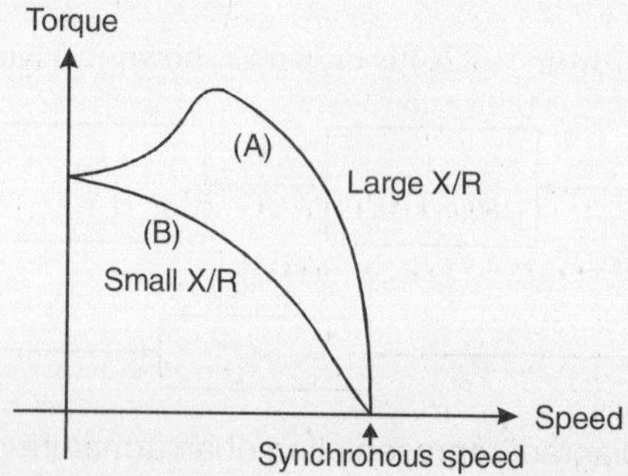

Figure 6.5 Torque–speed characteristics of an induction motor

Construction of AC servomotor

An AC servomotor is basically a two-phase induction motor with some special design features. The stator consists of two pole pairs (A-B and C-D) mounted on the inner periphery of the stator such that their axes are at an angle of 90° in space. Each pole pair carries a winding. One winding is called reference winding and the other is called control winding. The exciting current in the winding should have a phase displacement of 90°. The supply used to drive the motor is single phase, and therefore, a phase advancing capacitor is connected to one of the phases to produce a phase difference of 90°. The constructional features of an AC servomotor are shown in Figure 6.6.

The rotor construction is usually squirrel-cage or drag-cup type. The squirrel-cage rotor is made of laminations. The rotor bars are placed in the slots and short circuited at both ends by end rings. The diameter of the rotor is kept small in order to reduce inertia and to obtain good accelerating characteristics.

The drag-cup construction is employed for very low applications. In this type of construction, the rotor will be in the form of a hollow cylinder made of aluminium. The aluminium cylinder itself acts as a short-circuited rotor conductor (electrically both types of rotor are identical).

Working of a two-phase induction motor

Figure 6.7 shows a schematic diagram for balanced operation of the two-phase induction motor. The stator windings are excited by voltage of equal r.m.s. magnitude and 90° phase difference.

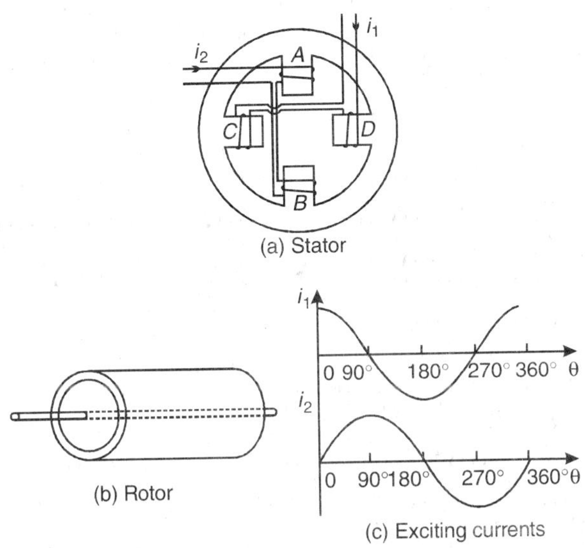

Figure 6.6 Simplified constructional features of a two-phase AC servomotor

This results in exciting currents i_1 and i_2 that are phase displaced by 90° and have equal r.m.s value. Therefore, their respective fields will be 90° apart in both time and space resulting in a magnetic field of constant magnitude rotating at synchronous speed. The direction of rotation depends upon phase relationship of the two currents (voltages). As the field sweeps over the rotor, voltages are induced in it producing current in the short-circuited rotor. The rotating magnetic field interacts with these currents producing a torque on the rotor in the direction of field of rotation, and therefore, the rotor starts moving in the same direction as that of the rotating magnetic field.

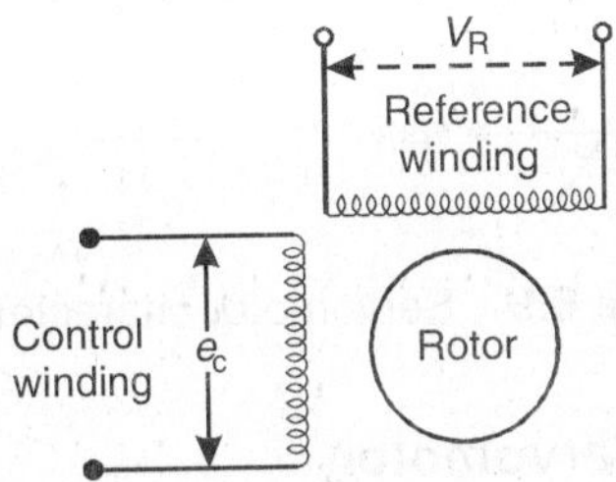

Figure 6.7 Schematic diagram of a two-phase induction motor

Working of an AC servomotor

Figure 6.8 shows a schematic diagram of an AC servomotor as a control system component. As shown in Figure 6.8, one of the phases, known as the reference winding, is excited by a constant voltage while the other phase known as the control winding is energised by a voltage, which is 90° out of phase with respect to the voltage of the reference winding. This 90° phase shift between the reference and the control winding voltage can also be obtained by connecting a capacitor in series with the reference winding or by adding a simple phase shifting network. The main field is supplied continuously to produce high torque, while the control winding is supplied with the controlling voltage, which is generally applied from a servoamplifier. By changing the control voltage, the motor output can be controlled. When the control voltage leads to the fixed voltage, shaft rotation is in one direction; however, when it lags the fixed voltage, the direction of rotation is reversed.

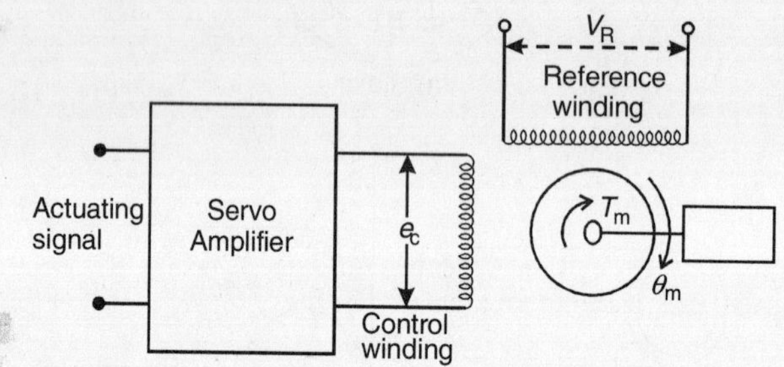

Figure 6.8 Schematic diagram of a two-phase servomotor

The torque–speed curves of a typical AC servomotor is plotted for a fixed reference phase voltage and with variable r.m.s. control voltage are shown in Figure 6.9. All these curves have a negative slope. Note that the curve for zero control voltage goes through the origin. This means that when the control phase voltage becomes zero, the motor develops a decelerating torque, and therefore, the motor stops. The curves show a large torque at zero speed. This is a requirement for the servomotor in order to provide rapid acceleration. The torque–speed curves of AC servomotors are nonlinear except in the low speed region.

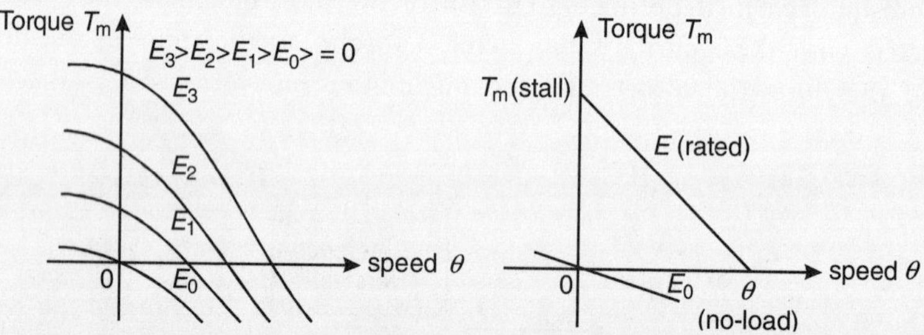

Figure 6.9 Servomotor characteristics

Transfer function of an AC servomotor

Let us consider the following variables:

T_m = Torque development by the motor in Nm
θ_m = Angular displacement of the shaft
J_m = Moment of inertia of the motor and load
m = Slope of the torque speed characteristics
K = Slope of the control-phase voltage versus torque characteristics

With reference to Figure 6.9, we can say that for speeds near zero, all the curves are straight lines parallel to the characteristics at rated input voltage ($e_c = E$) and are equally spaced for equal increments of the input voltage. Under this assumption, the torque developed by the motor is

$$T_{\mathrm{m}}(t) = Ke_{\mathrm{c}}(t) + m\frac{d\theta_{\mathrm{m}}(t)}{dt}. \tag{6.27}$$

The total torque produced by the motor is utilised in rotating the motor rotor with load and to overcome motor friction. Hence,

$$J_{\mathrm{m}}\frac{d^2\theta_{\mathrm{m}}(t)}{d^2t} + f_{\mathrm{m}}\frac{d\theta_{\mathrm{m}}(t)}{dt} = T_{\mathrm{m}}. \tag{6.28}$$

From (6.27) and (6.28), we have

$$J_{\mathrm{m}}\frac{d^2\theta_{\mathrm{m}}(t)}{d^2t} + f_{\mathrm{m}}\frac{d\theta_{\mathrm{m}}(t)}{dt} = Ke_{\mathrm{c}}(t) + m\frac{d\theta_{\mathrm{m}}(t)}{dt}. \tag{6.29}$$

By taking Laplace transforms on both sides and assuming that all the initial conditions are zero, we have

$$(s^2J_{\mathrm{m}} + sf_{\mathrm{m}})\theta_{\mathrm{m}}(s) = KE_{\mathrm{c}}(s) + sm\theta_{\mathrm{m}}(s)$$

$$\Rightarrow \quad (s^2J_{\mathrm{m}} + sf_{\mathrm{m}} - sm)\theta_{\mathrm{m}}(s) = KE_{\mathrm{c}}(s). \tag{6.30}$$

Therefore, the required transfer function is

$$\frac{\theta_{\mathrm{m}}(s)}{E_{\mathrm{c}}(s)} = \frac{K}{s^2J_{\mathrm{m}} + s(f_{\mathrm{m}} - m)} \tag{6.31}$$

$$= \frac{K}{s(f_{\mathrm{m}} - m)\left[1 + s\dfrac{J_{\mathrm{m}}}{f_{\mathrm{m}} - m}\right]}$$

$$= \frac{K}{(f_{\mathrm{m}} - m)}\frac{1}{s}\frac{1}{\left[1 + s\dfrac{J_{\mathrm{m}}}{f_{\mathrm{m}} - m}\right]}$$

$$= \frac{K_{\mathrm{m}}}{s(1 + s\tau_{\mathrm{m}})}.$$

$$\therefore \quad \frac{\theta_{\mathrm{m}}(s)}{E_{\mathrm{c}}(s)} = \frac{K_{\mathrm{m}}}{s(1 + s\tau_{\mathrm{m}})} \tag{6.32}$$

where $K_{\mathrm{m}} = \dfrac{K}{(f_{\mathrm{m}} - m)}$ is the motor gain constant and $\tau_{\mathrm{m}} = \dfrac{J_{\mathrm{m}}}{(f_{\mathrm{m}} - m)}$ is the motor time constant.

The block diagram of the motor from (6.32) is drawn as shown in Figure 6.10.

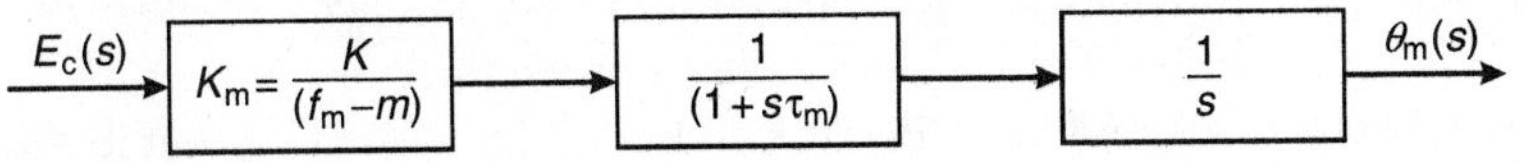

$$E_c(s) \rightarrow \boxed{K_m = \frac{K}{(f_m - m)}} \rightarrow \boxed{\frac{1}{(1 + s\tau_m)}} \rightarrow \boxed{\frac{1}{s}} \rightarrow \theta_m(s)$$

Figure 6.10 Block diagram representation of an AC motor

6.3 SYNCHROS

A synchro is an electromagnetic transducer commonly used to convert an angular position of a shaft into an electric signal. It is commonly known as selsyn or autosyn. It produces an output voltage depending upon the angular position of the rotor. It works on the principle of an induction motor. Synchros are small motors widely used for remote transmission of shaft position in AC servomechanisms. They can also be used as an error detector in position control systems.

6.3.1 Synchro Transmitter

The basic synchro is usually called a synchro transmitter. Its construction is similar to three-phase alternator. The basic structure consists of a wound rotor and wound stator. The stator is of laminated silicon steel and is slotted on the inner periphery to accommodate a balanced three-phase winding, which is usually of concentric coil type and is star connected. The rotor is a salient pole dumbbell-shaped magnetic structure with a single winding. A single-phase AC voltage is applied to the rotor winding through slip rings. The constructional features and a schematic diagram of synchro transmitter are shown in Figure 6.11.

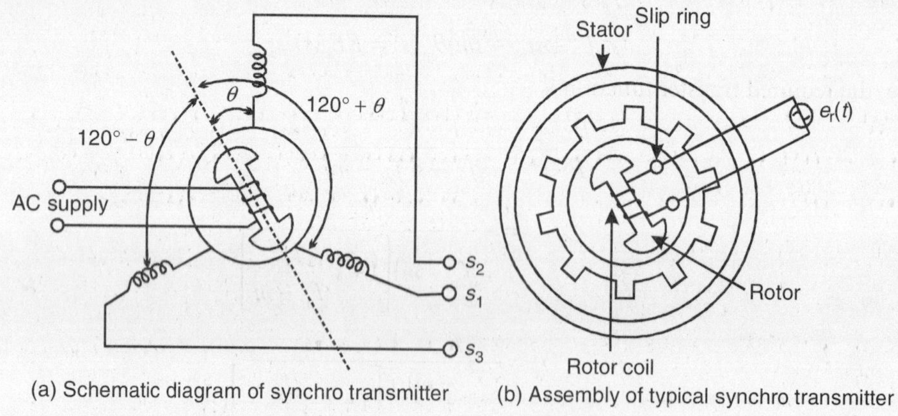

(a) Schematic diagram of synchro transmitter (b) Assembly of typical synchro transmitter

Figure 6.11 Synchro transmitter

Let an AC voltage

$$e_r(t) = E_r \sin \omega_c t \tag{6.33}$$

be applied to the rotor of the synchro transmitter, as shown in Figure 6.11(b). This applied voltage causes a flow of magnetising current in the rotor coil, which produces a sinusoidal time-varying flux directed along its axis and distributed nearly sinusoidally in the air gap along the stator periphery. Because of transformer action, voltages are induced in each of the stator coils. As the air gap flux is sinusoidally distributed, the flux linking with any stator coil is proportional to the cosine of the angle between the rotor and the stator coil axes and so is the voltage induced in the stator coil. The stator coil voltages are in time phase with each other. Thus, we see that the synchro transmitter acts like a single-phase transformer in which the rotor coil is the primary and the stator coils form the three secondaries.

Let e_{s1}, e_{s2} and e_{s3} be the voltages induced in stator coils S_1, S_2 and S_3 with respect to the neutral. Then, for the rotor position of the synchro transmitter as shown in Figure 6.11, where the rotor axis makes an angle θ with the axis of the stator coil S_2

$$\left.\begin{aligned} e_{s1} &= KE_r \sin \omega_c t.\cos(\theta + 120°) \\ e_{s2} &= KE_r \sin \omega_c t.\cos \theta \\ e_{s3} &= KE_r \sin \omega_c t.\cos(\theta - 120°) \end{aligned}\right\} \tag{6.34}$$

The three terminal voltages of the stator are

$$\left.\begin{aligned} e_{s1s2} &= e_{s1} - e_{s2} = \sqrt{3}KE_r \sin(\theta - 120°) \times \sin\omega_c t \\ e_{s2s3} &= e_{s2} - e_{s3} = \sqrt{3}KE_r \sin(\theta + 120°) \times \sin\omega_c t \\ e_{s3s1} &= e_{s3} - e_{s1} = \sqrt{3}KE_r \sin\theta \times \sin\omega_c t \end{aligned}\right\}. \tag{6.35}$$

When $\theta = 0$, from (6.34) and (6.35), it is seen that the maximum voltage is induced in stator coil S_2, while the terminal voltage e_{s3s1} is zero. This position of the rotor is defined as the 'electrical zero' of the transmitter and is used as reference for specifying the angular position of the rotor. Thus, it is seen that the input to the synchro transmitter is the angular position of its rotor shaft and the output is a set of three stator coils to coil single-phase voltages. The magnitudes of these voltages are functions of the shaft position. Therefore, by measuring and identifying the set of voltages at the stator terminals, it is possible to identify the angular position of the rotor.

6.3.2 Synchro Control Transformer

The constructional feature of a synchro control transformer (synchro receiver) is identical to that of a synchro transmitter except that rotor is cylindrically shaped so that the air gap flux is uniformly distributed around the rotor. This feature of the control transformer minimises changes in rotor impedance with the rotation of the shaft. Another distinguished feature is that the stator winding of the control transformer has high impedance per phase, which permits several control transformers to be fed from a single transmitter. The output of the synchro transmitter is applied to the stator windings of a synchro control transformer.

Working

The rotor shaft is connected to the load, the position of which has to be maintained at the desired value. Depending on the current position of the rotor and the applied e.m.f. on the stator, an e.m.f. is produced in the rotor winding. This e.m.f. can be measured and used to drive a motor so that the position of the load is corrected.

6.3.3 Synchro Error Detector

The synchro error detector is formed by the interconnection of the synchro transmitter and the synchro control transformer, as shown in Figure 6.12. The stator leads of the transmitter are directly connected to the stator leads of the control transformer. The angular position of the transmitter rotor is the reference input and the rotor is excited by a single-phase AC supply. The control transformer rotor is connected to a servomotor and to the shaft of the load, with its position indicating the desired output. Initially, the transmitter and control transformer rotor are assumed to be in their electrical zero position. The angular separation of both rotor axes in the position is 90°. The electrical zero position of a control transformer in a servo system is that position of its rotor for which the output voltage on the rotor winding is zero with the transmitter in its electrical zero position.

When the transmitter rotor is excited, the rotor flux is set up and e.m.fs are induced in the stator coils. These induced e.m.fs are impressed on the stator coils of the control transformer. The current in the stator coils sets up a flux in the control transformer. Due to similarity in the

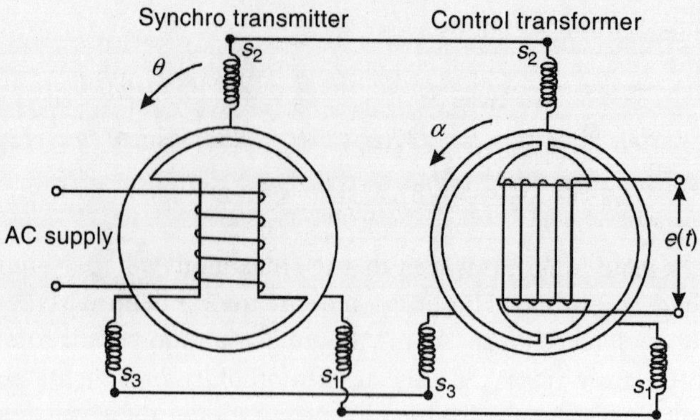

Figure 6.12 Synchro error detector

magnetic construction, the flux pattern produced in the two synchros will be the same, if all losses are neglected.

Let the rotor of the transmitter rotates through an angle θ from its electrical zero position in the direction indicated and let the control transformer rotate in the same direction through an angle α resulting in a net angular separation of $\phi = 90 - \theta + \alpha$ between the two rotors. Therefore, the voltage at the rotor terminals of the control transformer is

$$e(t) = K'E_r \cos(90 - \theta + \alpha)\sin\omega_c t$$
$$= K'E_r \sin(\theta - \alpha)\sin\omega_c t. \quad (6.36)$$

For small angular displacements between the two rotor positions, we have

$$e(t) = K'E_r(\theta - \alpha)\sin\omega_c t. \quad (6.37)$$

Thus, the synchro transmitter–control transformer pair acts as an error detector giving a voltage signal at the rotor terminals of the control transformer proportional to the angular difference between the shaft position of the transmitter and shaft position of the control transformer.

6.3.4 Applications of Synchros

The various applications of synchros are listed as follows:

1. Synchro systems were first used in the control system of the Panama Canal in the early 1900s to transmit lock gate and valve stem positions and water levels to the control desks.
2. Fire-control system designs developed during World War II used synchros extensively to transmit angular information from guns and sights to an analogue fire control computer and to transmit the desired gun position back to the gun location.
3. Smaller synchros are still used to remotely drive indicator gauges and as rotary position sensors for aircraft control surfaces, where the reliability of these rugged devices is needed. Digital devices like the rotary encoder have replaced synchros in most other applications.
4. Selsyn motors were widely used in motion picture equipment to synchronise movie cameras and sound recording equipment.
5. Large synchros were used on naval warships such as destroyers, to operate the steering gear from the wheel on the bridge.

6.4 TACHOMETERS OR TACHOGENERATORS

Tachometer is an electromechanical unit that generates an electrical output proportional to the speed of the shaft. In automatic control system, tachometer performs two main functions:

1. Stabilisation of system
2. Computation of closed loops in a control system

The following sections discusses the two types of tachometers.

6.4.1 DC Tachometer

It is sometimes necessary in a control system to provide a feedback voltage proportional to the speed of the shaft. In a DC servo mechanism, this can be achieved by using a DC tachometer, which is a permanent magnet DC generator. It contains a permanent magnet and an iron core rotor. No external supply voltage is required. The winding on the rotor is connected to commutator segments and the output voltage is taken across pair of brushes that ride on the commutator segments. DC tachometers provide visual speed readout of a rotating shaft. Such tachometers are directly coupled to a voltmeter, which is calibrated in r.p.m. The schematic diagram of DC tachometer is as shown in Figure 6.13.

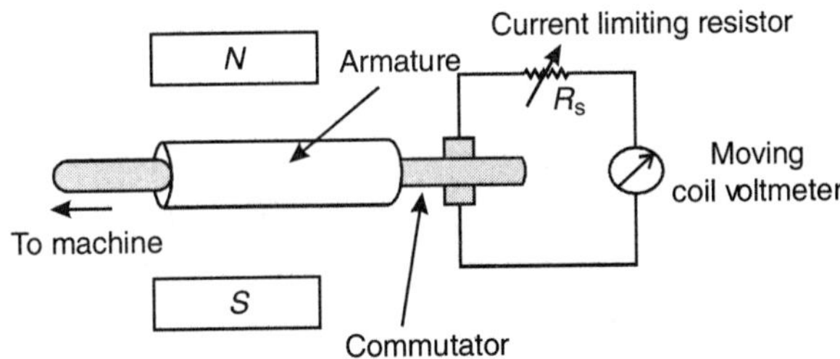

Figure 6.13 DC tachometer

The input to the DC tachometer is shaft speed, while the output is the voltage, which is proportional to its rotating speed.

$$\therefore \quad e_T = K_T \omega(t) \tag{6.38}$$

where K_T is the tachometer sensitivity, $\omega(t)$ is the angular speed of shaft and e_T is the generated output voltage.

Therefore, the transfer function of DC tachometer is

$$\frac{E_T(s)}{\omega(s)} = K_T. \tag{6.39}$$

A DC tachometer can be used on AC servomechanism by converting the DC output voltage to an AC voltage by using an inverter circuit.

Advantages

The advantages of DC tachometer are as follows:

1. Generated voltages are free from undesirable wave shapes and phase shifts.
2. Temperature is compensated.
3. There is no residual voltage present at zero speed.

Disadvantages

The disadvantages of DC tachometer are as follows:

1. Wear and tear of brushes.
2. It needs high torque.
3. Commutation sparking gives noise generated voltages.
4. A small ripple is present.

6.4.2 AC Tachometer

An AC tachometer is used in a control system to provide AC voltage feedback proportional to the speed of the shaft. This is a two-phase induction motor, as shown in Figure 6.14. One of the stator windings is used as the reference winding and the other one is used as the control winding. The reference winding is fed with a suitable AC voltage of constant frequency and magnitude. Therefore, a voltage of the same frequency is induced in the control winding. This output voltage is fed to the high input impedance circuit of an amplifier so that the control winding can be considered as open circuited.

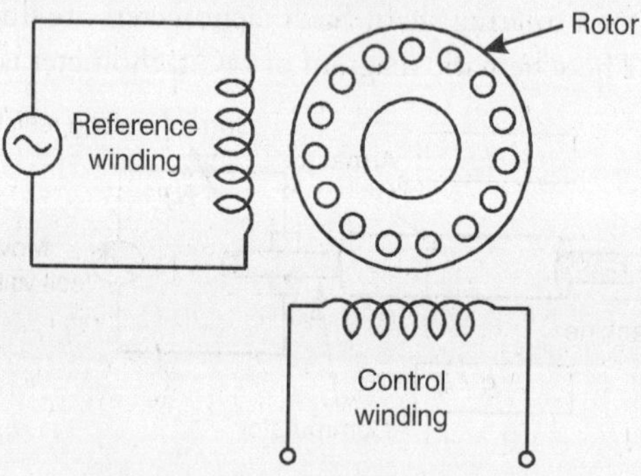

Figure 6.14 AC tachometer

When rotor shaft is rotated, the magnitude of sinusoidal output voltage is directly proportional to the speed of rotor. Thus, when the rotor shaft is stationary, the output voltage is zero. The polarity of the output voltage is determined by the direction of the rotation.

The transfer function of AC tachometer is given as

$$\therefore \quad e = K\frac{d\theta}{dt} \Rightarrow \frac{E(s)}{\theta(s)} = sK. \tag{6.40}$$

Advantages

The advantages of AC tachometer are as follows:

1. It is a brushless machine.
2. It provides speed feedback.
3. It reduces electromagnetic noise.
4. It reduces ripple also.

6.5 POTENTIOMETERS

A potentiometer, informally a pot, is a three-terminal resistor with a sliding contact that forms an adjustable voltage divider. If only two terminals are used, one end and the wiper, it acts as a variable resistor or rheostat. A potentiometer is essentially a voltage divider used for measuring electric potential (voltage). These are commonly used to control electrical devices like volume controls on audio equipment. Potentiometers can be used as position transducers; for example, in the mechanism used in a joystick. Potentiometers are rarely used to directly control significant power (more than a watt) since the power dissipated in the potentiometer would be comparable to the power in the controlled load.

Potentiometers comprise a resistive element, a sliding contact (wiper) that moves along the element making good electrical contact with one part of it, electrical terminals at each end of the element, a mechanism that moves the wiper from one end to the other and a housing containing the element and wiper, as shown in Figure 6.15.The symbolic representation of potentiometer is shown in Figure 6.16.

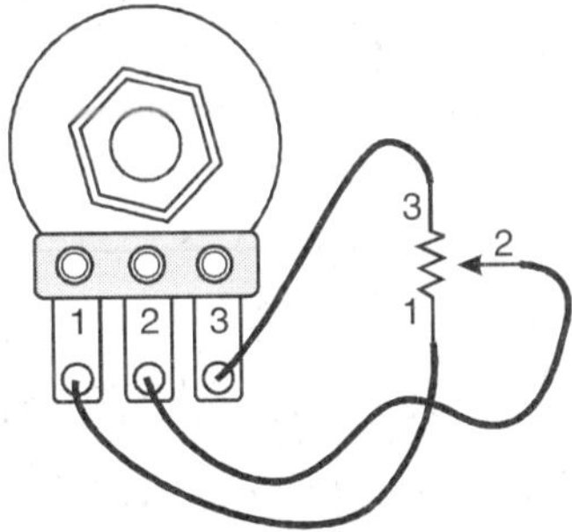

Figure 6.15 Construction details of a potentiometer

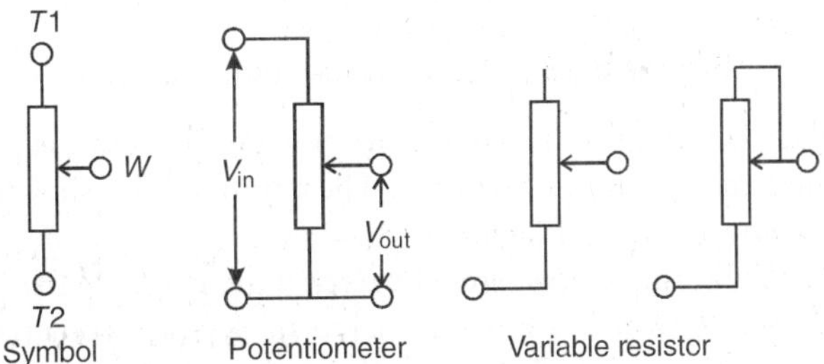

Figure 6.16 Potentiometer symbol

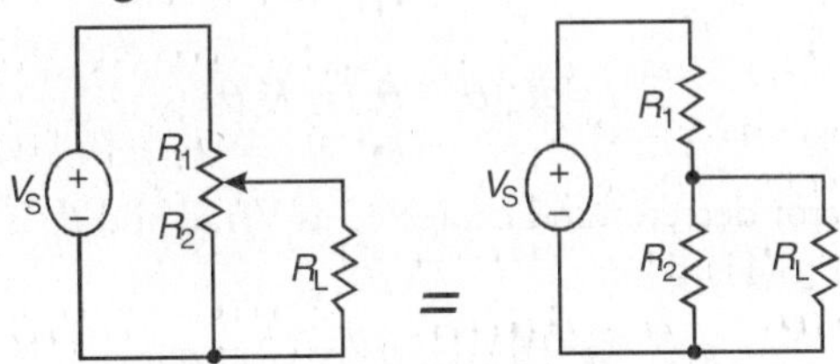

Figure 6.17 Potentiometer as a voltage divider

The most common use of a potentiometer is as a voltage divider as shown in Fgure 6.17, to obtain a manually adjustable output voltage at the slider (wiper) from a fixed input voltage applied across the two ends of the potentiometer.

The voltage across load can be calculated as

$$V_L = \frac{R_2 R_L}{R_1 R_L + R_2 R_L + R_1 R_2} V_s .$$

6.5.1 Potentiometer as an Error Detector

A potentiometer error detector consists of two similar potentiometers connected in parallel and supplied from a voltage source, as shown in Figure 6.18.

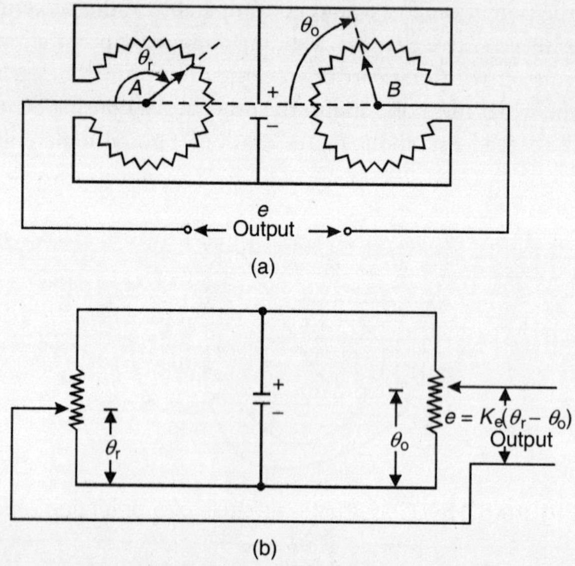

Figure 6.18 Potentiometer error detector

Figure 6.18(a) shows the input shaft of the control system connected to the potentiometer marked A and its setting fixed at desired angular position of θ_r. The output shaft is shown making an angle of θ_o. Figure 6.18(b) shows the arrangement of Figure 6.18(a) in a circuit diagram. Here, if angle θ_o equal to θ_r, then their difference, that is, the error is zero. However, if there is some error in the system response, then the difference of θ_o and θ_r, which is proportional to an electrical potential *e*, can be expressed as

$$e \;\alpha\; (\theta_r - \theta_0)$$

or

$$e = K_e(\theta_r - \theta_0) = K_e\theta_e \tag{6.41}$$

where K_e is the gain of the error detector and expressed as V/rad and θ_e is the angular error $(\theta_r - \theta_0)$.

The Laplace transform of (6.41) is

$$E(s) = K_e[\theta_r(s) - \theta_o(s)] = K_e\theta_e(s). \tag{6.42}$$

The block diagram representation is shown in Figure 6.19.

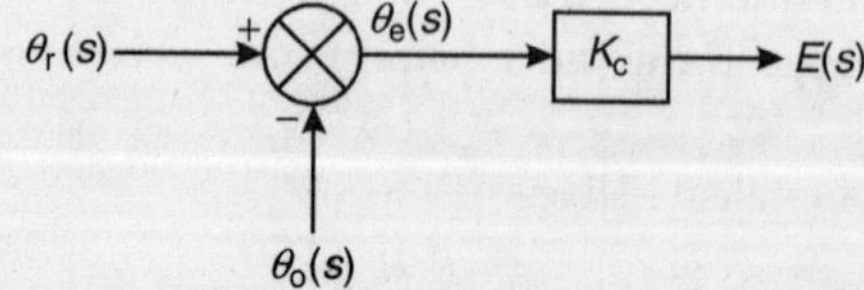

Figure 6.19 Block diagram representation of a potentiometer-type error detector

6.6 MAGNETIC AMPLIFIERS

The magnetic amplifier is an electromagnetic device for amplifying electrical signals. It was invented early in the 20th century and was used as an alternative to vacuum tube amplifiers, where robustness and high current capacity were required. It uses the principle of transformers along with the concept of magnetic saturation of the core. A magnetic amplifier is a device used for controlling the flow of power to a load by means of saturating a magnetic core. They are widely used in recent years for both high and low power applications.

Magnetic amplifier is a device consisting of combination of saturable reactors, rectifiers and conventional transformers used to secure control or amplification. In magnetic amplifiers, the load current is controlled by a DC magnetising current, which is very low as compared with load current. A large current value is controlled by a small current value, and hence such type of amplifiers are termed as current amplifiers. The load current is controlled by using a saturable reactor. The reactance of the reactor depends upon the DC control current. Thus, the load current is controlled by using magnetic property, and hence, the term magnetic amplifier. Magnetic materials suitable for magnetic amplifiers have a rectangular hysteresis loop with sharp saturation and low hysteresis losses. The excitation level is adjusted such that the flux swing in the core occurs between the knees of the *B–H* curve.

The basic principle of magnetic amplifier is to control the current through one coil by passing a direct current through another coil and both coils wound around a magnetic core material. In the three-limbed core shown in Figure 6.20, the outer limb windings are identical and are connected to AC supply. The coil wound on central limb is called as control winding and coil wound on outer limb is called as load winding. When there is no direct current, the effective flux in the control limb is zero and no voltage will be induced in the windings. The arrangement is called a transductor or saturable reactor.

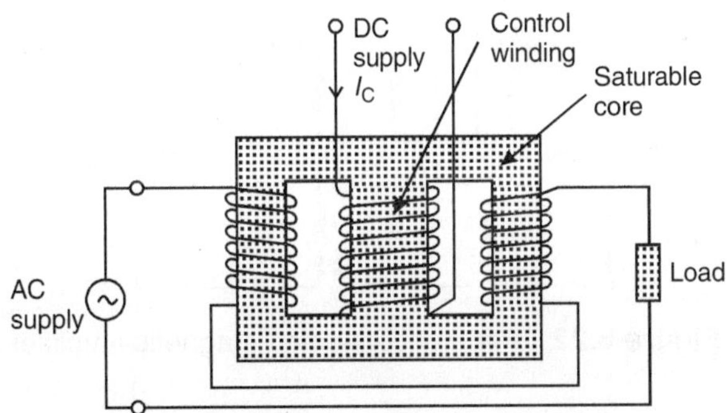

Figure 6.20 Basic magnetic amplifier

When a direct current is passed through the central limb winding, a flux is setup in the core. The impedance of two AC windings will change depending on the level at which the core flux is set by the control winding current. The alternating current through the load, which is connected in series with the AC winding can be controlled by direct current flowing through the control winding. The polarity of the input current does not affect the operation. By a suitable choice of the number of turns on the windings, the power required in the control winding (input) can be made small as compared to the power in the AC winding and the output.

The typical magnetic amplifier consists of two physically separate but similar transformer magnetic cores, each of which has two windings – a control winding and a load winding (or an AC winding). A small DC current from a low impedance source is fed into the series-connected control windings.

The AC windings may be connected either in series or in parallel, the configurations resulting in different types of magnetic amplifiers. The amount of control current fed into the control winding sets the point in the AC winding waveform at which either core will saturate. In saturation, the AC winding on the saturated core will go from a high impedance state ('off') into a very low impedance state ('on') – that is, the control current controls the voltage at which the magnetic amplifier switches 'on'. A relatively small DC current on the control winding is able to control or switch large AC currents on the AC windings. This results in current amplification.

In series-connected magnetic amplifier, the two reactors (load winding) are connected in series, as shown in Figure 6.21. The total reactance is twice and it is controlled by control winding. In such type of magnetic amplifiers, the current can be controlled right from near zero current. By increasing the control current, the reactance is slowly decreased, and hence, load current is increased. In parallel-connected magnetic amplifiers, the two reactors are connected in parallel, as shown in Figure 6.22. The total reactance in this case will be decreased. At zero control current, there will be appreciable load current.

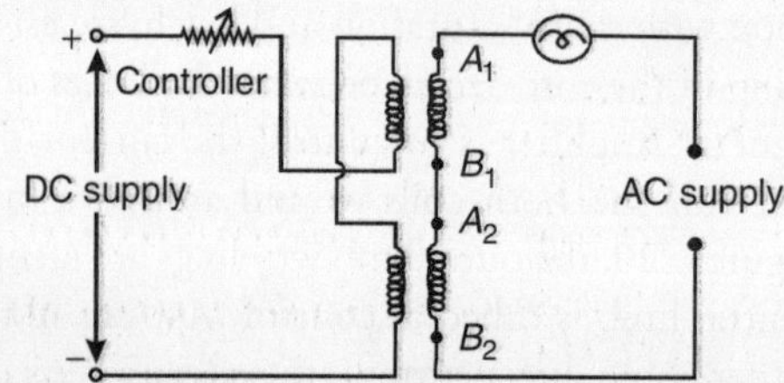

Figure 6.21 Series-connected magnetic amplifier

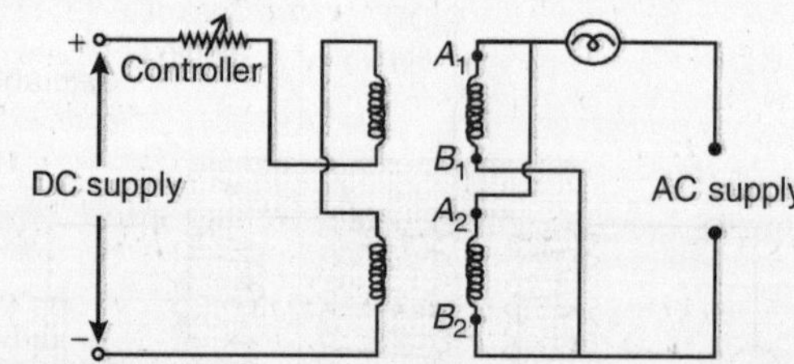

Figure 6.22 Parallel-connected magnetic amplifier

6.6.1 Applications of Magnetic Amplifier

The various applications of magnetic amplifier are listed as follows:

1. Magnetic amplifiers were important as modulation and control amplifiers in the early development of voice transmission by radio.
2. The ability to control large currents with small control power made magnetic amplifiers useful for controlling lighting circuits, stage lighting and lighting for advertisement signboards.
3. These are used extensively as the switching element in early switched-mode (SMPS) power supplies.
4. These are still used by some arc welders.
5. Magnetic amplifiers can be used for measuring high DC voltages without direct connection to the high voltage, and therefore, they are still used in high-voltage direct current (HVDC) technique.
6. These are used by locomotives to detect wheel slip.

REVIEW QUESTIONS

1. Explain the working of a DC servomotor with a neat sketch.
2. Explain the working principle of synchro receiver with a neat sketch.
3. Derive the transfer function of a DC motor with armature voltage control.
4. With the help of sketches, explain the construction and working principle of a synchro transmitter.
5. Derive the transfer function for the field-controlled DC motor with a neat sketch and explain the advantages of field-controlled DC motor over armature-controlled DC motor.
6. Derive the transfer function for armature-controlled DC motor with a neat diagram and explain the advantages of armature-controlled DC motor over field-controlled DC motor.
7. Explain about synchro transmitter and receiver and derive the transfer function for synchros.
8. Derive the transfer function of a DC motor with field control.
9. Explain the synchro error detector with a circuit diagram.
10. Derive the transfer function for AC servomotor. Explain the torque–speed characteristics of AC servomotor.
11. Explain the working principle and construction of AC servomotors.
12. What is the principle used in magnetic amplifiers?
13. For zero control current, why is load current more in parallel winding than in series winding?
14. Explain the construction and working principle of magnetic amplifiers.
15. Explain how potentiometer is used as error detector.

OBJECTIVE TYPE QUESTIONS

1. In servo applications, the motor used is
   (a) universal motor (b) synchronous motor
   (c) three-phase induction motor (d) all
2. Synchros are generally used as transmitters of
   (a) digital data (b) angular data (c) data logger (d) all
3. The torque–speed characteristics of a two-phase induction motor is largely affected by
   (a) voltage (b) speed (c) R/X ratio (d) X/R ratio
4. The stator of synchro is made up of
   (a) silicon steel (b) high carbon steel (c) cast iron (d) all
5. To minimise the loading effect of a potentiometer, we use
   (a) low impedance (b) very high impedance
   (c) low resistance (d) low inductance
6. In the field-controlled DC motor, the entire damping comes from the
   (a) armature resistance (b) back e.m.f.
   (c) motor friction and load (d) field resistance
7. The characteristic equation of an armature-controlled DC motor is
   (a) first order (b) second order (c) third order (d) fourth order
8. In a series DC servomotor, the following constitutes a drawback
   (a) High starting torque (b) Large current drawn
   (c) Poor speed regulation (d) All
9. A synchro transmitter receiver unit is a
   (a) two-phase AC device (b) three-phase AC device
   (c) DC device (d) single-phase AC device

10. In the position control systems, the device used for providing rate feedback voltage is called
(a) potentiometer (b) synchro transmitter
(c) synchro transformer (d) tachogenerator

11. The AC motor used in servo application is a
(a) single-phase induction motor (b) two-phase induction motor
(c) universal motor (d) synchronous motor

12. The most common use of the synchros is
(a) as an error detector (b) for the transmission of arithmetic data
(c) for transmission of angular data (d) for synchronisation

13. Synchro transmitter acts like
(a) an induction motor (b) a transformer
(c) an alternator (d) a rectifier

14. The following device may be used to control the position of a very small load
(a) an AC servomotor (b) a DC servomotor
(c) a synchro (d) None of these

15. The resolution of a potentiometer depends on
(a) type of contact (b) wire material
(c) size and width of the wire (d) none of these

16. The resolution of a potentiometer must be
(a) infinite (b) very high (c) medium (d) zero

ANSWERS

1. c **2.** c **3.** d **4.** a **5.** b **6.** c **7.** c **8.** d **9.** b **10.** d
11. b **12.** a **13.** b **14.** c **15.** c **16.** a

CHAPTER

7 Time Domain Analysis

Contents

7.1 INTRODUCTION

Most of the control systems use time as its independent variable and therefore, it is important to analyse the response given by the system for applied excitation, which is a function of time. The evaluation of the system is based on the analysis of such response. Stability analysis, system accuracy and complete evaluation is always based on the time response analysis. Thus, by definition, time response is the response given by the system, which is a function of time to the applied excitation.

In any practical system, output of the system takes some finite time to reach to its final value. This time varies from system to system and is dependent on different factors. Similarly, final value achieved by the output also depends on different factors such as friction, mass or inertia of moving elements, and some inherent nonlinearity.

Successfulness and accuracy of a system depend on the final value reached by the system output, which should be very close to what is desired from that system. While reaching to its final value, the output also has to be smooth without aberrations. The final state achieved by the output is called the steady-state, and the output variation within the time it takes to achieve the steady-state is called the transient response of the system.

7.2 CLASSIFICATION OF TIME RESPONSES

The response or behaviour of a system with respect to time is called time response. In the initial part of time response of a control system, a transient state appears and during the post-transient part, steady-state is achieved. Thus, the time response of any system is divided into two parts:

1. **Transient response:** It is defined as the part of the total response that goes to zero as time becomes very large.

   In this case, $\underset{t\to\infty}{\text{Lt}}\, c(t)=0$

   The transient part of time response reveals the nature of response (i.e., oscillatory or over-damped) and also gives an indication about its speed.

2. **Steady-state response:** It is defined as the part of the total response that attains a constant value as time approaches infinity (or) the part of the response that remains even after the transients have died out. The steady-state part of the time response reveals the accuracy of a system. Steady-state error is observed if the actual output does not exactly match with the input.

7.3 STANDARD TEST SIGNALS

1. **Step signal:** The step is a signal whose value changes from one level (usually zero) to another level 'A' in zero time. The mathematical representation of the step function is

$$r(t) = Au(t)$$

where

$$u(t)=1 \text{ for } t\geq 0$$

$$= 0 \text{ for } t<0$$

By taking Laplace transform, we get

$$R(s) = A/s \tag{7.1}$$

The graphical representation of a step signal is shown in Figure 7.1.

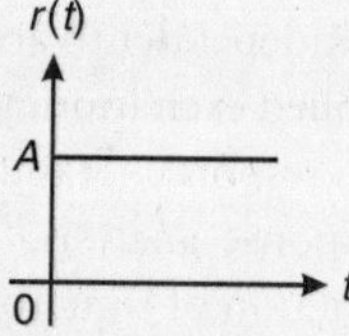

Figure 7.1 Step signal

2. **Ramp signal:** The ramp is a signal that starts at a value of zero and increases linearly with time. Mathematically,

$$r(t) = At \quad \text{for } t\geq 0$$

$$= 0 \quad \text{for } t<0$$

By taking Laplace transform, we get

$$R(s) = A/s^2 \tag{7.2}$$

The graphical representation of a ramp signal is shown in Figure 7.2.

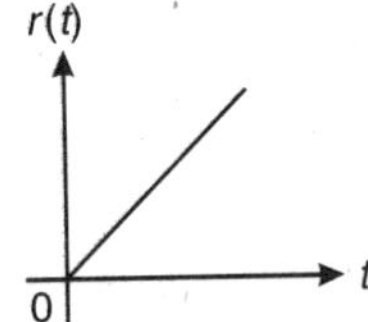

Figure 7.2 Ramp signal

From (7.1) and (7.2), it is seen that a ramp signal is integral of a step signal.

3. **Parabolic signal:** This is the input that is one degree faster than a ramp type of input, as shown in Figure 7.3.
The mathematical representation of this signal is

$$r(t) = At^2/2 \text{ for } t \geq 0$$

$$= 0 \text{ for } t < 0$$

By taking Laplace transform, we get

$$R(s) = A/s^3. \quad (7.3)$$

The graphical representation of a parabolic signal is shown in Figure 7.3.

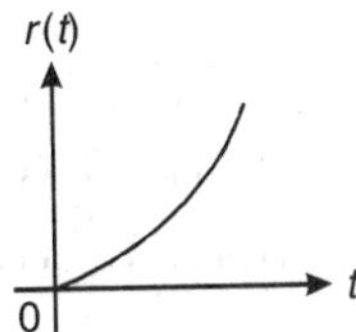

Figure 7.3 Parabolic signal

From (7.2) and (7.3), it is seen that a parabolic signal is integral of a ramp signal.

4. **Impulse signal:** A unit impulse is defined as a signal that has a zero value everywhere except at $t = 0$, where its magnitude is infinite. It is generally called the δ function and has the following property.

$$\delta(t) = 0; \ t \neq 0$$

$$\int_{-\varepsilon}^{\varepsilon} \delta(t)dt = 1 \text{ where } \varepsilon \to 0$$

Since a perfect impulse cannot be achieved in practice, it is usually approximated by a pulse of small width but unit area, as shown in Figure 7.4.

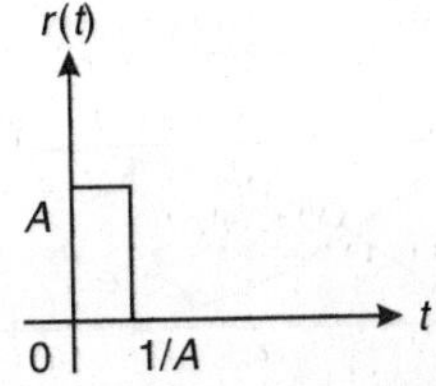

Figure 7.4 Impulse signal

Mathematically, an impulse function is the derivative of a step function

i.e.,
$$\delta(t) = \dot{u}(t)$$

The Laplace transform of a unit impulse is

$$L[\delta(t)] = 1 = R(s).$$

The impulse response of a system with transfer function $G(s) = \dfrac{C(s)}{R(s)}$ is given as

$$C(s) = G(s)R(s) = G(s)$$
$$\Rightarrow c(t) = L^{-1}\{G(s)\} = g(t)$$

Thus, the impulse response of a system indicated by $g(t)$ is the inverse Laplace transform of its transfer function. This is sometimes referred to as weighting function of the system using convolution integral:

$$c(t) = \int_0^t g(t-\tau)r(\tau)d\tau \tag{7.4}$$

7.4 TIME RESPONSE OF A FIRST-ORDER SYSTEM

Consider the first-order system of Figure 7.5 with unity feedback.

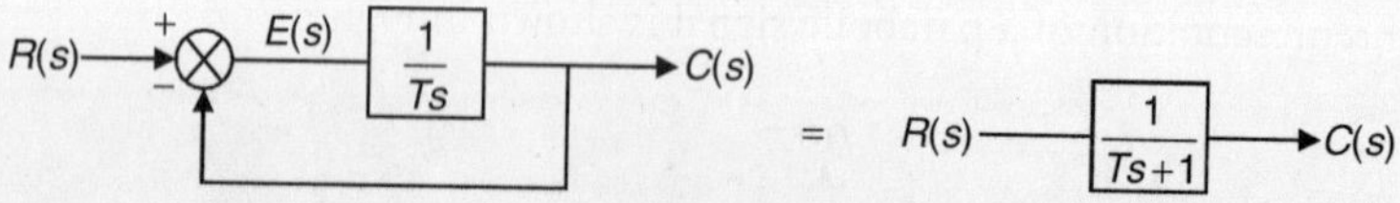

Figure 7.5 Block diagram of a first-order system

The transfer function of the system shown in Figure 7.5 is

$$\frac{C(s)}{R(s)} = \frac{1}{1+sT}.$$

Analyse the system response to various inputs.

1. **Response to unit step or displacement input:**
   For the unit step input, $R(s) = 1/s$
   The output response is given as
   $$C(s) = \frac{1}{s(1+sT)} = \frac{1}{s} - \frac{T}{1+sT}$$
   By taking the inverse Laplace transform, we get
   $$c(t) = 1 - e^{-t/T} \tag{7.5}$$
   which is plotted as shown in Figure 7.6.

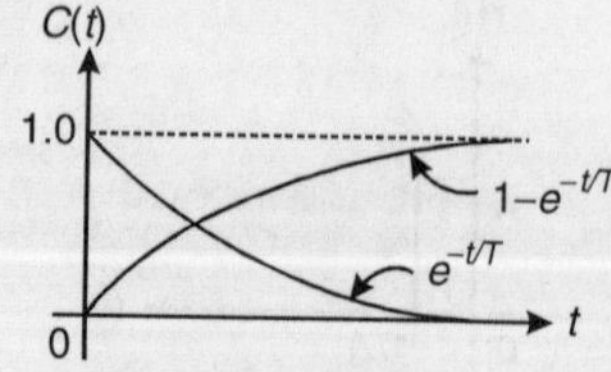

Figure 7.6 Unit step response of a first-order system

It is seen that the output rises exponentially from zero value to the final value of unity. The initial slope of the curve at $t = 0$ is given as

$$\left.\frac{dc}{dt}\right|_{t=0} = \left.\frac{1}{T}e^{-t/T}\right|_{t=0} = \frac{1}{T}$$

where T is known as the time constant of the system.

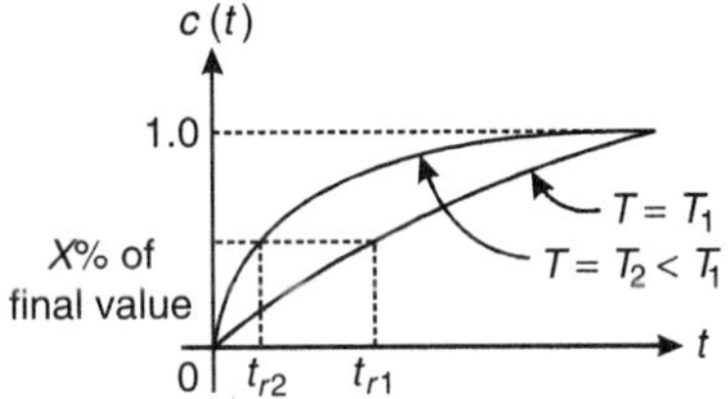

Figure 7.7 Effect of time constant on system response

The time constant is indicative of how fast the system tends to reach the final value. A large time constant corresponds to a sluggish system and a small time constant corresponds to a fast response, as shown in Figure 7.7.

The error response of the system is given as

$$e(t) = r(t) - e(t) = e^{-t/T}. \tag{7.6}$$

The steady-state error is given as

$$e_{ss} = \underset{t\to\infty}{\text{Lt}}\, e(t) = 0 \cdot \tag{7.7}$$

Thus, this system tracks the unit step input with zero steady-state error.

2. **Response to a unit ramp or velocity input:**
For the unit ramp input, $R(s) = 1/s2$
The output response is given as

$$C(s) = \frac{1}{s^2(1+sT)} = \frac{1}{s^2} - \frac{T}{s} + \frac{T^2}{1+sT}.$$

By taking the inverse Laplace transform, we get

$$c(t) = t - T\left(1 - e^{-t/T}\right). \tag{7.8}$$

The error response of the system is given as

$$e(t) = r(t) - c(t) = T(1 - e^{-t/T}). \tag{7.9}$$

The steady-state error is given as

$$e_{ss} = \underset{t\to\infty}{\text{Lt}}\, e(t) = T. \tag{7.10}$$

Thus, the first-order system under consideration will track the unit ramp input with a steady-state error T, which is equal to the time constant of the system, as shown in Figure 7.8. Therefore, reducing the system time constant not only improves its speed of response, but also reduces its steady-state error of a ramp input.

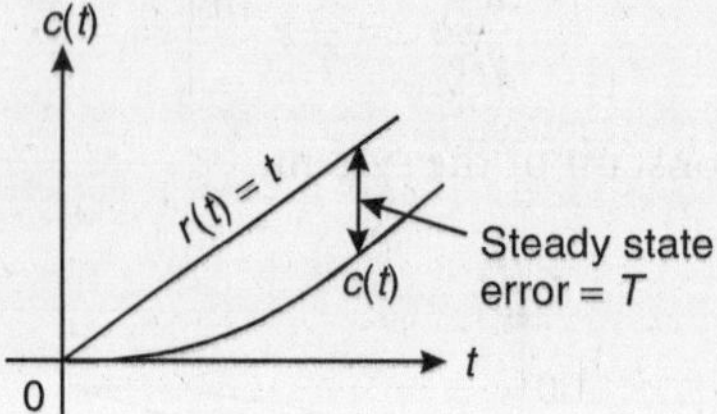

Figure 7.8 Unit ramp response of a first-order system

3. **Response to a unit parabolic or acceleration input**
For the unit parabolic input, $R(s) = 1/s^3$
The output response is given as

$$C(s) = \frac{1}{s^3(1+sT)}$$

$$c(t) = T^2 - Tt + \frac{t^2}{2} - T^2 e^{-t/T} \tag{7.11}$$

On differentiating (7.11), we get

$$\frac{dc(t)}{dt} = -T + t + Te^{-t/T} = t - T(1 - e^{-t/T}) \tag{7.12}$$

It is found that (7.12) is the same as (7.8), which is the response of first-order system to unit velocity input. Thus, subjecting the first-order system to a unit parabolic input does not give any additional information regarding transient behaviour of the system.
However, the error response of the system is given as

$$e(t) = r(t) - e(t) = -T^2 + Tt + T^2 e^{-t/T}. \tag{7.13}$$

The steady-state error is given as

$$e_{ss} = \underset{t\to\infty}{\text{Lt}}\ e(t) = \infty .$$

Thus, a first-order system has infinite steady-state error for a parabolic input.

Summarising the analysis of first-order system, we can say that the step input yields the desired information about the speed of transient response. It is observed that the speed of response is inversely proportional to the time constant T of the system. The ramp and the parabolic inputs do not give any additional information regarding the speed of response.

However, the steady-state errors are different for these three different inputs.

For a step input, e_{ss} = zero.

For a velocity input, $e_{ss} = T$ = finite

For an acceleration input, $e_{ss} = \infty$.

It is clear from the aforementioned discussion that it is sufficient to study the behaviour of any system to a unit step input for understanding its transient response and use the velocity input and acceleration input for understanding the steady-state behaviour of the system.

In general, transient response depends on

1. Poles of the closed-loop transfer function (t.f.).
2. Location of the poles of the closed-loop

Transient response is independent of the magnitude of the input applied. Any change in the magnitude of the selected input will not have any effect on the transient response of the system.

Problem-7.1: A system has the following t.f. $\frac{C(s)}{R(s)} = \frac{20}{s+10}$. Determine its unit impulse, step and ramp responses with zero initial conditions.

Solution:

Given that

$$\frac{C(s)}{R(s)} = \frac{20}{s+10}.$$

(a) **Unit impulse input:**

$$R(s) = 1,$$

$$\frac{C(s)}{R(s)} = \frac{20}{s+10}$$

$$C(s) = \frac{20}{s+10}R(s) = \frac{20}{s+10}$$

$$\Rightarrow \qquad c(t =) = L^{-1}\left[\frac{20}{s+10}\right] = 20e^{-10t}$$

(b) **Unit step input:**

$$R(s) = 1/s$$

$$C(s) = \frac{20}{s+10}R(s) = \frac{20}{s(s+10)} = \frac{2}{s} - \frac{2}{s+10}$$

$$\Rightarrow \qquad c(t =) = L^{-1}\left[\frac{2}{s} - \frac{2}{s+10}\right] = 2(1 - e^{-10t})$$

(c) **Unit ramp input:**

$$R(s) = \frac{1}{s^2}$$

$$C(s) = \frac{20}{s+10}R(s) = \frac{20}{s^2(s+10)} = \frac{2}{s^2} - \frac{0.2}{s} + \frac{0.2}{s+10}$$

$$\Rightarrow \qquad c(t =) = L^{-1}\left[\frac{2}{s^2} - \frac{0.2}{s} + \frac{0.2}{s+10}\right] = 2t - 0.2 + 0.2e^{-10t}$$

7.5 CHARACTERISTIC EQUATION

The closed-loop t.f. of a system is given as

$$\frac{C(s)}{R(s)} = \frac{G(s)}{1+G(s)H(s)}. \qquad (7.14)$$

The equation that gives poles of system is called characteristic equation, which is

$$1 + G(s)H(s) = 0. \tag{7.15}$$

For the first-order system, (7.15) is also first order in general, of the form

$$1 + Ts = 0 \tag{7.16}$$

$\therefore$ There is only one closed loop pole, That is, $s = -1/T$

Thus, the closed loop poles are the roots of characteristic equation. The value of closed loop pole appears as an exponential index in the transient output.

Consider two systems, one with closed loop pole at $s_1 = -2$, while other having closed loop pole at $s_2 = -4$. In the output of two systems, we will get e^{-2t} in the first system and e^{-4t} in the second system. Now, as $t \to \infty$, e^{-4t} will vanish earlier than e^{-2t}. Hence, the second system transients will vanish quickly and stabilise to its final value earlier than the first system. From this, as closed loop poles move in the left half of the *s*-plane away from the imaginary axis, transients die out more quickly; this makes the system more stable.

The pole–zero plot of the first-order system is as shown in Figure 7.9.

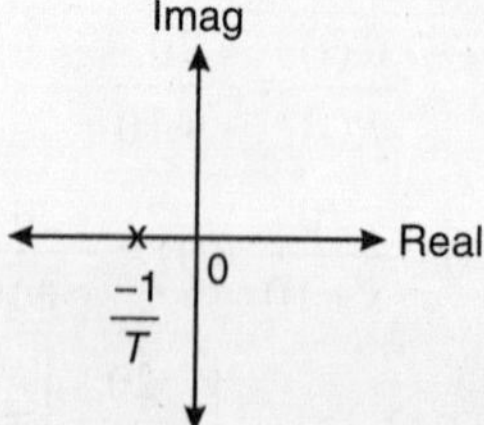

Figure 7.9 Pole–zero plot of first-order system

7.6 TIME RESPONSE OF A SECOND-ORDER SYSTEM

Consider a Type 1 second-order system, as shown in Figure 7.10. Since *G*(*s*) has one pole at the origin, it is a Type 1 system.

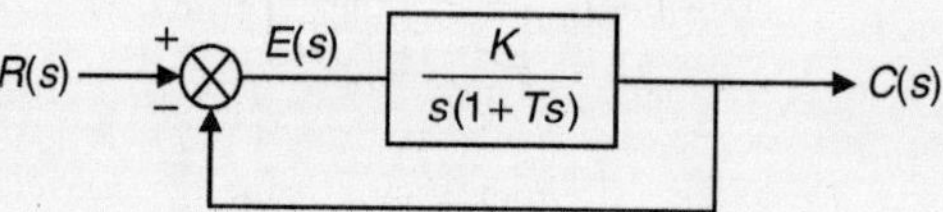

Figure 7.10 Second-order type 1 system

The closed-loop t.f. is given as

$$T(s) = \frac{C(s)}{R(s)} = \frac{K}{Ts^2 + s + K}. \tag{7.17}$$

It can also be written in the standard form as

$$T(s) = \frac{C(s)}{R(s)} = \frac{\omega_n^2}{s^2 + 2\xi\omega_n s + \omega_n^2}. \tag{7.18}$$

The characteristic equation of the system is

$$s^2 + 2\xi\omega_n s + \omega_n^2 = 0 \tag{7.19}$$

where ξ is the damping ratio and ω_n is the natural frequency of oscillations.

7.6.1 Damping Ratio

Every practical system takes finite time to reach to its steady-state, and during this period, it oscillates or increases exponentially. The behaviour of a system gets decided by the type of closed loop poles and locations of closed loop poles in the *s*-plane. The closed loop poles are dependent on the selection of the parameters of the system. Every system has a tendency that opposes the oscillatory behaviour of the system, called damping. Now, this tendency controls the type of closed loop poles, and hence, the nature of the response. This damping is measured by a factor or a ratio called damping ratio of the system. This factor gives us an indication of the extent of opposing behaviour faced by the oscillations, at the output. In some systems, it will be low; in which case system will oscillate, but slowly, that is, with damped frequency. In some systems, it may be very high so that the system output will not oscillate at all; further, it will be exponential and very slow that it will take very long time to reach the steady-state. This damping ratio explaining such behaviour is denoted by a Greek symbol ξ (zeta). Now, as this measures the opposition by the system to the oscillatory behaviour, if it is made zero $(\xi = 0)$, the system will oscillate with maximum frequency. As there is no opposition from the system, the system naturally and freely oscillates under such conditions. Hence, this frequency of oscillations under $\xi = 0$ condition is called natural frequency of oscillations of the system and its unit is rad/s. Equation (7.17) can be rewritten as

$$T(s) = \frac{\frac{K}{T}}{s^2 + \frac{l}{T}s + \frac{K}{T}} \tag{7.20}$$

where

$$\omega_n = \sqrt{\frac{K}{T}} = \text{natural frequency}$$

$$\xi = \frac{1}{2\sqrt{KT}} = \text{damping factor}$$

The poles of t.f. or the roots of the characteristic equation are given as

$$s^2 + 2\xi\omega_n s + \omega_n^2 = 0.$$

$$s_{1,2} = \frac{-2\xi\omega_n \pm \sqrt{4\xi^2\omega_n^2 - 4\omega_n^2}}{2}$$

For

$$\xi < 1$$

$$s_{1,2} = -\xi\omega_n \pm j\omega_n\sqrt{1-\xi^2} = -\xi\omega_n \pm j\omega_d \tag{7.21}$$

where $\omega_d = \omega_n\sqrt{1-\xi^2}$ is called the damped natural frequency of the system.

If $\xi > 1$, the two roots, s_1 and s_2 are real and we have an 'over-damped system'. If $\xi = 1$, the system is known as 'critically damped system'. The more common case of $\xi < 1$ is known as the 'under-damped system'.

1. **Response to a unit step input:**
   For a unit step input $R(s) = 1/s$.
   The output response is given as

$$C(s) = \frac{\omega_n^2}{s^2 + 2\xi\omega_n s + \omega_n^2} \times R(s) = \frac{\omega_n^2}{s^2 + 2\xi\omega_n s + \omega_n^2} \times \frac{1}{s} \quad (7.22)$$

$$s^2 + 2\xi\omega_n s + \omega_n^2 = (s + \xi\omega_n)^2 + \omega_n^2(1 - \xi^2)$$

Case (i): Under-damped ($0 < \xi < 1$)

$$C(s) = \frac{\omega_n^2}{s^2 + 2\xi\omega_n s + \omega_n^2} \times \frac{1}{s} = \frac{1}{s} - \frac{s + 2\xi\omega_n}{(s + \xi\omega_n)^2 + \omega_n^2(1 - \xi^2)} \quad (7.23)$$

Substitute

$$\omega_d = \omega_n\sqrt{1 - \xi^2}$$

$$C(s) = \frac{1}{s} - \frac{s + 2\xi\omega_n}{(s + \xi\omega_n)^2 + \omega_d^2}$$

$$= \frac{1}{s} - \frac{s + \xi\omega_n}{(s + \xi\omega_n)^2 + \omega_d^2} - \frac{\xi\omega_n}{\omega_d}\frac{\omega_d}{(s + \xi\omega_n)^2 + \omega_d^2}.$$

By taking inverse Laplace transforms on both sides, we get

$$c(t) = 1 - e^{-\xi\omega_n t}\cos\omega_d t - \frac{\xi\omega_n}{\omega_d}.e^{-\xi\omega_n t}\sin\omega_d t \quad (7.24)$$

$$= 1 - e^{-\xi\omega_n t}\left[\cos\omega_d t + \frac{\xi}{\sqrt{1 - \xi^2}}\sin\omega_d t\right]$$

$$= 1 - \frac{e^{-\xi\omega_n t}}{\sqrt{1 - \xi^2}}\left[\sqrt{1 - \xi^2}\cos\omega_d t + \xi\sin\omega_d t\right].$$

Substitute

$$\sin\theta = \sqrt{1 - \xi^2} \Rightarrow \cos\theta = \xi, \tan\theta = \frac{\sqrt{1 - \xi^2}}{\xi}$$

$$c(t) = 1 - \frac{e^{-\xi\omega_n t}}{\sqrt{1 - \xi^2}}\left[\sin\theta\cos\omega_d t + \cos\theta\sin\omega_d t\right]$$

$$= 1 - \frac{e^{-\xi\omega_n t}}{\sqrt{1 - \xi^2}}\sin(\omega_d t + \theta)$$

$$\therefore c(t) = 1 - \frac{e^{-\xi\omega_n t}}{\sqrt{1 - \xi^2}}\sin(\omega_d t + \theta) \quad (7.25)$$

where $\omega_d = \omega_n\sqrt{1 - \xi^2}$ and $\theta = \tan^{-1}\left[\frac{\sqrt{1 - \xi^2}}{\xi}\right]$

This response is plotted in Figure 7.11(a). The response is oscillatory and as $t \to \infty$, it approaches unity.

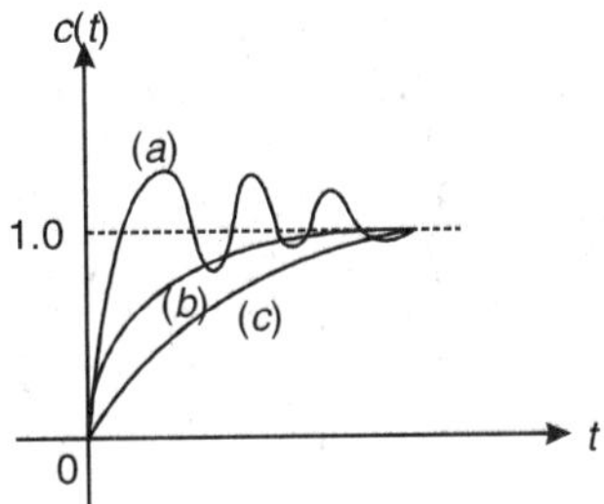

Figure 7.11 Step response of a second-order system

Case (ii): Critically damped ($\xi = 1$)

The output of the system is given as

$$C(s) = \frac{\omega_n^2}{s^2 + 2\xi\omega_n s + \omega_n^2} \times \frac{1}{s}$$

$$= \frac{\omega_n^2}{s(s^2 + 2\omega_n s + \omega_n^2)} = \frac{\omega_n^2}{s(s+\omega_n)^2}$$

$$C(s) = \frac{1}{s} - \frac{\omega_n}{(s+\omega_n)^2} - \frac{1}{(s+\omega_n)}.$$

By taking inverse Laplace transforms on both sides, we get

$$c(t) = 1 - \omega_n t e^{-\omega_n t} - e^{-\omega_n t} = 1 - e^{-\omega_n t}(1 + \omega_n t).$$

This response is plotted in Figure 7.11 (b)

Case (iii): Over-damped ($\xi > 1$)

The output of the system is given as

$$C(s) = \frac{\omega_n^2}{s^2 + 2\xi\omega_n s + \omega_n^2} \times \frac{1}{s}$$

$$= \frac{\omega_n^2}{s(s + \xi\omega_n + \omega_n\sqrt{\xi^2 - 1})(s + \xi\omega_n - \omega_n\sqrt{\xi^2 - 1})}$$

$$C(s) = \frac{1}{s} - \left[\frac{A}{(s + \xi\omega_n + \omega_n\sqrt{\xi^2 - 1})} + \frac{B}{(s + \xi\omega_n - \omega_n\sqrt{\xi^2 - 1})}\right].$$

By taking inverse Laplace transforms on both sides, we get

$$c(t) = 1 - \left[Ae^{-(\xi\omega_n + \omega_n\sqrt{\xi^2 - 1})} + Be^{-(\xi\omega_n - \omega_n\sqrt{\xi^2 - 1})}\right]. \tag{7.26}$$

This response is plotted in Figure 7.11 (c).

Case (iv): Un-damped ($\xi = 0$)

The output of the system is given as

$$C(s) = \frac{\omega_n^2}{s^2 + 2\xi\omega_n s + \omega_n^2} \times \frac{1}{s} = \frac{\omega_n^2}{s + (s^2 + \omega_n^2)} = \frac{1}{s} - \frac{s}{s^2 + \omega_n^2}$$

By taking inverse Laplace transforms on both sides, we get

$$c(t) = 1 - \cos\omega_n t\,. \tag{7.27}$$

This response is plotted in Figure 7.12.

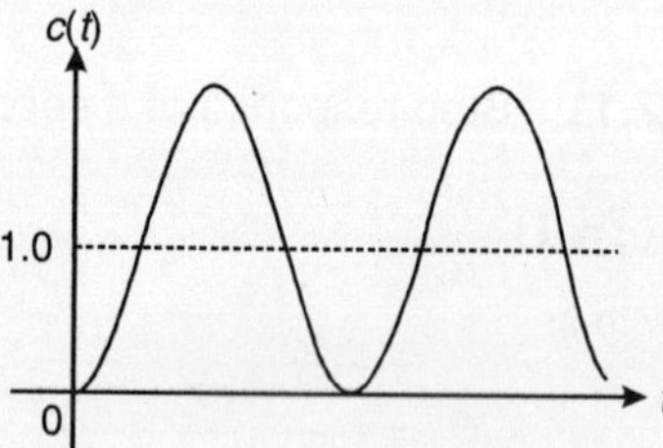

Figure 7.12 Step response of a second-order system for un-damped case

7.7 TIME DOMAIN SPECIFICATIONS OF SECOND-ORDER SYSTEMS

Higher order control systems usually have a pair of complex conjugate poles with damping less than one, which dominates over other poles. Therefore, the time response of the second- and higher order control systems to a step input is generally of damped oscillatory nature, as shown in Figure 7.13. It is observed from this figure that the step response has a number of overshoots and undershoots with respect to the final steady value. This type of step response is characterised by the following performance indices, which are usually evaluated in terms of the following qualities.

1. How fast the system moves to follow the input?
2. How oscillatory it is (indicative of damping)?
3. How long does it take to practically reach the final value?

In designing any system, certain design specifications are given based on the typical under-damped step response as shown as in Figure 7.13.

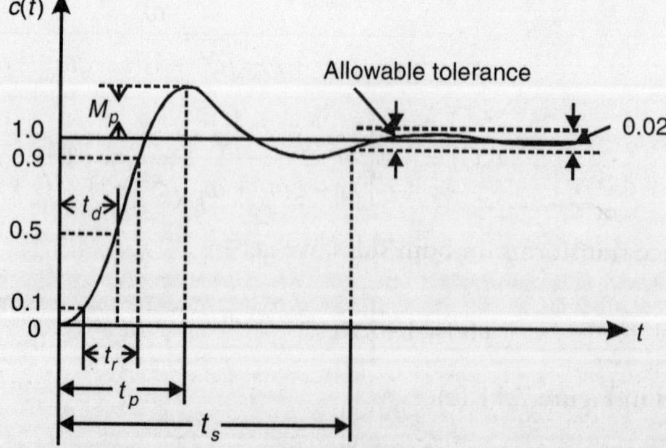

Figure 7.13 Time domain specifications of a second-order system

The various design specifications are

1. **Delay time (t_d)**
It is the time required for the response to step input to reach 50% of the final value in the first attempt.
From the definition,

$$c(t) = 1/2 \text{ at } t = t_d$$

$$\frac{1}{2} = 1 - \frac{e^{-\xi\omega_n t}}{\sqrt{1-\xi^2}}\sin(\omega_d t_d + \theta)$$

$$\frac{e^{-\xi\omega_n t}}{\sqrt{1-\xi^2}}\sin(\omega_d t_d + \theta) = \frac{1}{2} \tag{7.28}$$

The result of the above-mentioned equation is

$$t_d = \frac{1+0.7\xi}{\omega_n}. \tag{7.29}$$

2. **Rise time (t_r)**
It is the time required for the response to rise from 10% to 90% of the final value for over-damped systems and 0% to 100% of the final value for under-damped system. The rise time is reciprocal of the slope of the response at the instant and the response is equal to 50% of the final value.
From the definition,

$$c(t) = 1 \text{ at } t = t_r$$

$$1 = 1 - \frac{e^{-\xi\omega_n t}}{\sqrt{1-\xi^2}}.\sin(\omega_d t_r + \theta)$$

$$\Rightarrow \quad \frac{e^{-\xi\omega_n t}}{\sqrt{1-\xi^2}}\sin(\omega_d t_r + \theta) = 0$$

$$\Rightarrow \quad \sin(\omega_d t_r + \theta) = \pi$$

$$\Rightarrow \quad \omega_d t_r + \theta = n\pi$$

where $n = 1,2$
For the response in the first attempt, n is 1,

$$\therefore \quad \omega_d t_r + \theta = \pi$$

$$\Rightarrow \quad t_r = \frac{\pi - \theta}{\omega_d} = \frac{\pi - \tan^{-1}\left(\frac{\sqrt{1-\xi^2}}{\xi}\right)}{\omega_n\sqrt{1-\xi^2}} \tag{7.30}$$

3. **Peak time (t_p)**
It is the time required for the response to step input to reach the maximum or peak value of the response.

From the definition,

$$\frac{dc}{dt} = 0 \text{ at } t = t_p$$

$$\frac{d}{dt}[1 - \frac{e^{-\xi\omega_n t}}{\sqrt{1-\xi^2}}.\sin(\omega_d t_r + \theta)] = 0, \text{ at } t = t_p$$

$$-\xi\omega_n \times \frac{e^{-\xi\omega_n t_p}}{\sqrt{1-\xi^2}}\sin(\omega_d t_p + \theta) + \frac{e^{-\xi\omega_n t_p}}{\sqrt{1-\xi^2}}\omega_d \cos(\omega_d t_p + \theta) = 0$$

$$\xi\omega_n \sin(\omega_d t_p + \theta) - \omega_d \cos(\omega_d t_p + \theta) = 0$$

$$\Rightarrow \quad \frac{\sin(\omega_d t_p + \theta)}{\cos(\omega_d t_p + \theta)} = \frac{\omega_d}{\xi\omega_n}$$

$$\Rightarrow \quad \frac{\sin(\omega_d t_p + \theta)}{\cos(\omega_d t_p + \theta)} = \frac{\sqrt{1-\xi^2}}{\xi}$$

$$\Rightarrow \quad \tan(\omega_d t_p + \theta) = \frac{\sqrt{1-\xi^2}}{\xi}$$

$$\Rightarrow \quad \tan(\omega_d t_p + \theta) = \tan\theta = \tan(n\pi + \theta)$$

$$\Rightarrow \quad \omega_d t_p + \theta = n\pi + \theta$$

$$\Rightarrow \quad t_p = \frac{n\pi}{\omega_d}.$$

Since the peak time corresponds to the first overshoot, we get

$$t_p = \frac{\pi}{\omega_d} = \frac{\pi}{\omega_n\sqrt{1-\xi^2}}. \tag{7.31}$$

4. **Peak overshoot (M_p)**
It is the difference between the peak value of the response and the steady-state value. It is usually expressed in terms of the percent of the steady-state value. If t_p is the peak time, then percentage peak overshoot is given as

$$\%M_p = \left[\frac{c(t_p) - c(\infty)}{c(\infty)}\right] \times 100$$

$$\%M_p = \left[c(t_p) - 1\right] \times 100$$

$$= -\frac{e^{-\xi\omega_n t_p}}{\sqrt{1-\xi^2}}\sin(\omega_d t_p + \theta) \times 100.$$

However,
$$t_p = \frac{\pi}{\omega_d}$$

$$\%M_p = -\frac{e^{-\xi\omega_n t_p}}{\sqrt{1-\xi^2}}\sin(\pi+\theta)\times 100 = \frac{e^{-\xi\omega_n t_p}}{\sqrt{1-\xi^2}}\sin(\theta)\times 100 \quad (7.32)$$

We know that $\tan\theta = \frac{\sqrt{1-\xi^2}}{\xi}$ can be represented as shown in Figure 7.14.

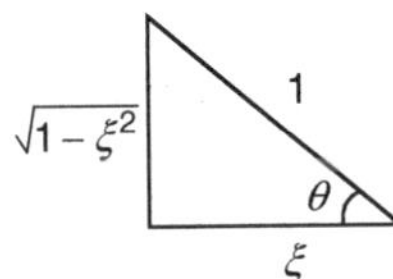

Figure 7.14 Relation between θ and ξ

From the above figure, we have

$$\sin\theta = \sqrt{1-\xi^2} \text{ and substituting this in } M_p$$

$$\%M_p = \frac{e^{-\xi\omega_n t_p}}{\sqrt{1-\xi^2}} \times \sqrt{1-\xi^2} \times 100$$

$$= e^{-\xi\omega_n \times \frac{\pi}{\omega_d}} \times 100$$

$$= e^{-\frac{\xi\pi}{\sqrt{1-\xi^2}}} \times 100 \quad (7.33)$$

5. **Setting time (t_s)**
   It is the time required for the response to reach and remain within a specified tolerance band, usually 2% to 5%, of the steady-state output.
   Thus, for a 2% tolerance band, we have

$$c(t)_{\text{at } t=t_s} = 0.98 = 1 - \frac{e^{-\xi\omega_n t_s}}{\sqrt{1-\xi^2}}$$

$$\Rightarrow \quad \frac{e^{-\xi\omega_n t_s}}{\sqrt{1-\xi^2}} = 0.02 \quad (7.34)$$

For low values of ξ, $\xi^2 << 1$, and therefore,

$$e^{-\xi\omega_n t_s} = 0.02$$

$$\Rightarrow \quad -\xi\omega_n t_s = \ln(0.02) = -3.912$$

$$\Rightarrow \quad t_s = \frac{4}{\xi\omega_n} \quad (7.35)$$

Similarly for 5% tolerance band, we get

$$c(t)_{\text{at } t=/t_s} = 0.95 = 1 - \frac{e^{-\xi\omega_n t_s}}{\sqrt{1-\xi^2}}$$

$$\Rightarrow \qquad \frac{e^{-\xi\omega_n t_n}}{\sqrt{1-\xi^2}} = 0.05. \tag{7.36}$$

For low values of ξ, $\xi^2 << 1$, and therefore,

$$e^{-\xi\omega_n t_s} = 0.05$$

$$-\xi\omega_n t_s = \ln(0.05) = -2.995$$

$$\Rightarrow \qquad t_s = \frac{3}{\xi\omega_n}. \tag{7.37}$$

6. **Steady-state error**

It is the difference between the desired output and the actual output as $t \to \infty$ or under steady-state conditions. The desired output is given by the reference input $r(t)$, and therefore,

$$e_{ss} = \underset{t\to\infty}{\text{Lt}}\left[r(t) - c(t)\right] \tag{7.38}$$

The steady-state error for a Type 1 second-order system considered for design specifications for a step input is obviously zero. Thus,

$$e_{ss} = \underset{t\to\infty}{\text{Lt}}[1 - c(t)] = 0 \tag{7.39}$$

and the steady-state error for a ramp input is

$$e_{ss} = \frac{2\xi}{\omega_n}. \tag{7.40}$$

The steady-state error reflects on the system accuracy. This error arises from the nature of inputs, type of system and nonlinearities of system components such as static friction and backlash. Since the actual input in a physical system is often a random signal, the steady-state errors are obtained for the standard test signals namely step, ramp and parabolic signals.

The various time domain specifications for a second-order control system are listed in Table 7.1

Table 7.1 Time domain specifications for a second-order system

S. No.	Time domain specification	Expression
01	Delay time	$t_d = \frac{1+0.7\xi}{\omega_n}$
02	Rise time	$t_r = \frac{\pi-\theta}{\omega_d} = \frac{\pi - \tan^{-1}\left(\frac{\sqrt{1-\xi^2}}{\xi}\right)}{\omega_n\sqrt{1-\xi^2}}$

(Continued)

Table 7.1 *(Continued)*

S. No.	Time domain specification	Expression
03	Peak time	$t_p = \frac{\pi}{\omega_d} = \frac{\pi}{\omega_n\sqrt{1-\xi^2}}$
04	Peak overshoot	$\%M_p = e^{-\frac{\xi\pi}{\sqrt{1-\xi^2}}} \times 100$
05	Setting time	$t_s = \frac{4}{\xi\omega_n}$ (2% tolerance band) $t_s = \frac{3}{\xi\omega_n}$ (5% tolerance band)
06	Steady-state error	$e_{ss} = 0$

Problem-7.2: A unity feedback system is characterised by an open-loop t.f. $G(s) = \frac{K}{s(s+10)}$. Determine the gain K, so that the system will have a damping ratio of 0.5. For this value of K, determine T_s, T_p and M_p for a unit step input.

Solution:

Given that

$$G(s) = \frac{K}{s(s+10)}, H(s) = 1.$$

$$\xi = 0.5$$

$$\frac{C(s)}{R(s)} = \frac{G(s)}{1+G(s)H(s)} = \frac{K/s(s+10)}{1+K/s(s+10)} = \frac{K}{s^2+10s+K}$$

The characteristic equation is $s^2 + 10s + K = 0$

By comparing the equation with $s^2 + 2\xi\omega_n s + \omega_n^2 = 0$, we get

$$\omega_n = \sqrt{K}, 2\xi\omega_n = 10 \Rightarrow \omega_n = \frac{10}{2\times 0.5} = 10$$

$$K = \omega_n^2 = 10^2 = 100\cdot$$

1. Settling time $T_s = \frac{4}{\xi\omega_n} = \frac{4}{0.5\times 10} = 0.8\,\text{s}$.

2. Peak time $T_p = \frac{\pi}{\omega_n\sqrt{1-\xi^2}} = \frac{\pi}{10\sqrt{1-0.25}} = 0.362$.

3. Peak overshoot $M_p = e^{-\frac{\xi\pi}{\sqrt{1-\xi^2}}} \times 100 = e^{-\frac{0.5\pi}{\sqrt{1-0.25}}} \times 100 = 16.3\%$.

***Problem-7.3*:** Measurement conducted on a servo mechanism shows the system response to be $c(t) = 1 + 0.2e^{-60t} - 1.2e^{-10t}$, when subjected to a unit-step input. Obtain the expression for closed-loop t.f., the damping ratio and un-damped natural frequency of oscillations.

***Solution*:**

Given that

$$r(t) = 1 \Rightarrow R(s) = 1/s$$

$$c(t) = 1 + 0.2e^{-60t} - 1.2e^{-10t}$$

$$\Rightarrow \qquad C(s) = \frac{1}{s} + \frac{0.2}{s+60} - \frac{1.2}{s+10}$$

$$= \frac{600}{s(s+60)(s+10)}$$

$$T(s) = \frac{C(s)}{R(s)} = \frac{600 \times s}{s(s+60)(s+10)} = \frac{600}{(s+10)(s+60)} = \frac{600}{s^2 + 70s + 600}$$

The characteristic equation is $s^2 + 70s + 600 = 0$.

By comparing the equation with $s^2 + 2\xi\omega_n s + \omega_n^2 = 0$, we get

$$\omega_n = \sqrt{600} = 24.5 \text{ rad/s}$$

$$2\xi\omega_n = 70 \Rightarrow \xi = \frac{70}{2 \times 24.5} = 1.43$$

***Problem-7.4*:** Find all the time domain specifications for a unity feedback control system whose open-loop t.f. is given as $G(s) = \frac{25}{s(s+6)}$.

***Solution*:**

Given that

$$G(s) = \frac{25}{s(s+6)} \quad \text{and } H(s) = 1$$

$$\frac{C(s)}{R(s)} = \frac{G(s)}{1 + G(s)H(s)} = \frac{25/s(s+6)}{1 + 25/s(s+6)} = \frac{25}{s^2 + 6s + 25}$$

The characteristic equation is $s^2 + 6s + 25 = 0$.

By comparing the equation with $s^2 + 2\xi\omega_n s + \omega_n^2 = 0$, we get

$$\omega_n = 5, \; 2\xi\omega_n = 6 \Rightarrow \xi = \frac{6}{2 \times 5} = 0.6$$

$$\omega_d = \omega_n\sqrt{1-\xi^2} = 5\sqrt{1-0.36} = 4\text{rad/s}$$

$$\theta = \tan^{-1}\left(\frac{\sqrt{1-\xi^2}}{\xi}\right) = \tan^{-1}\frac{\sqrt{1-0.36}}{0.6} = 53.12° = 0.92\text{rad} .$$

1. Rise time $T_r = \dfrac{\pi-\theta}{\omega_d} = \dfrac{\pi-0.92}{4} = 0.55\,\text{s}$.

2. Peak time $T_p = \dfrac{\pi}{\omega_d} = \dfrac{\pi}{4} = 0.785\,\text{s}$ ·

3. Delay time $T_d = \dfrac{1+0.7\xi}{\omega_n} = \dfrac{1+0.7\times0.6}{5} = 0.284\,\text{s}$.

4. Settling time $T_s = \dfrac{4}{\xi\omega_n} = \dfrac{4}{0.6\times5} = 1.33\,\text{s}$.

5. Peak overshoot $\%M_p = e^{-\frac{\xi\pi}{\sqrt{1-\xi^2}}}\times100 = e^{-\frac{0.6\pi}{\sqrt{1-0.36}}}\times100 = 9.5$

Problem-7.5: The forward path t.f. of a unity feedback control system is given as $G(s) = \dfrac{2}{s(s+3)}$. Obtain the expression for unit step response of the system.

Solution:

Given that

$$G(s) = \frac{2}{s(s+3)},\ H(s) = 1$$

The closed-loop t.f. is

$$\frac{C(s)}{R(s)} = \frac{G(s)}{1+G(s)H(s)} = \frac{2/s(s+3)}{1+2/s(s+3)} = \frac{2}{s^2+3s+2}$$

$$\Rightarrow \qquad \frac{C(s)}{R(s)} = \frac{2}{(s+1)(s+2)}$$

$$\Rightarrow \qquad C(s) = \frac{2}{(s+1)(s+2)}R(s).$$

As the input to the system is step input, $R(s) = 1/s$.

$$C(s) = \frac{2}{s(s+1)(s+2)} = \frac{1}{s} - \frac{2}{s+1} + \frac{1}{s+2}.$$

By taking inverse Laplace transforms on both sides, we get

$$c(t) = (1 - 2e^{-t} + e^{-2t})\cdot$$

Problem-7.6: A closed-loop servo is represented by the differential equation $\dfrac{d^2c}{dt^2} + 8\dfrac{dc}{dt} = 64e$, where c is the displacement of the output shaft, r is the displacement of the input shaft and $e = r - c$. Determine under-damped natural frequency, damping ratio and $\%M_p$ for unit step input.

Solution:
Given that

$$\frac{d^2c}{dt^2}+8\frac{dc}{dt}=64e=64(r-c)$$

By taking Laplace transforms on both sides with zero initial conditions, we get

$$s^2C(s)+8sC(s)=64[R(s)-C(s)]$$

$$(s^2+8s+64)C(s)=64R(s)$$

$$\Rightarrow \qquad \frac{C(s)}{R(s)}=\frac{64}{s^2+8s+64}.$$

The characteristic equation is $s^2+8s+64=0$.

By comparing the equation with $s^2+2\xi\omega_n s+\omega_n^2=0$, we get

$$\omega_n^2=64 \Rightarrow \omega_n=8\,\text{rad/s}$$

$$2\xi\omega_n=8 \Rightarrow \xi=0.5.$$

Peak overshoot, $\%M_p=e^{-\frac{\xi\pi}{\sqrt{1-\xi^2}}}\times 100=e^{-\frac{0.5\pi}{\sqrt{1-0.25}}}\times 100=16.3\%$

Problem-7.7: A second-order system is given as $\frac{C(s)}{R(s)}=\frac{25}{s^2+6s+25}$. Find its rise time, peak time, peak overshoot and settling time if subjected to unit-step input. Also calculate the expression for its output response.

Solution:

Given that

$$\frac{C(s)}{R(s)}=\frac{25}{s^2+6s+25}.$$

The characteristic equation is $s^2+6s+25=0$.

By comparing the equation with $s^2+2\xi\omega_n s+\omega_n^2=0$, we get

$$\omega_n^2=25 \Rightarrow \omega_n=5\,\text{rad/s}$$

$$2\xi\omega_n=6 \Rightarrow \xi=0.6$$

$$\omega_d=\omega_n\sqrt{1-\xi^2}=5\sqrt{1-0.36}=4\text{rad/s}$$

$$\theta=\tan^{-1}\frac{\sqrt{1-\xi^2}}{\xi}=\tan^{-1}\frac{\sqrt{1-0.36}}{0.6}=53.12°=0.92\text{rad}$$

1. Rise time $T_r=\frac{\pi-\theta}{\omega_d}=\frac{\pi-0.92}{4}=0.55\,\text{s}$

2. Peak time $T_p=\frac{\pi}{\omega_d}=\frac{\pi}{4}=0.785\,\text{s}.$

3. Settling time $T_s = \dfrac{4}{\xi\omega_n} = \dfrac{4}{0.6\times 5} = 1.33\,\text{s}.$

4. Peak overshoot $\%M_p = e^{-\frac{\xi\pi}{\sqrt{1-\xi^2}}}\times 100 = e^{-\frac{0.6\pi}{\sqrt{1-0.36}}}\times 100 = 9.5.$

5. The output response is given as

$$c(t) = 1 - \frac{e^{-\xi\omega_n t}}{\sqrt{1-\xi^2}}\sin(\omega_d t + \theta)$$

$$= 1 - \frac{e^{-3t}}{\sqrt{1-0.36}}\sin(4t + 53.12°)$$

$$= 1 - 1.563e^{-3t}\sin(4t + 53.12°)$$

7.8 STEADY-STATE ERROR AND ERROR CONSTANTS

Steady-state error is defined as the difference between the reference input and the steady-state output as time approaches infinity

i.e.,
$$e_{ss} = \underset{t\to\infty}{\text{Lt}}\ e(t) = \underset{t\to\infty}{\text{Lt}}\,[r(t) - c(t)].$$

It gives an idea about the accuracy of the designed system to the system designer. It may be caused in the system due to the nature of the input signal, type of system and non-linearity present in the system. As the steady-state error is an index of accuracy of a control system, it should be minimum as far as possible.

Consider a unit feedback system shown in Figure 7.15.

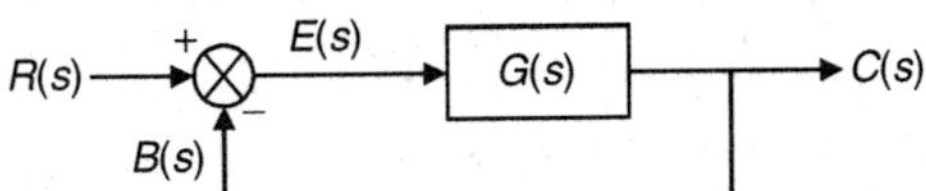

Figure 7.15 Unity feedback system

Here, input is $R(s)$, output is $C(s)$, feedback signal is $B(s)$ and the difference between the input and the output is the error signal $E(s)$.

From Figure 7.15, we get

$$\frac{C(s)}{R(s)} = \frac{G(s)}{1+G(s)}$$

$$C(s) = E(s)G(s).$$

$$\therefore \qquad E(s) = \frac{C(s)}{G(s)} = \frac{G(s)}{1+G(s)}.R(s)\times\frac{1}{G(s)} = \frac{1}{1+G(s)}.R(s).$$

The steady-state error e_{ss} may be found using the final value theorem as follows.

$$e_{ss} = \underset{t\to\infty}{\text{Lt}}\ e(t) = \underset{s\to 0}{\text{Lt}}\ sE(s) = \underset{s\to 0}{\text{Lt}} = \frac{sR(s)}{1+G(s)}. \qquad (7.41)$$

Equation (7.41) shows that the steady-state error depends upon input $R(s)$ and the forward t.f. $G(s)$. The expressions for steady-state error for various types of standard test signals are derived as follows:

1. **Unit-step or displacement input**
   For a unit-step input, we get

$$r(t) = u(t) = 1 \Rightarrow R(s) = 1/s$$

$$e_{ss} = \underset{s\to 0}{\text{Lt}} \frac{sR(s)}{1+G(s)} = \underset{s\to 0}{\text{Lt}} \frac{1}{1+G(s)} = \frac{1}{1+G(0)} = \frac{1}{1+K_p}$$

   where $K_p = \underset{s\to 0}{\text{Lt}}\, G(s) = G(0)$ is called 'position error constant'.

2. **Unit-ramp or velocity input**
   For a unit-step input, we have

$$r(t) = t \Rightarrow R(s) = 1/s^2$$

$$e_{ss} = \underset{s\to 0}{\text{Lt}} \frac{sR(s)}{1+G(s)} = \underset{s\to 0}{\text{Lt}} \frac{1}{s+sG(s)} = \underset{s\to 0}{\text{Lt}} \frac{1}{sG(s)} = \frac{1}{K_v}$$

   where $K_v = \underset{s\to 0}{\text{Lt}}\, sG(s)$ is called 'velocity error constant'.

3. **Unit parabolic or acceleration input**
   For a unit parabolic input, we have

$$r(t) = t^2/2 \Rightarrow R(s) = 1/s^3$$

$$e_{ss} = \underset{s\to 0}{\text{Lt}} \frac{sR(s)}{1+G(s)} = \underset{s\to 0}{\text{Lt}} \frac{1}{s^2+s^2G(s)} = \underset{s\to 0}{\text{Lt}} \frac{1}{s^2G(s)} = \frac{1}{K_a}$$

   where $K_a = \underset{s\to 0}{\text{Lt}}\, s^2G(s)$ is called 'parabolic error constant'.

Problem-7.8: Determine the error coefficients and static error for $G(s) = \dfrac{1}{s(s+1)(s+10)}$, $H(s) = s+2$.

Solution:

Given that

$$G(s) = \frac{1}{s(s+1)(s+10)}, \; H(s) = s+2$$

$$G(s)H(s) = \frac{s+2}{s(s+1)(s+10)}.$$

(a) Position error constant and corresponding steady-state error are

$$K_p = \underset{s\to 0}{\text{Lt}}\, G(s)H(s) = \underset{s\to 0}{\text{Lt}} \frac{s+2}{s(s+1)(s+10)} = \infty$$

$$e_{ss} = \frac{1}{1+K_p} = \frac{1}{1+\infty} = 0.$$

(b) Velocity error constant and corresponding steady-state error are

$$K_v = \underset{s\to 0}{\text{Lt}}\, sG(s)H(s) = \underset{s\to 0}{\text{Lt}} \frac{s(s+2)}{s(s+1)(s+10)} = \frac{2}{10} = 0.2$$

$$e_{ss} = \frac{1}{K_v} = \frac{1}{0.2} = 5\cdot$$

(c) Acceleration error constant and corresponding steady-state error are

$$K_a = \underset{s\to 0}{\text{Lt}}\, s^2G(s)H(s) = \underset{s\to 0}{\text{Lt}} \frac{s(s+2)}{(s+1)(s+10)} = 0$$

$$e_{ss} = \frac{1}{K_a} = \infty\cdot$$

Problem-7.9: The overall t.f. of a unity feedback control system is given as

$$\frac{C(s)}{R(s)} = \frac{10}{s^2+6s+10}\cdot$$

(a) Find K_p, K_v, K_a,
(b) Determine the steady-state error if the input is $r(t) = 1 + t + t^2$

Solution:
Given that

$$\frac{C(s)}{R(s)} = \frac{10}{s^2+6s+10}, H(s) = 1$$

$$\frac{C(s)}{R(s)} = \frac{G(s)}{1+G(s)H(s)} = \frac{G(s)}{1+G(s)} = \frac{10}{s^2+6s+10}$$

$$\Rightarrow \quad \frac{1+G(s)}{G(s)} = \frac{s^2+6s+10}{10}$$

$$\Rightarrow \quad \frac{1}{G(s)} = \frac{s^2+6s+10}{10} - 1 = \frac{s^2+6s}{10}$$

$$\Rightarrow \quad G(s) = \frac{10}{s^2+6s} = \frac{10}{s(s+6)}$$

(a) The various error constants are

$$K_p = \underset{s\to 0}{\text{Lt}}\, G(s)H(s) = \underset{s\to 0}{\text{Lt}} \frac{10}{s(s+10)} = \infty$$

$$K_v = \underset{s\to 0}{\text{Lt}}\, sG(s)H(s) = \underset{s\to 0}{\text{Lt}} \frac{10}{(s+6)} = 1.667$$

$$K_a = \underset{s\to 0}{\text{Lt}}\, s^2 G(s)H(s) = \underset{s\to 0}{\text{Lt}} \frac{10s}{(s+6)} = 0$$

(b)
$$r(t) = 1 + t + t^2 = A_1 + A_2 t + \frac{A_3}{2} t^2$$

The input is a combination of the three standard inputs. Therefore, let us calculate their magnitudes

$A_1 = 1,\quad A_2 = 1,\quad A_3 = 2$

$K_p = \infty,\quad K_v = 1.667,\quad K_a = 0$

$$e_{ss} = e_{ss1} + e_{ss2} + e_{ss3} = \frac{A_1}{1+K_P} + \frac{A_2}{K_V} + \frac{A_3}{K_a} = \frac{1}{1+\infty} + \frac{1}{1.667} + \frac{2}{0} = \infty$$

Problem-7.10: Consider a unity feedback system with a closed t.f. $\frac{C(s)}{R(s)} = \frac{ks+b}{s^2+as+b}$. Determine the open-loop t.f. $G(s)$. Show that the steady-state error with unit ramp input is given as $\frac{(a-k)}{b}$.

Solution:

Given that

$$\frac{C(s)}{R(s)} = \frac{ks+b}{s^2+as+b},\ H(s) = 1$$

The closed-loop t.f. is

$$\frac{C(s)}{R(s)} = \frac{G(s)}{1+G(s)H(s)} = \frac{G(s)}{1+G(s)} = \frac{ks+b}{s^2+as+b}$$

$$\Rightarrow \quad \frac{1+G(s)}{G(s)} = \frac{s^2+as+10}{ks+b}$$

$$\Rightarrow \quad \frac{1}{G(s)} = \frac{s^2+as+b}{ks+b} - 1 = \frac{s^2+as-ks}{ks+b}$$

$$\Rightarrow \quad G(s) = \frac{ks+b}{s^2+(a-k)s} = \frac{ks+b}{s[s+(a-k)]}.$$

Velocity error constant is

$$K_v = \underset{s\to 0}{\text{Lt}}\, sG(s)H(s) = \underset{s\to 0}{\text{Lt}}\, s\frac{ks+b}{s[s+(a-k)]} = \frac{b}{a-k}.$$

Steady-state error with velocity input is

$$e_{ss} = \frac{1}{K_v} = \frac{a-k}{b}.$$

Problem-7.11: A unity feedback system has $G(s) = \dfrac{180}{s(s+6)}$ and $r(t) = 4t$. Determine

(a) steady-state error
(b) value of K to reduce the error by 6%.

Solution:
Given that

$$G(s) = \frac{180}{s(s+6)}, \; H(s) = 1$$

$$r(t) = 4t \Rightarrow R(s) = 4/s^2$$

(a) Steady-state error is given as

$$e_{ss} = \underset{s\to 0}{\text{Lt}} \frac{sR(s)}{1+G(s)H(s)} = \underset{s\to 0}{\text{Lt}}\, s\frac{4/s^2}{1+180/s(s+6)} = \underset{s\to 0}{\text{Lt}} \frac{4(s+6)}{s^2+6s+180} = 24/180 = 0.134.$$

(b) Since new error is 6% less, we have

$$(e_{ss})_{new} = 0.94e_{ss} = 0.94 \times 0.134 = 0.12596$$

$$(e_{ss})_{new} = \underset{s\to 0}{\text{Lt}} \frac{sR(s)}{1+G(s)H(s)} = \underset{s\to 0}{\text{Lt}}\, s\frac{4/s^2}{1+K/s(s+6)}$$

$$= \underset{s\to 0}{\text{Lt}} \frac{4(s+6)}{s^2+6s+K} = 24/K = 0.12596$$

$$\Rightarrow \qquad K = \frac{24}{0.12596} = 190.35.$$

Problem-7.12: For a system $G(s)H(s) = \dfrac{K}{s^2(s+2)(s+3)}$, find the value of K to limit the steady-state error to 10 when the input to the system is $r(t) = 1 + 10t + \dfrac{40}{2}t^2$.

Solution:
Given that

$$r(t) = 1 + 10t + 40/2t^2$$

$$= A_1 + A_2t + A_3/2t^2$$

The input is a combination of the three standard inputs. Therefore, let us calculate their magnitudes.

$$A_1 = 1, A_2 = 10, A_3 = 40$$

$$K_p = \underset{s\to 0}{\text{Lt}}\, G(s)H(s) = \underset{s\to 0}{\text{Lt}} \frac{K}{s^2(s+2)(s+3)} = \infty$$

$$K_v = \underset{s\to 0}{\text{Lt}}\, sG(s)H(s) = \underset{s\to 0}{\text{Lt}} \frac{K}{s(s+2)(s+3)} = \infty$$

$$K_a = \underset{s\to 0}{\text{Lt}}\, s^2G(s)H(s) = \underset{s\to 0}{\text{Lt}} \frac{K}{(s+2)(s+3)} = K/6$$

$$e_{ss} = e_{ss1} + e_{ss2} + e_{ss3} = \frac{A_1}{1+K_P} + \frac{A_2}{K_V} + \frac{A_3}{K_a}$$

$$10 = \frac{1}{1+\infty} + \frac{10}{\infty} + \frac{40}{K/6}$$

$$\Rightarrow \qquad 10 = 240/K \Rightarrow K = 24$$

∴ The value of K is 24, and this value limits the steady-state error to 10.

7.9 STEADY-STATE ERROR FOR TYPE 0, TYPE 1 AND TYPE 2 SYSTEMS

The open-loop t.f. of a unity feedback system can be written in two standard forms:

1. Time constant form
2. Pole–zero form

In these two forms, $G(s)$ is given as

$$G(s) = k\frac{(1+sT_{z1})(1+sT_{z2})(1+sT_{z3})\cdots(1+sT_{zm})}{s^n(1+sT_{p1})(1+sT_{p2})(1+sT_{p3})\cdots(1+sT_{zn})}. \tag{7.42}$$

This form of representation is called as time constant form.

$$G(s) = k'\frac{(s+z_1)(s+z_2)(s+z_3)\cdots(s+z_m)}{s^n(s+p_1)(s+p_2)(s+p_3)\cdots(s+p_m)}. \tag{7.43}$$

This form of representation is called as pole-zero form.
The gain in the two forms are related by

$$k = k'\frac{\prod_{m} z_m}{\prod_{n} p_n}. \tag{7.44}$$

With the gain relation of (7.44) for the two forms of $G(s)$, it is sufficient to obtain steady-state errors in terms of the gains of any one of the forms. Using the time constant form, from (7.42), the term s^n in the denominator corresponds to the number of integrations in the system. As s tends to zero, this term, that is, the poles at the origin, dominate in determining the steady-state error.

Therefore, control systems are classified in accordance with the number of integrations in the open-loop t.f. $G(s)$, as described below. Hence, the steady-state error, which depends on $\underset{s\to 0}{\text{Lt}}\, G(s)$, $\underset{s\to 0}{\text{Lt}}\, sG(s)$ or $\underset{s\to 0}{\text{Lt}}\, s^2G(s)$, is dependent on the type of the system. Let us now obtain the steady-state error for various standard test signals for Type 0, Type 1 and Type 2 system.

1. **Type 0 system**
   If $n = 0$ in (7.42), then the system is of Type 0 system and the error constants and steady-state errors to various standard inputs are

$$\left.\begin{aligned}
K_p &= \underset{s\to 0}{\mathrm{Lt}}\, G(s) = \underset{s\to 0}{\mathrm{Lt}}\, k\frac{(1+sT_{z1})(1+sT_{z2})(1+sT_{z3})\cdots(1+sT_{zm})}{(1+sT_{p1})(1+sT_{p2})(1+sT_{p3})\cdots(1+sT_{pn})} = k \\
K_v &= \underset{s\to 0}{\mathrm{Lt}}\, sG(s) = \underset{s\to 0}{\mathrm{Lt}}\, k\frac{s(1+sT_{z1})(1+sT_{z2})(1+sT_{z3})\cdots(1+sT_{zm})}{(1+sT_{p1})(1+sT_{p2})(1+sT_{p3})\cdots(1+sT_{pn})} = 0 \\
K_a &= \underset{s\to 0}{\mathrm{Lt}}\, s^2G(s) = \underset{s\to 0}{\mathrm{Lt}}\, k\frac{s^2(1+sT_{z1})(1+sT_{z2})(1+sT_{z3})\cdots(1+sT_{zm})}{(1+sT_{p1})(1+sT_{p2})(1+sT_{p3})\cdots(1+sT_{pn})} = 0
\end{aligned}\right\}. \tag{7.45}$$

Steady-state errors for the various inputs are

$$\left.\begin{aligned}
e_{ss} &= \frac{1}{1+K_p} = \frac{1}{1+k} \\
e_{ss} &= \frac{1}{K_v} = \frac{1}{0} = \infty \\
e_{ss} &= \frac{1}{K_a} = \frac{1}{0} = \infty
\end{aligned}\right\} \tag{7.46}$$

2. **Type 1 system**
   If $n = 1$ in (7.42), then the system is of Type 1. The error constants and steady-state errors to various standard inputs are

$$\left.\begin{aligned}
K_p &= \underset{s\to 0}{\mathrm{Lt}}\, G(s) = \underset{s\to 0}{\mathrm{Lt}}\, k\frac{(1+sT_{z1})(1+sT_{z2})(1+sT_{z3})\cdots(1+sT_{zm})}{s(1+sT_{p1})(1+sT_{p2})(1+sT_{p3})\cdots(1+sT_{pn})} = \infty \\
K_v &= \underset{s\to 0}{\mathrm{Lt}}\, sG(s) = \underset{s\to 0}{\mathrm{Lt}}\, k\frac{(1+sT_{z1})(1+sT_{z2})(1+sT_{z3})\cdots(1+sT_{zm})}{(1+sT_{p1})(1+sT_{p2})(1+sT_{p3})\cdots(1+sT_{pn})} = k \\
K_a &= \underset{s\to 0}{\mathrm{Lt}}\, s^2G(s) = \underset{s\to 0}{\mathrm{Lt}}\, k\frac{s(1+sT_{z1})(1+sT_{z2})(1+sT_{z3})\cdots(1+sT_{zm})}{(1+sT_{p1})(1+sT_{p2})(1+sT_{p3})\cdots(1+sT_{pn})} = 0
\end{aligned}\right\}. \tag{7.47}$$

Steady-state errors for various inputs are

$$\left.\begin{aligned}
e_{ss} &= \frac{1}{1+K_p} = \frac{1}{1+\infty} = 0 \\
e_{ss} &= \frac{1}{K_v} = \frac{1}{k} \\
e_{ss} &= \frac{1}{K_a} = \frac{1}{0} = \infty
\end{aligned}\right\} \tag{7.48}$$

3. **Type 2 system**
   If $n = 2$ in (7.42), then the system is of Type 2 system and the error constants and steady-state errors to various standard inputs are

$$\left.\begin{aligned}K_p &= \underset{s\to 0}{\mathrm{Lt}}\, G(s) = \underset{s\to 0}{\mathrm{Lt}}\, k\frac{(1+sT_{z1})(1+sT_{z2})(1+sT_{z3})\cdots(1+sT_{zm})}{s^2(1+sT_{p1})(1+sT_{p2})(1+sT_{p3})\cdots(1+sT_{pn})} = \infty \\ K_v &= \underset{s\to 0}{\mathrm{Lt}}\, sG(s) = \underset{s\to 0}{\mathrm{Lt}}\, k\frac{(1+sT_{z1})(1+sT_{z2})(1+sT_{z3})\cdots(1+sT_{zm})}{s(1+sT_{p1})(1+sT_{p2})(1+sT_{p3})\cdots(1+sT_{pn})} = \infty \\ K_a &= \underset{s\to 0}{\mathrm{Lt}}\, s^2G(s) = \underset{s\to 0}{\mathrm{Lt}}\, k\frac{(1+sT_{z1})(1+sT_{z2})(1+sT_{z3})\cdots(1+sT_{zm})}{(1+sT_{p1})(1+sT_{p2})(1+sT_{p3})\cdots(1+sT_{pn})} = k\end{aligned}\right\}. \tag{7.49}$$

Steady-state errors for various inputs are

$$\left.\begin{aligned}e_{ss} &= \frac{1}{1+K_p} = \frac{1}{1+\infty} = 0 \\ e_{ss} &= \frac{1}{K_v} = \frac{1}{\infty} = 0 \\ e_{ss} &= \frac{1}{K_a} = \frac{1}{k}\end{aligned}\right\} \tag{7.50}$$

Thus,

$$\left.\begin{aligned}e_{ss} &= \frac{1}{1+K_p} \to \text{Type`0' system} \\ e_{ss} &= \frac{1}{K_v} \to \text{Type`1' system} \\ e_{ss} &= \frac{1}{K_a} \to \text{Type`2' system}\end{aligned}\right\} \tag{7.51}$$

The steady-state errors, for the various standard inputs for Type 0, Type 1 and Type 2 are summarised in Table 7.2.

Table 7.2 Steady-state error for various inputs

Standard input	**Steady-state errors e_{ss}**		
	Type 0 $K_p = \lim_{s\to 0} G(s)$	**Type 1** $K_v = \lim_{s\to 0} sG(s)$	**Type 2** $K_a = \lim_{s\to 0} s^2G(s)$
Unit step	$\frac{1}{1+k_p}$	0	0
Unit velocity	∞	$\frac{1}{K_v}$	0
Unit acceleration	∞	∞	$\frac{1}{K_a}$

From Table 7.2, as the type of the system and the number of integrations increase, more and more steady-state errors become zero. Hence, it may appear that it is better to design a system with more and more poles at the origin. However, if the type of the system is higher than 2, the systems tend to be more unstable and the dynamic errors tend to be larger.

7.9.1 Disadvantages of Static Error Coefficient Method

The disadvantages of static error coefficient method are as follows:

1. This method is applicable only to the three standard input signals and cannot give error coefficients for other signals.
2. Mathematical response of error is given as 0 or ∞; the method cannot give the exact value of the error.
3. This method cannot be used for unstable systems.

Problem-7.13: A unity feedback system has $G(s)=\dfrac{40(s+2)}{s(s+1)(s+4)}$. Determine

(a) Type of the system
(b) All the error coefficients
(c) Error for ramp input with magnitude 4

Solution:
Given that

$$G(s)=\frac{40(s+2)}{s(s+1)(s+4)},H(s)=1.$$

To determine the type of a system, arrange $G(s)H(s)$ in time constant form

$$G(s)=\frac{40(s+2)}{s(s+1)(s+4)}=\frac{20(1+0.5s)}{s(s+1)(1+0.25s)}.$$

By comparing the equation with the standard form, we get

$$G(s)H(s)=k\frac{(1+sT_{z1})(1+sT_{z2})(1+sT_{z3})\cdots(1+sT_{zm})}{s^n(1+sT_{p1})(1+sT_{p2})(1+sT_{p3})\cdots(1+sT_{pn})}.$$

(a) Type of the system = 1

(b) $K_p = \underset{s\to 0}{\text{Lt}}\, G(s)H(s) = \underset{s\to 0}{\text{Lt}}\, \dfrac{40(s+2)}{s(s+1)(s+4)} = \infty$

$$K_v = \underset{s\to 0}{\text{Lt}}\, sG(s)H(s) = \underset{s\to 0}{\text{Lt}}\, \frac{40(s+2)}{(s+1)(s+4)} = 20$$

$$K_a = \underset{s\to 0}{\text{Lt}}\, s^2G(s)H(s) = \underset{s\to 0}{\text{Lt}}\, \frac{40s(s+2)}{(s+1)(s+4)} = 0$$

(c) The steady-state error for ramp input is given as

$$e_{ss}=\frac{A}{K_v}=\frac{4}{20}=0.2.$$

***Problem-7.14*:** For a servo mechanism with the following open-loop t.f., explain what type of input signal gives rise to a constant steady-state error and calculate their values.

(a) $G(s) = \dfrac{20(s+2)}{s(s+1)(s+3)}$

(b) $G(s) = \dfrac{10}{(s+2)(s+3)}$

***Solution*:**

Given that

(a) $G(s) = \dfrac{20(s+2)}{s(s+1)(s+3)}$

Let us assume unity feedback system

$\therefore \quad H(s) = 1$

The open-loop system has a pole at the origin. Hence, it is a Type 1 system. In this system, the velocity (ramp) input will give a constant steady-state error. The steady-state error with unit velocity input is

$$e_{ss} = 1/K_v.$$

Velocity error constant is

$$K_v = \underset{s\to 0}{\text{Lt}}\, sG(s)H(s) = \underset{s\to 0}{\text{Lt}}\, s\frac{20(s+2)}{s(s+1)(s+3)} = \frac{20\times 2}{1\times 3} = 40/3.$$

Steady-state error is

$$e_{ss} = 1/K_v = 3/40 = 0.075.$$

(b) $G(s) = \dfrac{10}{(s+2)(s+3)}$

Let us assume the unity feedback system

$\therefore \; H(s) = 1.$

The open-loop system has no pole at the origin. Hence, it is a Type 0 system. In this system, the step input will give a constant steady-state error. The steady-state error with unit velocity input is

$$e_{ss} = \frac{1}{1+k_p}.$$

Position error constant is

$$K_p = \underset{s\to 0}{\text{Lt}}\, G(s)H(s) = \underset{s\to 0}{\text{Lt}}\, \frac{10}{(s+2)(s+3)} = \frac{10}{2\times 3} = 10/6 = 5/3.$$

Steady-state error is

$$e_{ss} = 1/(1+K_p) = 1/(1+5/3) = 0.375.$$

7.10 DESIGN SPECIFICATIONS OF A CONTROL SYSTEM

A second-order control system is required to satisfy three main specifications, namely, peak overshoot M_p, settling time t_s and steady-state accuracy. Peak overshoot is indicative of damping ξ in the system and for a given damping, settling time indicates the un-damped natural frequency of the system. The steady-state accuracy is specified by the steady-state error, which can be made to lie within the given limits by choosing an appropriate error constant K_p, K_v or K_a depending on the type of the system. If any other specifications like rise time or delay time are also specified, they must be specified consistent with the other specifications. Most control systems are designed to be under-damped with a damping factor lying between 0.3 and 0.7.

In a second-order system, the expressions for ξ, t_s and e_{ss} are given as

$$\xi = \frac{1}{2\sqrt{K_v \tau}} \tag{7.52}$$

$$t_s = \frac{4}{\xi \omega_n} \tag{7.53}$$

$$e_{ss} = \frac{1}{K_v} \tag{7.54}$$

and the only variables from these expressions are K_v and τ. Even if both of them are varied, we can satisfy only two of these specifications, namely, ξ, t_s and e_{ss}.

Generally, for a given system to which a suitable controller has to be designed, it means that the system time constant τ is fixed and the only variable available is the system gain K_v. By using a proportional controller, the gain can be adjusted to suit the requirement of steady-state accuracy. If K_v is adjusted for an allowable limit on steady-state error, this value of K_v is usually large enough to make the system damping considerably less, as given by the above-mentioned expression. Thus, the transient behaviours of the system are not satisfactory. Hence, suitable compensation schemes must be designed so that the dynamic response improves.

7.11 GENERALISED ERROR COEFFICIENTS OR DYNAMIC ERROR COEFFICIENTS OR ERROR SERIES

The main disadvantage of defining the steady-state error in terms of error constants is that only one of the constants is finite and non-zero for a particular system, while the other constants are either zero or infinity. If any error constant is zero, the steady-state error is infinity, but we do not have any clue as to how the error is approaching infinity. If the inputs are other than displacement, velocity or acceleration inputs, we can extend the concept of generalised error coefficients to include inputs, which can be represented by a polynomial.

Consider the system shown in Figure 7.16.

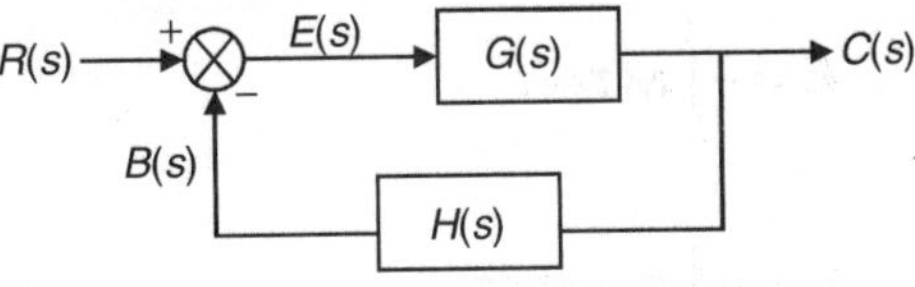

Figure 7.16 Feedback control system

From Figure 7.16, we get

$$\frac{C(s)}{R(s)} = \frac{G(s)}{1+G(s)H(s)}$$

$$C(s) = E(s)G(s)$$

$$\therefore \quad E(s) = \frac{C(s)}{G(s)} = \frac{G(s)}{1+G(s)H(s)}.R(s) \times \frac{1}{G(s)} = \frac{1}{1+G(s)H(s)}.R(s). \tag{7.55}$$

Equation (7.55) may be written as

$$E(s) = Y(s)R(s) \tag{7.56}$$

where

$$Y(s) = \frac{1}{1+G(s)H(s)}. \tag{7.57}$$

Using convolution theorem, (7.56) can be written as

$$e(t) = \int_0^t y(\tau)r(t-\tau)d\tau. \tag{7.58}$$

Using Taylor's series, we can write

$$r(t-\tau) = r(t) - \tau r'(t) + \frac{\tau^2}{2!}r''(t) - \frac{\tau^3}{3!}r'''(t) + \cdots \tag{7.59}$$

From (7.58) and (7.59), we get

$$e(t) = \int_0^t y(\tau)\left[r(t) - \tau r'(t) + \frac{\tau^2}{2!}r''(t) - \frac{\tau^3}{3!}r'''(t) + \cdots\right]d\tau$$

$$= r(t)\int_0^t y(\tau)d\tau - r'(t)\int_0^t \tau y(\tau)d\tau + r''(t)\int_0^t \frac{\tau^2}{2!}y(\tau)d\tau - \cdots.$$

Again

$$e_{ss} = \underset{t\to\infty}{\text{Lt}}\, e(t)$$

$$e_{ss} = \underset{t\to\infty}{\text{Lt}}\left[r(t)\int_0^t y(\tau)d\tau - r'(t)\int_0^t \tau y(\tau)d\tau + r''(t)\int_0^t \frac{\tau^2}{2!}y(\tau)d\tau - \cdots\right]$$

$$= k_0 r(t) + k_1 r'(t) + \frac{k_2}{2!}r''(t) + \cdots \tag{7.60}$$

where

$$k_0 = \int_0^\infty y(\tau)d\tau$$

$$k_1 = -\int_0^\infty \tau y(\tau)d\tau$$

$$k_2 = +\int_0^\infty \tau^2 y(\tau)d\tau$$

$$\vdots \qquad \vdots$$

$$k_n = (-1)^n \int_0^\infty \tau^n y(\tau)d\tau.$$

The coefficients k_0, k_1, k_2-----k_n are defined as generalised error coefficients. Equation (7.60) is known as generalised error series. It may be observed that the steady-state error is obtained as a function of time, in terms of generalised error coefficients, the steady-state part of the input and its derivatives. For a given t.f. $G(s)$, the error coefficients can be easily evaluated as follows:

Let $$y(t) = L^{-1}\left[Y(s)\right].$$

The Laplace transform of $y(t)$ is defined as

$$Y(s) = \int_0^{\infty} y(\tau)e^{-s\tau}d\tau. \tag{7.61}$$

By applying both sides as $\underset{s\to 0}{\text{Lt}}$

$$\underset{s\to 0}{\text{Lt}}\, Y(s) = \underset{s\to 0}{\text{Lt}}\left[\int_0^{\infty} y(\tau)e^{-s\tau}d\tau\right] = \int_0^{\infty} y(\tau)\,\underset{s\to 0}{\text{Lt}}\, e^{-s\tau}d\tau = \int_0^{\infty} y(\tau)d\tau = k_0$$

$$\Rightarrow \qquad k_0 = \underset{s\to 0}{\text{Lt}}\, Y(s). \tag{7.62}$$

By taking derivative of (7.61), we can write

$$\frac{dY(s)}{ds} = \int_0^{\infty} y(\tau)(-\tau)e^{-s\tau}d\tau$$

$$\underset{s\to 0}{\text{Lt}}\frac{dY(s)}{ds} = \underset{s\to 0}{\text{Lt}}\left[\int_0^{\infty} y(\tau)(-\tau)e^{-s\tau}d\tau\right]$$

$$= \int_0^{\infty} y(\tau)(-\tau)\,\underset{s\to 0}{\text{Lt}}\, e^{-s\tau}d\tau$$

$$= -\int_0^{\infty} \tau y(\tau)d\tau = k_1.$$

$$k_1 = \underset{s\to 0}{\text{Lt}}\left[\frac{dY(s)}{ds}\right] \tag{7.63}$$

$$\therefore \quad \left.\begin{aligned} k_0 &= \underset{s\to 0}{\text{Lt}}\, Y(s) \\ k_1 &= \underset{s\to 0}{\text{Lt}}\left[\frac{dY(s)}{ds}\right] \\ k_2 &= \underset{s\to 0}{\text{Lt}}\left[\frac{d^2Y(s)}{ds^2}\right] \\ &\;\;\vdots \\ k_n &= \underset{s\to 0}{\text{Lt}}\left[\frac{d^nY(s)}{ds^n}\right] \end{aligned}\right\} \tag{7.64}$$

The disadvantages of the static error coefficient method are eliminated in this method.

Problem-7.15: Evaluate the error series for a unity feedback system having a forward path t.f. $G(s)=\dfrac{50}{s(s+10)}$. Estimate the steady-state error of the system for the input $r(t)=1+2t+t^2$.

Solution:

Given that

$$G(s)=\frac{50}{s(s+10)},\ H(s)=1$$

$$Y(s)=\frac{1}{1+G(s)H(s)}=\frac{1}{1+\dfrac{50}{s(s+10)}}=\frac{s^2+10s}{s^2+10s+50}$$

$$k_0=\underset{s\to 0}{\text{Lt}}\,Y(s)=\underset{s\to 0}{\text{Lt}}\left[\frac{s^2+10s}{s^2+10s+50}\right]=0$$

$$\frac{dY(s)}{ds}=\frac{50(2s+10)}{(s^2+10s+50)^2}$$

$$k_1=\underset{s\to 0}{\text{Lt}}\left[\frac{50(2s+10)}{(s^2+10s+50)^2}\right]=\frac{50\times 10}{50^2}=0.2$$

$$\frac{d^2Y(s)}{ds^2}=\frac{(s^2+10s+50)^2 100-100(2s+10)^2(s^2+10s+50)}{(s^2+10s+50)^4}$$

$$=\frac{(s^2+10s+50)100-100(2s+10)^2}{(s^2+10s+50)^3}$$

$$k_2=\underset{s\to 0}{\text{Lt}}\left[\frac{d^2Y(s)}{ds^2}\right]=\underset{s\to 0}{\text{Lt}}\left[\frac{(s^2+10s+50)100-100(2s+10)^2}{(s^2+10s+50)^3}\right]$$

$$=\frac{50\times 100-100\times 10^2}{50^3}=-0.04.$$

Since the input is

$$r(t)=1+2t+t^2$$

$$r'(t)=2+2t,\ r''(t)=2.$$

Therefore, the error series is given as

$$e_{ss}(t)=k_0 r(t)+k_1 r'(t)+\frac{k_2}{2!}r''(t)+\cdots$$

$$=0+0.2(2+2t)-0.04\times 2$$

$$=0.4t+0.32.$$

Steady-state error is given as

$$e_{ss}=\underset{t\to\infty}{\text{Lt}}[0.4t+0.32]=\infty.$$

ADDITIONAL SOLVED PROBLEMS

ASP-1: A system is given by the differential equation $\frac{d^2y}{dt^2}+4\frac{dy}{dt}+8y=8x$, where y = output and x = input. Determine all the time domain specifications and obtain the output response for unit-step input.

Solution:

Given that the given differential equation is

$$\frac{d^2y}{dt^2}+4\frac{dy}{dt}+8y=8x.$$

By taking Laplace transforms on both sides with zero initial conditions, we get

$$s^2Y(s)+4sY(s)+8Y(s)=8X(s)$$

$$(s^2+4s+8)Y(s)=8X(s)$$

$$\Rightarrow \qquad \frac{Y(s)}{X(s)}=\frac{8}{s^2+4s+8}.$$

The characteristic equation is $s^2+4s+8=0$.

By comparing the equation with $s^2+2\xi\omega_n(s)+\omega_n^2=0$, we get

$$\omega_n^2=8 \Rightarrow \omega_n=2.83\ \text{rad/s}$$

$$2\xi\omega_n=4 \Rightarrow \xi=0.7067$$

$$\omega_d=\omega_n\sqrt{1-\xi^2}=2.83\sqrt{1-0.7067^2}=2\ \text{rad/s}.$$

$$\theta=\tan^{-1}\frac{\sqrt{1-\xi^2}}{\xi}=\tan^{-1}\frac{\sqrt{1-0.7067^2}}{0.7067}=45^\circ=\pi/4\ \text{rad}$$

1. Rise time $T_r=\frac{\pi-\theta}{\omega_d}=\frac{\pi-\pi/4}{2}=3\pi/8=1.178$ s.

2. Peak time $T_p=\frac{\pi}{\omega_d}=\frac{\pi}{2}=1.57$ s.

3. Settling time $T_s=\frac{4}{\xi\omega_n}=\frac{4}{0.7067\times 2.83}=2$ s.

4. Delay time $T_d=\frac{1+0.7\xi}{\omega_n}=\frac{1+0.7\times 0.7067}{2.83}=0.5284$ s.

5. Peak overshoot $\%M_p=e^{-\frac{\xi\pi}{\sqrt{1-\xi^2}}}\times 100=e^{-\frac{0.7067\pi}{\sqrt{1-0.5}}}\times 100=4.33\%$.

6. The output response is given as

$$c(t)=1-\frac{e^{-\xi\omega_n t}}{\sqrt{1-\xi^2}}\sin(\omega_d t+\theta)$$

$$= 1 - \frac{e^{-2t}}{\sqrt{1-0.5}} \sin(2t + 45^\circ)$$
$$= 1 - 1.41e^{-2t} \sin(2t + 45^\circ)$$

ASP-2: A closed-loop system has two complex conjugate poles at $s_1, s_2 = -2 \pm j1$. Determine the form of t.f. and the values of ω_n, T_p, T_r, T_s and M_p assuming standard second-order system.

Solution:

Given that a closed loop system has two complex conjugate poles at $s_1, s_2 = -2 \pm j1$.
The characteristic equation of the system is given as

$$(s + 2 + j)(s + 2 - j) = 0$$
$$s^2 + 4s + 5 = 0.$$

By comparing the equation with $s^2 + 2\xi\omega_n s + \omega_n^2 = 0$, we get

$$\omega_n^2 = 5 \Rightarrow \omega_n = 2.236 \text{ rad/s}$$
$$2\xi\omega_n = 4 \Rightarrow \xi = 0.8944$$
$$\omega_d = \omega_n\sqrt{1-\xi^2} = 2.236\sqrt{1-0.8944^2} = 1 \text{ rad/s}$$
$$\theta = \tan^{-1}\frac{\sqrt{1-\xi^2}}{\xi} = \tan^{-1}\frac{\sqrt{1-0.8944^2}}{0.8944} = 26.56^\circ = 04636 \text{ rad}$$

1. Rise time $T_r = \dfrac{\pi - \theta}{\omega_d} = \dfrac{\pi - 0.4636}{2} = 2.6779$

2. Peak time $T_p = \dfrac{\pi}{\omega_d} = \dfrac{\pi}{1} = 3.1415$ s.

3. Settling time $T_s = \dfrac{4}{\xi\omega_n} = \dfrac{4}{0.8944 \times 2.236} = 2$ s.

4. Delay time $T_d = \dfrac{1 + 0.7\xi}{\omega_n} = \dfrac{1 + 0.7 \times 0.8944}{2.236} = 0.7272$ s.

5. Peak overshoot $\%M_p = e^{-\frac{\xi\pi}{\sqrt{1-\xi^2}}} \times 100 = e^{-\frac{0.8944\pi}{\sqrt{1-0.8^2}}} \times 100 = 0.925\%$

ASP-3: Find the error coefficients for the system having the following block diagram.

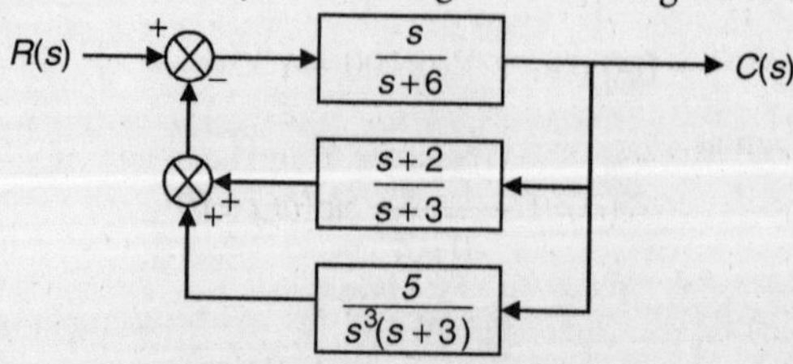

Solution:

The given block diagram can be reduced to

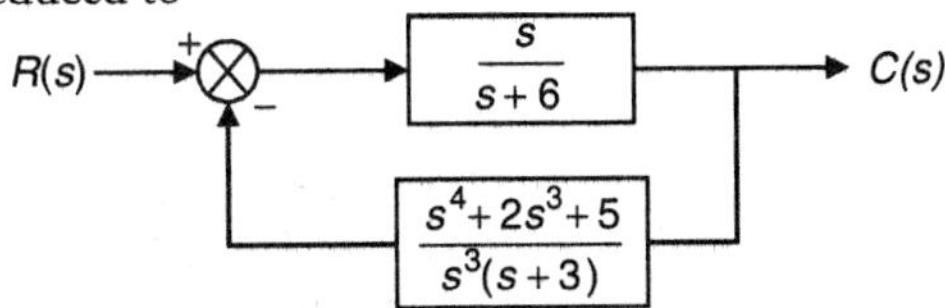

From the diagram, we get

$$G(s)H(s)=\frac{s(s^4+2s^3+5)}{s^3(s+3)(s+6)}=\frac{s^4+2s^3+5}{s^2(s+3)(s+6)}.$$

1. $K_p = \underset{s\to 0}{\text{Lt}}\, G(s)H(s) = \underset{s\to 0}{\text{Lt}}\, \frac{s^4+2s^3+5}{s^2(s+3)(s+6)} = \infty$

2. $K_p = \underset{s\to 0}{\text{Lt}}\, sG(s)H(s) = \underset{s\to 0}{\text{Lt}}\, \frac{s^4+2s^3+5}{s(s+3)(s+6)} = \infty$

3. $K_a = \underset{s\to 0}{\text{Lt}}\, s^2G(s)H(s) = \underset{s\to 0}{\text{Lt}}\, \frac{s^4+2s^3+5}{(s+3)(s+6)} = 5/18$

ASP-4: Find step, ramp and parabolic error coefficients and their corresponding steady-state error for unity feedback system having the t.f. $G(s)=\frac{14(s+3)}{s(s+5)(s^2+3s+2)}$.

Solution:

Given that

$$G(s)=\frac{14(s+3)}{s(s+5)(s^2+3s+2)},\ H(s) = 1.$$

1. The step error coefficient is

$$K_p = \underset{s\to 0}{\text{Lt}}\, G(s)H(s) = \underset{s\to 0}{\text{Lt}}\, \frac{14(s+3)}{s(s+5)(s^2+3s+2)} = \infty.$$

Steady-state error, $e_{ss} = \frac{1}{1+K_p} = 0$

2. The ramp error coefficient is

$$K_v = \underset{s\to 0}{\text{Lt}}\, sG(s)H(s) = \underset{s\to 0}{\text{Lt}} = \frac{14(s+3)}{(s+5)(s^2+3s+2)} = 14\times 3/10 = 4.2.$$

Steady-state error, $e_{ss} = \frac{1}{K_v} = \frac{1}{4.2} = 0.235$

3. The parabolic error coefficient is

$$K_a = \underset{s\to 0}{\text{Lt}}\, s^2G(s)H(s) = \underset{s\to 0}{\text{Lt}}\, \frac{14s(s+3)}{(s+5)(s^2+3s+2)} = 0.$$

Steady-state error, $e_{ss} = \frac{1}{K_a} = \infty$

ASP-5: The open-loop t.f. of a unity feedback system is $G(s)=\frac{K}{s(1+sT)}$.

1. Find by what factor the gain K can be reduced so that the overshoot is reduced from 75% to 25%.
2. Find by what factor the gain K can be reduced so that the damping ratio is increased from 0.1 to 0.6.

Solution:
Given that

$$G(s)=\frac{K}{s(1+sT)},\ H(s)=1.$$

The closed-loop t.f. is

$$\frac{C(s)}{R(s)}=\frac{G(s)}{1+G(s)H(s)}=\frac{K/s(1+sT)}{1+K/s(1+sT)}=\frac{K}{s^2T+s+K}$$

$$\Rightarrow \qquad \frac{C(s)}{R(s)}=\frac{K/T}{s^2+s/T+K/T}.$$

The characteristic equation of the system is

$$1+G(s)H(s)=0$$

$$s^2+\frac{1}{T}s+\frac{K}{T}=0\cdot$$

By comparing the equation with $s^2+2\xi\omega_n s+\omega_n{}^2=0$, we get

$$\omega_n{}^2=\frac{K}{T}\Rightarrow\omega_n=\sqrt{\frac{K}{T}}$$

$$2\xi\omega_n=\frac{1}{T}\Rightarrow\xi=\frac{1}{T\times 2\omega_n}=\frac{1}{2\sqrt{KT}}.$$

1. Let ξ_1 be the damping ratio when $\%M_p$ is 75.

$$\%M_p=e^{-\frac{\xi_1\pi}{\sqrt{1-\xi_1^2}}}\times 100=75\cdot$$

$$\Rightarrow \qquad \xi_1=0.091$$

Similarly,

$$\%M_p=e^{-\frac{\xi_2\pi}{\sqrt{1-\xi_2^2}}}\times 100=25$$

$$\Rightarrow \qquad \xi_2=0.4037$$

We know that

$$\xi=\frac{1}{2\sqrt{KT}}\Rightarrow\xi\alpha\frac{1}{\sqrt{K}}.$$

$$\Rightarrow \qquad \frac{\xi_1}{\xi_2}=\sqrt{\frac{K_2}{K_1}}$$

$$\Rightarrow \qquad \frac{K_2}{K_1}=\frac{\xi_1{}^2}{\xi_2{}^2}=\frac{0.091^2}{0.4037^2}=0.0508$$

$$\Rightarrow \qquad K_2=0.0508K_1=\frac{K_1}{20}\cdot$$

The gain must be multiplied by a factor 1/20 to reduce the overshoot from 75% to 25% .

2. Let when $\xi = \xi_1 = 0.1, K = K_1$

When $\xi = \xi_2 = 0.6, K = K_2$

$$\Rightarrow \qquad \frac{K_2}{K_1} = \frac{\xi_1^2}{\xi_2^2} = \frac{0.1^2}{0.6^2} = \frac{1}{36}$$

$$\Rightarrow \qquad K_2 = \frac{K_1}{36}$$

The gain must be multiplied by a factor 1/36 to increase the damping ratio from 0.1 to 0.6.

ASP-6: For a unity feedback system having $G(s) = \dfrac{K}{s(2+sT)}$, find

1. the factor by which gain K should be multiplied to increase the damping ratio from 0.15 to 0.6.
2. the factor by which time constant should be multiplied to reduce the damping ratio from 0.8 to 0.4.

Solution:

Given that

$$G(s) = \frac{K}{s(2+sT)}, \; H(s) = 1.$$

The closed-loop t.f. is

$$\frac{C(s)}{R(s)} = \frac{G(s)}{1+G(s)H(s)} = \frac{K/s(2+sT)}{1+K/s(2+sT)} = \frac{K}{s^2T+2s+K}$$

$$\Rightarrow \qquad \frac{C(s)}{R(s)} = \frac{K/T}{s^2+2s/T+K/T}.$$

The characteristic equation of the system is

$$1+G(s)H(s) = 0$$

$$s^2 + \frac{2}{T}s + \frac{K}{T} = 0.$$

By comparing the equation with $s^2 + 2\xi\omega_n s + \omega_n^2 = 0$, we get

$$\omega_n^2 = K/T \Rightarrow \omega_n = \sqrt{\frac{K}{T}}$$

$$\xi\omega_n = 1/T \Rightarrow \xi = \frac{1}{T\omega_n} = \frac{1}{\sqrt{KT}}$$

1. The new value of K

$$\xi = \frac{1}{\sqrt{KT}} \Rightarrow \xi \propto \frac{1}{\sqrt{K}}$$

$$\Rightarrow \qquad \frac{\xi_1}{\xi_2} = \sqrt{\frac{K_2}{K_1}}$$

$$\Rightarrow \qquad \frac{K_2}{K_1} = \frac{\xi_1^2}{\xi_2^2} = \frac{0.15^2}{0.6^2} = 0.0625 = 1/16$$

$$\Rightarrow \qquad K_2 = \frac{K_1}{16}.$$

The gain must be multiplied by factor 1/16 to increase the damping ratio from 0.15 to 0.6.

2. The new value of T is

$$\xi = \frac{1}{\sqrt{KT}} \Rightarrow \xi \propto \frac{1}{\sqrt{T}}$$

$$\Rightarrow \qquad \frac{\xi_1}{\xi_2} = \sqrt{\frac{T_2}{T_1}}$$

$$\Rightarrow \qquad \frac{T_2}{T_1} = \frac{\xi_1^2}{\xi_2^2} = \frac{0.8^2}{0.4^2} = 4$$

$$\Rightarrow \qquad T_2 = 4T_1.$$

The time constant T must be multiplied by a factor 4 to reduce ξ from 0.8 to 0.4.

ASP-7: For the system shown in the figure, obtain the closed-loop t.f., damping ratio, natural frequency and expression for the output response, if it is subjected to unit-step input.

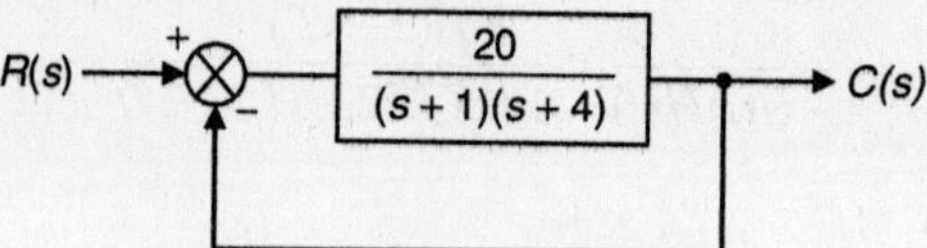

Solution:

From the given block diagram, we get

$$\frac{C(s)}{R(s)} = \frac{20/(s+1)(s+4)}{1+20/(s+1)(s+4)} = \frac{20}{s^2+5s+24}.$$

The characteristic equation is $s^2+5s+24=0$.

By comparing the equation with $s^2+2\xi\omega_n s+\omega_n^2=0$, we get

$$\omega_n^2 = 24 \Rightarrow \omega_n = 4.8989\,\text{rad/s}$$

$$2\xi\omega_n = 5 \Rightarrow \xi = 0.51031$$

$$\omega_d = \omega_n\sqrt{1-\xi^2} = 5\sqrt{1-0.51^2} = 4.2129\,\text{rad/s}$$

$$\theta = \tan^{-1}\frac{\sqrt{1-\xi^2}}{\xi} = \tan^{-1}\frac{\sqrt{1-0.51^2}}{0.51} = 59.012^0 = 1.03\,\text{rad}.$$

We know that

$$\frac{C(s)}{R(s)} = \frac{20}{s^2+5s+24} = \frac{20}{24}\frac{24}{s^2+5s+24}$$

$$C(s) = \frac{20}{24}\frac{24}{s^2+5s+24}R(s).$$

The output response is given as

$$c(t) = \frac{20}{24}\left[1 - \frac{e^{-\xi\omega_n t}}{\sqrt{1-\xi^2}}\sin(\omega_d t + \theta)\right]$$

$$= \frac{20}{24}\left[1 - \frac{e^{-2.5t}}{\sqrt{1-0.26}}\sin(4.2129t + 59.012^\circ)\right]$$

$$= \frac{20}{24}\left[1 - 1.1628e^{-2.5t}\sin(4.2129t - 59.012^\circ)\right].$$

ASP-8: A certain control system is described by the differential equation $\frac{d^2y}{dt^2} + 7\frac{dy}{dt} + 12y = 12x$, where y is the output and x is the input. Find its output response for unit step input.

Solution:

The given differential equation is

$$\frac{d^2y}{dt^2} + 7\frac{dy}{dt} + 12y = 12x.$$

By taking Laplace transforms on both sides with zero initial conditions, we get

$$s^2Y(s) + 7sY(s) + 12Y(s) = 12X(s)$$

$$(s^2 + 7s + 12)Y(s) = 12X(s)$$

$$\frac{Y(s)}{X(s)} = \frac{12}{s^2 + 7s + 12}.$$

The characteristic equation is $s^2 + 7s + 12 = 0$.

By comparing the equation with $s^2 + 2\xi\omega_n s + \omega_n^2 = 0$, we get

$$\omega_n^2 = 8 \Rightarrow \omega_n = 3.464\text{rad/s}$$

$$2\xi\omega_n = 4 \Rightarrow \xi = 1.010363.$$

As $\xi > 1$, the given system is over-damped. Hence, the output will not contain any oscillations. Hence, the standard expression for $c(t)$ cannot be used.

As per the given data, the input is in unit step, i.e., $X(s) = 1/s$

$$Y(s) = \frac{12}{s^2 + 7s + 12}X(s)$$

$$= \frac{12}{s(s+3)(s+4)} = \frac{A}{s} + \frac{B}{s+3} + \frac{C}{s+4}.$$

$$= \frac{1}{s} - \frac{4}{s+3} + \frac{3}{s+4}$$

By taking inverse Laplace transforms on both sides, we get

$$Y(t) = 1 - 4e^{-3t} + 3e^{-4t}.$$

ASP-9: For the feedback system, as shown in figure, determine the values of K_1 and K_2, so that peak overshoot of the system is 1.3% and the corresponding time is 1.06 s.

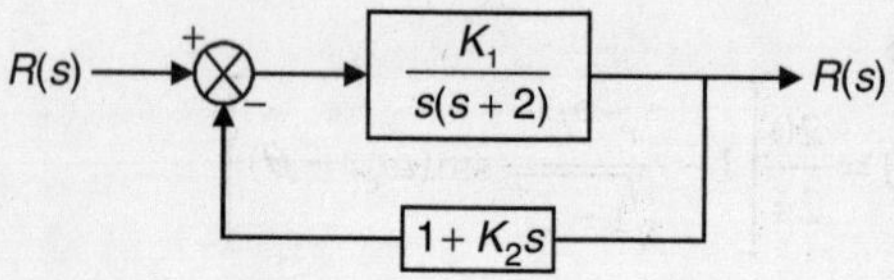

Solution:

From the given block diagram, $G(s)=\dfrac{K_1}{s(s+2)}, H(s)=1+sK_2$

The closed-loop t.f. is

$$\frac{C(s)}{R(s)}=\frac{G(s)}{1+G(s)H(s)}=\frac{K_1/s(s+2)}{1+K_1(1+sK_2)/s(s+2)}.$$

The characteristic equation of the system is

$$1+G(s)H(s)=0$$

$$s^2+(K_1K_2+2)s+K_1=0.$$

By comparing the equation with $s^2+2\xi\omega_n s+\omega_n{}^2=0$, we get

$$\omega_n{}^2=K_1 \Rightarrow \omega_n=\sqrt{K_1}$$

$$2\xi\omega_n=K_1K_2+2 \Rightarrow \xi=\frac{K_1K_2+2}{2\sqrt{K_1}} \tag{1}$$

$$\omega_d=\omega_n\sqrt{1-\xi^2}=\sqrt{K_1}\sqrt{1-\frac{(K_1K_2+2)^2}{4K_1}} \tag{2}$$

$$\%M_p=e^{-\frac{\xi\pi}{\sqrt{1-\xi^2}}}\times 100=1.3$$

$$\ln(0.013)=\frac{-\xi\pi}{\sqrt{1-\xi^2}}$$

$$\sqrt{1-\xi^2}=0.723\xi$$

$$\Rightarrow \qquad \xi=0.8103$$

Peak time, $$T_p=\frac{\pi}{\omega_d}=1.06 \Rightarrow \omega_d=2.96\text{rad/s} \tag{3}$$

$$\frac{K_1K_2+2}{2\sqrt{K_1}}=0.8103 \tag{4}$$

From (2) and (3), we get

$$\sqrt{K_1}\sqrt{1-\frac{(K_1K_2+2)^2}{4K_1}}=2.96$$

$$\sqrt{K_1}\sqrt{1-0.6566}=2.96.$$

$$\Rightarrow \qquad K_1 = 25.514$$

From (4), we can write

$$K_1K_2 + 2 = 1.6206\sqrt{K_1}\,.$$

$$\Rightarrow \qquad K_2 = 0.2424$$

The values of $K_1 = 25.514$ and $K_2 = 0.2424$

ASP-10: The step response of a second-order system is shown for an input of 2 $u(t)$. Determine the open-loop t.f. and the closed-loop t.f. Assume unity feedback.

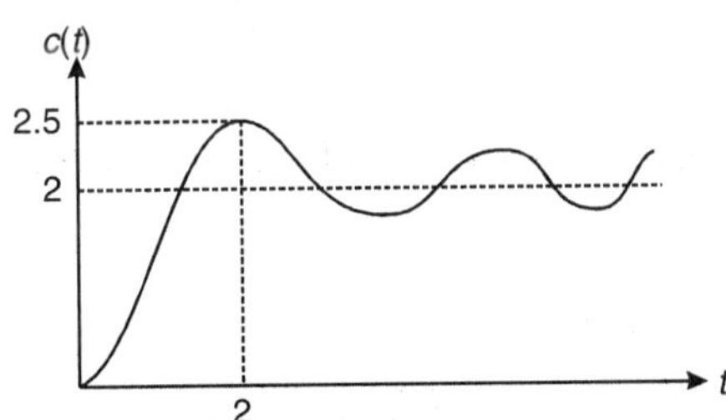

Solution:

From the given figure, $T_p = 2$ s

The overshoot for steady-state of 2 is

$$2.5 - 2 = 0.5$$

Hence, the % overshoot is $\%M_p = \dfrac{2.5-2}{2}\times 100 = 25$

$$\%M_p = e^{-\frac{\xi\pi}{\sqrt{1-\xi^2}}}\times 100 = 25$$

$$\Rightarrow \qquad \ln(0.25) = \frac{-\xi\pi}{\sqrt{1-\xi^2}}$$

$$\Rightarrow \qquad \sqrt{1-\xi^2} = 2.266\xi$$

$$\xi = 0.4037$$

$$T_p = \frac{\pi}{\omega_d} = \frac{\pi}{\omega_n\sqrt{1-\xi^2}}$$

$$\Rightarrow \qquad 2 = \frac{\pi}{\omega_n\sqrt{1-0.1629}} \Rightarrow \omega_n = 1.17169\,\text{rad/s}$$

∴ The closed-loop t.f. is

$$\frac{C(s)}{R(s)} = \frac{\omega_n^2}{s^2 + 2\xi\omega_n s + \omega_n^2} = \frac{1.7169^2}{s^2 + 2\times 0.4037\times 1.7169s + 1.7169^2}$$

$$\Rightarrow \qquad \frac{C(s)}{R(s)} = \frac{2.9478}{s^2 + 1.3862s + 2.9478}\,.$$

To get the open-loop t.f., divide numerator and denominator by $(s^2+1.3862s)$

$$\frac{C(s)}{R(s)}=\frac{2.9478/(s^2+1.3862s)}{1+2.9478/(s^2+1.3862s)}.$$

Compare the equation with

$$\frac{C(s)}{R(s)}=\frac{G(s)}{1+G(s)H(s)}$$

$$\Rightarrow \qquad G(s)=\frac{2.9478}{s^2+1.3862s}=\frac{2.9478}{s(s+1.3862)}$$

ASP-11: For a second-order system, $\xi=0.6$, $\omega=5\,\text{rad/s}$. Find the values of ω_d, T_r, T_p, T_s and M_p.

Solution:

Given that

$$\xi=0.6,\ \omega_n=5\,\text{rad/s}$$

$$\omega_d=\omega_n\sqrt{1-\xi^2}=5\sqrt{1-0.36}=4\,\text{rad/s}$$

$$\theta=\tan^{-1}\frac{\sqrt{1-\xi^2}}{\xi}=\tan^{-1}\frac{\sqrt{1-0.36}}{0.6}=53.12°=0.92\,\text{rad}$$

1. Rise time $T_r=\dfrac{\pi-\theta}{\omega_d}=\dfrac{\pi-0.92}{4}=0.55\,\text{s}.$

2. Peak time $T_p=\dfrac{\pi}{\omega_d}=\dfrac{\pi}{4}=0.785\,\text{s}.$

3. Delay time $T_d=\dfrac{1+0.7\xi}{\omega_n}=\dfrac{1+0.7\times0.6}{5}=0.284\,\text{s}.$

4. Settling time $T_s=\dfrac{4}{\xi\omega_n}=\dfrac{4}{0.6\times5}=1.33\,\text{s}.$

5. Peak overshoot $\%M_p=e^{-\frac{\xi\pi}{\sqrt{1-\xi^2}}}\times100=e^{-\frac{0.6\pi}{\sqrt{1-0.36}}}\times100=9.5\%.$

ASP-12: Consider the closed-loop system $\dfrac{C(s)}{R(s)}=\dfrac{\omega_n^2}{s^2+2\xi\omega_n s+\omega_n^2}$. Determine the values of ξ, ω_n, so that the system responds to a step input with approximately 5% overshoot and with a settling time of 2 s.

Solution:

The given closed-loop t.f. is

$$\frac{C(s)}{R(s)}=\frac{\omega_n^2}{s^2+2\xi\omega_n s+\omega_n^2}.$$

The given data is

Settling time $$T_s = \frac{4}{\xi\omega_n} = 2 \Rightarrow \xi\omega_n = 2$$

Peak overshoot $$\%M_p = e^{-\frac{\xi\pi}{\sqrt{1-\xi^2}}} \times 100 = 5$$

$$\Rightarrow \quad \frac{\xi\pi}{\sqrt{1-\xi^2}} = -\ln(0.05) = 3$$

$$\Rightarrow \quad \frac{\xi\pi}{\sqrt{1-\xi^2}} = 0.95 \Rightarrow \xi = 0.7237$$

$$\xi\omega_n = 2 \Rightarrow \omega_n = \frac{2}{\xi} = \frac{2}{0.7237} = 2.743\text{rad/s}.$$

ASP-13: A unity feedback system is characterised by the open-loop t.f. $G(s) = \dfrac{1}{s(1+0.5s)(1+0.2s)}$. Determine the steady-state error for unit-step, unit ramp and unit parabolic inputs. Further, determine the damping factor and natural frequency of the dominant roots.

Solution:
Given that

$$G(s) = \frac{1}{s(1+0.5s)(1+0.2s)}, \; H(s) = 1.$$

1. $$K_p = \underset{s\to 0}{\text{Lt}}\, G(s)H(s) = \underset{s\to 0}{\text{Lt}} \frac{1}{s(1+0.5s)(1+0.2s)} = \infty$$

$$e_{ss} = \frac{1}{1+K_p} = \frac{1}{1+\infty} = 0.$$

2. $$K_v = \underset{s\to 0}{\text{Lt}}\, sG(s)\; H(s) = \underset{s\to 0}{\text{Lt}} \frac{s}{s(1+0.5s)(1+0.2s)} = 1$$

$$e_{ss} = \frac{1}{K_v} = 1.$$

3. $$K_a = \underset{s\to 0}{\text{Lt}}\, s^2G(s)\; H(s) = \underset{s\to 0}{\text{Lt}} \frac{s}{s^2(1+0.5s)(1+0.2s)} = 0$$

$$e_{ss} = \frac{1}{K_a} = \infty.$$

The characteristic equation is given as

$$1+G(s)H(s)=0$$
$$1+s(1+0.5s)(1+0.2s)=0$$
$$0.1s^3+0.7s^2+s+1=0$$
$$s^3+7s^2+10s+10=0$$
$$(s+5.5)(s^2+1.5s+1.75\ )=0$$

$$s+5.5=0,\ s^2+1.5s+1.75=0$$

$$s^2+2\xi\omega_n s+\omega_n^{\ 2}=0$$

The dominant roots are $s=\dfrac{-1.5\pm\sqrt{1.5^2-4\times1.75}}{2}=-0.75\pm j0.433$

$$\omega_n=\sqrt{1.75}=1.32\text{rad/s}$$
$$2\xi\omega_n=1.5\Rightarrow\xi=0.568.$$

ASP-14: The unit-step response of a second-order linear system with zero initial state is given as $c(t)=1-1.25e^{-6t}\sin(8t+\tan^{-1}1.333)$. Determine the damping ratio, un-damped natural frequency of oscillations and peak overshoot.

Solution:
Given that

$$c(t)=1-1.25e^{-6t}\sin(8t+\tan^{-1}1.333).$$

By comparing the equation with the standard form, we get

$$c(t)=1-\frac{e^{-\xi\omega_n t}}{\sqrt{1-\omega_2}}\sin(\omega_d t+\theta)$$

$$\Rightarrow\qquad \xi\omega_n=6,\ \omega_d=8$$

$$\omega_d=\omega_n\sqrt{1-\xi^2}=8$$

$$\Rightarrow\qquad \omega_n^{\ 2}(1-\xi^2)=64$$

$$\Rightarrow\qquad \omega_n^{\ 2}-36=64\Rightarrow\omega_n^{\ 2}=100\Rightarrow\omega_n=10.$$

$$\xi=\frac{6}{\omega_n}=\frac{6}{10}=0.6$$

The peak overshoot is given as

$$\%M_p=e^{-\frac{\xi\pi}{\sqrt{1-\xi^2}}}\times100=e^{-\frac{0.6\pi}{\sqrt{1-0.36}}}\times100=9.4$$

ASP-15: The block diagram of a servo system is shown in the figure. Determine the characteristic equation of the system, and hence, calculate the following when a unit-step-input is given.

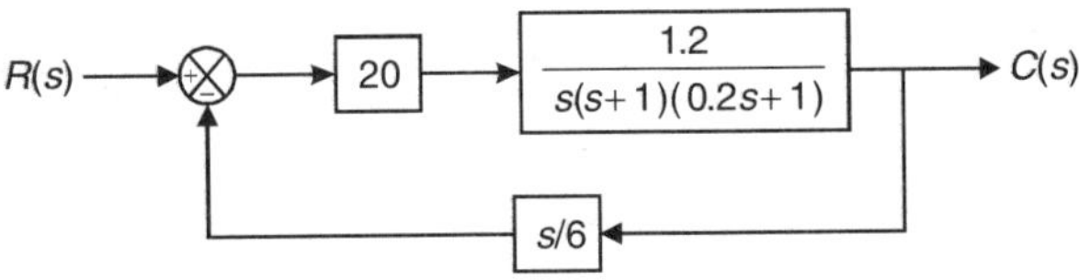

1. Un-damped frequency of oscillations.
2. Damped frequency of oscillations
3. Damping ratio and damping factor
4. Maximum overshoot
5. Time interval after which maximum and minimum will occur
6. Settling time
7. Number of cycles completed the output is settled within 2% and 5% of final value.

Solution:

From the given block diagram,

$$G(s) = \frac{20 \times 1.2}{s(s+1)(0.2s+1)} = \frac{24}{s(s+1)(0.2s+1)}, H(s) = \frac{s}{6}.$$

The characteristic equation of the system is

$$1 + G(s)H(s) = 0$$

$$1 + \frac{4}{(s+1)(0.2s+1)} = 0.$$

$$0.2s^2 + 1.2s + 1 + 4 = 0$$

$$s^2 + 6s + 25 = 0$$

By comparing the equation with $s^2 + 2\xi\omega_n s + \omega_n^2 = 0$, we get

1. $\omega_n^2 = 25 \Rightarrow \omega_n = 5\,\text{rad/s}.$

   $2\xi\omega_n = 6 \Rightarrow \xi = 0.6$

2. Damping frequency, $\omega_d = \omega_n\sqrt{1-\xi^2} = 5\sqrt{1-0.36} = 4\,\text{rad/s}.$
3. $2\xi\omega_n = 6 \Rightarrow \xi = 0.6.$

   Damping factor $\xi = 0.6.$

4. Maximum overshoot $\%M_p = e^{-\frac{\xi\pi}{\sqrt{1-\xi^2}}} \times 100 = e^{-\frac{0.6\pi}{\sqrt{1-0.36}}} \times 100 = 9.5\%.$

5. Number of cycles completed before settling within 2%.

$$2T_p = 2 \times \frac{\pi}{\omega_d} = \frac{2\pi}{4} = 1.57\,\text{s}.$$

6. Settling time (2%) $T_s = \dfrac{4}{\xi\omega_n} = \dfrac{4}{0.6 \times 5} = 1.33\,\text{s}.$

Settling time (5%) $T_s = \frac{3}{\xi\omega_n} = \frac{3}{0.6\times 5} = 1\,\text{s}.$

7. Number of cycles completed before settling within 2%

$$\frac{2\sqrt{1-\xi^2}}{\pi\xi} = \frac{2\times 0.8}{\pi\times 0.6} = 0.85 \text{ cycles.}$$

Number of cycles completed before settling within 5%

$$\frac{3\sqrt{1-\xi^2}}{\pi\xi} = \frac{3\times 0.8}{2\pi\times 0.6} = 0.6366 \text{ cycles.}$$

ASP-16: The open-loop t.f. of a unity feedback system $G(s) = \frac{K}{s(1+sT)}$. For this system, overshoot is reduced from 0.6 to 0.2 due to change in K only. Show that $\frac{TK_1 - 1}{TK_2 - 1} = 43.33,$ where K_1 and K_2 are values of K for 0.6 and 0.2 overshoot, respectively

Solution:

Given that

$$G(s) = \frac{K}{s(1+sT)},\ H(s) = 1.$$

The closed-loop t.f. is

$$\frac{C(s)}{R(s)} = \frac{G(s)}{1+G(s)H(s)} = \frac{K/s(1+sT)}{1+K/s(1+sT)} = \frac{K}{s^2T+s+K}$$

$$\Rightarrow \qquad \frac{C(s)}{R(s)} = \frac{K/T}{s^2+s/T+K/T}$$

The characteristic equation of the system is

$$1 + G(s)H(s) = 0$$

$$s^2 + \frac{1}{T}s + \frac{K}{T} = 0.$$

By comparing the equation with $s^2 + 2\xi\omega_n s + \omega_n^2 = 0$, we get

$$\omega_n^2 = \frac{K}{T} \Rightarrow \omega_n = \sqrt{\frac{K}{T}}$$

$$2\xi\omega_n = \frac{1}{T} \Rightarrow \xi = \frac{1}{T\times 2\omega_n} = \frac{1}{2\sqrt{KT}}.$$

Let ξ_1 be the damping ratio when M_p is 0.6.

$$\%M_p = e^{\frac{-\xi_1\pi}{\sqrt{1-\xi_1^2}}} \times 100 = 60$$

$\Rightarrow$ $$\xi_1 = 0.1602$$

Similarly,

$$\%M_p = e^{\frac{-\xi_2\pi}{\sqrt{1-\xi_2^2}}} \times 100 = 20$$

$\Rightarrow$ $$\xi_2 = 0.4559.$$

We know that

$$\xi = \frac{1}{2\sqrt{KT}} \Rightarrow \xi_1 = \frac{1}{2\sqrt{K_1T}},\ \xi_2 = \frac{1}{2\sqrt{K_2T}}$$

$$\xi_1 = \frac{1}{2\sqrt{K_1T}},\ \xi_2 = \frac{1}{2\sqrt{K_2T}}.$$

$\Rightarrow$ $$0.1602 = \frac{1}{2\sqrt{K_1T}},\ 0.4559 = \frac{1}{2\sqrt{K_2T}}$$

$\Rightarrow$ $$K_1T = 9.74,\ K_2T = 1.2028.$$

$\Rightarrow$ $$K_1T - 1 = 8.74,\ K_2T - 1 = 0.2028.$$

$\Rightarrow$ $$\frac{TK_1 - 1}{TK_2 - 1} = \frac{8.74}{0.2028} = 43.33.$$

REVIEW QUESTIONS

1. What is meant by time response? Explain about
   1. steady-state response and 2. transient response
2. Explain about the various test signals used in control system.
3. What is mean by steady-state error? Derive the expression for steady-state error.
4. Define the various time domain specifications for a standard second-order control system. Derive the expression for each of them.
5. How steady-state error of a control system is determined? How it can be reduced?
6. Explain the significance of generalised error series.
7. State how the type of a control system is determined. How does it affect the steady-state error?
8. Obtain the generalised error series and indicate the evaluation of dynamic error coefficients.
9. Define type and order of a control system. Explain each with an example.
10. Define steady-state response and steady-state error. How can you minimise the steady-state error.
11. How does the damping ratio affect the time response of a second-order system?
12. Obtain an expression for the unit-step response of a second-order system for an under-damped case.
13. Draw the transient response of a second-order system and define all the specifications for the under-damped case.
14. What are generalised error constants? State the advantages of generalised error coefficients.
15. Explain error constants K_p, K_v and K_a for Type 2 system.
16. Derive the static error constants and list the disadvantages.
17. Define the following systems and sketch their output waveforms for an unit step input:
    1. Under-damped system
    2. Un-damped system
    3. Over-damped system
    4. Critically damped system

18. For a first-order system, find out the output of the system when the input applied to the system is in unit step. Sketch the $r(t)$ and $c(t)$ and show the steady-state error.

EXERCISE PROBLEMS

1. Find the type and order of the following systems:

i. $G(s)H(s) = \dfrac{10(s+2)}{s(s^3+4s+20)}$

ii. $G(s)H(s) = \dfrac{20}{s^2(s^2+10s)}$

iii. $G(s)H(s) = \dfrac{(s^2+10)}{s(s+3)(s^2+2s+10)}$

iv. $G(s)H(s) = \dfrac{2s}{(1+0.1s)(1+0.5s)}$

2. A second-order control system is represented by t.f. $\dfrac{\theta(s)}{T(s)} = \dfrac{1}{Js^2 + fs + K}$, where θ is the proportional output and T is the input torque. When applied a step input of 10 Nm, we have the following results as $\%M_p = 6$, $t_p = 1$ s, $e_{ss} = 0.5$ rad. Now, determine the values of J, f and k.

3. For the given system as shown in the figure, find

i. peak overshoot for step input.
ii. settling time (2% tolerance) for step input.
iii. steady-state error for an input of $r(t) = 2 + 4t + 6t^2$.

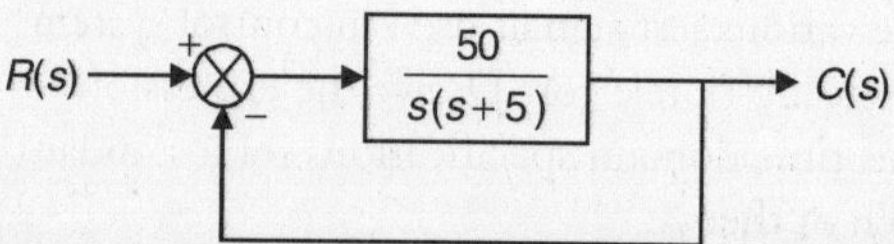

4. If the forward path of a unity feedback control system is given as $G(s) = 2/s(s+3)$, then find the expression for the unit-step response of the system.

5. For the given system as shown in the figure, find

i. system type and order
ii. error constants K_p, K_v and K_a.
iii. steady-state error for an input of $r(t) = 4u(t)$.

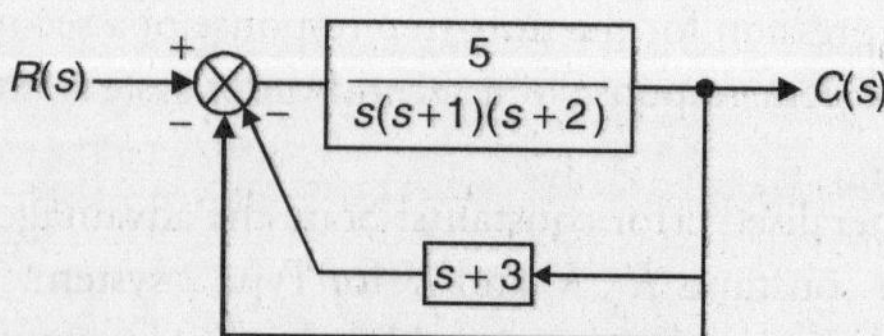

6. The open-loop t.f. of an unity feedback system is given as $G(s) = K/s(1 + sT)$. Find by what factor, K has to be reduced so that the overshoot is reduced from 60% to 20%.

7. Consider the closed loop system given as $\dfrac{C(s)}{R(s)} = \dfrac{\omega_n^2}{s^2 + 2\xi\omega_n s + \omega_n^2}$. Determine the values of ξ and ω_n so that the system responds to a step input with approximately 5% overshoot and a settling time of 2 s.

8. Using generalised error series, calculate the steady-state error of a unity feedback system given as $G(s) = 5/(1 + 0.5s)$ for the following inputs.
   i. $r(t) = 2t$
   ii. $r(t) = 1 + 2t + t^2/2$
   iii. $r(t) = 4$
9. Consider the system as shown in the figure. Determine the values of K_1 and K_2 so that the maximum overshoot in unit-step response is 2.5% and the peak time is 2 s.

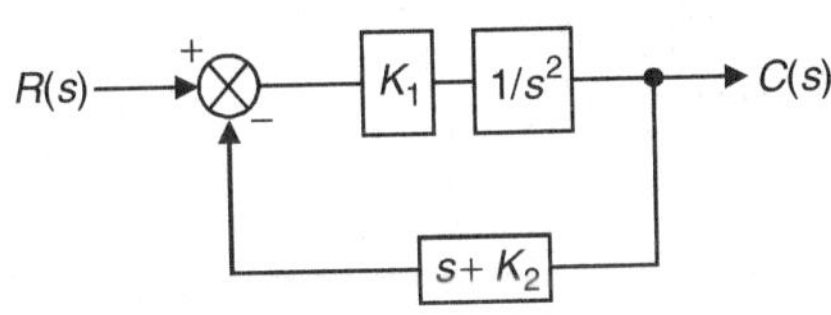

10. Consider the closed-loop system given as $\frac{C(s)}{R(s)} = \frac{50}{s^2 + s + 50}$. Determine the following step response specifications.
    i. Un-damped frequency.
    ii. Damped frequency.
    iii. Peak overshoot.
11. Find K to limit the steady-state error of a system for input $r(t) = 1 + 8t + 9t^2$ to 0.4 having
$$G(s)H(s) = \frac{K}{s^2(s+1)(s+2)}.$$
12. A system is given by differential equation $\frac{d^2y}{dt^2} + 5\frac{dy}{dt} + 16y = 16x$, where y is the output and x is the input. Determine all the time domain specifications and obtain output response for unit step input.
13. A unity feedback system is characterised by an OLTF $G(s) = \frac{K}{s(s+10)}$. Determine gain K so that the system will have a damping ratio of 0.5. For this value of K, determine settling time, peak time and peak overshoot.
14. Measurements conducted on a servo mechanism shows the system response to be $c(t) = 1 + 2e^{-40t} - 3e^{-10t}$ when subjected to a unit-step input. Obtain the expression for the closed-loop t.f.
15. The unit step response of a second-order control system is given as $c(t) = 1 - 1.25e^{-6t}\sin(8t - 53.12^0)$. Determine the damping ratio and undamped frequency of oscillations.
16. A unity feedback system is characterised by an open-loop transfer function $G(s) = \frac{1}{s(s+1)}$. The input to the system is described as $r(t) = 4 + 6t + 2t^2$. Determine the generalised error series and steady-state error.

ANSWERS

1. i. 1, 4 ii. 3, 4 iii. 1, 4 iv. 0, 2
2. $J = 1.13$, $f = 6.26$, $K = 20$
3. i. $M_p = 30.7\%$ ii. $t_s = 1.616$ s iii. $e_{ss} = \infty$
4. $c(t) = 1 - 2e^{-t} + e^{-2t}$
5. i. Type = 0, order = 3 ii. $K_p = \infty$, $K_v = 10$, $K_a = 0$ iii. $e_{ss} = 0$
6. $K_2 = K_1/8$

7. $\xi = 0.69$, $\omega_n = 2.89$ rad/s
8. i. $e_{ss} = 0.139 + 0.333t$ ii. $e_{ss} = 0.2928 + 0.4t + 0.0834t^2$ iii. $e_{ss} = 0.6666$
9. $K_1 = 3.69$, $K_2 = 1.59$
10. i. $\omega_n = 7.07$ rad/s ii. $\omega_d = 7.05$ rad/s iii. $M_p = 4.23\%$
11. $K = 90$
12. $t_d = 0.359$ s, $M_p = 8.09\%$, $t_s = 1.6$ s, $t_p = 1$ s, $c(t) = 1 - 1.28e^{-2.5t}\sin(3.12t + 51.32°)$
13. $K = 100$, $t_s = 0.8$ s, $t_p = 0.36$ s, $M_p = 16.3\%$
14. $T(s) = (50s + 400)/(s + 10)(s + 40))$
15. $\xi = 0.6$, $\omega_n = 10$ rad/s
16. $e(t) = 6+4t$, $e_{ss} = \infty$

OBJECTIVE TYPE QUESTIONS

1. If $M_p = 100\%$, then the damping ratio is________
   (a) 1 (b) 0 (c) 0.5 (d) infinite
2. The type and order of the system having t.f. $T(s) = \dfrac{s^2+1}{s(s^3+2s+1)}$ are
   (a) 1, 1 (b) 3, 3 (c) 1, 4 (d) None
3. The static error coefficient K_p is given as
   (a) $\underset{s\to 0}{\text{Lt}}\, G(s)H(s)$ (b) $\underset{s\to 0}{\text{Lt}}\, sG(s)H(s)$ (c) $\underset{s\to 0}{\text{Lt}}\, s^2G(s)H(s)$ (d) None
4. The velocity error coefficient K_v is given as
   (a) $\underset{s\to 0}{\text{Lt}}\, G(s)H(s)$ (b) $\underset{s\to 0}{\text{Lt}}\, sG(s)H(s)$ (c) $\underset{s\to 0}{\text{Lt}}\, s^2G(s)H(s)$ (d) None
5. The acceleration error coefficient K_a is given as
   (a) $\underset{s\to 0}{\text{Lt}}\, G(s)H(s)$ (b) $\underset{s\to 0}{\text{Lt}}\, sG(s)H(s)$ (c) $\underset{s\to 0}{\text{Lt}}\, s^2G(s)H(s)$ (d) None
6. If the response of a system to an impulse input is $5e^{-3t}$, then the t.f. of the system is ________
   (a) $\dfrac{5}{s-3}$ (b) $\dfrac{5}{s+3}$ (c) $5(s-3)$ (d) $5(s+3)$
7. When an unit step input is applied to a second-order system, then the peak time is given as ________
   (a) $\dfrac{\pi}{\omega_d}$ (b) $\dfrac{\pi}{\omega_n}$ (c) $\dfrac{2\pi}{\omega_d}$ (d) None
8. The ratio of settling to time constant for a 2% tolerance of an under-damped system is
   (a) 4 (b) 2 (c) 5 (d) 6
9. The steady-state error for a Type 2 system when subjected to unit-step input is
   (a) 1 (b) 0 (c) $1/K$ (d) infinite
10. The steady-state error for a Type 2 system when subjected to unit ramp input is
    (a) 1 (b) 0 (c) $1/K$ (d) infinite
11. The steady-state error for a Type 2 system when subjected to unit parabolic input is
    (a) 1 (b) 0 (c) $1/K$ (d) infinite
12. The static error coefficient for the given t.f. $G(s) = \dfrac{50}{(s+1)(s+2)}$ is
    (a) 50 (b) 25 (c) 5 (d) 2.5
13. The velocity error coefficient for the given t.f. $G(s) = \dfrac{50}{s(s+1)(s+2)}$ is0
    (a) 5 (b) 25 (c) 50 (d) 2.5

14. The static error coefficient for the given t.f. $G(s)=\dfrac{4(s^2+10s+100)}{(s+1)(s+2)}$ is
(a) 200 (b) 25 (c) 400 (d) None

15. The peak over-shoot due to the increase in damping factor
(a) Increases (b) Decreases (c) Remains constant (d) None

16. The ratio of damped frequency to natural frequency of the system having damping factor is ξ
(a) $\sqrt{1+\xi^2}$ (b) $\sqrt{1-\xi^2}$ (c) $\dfrac{1}{\sqrt{1-\xi^2}}$ (d) $\dfrac{1}{\sqrt{1+\xi^2}}$

17. Which of the following system is generally preferred?
(a) Under-damped (b) Over-damped (c) Critically damped (d) Un-damped

18. If an unity feedback has open-loop t.f. $G(s)=\dfrac{16}{s(s+4)}$, then
(a) $\omega_n = 4$ rad/s, $\xi = 0.5$ (b) $\omega_n = 1$ rad/s, $\xi = 0.25$
(c) $\omega_n = 2$ rad/s, $\xi = 0.65$ (d) None

19. A second control system with no zeros has its poles located at $-6 + j8$ and $-6 - j8$ in the s-plane. The un-damped natural frequency and the damping factor of the system are, respectively,
(a) 10 rad/s and 0.6 (b) 6 rad/s and 0.8
(c) 8 rad/s and 0.6 (d) None

20. The steady-state error is determined as the difference between the reference input and the system output as
(a) $t = t_p$ (b) $t = \infty$ (c) $t = 0$ (d) $t =$ time constant

21. A second control system is given by the differential equation $\dfrac{d^2x}{dt^2}+4\dfrac{dx}{dt}+16x=16y$, where x is the output and y is the input. Then, the damping ratio is
(a) 0.5 (b) 0.9 (c) 0.25 (d) 0.325

22. The value of damping ratio of 0.8 in the step response of a second-order system results in maximum overshoot of
(a) 1.5 (b) 0.9 (c) 5 (d) 3.25

23. If the impulse response of a control system is $5e^{-10t}$, then its step response is
(a) $0.5e^{-10t}$ (b) $5(1 - e^{-10t})$ (c) $10(1 - e^{-10t})$ (d) $0.5(1 - {}^{-10t})$

24. The steady-state error for a unity feedback system having open loop transfer function $G(s)=\dfrac{16}{s+4}$, when subjected to a unit-step input will be
(a) 4 (b) 1/5 (c) 16 (d) 1/16

25. In a critically damped system,
(a) oscillations are prominent (b) oscillations just disappear
(c) oscillations are sustained (d) no oscillations are observed

26. The transient response of a second-order control system is given as $c(t)=1-\dfrac{e^{-6t}}{\sqrt{1-\xi^2}}\sin(\omega_d t+\phi)$, the settling time for the 2% criterion will be
(a) 2/3 (b) 3/2 (c) 6 (d) cannot be stated

27. An increase in damping ratio
(a) decreases rise time (b) increases rise time
(c) remains same (d) cannot be stated

28. For the system $\frac{C(s)}{R(s)} = \frac{25}{s^2 + 6s + 25}$, the nature of the response will be
(a) under-damped (b) over-damped (c) critically damped (d) un-damped

29. In general, good time response demands
(a) less settling time (b) less overshoot
(c) smallest stead state error (d) all

30. The type of the t.f. denotes the number of
(a) poles at origin (b) zeros at origin (c) finite poles (d) poles at infinity

31. Which of the following t.f. represents the critically damped system?
(a) $\frac{1}{s^2 + 2s + 4}$ (b) $\frac{1}{s^2 + s + 2}$ (c) $\frac{1}{s^2 + s + 4}$ (d) $\frac{1}{s^2 + 4s + 4}$

32. Transient response of the system is basically because of
(a) coupling (b) forces (c) friction (d) stored energy

ANSWERS

1. b	**2.** c	**3.** a	**4.** b	**5.** c	**6.** b	**7.** a	**8.** a	**9.** b	**10.** b
11. c	**12.** b	**13.** b	**14.** a	**15.** b	**16.** b	**17.** a	**18.** a	**19.** a	**20.** b
21. a	**22.** a	**23.** d	**24.** b	**25.** b	**26.** a	**27.** a	**28.** a	**29.** d	**30.** a
31. d	**32.** d								

CHAPTER

8

Feedback Characteristics of Control Systems

Contents

8.1 INTRODUCTION

A system that maintains a prescribed relationship between the output and the reference input by comparing them and using the difference as a means of control is called feedback control system. Feedback exists whenever there is a closed sequence of cause-and-effect relationship. Feedback control systems are often referred to as closed-loop control systems.

The negative feedback results in better stability in steady-state and rejects any disturbance signals. It also has low sensitivity to parameter variations. Hence, negative feedback is preferred in closed loop systems.

The positive feedback increases the error signal and drives the output to instability. However, sometimes the positive feedback is used in minor loops in control systems to amplify certain internal signals or parameters.

The difference between the reference input and the controlled output is known as error in control systems. To reduce this error, feedback is used. Using feedback, not only the error is reduced but also the sensitivity of the system due to parameter variations and unwanted internal and external disturbances are decreased substantially. Since the parameters of a system have a tendency to vary under a variety of changing conditions, these parameter variations have adverse effect on system performance.

8.2 REDUCTION OF PARAMETER VARIATIONS BY USING FEEDBACK

One of the primary purposes of using feedback in control system is to reduce the sensitivity of the system to parameter variations. The parameters of a system may vary with age, changing environment, etc. Conceptually, sensitivity is a measure of the effectiveness of feedback in reducing the influence of these variations on system performance.

8.2.1 Sensitivity

The change in a variable due to variations in the parameters of control systems is expressed in terms of sensitivity. Let the variable in a control system that changes its value be A such as the output and this change is considered due to parameter variations of element K such as gain or feedback; then, the sensitivity of the control system is expressed as

$$\text{Sensitivity} = \frac{\%\,\text{change in } A}{\%\,\text{change in } K}.$$

In mathematical terms, sensitivity can be written as

$$S_K^A = \frac{\dfrac{\partial A}{A}\times 100}{\dfrac{\partial K}{K}\times 100} = \frac{\partial A}{\partial K}\times\frac{K}{A}. \tag{8.1}$$

The notation S_K^A denotes sensitivity of variable A with respect to parameter K.

Sensitivity of overall t.f. *T*(*s*) w.r.t forward path t.f. *G*(*s*)

The sensitivity function for the overall t.f. $T(S)$ w.r.t variations in $G(S)$ written as

$$S_G^T = \frac{\partial T}{\partial G}\times\frac{G}{T}.$$

1. Open-loop control system

R(s) → G(s) → C(s)

For the above-mentioned system, the t.f. can be written as

$$T(s) = \frac{C(s)}{R(s)} = G(s)$$

$$S_G^T = \frac{G}{T}\frac{\partial T}{\partial G} = \frac{G}{G}\times 1 = 1. \tag{8.2}$$

2. Closed-loop control system

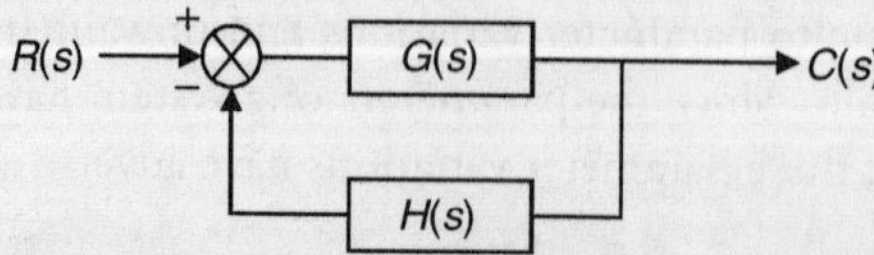

The overall t.f. for the abovementioned system is

$$T(s)=\frac{G(s)}{1+G(s)H(s)}$$

$$\frac{\partial T(s)}{\partial G(s)}=\frac{(1+GH)-1\times GH}{(1+GH)^2}=\frac{1}{(1+GH)^2}$$

$$S_G^T=\frac{G}{T}\frac{\partial T}{\partial G}$$

$$=\frac{G(s)}{G(s)/1+G(s)H(s)}\times\frac{1}{[1+G(s)H(s)]^2}$$

$$S_G^T=\frac{1}{1+G(s)H(s)} \tag{8.3}$$

By comparing (8.2) and (8.3), it is observed that the sensitivity of overall t.f. w.r.t forward path t.f. in the case of closed-loop control system is reduced by a factor $[1 + G(s)H(s)]$ as compared to that of open-loop control system.

Sensitivity of overall t.f. *T*(*s*) w.r.t feedback path t.f. *H*(*s*)

The overall t.f. is given as

$$T(s)=\frac{G(s)}{1+G(s)H(s)}$$

$$S_H^T=\frac{H}{T}\frac{\partial T}{\partial H}$$

$$\frac{\partial T(s)}{\partial H(s)}=\frac{\partial}{\partial H(s)}\left[\frac{G(s)}{1+G(s)H(s)}\right]$$

$$=G(s)\times\frac{-1}{[1+G(s)H(s)]^2}\times G(s)$$

$$=\frac{-[G(s)]^2}{[1+G(s)H(s)]^2}$$

$$S_H^T=\frac{H(s)}{\frac{G(s)}{1+G(s)H(s)}}\times\frac{-G(s)^2}{[1+G(s)H(s)]^2}$$

$$=\frac{-G(s)H(s)}{1+G(s)H(s)}. \tag{8.4}$$

By comparing (8.3) and (8.4), it is concluded that a closed-loop control system is more sensitive to variations in feedback path parameters than to variations in forward path parameters, and therefore, the specifications of feedback elements in a closed-loop control system should be more rigid as compared to that of forward path elements.

8.3 EFFECT OF FEEDBACK ON TIME CONSTANT OF A CONTROL SYSTEM

Let the t.f. of an open-loop control system $G(s)$ be given as

$$G(s) = \frac{k}{1+sT}.$$

If the applied input to the system be in unit steps, its response in open loop can be obtained as follows

$$\frac{C(s)}{R(s)} = \frac{k}{1+sT}$$

$$C(s) = \frac{kR(s)}{1+sT} = \frac{k}{s(1+sT)} = \frac{k/T}{s\left(s+\frac{1}{T}\right)} = k\left(\frac{1}{s} - \frac{1}{s+\frac{1}{T}}\right)$$

$$\Rightarrow \quad c(t) = k(1-e^{-t/T}). \tag{8.5}$$

In (8.5), T is the time constant of the open-loop system.

Introducing feedback in the system as shown in Figure 8.1, the overall t.f. of the system can be written as

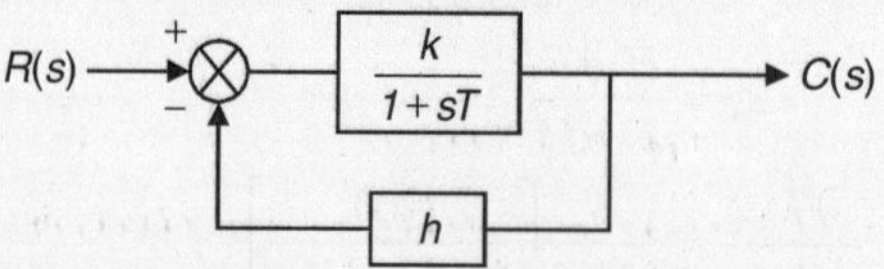

Figure 8.1 Effect of feedback on time constant

$$\frac{C(s)}{R(s)} = \frac{\frac{k}{1+sT}}{1+\frac{k}{1+sT}\times h} = \frac{k}{1+kh+sT} = \frac{k/T}{s+\left(\frac{1+kh}{T}\right)}. \tag{8.6}$$

Let the input of the closed-loop system be unit-step.

$$r(t) = u(t)$$

$$\Rightarrow \quad R(s) = \frac{1}{s}$$

$$\therefore \frac{C(s)}{R(s)} = \frac{k/T}{s+\left(\frac{1+kh}{T}\right)}$$

$$C(s)=\frac{k}{T}\times\frac{1}{\left(s+\frac{1+kh}{T}\right)}\times\frac{1}{s}$$

$$C(s)=\frac{k}{T}\left(\frac{T}{1+kh}\times\frac{1}{s}-\frac{T}{1+kh}\times\frac{1}{s+\frac{1+kh}{T}}\right)$$

$$C(s)=\frac{k}{1+kh}\left(\frac{1}{s}-\frac{1}{s+\frac{1+kh}{T}}\right)$$

By taking inverse Laplace transforms on both sides, we get

$$c(t)=\frac{k}{1+kh}\times\left(1-e^{-t\frac{(1+kh)}{T}}\right). \tag{8.7}$$

The new time constant of the feedback control system is $\frac{T}{1+kh}$. If the value of h is positive and $k > 1$, the new time constant is less than 1 and response is fast. Therefore, feedback improves the time response of a system.

8.4 EFFECT OF FEEDBACK ON OVERALL GAIN

For an open-loop system with overall t.f. $G(s)$, the gain of such system is $G(s)$.

Due to the introduction of feedback in such a system, the overall gain of the system becomes $\frac{G(s)}{1+G(s)H(s)}$. If the feedback is negative, gain $G(s)$ is reduced by a factor $\frac{1}{1+G(s)H(s)}$. Therefore, due to negative feedback, the overall gain reduces.

8.5 EFFECT OF FEEDBACK ON STABILITY (CONTROL OVER SYSTEM DYNAMICS BY USING FEEDBACK)

Let us consider a simple feedback system as shown in Figure 8.2

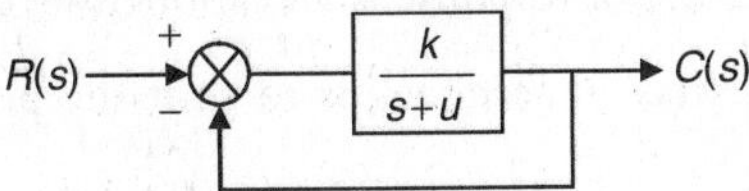

Figure 8.2 Feedback control system

The open-loop t.f. is given as

$$G(s)=\frac{k}{s+u}$$

The impulse response is given as $[\because R(s)=1]$

$$C(s) = G(s) = \frac{k}{s+u}$$

$$\Rightarrow \quad c(t) = ke^{-ut}. \tag{8.8}$$

For the feedback system, we have

$$\frac{C(s)}{R(s)} = \frac{\frac{k}{s+u}}{1+\frac{k}{s+u}} = \frac{k}{s+u+k}.$$

Since $R(s)=1$, for impulse input, the response is given as

$$C(s) = \frac{k}{s+u+k}$$

$$\Rightarrow \quad c(t) = k.e^{-(u+k)t}. \tag{8.9}$$

Figure 8.3 shows the pole–zero plot of both non-feedback and feedback systems.

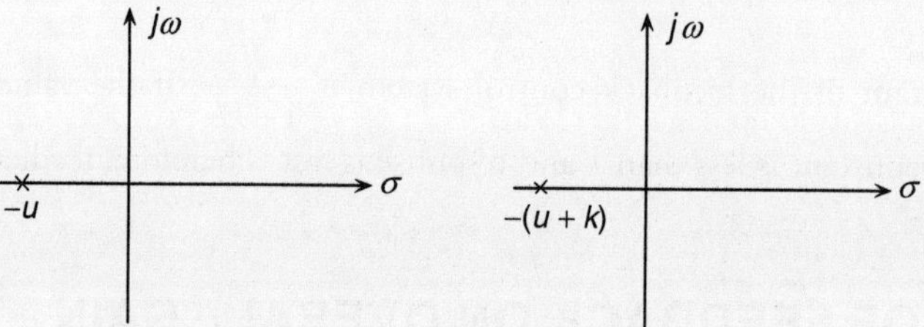

Figure 8.3 Pole–zero plots of (8.8) and (8.9)

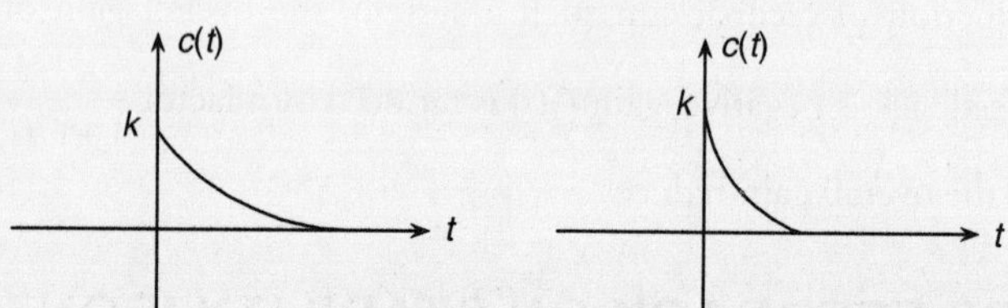

Figure 8.4 Impulse response of (a) Non-feedback and (b) Feedback system

The non-feedback system has the pole located at $s = -u$ of which the impulse response is shown in Figure 8.4. As we visualise, the nature of response is an exponential decay with time constant $T = \frac{1}{u}$. For the positive value of K, the effect of feedback is to shift the pole negatively to $s = -(u+k)$ and therefore, the time constant reduces to $\frac{1}{k+u}$. As k increases, the system dynamics becomes faster, andtherefore, the transient response decays more quickly, as shown in Figure 8.4(b). It should be noted that feedback controls the dynamics of the system by adjusting the location of its poles. It should also be noted that feedback increases the possibility of instability of systems, that is, a closed loop system may be unstable even though the open loop system is stable.

Thus, we are sure that feedback is a very useful technique to control the system dynamics.

8.6 EFFECT OF FEEDBACK ON DISTURBANCES

Feedback provided in control system partially eliminates the disturbance signal in the system. The disturbance signals cause the system to generate inaccurate output. Examples of disturbance signals are thermal noise in amplifier tubes, wind gusts on antenna of radar systems, high frequency noise in electronic application, distortion in non-linear systems. Disturbance may be introduced in the forward path, feedback path or output of the system.

1. **Disturbance in the forward path:** The disturbances in the forward path of control systems may be produced either due to the properties of elements present in the forward path or can be due to the surrounding condition changes of forward path elements. Figure 8.5 illustrates the disturbance signal $T_d(s)$ introduced in the forward path.

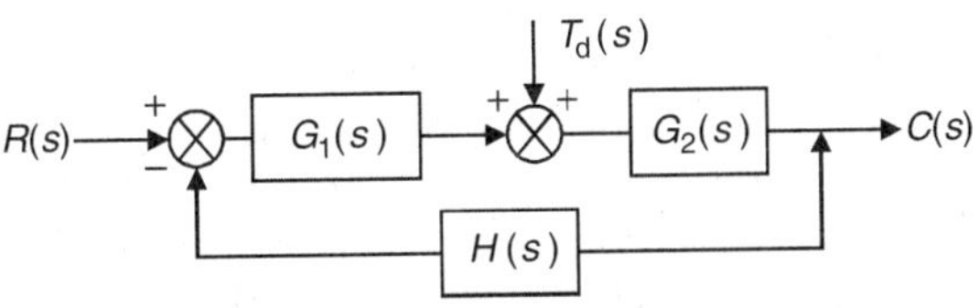

Figure 8.5 Disturbances in forward path

To get the effect of disturbance on the output, let us substitute $R(s) = 0$; therefore, Figure 8.5 becomes

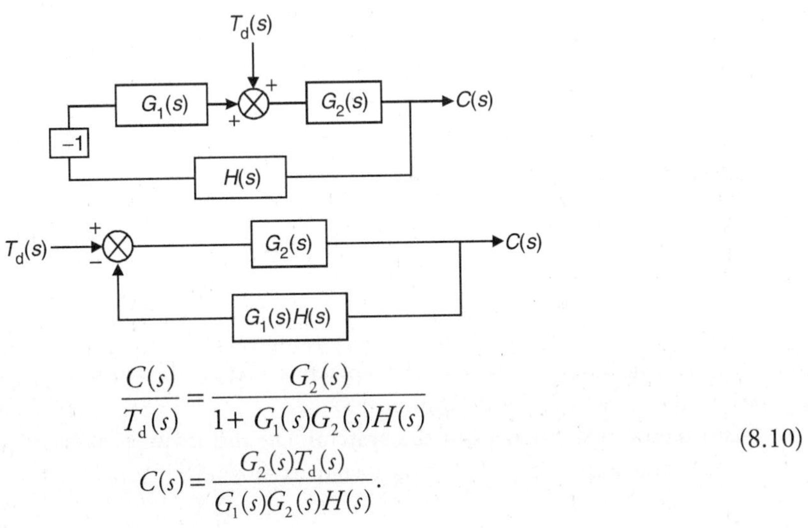

$$\frac{C(s)}{T_d(s)} = \frac{G_2(s)}{1 + G_1(s)G_2(s)H(s)} \tag{8.10}$$

$$C(s) = \frac{G_2(s)T_d(s)}{G_1(s)G_2(s)H(s)}.$$

If $G_1(s)G_2(s)H(s) >> 1$, then we have from (8.10),

$$C(s) = \frac{G_2(s)T_d(s)}{G_1(s)G_2(s)H(s)}$$

$$C(s) = \frac{T_d(s)}{G_1(s)H(s)}. \tag{8.11}$$

From (8.11), it is clear that if $G_1(s)$ is made very large, the effect of disturbance $T_d(s)$ on the output will be very small. Therefore, it is always desired to have large $G_1(s)$, so that the output may not change due to disturbances in the forward path of the system.

2. **Disturbance in the feedback path:** These disturbances are produced due to changes in the feedback path elements like sensing devices and illustrated as shown in Figure 8.6.

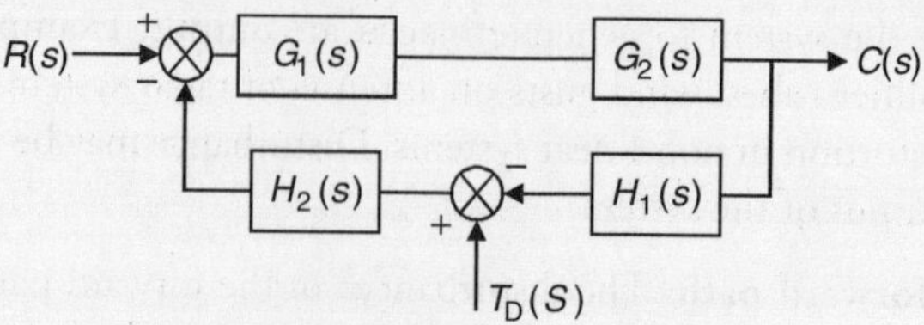

Figure 8.6 Disturbance in feedback loop

To get the effect of $T_d(s)$ on $C(s)$, let the input $R(s) = 0$. Therefore, the above figure can be redrawn as

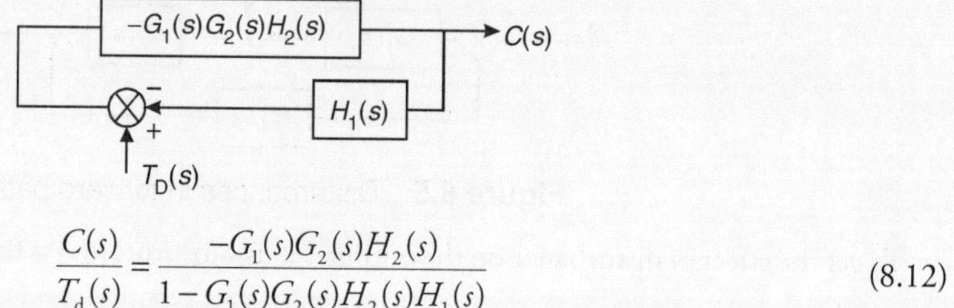

$$\frac{C(s)}{T_d(s)} = \frac{-G_1(s)G_2(s)H_2(s)}{1 - G_1(s)G_2(s)H_2(s)H_1(s)} \tag{8.12}$$

By assuming $G_1(s)G_2(s)H_1(s)H_2(s) >> 1$, we get

$$\frac{C(s)}{T_d(s)} = \frac{-G_1(s)G_2(s)H_2(s)}{-G_1(s)G_2(s)H_2(s)H_1(s)} = \frac{1}{H_1(s)}$$

$$C(s) = \frac{1}{H_1(s)} T_d(s). \tag{8.13}$$

Therefore, by making proper value of $H_1(s)$, the effect of disturbance in feedback path $T_d(s)$ can be reduced at the output of the system. In practice, $H_1(s)$is the optimum value for which the signal-to-noise ratio in the sensor is maximum.

3. **Disturbance at the output of the system:** The disturbance at the output of the system is introduced at the output terminals of the system as shown in Figure 8.7.

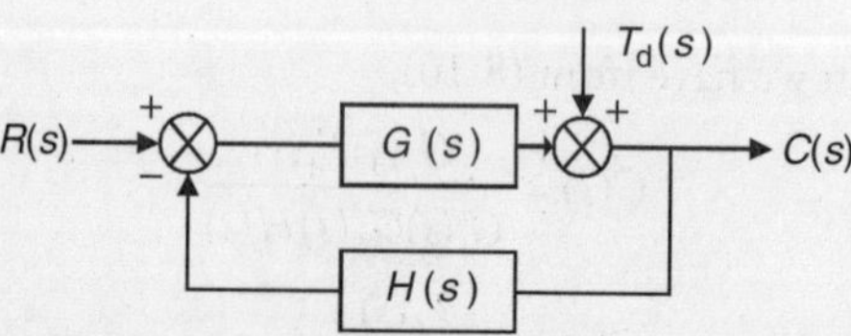

Figure 8.7 Disturbance at the output

To study the effect of disturbance on output only, let us substitute $R(s) = 0$, and the above figure can be redrawn as

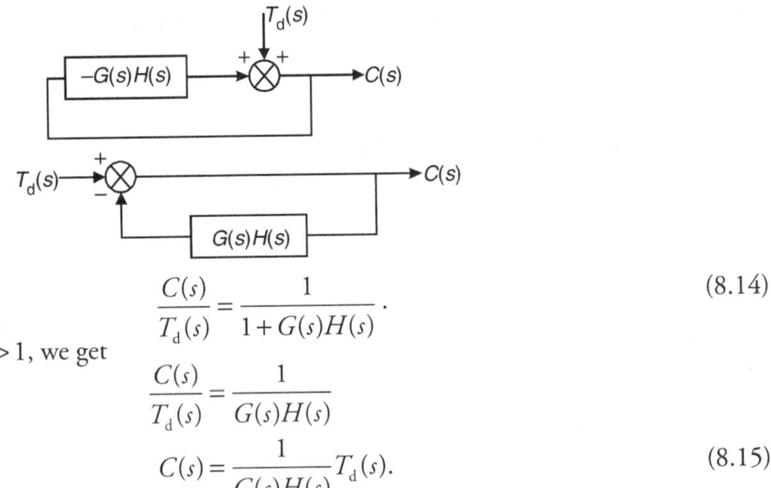

$$\frac{C(s)}{T_d(s)} = \frac{1}{1+G(s)H(s)}. \tag{8.14}$$

By assuming $G(s)H(s) >> 1$, we get

$$\frac{C(s)}{T_d(s)} = \frac{1}{G(s)H(s)}$$

$$C(s) = \frac{1}{G(s)H(s)} T_d(s). \tag{8.15}$$

Therefore, changing the value of $G(s)$ or $H(s)$ or both, the effect of disturbance on output can be reduced.

Problem-8.1: Determine the sensitivity of the closed loop system shown below at w = 1 rad/sw.r.t

(a) Forward path t.f.
(b) Feedback path t.f.

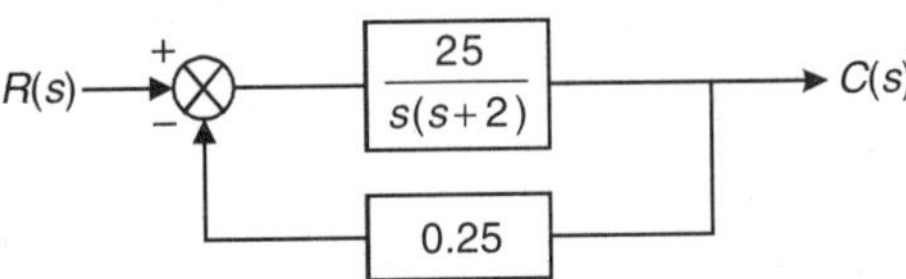

Solution:
From the above block diagram, we know that

$$G(s) = \frac{25}{s(s+2)}, H(s) = 0.25.$$

Sensitivity of closed loop system w.r.t forward path t.f.

The sensitivity of overall t.f. (T) w.r.t. forward path t.f. (G) is given as

$$S_G^T = \frac{1}{1+G(s)H(s)} = \frac{1}{1+\frac{25}{s(s+2)}\times 0.25} = \frac{s^2+2s}{s^2+2s+6.25}.$$

Let $\quad s = j\omega.$

$$S_G^T = \frac{-\omega^2+2j\omega}{-\omega^2+2j\omega+6.25}$$

Since $\quad \omega = 1\,\text{rad/s}$

$$S_G^T = \frac{-1+2j}{-1+2j+6.25} = \frac{-1+2j}{5.25+2j}$$

$\Rightarrow \quad |S_G^T| \approx 0.398.$

Sensitivity of closed loop system w.r.t feedback path t.f.

The sensitivity of overall t.f. (T) w.r.t. feedback path t.f. (H) is given as

$$S_{\mathrm{H}}^{T} = -\frac{G(s)H(s)}{1+G(s)H(s)} = -\frac{\dfrac{25}{s(s+2)}\times 0.25}{1+\dfrac{25}{s(s+2)}\times 0.25} = -\frac{6.25}{s^2+2s+6.25}.$$

Let $\qquad s = j\omega$

$$S_{\mathrm{H}}^{T} = \frac{-6.25}{-\omega^2+2j\omega+6.25}.$$

Since $\qquad \omega = 1\,\mathrm{rad/s}$,

$$S_{\mathrm{H}}^{T} = \frac{-6.25}{-1+2j+6.25} = \frac{-6.25}{5.25+2j}$$

$\Rightarrow \qquad |S_{\mathrm{G}}^{T}| \approx -1.11.$

***Problem-8.2*:** Determine the sensitivity of the closed-loop t.f. $T(s) = \dfrac{C(s)}{R(s)}$ for variations in parameter K at $\omega = 5\mathrm{rad/s}$. Assume that the normal value of K is 1.

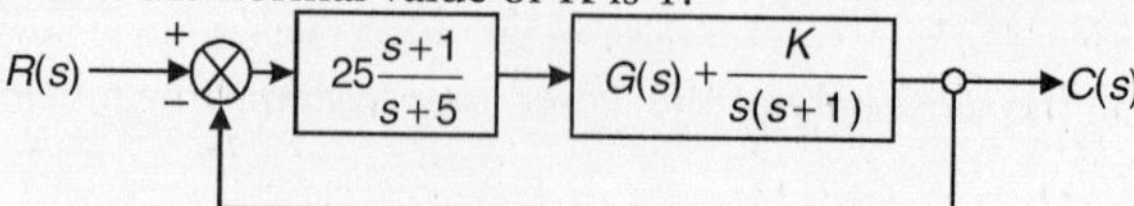

***Solution*:**

From the given figure, we have

$$G' = \frac{25K}{s\,(s+5)}$$

$$T(s) = \frac{C(s)}{R(s)} = \frac{G'(s)}{1+G'(s)} = \frac{25K}{s^2+5s+25K}.$$

$$\frac{\partial T}{\partial K} = \frac{(s^2+5s+25K)25 - 25K \times 25}{(s^2+5s+25K)^2}$$

$$= \frac{(s^2+5s)25}{(s^2+5s+25K)^2}$$

Therefore, $\qquad S_{\mathrm{K}}^{T} = \dfrac{\partial T}{\partial K}\times\dfrac{K}{T} = \dfrac{s(s+5)}{s^2+5s+25K}.$

Since the normal value of $K = 1$, we have

$$S_{\mathrm{K}}^{T} = \frac{s(s+5)}{s^2+5s+25}.$$

Let $\qquad s = j\omega.$

$$S_{\mathrm{K}}^{T} = \frac{-\omega^2+5j\omega}{-\omega^2+5j\omega+25}$$

Since $\qquad \omega = 5\,\mathrm{rad/s}$,

$$S_{\mathrm{K}}^{T} = \frac{-25+25j}{-25+25j+25} = -1+j$$

$\Rightarrow \qquad |S_{\mathrm{K}}^{T}| \approx 1.41.$

At a particular frequency $\omega = 5\mathrm{rad/s}$, the magnitude of sensitivity is $|S_{\mathrm{K}}^{T}| = 1.41$.

8.7 CONTROLLERS

A controller is a device introduced in the system to modify the error signal and to produce a control signal. The manner in which the controller produces the control signal is called control action. It can be introduced in the feedback or forward path of the system, which then controls the steady-state and transient response as per the requirement. The controller may be electrical, hydraulic or pneumatic, depending on the nature of the signal and system.

Depending on the control action provided, the controllers can be classified as follows:

1. Proportional (P) controller.
2. Proportional–derivative (PD) controller.
3. Proportional–integral (PI) controller.
4. Proportional–integral–derivative (PID) controller.
5. Rate feedback controller (output derivative controller).

8.7.1 Proportional Controller (P Controller)

In most of the practical systems, controller input is proportional to the error generated. Such systems are called proportional controllers (P-type controller).

Consider a second-order system where controller input is the error itself and the proportional constant is k = 1, as shown in Figure 8.8.

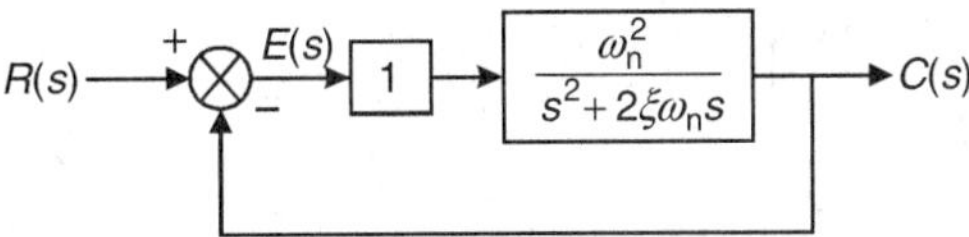

Figure 8.8 P-controller

From the above block diagram, we have

$$G(s) = \frac{\omega_n^2}{s(s+2\xi\omega_n)}, H(s) = 1$$

$$T(s) = \frac{C(s)}{R(s)} = \frac{\omega_n^2}{s^2 + 2\xi\omega_n s + \omega_n^2}. \quad (8.16)$$

To obtain steady-state error, we get

$$K_p = \underset{S\to 0}{\text{Lt}}\, G(s) = \infty$$

$$K_v = \underset{S\to 0}{\text{Lt}}\, sG(s) = \frac{\omega_n}{2\xi}.$$

Now, if transient response is to be improved, damping ratio must be changed.

In general, good time response demands

1. less settling time
2. less rise time
3. less overshoot
4. smallest steady-state error.

An increase in K_v results in an increase in the system gain and a decrease in steady-state error; however, due to high gain, settling time and peak overshoot also increase. This may lead to instability of the system. Therefore, a compromise is made to keep steady-state error and overshoot within acceptable limits.

8.7.2 Proportional–Derivative (PD) Controller

The PD controller produces an output signal consisting of two terms: one proportional to error signal and the other proportional to the derivative of the error signal.

$$\text{Input to the controller} = ke(t) + k_d \frac{de(t)}{dt}.$$

After taking Laplace transforms to the equation, we get

$$\text{Input to the controller} = kE(s) + k_d sE(s) = (k + sk_d)E(s).$$

Therefore, the t.f. of the PD controller $= k + sk_d$.

The block diagram representation of an unity feedback control system with PD controller is shown in Figure 8.9.

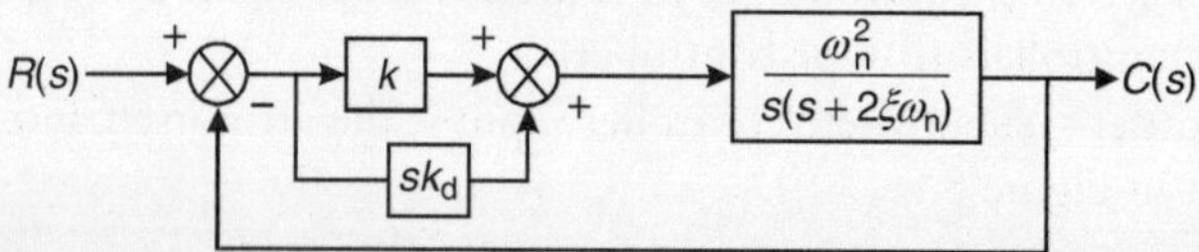

Figure 8.9 PD controller

By assuming $k = 1$, we get

$$G(s) = \frac{(1 + sk_d)\omega_n^2}{s(s + 2\xi\omega_n)}, H(s) = 1$$

$$T(s) = \frac{C(s)}{R(s)} = \frac{(1 + sk_d)\omega_n^2}{s^2 + s(2\xi\omega_n + \omega_n^2 k_d) + \omega_n^2}. \quad (8.17)$$

By comparing denominator with the standard form, we have

$$2\xi'\omega_n = 2\xi'\omega_n + \omega_n^2 k_d$$

$$\Rightarrow \quad \xi' = \xi + \frac{\omega_n k_d}{2}. \quad (8.18)$$

Because of this controller, damping ratio increases by a factor $\frac{\omega_n k_d}{2}$, while

$$K_p = \underset{s \to 0}{\text{Lt}}\, G(s) = \infty$$

$$K_v = \underset{s \to 0}{\text{Lt}}\, sG(s) = \frac{\omega_n}{2\xi}.$$

As there is no change in error coefficients, error also will remain same. Hence, PD controller has following effects on the system:

1. It increases damping ratio.
2. Type of the system remains unchanged.
3. ω_n for system remains unchanged.

4. It reduces peak overshot.
5. It reduces settling time.
6. Steady-state error remains unchanged.

In general, this controller improves the transient part without affecting the steady-state part.

8.7.3 Proportional–Integral (PI) Controller

The PI controller produces an output signal consisting of two terms: one proportional to the error signal and the other proportional to the integral of the error signal.

Input to the controller $= ke(t) + k_i \int e(t)dt$.

After taking Laplace transforms to the equation, we have

Input to the controller $= kE(s) + \frac{k_1}{s}E(s) = \left(k + \frac{k_i}{s}\right)E(s)$.

Therefore, the t.f. of the PI controller $= \left(k + \frac{k_i}{s}\right)$.

The block diagram representation of an unity feedback control system with PI controller is shown in Figure 8.10.

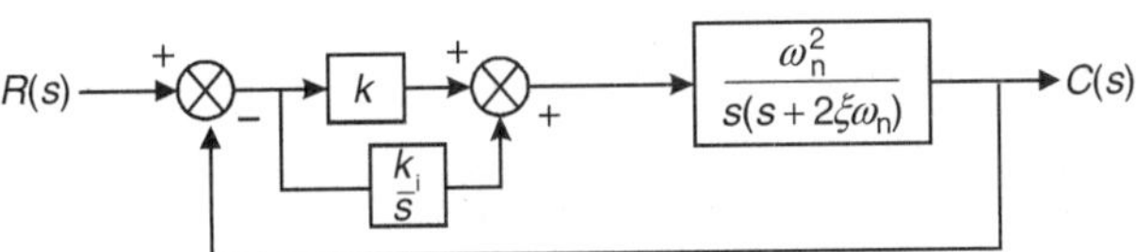

Figure 8.10 PI controller

By assuming $k = 1$, we get

$$G(s) = \frac{\left(1 + \frac{k_i}{s}\right)\omega_n^2}{s^2(s + 2\xi\omega_n)} = \frac{(k_i + s)\omega_n^2}{s^2(s + 2\xi\omega_n)}, H(s) = 1$$

$$T(s) = \frac{C(s)}{R(s)} = \frac{(k_i + s)\omega_n^2}{s^3 + 2\xi\omega_n s^2 + s\omega_n^2 + k_i\omega_n^2}. \tag{8.19}$$

This means that it becomes a third-order system.

As order increases by one, the system becomes relatively less stable as k_i must be designed in such a way that the system will remain in stable condition. A second-order system is always stable. Hence, transient response gets affected badly if controller is not designed properlyw.

$$K_p = \underset{s\to 0}{\text{Lt}}\, G(s) = \infty, \quad e_{ss} = 0$$

While

$$K_v = \underset{s\to 0}{\text{Lt}}\, sG(s) = \infty, \quad e_{ss} = 0$$

Hence, as type is increased by one, error becomes zero for ramp type of inputs. In other words, steady-state of the system gets improved and the system becomes more accurate in nature.

Hence, PI controller has the following effects:

1. It increases the 'order' of the system.
2. It increases the 'type' of the system.
3. Design of k_i must be proper to maintain stability of the system. Therefore, it makes the system relatively less stable.
4. Steady-state error reduces tremendously for same type of inputs.

In general, this controller improves steady-state part without affecting the transient part.

8.7.4 Proportional–Integral–Derivative (PID) Controller

An integral controller eliminates steady-state error due to a velocity input, but its effect on dynamic response is difficult to predict as the system order increases to three. Further, we have seen that a derivative term in the forward path improves the damping in the system. Hence, a suitable combination of integral and derivative controllers results in a proportional, integral and derivative controller, usually called PID controller.

The t.f. of the PID controller is given as

$$G_e(s) = k + sk_d + \frac{k_i}{s}.$$

The overall forward path t.f. is given by assuming $k = 1$

$$G_o(s) = \frac{\left(1 + sk_d + \frac{k_i}{s}\right)\omega_n^2}{s(s + 2\xi\omega_n)}$$

and the overall t.f. is given as

$$T(s) = \frac{C(s)}{R(s)} = \frac{(k_i + s + s^2 k_d)\omega_n^2}{s^3 + s^2\omega_n(2\xi + k_d\omega_n) + s\omega_n^2 + k_i\omega_n^2}. \tag{8.20}$$

Proper choice of k, k_d and k_i results in satisfactory transient and steady-state responses. The process of choosing proper k, k_d and k_i for a given system is known as tuning of a PID controller. The block diagram representation of an unity feedback control system with PID controller is shown in Figure 8.11.

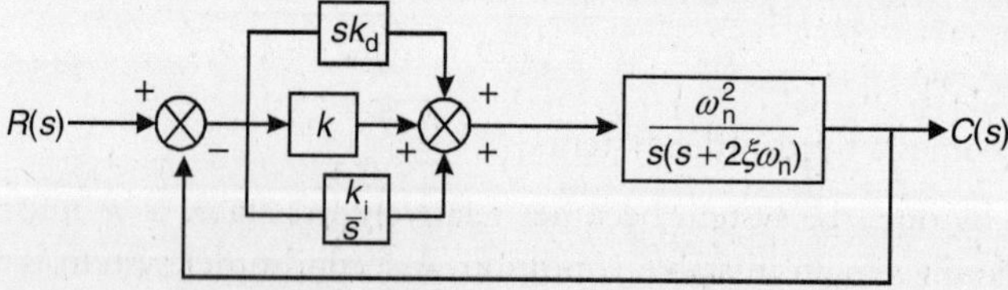

Figure 8.11 PID controller

8.7.5 Output Derivative Controller

This is achieved by feeding back the derivative of output signal internally using a tachogenerator and comparing with signal proportional to the error, as shown. This is called minor-loop feedback compensation.

Input to the controller $ke(t) - k_t \dfrac{dc(t)}{dt}$

After taking Laplace transforms of the above-mentioned equation, we have

Input to the controller $= kE(s) - sk_t C(s)$.

This can be realised as shown in Figure 8.12.

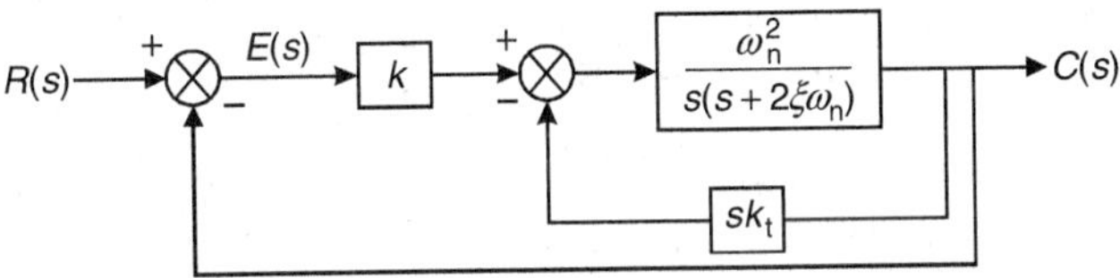

Figure 8.12 Output derivative controller

By assuming $k = 1$, let us study its effect on same system, which is considered earlier with

$$G(s) = \frac{\omega_n^2}{s(s+2\xi\omega_n)}.$$

Now, new

$$G(s) = \frac{\dfrac{\omega_n^2}{s(s+2\xi\omega_n)}}{1+\dfrac{\omega_n^2}{s(s+2\xi\omega_n)}sk_t} = \frac{\omega_n^2}{s(s+2\xi\omega_n + k_t\omega_n^2)}$$

and

$$T(s) = \frac{C(s)}{R(s)} = \frac{\omega_n^2}{s^2 + [2\xi\omega_n + \omega_n^2]s + K_t\omega_n^2} \quad . \tag{8.21}$$

Therefore, by comparing the denominator with the standard form, it is clear that there is no change in ω_n.

$$\therefore 2\xi'\,\omega_n = 2\xi\omega_n + \omega_n^2 k_t$$

$$\Rightarrow \qquad \xi' = \xi + \frac{\omega_n k_t}{2} \tag{8.22}$$

where ξ' is the new damping factor and ξ is the old damping factor.

Therefore, there is improvement in damping ratio, and hence, improvement in transient response when

$$K_p = \lim_{s\to 0} G(s)H(s) = \infty$$

$$K_v = \lim_{s\to 0} sG(s)H(s) = \lim_{s\to 0}\frac{s\omega_n^2}{s[s+2\xi\omega_n + k_t\omega_n^2]} = \frac{\omega_n}{2\xi + \omega_n k_t}.$$

As k_t is positive, K_v decreases; and as $e_{ss} \propto \dfrac{1}{K_v}$, the steady-state error increases.

Hence, this controller affects the steady state also. Hence, k_t must be designed in such a way that improvement in ξ will not cause a large increase in steady-state error. k_t should be such that error will be always within specified limits. However, for step input, as type of system is '1' before and after implementation of the controller, the error is always zero. Thus, we can say that the derivative output controller behaves exactly similar to the PD type of controller because of improvement in transient response of the system. However, the rise time for this controller is small, that is, response is fast in comparison with rate feedback controller because of addition of zero in the forward path.

Problem-8.3: The following figure shows the PD controller used for a system. Determine the value of T_d so that the system will be critically damped. Calculate its settling time.

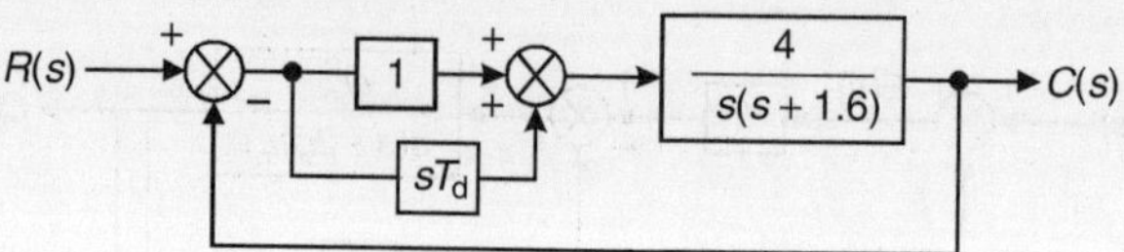

Solution:

From the given block diagram, we have

$$G(s) = \frac{(1+sT_d)4}{s(s+1.6)}, H(s) = 1.$$

The characteristic equation of the system is

$$1 + G(s)H(s) = 0$$

$$1 + \frac{(1+sT_d)4}{s(s+1.6)} \times 1 = 0.$$

$$s^2 + (1.6 + 4T_d)s + 4 = 0$$

By comparing the equation with $s^2 + 2\xi\omega_n s + \omega_n^2 = 0$, we have

$$\omega_n^2 = 4 \Rightarrow \omega_n = 2\,\text{rad/s}$$

$$2\xi\omega_n = 1.6 + 4T_d$$

$$\Rightarrow \quad \xi = \frac{1.6 + 4T_d}{4}.$$

However, as per the given data, the system is critically damped

$$\therefore \quad \xi = 1 = \frac{1.6 + 4T_d}{4} \Rightarrow T_d = 0.6.$$

Settling time $$T_s = \frac{4}{\xi\omega_n} = \frac{4}{1 \times 2} = 2\,\text{s}.$$

Problem-8.4: The feedback system is shown in the figure.

(a) In the absence of derivative feedback controller (K_0), find ξ and ω_n.
(b) Find K_0, if ξ is to be modified to 0.5 by the use of controller.

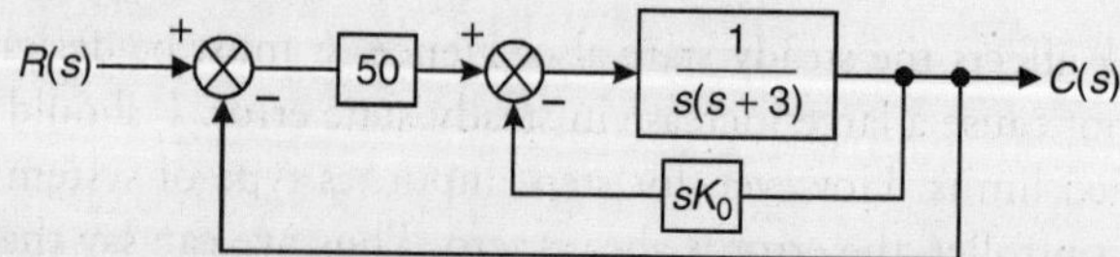

Solution:

(a) **In the absence of derivative feedback controller:**

From the given block diagram,

$$G(s) = \frac{50}{s(s+3)},\ H(s) = 1$$

The characteristic equation of the system is

$$1 + G(s)H(s) = 0\cdot$$

$$1 + \frac{50}{s(s+3)} = 0$$

$$s^2 + 3s + 50 = 0$$

By comparing the equation with $s^2 + 2\xi\omega_n s + \omega_n^2 = 0$, we have

$$\omega_n^2 = 50 \Rightarrow \omega_n = 7.071\,\text{rad/s}.$$

$$2\xi\omega_n = 3 \Rightarrow \xi = 0.2121$$

(b) **With the controller:**

From the given block diagram, we have

$$G(s) = 50 \times \frac{1/s(s+3)}{1 + sK_0/s(s+3)} = \frac{50}{s^2 + 3s + sK_0},\ H(s) = 1.$$

The characteristic equation of the system is

$$1 + G(s)H(s) = 0$$

$$1 + \frac{50}{s^2 + s(K_0 + 3)} \times 1 = 0$$

$$s^2 + (3 + K_0)s + 50 = 0$$

By comparing the equation with $s^2 + 2\xi\omega_n s + \omega_n^2 = 0$, we have

$$\omega_n^2 = 50 \Rightarrow \omega_n = 7.071\ \text{rad/s}$$

$$2\xi\omega_n = 3 + K_0$$

$$\Rightarrow \qquad \xi = \frac{3 + K_0}{2 \times 7.071}.$$

As per the given data, $\xi = 0.5$

$$\frac{3 + K_0}{2 \times 7.071} = 0.5 \Rightarrow K_0 = 4.071$$

ADDITIONAL SOLVED PROBLEMS

ASP-1: Determine the sensitivity of the overall gain for the system shown in the figure and calculate the value of K such that the sensitivity is 0.4 under steady-state conditions.

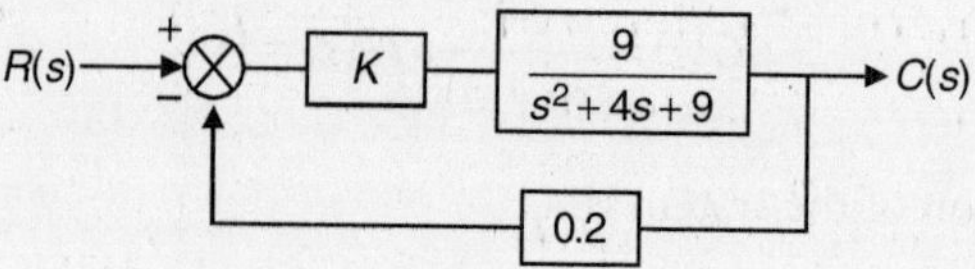

Solution:

$$G(s) = \frac{9K}{s^2 + 4s + 9}, \ H(s) = 0.2$$

$$S_G^T = \frac{1}{1 + G(s)H(s)} = 0.4$$

$$\Rightarrow \frac{1}{1 + \frac{9K}{s^2 + 4s + 9} \times 0.2} = 0.4$$

$$\Rightarrow \frac{s^2 + 4s + 9}{s^2 + 4s + 9 + 1.8K} = 0.4$$

Substitute $s = j\omega = j1$ (let $\omega = 1$ rad/s)

$$\Rightarrow \frac{-1 + j4 + 9}{-1 + j4 + 9 + 1.8K} = 0.4$$

$$\Rightarrow \frac{8 + j4}{(8 + 1.8K) + j4} = 0.4$$

$$\Rightarrow \frac{\sqrt{8^2 + 4^2}}{\sqrt{(8 + 1.8K)^2 + 16}} = 0.4$$

$$\Rightarrow (8 + 1.8K)^2 + 16 = \frac{80}{0.16} = 500.$$

$$(8 + 1.8K)^2 = 500 - 16 = 484$$

$$8 + 1.8K = 22$$

$$K = \frac{22 - 8}{1.8} = 7.8$$

ASP-2: For the feedback system as shown in the figure, the normal value of the process parameter K is 0.5. Find the sensitivity of the t.f. $T(s)$ w.r.t parameter $K = 0.5$ at $\omega = 2$ rad/s.

Solution:
The t.f. of the system is given as

$$T(s)=\frac{C(s)}{R(s)}=\frac{25\times\frac{(s+1)}{s+5}\times\frac{K}{s(s+1)}}{1+\frac{25(s+1)}{s+5}\times\frac{K}{s(s+1)}}$$

$$\Rightarrow \qquad T(s)=\frac{25K}{s^2+5s+25K}$$

$$S_K^T=\frac{K}{T}\frac{\partial T}{\partial K}$$

$$\frac{\partial T}{\partial K}=\frac{(s^2+5s\ +25K)25-25K\times 25}{(s^2+5s+25K^2)}$$

$$=\frac{25(s^2+5s)}{(s^2+5s+25K)^2}$$

$$S_K^T=\frac{K}{\frac{25K}{s^2+5s+25K}}\cdot\frac{25(s^2+5s)}{(s^2+5s+25K)^2}$$

$$=\frac{s^2+5s}{s^2+5s+25K}$$

Substitute $s=j\omega=j2,\ K=0.5$

$$S_K^T=\frac{-4+j10}{-4+j10+12.5}=\frac{-4+j10}{8.5+j10}=\frac{\sqrt{4^2+10^2}}{\sqrt{10^2+8.5^2}}=\frac{10.77}{13.12}=0.8206$$

$$\Rightarrow \qquad S_K^T=0.8206\cdot$$

ASP-3: A unity feedback system has an open-loop t.f. $G(s)=\dfrac{25}{s(s+8)}$. Determine its damping ratio, peak overshoot and time required to reach the peak output. Now, a derivative component having t.f. of $s/10$ is introduced in the system. Discuss its effect on the values obtained above.

Solution:
The open-loop t.f. is

$$G(s)=\frac{25}{s(s+8)}$$

$$\frac{C(s)}{R(s)}=\frac{G(s)}{1+G(s)H(s)}=\frac{25/s(s+8)}{1+25/s(s+8)}=\frac{25}{s^2+8s+25}.$$

The characteristics equation is $s^2+8s+25=0$

By comparing the equation with $s^2 + 2\xi\omega_n s + \omega_n^2 = 0$, we have

$$\omega_n = 5, 2\xi\omega_n = 8 \Rightarrow \xi = 8/2\times 5 = 0.8$$

$$\omega_d = \omega_n\sqrt{1-\xi^2} = 5\sqrt{1-0.64} = 3\text{rad/s}$$

Peak time $$T_p = \frac{\pi}{\omega_d} = \frac{\pi}{3} = 1.047\,\text{s}$$

Peak overshoot $\%M_p = e^{\frac{\xi\pi}{\sqrt{\sqrt{1-\xi^2}}}} \times 100 = e^{\frac{0.8\pi}{\sqrt{1-0.64}}} \times 100 = 1.52$

When the derivative control is introduced in the feedback, we get

$$H(s) = 1 + s/10 = (s+10)/10$$

$$\frac{C(s)}{R(s)} = \frac{G(s)}{1+G(s)H(s)} = \frac{25/s(s+8)}{1+25(s+10)/10s(s+8)} = \frac{25}{s^2+10.5s+25}.$$

The characteristics equation is $s^2 + 10.5s + 25 = 0$.

By comparing the equation with $s^2 + 2\xi\omega_n s + \omega_n^2 = 0$, we get

$$\omega_n = 5,\ 2\xi\omega_n = 10.5 \Rightarrow \xi = 10.5/2\times 5 = 1.05.$$

Since the system becomes overdamped, the peak overshoot will not be attained. Thus, the effect of derivative control in a unity feedback system can be enumerated as follows:

1. There is no change in natural frequency, but the damped frequency will change as per the damping ratio.
2. There is an increase in the value of damping ratio.
3. The time to reach the peak overshoot is increased.
4. The value of percentage peak overshoot is reduced

REVIEW QUESTIONS

1. What is feedback? Explain the effects of feedback.
2. What is sensitivity function? Explain it with respect to open loop and closed loop systems.
3. Explain the linearising effect of feedback.
4. Explain feedback characteristics of a closed-loop system.
5. Explain the effect of feedback on stability.
6. Explain the effect of feedback on bandwidth.
7. Explain reduction of parameter variations using feedback systems.
8. Why derivative controller is not used in control systems? What is the effect of PI controller on the system performance?
9. Discuss the advantages and disadvantages of proportional, proportional–derivative, proportional–integral and proportional–integral–derivative control system.
10. What are the types of controllers that are used in a closed-loop system? Explain them.

11. Discuss the effect on the performance of a second-order control system of
    (i) Proportional control
    (ii) Integral control
    (iii) Rate feedback control.

EXERCISE PROBLEMS

1. Find the magnitude of sensitivity $\left|S_K^T\right|$ at $\omega = 5$ rad/s, $K=1$ for the systems, as shown in Figures 1 and 2

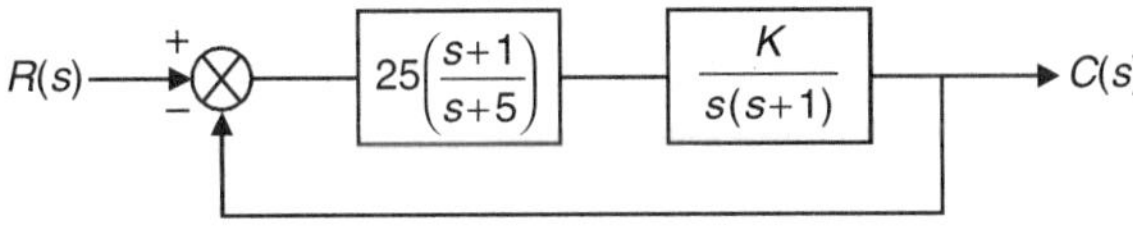

Figure 1

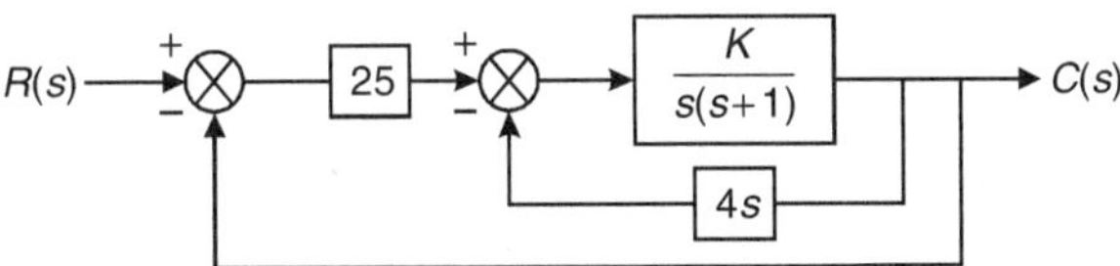

Figure 2

2. In the closed-loop system as shown in figure, find sensitivity $S_{G_1}^T$ and $S_{G_2}^T$.

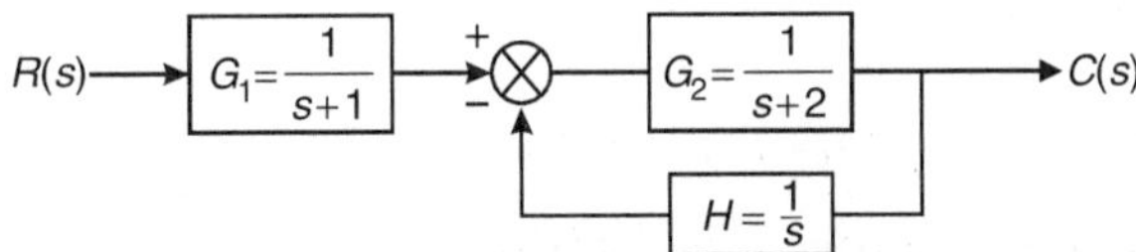

3. The system shown in figure uses a rate feedback controller. Determine the tachometer constant K_t so as to obtain the damping ration as 0.5. Calculate the corresponding ω_d, T_p, T_s and M_p.

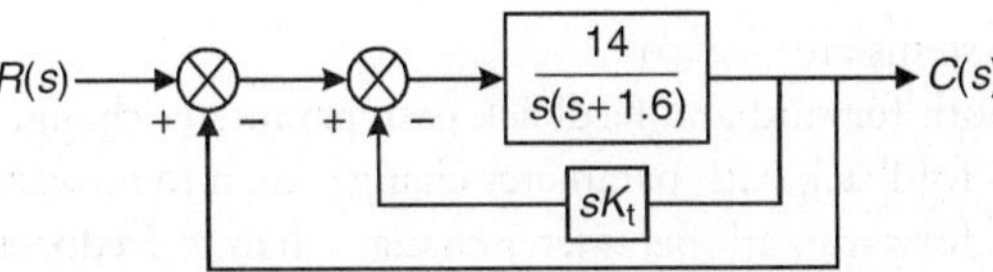

4. For the system as shown in the figure, determine the sensitivity of the closed-loop t.f. w.r.t. G, H and K. Assign $\omega = 2$ rad/s.

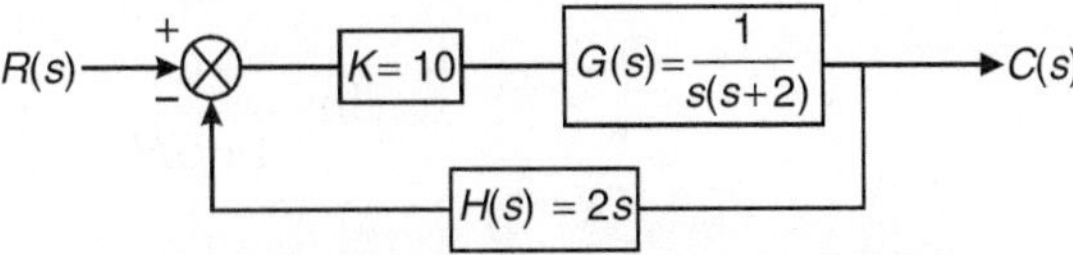

5. Calculate the sensitivity of the closed-loop system as shown in figure w.r.t. forward path t.f. and feedback path t.f. at $\omega = 1.3$ rad/s.

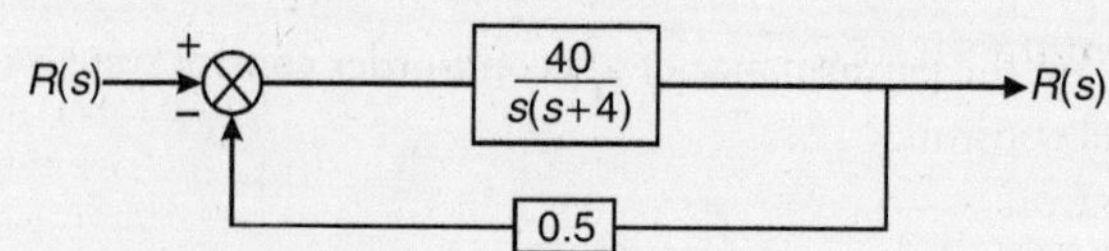

6. For the following system, determine M_p and T_s when excited by unit-step input. If for the same system, a PD controller having constant $T_d = 1/30$ is used in forward path, determine the new values of damping ration, M_p and T_s.

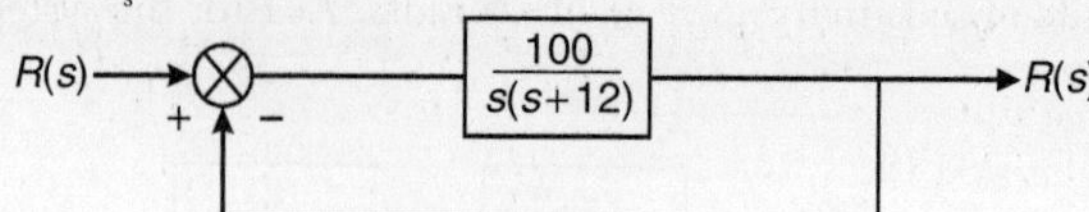

ANSWERS

1. $|S_K^T| = 1.414, |S_K^T| = 1.414$

2. $S_G^T = 1, S_G^T = \frac{(s^2+2s)}{s^2+2s+1}$

3. $K_t = 0.1529, \omega_d = 3.24$ rad/s, $T_p = 0.9695$ s, $T_s = 2.138$ s, $M_p = 16.3\%$

4. $S_G^T = 0.128,\ S_H^T = 0.905,\ S_K^T = 0.128$

5. $S_G^T = 0.287,\ S_H^T = -1.05$

6. (i) $\omega_d = 8$ rad/s, $M_p = 9.47\%$, $T_s = 0.667$ s
   (ii) $\xi = 0.7665$, $M_p = 2.353\%$, $T_s = 0.5218$ s

OBJECTIVE TYPE QUESTIONS

1. Feedback control systems are
   (a) insensitive to both forward and feedback path parameter changes
   (b) less sensitive to feedback path parameter changes than to forward path parameter changes
   (c) less sensitive to forward path parameter changes than to feedback path parameter changes
   (d) equally sensitive to both forward and feedback path parameter changes.

2. The sensitivity S_G^T of a system with the t.f. $T = \frac{G}{1+GH}$ is given as__________

   (a) $\frac{G}{1+GH}$ (b) $\frac{1}{1+GH}$

   (c) $\frac{-GH}{1+GH}$ (d) None

3. The sensitivity S_H^T of a system with the t.f. $T = \frac{G}{1+GH}$ is given as__________

(a) $\frac{G}{1+GH}$ (b) $\frac{1}{1+GH}$

(c) $\frac{-GH}{1+GH}$ (d) None

4. The main drawback of the feedback system is
(a) inaccuracy (b) inefficient
(c) insensitivity (d) instability

5. A system sensitivity can be expressed as
(a) the ratio of the change in the system t.f. to the change of process t.f. (or parameter) for a small incremental change
(b) product of the change in the system t.f. to the change of process t.f. (or parameter) for a small incremental change
(c) the effect of change on performance of the system as a consequence of change of parameters
(d) a number that signifies the sensitivity with respect to the response of the system to any change

6. The system sensitivity to parameter variations
(a) increases (b) decreases
(c) becomes zero (d) none

7. Which of the following system provides excellent transient and steady-state response?
(a) P controller (b) PI controller
(c) PD controller (d) PID controller

8. PI controller, in general, improves
(a) transient response (b) steady-state response
(c) both (d) None

9. PD controller, in general, improves
(a) transient response (b) steady-state response
(c) both (d) None

10. The effect of introducing the feedback can be expressed as
(a) increasing sensitivity, improving transient response and removing all possibilities of the system becoming unstable
(b) reducing sensitivity, empowering transient response and increasing the effect of transient response
(c) reducing sensitivity, improving transient response, minimising effect of disturbance signals, reducing the gain of the system and introducing the possibility of instability
(d) increasing sensitivity, improving transient response, minimising effects of disturbance signals, increasing the gain of the systems and introducing the possibility of instability.

11. The rate feedback controller improves
(a) transient response (b) steady-state response
(c) both (d) None

12. Due to feedback, the steady-state error
(a) gets increased (b) gets decreased
(c) remains constant (d) remains independent of parameter variations

ANSWERS

1. c **2.** b **3.** c **4.** c **5.** c **6.** b **7.** d **8.** b **9.** a **10.** c
11. a **12.** b

CHAPTER

9 Stability of Systems

Contents

9.1 INTRODUCTION

The term 'stability' refers to the stable working condition of a control system. Every working system is designed to be stable. In a stable system, the response or output is predictable, finite and stable for a given input. Stability is an important property that a system is required to possess. It is not only essential to design a system with a desired response but also necessary to design a system that is stable.

9.2 CONCEPT OF STABILITY

1. **Stable system:** In general, there are two concepts of stability. The first concept involves the Bounded-Input Bounded-Output (BIBO) stability. According to this concept, when the system is excited by a bounded input, the output is also bounded and controllable. According to the second concept, when the system does not have an input, the output must end to zero irrespective of the initial conditions.

2. **Unstable system:** A linear time-invariant (LTI) system is said to be unstable, if

    1. For a bounded input, it produces unbounded output.
    2. In the absence of input, output does not return to zero, that is, it shows certain output without input.

3. **Conditionally stable system:** A linear time-invariant system is said to be conditionally stable for a certain condition of a particular parameter of the system, and its output is a bounded one.

The system is said to be unstable when that condition is violated and the output becomes unbounded. In other words, the stability of system depends on the condition of parameter of the system. Such a system is called conditionally stable system.

4. **Critically stable or marginally stable systems:** An LTI system is marginally stable, if the natural response remains constant or oscillates as time approaches infinity.
5. **Asymptotic stability:** Some systems, in practice, may get driven by the initial conditions, without any input applied to it. If the zero input response of a system subjected to finite initial conditions reaches zero as time t approaches infinity, the system is said to be zero input stable or asymptoticallystable.

   For example, let us consider a series RC circuit with capacitor initially charged to some voltage. This initial voltage is enough to start the operation of the system; further, without any external input, this voltage drives the current till the capacitor gets fully discharged. The stability related to such a system under zero input conditions but operated under initial conditions is called zero input stability of the system. However, the current through RC circuit reduces to zero as the capacitor gets fully discharged.
6. **Relative stability:** The system is said to be relatively more stable if its settling time is less than that of the other system. The settling time is inversely proportional to real parts of the roots. The settling time will be large for the roots located near the $j\omega$-axis. As roots or pair of complex conjugate roots move away from $j\omega$-axis, that is, towards the left half of the s-plane, settling time becomes less or small and the system becomes more and more stable. Hence, the relative stability of systems improve, as the closed loop-poles move away from the imaginary axis in the left half of s-plane.

9.3 EFFECT OF LOCATION OF POLES ON STABILITY

The location of poles has a direct effect on stability. The entire s-plane is taken into account and it is divided as follows

1. **Left Half-Plane (LHP):** The system is stable when the poles are located on
   (a) real axis and simple
   (b) real axis and multiple
   (c) complex conjugates and simple
2. **$j\omega$-axis:** The poles are located
   (a) on $j\omega$-axis and simple
   (b) on $j\omega$-axis and multiple
   (c) at origin
   (d) at origin and multiple

   Here, if the pole has zero real part and is not repeated, the system is said to be stable.
3. **Right Half-Plane (RHP):** In this case, the system is unstable when the poles are located
   (a) on real axis and simple
   (b) on real axis and multiple
   (c) complex conjugates and simple

S.No	T.F and location of poles	Impulse response	Stability condition
1	$G(s)=\frac{K}{s+a}$ 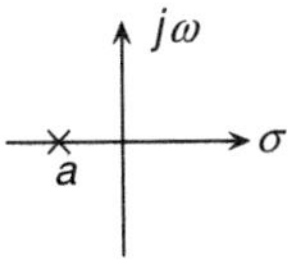 $j\omega$, σ, a Root on positive real axis	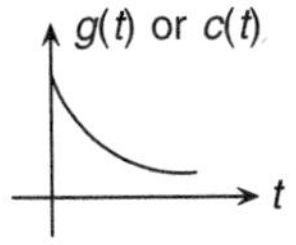 $c(t)=g(t)=L^{-1}\left(\frac{K}{s+a}\right)$ $=Ke^{-at}$ $g(t)$ or $c(t)$, t Response is exponentially decaying	Absolutely stable
2	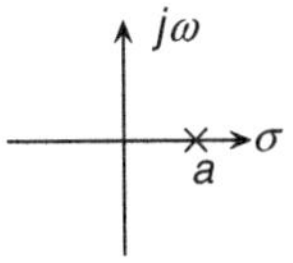 $G(s)=\frac{K}{s-a}$ $j\omega$, σ, a Root on positive real axis	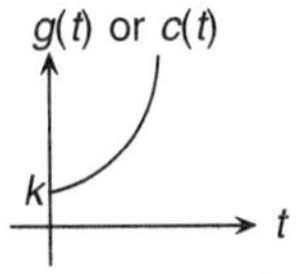 $c(t)=g(t)=L^{-1}\left(\frac{K}{s-a}\right)$ $=Ke^{at}$ $g(t)$ or $c(t)$, k, t Response is exponentially increasing	Unstable
3	$G(s)=\frac{K}{s+a+jb}+\frac{K}{s+a-jb}$ $j\omega$, jb, σ, $-a$, $-jb$ 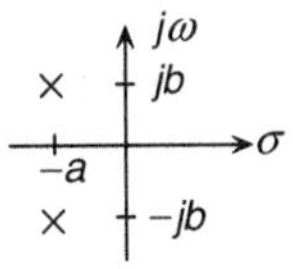 Complex conjugate root on the left half of *s*-plane	$g(t)=L^{-1}\left(\frac{K}{s+a+jb}+\frac{K}{s+a-jb}\right)$ $=Ke^{-(a+jb)t}+Ke^{-(a-jb)t}$ $=2Ke^{-at}\cos bt$ 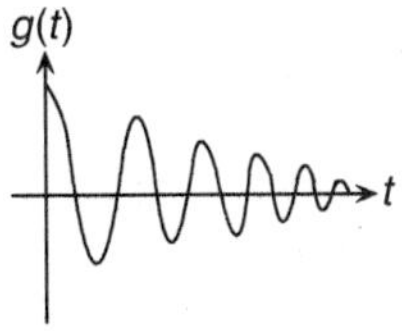 $g(t)$, t The response is damped sinusoidal (i.e damped oscillatory)	Absolutely stable
4	$G(s)=\frac{K}{s-a+jb}+\frac{K}{s-a-jb}$ $j\omega$, jb, σ, a, $-jb$ 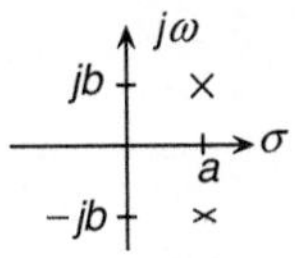 Complex conjugate root on the right half of *s*-plane	$g(t)=L^{-1}\left(\frac{K}{s-a+jb}+\frac{K}{s-a-jb}\right)$ $=Ke^{-(a+jb)t}+Ke^{-(a-jb)t}$ $=2Ke^{at}\cos bt$ 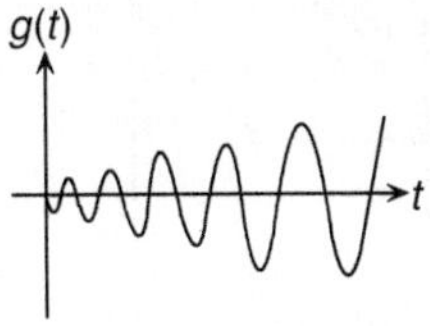 $g(t)$, t The response is exponentially increasing sinusoidal	Unstable

(*Continued*)

(*Continued*)

5	$G(s)=\frac{K}{s+jb}+\frac{K}{s-jb}$ $j\omega$, jb, σ, $-jb$ 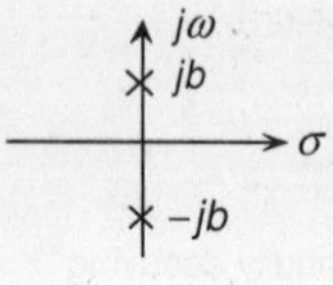 Single pair of roots on the $j\omega$-axis	$g(t)=L^{-1}\left(\frac{K}{s+jb}+\frac{K}{s-jb}\right)$ $=Ke^{-jbt}+Ke^{jbt}$ $=2Ke^{at}\cos bt$ $=2K\sin(bt+90)$ 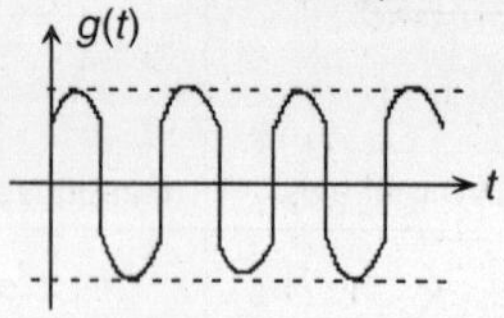 $g(t)$, t The response is oscillatory	Marginally or critically stable
6	$G(s)=\frac{K}{(s+jb)^2}+\frac{K}{(s-jb)^2}$ 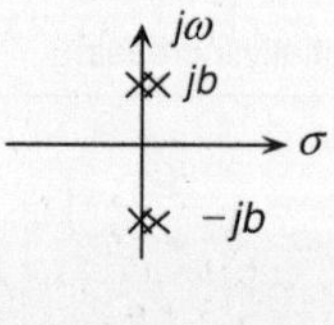 $j\omega$, jb, σ, $-jb$ Double pair of roots on the $j\omega$-axis	$g(t)=L^{-1}\left(\frac{K}{(s+jb)^2}+\frac{K}{(s-jb)^2}\right)$ $=Kte^{-jbt}+Kte^{jbt}$ $=2Kt\cos bt$ $=2Kt\sin(bt+90)$ 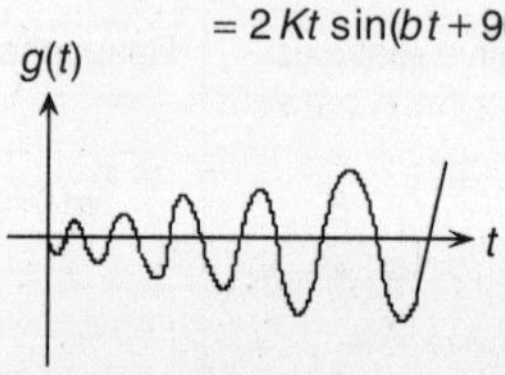 $g(t)$, t Response is linearly increasing sinusoidal	Unstable
7	$G(s)=\frac{K}{S}$ 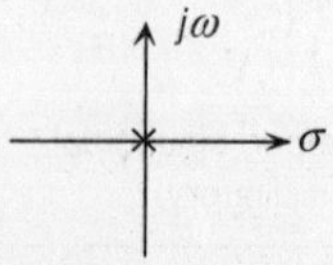 $j\omega$, σ Single root at origin	$g(t)=L^{-1}\left(\frac{K}{S}\right)$ $=K$ 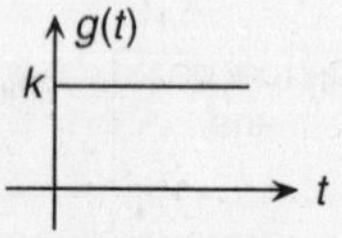 $g(t)$, k, t The response is constant	Absolutely stable
8	$G(s)=\frac{K}{S^2}$ 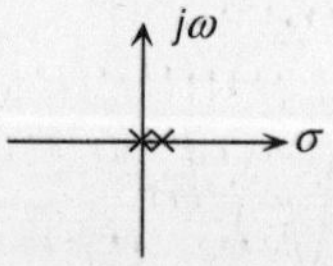 $j\omega$, σ Double root at origin	$g(t)=L^{-1}\left(\frac{K}{S^2}\right)$ $=Kt$ 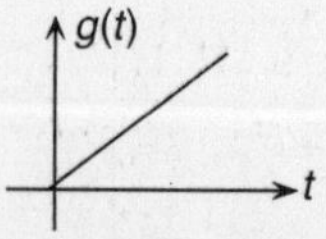 $g(t)$, t The response will linearly increase with time	Unstable

Figure 9.1 Roots in *s*-plane and its impulse response and stability condition

This figure depicts the roots in the *s*-plane as well as its impulse response and stability analysis. From the definition of stability, the impulse response must tend to zero (or constant) as time approaches infinity. Therefore, from Figure 9.1, the following conclusions can be made.

1. For all the poles located in the left half of the *s*-plane, the impulse response tends to zero and the system is in stable state.
2. For all the poles located on the $j\omega$ -axis and simple, the system is marginally stable or limitedly stable.
3. For all the poles located in the right half of the *s*-plane, the impulse response does not tend to zero and it increases with time; therefore, the output becomes unbounded. Hence, the system is unstable.

Therefore, the roots of the characteristic equation have the following conditions

1. When the roots have a negative real part, system is stable.
2. When the roots have a zero real part and the pole is not repeated, the system is marginally or limitedly stable.
3. When the roots have a positive real part, the system is unstable.

Figure 9.2 depicts all the above-mentioned conditions.

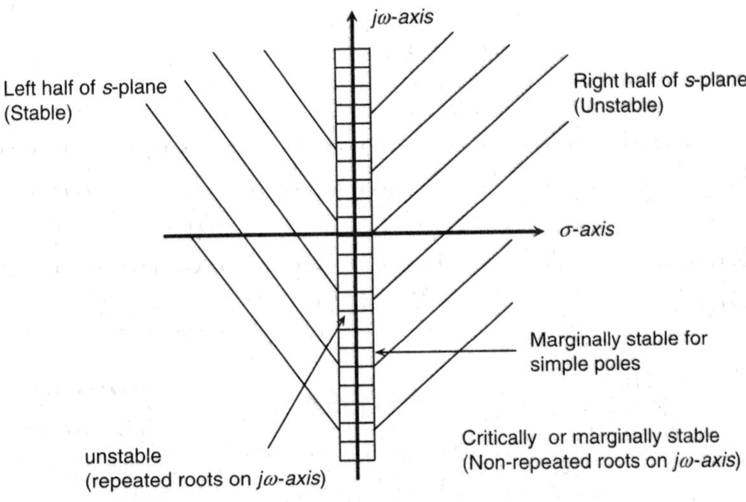

Figure 9.2 Stability analysis using *s*-plane

***Problem-9.1*:** Determine whether the following systems are stable, marginally stable or unstable.

(a) $-2, -4$ (b) $-3, 3, -2j, 2j$ (c) $-3, 4$ (d) $-4, 0$
(e) $-2 + j, -2 - j$ (f) $-2 + j4, -2 - j4, -2$

***Solution*:**

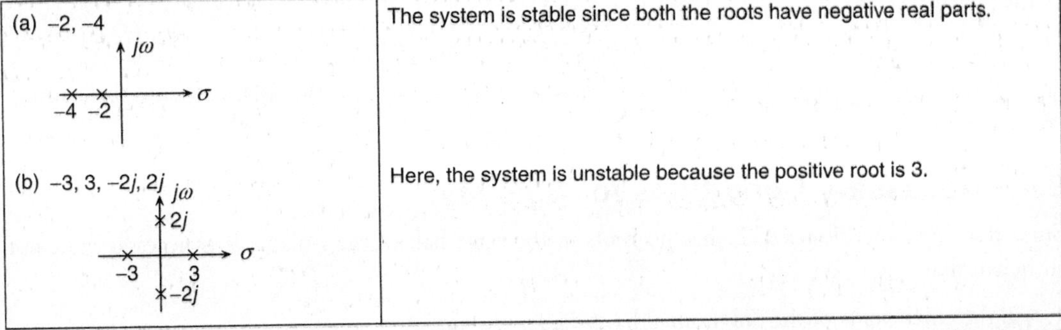

(a) −2, −4	The system is stable since both the roots have negative real parts.
(b) −3, 3, −2*j*, 2*j*	Here, the system is unstable because the positive root is 3.

(Continued)

(*Continued*)

(c) −3, 4 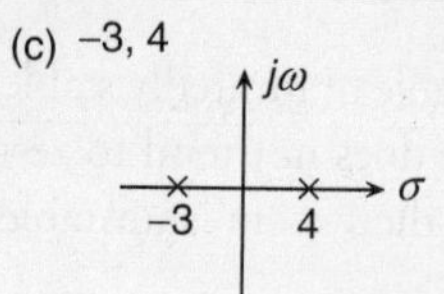	The system is unstable since one root has a positive real part.
(d) −4, 0	The system is marginally stable because the roots are non-positive.
(e) $-2+j, -2-j$	Since the positive real parts are only considered, the system is stable.
(f) $-2+j4, -2-j4, -2$	Since only negative real parts are considered, the system is stable.

9.4 ROUTH–HURWITZ (*R–H*) CRITERION

The *R–H* criterion provides a method for determining the location of poles of the characteristic equation with respect to the left half and right half of the *s*-plane without solving the equation. The transfer function (t.f.) of any closed-loop system can be represented as

$$\frac{C(s)}{R(s)} = \frac{b_0 s^m + b_1 s^{m-1} + \ldots b_m}{a_0 s^n + a_1 s^{n-1} + \ldots a_n} = \frac{B(s)}{F(s)}$$

where *a* and *b* are constants.

To find closed loop poles, we equate $F(s) = 0$. This equation is called characteristic equation (C.E.) of the system

i.e.,
$$F(s) = a_0 s^n + a_1 s^{n-1} + a_2 s^{n-2} + \ldots a_n = 0.$$

Thus, the roots of the C.E. are the closed-loop poles of the system; this factor decides the stability of the system.

9.4.1 Necessary Conditions for Stability

Since the above-mentioned C.E. has no roots in the right half of the *s*-plane, it is necessary but not sufficient that

1. All the coefficients of the polynomial have the same sign.

2. None of the coefficient vanishes, that is, all powers of s must be present in the descending order from n to '0'. When all the coefficients are positive, the system is not necessarily stable. Even though the coefficients are positive, some of the roots may lie on the right half of the s-plane or on the imaginary axis. In order for all the roots to have negative real parts, it is necessary but not sufficient that all the coefficients of the C.E. should be positive. If all the coefficients of the C.E. are positive, then the system may be stable and one should proceed further to examine the sufficient conditions of stability.

9.4.2 HURWITZ STABILITY CRITERION

The C.E. is given as

$$a_0 s^n + a_1 s^{n-1} + a_2 s^{n-2} + \dots\dots\dots\dots + a_n = 0.$$

Hurwitz gave the sufficient condition for having all the roots of the C.E. in the left half of the s-plane as if and only if all the n sub-determinants D_n, where $n = 1, 2\ldots\ldots n$, of Hurwitz determinant (H) are positive.

The Hurwitz determinant is given as

$$H = \begin{bmatrix} a_1 & a_0 & 0 & 0 & \cdots & 0 \\ a_3 & a_2 & a_1 & a_0 & \cdots & 0 \\ a_5 & a_4 & a_3 & a_2 & \cdots & 0 \\ \vdots & \vdots & \vdots & \vdots & & \\ a_{2n-1} & a_{2n-2} & a_{2n-3} & a_{2n-4} & & \end{bmatrix}$$

$$D_1 = |a_1| > 0$$

$$D_2 = \begin{vmatrix} a_1 & a_0 \\ a_3 & a_2 \end{vmatrix} > 0$$

$$D_3 = \begin{vmatrix} a_1 & a_0 & 0 \\ a_3 & a_2 & a_1 \\ a_5 & a_4 & a_3 \end{vmatrix} > 0$$

$$D_n = |H| > 0$$

that is, all the sub-determinants must be positive for a stable system.

The disadvantages of Hurwitz stability criterion are as follows:

1. It is very complicated and time consuming for solving higher-order systems.
2. This method is unable to find the exact number of poles located in the right half of the s-plane.
3. It is very difficult to predict marginal stability.

Due to the above-mentioned disadvantages, the method suggested by Routh, known as Routh's method, is used. It is also known as the Routh–Hurwitz (R–H) method.

Problem-9.2: Consider a fourth-order system with the C.E. $S^4 + 8S^3 + 18S^2 + 16^S + 5 = 0$, State whether the system is stable or not.

***Solution*:**

The Hurwitz arrangement is given as

$$H = \begin{bmatrix} 8 & 1 & 0 & 0 \\ 16 & 18 & 8 & 1 \\ 0 & 5 & 16 & 18 \\ 0 & 0 & 0 & 5 \end{bmatrix}$$

$$D_1 = |8| > 0$$

$$D_2 = \begin{bmatrix} 8 & 1 \\ 16 & 18 \end{bmatrix} = 128 > 0$$

$$D_3 = \begin{vmatrix} 8 & 1 & 0 \\ 16 & 18 & 8 \\ 0 & 5 & 16 \end{vmatrix} = 1.728 > 0$$

$$D_4 = |H| = 5D_3 > 0$$

Therefore, the system is stable.

9.4.3 Routh–Hurwitz Stability Criterion

In this criterion, the coefficients are arranged in an array, which is known as Routh's array. The C.E. of the system is given as

$$F(s) = a_0 s^n + a_1 s^{n-1} + a_2 s^{n-2} + \ldots\ldots\ldots + a_n = 0.$$

The Routh's array is given as

$$\begin{array}{c|ccccc} s^n & a_0 & a_2 & a_4 & a_6 & \cdots \\ s^{n-1} & a_1 & a_3 & a_5 & a_7 & \cdots \\ s^{n-2} & b_1 & b_2 & b_3 & b_4 & \\ s^{n-3} & c_1 & c_2 & c_3 & \cdots & \cdots \\ \vdots & \vdots & \vdots & \vdots & \vdots & \\ s^0 & a_n & & & & \end{array}$$

The coefficients for the first two rows are written directly from the C.E.. From these two rows, the next rows can be obtained as follows:

$$b_1 = \frac{a_1 a_2 - a_0 a_3}{a_1},\ b_2 = \frac{a_1 a_4 - a_0 a_5}{a_1},\ b_3 = \frac{a_1 a_6 - a_0 a_7}{a_1}.$$

From the second and third row, the fourth row can be determined as

$$c_1 = \frac{b_1 a_3 - a_1 b_2}{b_1},\ c_2 = \frac{b_1 a_5 - a_1 b_3}{b_1}.$$

This process is to be continued until the coefficient for s^0 is obtained, which will be a_n. From this array, the stability of system can be predicted.

The stability of the system can be predicted from this array as follows:

1. All elements in the first column of Routh's array should be positive (i.e., should have same sign) and there should not be any sign change. This is a necessary and sufficient condition for the system to be stable.
2. Any change of sign in the first column of Routh's array indicates that
   (a) the system is unstable.
   (b) the number of sign changes gives the number of roots lying on the right half of the s-plane.

Problem-9.3: Examine the stability of the given equations using Routh's stability criterion:

(a) $s^3 + 6s^2 + 11s + 6 = 0$
(b) $s^3 + 4s^2 + s + 16 = 0$

Solution:

(a) $s^3 + 6s^2 + 11s + 6$

$$\begin{array}{c|cc} s^3 & 1 & 11 \\ s^2 & 6 & 6 \\ s^1 & \dfrac{11\times 6-6}{6}=10 & 0 \\ s^0 & 6 & \end{array}$$

As there is no sign change in the first column, the given system is stable.

(b) $s^3 + 4s^2 + s + 16 = 0$

$$\begin{array}{c|cc} s^3 & 1 & 1 \\ s^2 & 4 & 16 \quad \leftarrow \text{First sign change} \\ s^1 & \dfrac{4-16}{4}=-3 & 0 \quad \leftarrow \text{Second sign change.} \\ s^0 & 16 & \end{array}$$

As there are two sign changes, the system is unstable. There are two roots in the right half of the s-plane.

9.4.4 Difficulties in *R–H* Stability Criterion

The following two difficulties are generally encountered while forming Routh's array.

A row of all zeros

When all the elements in any row of the Routh's array are zero because of a zero row in the array, the Routh's test fails. To overcome this difficulty, the following method is adopted.

1. Determine the auxiliary polynomial $A(s)$
2. Differentiate the auxiliary polynomial with respect to s to get $\dfrac{dA(s)}{ds}$
3. The row of zeros is replaced with coefficients of $\dfrac{dA(s)}{ds}$
4. Continue the construction of the array in usual manner and determine the stability.

First element in a row is zero

When the first element in any row of the Routh's array is zero, while rest of the row has at least one non-zero term, the terms in the next row become infinite and Routh's test fails because of this zero term.

The following methods are adopted to overcome this difficulty:

1. Substitute a small positive number $\in$ for the zero and we proceed to evaluate the rest of the Routh's array, and then, examine the signs of the first column of Routh's array by substituting $\in \rightarrow 0$
2. Modify the original C.E. by replacing s with $\frac{1}{z}$. We apply the Routh's test on the modified equation, which is now given in terms of z. Here, the number of z roots with positive real parts are the same as the number of s roots with positive real parts.

Problem-9.4: Examine the stability of the following systems using *R–H* stability criterion:

(a) $s^5 + s^4 + 3s^3 + 3s^2 + 4s + 8 = 0$
(b) $s^6 + 2s^5 + 8s^4 + 12s^3 + 20s^2 + 16s + 16 = 0$

Solution:

(a) The C.E. is

$$s^5 + s^4 + 3s^3 + 3s^2 + 4s + 8 = 0.$$

$$\begin{array}{c|ccc} s^5 & 1 & 3 & 4 \\ s^4 & 1 & 3 & 8 \\ s^3 & 0 & -4 & 0 \\ s^2 & \infty & & \end{array} \leftarrow \text{Routh's test failed}$$

To remove the above-mentioned difficulty, the following methods are adopted.

Method 1:
Substitute $s = \frac{1}{z}$, then C.E. becomes

$$8z^5 + 4z^4 + 3z^3 + 3z^3 + 3z^2 + z + 1 = 0$$

$$\begin{array}{c|ccc} z^5 & 8 & 3 & 1 \\ z^4 & 4 & 3 & 1 \\ z^3 & -3 & -1 & 0 \\ z^2 & 1.67 & 1 & 0 \\ z^1 & 0.8 & 0 & 0 \\ z^0 & 1 & 0 & 0 \end{array}$$

There are two sign changes in the first column of Routh's array, and hence, the given system is unstable.

Method 2:
Replace 0 by $\in$ (a small number) and complete the array with $\in$. Then, examine the sign change by taking $\in \rightarrow 0$.

The above-mentioned problem is solved as follows:

$$\begin{array}{c|ccc} s^5 & 1 & 3 & 4 \\ s^4 & 1 & 3 & 8 \\ s^3 & \in & -4 & 0 \\ s^2 & \dfrac{3\in+4}{\in} & 8 & 0 \\ s^1 & \dfrac{\dfrac{-4(3\in+4)}{\in}-8\in}{\dfrac{3\in+4}{\in}} & 0 & 0 \\ s^0 & 8 & 0 & 0 \end{array}$$

$$\underset{e\to 0}{\text{Lt}}\ \frac{3\in+4}{\in}=\infty$$

$$\underset{e\to 0}{\text{Lt}}\ \frac{-4\left(\dfrac{3\in+4}{\in}\right)-8\in}{\dfrac{3\in+4}{\in}}=\underset{e\to 0}{\text{Lt}}\ \frac{-4(3\in+4)-8\in^2}{3\in+4}=-4.$$

Routh's array becomes

$$\begin{array}{c|ccc} s^5 & 1 & 3 & 4 \\ s^4 & 1 & 3 & 8 \\ s^3 & \in & -4 & 0 \\ s^2 & \infty & 8 & 0 \\ s^1 & -4 & 0 & 0 \\ s^0 & 8 & & \end{array}$$

There are two sign changes in the first column, and hence, the system is unstable.

(b) $s^6 + 2s^5 + 8s^4 + 12s^3 + 20s^2 + 16s + 16 = 0$

The Routh's array is

$$\begin{array}{c|cccc} s^6 & 1 & 8 & 20 & 16 \\ s^5 & 2 & 12 & 16 & 0 \\ s^4 & 2 & 12 & 16 & \\ s^3 & 0 & 0 & & \\ s^2 & & & & \end{array}$$

$\rightarrow$ Routh's test failed

To eliminate the above-mentioned difficulty, form an equation by using the coefficients of the array that is just above the row of zeros. This equation is known as auxiliary equation.

$$\therefore A(s) = 2s^4 + 12s^2 + 16$$

$$\frac{dA(s)}{ds} = 8s^3 + 24s.$$

The above-mentioned Routh's array becomes

$$\begin{array}{c|cccc} s^6 & 1 & 8 & 20 & 16 \\ s^5 & 2 & 12 & 16 & 0 \\ s^4 & 2 & 12 & 16 & \\ s^3 & 8 & 24 & & \\ s^2 & 6 & 16 & & \\ s^1 & 2.67 & & & \\ s^0 & 16 & & & \end{array}$$

There is no sign change in the first column, and hence, the given system is stable.

9.4.5 Determining the Range of Values of *K*

In practical systems, an amplifier of gain k is introduced in the feed-forward path of the system.

The closed-loop t.f. is

$$\frac{C(s)}{R(s)} = \frac{kG(s)}{1 + kG(s)H(s)}.$$

Hence, C.E. is

$$F(s) = 1 + kG(s)H(s) = 0.$$

Therefore, the unknown k appears in C.E.. In such cases, Routh's array should be constructed in terms of k, and then, the range of values of k can be obtained in such a way that it will not produce any sign change in the first column of the Routh's array. Hence, it is possible to obtain the range of values of k for absolute stability of the system using Routh's criterion. Such a system where stability depends on the condition of parameter k is called a conditionally stable system.

Problem-9.5: Determine the range of values of k so that the system having the following C.E. will be stable.

$$s(s^2 + 2s + 3)(s + 2) + k = 0$$

Solution:

The given C.E. is

$$s(s^2 + 2s + 3)(s + 2) + k = 0$$

$$s^4 + 4s^3 + 7s^2 + 6s + k = 0.$$

The Routh's array is

$$\begin{array}{c|ccc} s^4 & 1 & 7 & k \\ s^3 & 4 & 6 & 0 \\ s^2 & 5.5 & k & 0 \\ s^1 & \frac{33-4k}{5.5} & 0 & \\ s^0 & k & & \end{array}$$

The system will be stable, if

$$k > 0, \ \frac{33 - 4k}{5.5} > 0.$$

$$\Rightarrow 33 - 4k > 0 \Rightarrow k < 8.25$$

$\therefore$ Range of k is $0 < k < 8.25$.

9.4.6 Limitation of *R–H* Stability Criterion

The following are the limitations of *R–H* stability criterion:

1. This method is used only if C.E. is algebraic and all the coefficients are real.
2. This method gives information only on the absolute stability of the system, but does not give any indication of the relative stability.
3. It gives information only about the number of roots lying on the right-hand side of the *s*-plane, and the values of the roots that are not available. It cannot distinguish real and complex roots.

ADDITIONAL SOLVED PROBLEMS

ASP-1: Apply *R–H* criterion for equation $S^4 + 5S^3 + 2S^2 + 3S + 2 = 0$ to determine the stability and find the number of roots lying in the right half of the *s*-plane.

Solution:

The C.E. is $S^4 + 5S^3 + 2S^2 + 3S + 2 = 0$.

The Routh's array is

S^4	1	2	2
S^3	5	3	0
S^2	1·4	2	0
S^1	−4·14	0	0
S^0	2	0	0

There are two sign changes, and hence, the system is unstable. There are two roots in the right half of the *s*-plane.

ASP-2: The C.E. of the two systems are given as follows:

(a) $S^4 + 21S^3 + 21S^2 + 36S + 20 = 0$
(b) $S^5 + 6S^4 + 3S^3 + 2S^2 + S + 1 = 0$

Find whether the systems are stable or not using *R–H* criterion.

Solution:

(a) The C.E. is $S^4 + 21S^3 + 21S^2 + 36S + 20 = 0$.

The Routh's array is

S^4	1	21	20
S^3	21	36	0
S^2	19.28	20	0
S^1	13.216	0	0
S^0	20	0	0

There is no sign change in the first column, and hence, the given system is stable.

(b) The C.E. is $S^5 + 6S^4 + 3S^3 + 2S^2 + S + 1 = 0$.

The Routh's array is

$$\begin{array}{c|ccc} S^5 & 1 & 3 & 1 \\ S^4 & 6 & 2 & 1 \\ S^3 & 2.67 & 0.833 & 0 \\ S^2 & 0.128 & 1 & 0 \\ S^1 & -20 & 0 & 0 \\ S^0 & 1 & 0 & 0 \end{array}$$

There are two sign changes in the first column, and hence, the given system is unstable.

ASP-3: The open-loop t.f. of a unity feedback control system is given as $G(s) = \dfrac{K}{S(1+ST_1)(1+ST_2)}$. By applying *R–H* stability criterion, determine the value of K in terms of T_1 and T_2 for the system to be stable.

Solution:

The C.E. of the system is

$$1 + G(s)H(s) = 0$$

$$1 + \frac{K}{S(1+ST_1)(1+ST_2)} = 0$$

$$S^3 T_1 T_2 + S^2(T_1 + T_2) + S + K = 0.$$

The Routh's array is

$$\begin{array}{c|cc} S^3 & T_1T_2 & 1 \\ S^2 & T_1 + T_2 & K \\ S^1 & \dfrac{(T_1 + T_2 - KT_1T_2)}{T_1 + T_2} & 0 \\ S^0 & K & 0 \end{array}$$

For the stable system, $K > 0$

$$\frac{(T_1 + T_2 - KT_1T_2)}{T_1 + T_2} > 0$$

$$T_1 + T_2 - KT_1T_2 > 0$$

$$T_1 + T_2 < KT_1T_2$$

$$K < (T_1 + T_2)/T_1T_2 < 1/T_1 + 1/T_2.$$

The range of K is $0 < K < (1/T_1 + 1/T_2)$.

ASP-4: Examine the stability of the following systems using *R–H* criterion.

(a) $S^5 + 2S^4 + 4S^3 + 8S^2 + 3S + 1 = 0$

(b) $S^5 + 2S^4 + 2S^3 + 4S^2 + 4S + 8 = 0$

Solution:

(a) The C.E. is $S^5 + 2S^4 + 4S^3 + 8S^2 + 3S + 1 = 0$.

$$
\begin{array}{c|ccc}
S^5 & 1 & 4 & 3 \\
S^4 & 2 & 8 & 1 \\
S^3 & 0 & 2.5 & 0 \\
S^2 & \infty & &
\end{array}
\quad \rightarrow \text{Routh's test failed}
$$

To remove the above-mentioned difficulty, the following method is adopted.

When we substitute $S = 1/Z$, then the C.E. becomes

$$Z^5 + 3Z^4 + 8Z^3 + 4Z^2 + 2Z + 1 = 0$$

$$
\begin{array}{c|ccc}
Z^5 & 1 & 8 & 2 \\
Z^4 & 3 & 4 & 1 \\
Z^3 & 6.67 & 1.67 & 0 \\
Z^2 & 3.25 & 1 & 0 \\
Z^1 & -0.382 & 0 & 0 \\
Z^0 & 1 & 0 & 0
\end{array}
$$

There are two sign changes in the first column of Routh's array, and hence, the given system is unstable.

(b) The C.E. is $S^5 + 2S^4 + 2S^3 + 4S^2 + 4S + 8 = 0$.

$$
\begin{array}{c|ccc}
S^5 & 1 & 2 & 4 \\
S^4 & 2 & 4 & 8 \\
S^3 & 0 & 0 & 0
\end{array}
\quad \rightarrow \text{Routh's test failed}
$$

To eliminate the above-mentioned difficulty, we form an equation by using the coefficients of the array, which is just above the row of zeros. The equation is known as auxiliary equation.

$$A(s) = 2S^4 + 4S^2 + 8$$

$$\frac{dA}{dS} = 8S^3 + 8S$$

$$
\begin{array}{c|ccc}
S^5 & 1 & 2 & 4 \\
S^4 & 2 & 4 & 8 \\
S^3 & 8 & 8 & 0 \\
S^2 & 2 & 8 & 0 \\
S^1 & -24 & 0 & 0 \\
S^0 & 8 & 0 & 0
\end{array}
$$

There are two sign changes in the first column of Routh's array, and hence, the system is unstable.

ASP-5: For a unity feedback system having forward path t.f.

$$G(s) = \frac{K}{S(1+0.6S)(1+0.4S)}.$$

Determine

(a) The range of values of K

(b) Marginal value of K

(c) Frequency of sustained oscillations

Solution:
The C.E. of the system is

$$1 + G(s)H(s) = 0$$

$$1 + \frac{K}{S(1+0.6S)(1+0.4S)} = 0$$

$$0.24S^3 + S^2 + S + K = 0.$$

The Routh's array is

$$\begin{array}{c|cc} S^3 & 0.24 & 1 \\ S^2 & 1 & K \\ S^1 & 1-0.24K & 0 \\ S^0 & K & 0 \end{array}$$

For a stable system,

$$K > 0, 1 - 0.24K > 0$$

$$K > 0,\ K < 4.167.$$

(a) The range of K is $0 < K < 4.167$.
(b) For a marginally stable value, we have

$$1 - 0.24K = 0 \Rightarrow K = 4.167.$$

(c) The Routh's array becomes

$$\begin{array}{c|cc} S^3 & 0.24 & 1 \\ S^2 & 1 & K \\ S^1 & 0 & 0 \end{array}$$

The auxiliary equation is

$$A(s) = S^2 + K = 0$$

$$S^2 = -4.167$$

$$S = \pm j2.04 = \pm j\omega.$$

The frequency of sustained oscillations = 2.04 rad/s.

ASP-6: Construct Routh's array and determine the stability of the system whose C.E. is $S^6 + 2S^5 + 8S^4 + 12S^3 + 20S^2 + 16S + 16 = 0$. Further, determine the number of roots lying on the right half of s-plane, left half of s-plane, and on imaginary axis.
Solution:
The C.E. is $S^6 + 2S^5 + 8S^4 + 12S^3 + 20S^2 + 16S + 16 = 0$.
The Routh's array is

$$\begin{array}{c|cccc} S^6 & 1 & 8 & 20 & 16 \\ S^5 & 2 & 12 & 16 & 0 \\ S^4 & 2 & 12 & 16 & 0 \\ S^3 & 0 & 0 & 0 & 0 \end{array} \rightarrow \text{Routh's test failed}$$

To eliminate the above difficulty, form an equation by using the coefficients of the array, which is just above the row of zeros. The equation is known as auxiliary equation.

$$A(s) = 2S^4 + 12S^2 + 16 = S^4 + 6S^2 + 8$$

$$\frac{dA}{dS} = 4S^3 + 12S.$$

Now, the Routh's array becomes

$$\begin{array}{c|cccc} S^6 & 1 & 3 & 20 & 16 \\ S^5 & 1 & 6 & 8 & 0 \\ S^4 & 1 & 6 & 8 & 0 \\ S^3 & 1 & 3 & 0 & 0 \\ S^2 & 3 & 8 & 0 & 0 \\ S^1 & 0.33 & 0 & 0 & 0 \\ S^0 & 8 & 0 & 0 & 0 \end{array}$$

The auxiliary polynomial is

$$2S^4 + 12S^2 + 16 = 0$$

$$S^4 + 6S^2 + 8 = 0.$$

Let $S^2 = x$, then $x^2 + 6x + 8 = 0$

$$\Rightarrow \qquad x = -2,\ x = -4.$$

The roots of the auxiliary polynomial are $S = \pm\sqrt{x} = \pm\sqrt{2}$ and $\pm\sqrt{4}$

$$= j\sqrt{2}, -j\sqrt{2}, j2, -j2.$$

The roots of the auxiliary polynomial are also roots of C.E.. Hence, four roots are lying on imaginary axis and the remaining two are lying on the left half of the *s*-plane and the given system is marginally or limitedly stable.

ASP-7: A feedback system has an open-loop t.f. of $G(s)H(s) = \dfrac{Ke^{-S}}{S(S^2+5S+9)}$. Using the *R–H* criterion, determine the maximum value of *K* for the closed-loop system to be stable.

Solution:

By assuming $e^{-s} = 1 - S$.

The C.E. of the system is

$$1 + G(s)H(s) = 0$$

$$1 + \frac{Ke^{-S}}{S(S^2+5S+9)} = 0$$

$$1 + \frac{K(1-S)}{S(S^2+5S+9)} = 0$$

$$S^3 + 5S^2 + (9 - K)S + K = 0.$$

The Routh's array is

$$\begin{array}{c|cc} S^3 & 1 & 9-K \\ S^2 & 5 & K \\ S^1 & \dfrac{45-6K}{5} & 0 \\ S^0 & K & 0 \end{array}$$

For stable system, $K > 0$, $\dfrac{45-6K}{5} > 0 \Rightarrow K < 7.5$

$K > 0$, $K < 7.5$; the range of K is $0 < K < 7.5$.

ASP-8: For the system having C.E.$2S^4 + 4S^2 + 1 = 0$, find the:

(a) number of roots in the left half of the s-plane.
(b) number of roots in the right half of the s-plane.
(c) number of roots on the imaginary axis.

Here, we use the *R–H* stability criterion.
Solution:
The C.E. of the system is

$$2S^4 + 4S^2 + 1 = 0$$

$$2S^4 + 0.S^3 + 4S^2 + 0.S + 1 = 0.$$

The Routh's array is

$$\begin{array}{c|ccc} S^4 & 2 & 4 & 1 \\ S^3 & 0 & 0 & 0 \rightarrow \text{Routh's test failed} \end{array}$$

The auxiliary equation is

$$A(s) = 2S^4 + 4S^2 + 1$$

$$\frac{dA(s)}{dS} = 8S^3 + 8S.$$

Now, the Routh's table becomes

$$\begin{array}{c|ccc} S^4 & 2 & 4 & 1 \\ S^3 & 8 & 8 & 0 \\ S^2 & 2 & 1 & 0 \\ S^1 & 4 & 0 & 0 \\ S^0 & 1 & 0 & 0 \end{array}$$

There is no sign change in the first column, and hence, the given system is stable; further, there are no roots lying on the right half of the s-plane.

Again, solving $A(s) = 2S^4 + 4S^2 + 1.$

$$S^2 = \frac{-4 \pm \sqrt{16-8}}{4} = \frac{-4 \pm 2\sqrt{2}}{4} \Rightarrow S = \pm j0.927, \ \pm j1.7732$$

Therefore,

(a) The number of roots in the left half of the s-plane = 0.
(b) The number of roots in the right half of the s-plane = 0.
(c) The number of roots on the imaginary axis = 4.

ASP-9: A given system oscillates with frequency 2 rad/s. Find the values of K_{mar} and P. The number of poles is on the right-hand side.

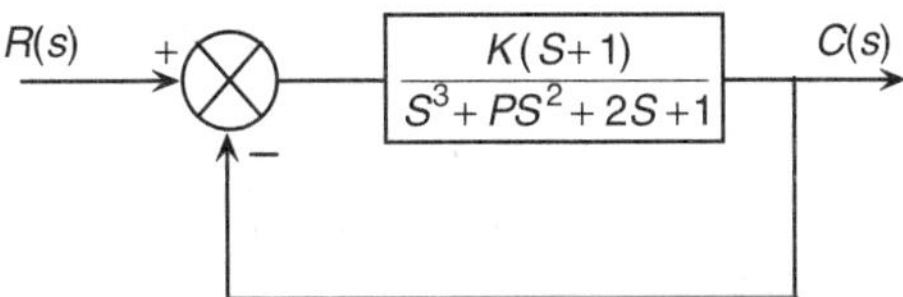

***Solution*:**

As the system oscillates, it is marginally stable, and the value of K, at this situation, is the marginal value of K. As the system is marginally stable, there must be a row of zeros occurring in Routh's array.

The C.E. of the system is

$$1 + G(s)H(s) = 0$$

$$1+\frac{K(S+1)}{S^3+PS^2+2S+1}=0$$

$$S^3 + PS^2 + (2 + K)S + (1 + K) = 0$$

$$\begin{array}{c|cc} S^3 & 1 & 2+K \\ S^2 & P & 1+K \\ S^1 & \dfrac{P(2+K)-(1-K)}{P} & 0 \\ S^0 & 1+K & 0 \end{array}$$

At marginal value of K, we have

$$(2 + K)P - (1 + K) = 0 \rightarrow P=\frac{1+K}{K+2}. \tag{1}$$

Now, at this value, the auxiliary equation is

$$A(s) = PS^2 + K + 1 = 0$$

$$S^2=-\frac{(1+K)}{P}\Rightarrow S=\pm j\sqrt{\frac{K+1}{P}}=\pm j\omega\Rightarrow\omega=\sqrt{\frac{K+1}{P}}$$

As per the given data, ω = 2 rad/s.

$\therefore$
$$\sqrt{\frac{K+1}{P}}=2\Rightarrow\frac{K+1}{4}=P \tag{2}$$

From Equation (1) and (2),
$$\frac{K+1}{4}=\frac{1+K}{K+2}\Rightarrow K_{mar}=2$$

$$P=\frac{K_{mar}+1}{4}=\frac{3}{4}=0.75.$$

ASP-10: The closed-loop t.f. of a system is given as $\frac{C(s)}{R(s)} = \frac{k}{S(1+ST)(1+0.5S)+50}$.Find the value of T such that the system is driven on to the verge of instability and the resulting frequency of oscillation.

Solution:

The closed-loop t.f. of a system is given as

$$\frac{C(s)}{R(s)} = \frac{k}{S(1+ST)(1+0.5S)+50}.$$

The C.E. of the system is given as

$$S(1+ST)(1+0.5S)+50=0$$

$$TS^3+(1+2T)S^2+2S+100=0$$

$$\begin{array}{c|cc} S^3 & T & 2 \\ S^2 & 1+2T & 100 \\ S^1 & \frac{2-96T}{1+2T} & 0 \\ S^0 & 100 & 0 \end{array}$$

For a stable system, $T > 0$, $T > -1/2$, $\frac{2-96T}{1+2T} > 0 \Rightarrow T < 1/48$.

The system becomes unstable if $T = 1/48$. If we substitute $T = 1/48$, we will get a zero row in the Routh's array. Since the zero row occurs on the odd powers of S^1, the S^1 row will be a zero row.

1. The value of T such that the system is driven on to the verge of instability = 1/48.
   The auxiliary equation is

$$1.04S^2+100=0$$

$$S=\pm j9.8=\pm j\omega.$$

2. The angular frequency of oscillations = 9.8 rad/s.

ASP-11: Determine the values of k and b so that the system whose open t.f. is $G(s) = \frac{k(S+1)}{S^3+bS^2+3S+1}$ oscillates at a frequency of oscillations of 2 rad/s.
(Let us assume unity feedback)

Solution:

The C.E. of the system is

$$1+G(s)H(s)=0$$

$$1+\frac{k(S+1)}{S^3+bS^2+3S+1}=0$$

$$S^3+bS^2+(K+3)S+K+1=0.$$

The Routh's array is

$$\begin{array}{c|cc} S^3 & 1 & K+3 \\ S^2 & b & K+1 \\ S^1 & \frac{b(K+3)-(K+1)}{b} & 0 \\ S^0 & K+1 & 0 \end{array}$$

The system will have sustained oscillations when

$$\frac{b(K+3)-(K+1)}{b}=0 \Rightarrow b=\frac{K+1}{K+3}. \quad (1)$$

The auxiliary equation is

$$bS^2+(K+1)=0$$

$$S^2=-\frac{(K+1)}{b} \Rightarrow S=\pm j\sqrt{\frac{K+1}{b}} \Rightarrow \omega=\sqrt{\frac{K+1}{b}}.$$

As per the given data, we have

$$\sqrt{\frac{K+1}{b}}=2 \Rightarrow \frac{(K+1)}{b}=4 \Rightarrow b=\frac{K+1}{4}. \quad (2)$$

From Equations (1) and (2), we get

$$\frac{K+1}{K+3}=\frac{K+1}{4} \Rightarrow K+3=4 \Rightarrow K=1$$

$$b=\frac{K+1}{4}=\frac{2}{4}=0.5.$$

The values of K and b are $K=1$, $b=0.5$.

ASP-12: A system has $G(s)=\dfrac{2}{S(S+1)(S+2)}, H(S)=1$. With R–H stability criterion, determine its relative stability about the line $S=-1$.

Solution:

To determine stability about $S=-1$ means to find whether there is any root of C.E. to the right of $S=-1$. The C.E. of the system is

$$1+G(s)H(s)=0$$

$$1+\frac{2}{S(S+1)(S+2)}=0$$

$$S^3+3S^2+2S+2=0.$$

Now, to get the new Routh's array w.r.t. line $S=-1$, replace S with $(Z-1)$.
i.e.,

$$(Z-1)^3+3(Z-1)^2+2(Z-1)+2=0$$

$$Z^3-Z+2=0$$

$$\begin{array}{c|cc} Z^3 & 1 & -1 \\ Z^2 & 0 & 2 \end{array}$$

Due to the presence of zero in the first column of the Routh's array, the Routh's test fails. This difficulty can be overcome by replacing 0 with small positive number (ε).

$$\begin{array}{c|cc} Z^3 & 1 & -1 \\ Z^2 & 0(\varepsilon) & 2 \\ Z^1 & -(\varepsilon+2)/\varepsilon & 0 \\ Z^0 & 2 & 0 \end{array}$$

$$\lim_{z\to 0}\frac{-\varepsilon-2}{\varepsilon} \to -\infty, \text{i.e., negative}$$

There are two sign changes; that is, two roots are located to the right of $S = -1$ line, and therefore, about $S = -1$, it is not stable.

ASP-13: Using *R–H* stability criterion, determine the stability of the following systems:

(a) Its loop t.f. has poles at $S = 0$, $S = -1$, $S = -3$ and zero at $S = -5$; gain K of forward path is 10.
(b) It is a Type 1 system with an error constant of 10 S^{-1} and poles at $S = -3$ and $S = -6$.

Solution:
(a) Loop t.f. has poles at $S = 0$, $S = -1$, $S = -3$ and zero at $S = -5$; gain K of forward path is 10, then t.f is

$$G(s)H(s) = \frac{10(S+5)}{S(S+1)(S+3)}.$$

The C.E. of the system is

$$1 + G(S)H(S) = 0$$

$$1 + \frac{10(S+5)}{S(S+1)(S+3)} = 0$$

$$S^3 + 4S^2 + 13S + 50 = 0$$

$$\begin{array}{c|cc} S^3 & 1 & 13 \\ S^2 & 4 & 50 \\ S^1 & 0.2 & 0 \\ S^0 & 50 & 0 \end{array}$$

There is no sign change in the first column of Routh's array, and hence, the system is absolutely stable.
(b) Type 1, $K_v = 10\ S^{-1}$, poles at $S = -3$ and $S = -6$

Then, the t.f. is $$G(S)H(S) = \frac{K}{S(S+3)(S+6)}.$$

Now, $$K_V = \lim_{S\to 0} G(s)H(s) = \lim_{S\to 0} \frac{K.S}{S(S+3)(S+6)} = K/18 .$$

Given that $$K/18 = 10 \Rightarrow K = 180$$

$$\therefore \quad G(s)H(s) = \frac{180}{S(S+6)(S+3)}.$$

The C.E. of the system is $1 + G(S)H(S) = 0$.

$$1 + \frac{180}{S(S+6)(S+3)} = 0$$

$$S^3 + 9S^2 + 18S + 180 = 0$$

$$\begin{array}{c|cc} S^3 & 1 & 18 \\ S^2 & 9 & 180 \\ S^1 & -2 & 0 \\ S^0 & 180 & 0 \end{array}$$

There are two sign changes in the first column, and hence, two roots are located in RHS of the *s*-plane. Hence, the system is unstable.

REVIEW QUESTIONS

1. Define the following terms
   (a) absolute stability
   (b) marginal stability
   (c) conditional stability
   (d) critically stable system
2. According to *R–H* stability criteria, how can you analyse the stability of the control system?
3. State the advantages and limitations of *R–H* stability criterion.
4. What are the necessary conditions to have all the roots of the C.E. in the left half of the *s*-plane?
5. What are the difficulties in *R–H* stability criterion? Explain, how you can overcome them.
6. Explain the significance of auxiliary equation.
7. How the roots of the C.E. are related to stability?
8. What is the relation between the stability and the coefficients of characteristic polynomial?

EXERCISE PROBLEMS

1. The C.E. of two systems are given as
   (a) $S^4 + 2S^3 + 3S^2 + 8S + 2 = 0$
   (b) $S^5 + 3S^4 + 7S^3 + 20S^2 + 6S + 15 = 0$
   Find whether the systems are stable or not using *R–H* criterion.
2. Investigate the stability of the closed loop system for the following C.Es using *R–H* stability criterion.
   (a) $S^5 + S^4 + 3S^3 + 3S^2 + 6S + 4 = 0$
   (b) $S^6 + S^5 + 6S^4 + 5S^3 + 10S^2 + 5S + 5 = 0$
3. Using *R–H* stability criterion, determine how many roots of the C.E. are to the right of the line $S = -1$ in the *S*-plane. The C.E is $S^2 + 5S + 5 = 0$.
4. For the unity feedback system with forward path t.f. $G(s) = \dfrac{K(s+20)}{s(s+2)(s+3)}$, find the range of *K* to make the system stable.
5. For the following system as shown in the figure:

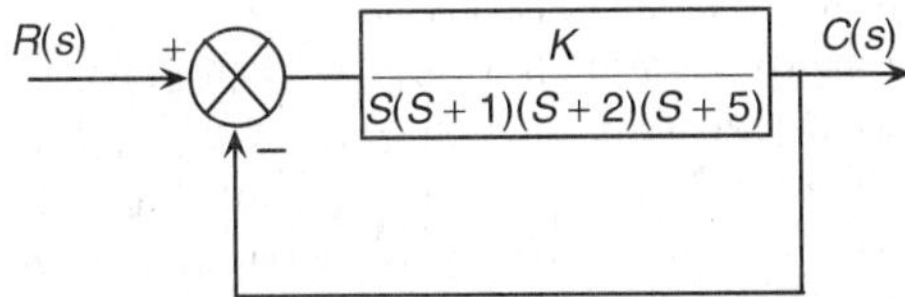

   (a) Find the range of *K* for stability.
   (b) Find the actual location of closed loop poles when the system is marginally stable.
6. The open-loop t.f. of a unity feedback system is given as $G(s) = \dfrac{K}{s(1+0.4s)(1+0.25s)}$. Find the restriction on *K*, so that the closed loop system is absolutely stable.
7. Determine whether the following systems are stable or unstable or marginally stable.
   (a) $-4, 3$
   (b) $-3 + 2j, -3 - 2j$
   (c) $-2 + 2j, -2 - 2j, -4j, 4j$
   (d) $-3, -5$
8. A unity feedback system has an open-loop t.f. of $G(s)H(s) = \dfrac{Ks^{-s}}{s(s^2+5s+9)}$. Find the range of *K* for the closed loop system to be stable.
9. For a unity feedback system having forward path t.f. $G(s) = \dfrac{K(s+4)}{s(s+1)(s+2)}$, determine the
   (a) range of values of *K*.
   (b) value of *K* that makes the system to oscillate.
   (c) frequency of sustained oscillations.

10. If a unity feedback system having $G(s)=\dfrac{5}{s(s^2+Ps+K)}$ is marginally stable and oscillates with a frequency 5 rad/s, then calculate P and K.

ANSWERS

1. (a) Unstable (b) Stable
2. (a) Unstable (b) Limitedly stable
3. No root is to the right of the s-plane
4. $0 < K < 2$
5. (a) $0 < K < 19.7$ (b) $s = +j1.12, -j1.12, -4.5, -3.5$
6. $0 < K < 6.5$
7. (a) Unstable (b) Stable (c) Marginally stable (d) Stable
8. $0 < K < 7.5$
9. (a) $0 < K < 6$ (b) $K = 6$ (c) $\omega = 2.83$ rad/s
10. $P = 0.2$, $K = 25$

OBJECTIVE TYPE QUESTIONS

1. The number of the roots of the C.E. $S^4+2S^3+3S^2+8S+2=0$ that lie in the right half of the s-plane is
   (a) 1 (b) 2 (c) 3 (d) 0
2. The control system whose poles are given as −2, −3 and +4 is
   (a) stable (b) unstable (c) marginally stable (d) none
3. The control system whose poles are given as $-2+j2, -2-j2, -j2, j2$ is
   (a) stable (b) unstable (c) marginally stable (d) None
4. The measure of relative stability is
   (a) peak overshoot (b) damping ratio (c) GM and PM (d) all
5. The number of sign changes in the entries of the first column of Routh's array denotes
   (a) the number of poles in RHP (b) the number of poles in LHP
   (c) the number of zeros in RHP (d) the number of zeros in LHP
6. If poles of the control system lie on the imaginary axis, then the system is
   (a) stable (b) unstable
   (c) marginally stable (d) conditionally stable
7. The necessary condition for the system to be stable is
   (a) all the coefficients of the characteristic polynomial should be positive
   (b) all the coefficients of the characteristic polynomial should be negative
   (c) some of the coefficients are positive and some of the coefficients are negative
   (d) all the above.
8. If the C.E of the system is $S^3+3S^2+3S+K=0$, then the system is stable when the value of K is
   (a) 2 (b) 9 (c) 6 (d) 4
9. How many roots of the C.E. $S^5+S^4+2S^3+2S^2+3S+15=0$ lie on the left half of the s-plane?
   (a) 1 (b) 2 (c) 3 (d) 4
10. If the first column elements in the Routh table are 1, 4, 3, −2 and 1, then
    (a) there is one root in the left half of the s-plane
    (b) there are two roots in the left half of the s-plane

(c) there is one root in the right half of the s-plane
(d) there are two roots in the right half of the s-plane.

11. If the elements in the first column of the Routh table are 1,0.5,4,3 and −1, then
(a) there is one root in the left half of the s-plane
(b) there are two roots in the left half of the s-plane
(c) there is one root in the right half of the s-plane
(d) there are two roots in the right half of the s-plane.

12. If the open-loop t.f. of the unity feedback control system is given as $G(s)=\dfrac{K}{s(1+sT_1)(1+sT_2)}$, then the system is stable when the value of K is
(a) $0<K<T_1T_2$
(b) $0<K<(1/T_1+1/T_2)$
(c) $0<K<T_1+T_2$
(d) All

13. If the open-loop t.f. of the unity feedback control system is given as $G(s)=\dfrac{K}{s(1+0.4s)(s+0.25s)}$, then the system is stable when the maximum value of K is
(a) 6
(b) 2.5
(c) 6.5
(d) 9.5

14. The Routh's array is given as

$$\begin{array}{l|lll} S^4 & 1 & 5 & 4 \\ S^3 & 1 & 4 & 0 \\ S^2 & 1 & 4 & 0 \\ S^1 & 0 & 0 & 0 \end{array}$$

The auxiliary equation of the system is
(a) $S^4+5S^2+4=0$
(b) $S^3+4S=0$
(c) $S^2+4=0$
(d) None

15. The *R–H* stability criterion gives
(a) absolute stability
(b) relative stability
(c) gain margin
(d) resonant frequency

16. Any sign change in the first column of the Routh's array indicates
(a) the system is unstable
(b) the number of roots lying on the right half of the s-plane
(c) Both (a) and (b).
(d) cannot be determined.

17. If there are repeated roots of C.E. on the $j\omega$-axis, the system is
(a) stable
(b) unstable
(c) oscillatory
(d) conditionally stable

18. If any root of the C.E. has a positive real part, then the system is
(a) stable
(b) unstable
(c) oscillatory
(d) conditionally stable

ANSWERS

1. b	**2.** b	**3.** c	**4.** d	**5.** a	**6.** c	**7.** a	**8.** b	**9.** b
10. d	**11.** c	**12.** b	**13.** c	**14.** c	**15.** a	**16.** c	**17.** b	**18.** b

CHAPTER

10 Root Locus Method

Contents

10.1 INTRODUCTION

The stability as well as the transient response of any closed-loop system depends on the location of closed-loop poles in the s-plane. If some parameters of the closed-loop system are varied, there is a movement of its poles. The cognisance of such movement of the poles with small variations of the parameter is remarkably useful in the design of the closed-loop system. In 1948, W.R. Evans introduced the root locus technique, which provides the knowledge of such movement of poles by using a graphical method in the s-plane. The root locus method is the graphical method for obtaining the roots of the characteristic equation (C.E.) in the s-plane, thereby investigating the system response performance. In this method, the movement of pole in the s-plane is sketched when a particular parameter, usually the gain of the system, is varied from $-\infty$ to $+\infty$. By using the root locus method, the designer can predict the effects on the location of the closed-loop poles by varying the gain value or adding open-loop poles or open-loop zeros.

10.2 BASIC CONCEPT

In general, the C.E. of a closed-loop system is given as

$$1+G(s)H(s)=0.$$

If $G(s) = KG'(s)$, where K is the gain of the amplifier in forward path or also called system gain, then

$$1 + KG'(s)H(s) = 0$$

This equation contains K as a variable parameter. The closed-loop poles, that is, the roots of the above-mentioned equation are now dependent on the values of K. Now, if gain K is varied from $-\infty$ to $+\infty$, then for each value of K will get separate set of location of the roots of the C.E. If all such locations are joined, the resulting locus is called root locus. Therefore, we can define root locus as the locus of the closed-loop poles obtained when system gain K is varied from $-\infty$ to $+\infty$. When K is varied from 0 to $+\infty$, the plot obtained is called direct root locus. On the other hand, when K is varied from $-\infty$ to 0, the plot obtained is called inverse root locus. However, generally, the term root locus is used in the sense of direct root locus.

10.3 ANGLE AND MAGNITUDE CRITERION

The C.E. of the closed-loop transfer function (CLTF) is given as

$$1 + G(s)H(s) = 0$$

$$G(s)H(s) = -1$$

$$G(s)H(s) = 1\angle \pm 180° \quad (10.1)$$

$$\therefore \quad \text{Magnitude} = |G(s)H(s)| = 1 \quad (10.2)$$

$$\text{Phase angle} = \angle G(s)H(s) = \pm(2q+1)180° \quad (10.3)$$

where q is a positive integer.

Actually, in $-1 + j0 = 1\angle \pm 180°$, the point $-1 + j0$ is a point on the negative real axis, and it can be traced as magnitude 1 at an angle ± 180, ± 540, 900 ……$(2q + 1)180°$. Any point on the root locus must satisfy either the magnitude condition or the angle condition. Since K is not known in the function $G(s)\ H(s)$, it is difficult to get $|G(s)H(s)| = 1$ at any point in the s-plane, and this condition is unsuitable to check the existence of any point on the root locus. If a point on the root locus is a function of $G(s)\ H(s)$, it helps to calculate the value of K.

Problem-10.1: Check whether the point $s = -3 + j5$ lies on the root locus of a system having $G(s)H(s) = \dfrac{K}{(s+1)(s+5)}$. Determine the corresponding value of K.

Solution:

Given that

$$G(s)H(s) = \frac{K}{(s+1)(s+5)}$$

$$G(s)H(s)|_{s=-3+j5} = \frac{K}{(-3+j5+1)(-3+j5+5)} = \frac{K}{(-2+j5)(2+j5)}$$

$$= \frac{K}{(5.3851\angle 111.8°) \times 5.3851\angle 68.2°} = \frac{K}{29\angle 180°} = \frac{K}{29}\angle -180$$

$$\therefore \quad \angle G(s)H(s)|_{s=-3+j5} = -180°$$

Since the angle condition is satisfied, point $-3 + j5$ lies on the root locus of the given system.

To find the value of K,

$$\left|G(s)H(s)\right|_{s=-3+j5} = 1 \Rightarrow \frac{K}{29} = 1 \Rightarrow K = 29.$$

***Problem-10.2*:** Check whether the points $(-1+j)$and $(-3+j)$ lie on the root locus of a system given as $G(s)H(s) = \dfrac{K}{(s+1)(s+2)}$. Use the angle condition.

***Solution*:**

1. The open loop t.f. (OLTF) is

$$G(s)H(s) = \frac{K}{(s+1)(s+2)}$$

$$G(s)H(s)\Big|_{s=-1+j} = \frac{K}{(-1+j+1)(-1+j+2)} = \frac{K}{(1+j)j} = \frac{K}{\sqrt{2}}\angle{-135^\circ}$$

$$\therefore \qquad \angle G(s)H(s)\Big|_{s=-1+j} \neq -180^\circ.$$

   Since the angle condition is not satisfied, point $(-1+j)$ does not lie on the root locus of the given system.

2. The OLTF is

$$G(s)H(s) = \frac{K}{(s+1)(s+2)}$$

$$G(s)H(s)\Big|_{s=-3+j} = \frac{K}{(-3+j+1)(-3+j+2)} = \frac{K}{(1-3j)} = \frac{K}{3.162}\angle 71.56^\circ$$

$$\therefore \qquad \angle G(s)H(s)\Big|_{s=-3+j} \neq -180^\circ.$$

   Since the angle condition is not satisfied, point $(-3+j)$ does not lie on the root locus of the given system.

10.4 RELATION BETWEEN POLES AND ZEROS OF OLTF WITH POLES OF CLTF

To discuss the relation between poles and zeros of open-loop transfer function (OLTF) with poles of closed-loop transfer function (CLTF), let us consider

$$G(s)H(s) = \frac{KN(s)}{D(s)}. \tag{10.4}$$

where K is the gain of the OLTF, $N(s)$ is the numerator polynomial in s and $D(s)$ is the denominator polynomial in s.

Case (a): OLTF zeros can be obtained by taking

$$G(s)H(s) = 0$$

or

$$\frac{KN(s)}{D(s)} = 0$$

$$\therefore \quad N(s) = 0 [\because K \neq 0].$$

Therefore, OLTF zero can be obtained from

$$N(s) = 0. \tag{10.5}$$

Case (b): OLTF poles can be obtained by taking

$$G(s)H(s) = \infty$$

or

$$\frac{KN(s)}{D(s)} = \infty$$

$$\therefore \quad D(s) = 0.$$

Therefore, OLTF zero can be obtained from

$$D(s) = 0. \tag{10.6}$$

Case (c): CLTF poles can be obtained by taking

$$1 + G(s)H(s) = 0$$

or

$$1 + \frac{KN(s)}{D(s)} = 0$$

or

$$D(s) + KN(s) = 0.$$

1. If $K = 0$, then $D(s) = 0$. Then, it is same as (10.6) and gives OLTF poles. Therefore, closed-loop poles are same as open-loop poles for $K = 0$.
2. If $K = \infty$, $D(s) + KN(s) = N(s)\left[K + \frac{D(s)}{N(s)}\right] = 0 \Rightarrow N(s) = 0$. It is same as (10.5) and gives OLTF zeros. Therefore, closed-loop poles are same as open-loop zeros for $K = \infty$. Thus, the root locus starts from open-loop pole at $K = 0$ and ends at open-loop zero at $K = \infty$.

10.5 PROPERTIES OF ROOT LOCUS

The various properties of root locus are listed as follows:

1. The root locus starts from an open-loop pole, that is, at $K = 0$ and terminates at an open-loop zero, that is, $K = \infty$.
2. The number of separate root locus branches are equal to either the number of open-loop poles or the number of open-loop zeros, whichever is greater.
3. A section of root locus lies on the real axis if the total number of open-loop poles and zeros to the right of the section is odd.
4. The root locus is symmetrical with respect to real axis.
5. At some point, on the root locus between two open-loop poles, the root locus breaks away.
6. For higher values of K, the root locus can be approximated by asymptotic lines and these asymptotic lines intersect at a point on the real axis.
7. The asymptotic lines make an angle with negative real axis at an intersection point (centre of gravity).
8. If the root locus intersects the imaginary axis, then the points of intersection are conjugate.
9. From the open-loop complex pole, the root locus departs by making an angle with the horizontal line.

10.6 PROCEDURE FOR CONSTRUCTING ROOT LOCUS

The following gives the procedure for constructing the root locus of the loop t.fs., when gain K is varied from 0 to $+\infty$.

1. **Symmetry:** The root locus plot is always symmetrical with respect to the real axis in s-plane.
2. **Starting and Ending Points:** The root locus originates from an open-loop pole, i.e., at $K = 0$ and terminates at an open-loop zero, i.e., at $K = \infty$.
3. **Number of Loci:** The number of separate root loci (N) depends upon the number of poles (P) and zeros (Z) of the loop t.f.

$$N = Z \quad \text{for } Z > P$$

$$N = P \quad \text{for } P > Z$$

   where Z is the number of finite zeros of $G(s)H(s)$ and
   P is the number of finite poles of $G(s)H(s)$

   Thus, the number of separate root locus is equal to the number of poles or zeros, whichever is greater.
4. **Existence on real axis:** Some of the loci will lie on the real axis. A point on the real axis will lie on the real axis if and only if the sum of OLTF poles and zeros to the right of the point is odd.
5. **Number of Asymptotic lines:** Asymptote is defined as a line on which the root locus touches at infinity. For the function $G(s)H(s)$ having P finite poles and Z finite zeros, the number of asymptotes is given as $n = (P - Z)$.
6. **Angle of asymptotes:** Since the number of poles is greater than the number of zeros, i.e., $P > Z$, $P - Z$ branches will move to infinity, and these branches move along the asymptote. For root locus, the angle of asymptote is given as

$$\varphi = \frac{(2q+1)\pi}{P-Z}$$

   where q is the positive integer having values 0, 1, 2......up to $q = (P - Z)$.
7. **Centre of asymptote or centroid:** For the function $G(s)H(s)$ having P finite poles and Z finite zeros, the number of asymptotes is $n = (P - Z)$. These asymptotes intersect on the real axis in s-plane at a point called centroid and is given as

$$\sigma = \frac{\Sigma \text{ poles of } G(s)H(s) - \Sigma \text{ zero of } G(s)H(s)}{P-Z}.$$

8. **Break-away or break-in point:** Break-away point is defined as the point at which root locus comes out of the real axis and break-in point is defined as a point at which root locus enters the real axis. The break-away or break-in points are the points on the root locus at which multiple roots of the C.E. occur. The following are the steps to determine the break-away or break-in points:
   (a) Find the C.E. of the system, i.e., $1 + G(s)H(s) = 0$.
   (b) Write K in terms of s, i.e., $K = f(s)$.
   (c) Derive $\frac{dK}{ds}$ and put $\frac{dK}{ds} = \mathbf{0}$.
   (d) The roots of the equation $\frac{dK}{ds} = \mathbf{0}$ are the break-away or break-in points.
   If the value of K is positive for any root of $\frac{dK}{ds} = \mathbf{0}$, the root(s) is (are) valid break-away or break-in point(s).

9. **Intersection of root locus with imaginary axis:** The point of intersection of root locus with the imaginary axis in the s-plane can be determined by applying the Routh's criterion to the C.E. of the system, which gives the critical values of K at the intersection point. The following are the steps to determine the intersection of root locus with imaginary axis:
   (a) Find the C.E. of the system, i.e., $1+G(s)H(s)=0$.
   (b) Develop Routh's array in terms of K.
   (c) Find K_{mar} that creates one of the rows of the Routh's array as a row of zeros.
   (d) Frame the auxiliary equation $A(s) = \mathbf{0}$ with the help of the coefficients of array just above the row of zeros.
   (e) The roots of auxiliary equation $A(s) = \mathbf{0}$ for $K = K_{mar}$ gives the intersection points of the root locus with imaginary axis.
10. **Angle of departure or arrival:** The root locus leaves from a complex pole and arrives at a complex zero. These two angles are known as angle of departure and angle of arrival, respectively. The angles of departure and arrival of root locus at the complex poles and zeros, respectively, depend upon the contribution of angles due to other poles and zeros of the loop t.f. $G(s)H(s)$.
    The angle of departure is given as
    $$\phi_d = 180° - \phi_p + \phi_z.$$
    The angle of arrival is given as
    $$\phi_a = 180° + \phi_p - \phi_z,$$
    where ϕ_p is the sum of the angles due to other poles and ϕ_z is the sum of the angles due to other zeros.
11. **Value of K at a point on the root locus:** The value of K at a point s_1 on the root locus is determined by measuring the vectors from the poles and zeros of loop t.f. to point s_1 on the root and is given as
    $$K = \frac{1}{|G(s)H(s)|} = \frac{\underset{j-1}{\overset{n+m}{\pi}}|s_1 + P_j|}{\underset{i-1}{\overset{n}{\pi}}|s_1 + Z_i|} = \frac{\text{Product of all vector lengths from poles of } G(s)H(s) \text{ to } s_1}{\text{Product of all vector lengths from zeros of } G(s)H(s) \text{ to } s_1}.$$

***Problem-10.3*:** Calculate the angle of asymptotes and the centroid for the system having $G(s)H(s) = \dfrac{K(s+3)}{s(s+2)(s+4)(s+5)}$.

***Solution*:**
Given that
$$G(s)H(s) = \frac{K(s+3)}{s(s+2)(s+4)(s+5)}.$$

The poles are given as
$$s(s+2)(s+5)(s+4) = 0$$
$$s = 0,\ s = -2,\ s = -5,\ s = -4.$$

The zeros are given as
$$s = -3$$
$$\text{Number of poles} = P = 4$$
$$\text{Number of zeros} = Z = 1$$

$$\text{Number of root locus branches} = N = P = 4$$
$$\text{Number of asymptotic lines} = n = P - Z = 3$$

1. The angle of asymptotic lines with negative real axis, we get

$$\phi = \frac{(2q+1)180°}{P-Z}, \text{ where } q = 0, 1, 2$$

$$= 1 \times 180°/3, 3 \times 180°/3, 5 \times 180°/3$$
$$= 60°, 180°, 300°.$$

2. The centroid is given as

$$\text{Centroid} = \frac{\text{Sum of poles} - \text{Sum of zeros}}{P-Z} = \frac{(0-2-4-5)(-3)}{4-1} = -2.667.$$

Problem-10.4: For $G(s)H(s) = \dfrac{K}{s(s+1)(s+3)}$, find the intersection point of the root locus with the $j\omega$-axis.

Solution:

Given that

$$G(s)H(s) = \frac{K}{s(s+1)(s+3)}.$$

The intersection point of the root locus branches with imaginary axis can be determined by using *R–H* stability criterion.

The C.E. of the system is given as

$$1 + G(s)H(s) = 0$$

$$1 + \frac{K}{s(s+1)(s+3)} = 0$$

$$K = -s(s+1)(s+3)$$

$$s^3 + 4s^2 + 3s + K = 0$$

s^3	1	3
s^2	4	K
s^1	$\dfrac{(12-K)}{4}$	0
s^0	K	0

For the stable system, $K > 0, (12-K)/4 > 0$

$$\Rightarrow K > 0, K < 12$$

For the stable system, the maximum value of K is 12. For $K > 12$, the roots lie on the RHS of the s-plane, and hence, $K = 12$ is the value for marginal stability.

The auxiliary equation is

$$4s^2 + K = 0$$

$$\Rightarrow \quad 4s^2 + 12 = 0$$

$$\Rightarrow \quad s = \pm j\,1.73.$$

Therefore, the intersection point of the root locus branches with the imaginary axis is $\pm j1.73$.

Problem-10.5: Sketch the root locus plot for the systems whose OLTF is given as $G(s)H(s)=\dfrac{K}{s(s+1)(s+3)(s+4)}$.

Solution:
Given that

$$G(s)H(s)=\frac{K}{s(s+1)(s+3)(s+4)}.$$

1. Root locus is symmetrical about real axis.
2. The root locus plot starts from a pole at $K=0$ and terminates at an open-loop zero $K=\infty$.
3. Number of poles $= P = 4$
   Number of zeros $= Z = 0$
   Number of root locus branches $= N = P = 4$
   Number of asymptotic lines $= n = P - Z = 4$
4. The angle of asymptotic lines with positive real axis

$$\phi=\frac{(2q+1)180^\circ}{P-Z},\ q=0,1,2,3$$

$$=180^\circ/4, 3\times 180^\circ/4, 5\times 180^\circ/4, 7\times 180^\circ/4$$

$$=45^\circ,\ 135^\circ,\ 225^\circ,\ 315^\circ.$$

5. The centroid is given as

$$\text{Centroid}=\frac{\text{Sum of poles}-\text{Sum of zeros}}{P-Z}=\frac{-8}{4}=-2$$

6. **Break-away points:** The C.E. is given as

$$1+G(s)\,H(s)=0$$

$$1+\frac{K}{s(s+1)(s+3)(s+4)}=0$$

$$K=-s(s+1)\,(s+3)\,(s+4)$$

$$K=-(s^4+8s^3+19s^2+12s)$$

$$\frac{dK}{ds}=0\Rightarrow 4s^3+24s^2+38s+12=0$$

$$\Rightarrow s_1=-0.418,\ s_2=-2,\ s_3=-3.581.$$

The valid break-away points are $b_1=-0.418$ and $b_2=-3.581$ (since K is positive for s_1 and s_3).

7. **Intersection points of the root locus branches with imaginary axis:** The C.E. is given as

$$1+G(s)\,H(s)=0.$$

$$1+\frac{K}{s(s+1)(s+3)(s+4)}=0$$

$$K = -s(s+1)(s+3)(s+4)$$

$$s^4 + 8s^3 + 19s^2 + 12s + K = 0$$

s^4	1	19	K
s^3	8	12	0
s^2	17.5	K	0
s^1	$\frac{(210-8K)}{17.5}$	0	0
s^0	K	0	0

For a stable system, $(210 - 8K)/17.5 > 0$

$$\Rightarrow K < 26.25.$$

For a stable system, the maximum value of K is 26.25. For $K > 26.25$, the roots lie on the RHS of the s-plane, and hence, $K = 26.25$ is the value for marginal stability.

The auxiliary equation of the system is

$$17.5s^2 + K = 0$$

$$17.5s^2 + 26.25 = 0$$

$$\Rightarrow \qquad s = \pm j\,1.224.$$

The complete root locus plot for the above-mentioned system is as shown in Figure (a).

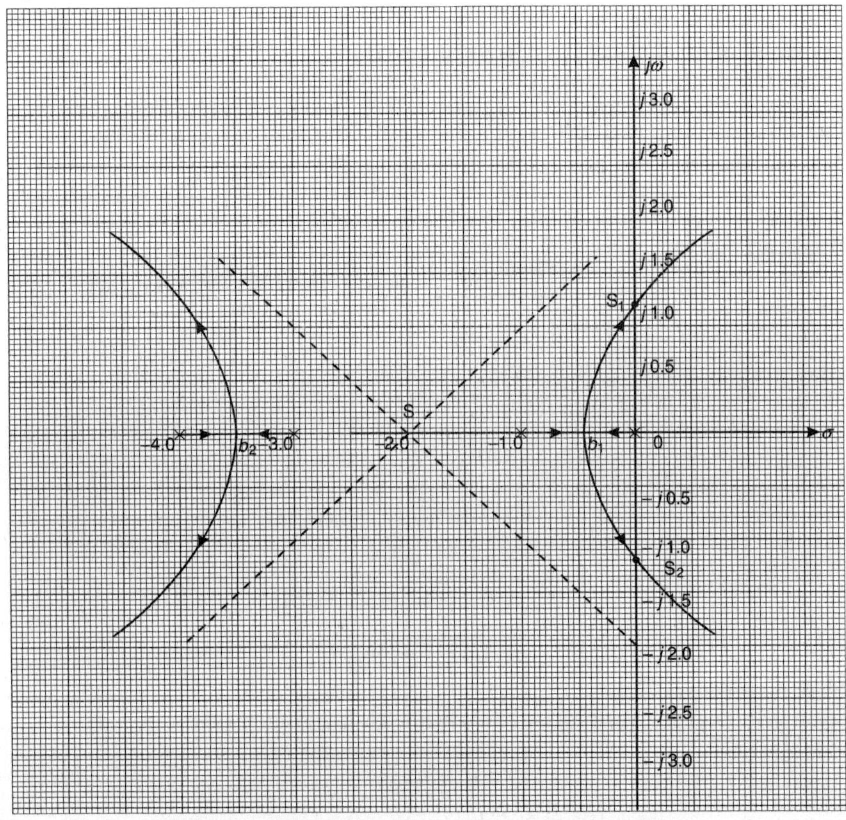

Figure (a) Root locus plot for Problem 10.5

***Problem-10.6*:** The OLTF of a control system is given as $G(s)H(s) = \dfrac{K}{s(s+6)(s^2+4s+13)}$. Sketch the root locus plot and determine

1. the break-away points
2. the angle of departure from complex poles
3. the stability condition

***Solution*:**

Given that

$$G(s)H(s) = \frac{K}{s(s+6)(s^2+4s+13)}.$$

1. Root locus is symmetrical about real axis.
2. The root locus plot starts from a pole at $K = 0$ and terminates at an open-loop zero, i.e., at $K = \infty$.
3. The poles are given as
   $s = 0,\ s = -6,\ s^2 + 4s + 13 = 0$
   $P_1 = 0,\ P_2 = -6,\ P_3 = -2 + j3,\ P_4 = -2 - j3$
   Number of poles = $P = 4$
   Number of zeros = $Z = 0$
   Number of root locus branches = $N = P = 4$
   Number of asymptotic lines = $n = P - Z = 4$
4. The angle of asymptotic lines with negative real axis
   $$\phi = \frac{(2q+1)180°}{P-Z},\ q = 0, 1, 2, 3$$
   $$= 180°/4,\ 3\times 180°/4,\ 5\times 180°/4,\ 7\times 180°/4$$
   $$= 45°,\ 135°,\ 225°,\ 315°.$$
5. The centroid is given as
   $$\text{Centroid} = \frac{\text{Sum of poles} - \text{Sum of zeros}}{P-Z} = \frac{(0-6-2+j3-2-j3)-(0)}{4-0} = -2.5.$$
6. **Break-away points:** The C.E. is given as
   $$1 + G(s)H(s) = 0$$
   $$1 + \frac{K}{s(s+6)(s^2+4s+13)} = 0$$
   $$K = -s(s+6)(s^2+4s+13)$$
   $$K = -(s^4 + 10s^3 + 37s^2 + 78s)$$
   $$\frac{dK}{ds} = 0 \Rightarrow 4s^3 + 30s^2 + 74s + 78 = 0$$
   $\Rightarrow\quad s_1 = -4.2,\ s_2 = -1.65 + j1.38,\ s_3 = -1.65 - j1.38.$
   The valid break-away point is $b_1 = -4.2$ (since K is positive for s_1).
7. **Intersection points of the root locus branches with imaginary axis:**
   The C.E. is given as
   $$1 + G(s)H(s) = 0$$
   $$s^4 + 10s^3 + 37s^2 + 78s + K = 0$$

s^4	1	37
s^3	10	78
s^2	29.2	K
s^1	$78-0.34K$	0
s^0	K	0

For the stable system, $K > 0,\ 78 - 0.34\,K > 0$

$K > 0,\ K < 228.4.$

For the stable system, the maximum value of K is 228.4. For $K > 228.4$, the roots lie on the RHS of the s-plane; hence, $K = 228.4$ is the value for marginal stability.

The auxiliary equation is $29.2s^2 + K = 0 \Rightarrow s = \pm j2.8.$

8. **Angle of departure:**

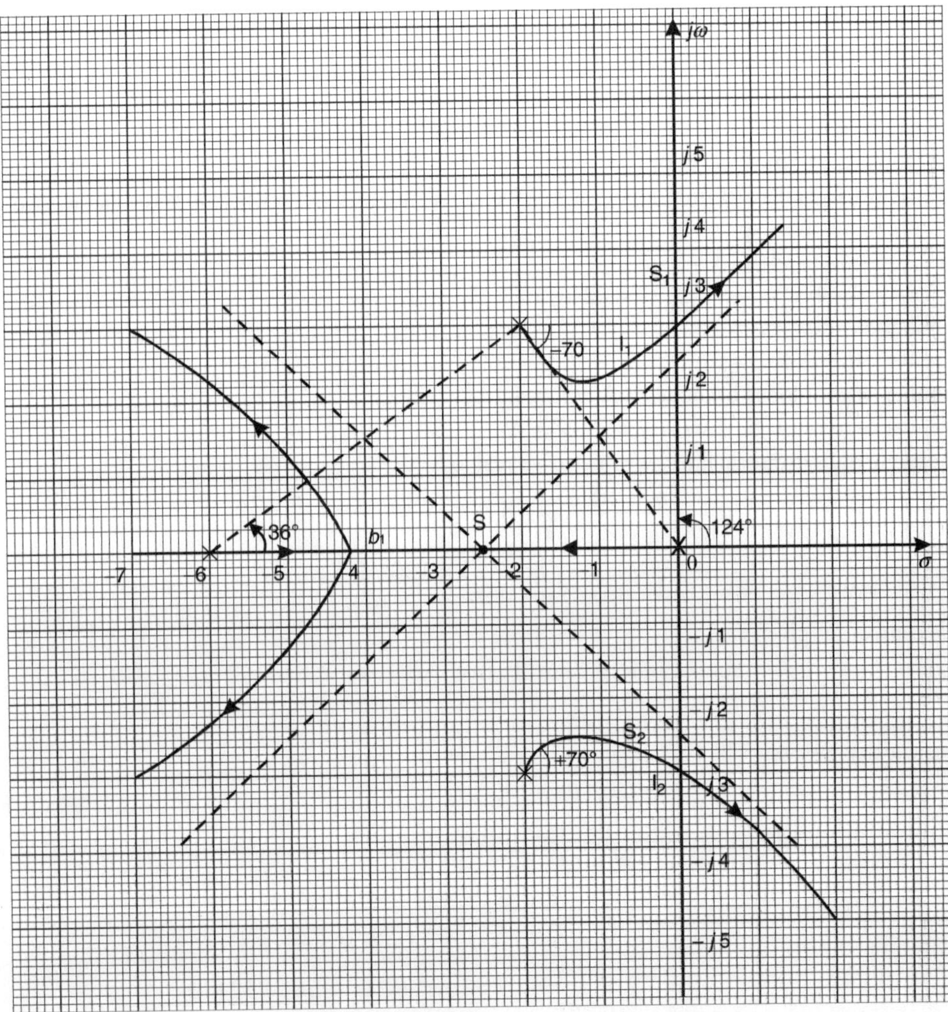

Figure (b) Root locus plot for Problem 10.6

$$\phi_{d1=(-2+j3)} = 180° - (\phi_{p1} + \phi_{p2} + \phi_{p4})$$
$$= 180° - (124° + 36° + 90°) = -70°$$
$$\phi_{d2=(-2-j3)} = +70°$$

The complete root locus plot is as shown in Figure (b).

(a) The break-away point is $B_1 = \mathbf{-4.2}$

(b) The angle of departure from complex pole is

$$\phi_{d1=(-2+j3)} = -70°$$

$$\phi_{d2=(-2-j3)} = +70°$$

(c) The system is stable if $\mathbf{0 < K < 228.4}$.

***Problem-10.7*:** Sketch the root locus plot of the system whose OLTF is given as

$$G(s)H(s) = \frac{K}{s(s+4)(s^2+4s+13)}.$$

***Solution*:**

Given that

$$G(s)H(s) = \frac{K}{s(s+4)(s^2+4s+13)}.$$

1. Root locus is symmetrical about real axis.
2. The root locus plot starts from a pole at $K = 0$ and terminates at an open-loop zero, that is, at $K = \infty$.
3. The poles are given as
   $s = 0, s = -4, s^2 + 4s + 13 = 0$
   $P_1 = 0, P_2 = -4, P_3 = -2 + j3, P_4 = -2 - j3$
   Number of poles = $P = 4$
   Number of zeros = $Z = 0$
   Number of root locus branches = $N = P = 4$
   Number of asymptotic lines = $n = P - Z = 4$
4. The angle of asymptotic lines with negative real axis
   $$\phi = \frac{(2q+1)180°}{P-Z},\ q = 0, 1, 2, 3$$
   $$= 180°/4, 3 \times 180°/4, 5 \times 180°/4, 7 \times 180°/4$$
   $$= 45°, 135°, 225°, 315°.$$
5. The centroid is given as
   $$\text{Centroid} = \frac{\text{Sum of poles} - \text{Sum of zeros}}{P-Z} = \frac{(0-4-2+j3-2-j3)-(0)}{4-0} = -2$$
6. **Break-away points:** The C.E. is given as
   $$1 + G(s)H(s) = 0$$
   $$1 + \frac{K}{s(s+4)(s^2+4s+13)} = 0$$
   $$K = -s(s+4)(s^2+4s+13)$$
   $$K = -(s^4 + 8s^3 + 29s^2 + 52s)$$
   $$\frac{dK}{ds} = 0 \Rightarrow \quad 4s^3 + 24s^2 + 58s + 52 = 0$$
   $$\Rightarrow \quad s_1 = -2, s_2 = -2 + j1.58, s_3 = -2 - j1.58.$$

The valid break-away point are $b_1 = -2, b_2 = -2 + j1.58, b_3 = -2 - j1.58$.

7. **Intersection points of the root locus branches with imaginary axis:** The C.E. is given as

$$1+G(s)H(s) = 0$$

$$s^4 + 8s^3 + 29s^2 + 52s + K = 0$$

s^4	1	29	K
s^3	8	52	0
s^2	22.5	K	0
s^1	$52 - 0.35K$	0	0
s^0	K	0	0

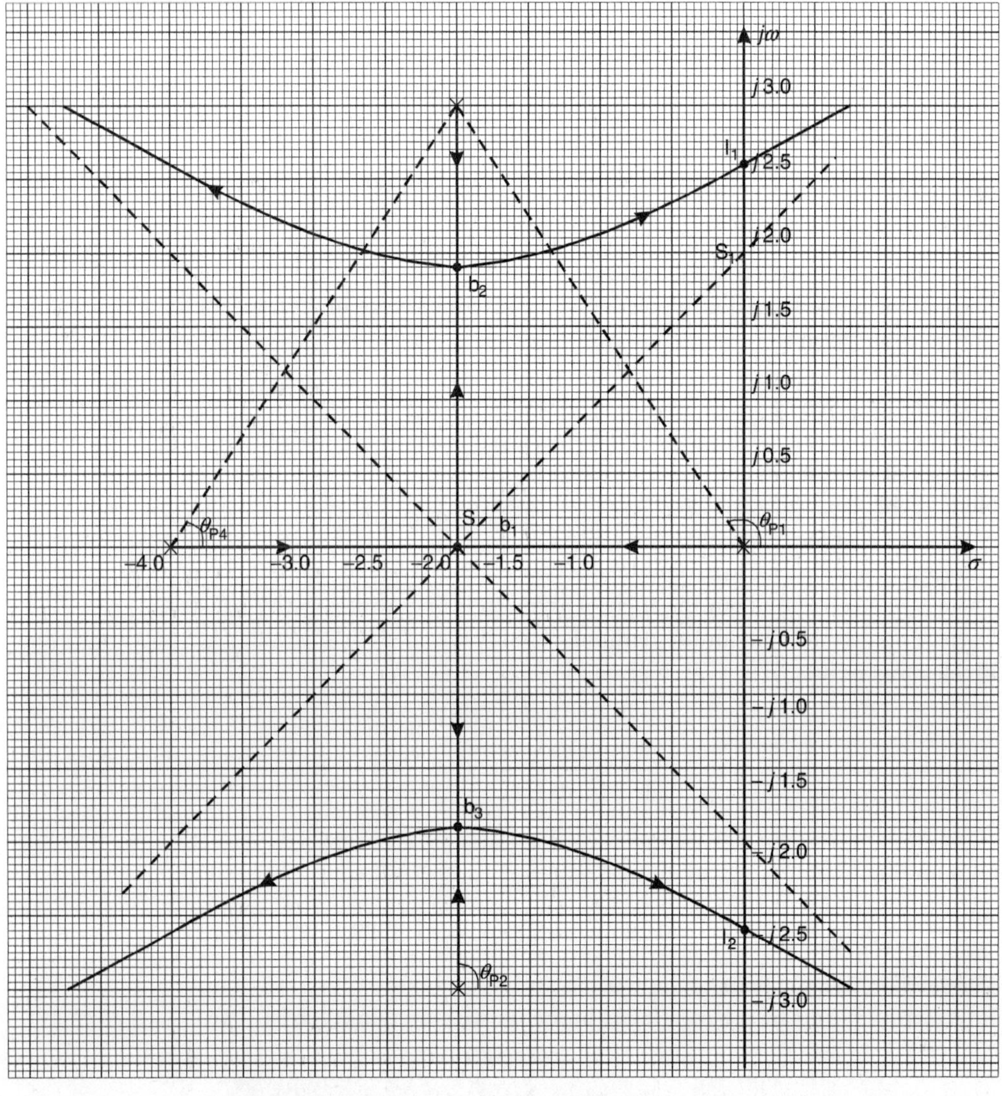

Figure (c) Root locus plot for Problem 10.7

For a stable system, $K > 0$, $52 - 0.35K > 0$

$$\Rightarrow \quad K > 0, K < 148.6$$

For a stable system, the maximum value of K is 148.6. For $K > 148.6$, the roots lie on the RHS of the s-plane, and hence, $K = 148.6$ is the value for marginal stability.
The auxiliary equation is

$$22.5s^2 + K = 0$$

$$22.5s^2 + 148.6 = 0$$

$$\Rightarrow \quad s = \pm j2.56.$$

8. **Angle of departure:**

$$\phi_{d1=(-2+j3)} = 180^\circ - (\phi_{p1} + \phi_{p2} + \phi_{p4})$$

$$= 180^\circ - (125^\circ + 90^\circ + 55^\circ) = -90^\circ$$

Similarly,

$$\phi_{d2=(-2-j3)} = +90^\circ$$

The complete root locus plot is as shown in Figure (c).

Problem-10.8: Sketch the root locus plot of the system whose OLTF is given as $G(s)H(s) = \dfrac{K(s+1)(s+2)}{(s+0.1)(s-1)}$.

Solution:

1. Root locus is symmetrical about real axis.
2. The root locus plot starts from a pole at $K = 0$ and terminates at an open-loop zero, that is, at $K = \infty$.
3. The poles are given as
   $s = 0.1, s = 1.$
   The zeros are given as
   $s = -1, s = -2$
   Number of poles $= P = 2$
   Number of zeros $= Z = 2$
   Number of root locus branches $= N = P = 2$
   Number of a symptotic lines $= n = P - Z = 0$
4. There is no need to find out the angle of asymptotic lines with negative real axis.
5. There is no need to find out the centroid.
6. **Break-away points:** The C.E. is given as

$$1 + G(s)H(s) = 0$$

$$1 + \frac{K(s+1)(s+2)}{(s+0.1)(s-1)} = 0$$

$$K = \frac{-(s+0.1)(s-1)}{(s+1)(s+2)} = \frac{s^2 - 0.9s - 0.1}{s^2 + 3s + 2}$$

$$\frac{dK}{ds} = 0 \Rightarrow 3.9s^2 + 4.2s - 1.5 = 0 \Rightarrow s_1 = +0.282,\ s_2 = -1.3589.$$

The valid break-away point are $b_1 = 0.282$ and $b_2 = -1.3589$.

7. **Intersection points of the root locus branches with imaginary axis:** The C.E. is given as

$$1 + G(s)H(s) = 0$$

$$s^2 - 0.9s - 0.1 + K(s^2 + 3s + 2) = 0$$

$$(1+K)s^2+(3K-0.9)s+2K-0.1=0$$

$$\begin{array}{c|cc} s^2 & 1+K & 2K-0.1 \\ s^1 & 3K-0.9 & 0 \\ s^0 & 2K-0.1 & 0 \end{array}$$

For the stable system, $3K-0.9>0$

$$K>0.3.$$

For the stable system, the maximum values of K is 0.3. For $K>0.3$, the roots lie on the RHS of the s-plane, and hence, $K=0.3$ is the value for marginal stability.

The auxiliary equation is

$$(1+K)s^2+2K-0.1=0$$

$$1.3s^2+0.5=0$$

$$s=\pm j0.62$$

The complete root locus plot is as shown in Figure (d).

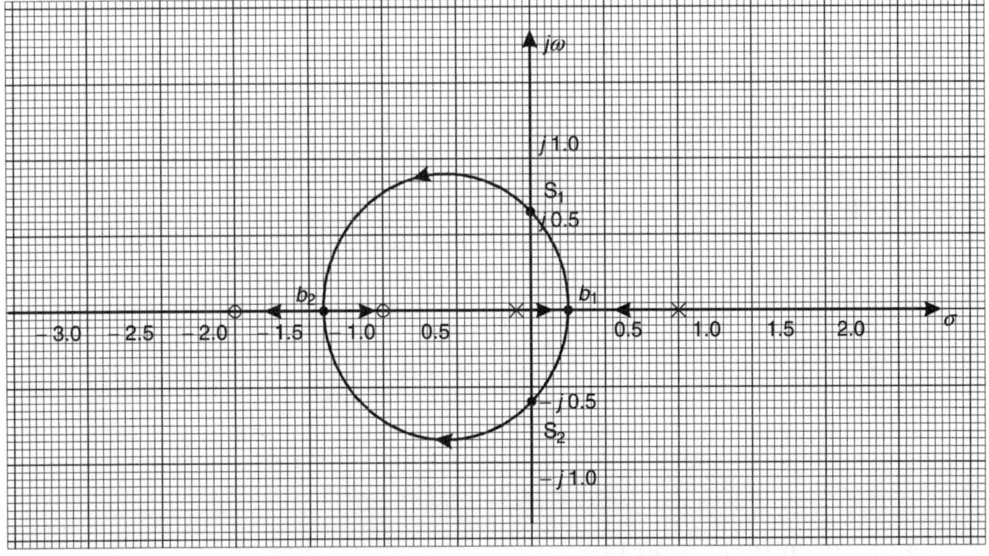

Figure (d) Root locus plot for Problem 10.8

Problem-10.9: The C.E. of the system is given as $s^3+2s^2+(K+1)s+3K=0$. Sketch the complete root locus.

Solution: Given that the C.E. of the system is

$$1+G(s)H(s)=0$$

$$s^3+2s^2+(K+1)s+3K=0$$

$$s^3+2s^2+s+(s+3)K=0$$

$$1+\frac{K(s+3)}{s(s^2+2s+1)}=0$$

$$\therefore \qquad G(s)H(s)\frac{K(s+3)}{s(s^2+2s+1)} = \frac{K(s+3)}{s(s+1)^2}.$$

1. Root locus is symmetrical about real axis.
2. The root locus plot starts from a pole at $K = 0$ and terminates at an open-loop zero, i.e., at $K = \infty$.
3. The poles are given as
   $s = 0, s = -1, s = -1$
   The zeros are given as
   $s = -3$
   Number of poles $= P = 3$
   Number of zeros $= Z = 1$
   Number of root locus branches $= N = P = 3$
   Number of asymptotic lines $= n = P - Z = 2$
4. The angle of asymptotic lines with negative real axis
   $$\phi = \frac{(2q+1)180°}{P-Z}, \text{ where } q = 0, 1$$
   $$= 180°/2,\ 3 \times 180°/2$$
   $$= 90°,\ 270°.$$
5. The centroid is given as
   $$\text{Centroid} = \frac{\text{Sum of poles} - \text{Sum of zeros}}{P-Z} = \frac{(0-1-1)-(-3)}{3-1} = 0.5.$$
6. **Break-away points:** The C.E. of the system is
   $$1 + G(s)H(s) = 0$$
   $$s^3 + 2s^2 + (K+1)s + 3K = 0$$
   $$1 + \frac{K(s+3)}{s(s^2+2s+1)} = 0$$
   $$K = \frac{-(s^3+2s^2+s)}{s+3}$$
   $$\frac{dK}{ds} = 0 \Rightarrow 2s^3 + 11s^2 + 12s + 3 = 0$$
   $\Rightarrow \quad s_1 = -1, s_2 = -0.3625, s_3 = -4.1375$
   The valid break-away points are $b_1 = -0.3625$, $b_2 = -1$ and $b_3 = -4.1375$.
7. **Intersection point of the root locus branches with imaginary axis**
   The C.E. of the system is
   $$1 + G(s)H(s) = 0$$
   $$s^3 + 2s^2 + (K+1)s + 3K = 0$$
   $$\begin{array}{c|cc} s^3 & 1 & 1+K \\ s^2 & 2 & 3K \\ s^1 & (2-K)/2 & 0 \\ s^0 & 3K & 0 \end{array}$$

For a stable system, $\quad K > 0, (2-K)/2 > 0$

$$\Rightarrow \quad K > 0, K < 2$$

For a stable system, the maximum value of K is 2. For $K > 2$, the roots lie on the RHS of the s-plane, and hence, $K = 2$ is the value for marginal stability.

The auxiliary equation is

$$2s^2 + 3K = 0$$

$$2s^2 + 6 = 0$$

$$\Rightarrow \quad s = \pm j1.732.$$

8. **Angle of departure:** No need to find out angle of departure.
   The complete root locus plot is as shown in the Figure (e).

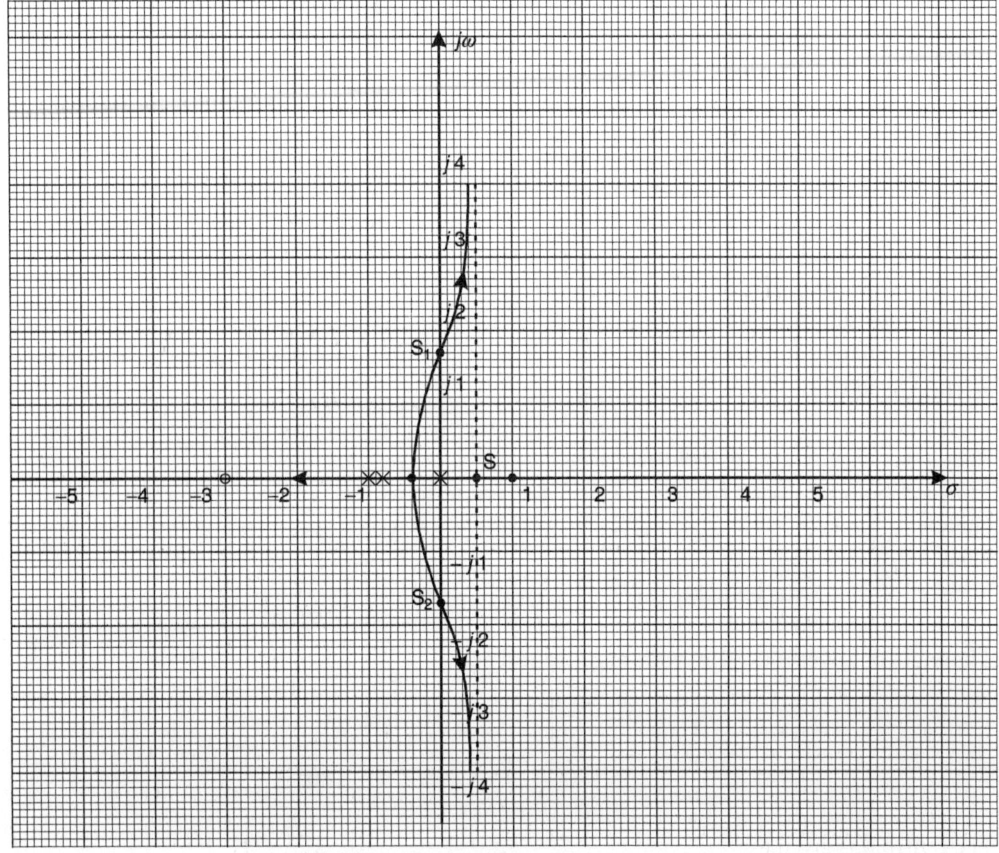

Figure (e) Root locus plot for Problem 10.9

10.7 GAIN MARGIN AND PHASE MARGIN FROM ROOT LOCUS

10.7.1 Gain Margin

The ratio of K at imaginary crossover (i.e., stability boundary) to the design value of K of the system is called the value of gain margin.

$$\text{Gain margin(GM)} = \frac{\text{Value of } K \text{ at imaginary crossover}}{\text{Design value of } K}.$$

If the root locus does not cross the $j\omega$-axis, gain is ∞. Gain margin indicates the maximum gain that can be multiplied for the system to be just on the verge of instability.

10.7.2 Phase Margin

Phase margin (PM) can be determined for a given value of K as follows:

1. Calculate ω, for which $|G(j\omega)H(j\omega)| = 1$ for the given value of K by using trial-and-error method.
2. Calculate Arg$[G(j\omega)H(j\omega)]$.
3. Phase margin $180° + \text{Arg}[G(jw)H(j\omega)]$.

10.8 DAMPING RATIO FROM ROOT LOCUS

To determine the open-loop gain k for a given value of damping ratio (ξ) of the dominant roots or vice versa, a damping line making an angle of $\theta = \cos^{-1}\xi$ with negative real axis is drawn for the specified damping. In the region where the rough root locus sketch intersects the (ξ) line, by trial-and-error procedure, the point of intersection is determined along the (ξ) line. This point is the desired root, at which the open-loop gain is calculated.

***Problem-10.10*:** Sketch the root locus plot for the system whose OLTF is given as $G(s)H(s) = \dfrac{K}{s(s+1)(s+3)}$. Determine

1. The value of K for $\xi = 0.5$
2. The gain margin for $K = 6$
3. The phase margin for $K = 6$

***Solution*:** Given that

$$G(s)H(s) = \frac{K}{s(s+1)(s+3)}.$$

1. Root locus is symmetrical about real axis.
2. The root locus plot starts from a pole at $K = 0$ and terminates at an open-loop zero $K = \infty$.
3. Number of poles = $P = 3$
   Number of zeros = $Z = 0$
   Number of root locus branches = $N = P = 3$
   Number of asymptotic lines = $n = P - Z = 3$
4. The angle of asymptotic lines with negative real axis

$$\phi = \frac{(2q+1)180°}{P-Z}, q = 0, 1, 2$$

$$= 180°/3, 3 \times 180°/3, 5 \times 180°/3$$
$$= 60°, 180°, 300°.$$

5. The centroid is given as

$$\text{Centroid} = \frac{\text{Sum of poles}}{P-Z} = -\frac{4}{3} = -1.333$$

6. **Break-away points:** The C.E. is given as

$$1 + G(s)H(s) = 0$$

$$K = -s(s+1)(s+3)$$
$$K = -(s^3 + 4s^2 + 3s)$$
$$\frac{dK}{ds} = 0 \Rightarrow 3s^2 + 8s + 3 = 0 \Rightarrow s_1 = -0.45, s_2 = -2.21.$$

The valid break-away point is $b_1 = -0.45$ (since K is positive for s_1).

7. **Intersection points of the root locus branches with imaginary axis:** The C.E. is given as $1 + G(s)H(s) = 0$

$$1+\frac{K}{s(s+1)(s+3)}=0$$

$$K = -s(s+1)(s+3)$$

$$s^3 + 4s^2 + 3s + K = 0$$

$$\begin{array}{c|cc} s^3 & 1 & 3 \\ s^2 & 4 & K \\ s^1 & (12-K)/4 & 0 \\ s^0 & K & 0 \end{array}$$

For a stable system, $K > 0$, $(12 - K)/4 > 0$

$\Rightarrow$ $K > 0, K < 12$

For a stable system, the maximum value of K is 12. For $K > 12$, the roots lie on the RHS of the s-plane, and hence, $K = 12$ is the value for marginal stability.

The auxiliary equation is

$$4s^2 + K = 0$$

$$4s^2 + 12 = 0$$

$\Rightarrow$ $s = \pm j1.73.$

The complete root locus plot for the abovementioned system is as shown in Figure (f).

1. **Value of K for $\xi = 0.5$.**
   Since $\xi = 0.5 \Rightarrow \cos\theta = \xi = 0.5 \Rightarrow \theta = 60°$.
   From the origin, a line at an angle $\theta = 60°$ is drawn, which intersects the root locus plot at $s = -0.35 + j0.6$. As this point lies on the root locus, the following equation is satisfied.

$$|G(s)H(s)| = 1$$

$$\left|\frac{K}{s(s+1)(s+3)}\right| = 1$$

$$\left|\frac{K}{(-0.35+j0.6)(-0.35+j0.6+1)(-0.35+j0.6+3)}\right| = 1$$

$\Rightarrow$ $$\frac{K}{1.6} = 1 \Rightarrow K = 1.6.$$

2. **Gain margin:**
   The desired value of $K = 6$ and the value for marginal stability as obtained from root locus is $K = 12$.

$$\therefore \text{Gain margin} = \frac{K_{\text{marginal stability}}}{K_{\text{desired}}} = \frac{12}{6} = 2$$

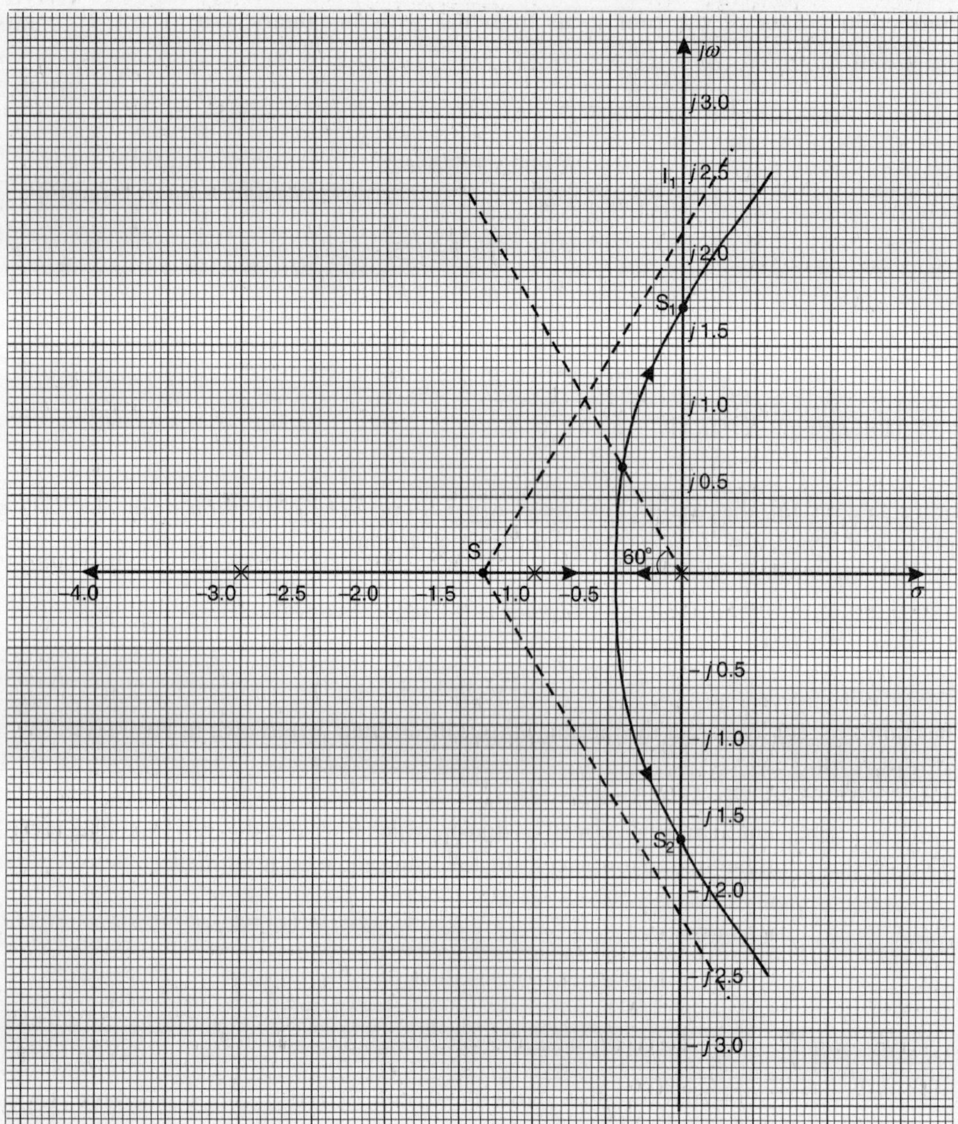

Figure (f) Root locus plot for Problem 10.10

3. **Phase margin:**

   To determine phase margin for a given value of $K = 6$, locate a point $j\omega_1$ on the imaginary axis of the root locus plot such that $|G(j\omega_1)H(j\omega_1)| = 1$; thus, ω_1 is the gain crossover frequency (GCF). The point $j\omega_1$ is located by trial-and-error method.
   Substitute $s = j\omega$, $K = 6$

$$|G(j\omega)H(j\omega)| = \left|\frac{6}{j\omega(1+j\omega)(3+j\omega)}\right| = 1$$

$\Rightarrow$
$$\frac{6}{\omega\sqrt{1+\omega^2}\sqrt{9+\omega^2}} = 1$$

$\Rightarrow$
$$\omega\sqrt{1+\omega^2}\sqrt{9+\omega^2} - 6 = 0$$

Let
$$f(\omega_1) = \omega_1\sqrt{1+\omega_1^2}\sqrt{9+\omega_1^2} - 6$$

$$f(1) = -1.5278, f(1.4) = 1.97$$

Since there is a change of sign, root lies between 1 and 1.4.

$f(1.2) = 0$, and hence, $\omega_1 = 1.2$ rad/s is the GCF.

Phase margin = $180 + (-\phi_{p1} - \phi_{p2} - \phi_{p3}) = 180° - 90° - 50.2° - 21.8° = 18°$.

10.9 OTHER PROPERTIES OF ROOT LOCUS

The properties of the root locus w.r.t. the open-loop pole–zero configuration as described, below are necessary in understanding the system performance.

10.9.1 Addition of Open-loop Poles

The addition of a pole to the OLTF pulls the root locus to the right, thereby making the system less stable and slow in response.

1. Consider a system of OLTF

$$G(s)H(s) = \frac{k}{s(s+a)} \tag{10.7}$$

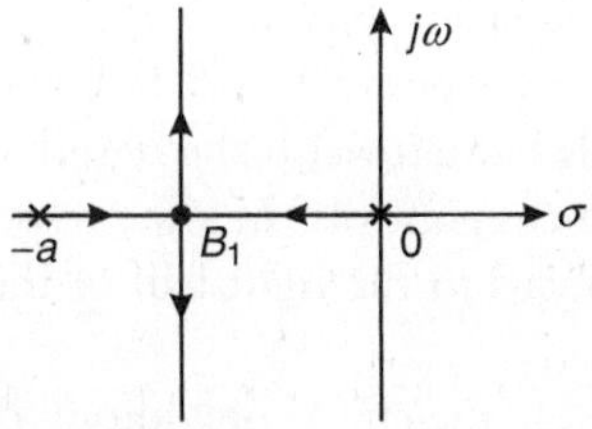

Figure 10.1 Root locus for $G(s)H(s) = k/s(s + a)$

The root locus of (10.7) is sketched in Figure 10.1 having the break-away point B_1.

2. When a simple pole at $s = -b$ is added to the OLTF system, (10.7) reduces to

$$G(s)H(s) = \frac{k}{s(s+a)(s+b)}. \tag{10.8}$$

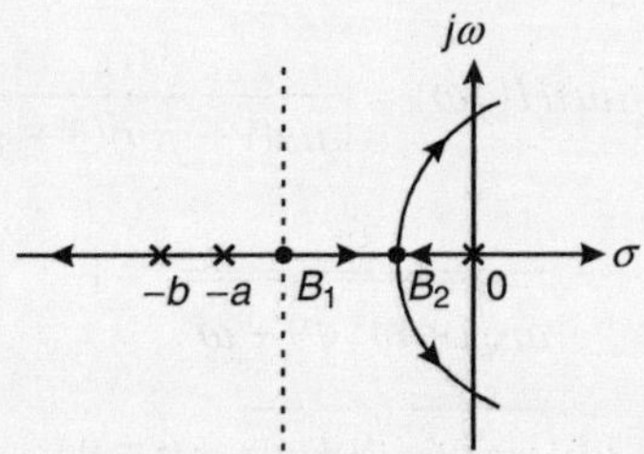

Figure 10.2 Root Locus for $G(s)H(s) = k/s(s + a)(s + b)$

The root locus of (10.8) is sketched in Figure 10.2. It is noted from this figure that by adding the pole, the break-away point has shifted from B_1 to B_2 and the root locus has bent towards the right half of the s-plane.

3. If one more pole at $s = -c$ is added to the OLTF, (10.8) changes to

$$G(s)H(s) = \frac{k}{s(s+a)(s+b)(s+c)}. \tag{10.9}$$

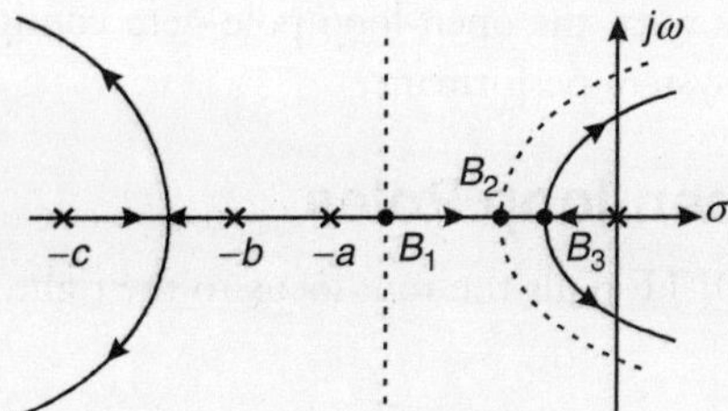

Figure 10.3 Root Locus for $G(s)H(s) = k/s(s + a)(s + b)(s + c)$

The root locus of (10.9) is sketched in Figure 10.3. It is noted from this figure that the addition of one more pole pulls the break-away point B_2 to B_3 to further towards the right half of the s-plane and the conjugate root locus bends more towards the right half of the s-plane.

Since the conjugate root locus controls the time response of the system, the addition of a pole to the OLTF pulls the root locus to the right half of the s-plane, thereby making the system less stable.

In summary, the following are the effects of addition of poles to the OLTF.

1. There is change in shape of the root locus and it shifts towards the imaginary axis.
2. The intercept on the $j\omega$-axis occurs for a lower value of k because of the asymptote's angle being lower down.
3. System becomes oscillatory.
4. Gain margin and relative stability decreases.
5. There is a reduction in the range of k.
6. A sluggish response can be changed to a quick response for the artful introduction of a pole.
7. Settling time increases.

10.9.2 Addition of Open-loop Zeros

The addition of a zero to the system produces the effect of pulling the root locus to the left half of the s-plane, thereby making the system more stable and fast in response.

1. Consider the system whose OLTF is

$$G(s)H(s) = \frac{k}{s(s+a)}. \tag{10.10}$$

The root locus of (10.10) is drawn as shown in Figure 10.4 with a break-away point B_1.

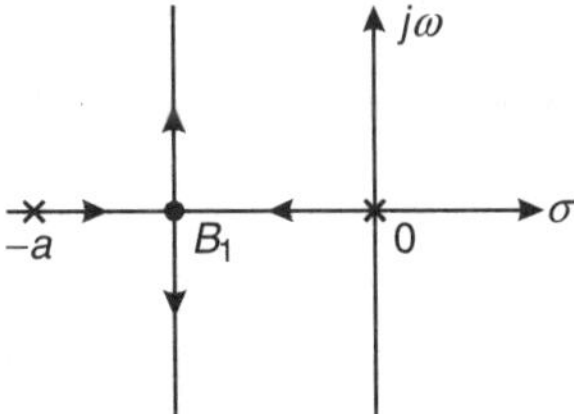

Figure 10.4 Root Locus for $G(s)H(s) = k/s(s+a)$

2. On adding a simple zero at $s = -b$ to the OLTF of the system (i.e., given as in 10.10), we have

$$G(s)H(s) = \frac{k(s+b)}{k(s+a)}. \tag{10.11}$$

The root locus for (10.11) is shown in Figure 10.5. From this figure, it is noted that the break-away point B_1 has shifted to the left from B_1 to B_2 and the root locus has bent towards the left half of the s-plane.

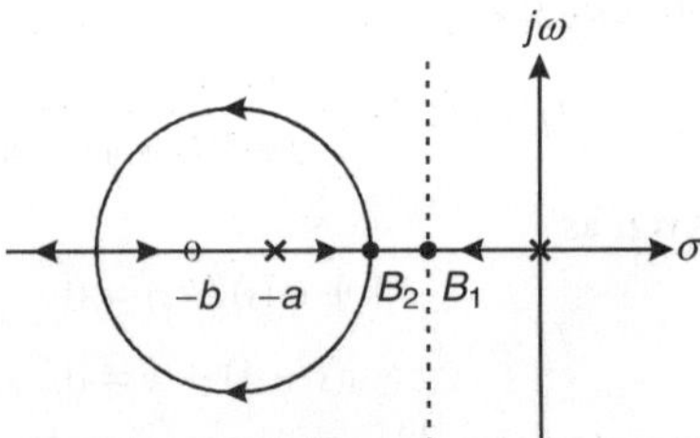

Figure 10.5 Root Locus for $G(s)H(s) = k(s+b)/s(s+a)$

3. If a pair of complex conjugate zero is added to the OLTF of the system, (10.10) changes to

$$G(s)H(s) = \frac{k(s^2+cs+b)}{s(s+a)}. \tag{10.12}$$

The root locus of (10.12) is plotted in Figure (10.6), from which it is noted that by adding these zeros the break-away point has shifted to the left and the root locus has bent towards the left half of the s-plane.

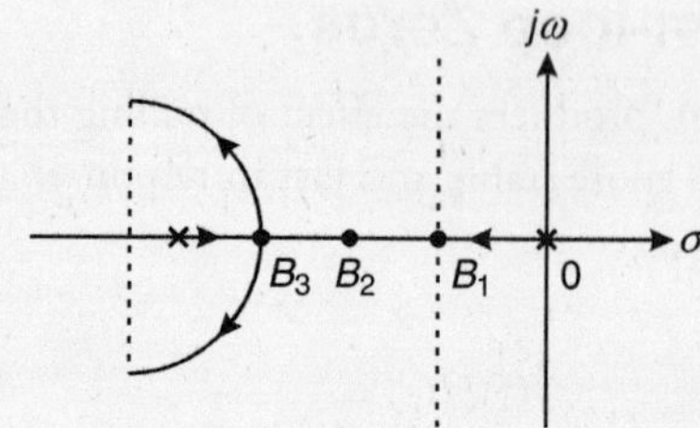

Figure 10.6 Root Locus for $G(s)H(s) = k(s^2 + cs + b)/s(s + a)$

In summary, the following are the effects of addition of zeros to the OLTF

1. There is change in the shape of the root locus and it shifts to the left half of the *s*-plane.
2. Stability of the system is enhanced.
3. Range of *k* increases.
4. Settling time speeds up.

10.10 CANCELLATION OF POLES OF *G*(*s*) AND ZEROS OF *H*(*s*)

Let us consider

$$G(s) = \frac{k}{s(s+3)(s+4)},\ H(s) = s+3$$

$$\text{t.f.} = \frac{C(s)}{R(s)} = \frac{G(s)}{1+G(s)H(s)} = \frac{\dfrac{k}{s(s+3)(s+4)}}{1+\dfrac{k(s+3)}{s(s+3)(s+4)}} = \frac{k}{s(s+3)[s(s+4)+k]}.$$

In root locus, $G(s)H(s)$ is considered as

$$G(s)H(s) = \frac{k(s+3)}{s(s+3)(s+4)} = \frac{k}{s(s+4)}$$

The closed-loop poles are given as

$$1 + G(s)H(s) = 0$$

$$s(s + 4) + k = 0.$$

The term $(s + 3)$ in the numerator and denominator cancel each other. Again $(s + 3)$ is a closed-loop pole, and therefore, the equation is wrong. In the above-mentioned case, the cancelled pole of $G(s)$ must be added as a closed-loop pole. If zero of $G(s)$ and pole of $H(s)$ are cancelled, no correction is required.

Problem-10.11: Sketch the root locus plot for the system having $G(s) = \dfrac{K}{s+1}$, $H(s) = \dfrac{s+1}{s^2+4s+5}$.

Solution:

In this problem, cancellation of poles of $G(s)$ at $s = -1$ and zero of $H(s)$ at $s = -1$ takes place. Therefore, as a closed-loop pole, at $s = -1$ must be added.

$$G(s)H(s) = \frac{K}{(s+1)(s^2+4s+5)}.$$

1. Root locus is symmetrical about real axis.
2. The root locus plot starts from a pole at $K = 0$ and terminates at an open-loop zero, that is, at $K = \infty$.
3. The poles are given as
   $s = -1$, $s^2 + 4s + 5 = 0$
   $P_1 = -1$, $P_2 = -2 + j$, $P_3 = -2 - j$
   Number of poles = $P = 3$
   Number of zeros = $Z = 0$
   Number of root locus branches = $N = P = 3$
   Number of asymptotic lines = $n = P - Z = 3$
4. The angle of asymptotic lines with negative real axis

$$\phi = \frac{(2q+1)\,180^\circ}{P-Z}, \text{ where } q = 0, 1, 2$$

$$= 180^\circ/3,\ 3\times 180^\circ/3,\ 5\times 180^\circ/3$$

$$= 60^\circ,\ 180^\circ,\ 300^\circ.$$

5. The centroid is given as

$$\text{Centroid} = \frac{\text{Sum of poles} - \text{Sum of zeros}}{P-Z} = \frac{(-1-2+j-2-j)-(0)}{3-0} = -\frac{5}{3} = -1.667$$

6. **Break-away points:**
   The C.E. is given as

$$1 + G(s)H(s) = 0$$

$$1 + \frac{K}{(s+1)(s^2+4s+5)} = 0$$

$$K = -(s+1)(s^2+4s+5)$$

$$\frac{dK}{ds} = 0 \Rightarrow 3s^2 + 10s + 9 = 0$$

$$\Rightarrow \quad s_1 = -1.667 + j0.4707,\ s_2 = -1.667 - j0.4707.$$

   The points are not on the root locus. Therefore, there are no break-away points.
7. **Intersection points of the root locus branches with imaginary axis:**
   The C.E. is given as

$$1 + G(s)H(s) = 0$$

$$s^3 + 5s^2 + 9s + 5 + K = 0$$

s^3	1	9
s^2	5	$5+K$
s^1	$(40-K)/5$	0
s^0	$K+5$	0

For a stable system $\quad K > 0,\ (40-K)/5 > 0$

$\Rightarrow \quad K > 0,\ K < 40$

For a stable system, the maximum value of K is 40. For $K > 40$, the roots lie on the RHS of the s-plane, and hence, $K = 40$ is the value for marginal stability.

The auxiliary equation is $5s^2 + 5 + K = 0$

$$5s^2 + 45 = 0$$

$$\Rightarrow \qquad s = \pm j3.$$

8. **Angle of departure:**

$$\phi_{d1=(-2+j)} = 180° - (\phi_{p3} + \phi_{p4})$$

$$= 180° - (135° + 90°) = -45°$$

$$\phi_{d2=(-2-j)} = +45°.$$

The complete root locus plot for the above-mentioned system is shown in Figure (g).

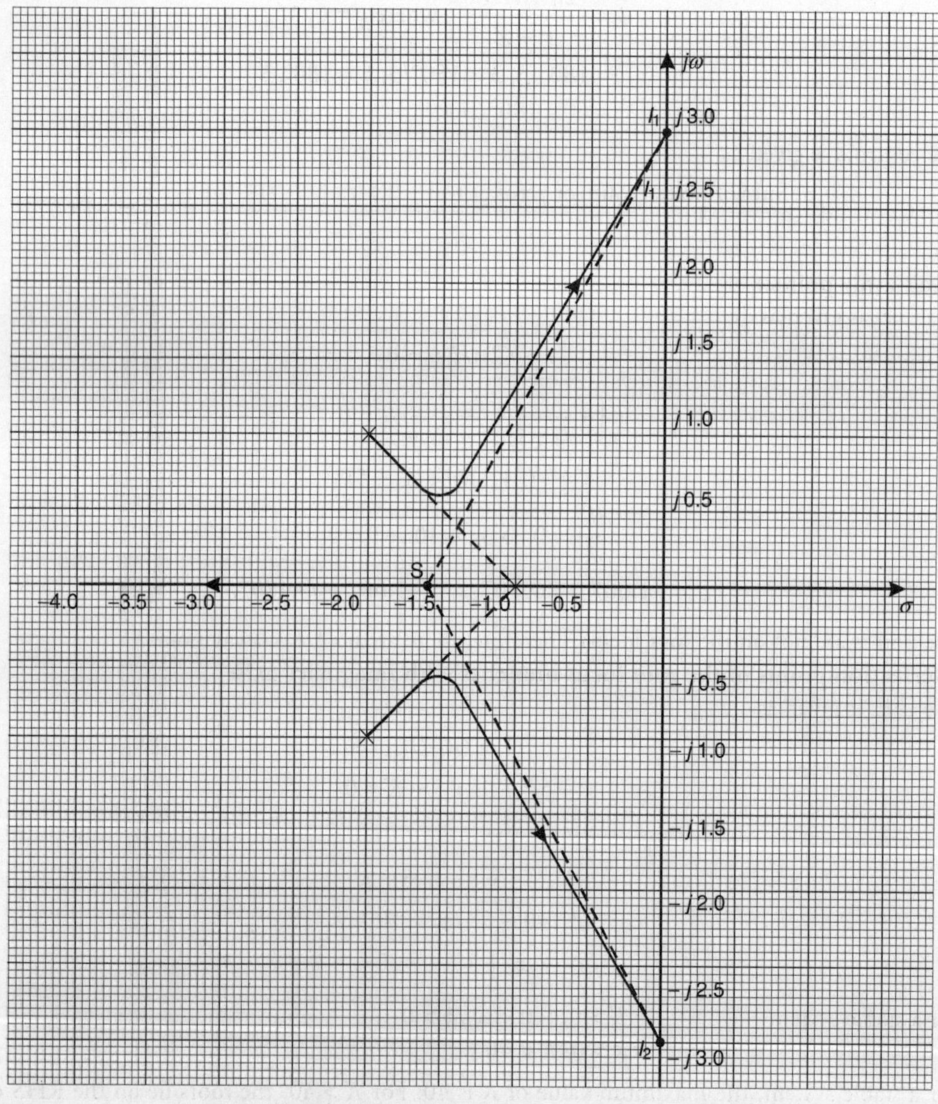

Figure (g) Root locus plot for Problem 10.11

ADDITIONAL SOLVED PROBLEMS

ASP-1: Calculate the angle of asymptotes and the centroid for the system having $G(s)H(s)=\dfrac{K(s+3)}{s(s+2)(s+4)(s+5)}$.

Solution:

Given that

$$G(s)H(s)=\frac{K(s+3)}{s(s+2)(s+4)(s+5)}.$$

The poles are given as

$$s(s+2)(s+5)(s+4)=0$$

$$s=0,\ s=-2,\ s=-5,\ s=-4.$$

The zeros are given as

$$s=-3$$

Number of poles = $P=4$.
Number of zeros = $Z=1$.
Number of root locus branches = $N=P=4$.
Number of asymptotic lines = $n=P-Z=3$.

1. The angle of asymptotic lines with negative real axis

$$\phi=\frac{(2q+1)180^\circ}{P-Z},\ \text{where } q=0,\ 1,\ 2$$

$$=1\times 180^\circ/3,\ 3\times 180^\circ/3,\ 5\times 180^\circ/3$$

$$=60^\circ,\ 180^\circ,\ 300^\circ.$$

2. The centroid is given as

$$\text{Centroid}=\frac{\text{Sum of poles}-\text{Sum of zeros}}{P-Z}=\frac{(0-2-4-5)-(-3)}{4-1}=-2.667.$$

ASP-2: Sketch the root locus plot of the system whose OLTF is given as $G(s)H(s)=\dfrac{K}{s(s^2+8s+32)}$.

Solution:

Given that

$$G(s)H(s)=\frac{K}{s(s^2+8s+32)}.$$

1. Root locus is symmetrical about real axis.
2. The root locus plot starts from a pole at $K=0$ and terminates at an open-loop zero, that is, at $K=\infty$.
3. The poles are given as
   $s=0,\ s^2+8s+32=0$
   $P_1=0,\ P_2=-4+j4,\ P_3=-4-j4$
   Number of poles = $P=3$
   Number of zeros = $Z=0$

Number of root locus branches = $N = P = 3$
Number of asymptotic lines = $n = P - Z = 3$.

4. The angle of asymptotic lines with negative real axis

$$\phi = \frac{(2q+1)180°}{P-Z}, \text{ where } q = 0, 1, 2$$

$$= 180°/3,\ 3 \times 180°/3,\ 5 \times 180°/3$$

$$= 60°,\ 180°,\ 300°.$$

5. The centroid is given as

$$\text{Centroid} = \frac{\text{Sum of poles} - \text{Sum of zeros}}{P-Z} = \frac{(0-4+j4-4-j4)-0}{3-0} = -\frac{8}{3} = -2.667.$$

6. **Break-away points:**
The C.E. is given as

$$1 + G(s)H(s) = 0$$

$$1 + \frac{K}{s(s^2+8s+32)} = 0$$

$$K = -s(s^2 + 8s + 32)$$

$$K = -(s^3 + 8s^2 + 32s)$$

$$\frac{dK}{ds} = 0 \Rightarrow 3s^2 + 16s + 32 = 0$$

$$\Rightarrow \quad s_1 = 2.667 + j1.88,\ s_2 = -2.667 - j1.88.$$

The points are not on the root locus. Therefore, there are no break-away points.

7. **Intersection points of the root locus branches with imaginary axis:**
The C.E. is given as

$$1 + G(s)H(s) = 0$$

$$s^3 + 8s^2 + 32s + K = 0$$

s^3	1	32
s^2	8	K
s^1	$(256-K)/8$	0
s^0	K	0

For a stable system, $K > 0$, $(256 - K)/8 > 0$

$$\Rightarrow \quad K > 0,\ K < 256$$

For a stable system, the maximum value of K is 256. For $K > 256$, the roots lie on the RHS of the s-plane, and hence, $K = 256$ is the value for marginal stability.
The auxiliary equation is

$$8s^2 + K = 0$$

$$8s^2 + 256 = 0$$

$$\Rightarrow \quad s = \pm j5.656.$$

8. **Angle of departure:**

$$\phi_{d1=(-4+j4)} = 180° - (\phi_{p3})$$

$$= 180° - (225°) = -45°$$

$$\phi_{d2=(-4-j4)} = +45° .$$

The complete root locus plot for the above-mentioned system is as shown in Figure (h).

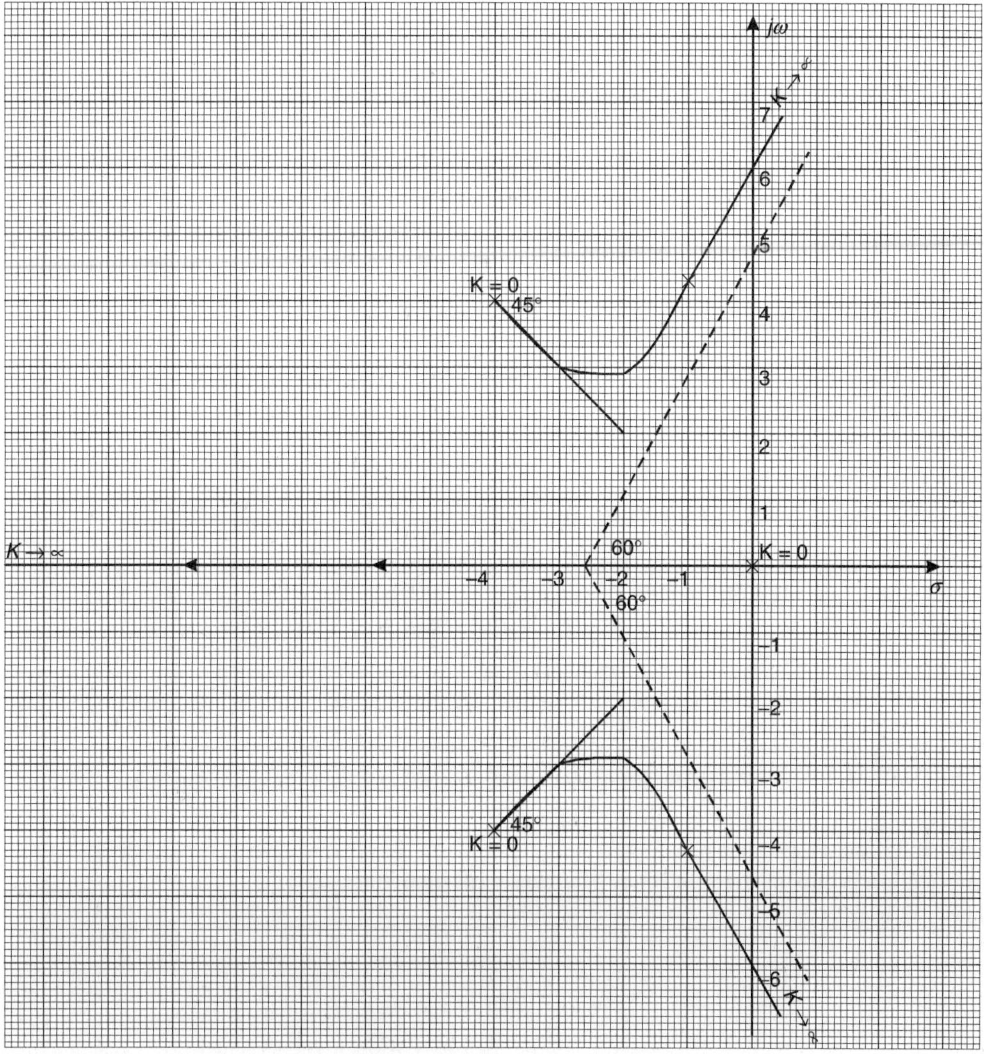

Figure (h) Root locus plot for ASP-2

ASP-3: Sketch the root locus plot of the system whose OLTF is given as $G(s)H(s) = \dfrac{K}{s(s+2)(s^2+2s+2)}$.

Solution:

Given that

$$G(s)H(s) = \frac{K}{s(s+2)(s^2+2s+2)}.$$

1. Root locus is symmetrical about the real axis.
2. The root locus plot starts from a pole at $K = 0$ and terminates at an open-loop zero, that is, at $k = \infty$.

3. The poles are given as
   $s = 0,\ s = -2,\ s^2 + 2s + 2 = 0$
   $P_1 = 0,\ P_2 = -2,\ P_3 = -1 + j1,\ P_4 = -1 - j1$
   Number of poles = $P = 4$
   Number of zeros = $Z = 0$
   Number of root locus branches = $N = P = 4$
   Number of asymptotic lines = $n = P - Z = 4$
4. The angle of asymptotic lines with negative real axis

$$\phi = \frac{(2q+1)180^\circ}{P-Z},\ \text{where } q = 0, 1, 2, 3$$

$$= 180^\circ/4,\ 3 \times 180^\circ/4,\ 5 \times 180^\circ/4,\ 7 \times 180^\circ/4$$

$$= 45^\circ,\ 135^\circ,\ 225^\circ,\ 315^\circ.$$

5. The centroid is given as

$$\text{Centroid} = \frac{\text{Sum of poles} - \text{Sum of zeros}}{P-Z} = \frac{(0-2-1+j1-1-j1)-(0)}{4-0} = -1.$$

6. **Break-away points:**
   The C.E. is given as

$$1 + G(s)H(s) = 0$$

$$1 + \frac{K}{s(s+2)(s^2+2s+2)} = 0$$

$$K = -s(s+2)(s^2+2s+2)$$

$$K = -s(s^4 + 4s^3 + 6s^2 + 4s)$$

$$\frac{dK}{ds} = 0 \Rightarrow 4s^3 + 12s^2 + 12s + 4 = 0$$

$$\Rightarrow \quad s^3 + 3s^2 + 3s + 1 = 0$$

$$\Rightarrow \quad (s+1)^3 = 0 \Rightarrow s = -1$$

   The valid break-away point is $b_1 = -1$.
7. **Intersection points of the root locus branches with imaginary axis:**
   The C.E. is given as

$$1 + G(s)H(s) = 0$$

$$s^4 + 4s^3 + 6s^2 + 4s + K = 0$$

s^4	1	6	K
s^3	4	4	0
s^2	5	K	0
s^1	$(20-4K)/5$	0	0
s^0	K	0	0

For a stable system, $\quad K > 0,\ (20 - 4K)/4 > 0$

$\Rightarrow \quad K > 0,\ K < 5$

For a stable system, the maximum value of K is 5. For $K > 5$, the roots lie on the RHS of the s-plane, and hence, $K = 5$ is the value for marginal stability.

The auxiliary equation is

$$5s^2 + K = 0$$

$$\Rightarrow \qquad 5s^2 + 5 = 0$$

$$\Rightarrow \qquad s = \pm j1$$

8. **Angle of departure:**

$$\phi_{d1=(-1+j1)} = 180° - (\phi_{p1} + \phi_{p2} + \phi_{p4})$$

$$= 180° - (135° + 45° + 90°) = -90°$$

$$\phi_{d2=(-1-j1)} = +90°$$

The complete root locus plot is as shown in Figure (i).

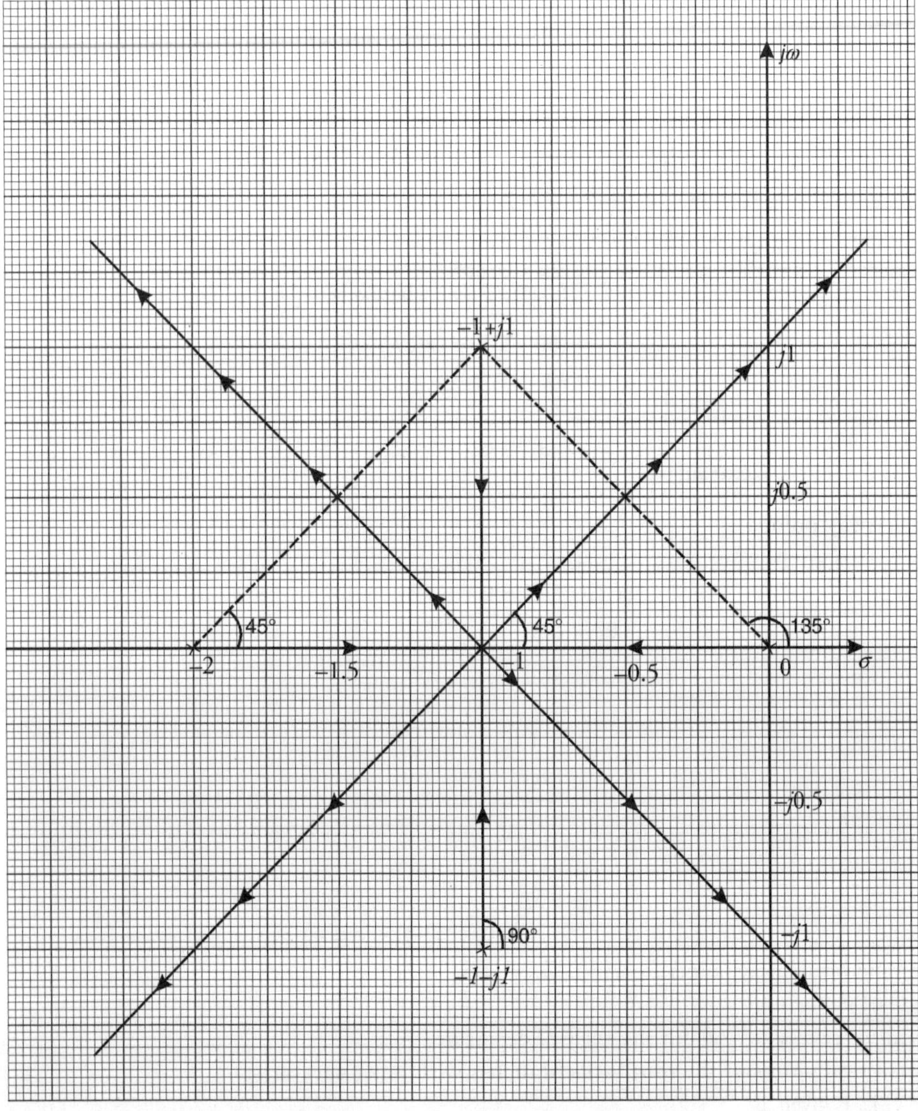

Figure (i) Root locus plot for ASP-3

ASP-4: Sketch the root locus plot for a unity feedback system whose OLTF is given as $G(s)=\dfrac{K(s+0.5)}{s^2(s+4.5)}$.

Solution:

Given that

$$G(s)=\frac{K(s+0.5)}{s^2(s+4.5)}.$$

1. Root locus is symmetrical about real axis.
2. The root locus plot starts from a pole at $K=0$ and terminates at an open-loop zero, i.e., at $K=\infty$.
3. The poles are given as
   $s=0,\ s=0,\ s=-4.5$
   $P_1=0,\ P_2=0,\ P_3=-4.5$
   The zeros are given as
   $s=-0.5$
   Number of poles $= P = 3$
   Number of zeros $= Z = 1$
   Number of root locus branches $= N = P = 3$
   Number of asymptotic lines $= n = P - Z = 2$
4. The angle of asymptotic lines with negative real axis

$$\phi=\frac{(2q+1)180°}{P-Z},\quad \text{where } q=0,1$$
$$=180°/2,\ 3\times 180°/2$$
$$=90°,\ 270°.$$

5. The centroid is given as

$$\text{Centroid}=\frac{\text{Sum of poles}-\text{Sum of zeros}}{P-Z}=\frac{-4.5-(-0.5)}{2}=-2.$$

6. **Break-away points:**
   The C.E. is given as

$$1+G(s)H(s)=0$$
$$1+\frac{K(s+0.5)}{s^2(s+4.5)}=0$$
$$\Rightarrow \quad K=-\frac{s^2(s+4.5)}{s+0.5}$$
$$\frac{dK}{ds}=0 \Rightarrow 2s^3+6s^2+4.5s=0$$
$$\Rightarrow \quad s(s^2+3s+2.25)=0$$
$$\Rightarrow \quad s(s+1.5)^2=0$$
$$\Rightarrow \quad s_1=0,\ s_2=-1.5,\ s_3=-1.5.$$

   The valid break-away points are $b_1=0,\ b_2=-1.5$.
7. **Angle of departure:**
   It is not necessary to find out the angle of departure.
8. **Intersection points of the root locus branches with imaginary axis:**
   The C.E. is given as

$$1+G(s)H(s)=0$$

$$1+\frac{K(s+0.5)}{s^2(s+4.5)}=0$$

$\Rightarrow$ $$s^3+4.5s^2+Ks+0.5K=0$$

s^3	1	K	0
s^2	4.5	$0.5K$	0
s^1	$0.88K$	0	0
s^0	$0.5K$	0	0

For a stable system, $0.88K > 0,\ 0.5K > 0$

$\Rightarrow$ $K > 0.$

The range of values of K is $0 < K < \infty$.

The complete root locus plot is as shown in Figure (j).

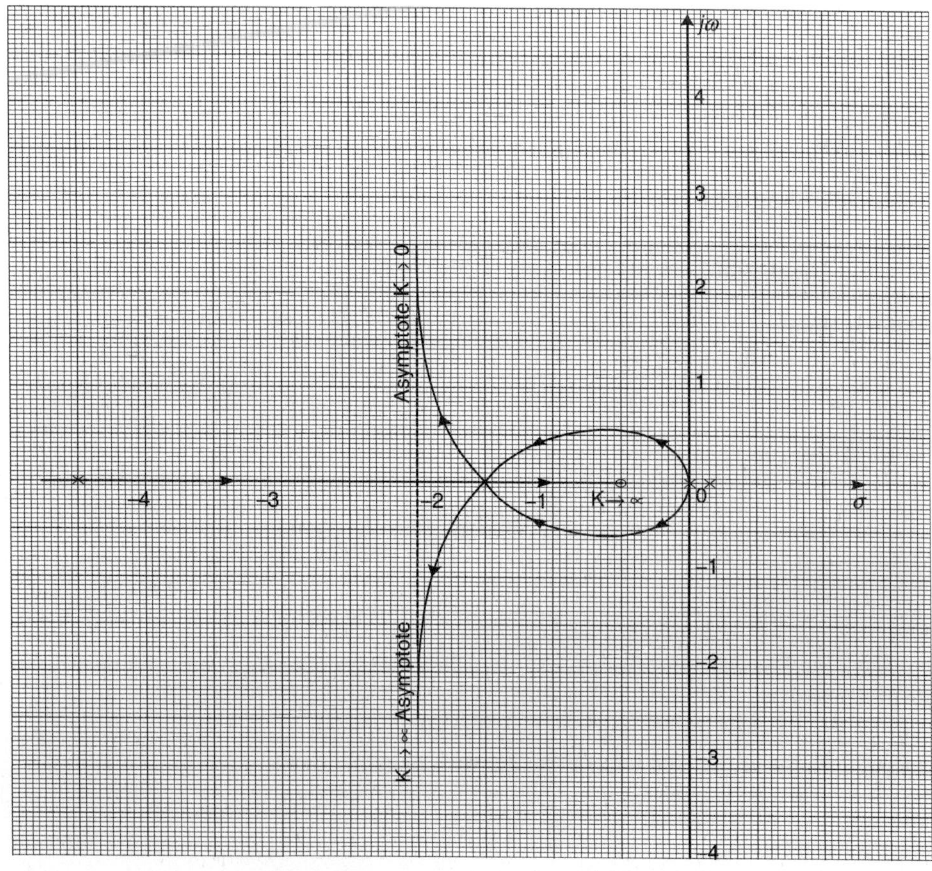

Figure (j) Root locus plot for ASP-4

ASP-5: Sketch the root locus plot for a unity feedback system whose OLTF is given as $G(s)=\frac{K(s+1)}{(s-1)(s+2)(s+4)}$. Find the range of K for which the system is stable.

Solution:

Given that

$$G(s) = \frac{K(s+1)}{(s-1)(s+2)(s+4)}.$$

1. Root locus is symmetrical about the real axis.
2. The root locus plot starts from a pole at $K = 0$ and terminates at an open-loop zero, that is, at $K = \infty$.
3. The poles are given as
   $s = 1, s = -2, s = -4$
   $s = -1$
   Number of poles = $P = 3$
   Number of zeros = $Z = 1$
   Number of root locus branches = $N = P = 3$
   Number of asymptotic lines = $n = P - Z = 2$
4. The angle of asymptotic lines with negative real axis
$$\phi = \frac{(2q+1)180°}{P-Z}, \text{ where } q = 0, 1$$
$$= 180°/2, 3 \times 180°/2$$
$$= 90°, 270°$$
5. The centroid is given as
$$\text{Centroid} = \frac{\text{Sum of poles} - \text{Sum of zeros}}{P-Z} = \frac{(1-2-4)-(-1)}{2} = -2.$$
6. **Break-away points:**
   The C.E. of the system is given as
$$1 + G(s)H(s) = 0$$
$$1 + \frac{K(s+1)}{(s-1)(s+2)(s+4)} = 0.$$
$$K = -\frac{(s^3 + 5s^2 + 2s - 8)}{s+1}$$
$$\frac{dK}{ds} = 0 \Rightarrow 2s^3 + 8s^2 + 10s + 10 = 0 \Rightarrow s^3 + 4s^2 + 5s + 5 = 0.$$
   After solving the above-mentioned equation, the valid break-away points is $b_1 = -2.863$
7. **Angle of departure:**
   It is not necessary to find out the angle of departure.
8. **Intersection points of the root locus branches with imaginary axis:**
   The C.E. of the system is given as
$$1 + G(s)H(s) = 0$$
$$1 + \frac{K(s+1)}{(s-1)(s+2)(s+4)} = 0$$
$$s^3 + 5s^2 + 2s - 8 = -K(s + 1)$$
$$s^3 + 5s^2 + (2 + K)s - 8 + K = 0$$

s^3	1	$2+K$
s^2	5	$-8+K$
s^1	$\frac{4K+18}{5}$	0
s^0	$-8+K$	0

For a stable system, $-8 + K > 0$, $2 + 6K > 0$

$\Rightarrow$ $K > 8$, $K > -1/3$

The range of K for the system being stable is $8 < K < \infty$. The asymptotes are 90° and 270° so that the locus do not cross the $j\omega$-axis.

The complete root locus plot is as shown in Figure (k).

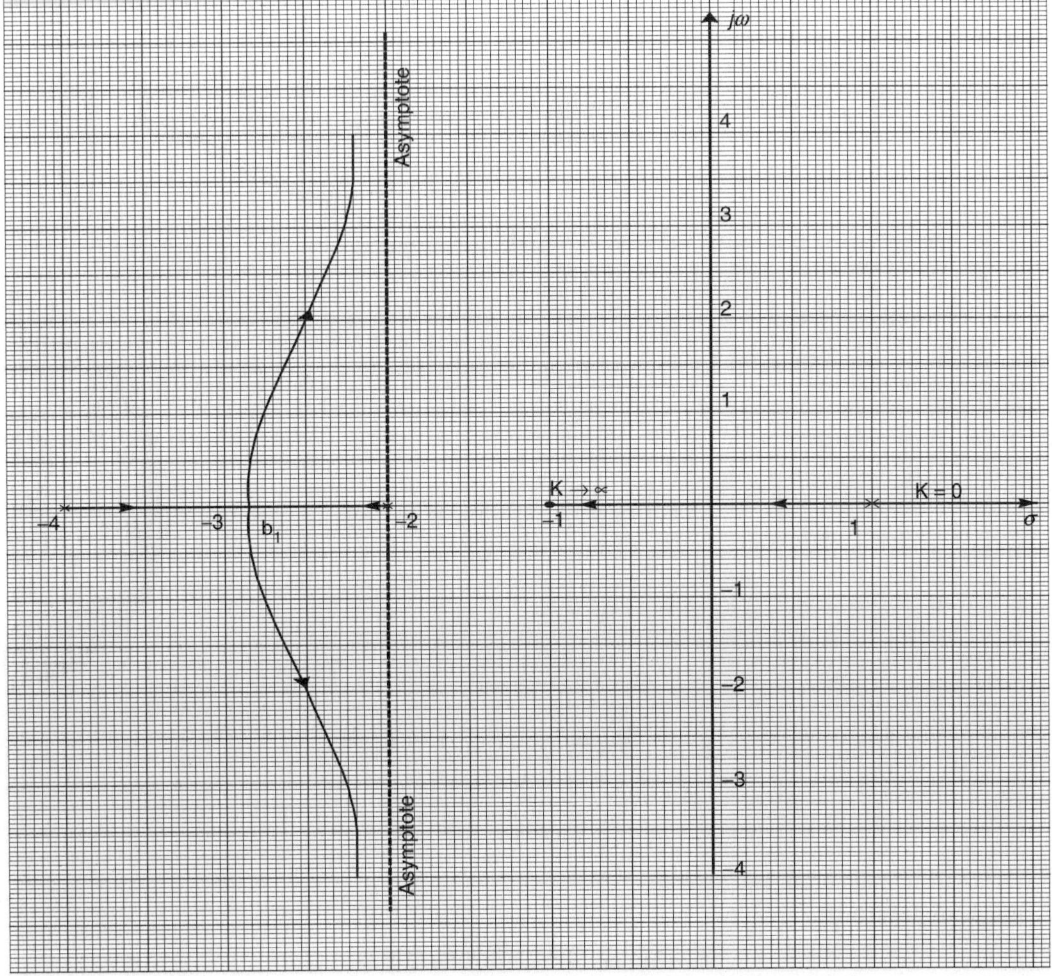

Figure (k) Root locus plot for ASP-5

ASP-6: Sketch the root locus plot of a unity feedback system whose OLTF is $G(s) = \dfrac{K(s+9)}{s(s^2+4s+11)}$.

Solution:

Given that

$$G(s)=\frac{K(s+9)}{s(s^2+4s+11)}.$$

1. Root locus is symmetrical about real axis.
2. The root locus plot starts from a pole at $K = 0$ and terminates at an open-loop zero, that is, at $K = \infty$.
3. The poles are given as
   $s = 0, s = -2 + j2.64, s = -2 - j2.64$
   The zeros are given as
   $s = -9$
   Number of poles = $P = 3$
   Number of zeros = $Z = 1$
   Number of root locus branches = $N = P = 3$
   Number of asymptotic lines = $n = P - Z = 2$
4. The angle of asymptotic lines with negative real axis

$$\phi=\frac{(2q+1)180°}{P-Z}\text{, where } q = 0, 1$$

$$= 180°/2,\ 3 \times 180°/2$$

$$= 90°,\ 270°.$$

5. The centroid is given as

$$\text{Centroid}=\frac{\text{Sum of poles}-\text{Sum of zeros}}{P-Z}=\frac{(0-2+j2.64-2-j2.64)-(-9)}{3-1}=2.5.$$

6. **Break-away points:**
   The C.E. of the system is given as

$$1 + G(s)H(s) = 0$$

$$1+\frac{K(s+9)}{s(s^2+4s+11)}=0$$

$$K=\frac{s(s^2+4s+11)}{s+9}$$

$$\frac{dK}{ds}=0 \Rightarrow 2s^3+31s^2+61s=0$$

$$\Rightarrow \quad s(s^2 + 15.5s + 30.5) = 0$$

$$\Rightarrow \quad s = 0, s = -13.157, s = -2.313.$$

   There are no valid break-away points.
7. **Angle of departure:**

$$\phi_{d1=(-2+j2.64)} = 180° - (\phi_{p1} + \phi_{p3}) + \phi_{z1}$$

$$= 180° - (127.1 + 90°) + 20.7° = 16.4°$$

$$\phi_{d2=(-2-j2.64)} = +16.4°.$$

8. **Intersection points of the root locus branches with imaginary axis:**

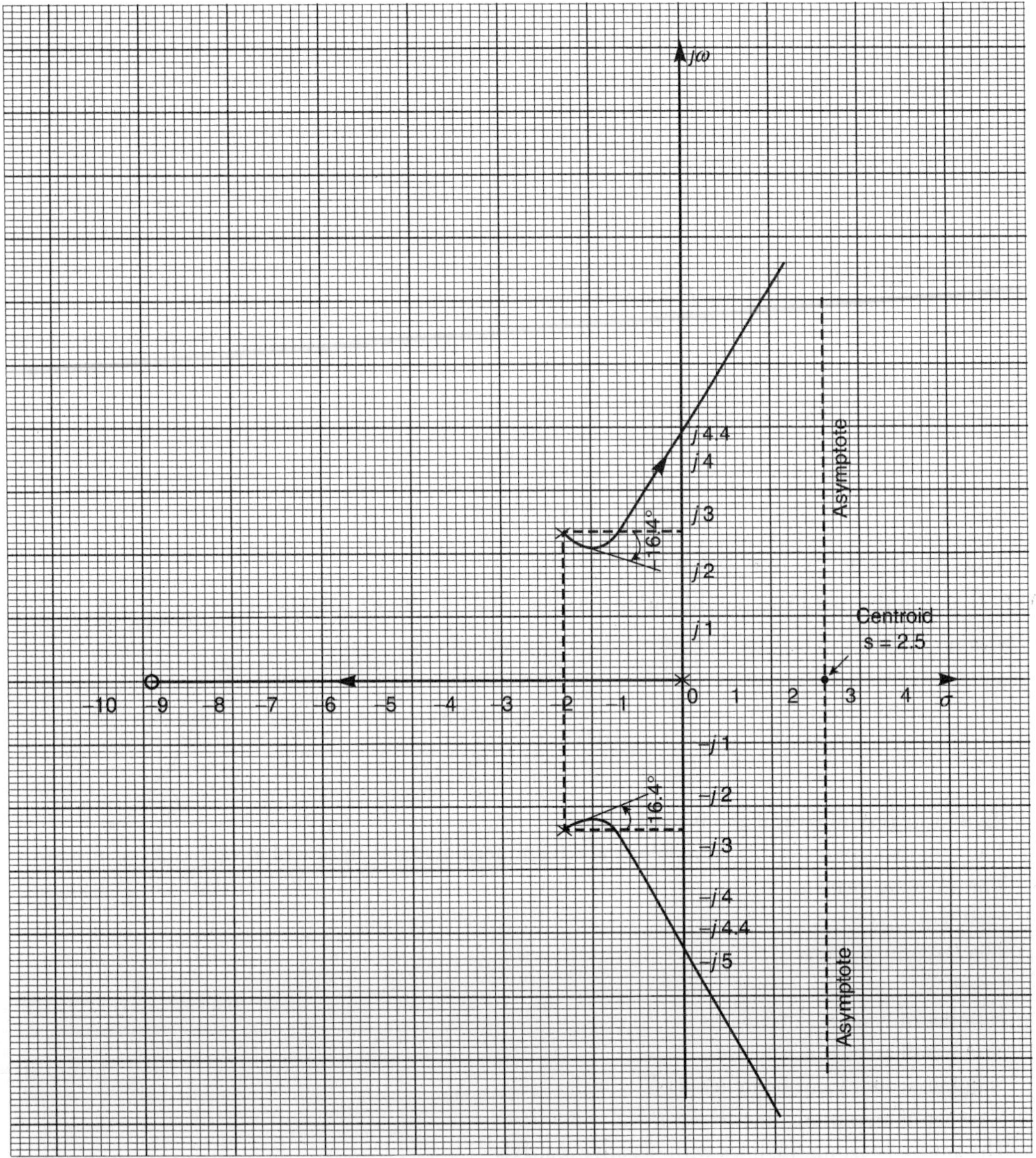

Figure (I) Root locus plot for ASP-6

The C.E. of the system is given as

$$1+G(s)H(s) = 0$$

$$1+\frac{K(s+9)}{s(s^2+4s+11)} = 0$$

$$\Rightarrow \qquad s(s^2 + 4s + 11) + K(s + 9) = 0$$

$$\Rightarrow \qquad s^3 + 4s^2 + (11 + K)s + 9K = 0$$

s^3	1	$11+K$	0
s^2	4	$9K$	0
s^1	$(44-5K)/4$	0	0
s^0	$9K$	0	0

For a stable system, $9K > 0,\ (44 - 5K/4) > 0$

$$\Rightarrow \qquad K > 0,\ K < 8.8.$$

For a stable system, the maximum value of K is 8.8. For $K > 8.8$, the roots lie on the RHS of the s-plane, and hence, $K = 8.8$ is the value for marginal stability.
The auxiliary equation is $4s^2 + 9K = 0$.

$$\Rightarrow \qquad 4s^2 + 79.2 = 0$$

$$\Rightarrow \qquad s = \pm j4.45$$

The complete root locus plot is as shown in Figure (l).

ASP-7: Sketch the root locus plot of a unity feedback system whose OLTF is $G(s) = \dfrac{K(s+1.5)}{s(s+1)(s+5)}$.

Solution:

Given that

$$G(s) = \frac{K(s+1.5)}{s(s+1)(s+5)}.$$

1. Root locus is symmetrical about real axis.
2. The root locus plot starts from a pole at $K = 0$ and terminates at an open-loop zero, i.e., at $K = \infty$.
3. The poles are given as
   $s = 0,\ s = -1,\ s = -5.$
   The zeros are given as
   $s = -1.5$
   Number of poles $= P = 3$
   Number of zeros $= Z = 1$
   Number of root locus branches $= N = P = 3$
   Number of asymptotic lines $= n = P - Z = 2$
4. The angle of asymptotic lines with negative real axis

$$\phi = \frac{(2q+1)180^\circ}{P-Z}, \text{ where } q = 0, 1$$

$$= 180^\circ/2,\ 3 \times 180^\circ/2$$

$$= 90^\circ,\ 270^\circ.$$

5. The centroid is given as

$$\text{Centroid} = \frac{\text{Sum of poles} - \text{Sum of zeros}}{P - Z} = \frac{(0-1-5)-(-1.5)}{3-1} = -2.25$$

6. **Break-away points:**
   The C.E. of the system is given as

$$1 + G(s)H(s) = 0$$

$$1 + \frac{K(s+1.5)}{s(s+1)(s+5)} = 0$$

$$K = \frac{s(s^2+6s+5)}{s+1.5}$$

$$\frac{dK}{ds} = 0 \Rightarrow s^3 + 5.25s^2 + 9s + 3.75 = 0$$

$$\Rightarrow \qquad s = 0.6,\ s = 2.3 + j0.89,\ s = -2.3 - j0.89.$$

   The valid break-away point is $b_1 = -0.6$

7. **Angle of departure:**
   Since there is no complex pole and zero, we need not find the angle of departure or arrival.

8. **Intersection points of the root locus branches with imaginary axis:**
   The C.E. of the system is given as

$$1 + G(s)H(s) = 0$$

$$1 + \frac{K(s+1.5)}{s(s+1)(s+5)} = 0$$

$$\Rightarrow \qquad s(s^2 + 6s + 5) + K(s + 1.5) = 0$$

$$\Rightarrow \qquad s^3 + 6s^2 + (5 + K)s + 1.5K = 0$$

s^3	1	$5+K$	0
s^2	6	$1.5K$	0
s^1	$(30+4.5)/6$	0	0
s^0	$1.5K$	0	0

   For a stable system, $1.5K > 0$, $(30 + 4.5K)/6 > 0$

$$\Rightarrow \qquad K > 0,\ K < -6.67$$

   Since the value of K is negative, there is no crossing point on the imaginary axis. In other words for any positive value of K, the root locus will not cross the imaginary axis.
   The complete root locus plot is as shown in Figure (m).

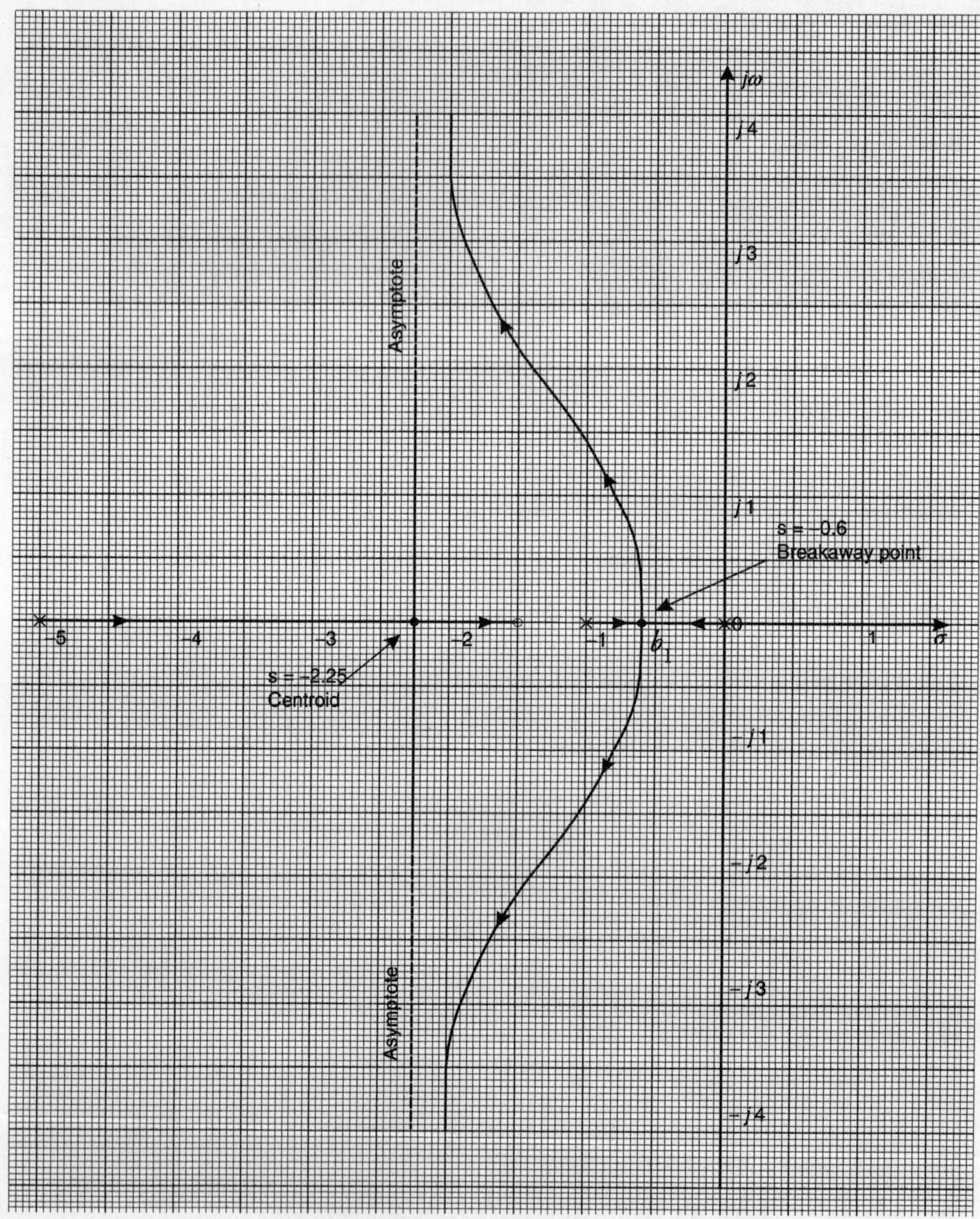

Figure (m) Root locus plot for ASP-7

ASP-8: Sketch the root locus plot of a unity feedback system whose OLTF is $G(s) = \dfrac{s}{(s^2+4)(s+2)}$.

Solution:

Given that

$$G(s) = \frac{s}{(s^2+4)(s+2)}.$$

1. Root locus is symmetrical about real axis.
2. The root locus plot starts from a pole at $K = 0$ and terminates at an open-loop zero, that is, at $K = \infty$.
3. The poles are given as
   $s = -2, s^2 + 4 = 0$
   $P_1 = -2, P_2 = j2, P_3 = -j2,$
   The zeros are given as
   $s = 0$
   Number of poles = $P = 3$
   Number of zeros = $Z = 1$
   Number of root locus branches = $N = P = 3$
   Number of asymptotic lines = $n = P - Z = 2$
4. The angle of asymptotic lines with negative real axis

$$\phi = \frac{(2q+1)180°}{P-Z}, \text{ where } q = 0, 1$$
$$= 180°/2,\ 3 \times 180°/2$$
$$= 90°,\ 270°.$$

5. The centroid is given as

$$\text{Centroid} = \frac{\text{Sum of poles} - \text{Sum of zeros}}{P-Z} = \frac{(0-2-j2+j2)-(0)}{3-1} = -1.$$

   The complete root locus is as shown in Figure (n).

ASP-9: Sketch the root locus plot of the system whose OLTF is given as $G(s)H(s) = \dfrac{K}{s(s+4)(s^2+8s+32)}$.

Hence, find the value of K so that the system has a damping factor 0.707.

Solution:

Given that

$$G(s)H(s) = \frac{K}{s(s+4)(s^2+8s+32)}.$$

1. Root locus is symmetrical about real axis.
2. The root locus plot starts from a pole at $K = 0$ and terminates at an open-loop zero, that is, at $K = \infty$.
3. The poles are given as
   $s = 0, s = -4, s^2 + 8s + 32 = 0$
   $P_1 = 0, P_2 = -4, P_3 = -4 + j4, P_4 = -4 - j4,$
   Number of poles = $P = 4$
   Number of zeros = $Z = 0$
   Number of root locus branches = $N = P = 4$
   Number of asymptotic lines = $n = P - Z = 4$
4. The angle of asymptotic lines with negative real axis

$$\phi = \frac{(2q+1)180°}{P-Z}, \text{ where } q = 0, 1, 2, 3$$
$$= 180°/4,\ 3 \times 180°/4,\ 5 \times 180°/4,\ 7 \times 180°/4$$
$$= 45°,\ 135°,\ 225°,\ 315°.$$

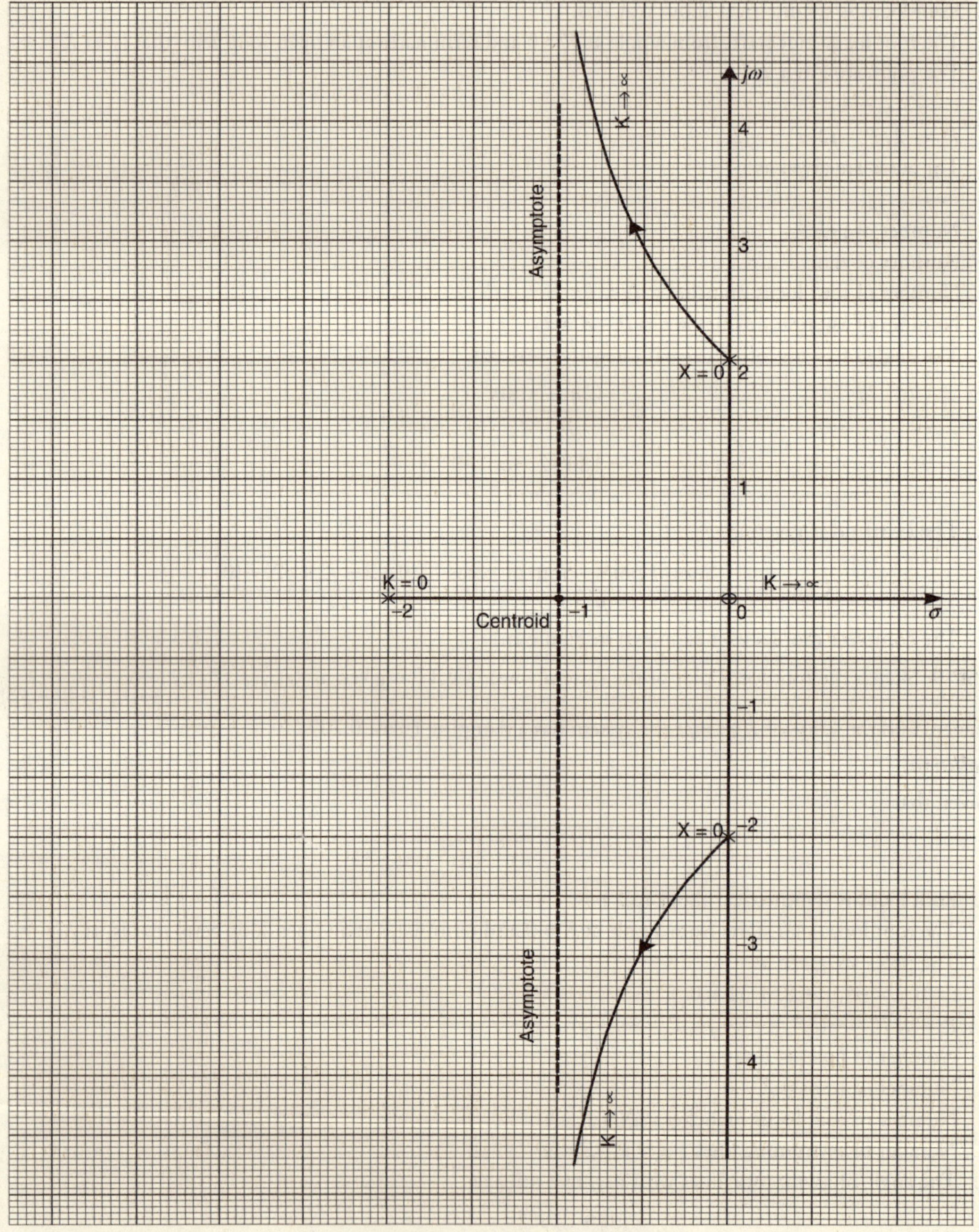

Figure (n) Root locus for ASP-8

5. The centroid is given as

$$\text{Centroid} = \frac{\text{Sum of poles} - \text{Sum of zeros}}{P - Z} = \frac{(0 - 4 - 4 + j4 - 4 - j4) - (0)}{4 - 0} = -3.$$

The complete root locus is as shown in Figure (n).

6. **Break-away points:**

The C.E. of the system is given as

$$1 + G(s)H(s) = 0$$

$$1+\frac{K}{s(s+4)(s^2+8s+32)}=0$$

$$K = -s(s+4)(s^2+8s+32)$$

$$K = -(s^4 + 12s^3 + 64s^2 + 128s)$$

$$\frac{dK}{ds} = 0 \Rightarrow 4s^3 + 6s^2 + 128s + 128 = 0$$

$$\Rightarrow \quad s^3 + 9s^2 + 32s + 32 = 0.$$

After solving the above-mentioned equation, the valid break-away point is $b_1 = -1.57$.

7. **Intersection points of the root locus branches with imaginary axis:**

The C.E. of the system is given as

$$1 + G(s)H(s) = 0$$

$$s^4 + 12s^3 + 64s^2 + 128s + K = 0$$

s^4	1	64	K
s^3	12	128	0
s^2	53.33	K	0
s^1	$128-0.225K$	0	0
s^0	K	0	0

For a stable system, $K > 0$, $128 - 0.225K > 0$

$$\Rightarrow \quad 0 < K < 569.$$

For a stable system, the maximum value of K is 569. For $K > 569$, the roots lie on the RHS of the s-plane, and hence, $K = 569$ is the value for marginal stability.

The auxiliary equation is $\quad 53.33s^2 + K = 0$

$$\Rightarrow \quad 53.33s^2 + 569 = 0$$

$$\Rightarrow \quad s = \pm j\,3.266.$$

8. **Angle of departure:**

$$\phi_{p3=(-4+j4)} = 180° - (\phi_{p1} + \phi_{p2} + \phi_{p4})$$

$$= 180° - (135° + 90° + 90°) = -135°$$

$$\phi_{p4=(-4-j4)} = +135°$$

The complete root locus plot is as shown in Figure (o).

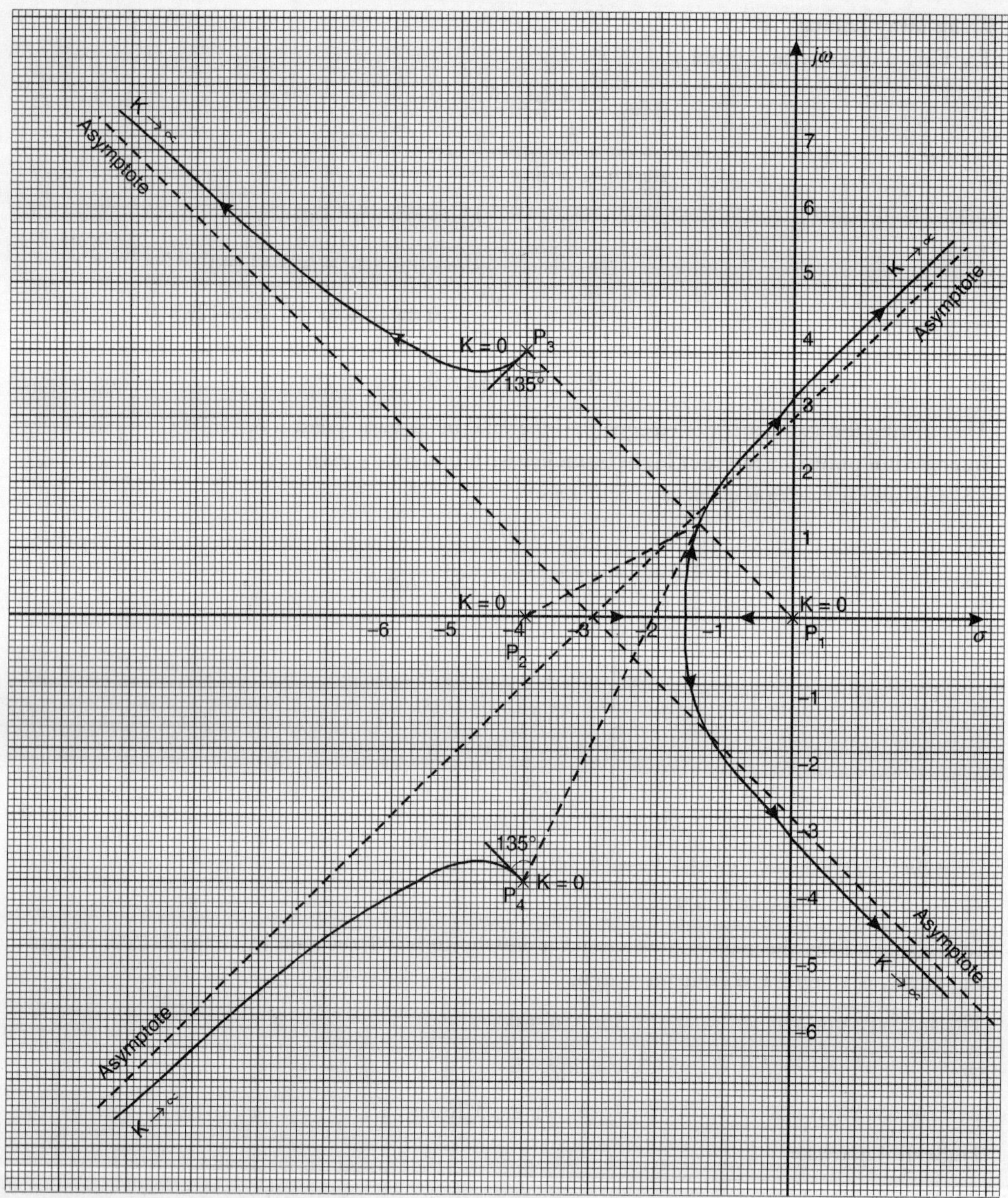

Figure (o) Root locus plot for ASP-9

Since $\xi = 0.707 \Rightarrow \cos\theta = \xi = 0.707 \Rightarrow \theta = 45°$

From the origin, a line at an angle $\theta = 45°$ is drawn, which intersects the root locus plot at $s = -1.35 + j1.35$. As this point lies on the root locus, the following condition is satisfied.

$$|G(s)H(s)| = 1$$

$$\left|\frac{K}{s(s+4)(s^2+8s+32)}\right| = 1$$

$$\left|\frac{K}{(-1.35+j1.35)(-1.35+j1.35+4)[(-1.35+j1.35)^2+8(-1.35+j1.35)+32]}\right|=1$$

$$\Rightarrow \qquad \frac{k}{130}=1\Rightarrow K=130\,.$$

ASP-10: The OLTF of a unity feedback system is $G(s)=\dfrac{K(s+a)}{s(s+b)}$.

1. Prove that break-away and break-in points will exist only when $|a|>|b|$.
2. Prove that the complex points on the root locus form a circle with centre $(-a, 0)$ and radius $\sqrt{a^2-ab}$.

Solution:

Given that

$$G(s)H(s)=\frac{K(s+a)}{s(s+b)}.$$

There are two open-loop poles, and hence, the number of branches are 2 and start at $P_1 = 0$ and $P_2 = -b$, respectively. One of the branches terminates at infinity and other terminates at zero $Z_1 = -a$.

1. The C.E. is given as

$$1 + G(s)H(s) = 0$$

$$1+\frac{K(s+a)}{s(s+b)}=0$$

$$\Rightarrow \qquad K=-\frac{s(s+b)}{s+a} \tag{1}$$

$$\frac{dK}{ds}=0 \Rightarrow s^2+2as+ab=0$$

$$\Rightarrow \qquad s=-a\pm\sqrt{a^2-ab}. \tag{2}$$

Substituting the value of s in (1), we will get

$$K=-\frac{\left[-a+\sqrt{a(a-b)}\right]^2+b\left[-a+\sqrt{a(a-b)}\right]}{-a+\sqrt{a(a-b)}+a}$$

$$=\frac{a^2+a(a+b)-2a\sqrt{a(a-b)}+\left[-ab+\sqrt{a(a-b}\right]}{\sqrt{a(a-b)}}$$

$$=-\left[2\sqrt{a(a-b)}+b-2a\right].$$

From the above-mentioned equation of K, it is clear that for all values of a and b such that $|a|>|b|$, K is positive. Therefore, the break-away or break-in points will exists only when $|a|>|b|$.

1. If $|a|>|b|$, then break-away and break-in points exist. Let us choose a test point in the complex plane $s = x + jy$ that lies on the root locus. This point should satisfy angle criterion so that the point is a part of the root locus.

$$G(s)H(s)_{s-x+jy}=\frac{K(x+jy+a)}{(x+jy)(x+jy+b)}.$$

The phase angle is given as

$$\angle G(s)H(s)_{s=-x+jy} = \tan^{-1}\frac{y}{x+a} - \tan^{-1}\frac{y}{x} - \tan^{-1}\frac{y}{x+b} = -180°$$

$$180 + \tan^{-1}\frac{y}{x+a} = \tan^{-1}\frac{y}{x} + \tan^{-1}\frac{y}{x+b}$$

By taking 'tan' on both sides, we have

$$\frac{\frac{y}{x} + \frac{y}{x+b}}{1 - \frac{y}{x} \times \frac{y}{x+b}} = \frac{\tan 180 + \frac{y}{x+a}}{1 - \tan 180 \times \frac{y}{x+a}}.$$

After simplifying the above-mentioned equation, we have

$$x^2 + 2xa + y^2 = -ab$$

By adding both sides a^2, we get

$$(x+a)^2 + y^2 = (a^2 - ab)$$

$$(x+a)^2 + (y+0)^2 = \left[\sqrt{a^2 - ab}\right]^2.$$

This is equation of a circle with centre $(-a, 0)$ and radius $\sqrt{a^2 - ab}$. Hence, it is proved.

REVIEW QUESTIONS

1. What do you understand by root locus plot?
2. Explain the constructional procedure of root locus plot.
3. Write a short note on angle and magnitude condition of root locus.
4. Explain the terms:
   (i) asymptotes (ii) centroid
   (iii) break-away points (iv) angle of departure
5. Define the term 'root locus' and state the rule for finding out the root locus on the real axis.
6. Explain the method of calculating the break-away points.
7. What is a root locus plot? Explain with a suitable example.
8. What are the features of root locus plot?
9. Show that the break-away and break-in points, if any, on the real axis for the root locus for $G(s)H(s) = \frac{KN(s)}{D(s)}$, where $N(s)$ and $D(s)$ are rational polynomials in S, and that they can be obtained by solving the equation $\frac{dK}{ds} = 0$.
10. Explain the stability of the system from the root locus plot in the following situations with suitable examples:
    (a) addition of open-loop poles.
    (b) addition of open-loop zeros.
11. What is break-away and break-in point? How are they determined?
12. What is centroid? How the centroid is calculated?
13. Open-loop t.f. of a unity feedback system is $G(s) = \frac{K(s+a)}{s(s+b)}$
    (a) Prove that break-away and break-in points will exist only when $|a| > |b|$
    (b) Prove that the complex points on the root locus form a circle with centre $(-a, 0)$ and radius $\sqrt{a^2 - ab}$

EXERCISE PROBLEMS

1. Check whether the point $s = -3 + j5$ lies on the root locus of a system having $G(s)H(s) = \frac{K}{(s+1)(s+5)}$. Determine the corresponding value of K.
2. Sketch the root locus for the system having the OLTF, $G(s)H(s) = \frac{K(s+4)(s+5)}{(s+1)(s+3)}$.
3. A unity feedback system has an OLTF, $G(s) = \frac{K}{s^2(s+2)}$.
   1. By sketching a root locus plot, show that the system is unstable for all values of K.
   2. Add a zero to $s = -a (0 \le a \le 2)$ and show that the addition of zero stabilises the system.
4. A feedback control system has an OLTF $G(s)H(s) = \frac{K}{s(s+3)(s^2+2s+2)}$. Find the root locus as K is varying from 0 to ∞.
5. Sketch the root locus plot for all the values of K from 0 to ∞ for a negative feedback system whose OLTF is given as $G(s)H(s) = \frac{K(s+1)}{s^2(s+9)}$?
6. Sketch the root locus plot for all the values of K from 0 to ∞ for a negative feedback system whose OLTF $G(s)H(s) = \frac{K(s^2-4s+8)}{(s+1)(s-0.5)}$?
7. Using the root locus method, find gain (K) for the system having OLTF as $G(s)H(s) = \frac{K(s+2)(s+4)}{(s+1)(s+3)(s+5)}$ when $\xi = 0.246$.
8. Consider a negative feedback system characterised by $G(s)H(s) = \frac{K(s+1)}{s(s+2)(s+3)(s+4)}$. Sketch the root locus plot for all values of K ranging from 0 to ∞. Further, find gain K for which the CLTF will have a pole on the real axis at -0.5.
9. Sketch the root locus plot for a closed-loop system having an OLTF $G(s)H(s) = \frac{K(s+1)(s+2)}{s(s-1)}$ for all values of K ranging from 0 to ∞. Find the range of values of K for closed-loop stability.

ANSWERS

1. Since $\phi = -180°$, the point $-3 + j5$ lies on the root locus, $K = 29$.
7. $K = 33$
8. For closed-loop stability, $0 < K < 141$.
9. For closed-loop stability, $0.33 < K < \infty$.

OBJECTIVE TYPE QUESTIONS

1. Intersection of root locus branches with imaginary axis can be determined by
   (a) Nyquist criterion (b) R–H criterion
   (c) Polar plot (d) Bode plot
2. The break-away point is determined by solving the equation
   (a) $\frac{dK}{ds} = 0$ (b) $1 + G(s)H(s) = 0$
   (c) $\int \frac{dK}{ds} = 0$ (d) $G(s)\,H(s) = 0$
3. The angle of asymptotic lines making with negative real axis is given as

(a) $\dfrac{2k\pi}{P-Z}$ (b) $\dfrac{(2k+1)\pi}{P-Z}$ (c) $\dfrac{(k+1)\pi}{P-Z}$ (d) $\dfrac{(2k+1)\pi}{P-Z}\cdot\dfrac{\pi}{2}$

4. When the root locus separates at a point between two open-loop poles, the point is called
(a) critical point (b) crossover point (c) centroid (d) break-away point

5. If the root locus branches cross the imaginary axis, the system becomes
(a) marginally stable (b) conditionally stable
(c) unstable (d) None

6. The intersection of the root locus branch with the imaginary axis can determined by using the equation
(a) $G(s)\,H(s) = 0$ (b) $1 - G(s)\,H(s) = 0$
(c) $1 + G(s)\,H(s) = 0$ (d) $\dfrac{dK}{ds} = 0$

7. The asymptotes to root locus branches intersect on the real axis at a point given as
(a) $\dfrac{\sum \text{Poles} - \sum \text{Zeros}}{(P-Z)}$ (b) $\dfrac{\sum \text{Poles} + \sum \text{Zeros}}{(P-Z)}$
(c) $\dfrac{\sum \text{Poles} - \sum \text{Zeros}}{(P+Z)}$ (d) None

8. In loop t.f., open-loop zeros are the same as
(a) zeros of $G(s)H(s)$ (b) zeros of $1 + G(s)\,H(s)$
(c) poles of $G(s)H(s)$ (d) poles of $1 + G(s)H(s)$

9. The value of K at which the root locus crosses the imaginary axis makes the system
(a) stable (b) under-damped
(c) marginally stable (d) unstable

10. The starting points of the root locus are
(a) closed-loop poles (b) closed-loop zeros
(c) open-loop poles (d) open-loop zeros

11. The ending points of the root locus are
(a) closed-loop poles (b) closed-loop zeros
(c) open-loop poles (d) open-loop zeros

12. The number of root loci branches for the OLTF $G(s)H(s) = \dfrac{K(s+5)(s+6)}{s(s+1)(s+3)(s^2+5s+6)}$ is
(a) 2 (b) 3 (c) 4 (d) 5

13. If the root locus branches do not cross the imaginary axes, the system is
(a) conditionally stable (b) inherently stable
(c) marginally stable (d) unstable

14. A unity feedback system has an OLTF $G(s)H(s) = \dfrac{K}{s(s^2+4s+13)}$. The centroid of the asymptotes of the root locus plot lies at
(a) –4 (b) –4/3 (c) –13 (d) –2

15. The OLTF of an unity feedback is given as $G(s) = \dfrac{5}{s(s+1)(s+3)(s+4)}$. The number of asymptotes of the root locus plot that tend to infinity is given as
(a) 1 (b) 2 (c) 3 (d) 4

16. The OLTF of an unity feedback system is given as $G(s) = \dfrac{K}{s(s+1)(s+3)}$. The break-away point of the root locus plot is given as
(a) –0.45 (b) –0.523 (c) –0.7 (d) None

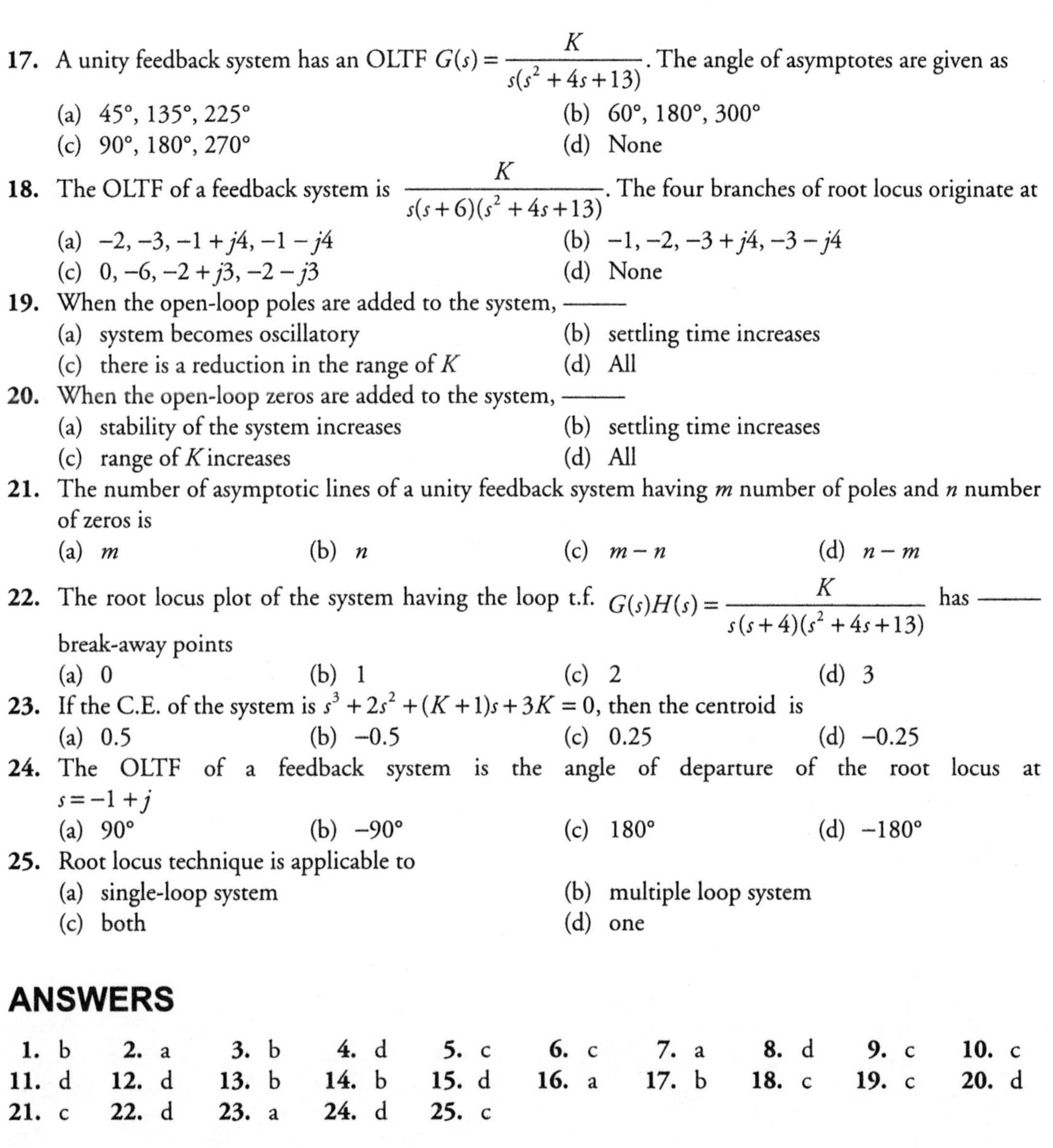

17. A unity feedback system has an OLTF $G(s) = \dfrac{K}{s(s^2+4s+13)}$. The angle of asymptotes are given as
   (a) 45°, 135°, 225°
   (b) 60°, 180°, 300°
   (c) 90°, 180°, 270°
   (d) None

18. The OLTF of a feedback system is $\dfrac{K}{s(s+6)(s^2+4s+13)}$. The four branches of root locus originate at
   (a) $-2, -3, -1+j4, -1-j4$
   (b) $-1, -2, -3+j4, -3-j4$
   (c) $0, -6, -2+j3, -2-j3$
   (d) None

19. When the open-loop poles are added to the system, ———
   (a) system becomes oscillatory
   (b) settling time increases
   (c) there is a reduction in the range of K
   (d) All

20. When the open-loop zeros are added to the system, ———
   (a) stability of the system increases
   (b) settling time increases
   (c) range of K increases
   (d) All

21. The number of asymptotic lines of a unity feedback system having m number of poles and n number of zeros is
   (a) m (b) n (c) $m-n$ (d) $n-m$

22. The root locus plot of the system having the loop t.f. $G(s)H(s) = \dfrac{K}{s(s+4)(s^2+4s+13)}$ has ——— break-away points
   (a) 0 (b) 1 (c) 2 (d) 3

23. If the C.E. of the system is $s^3+2s^2+(K+1)s+3K=0$, then the centroid is
   (a) 0.5 (b) −0.5 (c) 0.25 (d) −0.25

24. The OLTF of a feedback system is the angle of departure of the root locus at $s=-1+j$
   (a) 90° (b) −90° (c) 180° (d) −180°

25. Root locus technique is applicable to
   (a) single-loop system
   (b) multiple loop system
   (c) both
   (d) one

ANSWERS

1. b	**2.** a	**3.** b	**4.** d	**5.** c	**6.** c	**7.** a	**8.** d	**9.** c	**10.** c
11. d	**12.** d	**13.** b	**14.** b	**15.** d	**16.** a	**17.** b	**18.** c	**19.** c	**20.** d
21. c	**22.** d	**23.** a	**24.** d	**25.** c					

CHAPTER 11

Frequency Domain Analysis–Bode Plot

Contents

11.1 INTRODUCTION

Frequency response means the steady-state response of a system to a sinusoidal input. The magnitude and phase relationship between the sinusoidal input and the steady-state output of a system is known as frequency response. The frequency of the input signal over a certain range is varied to study the resulting frequency response. In the system t.f. $T(s)$, if s is replaced by $j\omega$, then the resulting t.f. $T(j\omega)$ is called sinusoidal t.f. The frequency response of the system can be directly obtained from the sinusoidal t.f. $T(j\omega)$ of the system. The magnitude and phase of $T(j\omega)$ are functions of frequency and can be evaluated for various values of frequencies. The frequency response can be evaluated for both open-loop and closed-loop systems. Frequency response methods are simple to use and offer many advantages when compared to time-domain methods.

11.2 ADVANTAGES AND LIMITATIONS OF FREQUENCY RESPONSE

11.2.1 Advantages

1. The frequency response of stable open-loop system can be obtained experimentally, without the knowledge of the t.f.

2. Frequency response tests are, in general, simple and can be made accurately by the use of readily available sinusoidal signal generators and precise measuring equipments.
3. The t.fs. of complicated components can be determined experimentally by frequency response tests.
4. A system may be designed so that the effects of undesirable noise are negligible and such analysis and design can be extended to certain non-linear control systems.
5. The apparatus required for obtaining frequency response is simple, inexpensive and easy to use.
6. For difficult cases, like conditionally stable systems, Nyquist plot is probably the only method to analyse stability.

11.2.2 Limitations

1. These methods can be applied only to linear systems. When these methods are applied to systems possessing some degree of non-linearity, the results of analysis and design are not exact. Hence, seasoned professionals are not satisfied by such methods since the evaluation of the results require experience and judgement regarding the degree of non-linearity.
2. For an existing system, frequency response is obtained only if the time constants are up to a few minutes. However, if the time constants are a few hours or days, then the methods are not convenient to apply for practical testing.

11.3 FREQUENCY RESPONSE TO SINUSOIDAL INPUT

The steady-state response of the system to a purely sinusoidal input is defined as frequency response of a system. In such a method, frequency of the input signal is to be varied over a certain range and the resulting response of a system is to be studied. Such a response is called frequency response. Let us consider a linear time-invariant system with its t.f., as shown in Figure 11.1. The input $r(t)$ is sinusoidal and is given as

r(t) → G(s) → c(t)
R(s) C(s)

Figure 11.1 Linear time-invariant systems

$$r(t) = A\sin\omega t$$

$$R(s) = \frac{A\omega}{s^2 + \omega^2}$$

Let $G(s)$ be the t.f. = ratio of two polynomials in s, namely

$$\frac{p(s)}{q(s)} = \frac{p(s)}{(s+s_1)(s+s_2)\cdots(s+s_n)}. \tag{11.1}$$

The Laplace-transformed output is given as

$$C(s) = R(s).G(s)$$

$$= \frac{A\omega}{(s^2+\omega^2)}\frac{p(s)}{(s+s_1)(s+s_2)\cdots(s+s_n)}.$$

By assuming system to be in stable state, we get

$$C(s) = \frac{a}{s+j\omega} + \frac{\bar{a}}{s-j\omega} + \frac{b_1}{s+s_1} + \cdots + \frac{b_n}{s+s_n} \tag{11.2}$$

where a is a constant and $\bar{a}$ is complex conjugate of a.
By taking inverse Laplace transform, we have

$$c(t) = ae^{-j\omega t} + \bar{a}e^{j\omega t} + b_1 e^{-s_1 t} + \cdots + b_n e^{-s_n t}. \tag{11.3}$$

For a stable system, $-s_1, -s_2 \ldots -s_n$, have negative real parts. Therefore, as s approaches infinity, the terms $e^{-s_1 t}, e^{-s_2 t} \ldots e^{-s_n t}$ approach zero. Thus, all the terms on the right-hand side of this equation, except the first two, drop out at steady state. If $C(s)$ involves multiple poles s_j of multiplicity m_j, then $c(t)$ will involve terms like $t^{h_j} e^{-s_j t}$ $(h_j = 0, 1, 2 \ldots m_j - 1)$. Since the real parts of the $-s_j$ are negative for stable system, the terms $t^{h_j} e^{-s_j t}$ approach zero as t approaches infinity.

Thus, regardless of whether or not the system is of the distinct pole type, the steady-state response becomes

$$c_{ss}(t) = ae^{-j\omega t} + \bar{a}e^{j\omega t}$$

where constant a can be evaluated as

$$a = G(s) \times \frac{A\omega}{s^2 + \omega^2}(s + j\omega)\Big|_{s=-j\omega} = \frac{-AG(-j\omega)}{2j}$$

$$\bar{a} = G(s) \times \frac{A\omega}{s^2 + \omega^2}(s + j\omega)\Big|_{s=+j\omega} = \frac{+AG(+j\omega)}{2j}.$$

Since $G(j\omega)$ is a complex quantity, it can be written in the following form:

$$G(j\omega) = |G(j\omega)|e^{j\phi}$$

where $G(j\omega)$ represents the magnitude and is given as

$$|G(j\omega)| = \sqrt{(\text{Real part})^2 + (\text{Imaginary part})^2} \quad \text{and}$$

ϕ represents the angle of $G(j\omega)$ and is given as

$$\phi = \angle G(j\omega) = \tan^{-1}\frac{[\text{Imaginary part of } G(j\omega)]}{[\text{real part of } G(j\omega)]}.$$

The angle ϕ may be negative, positive or zero. Similarly, we obtain the following expression for $G(j\omega)$.

$$G(-j\omega) = |G(-j\omega)|e^{-j\phi} = |G(j\omega)|e^{-j\phi}$$

$$\therefore \quad c_{ss}(t) = \frac{-A|G(-j\omega)|}{2j}e^{-j\phi}.e^{-j\omega t} + \frac{A|G(j\omega)|}{2j}e^{j\phi}.e^{j\omega t}$$

$$= A|G(j\omega)|\frac{e^{j(\omega t+\phi)} - e^{-j(\omega t+\phi)}}{2j}.$$

Now,

$$e^{j\theta} = \cos\theta + j\sin\theta$$

$$e^{-j\theta} = \cos\theta - j\sin\theta$$

$$\therefore \quad \cos\theta = \frac{e^{j\theta} + e^{-j\theta}}{2} \text{ and } \sin\theta = \frac{e^{j\theta} - e^{-j\theta}}{2j}$$

$$\therefore \quad c_{ss}(t) = A|G(j\omega)|\sin(\omega t + \phi). \tag{11.4}$$

From the above-mentioned equation, we see that a stable linear time-invariant system subjected to a sinusoidal input will, at steady state, have a sinusoidal output of the same frequency as the input. However, the amplitude, that is, magnitude and phase angle of the output is different from that of the input.

Magnitude of output = product of magnitude of input and $G(j\omega)$ and phase angle difference $\phi = \angle G(j\omega)$

The sinusoidal t.f. $G(j\omega)$ is a complex quantity and can be represented as magnitude and phase angle with frequency ω as a variable parameter. Such frequency domain t.f. can be obtained by substituting $j\omega$ for s in the t.f. $G(s)$ of the system.

$$G(j\omega) = G(s)\big|_{s=j\omega} = \text{frequency domain t.f.}$$

Therefore, to get frequency response means to sketch the variation in magnitude and phase angle of $G(j\omega)$, when ω is varied from 0 to ∞.

$$G(j\omega) = M\angle\phi \tag{11.5}$$

where M is the magnitude, ϕ is the phase angle and ω is the input frequency.

11.4 FREQUENCY DOMAIN SPECIFICATIONS

The basic objective of the control system design is to meet the performance specifications. These specifications are the constraints or limitations imposed on the mathematical functions, describing the system's characteristics.

Let us consider a general frequency response of a system, as shown in Figure 11.2.

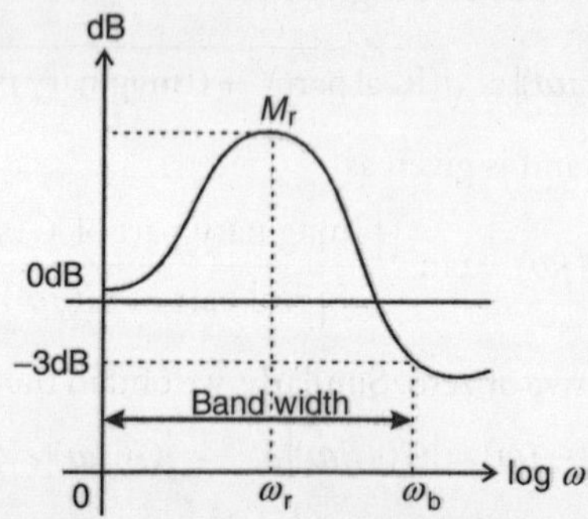

Figure 11.2 Plot of M_r versus ω

1. **Bandwidth (ω_b):** It is defined as the range of frequencies over which the system will respond satisfactorily. It can also be defined as a range of frequencies in which the magnitude response is almost flat in nature. Therefore, it can be considered as a range of frequencies over the magnitude of closed-loop response, that is, $\left|\dfrac{C(j\omega)}{R(j\omega)}\right|$ does not drop by more than 3 dB from its zero frequency value. Bandwidth indicates the speed of the response. It is inversely proportional to rise time. Therefore, large bandwidth means small rise time and fast response.

   It is given as

$$\omega_b = \omega_n[1 - 2\xi^2 + \sqrt{1 + (1 - 2\xi^2)^2}\,]^{1/2}.$$

2. **Cut-off frequency:** The frequency at which the magnitude of the closed-loop response is 3 dB down from its zero frequency value is called cut-off frequency. The range 0 to ω_b is the bandwidth of the system whose frequency response is as shown in Figure 11.2.
3. **Cut-off rate:** The slope of the resultant magnitude curve near the cut-off frequency is called cut-off rate.
4. **Resonant peak (M_r):** It is the maximum value of magnitude of the closed-loop frequency, as shown in Figure 11.2. The larger the value of resonant peak the more is the value of peak overshoot of the system for step input.
   It is given as
   $$M_r = \frac{1}{2\xi\sqrt{1-\xi^2}}.$$
5. **Resonant frequency (ω_r):** The frequency at which resonant peak M_r occurs in closed-loop frequency response is called resonant frequency. It is inversely proportional to the rise time.
   It is given as
   $$\omega_r = \omega_n\sqrt{1-2\xi^2}.$$

   These are the general specifications of the frequency response. The conditions that are most important from the stability and relative stability analysis viewpoint are discussed below.
6. **Gain crossover frequency(ω_{gc}):** The frequency at which magnitude of $G(j\omega)H(j\omega)$ is unity (i.e., 1) is called gain crossover frequency (i.e., $20 \log 10^1 = 0$ dB).
7. **Phase crossover frequency (ω_{pc}):** The frequency at which phase angle of $G(j\omega)H(j\omega)$ is $-180°$ is called phase crossover frequency.
8. **Gain margin (GM):** It can be defined as reciprocal of the magnitude of the $G(j\omega)H(j\omega)$ measured in phase crossover frequency.
   $$\therefore \quad \text{GM} = \frac{1}{\left|G(j\omega)H(j\omega)\right|_{\text{at } \omega=\omega_{pc}}}$$
9. **Phase margin (PM):** It can be defined as the amount of additional phase lag at the gain crossover frequency required to bring the system to the verge of instability.
   $$\therefore \quad \text{PM} = 180° + \angle G(j\omega)H(j\omega)_{\text{at } \omega=\omega_{gc}}$$

11.5 ESTIMATION OF FREQUENCY DOMAIN SPECIFICATIONS

1. **Resonant Peak (M_r)**
   Let us consider the closed-loop t.f. of a second-order system:
   $$T(s) = \frac{C(s)}{R(s)} = \frac{\omega_n^2}{s^2 + 2\xi\omega_n s + \omega_n^2}.$$
   The sinusoidal t.f. is obtained by substituting $s = j\omega$
   $$\therefore \quad T(j\omega) = \frac{C(j\omega)}{R(j\omega)} = \frac{\omega_n^2}{(j\omega)^2 + 2j\xi\omega_n\omega + \omega_n^2}$$
   $$= \frac{\omega_n^2}{-\omega^2 + j2\xi\omega_n\omega + \omega_n^2}$$

$$= \frac{\omega_n^2}{\omega_n^2\left(-\frac{\omega^2}{\omega_n^2} + j2\omega\xi\frac{\omega_n}{\omega_n^2} + 1\right)}$$

$$= \frac{1}{1 - \left(\frac{\omega}{\omega_n}\right)^2 + j2\xi\frac{\omega}{\omega_n}}. \tag{11.6}$$

Let the normalised frequency $u = \frac{\omega}{\omega_n}$

$$\therefore \qquad T(j\omega) = \frac{1}{1 - u^2 + j2\xi u}.$$

$$= \frac{1}{\sqrt{(1-u^2)^2 + 4\xi^2 u^2}\angle\tan^{-1}\left(\frac{2\xi u}{1-u^2}\right)}$$

$$= \frac{1}{\sqrt{(1-u^2)^2 + 4\xi^2 u^2}}\angle -\tan^{-1}\left(\frac{2\xi u}{1-u^2}\right)$$

$$= |M|\angle\phi. \tag{11.7}$$

where M is the magnitude of closed-loop t.f.

$$= |T(j\omega)|$$

$$= \frac{1}{\sqrt{(1-u^2)^2 + 4\xi^2 u^2}}$$

$$= [(1-u^2)^2 + 4\xi^2 u^2]^{-1/2}$$

where ϕ is the phase angle of the closed-loop t.f.

$$= \angle T(j\omega)$$

$$= -\tan^{-1}\left(\frac{2\xi u}{1-u^2}\right).$$

The resonant peak is the maximum value of M. The condition for the maximum value of M can be obtained by differentiating the equation of M with respect to u and substituting $\frac{dM}{du} = 0$ when $u = u_r$.

Further, $u_r = \frac{\omega_r}{\omega_n}$ = normalised resonant frequency.

$$\frac{dM}{du} = -\frac{1}{2}[(1-u^2)^2 + 4\xi^2 u^2]^{-3/2}[2(1-u^2)(-2u) + 8\xi^2 u]$$

$$= \frac{4u(1-u^2) - 8\xi^2 u}{2[(1-u^2)^2 + 4\xi^2 u^2]^{3/2}}$$

$$\left.\frac{dM}{du}\right|_{u=u_r} = \frac{4u_r(1-u_r^2) - 8\xi^2 u_r}{2[(1-u_r^2)^2 + 4\xi^2 u_r^2]^{3/2}} = 0$$

$$\Rightarrow \qquad 4u_r(1-u_r^2)-8\xi^2u_r=0$$

$$1-u_r^2=2\xi^2$$

$$u_r^2=1-2\xi^2$$

$$u_r=\sqrt{1-2\xi^2}$$

The resonant peak occurs when $u_r=\sqrt{1-2\xi^2}$. (11.8)

By substituting this condition in magnitude M, we can get the resonant peak as

$$M_r=\left.\frac{1}{[(1-u^2)^2+4\xi^2u^2]^{1/2}}\right|_{u=u_r=\sqrt{1-2\xi^2}}$$

$$=\frac{1}{[(1-1+2\xi^2)^2+4\xi^2(1-2\xi^2)]^{1/2}}$$

$$=\frac{1}{[4\xi^4+4\xi^2-8\xi^4]^{1/2}}$$

$$=\frac{1}{[4\xi^2(1-\xi^2)]^{1/2}}$$

$$=\frac{1}{2\xi\sqrt{1-\xi^2}}$$

$$M_r=\frac{1}{2\xi\sqrt{1-\xi^2}}. \qquad (11.9)$$

2. **Resonant frequency (ω_r)**

   The normalised resonant frequency is

$$u_r=\frac{\omega_r}{\omega_n}=\sqrt{1-2\xi^2}$$

$$\Rightarrow \qquad \omega_r=\omega_n\sqrt{1-2\xi^2}$$

Resonant frequency, $\qquad \omega_r=\omega_n\sqrt{1-2\xi^2}$. (11.10)

3. **Bandwidth (ω_b)**

   Let the normalised bandwidth $u_b=\dfrac{\omega_b}{\omega_n}$.

   When $u=u_b$, the magnitude M of the closed-loop system is $\dfrac{1}{\sqrt{2}}$. Therefore, by substituting $u=u_b$ in the magnitude M and equating it to $\dfrac{1}{\sqrt{2}}$; then

$$M=\frac{1}{[(1-u_b^2)^2+4\xi^2u_b^2]^{1/2}}=\frac{1}{\sqrt{2}}$$

$$1+u_b^4-2u_b^2+4\xi^2u_b^2-2=0$$

$$u_b^4-2u_b^2(1-2\xi^2)-1=0.$$

Let $\qquad k=u_b^2 \Rightarrow u_b=\sqrt{k}$

$$\therefore \qquad k^2-2k(1-2\xi^2)-1=0$$

$$k = \frac{2(1-2\xi^2) \pm \sqrt{4(1-2\xi^2)^2 + 4}}{2}$$

$$= (1-2\xi^2) \pm \sqrt{(1-2\xi^2)^2 + 1}$$

By taking only positive sign, we get

$$k = (1-2\xi^2) + \sqrt{(1-2\xi^2)^2 + 1}$$

$$u_b = \sqrt{k} = \left[(1-2\xi^2) + \sqrt{(1-2\xi^2)^2 + 1}\right]^{1/2}.$$

We know that $u_b = \dfrac{\omega_b}{\omega_n}$

$$\Rightarrow \quad \omega_b = \omega_n \left[1 - 2\xi^2 + \sqrt{1 + (1-2\xi^2)^2}\right]^{1/2} = \text{bandwidth.} \tag{11.11}$$

4. **Phase margin (*r*)**
   The open-loop t.f. of a second-order system is

$$G(s) = \frac{\omega_n^{\ 2}}{s(s + 2\xi\omega_n)}.$$

The sinusoidal t.f. is obtained by substituting $s = j\omega$

$$G(j\omega) = \frac{\omega_n^{\ 2}}{j\omega(j\omega + 2\xi\omega_n)}$$

$$= \frac{\omega_n^{\ 2}}{-\omega^2 + 2j\omega\xi\omega_n}$$

$$= \frac{\omega_n^{\ 2}}{\omega_n^2\left(-\dfrac{\omega^2}{\omega_n^2} + 2j\xi\dfrac{\omega}{\omega_n}\right)}$$

$$= \frac{1}{-\left(\dfrac{\omega}{\omega_n}\right)^2 + 2j\xi\dfrac{\omega}{\omega_n}}$$

$$= \frac{1}{-u^2 + 2j\xi u}$$

$$= \frac{1}{\sqrt{u^4 + 4\xi^2u^2}\angle\tan^{-1}\left(\dfrac{-2\xi}{u}\right)}$$

$$= [u^4 + 4\xi^2u^2]^{-1/2}\angle\tan^{-1}\left(\frac{+2\xi}{u}\right). \tag{11.12}$$

$$\text{Magnitude} = |G(j\omega)| = [u^4 + 4\xi^2u^2]^{-1/2}$$

$$\text{Phase angle} = |G(j\omega)| = \tan^{-1}\left(\frac{+2\xi}{u}\right).$$

At the gain crossover frequency ω_{gcf}, the magnitude of $G(j\omega)$ is unity. Let the normalised gain crossover frequency be $u_{gcf} = \frac{\omega_{gcf}}{\omega_n}$. By substituting u by u_{gcf} in the equation for magnitude of $G(j\omega)$ and equating to unity, we get

i.e., at $u = u_{gcf}, |G(j\omega)| = 1$

$$\Rightarrow \left[u_{gcf}^4 + 4\xi^2 u_{gcf}^2\right]^{1/2} = 1$$

$$u_{gcf}^4 + 4\xi^2 u_{gcf}^2 = 1.$$

Let $k = u_{gcf}^2 \Rightarrow u_{gcf} = \sqrt{k}$

$$k^2 + 4\xi^2 k - 1 = 0$$

$$k = \frac{-4\xi^2 \pm \sqrt{16\xi^4 + 4}}{2}$$

$$= -2\xi^2 \pm \sqrt{4\xi^4 + 1}$$

By taking the positive sign, we get

$$u_{gcf} = \sqrt{k}$$

$$= \left[-2\xi^2 + \sqrt{4\xi^4 + 1}\right]^{1/2}. \tag{11.13}$$

Phase margin, $r = 180° + \angle G(j\omega)\Big|_{\omega=\omega_{gcf}, u=u_{gcf}}$

$$= 180° + \tan^{-1}\left(\frac{2\xi}{u_{gcf}}\right)$$

$$= 180° + \tan^{-1}\frac{2\xi}{\left[-2\xi^2 + \sqrt{4\xi^4 + 1}\right]^{1/2}}.$$

$\therefore$ Gain crossover frequency, $\omega_{gcf} = \omega_n\left[-2\xi^2 + \sqrt{4\xi^4 + 1}\right]^{1/2}$. (11.14)

Phase margin, $r = 180° + \tan^{-1}\left[\frac{2\xi}{\left[-2\xi^2 + \sqrt{4\xi^4 + 1}\right]^{1/2}}\right]$. (11.15)

Note: The gain margin of a second-order system is infinite.

Problem-11.1: Determine the frequency domain specifications of the system having $\frac{C(s)}{R(s)} = \frac{5}{s^2 + 2s + 5}$.

Solution:

The C.E. of the system is $s^2 + 2s + 5$.

By comparing the equation with $s^2 + 2\xi\omega_n s + \omega_n^2 = 0$, we get

$$\omega_n = \sqrt{5}, 2\xi\omega_n = 2 \Rightarrow \xi = 2/2 \times \sqrt{5} = 0.45.$$

(a) Resonant frequency: $\omega_r = \omega_n\sqrt{1 - 2\xi^2} = \sqrt{5}\sqrt{1 - 2 \times 0.45^2}$

$$= 1.725\,\text{rad/s}.$$

(b) Resonant peak:

$$M_r = \frac{1}{2\xi\sqrt{1-\xi^2}} = \frac{1}{2\times0.45\sqrt{1-0.45^2}} = 1.24.$$

(c) Bandwidth:

$$\omega_b = \omega_n\left[1-2\xi^2+\sqrt{1+(1-2\xi^2)^2}\right]^{1/2}$$

$$= \sqrt{5}\left[1-2\times(0.45^2)+\sqrt{1+(1-2\times0.45^2)^2}\right]^{1/2}$$

$$= 3.01\,\text{rad/s}$$

Problem-11.2: Find the frequency domain specifications with a unity feedback system having $G(s) = \dfrac{225}{s(s+6)}$.

Solution:

The t.f. of the system is given as

$$\frac{C(s)}{R(s)} = \frac{G(s)}{1+G(s)H(s)} = \frac{225/s(s+6)}{1+225/s(s+6)} = \frac{225}{s^2+6s+225}.$$

The C.E. of the system is $s^2+6s+225$.

By comparing the equation with $s^2+2\xi\omega_n s+\omega_n{}^2=0$.

$$\omega_n = 15,\ 2\xi\omega_n = 6 \Rightarrow \xi = 6/2\times15 = 0.2$$

(a) Resonant frequency:

$$\omega_r = \omega_n\sqrt{1-2\xi^2} = 15\sqrt{1-2\times0.2^2}$$

$$= 14.4\text{rad/s}.$$

(b) Resonant peak:

$$M_r = \frac{1}{2\xi\sqrt{1-\xi^2}} = \frac{1}{2\times0.2\sqrt{1-0.2^2}} = 2.53.$$

(c) Bandwidth:

$$\omega_b = \omega_n\left[(1-2\xi^2)+\sqrt{1+(1-2\xi^2)^2}\right]^{1/2}$$

$$= 15\left[(1-2\times(0.2)^2)+\sqrt{1+(1-2\times0.2^2)^2}\right]^{1/2}$$

$$= 34.185\,\text{rad/s}.$$

11.6 FREQUENCY DOMAIN METHODS

The variations in M and ϕ of $G(j\omega)$ as ω, the input frequency is varied from 0 to ∞, can be obtained by number of methods. The commonly used method to sketch such frequency response for the given systems are

1. Bode plot or logarithmic plot
2. Polar plot
3. Nyquist plot
4. M–ϕ plot or log magnitude versus phase plot

11.7 LOGARITHMIC PLOT OR BODE PLOT

A sinusoidal t.f. may be represented by two separate plots: magnitude versus frequency in logarithmic scale and phase angle versus frequency in logarithmic scale as shown in Fig.11.3. The magnitude can be expressed in its logarithmic values by finding 20log $G(j\omega)H(j\omega)$, which has a unit decibel denoted by dB, and they are to be plotted against $\log_{10}\omega$. In the phase angle plot, angle of $G(j\omega)H(j\omega)$ is to be expressed in degrees, which is to be plotted against $\log_{10}\omega$.

To sketch the magnitude in dB and phase angle in degrees against $\log_{10}\omega$, logarithmic scale is used, which is available on semi-log graph paper. In such a paper, the x-axis is divided into a logarithmic scale, which is a non-linear one, while y-axis is divided into linear scale, and hence, it is called semi-log paper.

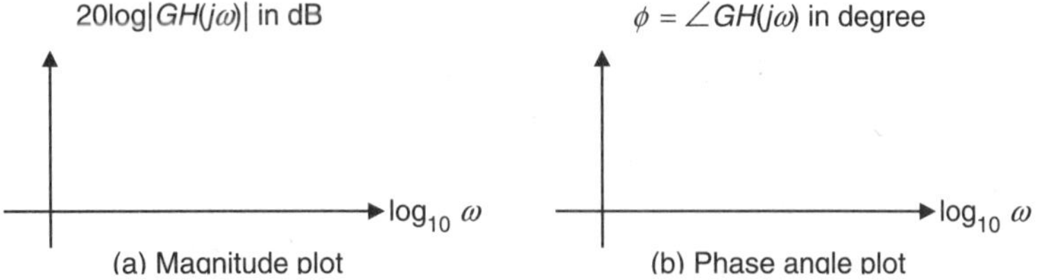

Figure 11.3 Bode plot

The distance between 1 and 10 is called a decade because log 1= 0 and log 10 = 1 and this distance is divided into 10 parts according to logarithmic scale, that is, log 2, log 3, -------------. Hence, it is not necessary to find logarithmic value of ω while plotting on x-axis but the logarithmic scale available takes care of the logarithmic value of ω. The advantage of this scale is in the wide range of frequencies and can be accommodated on a single paper. As log 0 = $-\infty$, it is obvious that x-axis cannot be calibrated from 0 but as per requirement, the smallest frequency may be selected as starting frequency such as 0.01 and 0.1. This hardly affects the result of the frequency response.

Consider
$$G(s)H(s) = k'\frac{s^m(s+z_1)(s+z_2)(s+z_3)\ldots}{s^n(s+p_1)(s+p_2)(s+p_3)\ldots}. \quad (11.16)$$

This form is not useful to sketch the Bode plot because either s^m or s^n will be present at a time but not both. Hence, it is necessary to rewrite $G(s)H(s)$ in the time constant form as

$$G(s)H(s) = k\frac{s^m(1+sT_{z1})(1+sT_{z2})(1+sT_{z3})\ldots}{s^n(1+sTP_1)(1+sTP_2)(1+sTP_3)\ldots}.$$

where
$$k = k'\frac{z_1z_2\ldots}{p_1p_2\ldots}.$$

Again, either s^m or s^n is present but not both.

The standard time constant form can be denoted as

$$G(s)H(s) = k\frac{(1+sT_{z1})(1+sT_{z2})(1+sT_{z3})\ldots}{s^n(1+sT_{p1})(1+sT_{p2})(1+sT_{p3})\ldots}. \quad (11.17)$$

where k is the resultant system gain, n is the type of the system and $T_{p1}, T_{p2}\ldots, T_{z1}, T_{z2}, \ldots$ are the time constants of different poles and zeros.

Each of the factors involved in (11.17) will contribute to magnitude and phase angle variations of $G(j\omega)H(j\omega)$ in frequency domain. The transfer of frequency domain can be obtained by substituting $s = j\omega$ in the above-mentioned expression.

11.8 BASIC FACTORS OF $G(j\omega)H(j\omega)$

1. **System gain k**

$$G(s)H(s) = k = k + j0$$

$$|G(j\omega)H(j\omega)| = \sqrt{k^2 + 0} = k$$

$$20\log|G(j\omega)H(j\omega)| = 20\log_{10} k\,\text{dB}$$

As gain k is constant, $20\log_{10}k$ is always constant though ω is varied from 0 to ∞. Therefore, its magnitude plot will be a straight line parallel to x-axis as shown in Fig.11.4.

$$\text{Phase angle } \phi = \tan^{-1}\left[\frac{\text{Imaginary part}}{\text{Real part}}\right]$$

$$= \tan^{-1}\left(\frac{0}{k}\right) = 0°.$$

Figure 11.4 Bode plot of constant '*k*'

Therefore, it does not affect the phase angle plot as its contribution to phase angle plot is 0°.

2. **Poles or zeros at the origin ($s^{\pm n}$)**

Let us consider for simplicity single pole at the origin.

$$G(s)H(s) = \frac{1}{s}$$

$$\therefore \quad G(j\omega)H(j\omega) = \frac{1}{j\omega} = \frac{1}{0 + j\omega}$$

$$|G(j\omega)H(j\omega)| = \frac{1}{\sqrt{0 + \omega^2}} = \frac{1}{\omega}.$$

$$\text{Magnitude in dB} = 20\log|G(j\omega)H(j\omega)|$$

$$= 20\log\left(\frac{1}{\omega}\right) = 20\log(\omega^{-1})$$

$$= -20\log\omega.$$

Thus, this equation is similar to $y = mx$, that is, 1 pole at the origin contributes to the magnitude plot according to $-20\log\omega$, that is, according to the straight line of slope -20 as shown in Fig.11.5.

$$\text{Magnitude in dB} = 20 \text{ if } \omega = 0.1$$
$$= 0 \text{ if } \omega = 1.0$$
$$= -20 \text{ if } \omega = 10.0$$
$$= -40 \text{ if } \omega = 100.0$$

Hence, the slope of magnitude plot of 1 pole at the origin is called -20 dB/dec. Now, at $\omega = 1$, $|G(j\omega)H(j\omega)| = 0$ dB. In other words, this line intersects the reference 0 dB line at $\omega = 1$ and at $\omega = 0.1$, it has magnitude of +20 dB, while at $\omega = 10$, it has magnitude of -20 dB.

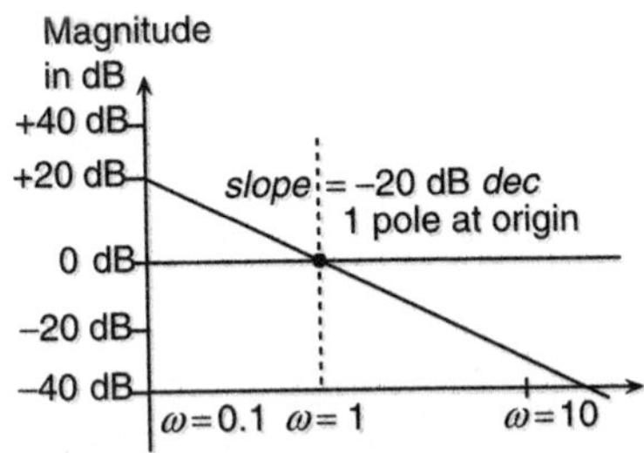

Figure 11.5 Bode plot of pole at origin

Consider two poles at origin

$$G(s)H(s) = \frac{1}{s^2}$$
$$= \frac{1}{j\omega} \cdot \frac{1}{j\omega}$$
$$|G(j\omega)H(j\omega)| = \frac{1}{\omega^2}$$
$$20\log|G(j\omega)H(j\omega)| = 20\log\left(\frac{1}{\omega^2}\right)\text{dB}$$
$$= 20\log(\omega^{-2})\text{dB}$$
$$= -40\log\omega.$$

Therefore, it is a straight line of slope -40 dB/dec.

In logarithmic plot, multiplication gets replaced by addition. At $\omega = 1, |G(j\omega)H(j\omega)| = 0\text{dB}$, that is., this line has slope -40 dB/dec and it intersects 0 dB line at $\omega = 1$.

Similarly. for n number of poles at the origin, we have

$$G(s)H(s) = \frac{1}{s^n}$$
$$|G(j\omega)H(j\omega)| = \frac{1}{\omega^n}$$

$$|G(j\omega)H(j\omega)| \text{ in dB} = 20\log(\omega)^{-n} = -20n\log\omega$$

Therefore, this is a straight line of slope $-20 \times n$ dB/s and intersecting 0 dB line at $\omega = 1$ as shown in Figure 11.6.

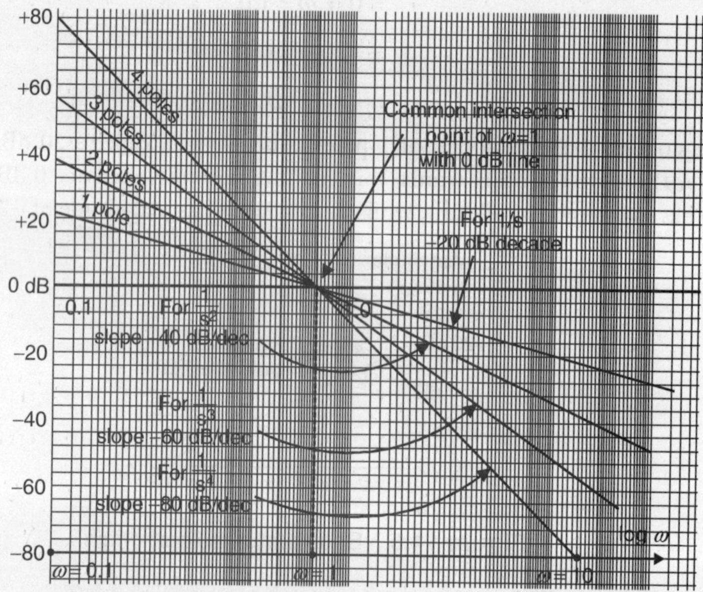

Figure 11.6 Bode plot of multiple poles at origin

For a zero at the origin, we have

$$G(s)H(s) = s$$
$$G(j\omega)H(j\omega) = 0 + j\omega$$

$$\text{magnitude in dB} = 20 \log \omega \text{ dB.}$$

This is the equation of a straight line whose slope is +20 dB/dec. The only change is the sign of the slope; for pole, it is −20 dB/dec, while for zero, it is +20 dB/dec. However, for both, intersection of line with 0 dB occurs at $\omega = 1$ only.

In general form, the number of zero at the origin,

$$G(s)H(s) = s^p$$
$$G(j\omega)H(j\omega) = (j\omega)^p$$
$$|G(j\omega)H(j\omega)| = \omega^p$$

$$\text{magnitude in dB} = 20 \times p \log \omega$$

$$\text{slope} = 20 \times p \text{ dB/dec}$$

Therefore, it gives family of lines with slopes as 20, 40.........20p dB/dec passing through intersection point of $\omega = 1$ with 0 dB line, as shown in Figure11.7.

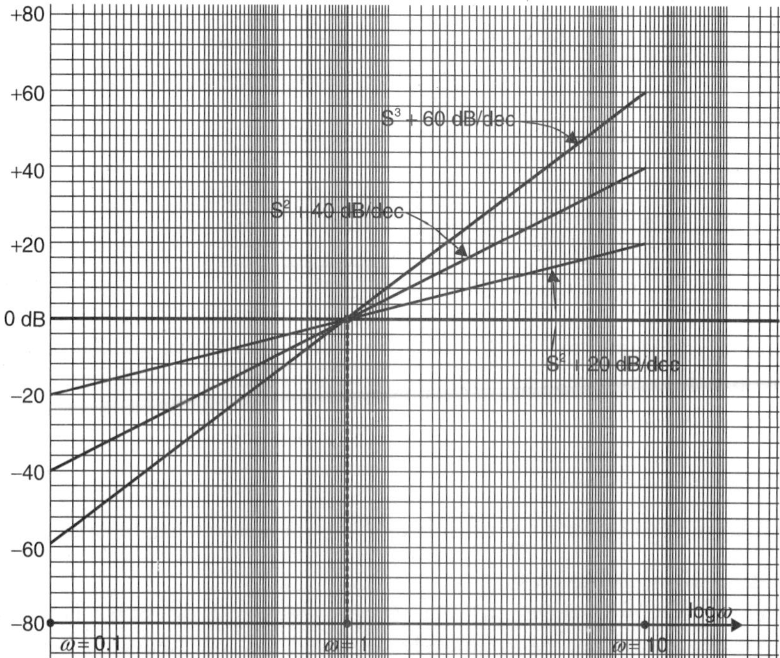

Figure 11.7 Bode plot of multiple zeros at origin

Phase angle plot: consider 1 pole at origin

$$G(s)H(s) = \frac{1}{s}$$

$$G(j\omega)H(j\omega) = \frac{1}{j\omega}$$

Phase angle $$G(j\omega)H(j\omega) = \frac{1}{\angle j\omega} = \frac{0^\circ}{90^\circ} = -90^\circ.$$

This is independent of ω. Therefore, phase angle plot of pole at origin is a line parallel to *X*-axis contributing –90° to phase angle.

For two poles at origin,

$$G(s)H(s) = \frac{1}{s^2}$$

$$G(j\omega)H(j\omega) = \frac{1}{\omega^2} = \frac{0^\circ}{90^\circ . 90^\circ} = -180^\circ.$$

Angle gets added to each other.

In general, *p* number of poles at the origin contribute to (–90° *p*) angle to overall phase angle plot.

3. **Simple pole or zero (first-order factors)**

   Consider

$$G(s)H(s) = \frac{1}{(1 + sT)}.$$

$$|G(j\omega)H(j\omega)| = \frac{1}{\sqrt{1+(\omega T)^2}} = \left[\sqrt{1+(\omega T)^2}\right]^{-1}$$

$$\text{Magnitude in dB} = 20\log|G(j\omega)H(j\omega)|$$

$$= 20\log\left[\sqrt{1+(\omega T)^2}\right]^{-1}$$

$$= -20\log\sqrt{1+(\omega T)^2}\text{dB}$$

Now, instead of sketching the magnitude plot exactly according to this equation, we can approximate this into two regions. We can draw a straight line approximation to the magnitude plot, and then, by applying correction factors can modify it to an accurate one if required.

The approximation is as follows

1. For low frequency range, $\omega << \frac{1}{T}$, that is, $\omega^2T^2 << 1$; hence, it can be neglected.
   Magnitude in dB = –20log 1 = 0.
   Therefore, for low frequencies, it is a straight line of 0 dB only. Thus, the contribution by such a factor can be completely neglected for low frequency range as it is very small.
2. For high frequency range, $\omega >> \frac{1}{T}$

$$\text{i.e., } \omega^2T^2 >> 1$$

$$\text{Magnitude in dB} = -20\log\omega T$$

In other words, it is a straight line of slope –20 dB/dec. Thus, for low frequency, magnitude plot is 0 dB line, while for high frequency, it is a straight line of slope –20 dB/dec. It is necessary to find a frequency ω_c, which separates the frequency range into two: low and high. This is the value of frequency at which straight line of slope –20 dB/dec in high frequency range intersects with 0 dB line in low frequency zone. Therefore, such an intersection takes place when

$$20\log\omega T = 0\text{ dB}$$

i.e.,

$$\omega T = 1$$

or $\omega = \frac{1}{T}$, which is called as corner frequency denoted as ω_c.

The magnitude plot discussed above is called asymptotic magnitude plot, is as shown in Figure 11.8.

The actual equation of magnitude plot (in dB) is $-20\log\sqrt{1+\omega^2T^2}$. Now, by approximation, at $\omega = \omega_c = \frac{1}{T}$, magnitude in dB = 0. However, actually, it can be calculated as actual magnitude (in dB), that is, $-20\log\sqrt{1+\omega_c^2T^2}$.

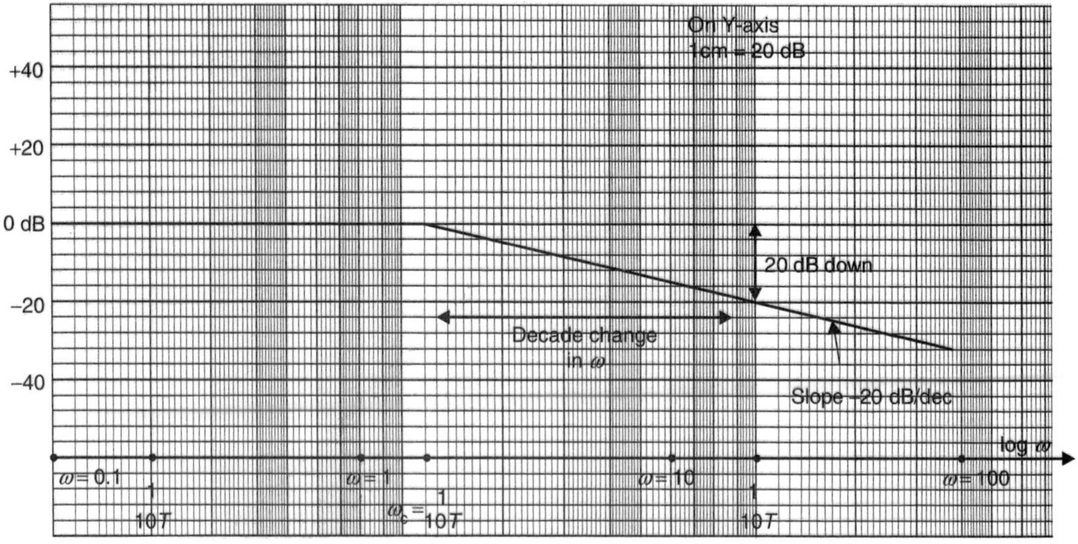

Figure 11.8 Bode plot of a simple pole

$$= -20\log\sqrt{1+\frac{1}{T^2}.T^2}$$

$$= -20\log\sqrt{2} = -3\text{ dB}$$

Similarly, at $\omega = \frac{2}{T}$

Actual magnitude (in dB) = $20\log\sqrt{1+\frac{4}{T^2}.T^2}$

$$= -20\log\sqrt{5} = -7\text{dB}$$

However, approximate magnitude (in dB) at $\omega = \frac{2}{T}$ is

$$= -20\log\omega T = -20\log\times\frac{2}{T}\times T = -6\text{ dB.}$$

$\therefore$ Correction at $\omega = \frac{2}{T} = -1\text{dB}$ down

While at $\omega = \frac{1}{2T}$,

actual magnitude (in dB) is $-20\log\sqrt{1+\frac{1}{4T^2}.T^2}$

$$= -20\log\sqrt{\frac{5}{4}} = -1\text{dB}.$$

Further, the value is approximately 0 dB as $\omega = \dfrac{1}{2T}$ is less than ω_c, that is, in the low frequency range.

Correction at $\omega = \dfrac{1}{2T} = -1$ dB (i.e., 1 dB down).

General error values are

For
$$\omega = \frac{\omega_c}{2} \rightarrow \text{error} = 1 \text{ dB down.}$$

$$\omega = \omega_c \rightarrow \text{error} = 3 \text{ dB down}$$

$$\omega = 2\omega_c \rightarrow \text{error} = 1 \text{ dB down}$$

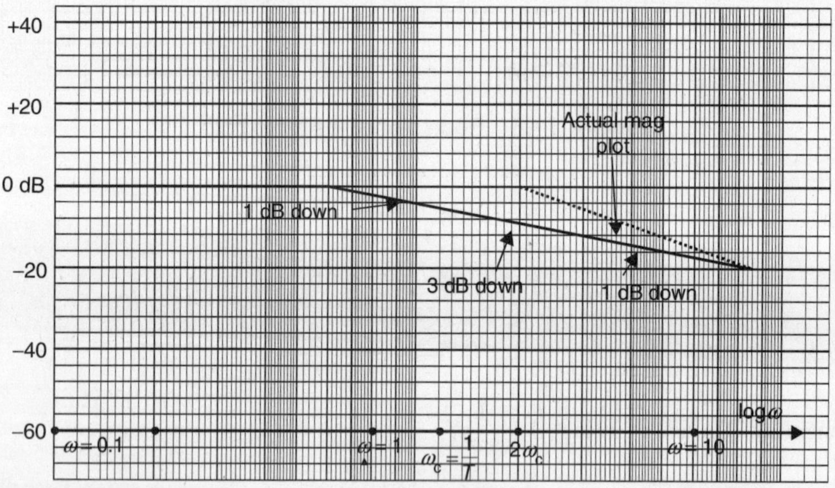

Figure 11.9 Bode plot of a simple pole showing correction factor

By applying these errors, the actual magnitude plot may be as shown in Figure 11.9.

For a simple zero, that is, first-order zero, we have

$$G(s)H(s) = (1 + Ts)$$

$$G(j\omega)H(j\omega) = (1 + j\omega T)$$

$$\text{Magnitude (in dB)} = 20\log|G(j\omega)H(j\omega)| = 20\log\sqrt{1+\omega^2T^2}$$

Hence, all analyses are applicable for a simple zero with a change in sign of slope, as shown in Figure 11.10.

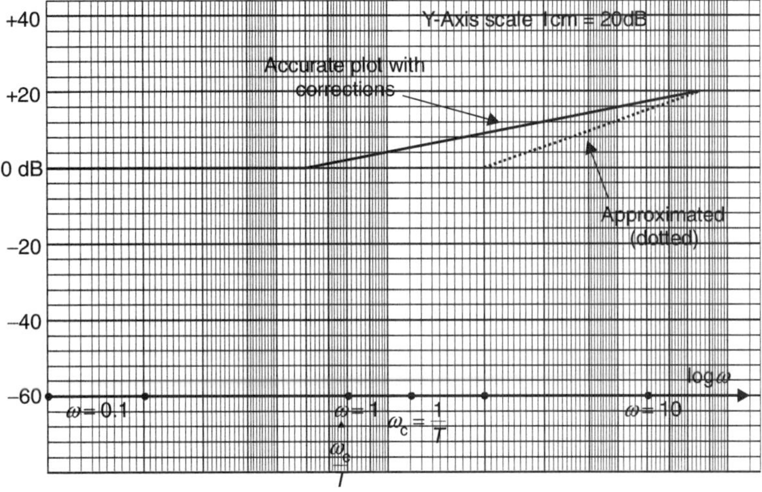

Figure 11.10 Bode plot of a simple zero

Phase angle plot:

Consider a simple pole

$$G(s)H(s) = \frac{1}{1+Ts}$$

$$G(j\omega)H(j\omega) = \frac{1}{1+j\omega T}$$

$$\angle G(j\omega)H(j\omega) = \frac{0^\circ}{\tan^{-1}\left(\frac{\omega T}{1}\right)} = -\tan^{-1}\omega T$$

While for a simple zero, we have

$$G(s)H(s) = (1+Ts)$$

$$G(j\omega)H(j\omega) = (1+j\omega T)$$

$$G(j\omega)H(j\omega) = \tan^{-1}\left(\frac{\omega T}{1}\right) = \tan^{-1}\omega T.$$

Therefore, the shape remains the same. Only sign of the angles change from negative to positive when factor changes from pole to zero. Such a plot is to be constructed by actually calculating angles for different frequencies. Table 11.1 lists the values.

Table 11.1 Phase angle for different values of ω

ω	**$\pm\tan^{-1}$ or (+ for zero, – for pole)**
$0.1\omega_c = \frac{1}{10T}$	±5.71°
$0.5\omega_c = \frac{1}{2T}$	±26.6°

(Continued)

(*Continued*)

ω	$\pm\tan^{-1}$ or (+ for zero, – for pole)
$\omega_c = \frac{1}{T}$	±45°
$2\omega_c = \frac{2}{T}$	±63.4°
$10\omega_c = \frac{10}{T}$	±84.3°

The data in the table can be plotted as shown in Figure 11.11.

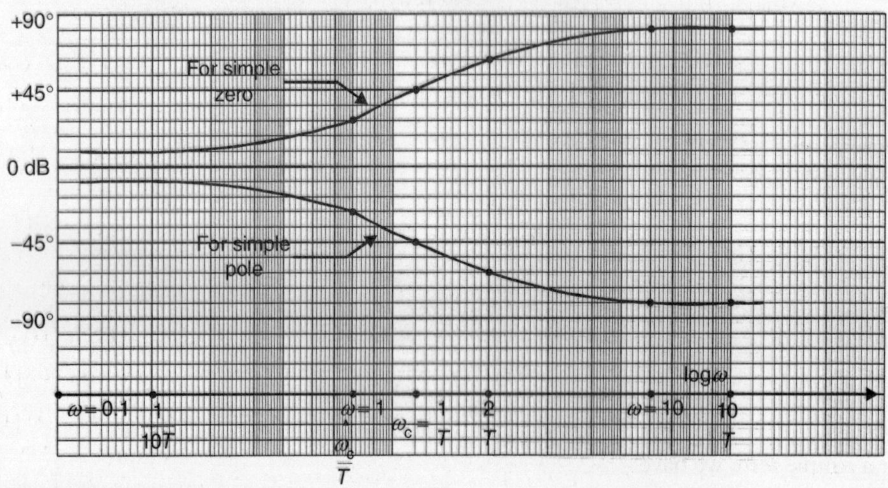

Figure 11.11 Actual Bode plot of a simple pole/zero

3. **Quadratic factors of the form**

$$\left(1+\left(\frac{2\xi}{\omega_n}\right)s+\frac{s^2}{\omega_n^2}\right)^{\pm 1} \simeq \left[1+2\xi j\left(\frac{\omega}{\omega_n}\right)+j^2\left(\frac{\omega}{\omega_n}\right)^2\right]^{\pm 1}$$

where ω is variable and ω_n is the constant for that factor.

$$G(j\omega)H(j\omega)=\frac{1}{1+2\xi j\left(\frac{\omega}{\omega_n}\right)-\left(\frac{\omega}{\omega_n}\right)^2} \quad \text{as } j^2=-1$$

$$=\frac{1}{\left[1-\left(\frac{\omega}{\omega_n}\right)^2\right]+j2\xi\left(\frac{\omega}{\omega_n}\right)}$$

$$\therefore \qquad j\omega) = \frac{1}{\sqrt{\left[1-\left(\frac{\omega}{\omega_n}\right)^2\right]^2 + 4\xi^2\left(\frac{\omega}{\omega_n}\right)^2}}$$

$$\therefore \qquad \text{Magnitude in dB} = 20\log\frac{1}{\sqrt{\left[1-\left(\frac{\omega}{\omega_n}\right)^2\right]^2 + 4\xi^2\left(\frac{\omega}{\omega_n}\right)^2}}$$

$$= 20\log\left[\sqrt{\left[1-\left(\frac{\omega}{\omega_n}\right)^2\right]^2 + 4\xi^2\left(\frac{\omega}{\omega_n}\right)^2}\right]^{-1}$$

$$\therefore \qquad \text{Magnitude in dB} = 20\log\left[\sqrt{\left[1-\left(\frac{\omega}{\omega_n}\right)^2\right]^2 + 4\xi^2\left(\frac{\omega}{\omega_n}\right)^2}\right]^{-1}.$$

For low frequencies, $\omega << \omega_n \; \therefore \left(\frac{\omega}{\omega_n}\right)^2 << 1.$

Magnitude in dB $= -20\log 1 = 0\,\text{dB}$

Thus, similar to simple pole, quadratic pole is also negligible until its corner frequency occurs.

For high frequency,

$\omega > \omega_n$, $\qquad 4\xi^2\left(\frac{\omega}{\omega_n}\right)^2 << \left(\frac{\omega}{\omega_n}\right)^4$, as ξ is very low.

$$\text{Magnitude in dB} = -20\log\sqrt{\left[\left(\frac{\omega}{\omega_n}\right)^2\right]^2}$$

$$= -20\log\left(\frac{\omega}{\omega_n}\right)^2 = -40\log\left(\frac{\omega}{\omega_n}\right).$$

$$= -40\log\omega + 40\log\omega_n$$

This is the equation of straight line of slope –40 dB/dec. Hence, general magnitude plot for quadratic factor is 0 dB line till corner frequency, and then, a straight line of slope –40 dB/dec, as shown in Figure 11.12.

To find corner frequency ω_c, we have

$$-40\log_n\frac{\omega}{\omega_n} = 0\text{dB}$$

$$\frac{\omega}{\omega_n} = 1$$

$$\therefore \qquad \omega_c = \omega_n.$$

Therefore, ω_n is the corner frequency for such a factor.

However, the above asymptotic plot is not very accurate as the error for the quadratic factor not only depends on ω but also on the value of ξ damping ratio.

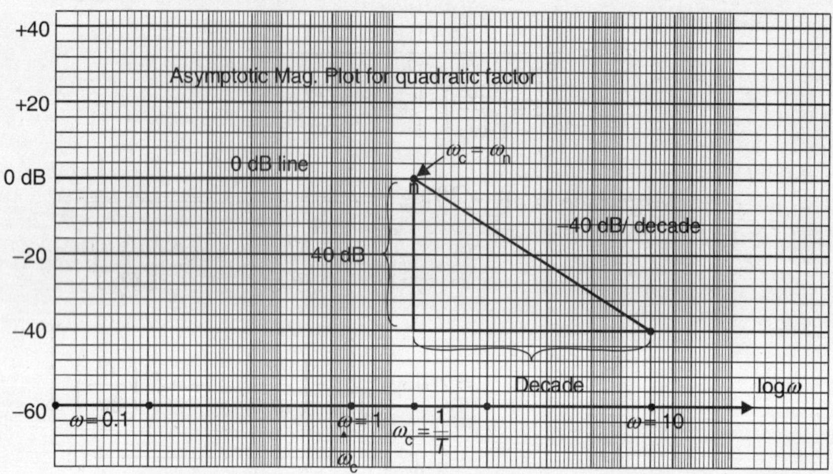

Figure 11.12 Bode plot of a quadratic pole

Let us see the effect of variation of ξ on the magnitude plot.

$$\text{Actual magnitude in dB} = -20\log\sqrt{\left(1-\left(\frac{\omega}{\omega_n}\right)^2\right)^2 + 4\xi^2\left(\frac{\omega}{\omega_n}\right)^2}.$$

$$\text{Now, at } \omega = \omega_n \text{, i.e, } \frac{\omega}{\omega_n} = 1.$$

$$\text{Actual magnitude in dB} = -20\log\sqrt{4\xi^2}.$$

Table 11.2 shows the corresponding error values for various values of ξ.

Table 11.2 Error for different values of ξ

ξ	**Accurate magnitude (in dB)**	**Approximate magnitude (in dB)**	**Error for quadratic pole**
0.1	+13.97	0	+13.97 dB up
0.2	+7.95	0	+7.95 dB up
0.3	+4.43	0	+4.43 dB up
0.4	+1.93	0	+1.93 dB up
0.7	−2.92	0	2.92 dB up
0.9	−5.10	0	5.1 dB down
1	6.02	0	6 dB down

Further, as errors are calculated, the magnitude plot for quadratic pole gets modified as shown in Figure 11.13. Hence, it is necessary to modify magnitude plot. We can use Table 11.2 to decide the correction for given ξ or find the correction using the formula.

$$\text{Correction} = -20\log 2\xi \text{ dB at } \omega = \omega_n \text{ of pole.}$$

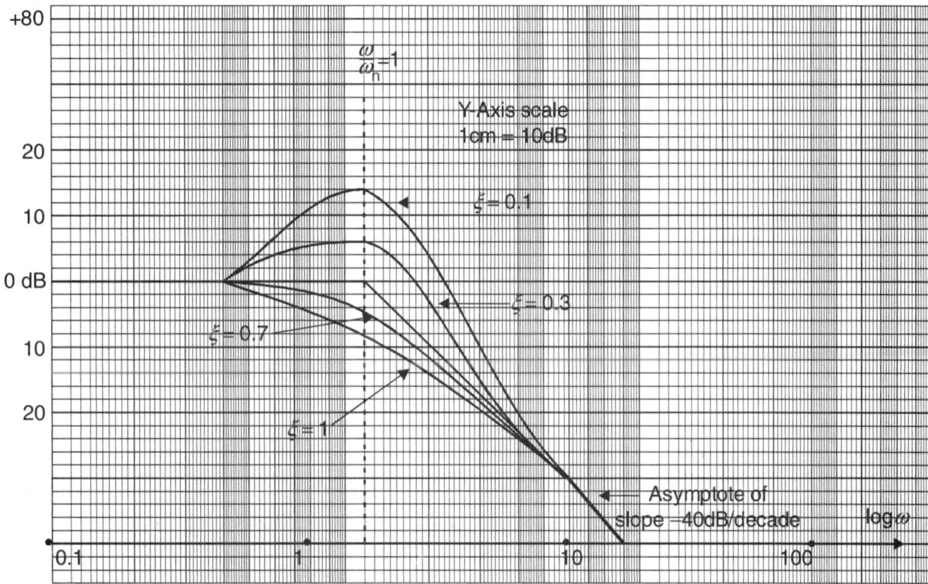

Figure 11.13 Exact Bode plot of a quadratic pole for different values of ξ

Here, the positive correction is upwards and negative correction is downwards.

The magnitude plot for a quadratic zero can be obtained by reversing the sign of the slope of basic asymptote and the signs of the corrections at corner frequency for various values of ξ, as shown in Figure 11.14.

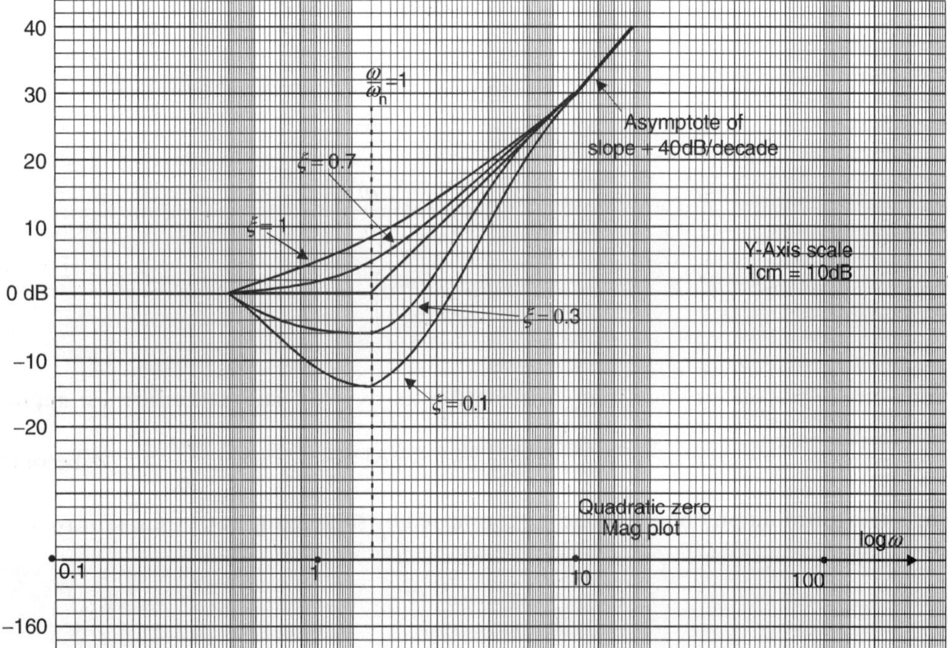

Figure 11.14 Exact Bode plot of a quadratic zero for different values of ξ

Phase angle

$$\angle G(j\omega)H(j\omega) = \frac{0^\circ}{\tan^{-1}\left\{\frac{2\xi\left(\frac{\omega}{\omega_n}\right)}{1-\left(\frac{\omega}{\omega_n}\right)^2}\right\}}.$$

$$\angle GH(j\omega) = -\tan^{-1}\frac{2\xi\left(\frac{\omega}{\omega_n}\right)}{1-\left(\frac{\omega}{\omega_n}\right)^2}$$

Table 11.3 shows the values for $\xi = 0.3$.

Table 11.3 Phase angle for different values of ω/ω_n

ω/ω_n	ϕ
0.1	−3.46°
0.5	−21.6°
1.0	−90.0°
2.0	+21.8 − 180° = −158.19°
4.0	+10.09 − 180° = −170.9°
10.0	+3.46 − 180° = −176.53°
⋮	⋮
∞	−180°

$$\phi = -\tan^{-1}\left\{\frac{2\times 0.3\times\left(\frac{\omega}{\omega_n}\right)}{1-\left(\frac{\omega}{\omega_n}\right)^2}\right\}.$$

Therefore, at $\frac{\omega}{\omega_n} = 1$ (i.e., $\omega = \omega_c = \omega_n$), it contributes to −90°, and hence, it must approach to −180° as $\frac{\omega}{\omega_n} \to \infty$. However, according to above-mentioned formula when $\frac{\omega}{\omega_n} > 1$, ϕ becomes positive, the angle contribution obtained must be considered by subtracting 180° from positive ϕ.

At $$\frac{\omega}{\omega_n} = 2,$$

we have $\phi = -\tan^{-1}(-0.4) = -(-21.8) = +21.8$

However, the actual angle contribution must be considered by applying correction of −180°, i.e., 21.8 − 180°= −158.19°. This happens because the behaviour of $\tan^{-1}$ function for the complex quantities with real part negative or imaginary part negative cannot be identified on calculator by using the abovementioned formula. For quadratic zero, the sign of the angle should be made positive

11.9 RENEWED METHOD OF DRAWING BODE PLOT

1. Adjust all the factors in the form of $(1 + Ts)$, and then, find the new value of k.
2. Calculate all the corner frequencies and find the power of s in the numerator as well as in the denominator. Numerator power of s is m, denominator power of s is n, identify m and n from the given t.f.
3. Calculate the starting point equal to $20 \log k - (m - n)\, 20$.
   Also calculate the starting slope = $(m - n)\, 20$.
4. Make the initial mark and set the initial frequency as $\omega = 0.1$. Successively, for each decade, mark ω=1, 10, 100, 1000, etc. Draw '0' dB line on the semi-log sheet. Fix the calculated starting point and draw vertical dotted lines for all the corner frequencies of its identity.
5. From the starting point, with starting slope, draw magnitude plot up to first corner frequency.
6. Check whether first corner frequency term is in numerator or in denominator. If the term is in numerator, the factor contributes to +20 dB slope. If it is in denominator, it contributes to −20 dB slope. Calculate new slope equal to existing slope ± contributing slope. From the first corner frequency point, draw further graph with new slope until it cuts next corner frequency term. Therefore, between two corner frequencies, the slope remains the same. It changes only at corner frequency. Thus, we calculate a new slope at each corner frequency, and we continue to draw the total magnitude plot at the end of the last corner frequency and beyond that value.

11.9.1 How to Draw Required Slope Line on Semi-log Paper from the Given Point

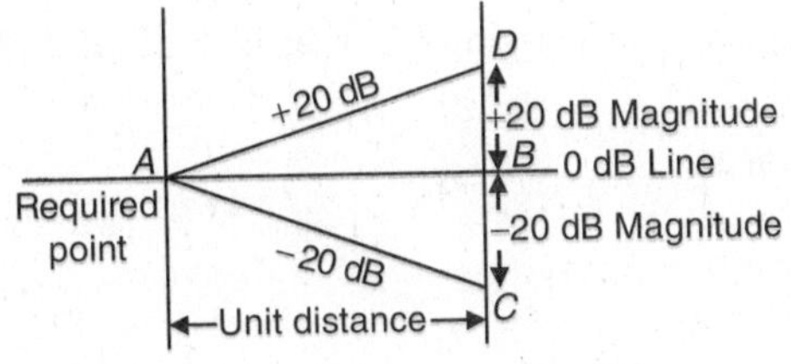

Note: In semi-log paper, a decade distance is a unit distance.
Slope = Opposite side/Adjacent side
If adjacent side = 1, then slope = opposite side.

Let us assume that from point A in above figure, it is required to draw ±20 dB slope line. From point A, move further unit distance horizontally equal to one decade distance, mark it as B. At point B, move vertically upward 20 dB magnitude for +20 dB slope and mark it as D. From point B, go vertically downwards −20 dB magnitude for −20 dB slope line and mark it as C. Further, hold the foot rule between point A and C or between A and D; now, draw a straight line, and this gives required slope line.

11.10 STABILITY CONSIDERATION FROM BODE PLOT

1. **Gain margin (GM)**

It can be defined as reciprocal of the magnitude of $G(j\omega)H(j\omega)$ measured in phase crossover frequency.

$$\text{GM} = \frac{1}{|G(j\omega)H(j\omega)|_{\text{at } \omega=\omega_{pc}}}.$$

Phase crossover frequency (ω_{pc})

The frequency at which phase angle of $G(j\omega)H(j\omega)$ is $-180°$ is called phase crossover frequency.

2. **Phase margin (PM)**

It can be defined as the amount of additional phase lag at the gain crossover frequency required to bring the system to the verge of instability

$$\text{PM} = 180° + \angle G(j\omega)H(j\omega)_{\text{at } \omega=\omega_{gc}}.$$

Gain crossover frequency (ω_{gc})

The frequency at which magnitude of $G(j\omega)\ H(j\omega)$ is unity, i.e., 1, is called gain crossover frequency (i.e., 20 log10' = 0 dB).

Bode plot showing GM, PM, ω_{gc} and ω_{pc} is as shown in Figure 11.15.

3. **Stability analysis from Bode plot**

The system is said to be stable when PM and GM are positive, while it is said to be unstable when both PM and GM are negative. Now, when the system is on the verge of instability, that is, marginally stable in nature, then both GM and PM are zero. This is possible when

$$\omega_{gc} = \omega_{pc}.$$

For stable system, $\omega_{gc} < \omega_{pc}$, i.e., GM and PM are positive.
For unstable system, $\omega_{gc} > \omega_{pc}$, i.e., GM and PM are negative.
For marginally stable system, $\omega_{gc} = \omega_{pc}$, i.e., GM and PM are zero.
If GM = ∞, system is absolutely stable and when GM = −∞, the system is inherently unstable.
As a thumb rule, an adequate gain margin of 2, that is, 6 dB and phase margin of about 30° are generally considered good.

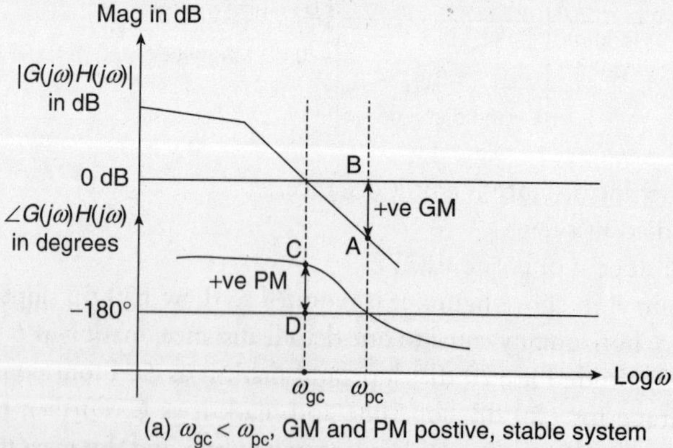

(a) $\omega_{gc} < \omega_{pc}$, GM and PM postive, stable system

Figure 11.15 Bode Plot Showing PM, GM, ω_{gc} and ω_{pc}

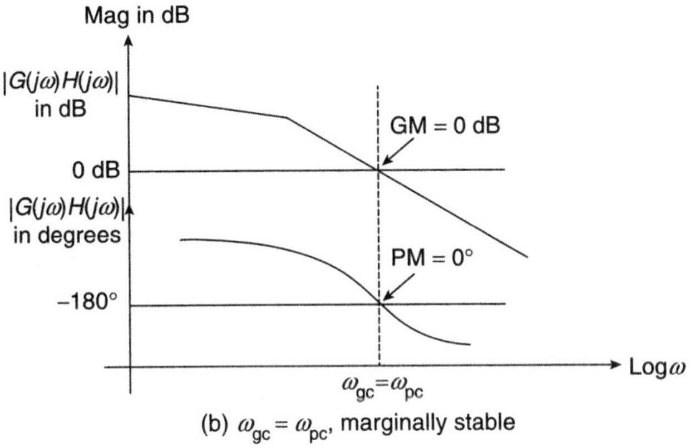

(b) $\omega_{gc} = \omega_{pc}$, marginally stable

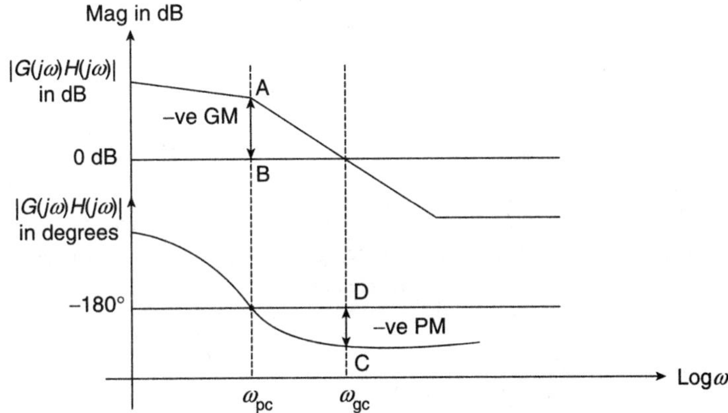

(c) $\omega_{gc} > \omega_{pc}$, GM and PM negative, unstable system

Figure 11.15 (*Continued*)

Problem-11.3: A unity feedback control system has the t.f. $G(s) = \dfrac{80}{s(s+2)(s+20)}$. Draw the Bode plot, and hence, determine GM, PM, ω_{gc} and ω_{pc}; further, comment on the stability.

Solution:

The open-loop t.f. is given as

$$G(s)H(s) = \frac{80}{s(s+2)(s+20)} = \frac{80}{s2(1+s/2)20(1+s/20)} = \frac{2}{s(1+0.5s)(1+0.05s)}.$$

To get the sinusoidal t.f., substitute $s = j\omega$ in $G(s)H(s)$.

$$G(j\omega)H(j\omega) = \frac{2}{j\omega(1+j0.5\omega)(1+j0.05\omega)}.$$

Phase angle is given as

$$\phi = \angle G(j\omega)H(j\omega) = -90 - \tan^{-1} 0.5\omega - \tan^{-1} 0.05\omega \; .$$

Phase plot

ω	−90	$-\tan^{-1} 0.5\omega$	$-\tan^{-1} 0.05\omega$	ϕ
0	−90	0	0	−90
0.2	−90	−5.7	−0.57	−96.27
2	−90	−4.5	−5.7	−140.7
8	−90	−75.96	−21.8	−187.76
10	−90	−78.69	−26.56	−195.27
20	−90	−84.28	−45.0	−219.28
40	−90	−87.13	−63.43	−240.58

Magnitude plot

The type of the given t.f. is 1 i.e., $m = 0$, $n = 1$ and hence, we have

1. Initial slope of the Bode plot = $(m-n)20 = -20$ dB/dec
2. Gain value in dB or intersection point on dB axis

$$= 20\log k - (m-n)20 = 20 + 20 \log k = 20 + 20 \log 2 = 26 \text{ dB.}$$

The corner frequencies are 2, 20 rad/s

Factor	Corner frequency in rad/s	Slope in dB/dec	Change in slope in dB/dec	Asymptotic log-magnitude characteristic
2	–	–	–	Straight line of constant slope of 0 dB/s starting from 20 log 2 = 6 dB point.
$\frac{1}{j\omega}$	None	−20	–	Straight line of constant slope −20 dB/dec passing through $\omega = 1$
$\frac{1}{1+j0.5\omega}$	2	−20	−20 − 20 = −40	Straight line of constant slope −20 dB/dec originating at $\omega = 2$
$\frac{1}{1+j0.05\omega}$	20	−20	−40 − 20 = −60	Straight line of constant slope −20 dB/dec originating at $\omega = 20$

The Bode plot is as shown in Figure (a).
From the Bode plot as shown in the figure, we have

1. Gain crossover frequency, $\omega_{gc} = 2.1$ rad/s
2. Phase crossover frequency, $\omega_{pc} = 6.35$ rad/s
3. Gain margin, GM = +21 dB
4. Phase margin, PM = +38°

Both GM and PM are positive, and hence, the system is stable.

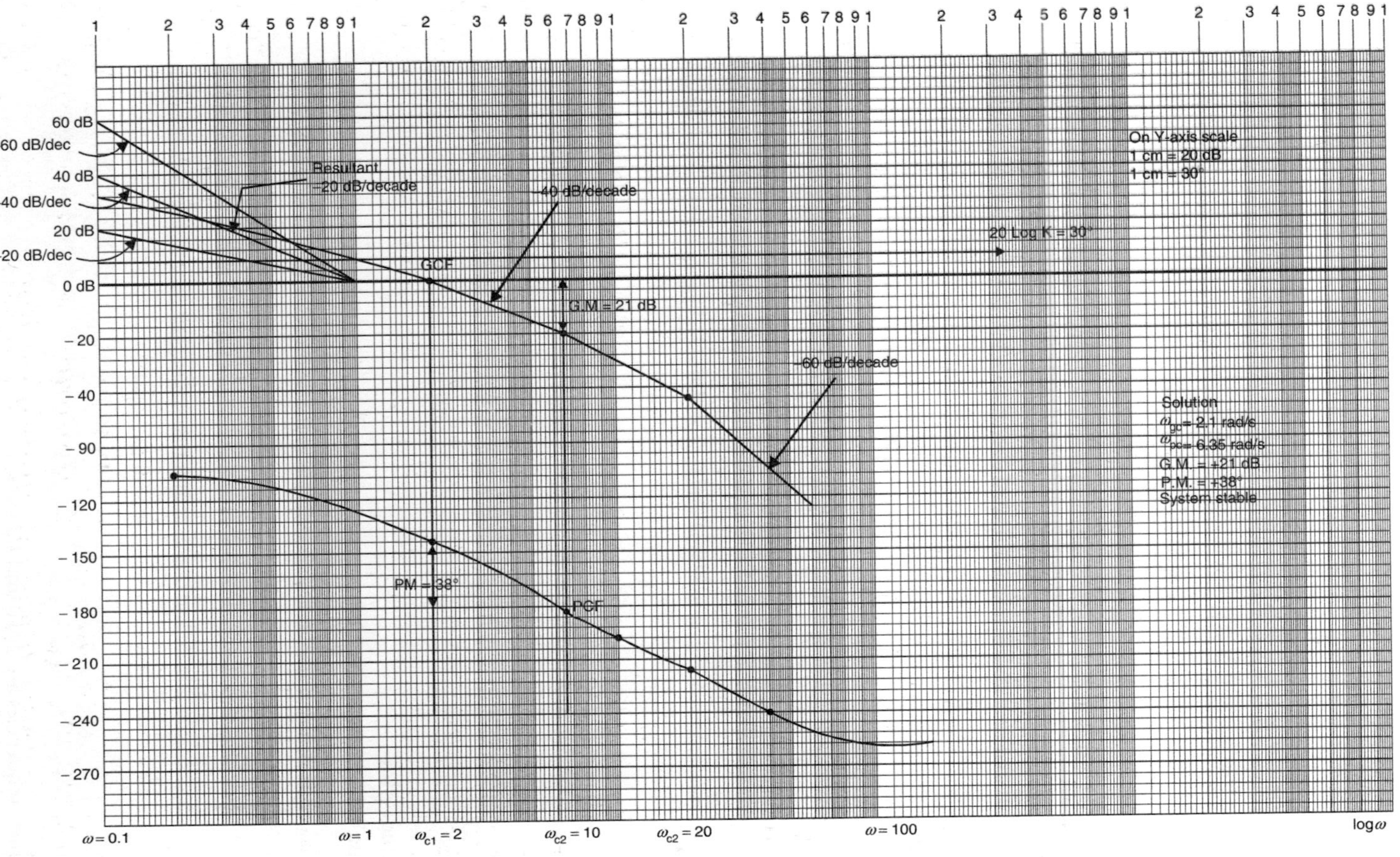

Figure (a) Bode plot for Problem 11.3

Problem-11.4: For a unity feedback system, $G(s)=\dfrac{800(s+2)}{s^2(s+10)(s+40)}$. Sketch the Bode plot, and comment on the stability.

Solution:
The open-loop t.f. is given as

$$G(s)=\frac{800(s+2)}{s^2(s+10)(s+40)}$$

$$=\frac{800\times 2(1+s/2)}{s^2 10(1+s/10)40(1+s/40)}=\frac{4(1+0.5s)}{s^2(1+0.1s)(1+0.025s)}.$$

To get sinusoidal t.f., substitute $s=j\omega$ in $G(s)H(s)$.

$$\therefore \qquad G(j\omega)H(j\omega)=\frac{4(1+j0.5\omega)}{(j\omega)^2(1+j0.1\omega)(1+j0.025\omega)}$$

The phase angle is given as

$$\phi=\angle G(j\omega)H(j\omega)=-180-\tan^{-1}0.1\omega-\tan^{-1}0.025\omega+\tan^{-1}0.5\omega.$$

Phase plot

ω	–180	$\tan^{-1}0.5\omega$	$-\tan^{-1}0.1\omega$	$-\tan^{-1}0.025\omega$	ϕ
0	–180	0	0	0	–180
0.2	–180	57	–1.14	–0.28	–175.7
2	–180	45	–11.5	–2.86	–149.16
10	–180	78.6	–45.3	–14.03	–160.43
20	–180	84.28	–63.43	–26.56	–185.7
50	–180	87.7	–78.6	–51.3	–222
100	–180	88.9	–84.28	–68.2	–243.5
α	–180	90	–90	–90	–270

Magnitude plot
The type of the given t.f. is 2 i.e., $m=0$, $n=2$ and hence

1. Initial slope of the Bode plot = $(m–n)20$ = –40 dB/dec
2. Gain value in dB or intersection point on dB axis is given as

$$=20\log k-(m-n)20$$
$$=20\log k-(0-2)20$$
$$=20\log 4+40$$
$$=+40+12$$
$$=52\,\text{dB}$$

The corner frequencies are 2, 10 and 40 rad/s

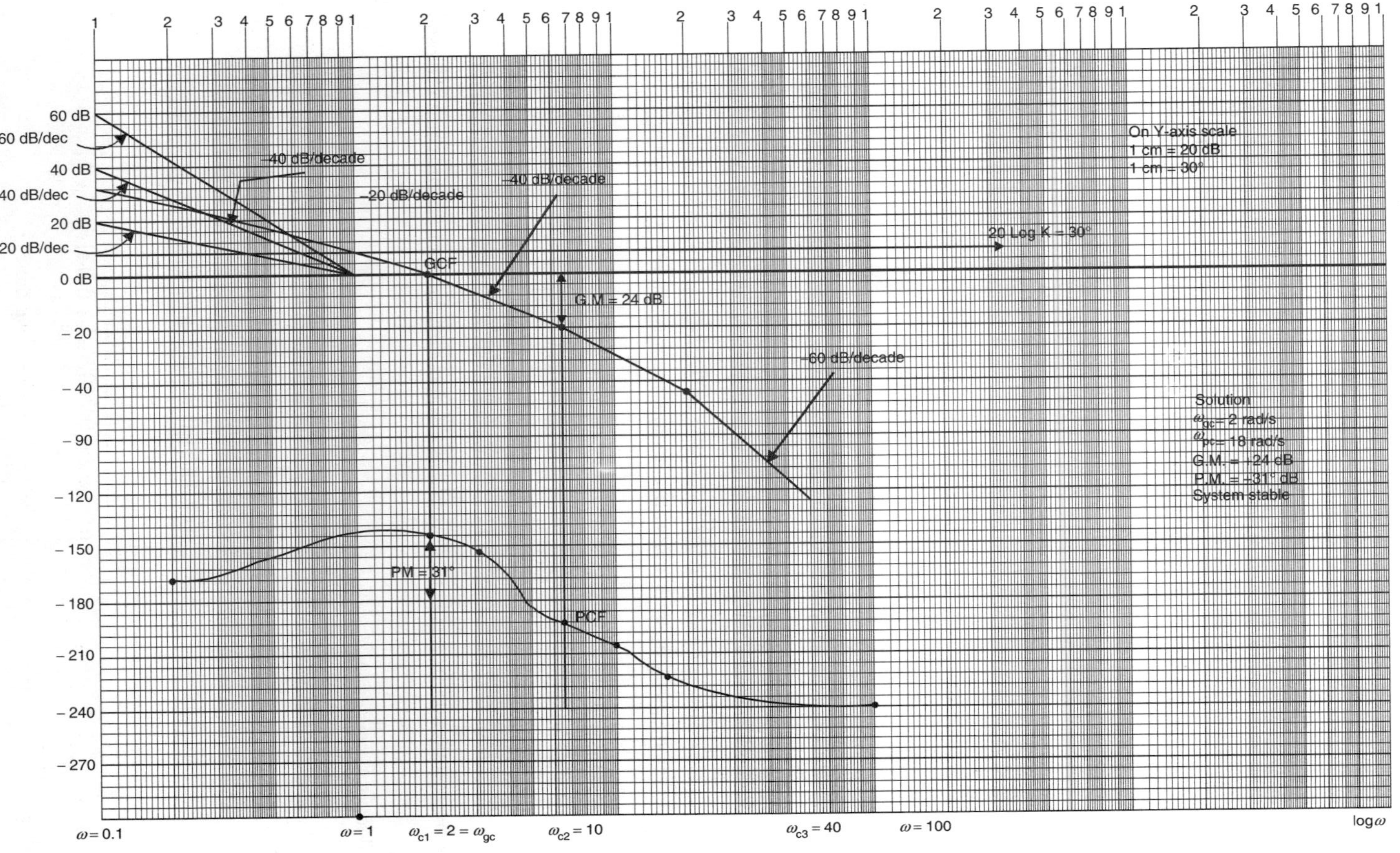

Figure (b) Bode plot for Problem 11.4

Factor	Corner frequency in rad/s	Slope in dB/dec	Change in slope in dB/dec	Asymptotic log-magnitude characteristic
4	-	0	-	Straight line of constant slope of 0 dB/s starting from 20 log k = 20 log 4 = 12 dB point
$\frac{1}{(j\omega)^2}$	-	–40	–40	Straight line of constant slope – 40 dB/dec passing through $\omega = 1$
$1 + j0.5\omega$	2	+20	–40 + 20 = –20	Straight line of constant slope + 20dB/dec originating at $\omega = 2$
$\frac{1}{1 + j0.1\omega}$	10	–20	–20 – 20 = –40	Straight line of constant slope – 20 dB/dec originating at $\omega = 10$
$\frac{1}{1 + j0.025\omega}$	40	–20	–40 – 20 = –60	Straight line of constant slope –20 dB/dec originating at $\omega = 40$

The Bode plot is as shown Figure (b).

From the Bode plot, we have

$$\text{GM} = +24 \text{ dB}, \text{ PM} = +31°$$

Both GM and PM are positive, and hence, the given system is stable.

Problem-11.5: For a particular unity feedback system, we have

$$G(s) = \frac{242(s+5)}{s(s+1)(s^2+5s+121)}.$$

Sketch the Bode plot, and hence, find ω_{gc}, ω_{pc}, GM and PM. Further, comment on the stability.

Solution:

The open-loop t.f. is given as

$$G(s)H(s) = \frac{242(s+5)}{s(s+1)(s^2+5s+121)}$$

$$= \frac{242(1+s/5)\times 5}{s(s+1)121\left(1+\frac{5}{121}s+\frac{s^2}{121}\right)}.$$

$$= \frac{10(1+0.2s)}{s(1+s)(1+0.00826s^2+0.041s)}$$

By comparing $\frac{1}{s^2+5s+121} = \frac{1}{s^2+2\xi\omega_n s+\omega_n^2}$, we have

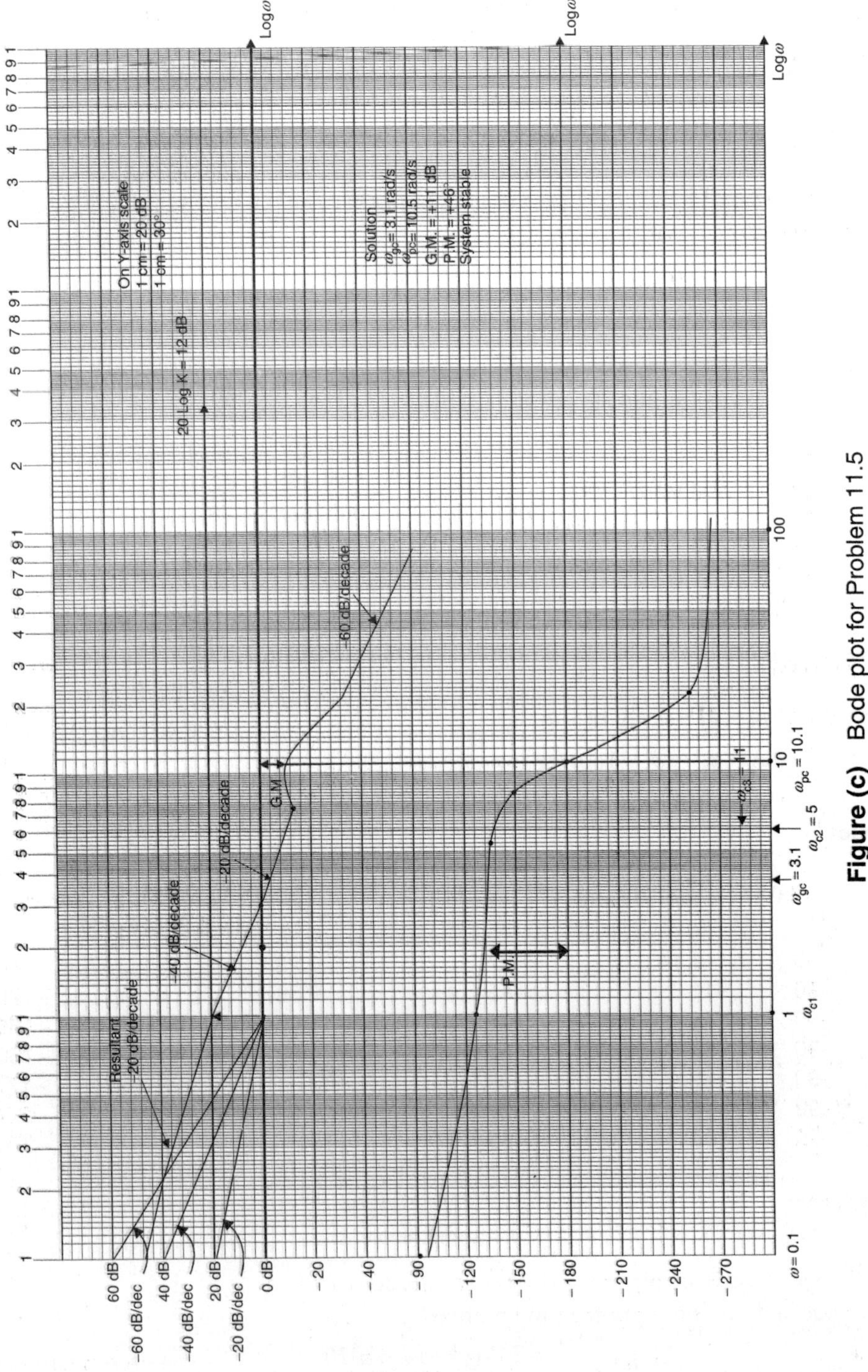

Figure (c) Bode plot for Problem 11.5

$$\omega_n^2 = 121 \Rightarrow \omega_n = 11$$

$$2\xi\omega_n = 5 \Rightarrow \xi = \frac{5}{2\times 11} = 0.22$$

Correction can be calculated precisely as

$$\begin{aligned} &= -20\log\sqrt{4\xi^2} \\ &= -20\log 2\xi \\ &= -20\log 2\times 0.22. \\ &= 7.13\text{dB} \end{aligned}$$

To get sinusoidal t.f., substitute $s = j\omega$ in $G(s)H(s)$.

$$G(j\omega)H(j\omega) = \frac{10(1+j0.2\omega)}{j\omega(1+j\omega)(1-0.00826\omega^2)+j0.041\omega}.$$

The phase angle is given as

$$\begin{aligned} \phi = \angle G(j\omega)H(j\omega) &= -90 + \tan^{-1}0.2\omega - \tan^{-1}\omega - \tan^{-1}\left(\frac{0.041}{1-10.00826\omega^2}\right) \text{for } \omega \le \omega_n \\ &= -90 + \phi_1 - \phi_2 - \phi_3 \text{ for } \omega \le \omega_n \\ &= -90 + \phi_1 - \phi_2 - (\phi_3 + 180) \text{ for } \omega > \omega_n \end{aligned}$$

Phase plot

ω	**–90**	$+\tan^{-1}0.5\omega(\phi_1)$	$-\tan^{-1}\omega(\phi_2)$	$-\tan^{-1}\left(\frac{0.041}{1-0.00826\omega^2}\right)$	ϕ
0.1	-90	+1.14	-57	-0.23	-94.7
1	-90	+11.3	-45	-2.36	-126
5	-90	+45	-78.6	-14.4	-138
8	-90	+58	-82.6	-34.8	-150
10	-90	+63.4	-84.2	-67	-177.8
20	-90	+76	-87.13	+19.5 – 180 = –160.4	-261.63
∞	-90	+90	-90	-180	-270

Magnitude plot

The type of the given t.f. is 1, i.e., $m = 0$, $n = 1$ and hence,

1. Initial slope of the Bode plot = $(m-n)20$ = –20 dB/dec
2. Gain value in dB or intersection point on dB axis

$$\begin{aligned} &= 20\log k - (m-n)20 \\ &= 20\log 10 + 20 \\ &= 40\text{dB}. \end{aligned}$$

The corner frequencies are 1, 5 and 11 rad/s.

Factor	Corner frequency in rad/s	Slope in dB/dec	Change in slope in dB/dec	Asymptotic log-magnitude characteristic
10	-	0	-	Straight line of constant slope of 0 dB/s starting from 20 log k = 20 log 10 = 12 dB point
$\frac{1}{j\omega}$	-	−20	0 − 20 = −20	Straight line of constant slope −20 dB/dec passing through $\omega = 1$
$\frac{1}{1+j\omega}$	1	−20	−20 − 20 = −40	Straight line of constant slope −20 dB/dec originating at $\omega = 1$
$1+j0.2\omega$	5	+20	−40 + 20 = −20	Straight line of constant slope +20 dB/dec originating at $\omega = 5$
$\frac{1}{1-j0.0088\omega^2 + j.0.041\omega}$	11	−40	−20 − 40 = −60	Straight line of constant slope 0 dB/dec for $\omega \le 11$ and straight line of slope −40 dB/dec for $\omega > 11$

At $\omega = 11$, quadratic pole with $\xi = 0.22$ occurs, and therefore, it contributes −40 dB/dec individually; the result will have slope −20 − 40 = −60 dB/dec exhibiting +7.95 dB upwards shift at $\omega = 11$ and will continue till ∞ as there is no other factor. The Bode plot is as shown in Figure (c).

From the Bode plot, we have

1. Gain crossover frequency, ω_{gc} = 3.1 rad/s.
2. Phase crossover frequency, ω_{pc} = 10.1 rad/s.
3. Gain margin, GM = +11 dB.
4. Phase margin, PM = +46°.

Both GM and PM are positive, and hence, the given system is stable.

Problem-11.6: A system has $G(s)H(s) = \frac{k}{s(1+s)(1+0.1s)(1+0.01s)}$.

Determine the value of K so that

1. GM = 10 dB 2. PM is 25°

Solution:

The open-loop t.f. of the given system, by assuming $k = 1$, is

$$G(s)H(s) = \frac{1}{s(1+s)(1+0.1s)(1+0.01s)}.$$

To get the sinusoidal t.f., substitute $s = j\omega$ in $G(s)H(s)$.

$$G(j\omega)H(j\omega) = \frac{1}{j\omega(1+j\omega)(1+0.1j\omega)(1+0.01j\omega)}.$$

The phase angle is given as

$$\phi = \angle G(j\omega)H(j\omega) = -90 + \tan^{-1}\omega - \tan^{-1}0.1\omega - \tan^{-1}0.01\omega.$$

Phase plot

ω	−90	$-\tan^{-1}\omega$	$-\tan^{-1}0.1\omega$	$-\tan^{-1}0.01\omega$	ϕ
0.1	−90	−5.7	−0.57	−0.057	−96.32
1	−90	−45	−5.7	−0.57	−141.27
2	−90	−63.43	−11.3	−1.14	−165.87
5	−90	−78.69	−26.56	−2.86	−198.11
10	−90	−84.28	−45	−57	−225
50	−90	−88.85	−78.69	−26.56	−284.1
100	−90	−89.42	−84.28	−45	−308.7

Magnitude plot

The type of the given t.f. is 1, i.e., $m = 0$, $n = 1$ and hence,

1. The initial slope = $(m–n)20$ = −20 dB/s
2. Gain value in dB or intersection point on dB axis
   = $20 \log k - (m - n)20$
   = $20 \log k + 20$
   = 20 dB.

The corner frequencies are 1, 10 and 100.rad/s

Factor	Corner Frequency in rad/s	Slope in dB/s	Change in slope in dB/s	Asymptotic log-magnitude characteristic
$\frac{1}{j\omega}$	-	−20	−20	Straight line of constant slope of −20 dB/s passing through ω= 1
$\frac{1}{1+j\omega}$	1	−20	−20 − 20 = 40	Straight line of constant slope −20 dB/dec originating at ω= 1.
$\frac{1}{1+j0.1\omega}$	10	−20	−40 − 20 = −60	Straight line of const. slope −20 dB/dec originating at ω= 10.
$\frac{1}{1+j0.01\omega}$	100	−20	−60 − 20 = −80	Straight line of constant slope −20 dB/dec originating at ω= 100.

Sketch the Bode plot and adjust the shift to meet the desired specifications one by one, as shown in Figure (d).

Now, *pcf* = 3.1 rad/s and it is independent of *k*; it intersects the plot at point *C*, which is without *k*. Point *C* is 18 dB down w.r.t 0 dB line. Therefore, under such condition, GM is 18 dB w.r.t 0 dB line. However, we want it to be 10 dB, i.e., $|G(j\omega)H(j\omega)|_{\omega-\omega_{pc}} = -10$ dB so that GM = +10 dB, i.e., point *C* should be

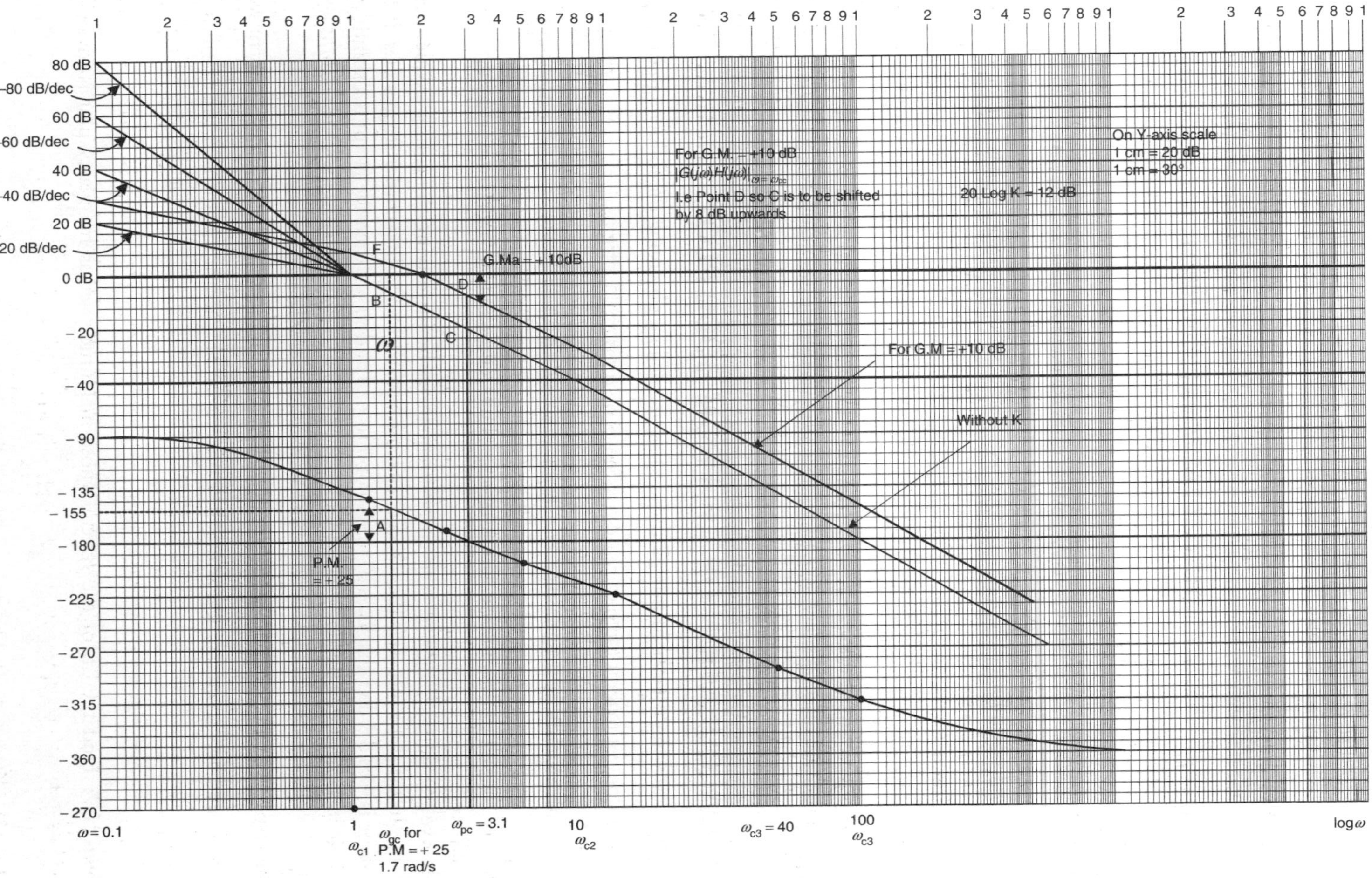

Figure (d) Bode plot for Problem 11.6

10 dB down w.r.t. 0 dB line. Hence, C should be shifted upwards by 8 dB to point D as shown to get GM = 10 dB. Shift CD = +8 dB = 20 log K, and upward shift is positive.

For GM = −10 dB, $k = 2.51$.

Now, $\text{PM} = \angle G(j\omega)H(j\omega)|_{\omega=\omega_{gc}} + 180°$

$$25° = \angle G(j\omega)H(j\omega)|_{\omega=\omega_{gc}} + 180°$$

Therefore, to get $\quad \text{PM} = +25°$

$$\angle G(j\omega)H(j\omega)|_{\omega=\omega_{pc}} = 25° - 180° = -155°.$$

So, we have to find a point on the phase angle plot that has an angle −155°, that is, the point A shown on the graph and corresponding frequency ω must be $\omega_{gc} = 1.7$ rad/s. Now, this intersects magnitude plot without k at point B and it should be at E to get 1.7 as ω_{gc} and PM is +25°. Therefore, shift point B to E is again +8 dB, and hence, in this problem for PM = +25° also 20 log k = 8 dB $\Rightarrow$ $k = 2.51$.

In this problem, single value of $K = 2.51$ satisfies both GM = 10 dB and PM = +25°. Note that this may not be the case for all problems. This is a rare case. For other problems, two separate k values must be calculated for GM and PM specifications.

11.11 CALCULATION OF T.F. FROM BODE PLOT

In this case, knowing the magnitude plot, the t.f. is decided by observing the following points.

1. Starting slope of magnitude plot represents poles or zeros at the origin.
   If starting slope is −20 dB/dec, there is 1 pole at origin; if it is −40 dB/dec and 0, there are 2 poles and no pole at origin, respectively. If it is +20 dB/dec, there is 1 zero at origin and so on.
2. Observe the shift in magnitude plot at $\omega = 1$, which represents 20 log k from which the value of k is decided.
3. Observe the slope changes at each corner frequencies by identifying the corner frequencies. The change in the slope indicates the respective factor with corresponding corner frequencies. If change in slope is −20 dB/dec, that is, −20 to −40 or 0 to −20 and so on, then the factor is a simple pole. If change in slope is +20 dB/dec, that is, 0 to +20 or −40 to −20 and so on, then the factor is a simple zero.
4. The respective time constant is reciprocal of the corner frequency, that is, $\omega_c = \frac{1}{T}$; then, we have $T = \frac{1}{\omega_c}$ and hence, we can determine the different factors present to construct the t.f. of the given system.

Problem-11.7: Find the t.f. of the system whose magnitude versus frequency plot is given in the figure.

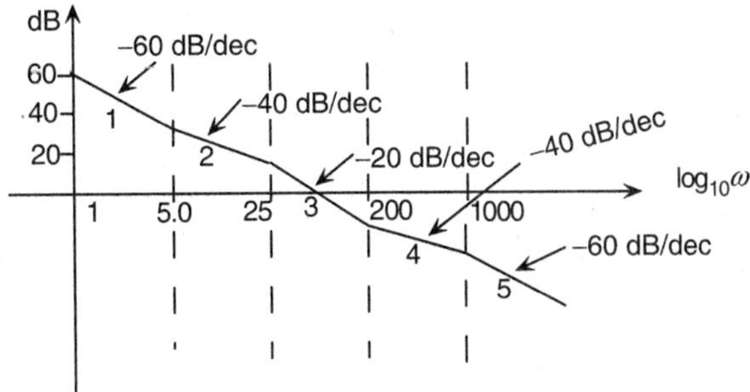

Solution:

For Part 1, at ω=1.0, dB = 60, and the initial slope of the curve is –60 dB/dec, which indicates the presence of triple pole at origin.

$$\therefore \qquad \text{t.f}_1 = \frac{k}{s^3}$$

$$\text{dB} = 20\log k - 60\log\omega$$

At $\omega = 1, 60 = 20\log_{10}^{k} - 60\log 1$

$$60 = 20\log_{10}^{k}$$

$$k = 10^3 = 1000$$

$$\therefore \qquad \text{t.f}_1 = \frac{1000}{s^3}$$

The corner frequencies are 5, 25, 200 and 1000.

At ω = 5, the slope of the curve changes from –60 dB/dec to –40 dB/dec, there is a zero at ω= 5.

For Parts 1 and 2, we can write

$$\therefore \qquad \text{t.f}_{12} = \frac{1000}{s^3}\left(1+\frac{s}{5}\right).$$

At ω= 25, the slope of the curve changes from –40 dB/dec to –20 dB/dec, there is a zero at ω= 25.

For Parts 1, 2 and 3, we can write

$$\text{t.f}_{123} = \frac{1000}{s^3}\left(1+\frac{s}{5}\right)\left(1+\frac{s}{25}\right)$$

At ω= 200, the slope of the curve changes from –20 dB/dec to –40 dB/dec, there is a pole at ω= 200.

For Parts 1, 2, 3 and 4, we can write

$$\text{t.f}_{1234} = \frac{1000(1+s/5)(1+s/25)}{s^3(1+s/200)}.$$

At ω= 1000, the slope of the curve changes from –40 dB/dec to –60 dB/dec, there is a pole at ω= 1000. Hence, the system has t.f. as follows

$$\text{t.f} = G(s) = \frac{1000(1+s/5)(1+s/25)}{s^3(1+s/200)(1+s/1000)}$$

$$= \frac{1000(1+0.2s)(1+0.04s)}{s^3(1+0.005s)(1+0.001s)}$$

Problem-11.8: Find the t.f. of the system whose Bode's diagram is given as follows:

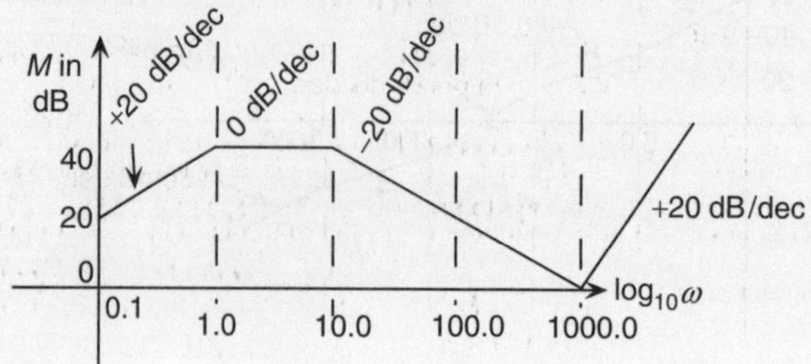

Solution:

The corner frequencies are at $\omega = 1.0$, $\omega = 10.0$ and $\omega = 1000.0$ rad/s.

$\omega = 1.0$ rad/s ------ corresponds to a pole.

$\omega = 10.0$ rad/s------ corresponds to a pole.

$\omega = 100.0$ rad/s------ corresponds to a zero of second order.

At $\omega = 1.0$ rad/s the magnitude $M = 40$ dB

$$40 = 20\log_{10} k + 20\log_{10} \omega$$

At $\omega = 1$,

$$40 = 20\log_{10} k$$

$\therefore$

$$k = 100.$$

Therefore, the t.f. is

$$G(s) = \frac{100s(1+s/1000)^2}{(1+s/10)(1+s)} = \frac{100s(1+0.001s)^2}{(1+0.1s)(1+s)}.$$

Problem-11.9: Find the closed-loop t.f. of a system from the Bode's plot given in the figure. The plot represents $|G(j\omega)H(j\omega)|$ in dB along the y-axis with frequency ω on the x-axis. Assume unity feedback.

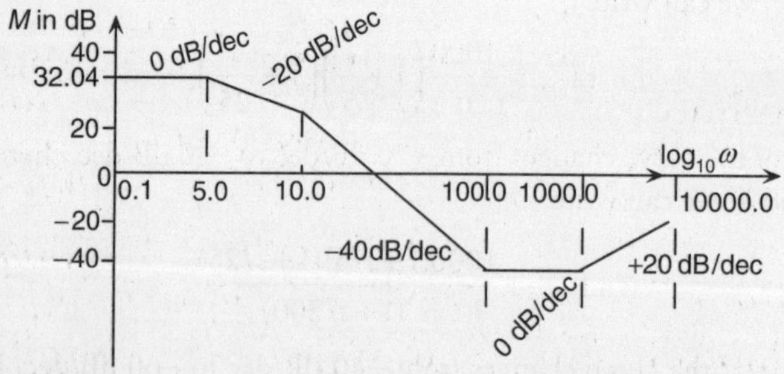

Solution:

The corner frequencies are at $\omega = 5$ rad/s, 10 rad/s, 100 rad/s and 1000 rad/s.

$\omega = 5$ and 10 rad/s ------ correspond to poles.

$\omega = 100$ and 1000.0 rad/s ------ correspond to zero.

$\omega = 100$ rad/s ------ double zero.

$\omega = 1000.0$ rad/s ------ simple zero since slope changes to 0 from −40 dB/dec and 0 dB to 20 dB/dec, respectively.

$$G(s)H(s) = \frac{k\ (1+0.01s\ )^2(1+0.001s)}{(1+0.2s)(1+0.1s)}$$

To evaluate *k*

At $\omega = 0.1$ rad/s, the contribution from the poles and zeros is negligible, and hence, we have

$$20\log|G(j\omega)H(j\omega)| = 20\log_{10} k = 32.04\text{dB}$$
$$\log_{10} k = 1.6020$$
$$k = 40$$

$$\therefore G(s)H(s) = \frac{40(1+0.01s)^2(1+0.001s)}{(1+0.2s)(1+0.1s)}.$$

The closed-loop t.f. is

$$\frac{G(s)H(s)}{1+G(s)H(s)} = \frac{40(1+0.01s^2)(1+0.001s)}{(1+0.2s)(1+0.01s)+40(1+0.01s^2)(1+0.001s)}.$$

11.12 ADVANTAGES OF BODE PLOT

The various advantages of Bode plot are

1. It shows both high and low frequency characteristics of a t.f. in a single diagram.
2. The Bode plot can be easily constructed using some valid approximations.
3. The relative stability of the system can be studied by calculating gain margin and phase margin.
4. It is possible to determine various other frequency domain specifications such as cut-off frequency and bandwidth.
5. Transfer function of the system can be obtained from Bode plot.
6. Using Bode plot, it is possible to compensate the system to obtain the desired response.
7. The value of system gain *k* can be designed for required specifications of GM and PM from Bode plot.
8. Experimentally, it is possible to draw a Bode plot without knowing the t.f.

ADDITIONAL SOLVED PROBLEMS

ASP-1: Sketch the Bode plot and determine the gain crossover frequency (ω_{gc}), phase crossover frequency (ω_{pc}), gain margin (GM), and phase margin (PM). Find the stability of the system for the given t.f.

$$G(s) = \frac{10}{s(s+1)(1+0.02s)}.$$

Solution:

Method 1 The open-loop t.f. is given as

$$G(s) = \frac{10}{s(s+1)(1+0.02s)}$$

To get the sinusoidal t.f., substitute $s = j\omega$ in $G(s)$.

$$G(j\omega) = \frac{10}{j\omega(j\omega+1)(1+0.02j\omega)}.$$

Phase angle $\phi = -90 - \tan^{-1}\omega - \tan^{-1}0.02\omega$

Phase plot

Frequency (ω)	Phase angle (ϕ)
0	-90
0.1	-95.82
1	-136.0
10	-185.6
15	-192.884
50	-223.8

Magnitude plot

The type of the gain t.f. is 1, i.e., $m = 0$, $n = 1$ and hence, we have

1. Initial slope = $(m-n)$ 20 = – 20 dB/dec
2. Gain value in dB or intersection point on dB axis

$$= 20\log k - (m-n)\,20 = 20 + 20\log k = 20 + 20\log 10 = 40\text{dB}$$

The corner frequencies are 1 and 50 rad/s

Factor	Corner frequency in rad/s	Slope in dB/s	Change in slope in dB/s	Asymptotic log-magnitude characteristic
$\frac{1}{j\omega}$	None	-20	-20	Straight line of constant slope -20 dB/dec originating at $\omega = 1$.
$\frac{1}{1+j\omega}$	2	-20	$-20 - 20 = -40$	Straight line of constant slope -20 dB/dec originating at $\omega = 2$.
$\frac{1}{1+j0.02\omega}$	20	-20	$-40 - 20 = -60$	Straight line of constant slope -20 dB/dec originating at $\omega = 20$.

From the Bode plot, as shown in the figure, we have

1. Gain crossover frequency = 3.2 rad/s
2. Phase crossover frequency = 8 rad/s
3. Gain margin = 15 dB
4. Phase margin = 17°

Stability Analysis

Both GM and PM are positive, and hence, the system is stable (or) if GCF < PCF, the system is stable.

Method 2 Let us consider using the renewed method of plotting Bode's plot directly from the resultant plot.

Consider
$$G(s) = \frac{10}{s(1+s)(1+0.02s)}.$$

Substitute $s = j\omega$,

$$G(j\omega) = \frac{10}{j\omega(1+j\omega)(1+j0.02\omega)}.$$

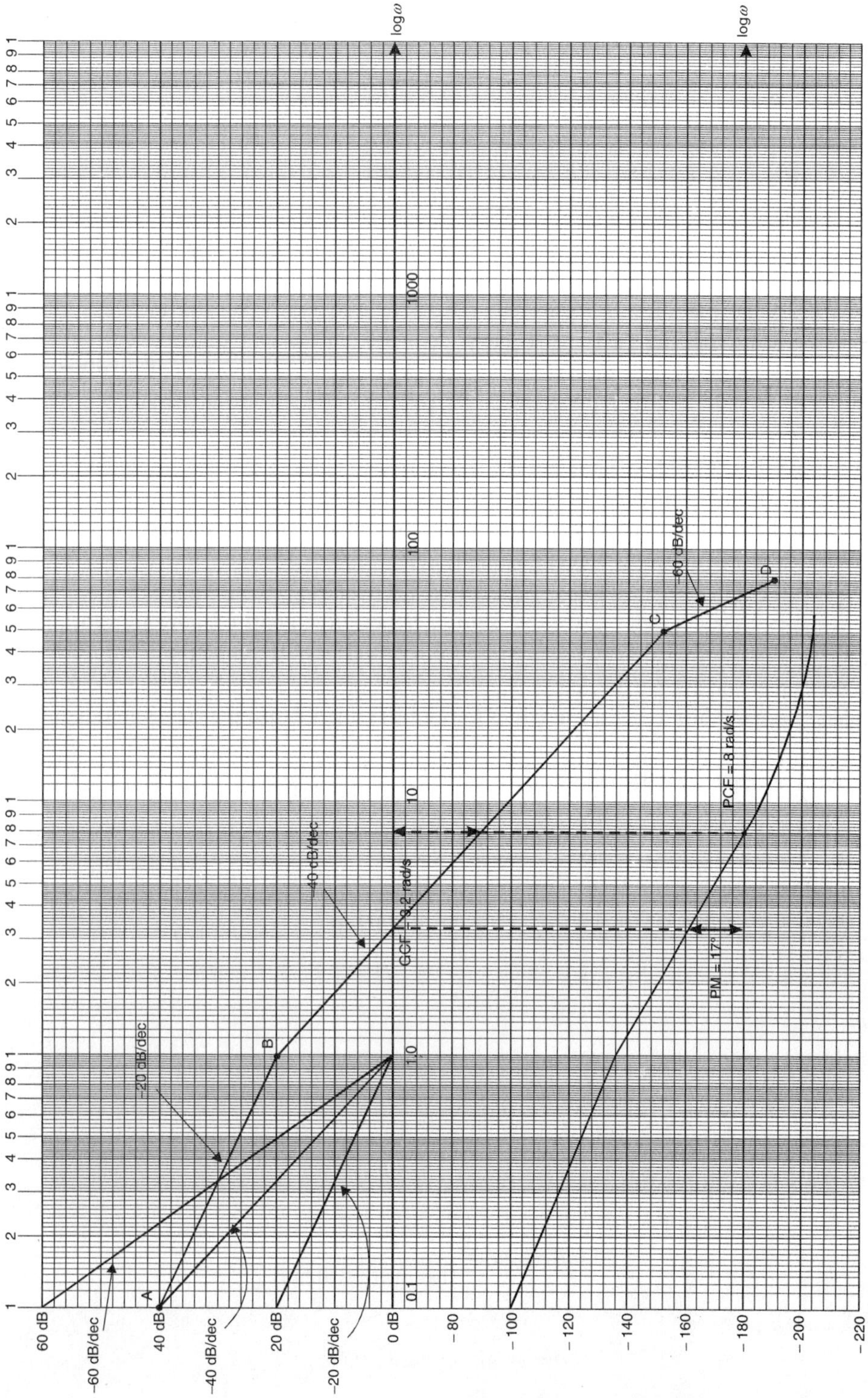

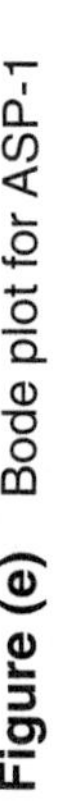

Figure (e) Bode plot for ASP-1

The corner frequencies are at 1, 50.
The starting point for the magnitude plot is calculated as

$$= 20\log k - (m-n)20.$$

$$= 20\log 10 + 20$$

$$= 20 + 20 = 40 \text{ dB}$$

The magnitude is

$$20\log|G(j\omega)| = 20\log 10 - 20\log\omega - 20\log\sqrt{1+\omega^2} - 20\log\sqrt{1+(0.020)^2}$$

1. The starting point is 40 dB with a slope of –20 dB/dec as there is a pole due to s^{-1} factor in the given t.f. Let us assume that this as point *A*. From point *A*, move a unit distance horizontally equal to 1 dec, go vertically downwards till 20 dB magnitude, draw a line from point *A* till 20 dB magnitude for –20 dB/dec slope line, as there is a pole due to the factor s^{-1} up to the corner frequency 1.0 rad/s. Let us assume this point as *B*, as shown in Figure (e).
2. From this corner frequency 1.0 rad/s, as the next corner frequency 50 rad/s corresponds to a pole, the resultant slope is –20 – 20 = –40 dB/dec. Move again a decade distance horizontally from this point *B* and go vertically downwards up to 40 dB magnitude for –40 dB/dec slope line and join *B* with this point till the next corner frequency is 50 rad/s; let us assume it as *C*.
3. For the corner frequencies greater than 50 rad/s, as this frequency is contributed by a pole in the above t.f., the resultant slope is –40 – 20 = –60 dB/dec. Hence, from point *C*, move a decade distance horizontally and go vertically downwards for 60 dB magnitude. Draw a line from *C* to reach this 60 dB magnitude point and assume this as *D*.
4. The phase angle is drawn on similar grounds for various values of ω in the same semi-log sheet, so as to find
   gain crossover frequency = 3.2 rad/s
   phase crossover frequency = 8 rad/s
   gain margin = 15 dB
   phase margin = 17°

ASP-2: Draw the Bode plot for the system having open-loop t.f.

$$G(s) = \frac{50}{(s+1)(s+2)}, H(s) = 1.$$

Solution:

The open-loop t.f. is given as

$$G(s)H(s) = \frac{50}{(s+1)(s+2)} = \frac{50}{2(1+s)(1+s/2)} = \frac{25}{(1+s)(1+0.5s)}$$

To get the sinusoidal t.f., substitute $s = j\omega$ in $G(s)H(s)$.

$$G(j\omega)H(j\omega) = \frac{25}{(1+j\omega)(1+j0.5\omega)}.$$

The phase angle is given as

$$\phi = -\tan^{-1}\omega - \tan^{-1}0.5\omega.$$

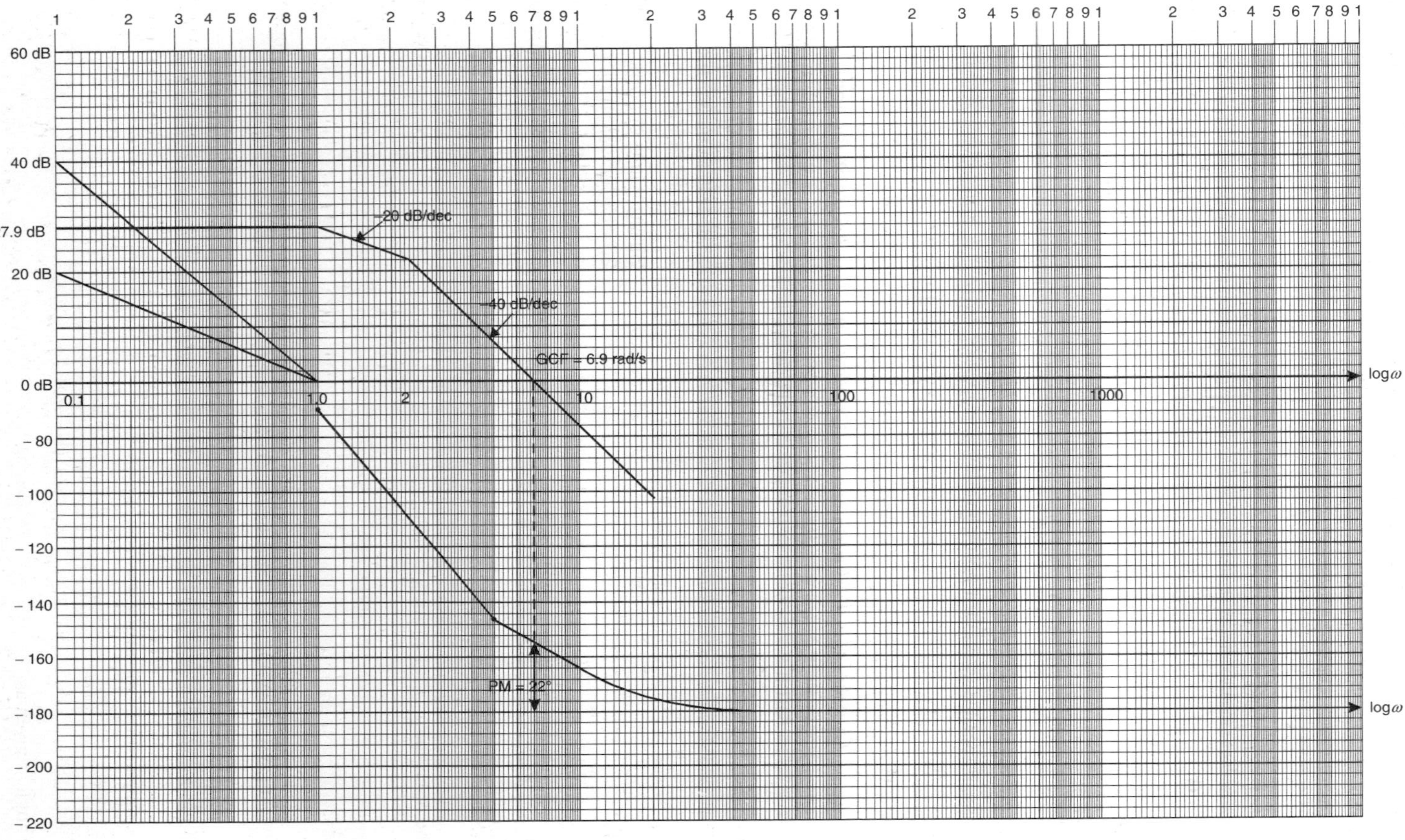

Figure (f) Bode plot for ASP-2

Phase plot

Frequency (ω)	Phase angle (ϕ)
0.1	−8.57
1	−71.56
2	−108.43
10	−162.9
25	−173.13
75	−177.7
100	−178.2

Magnitude plot

The type of the given t.f. is 0, i.e., $m = 0$, $n = 0$ and hence,

1. Initial slope = 0 dB/dec
2. Gain value in dB or intersection point on dB axis

$$20 \log k = 20 \log 25 = 27.9\text{dB}$$

The corner frequencies are 1 and 2, as shown in Figure (f).

Factor	Corner frequency in rad/s	Slope in dB/dec	Change in slope in dB/dec	Asymptotic log-magnitude characteristic
25	-	0	-	Straight line of constant slope of 0 dB/dec starting from 20 log 25 = 27.9 dB point
$\dfrac{1}{1+j\omega}$	$\omega_1 = 1$	−20	0 − 20 = −20	Straight line of constant slope 20 dB/dec originating at $\omega_1 = 1$ rad/s
$\dfrac{1}{1+j0.5\omega}$	$\omega_2 = 2$	−20	−20 − 20 = −40	Straight line of constant slope 20 dB/dec originating at $\omega_2 = 2$ rad/s

ASP-3: The open-loop t.f. of a unity feedback system is

$$G(s) = \frac{2(s+0.25)}{s^2(s+1)(s+0.5)} = \frac{1+4s}{s^2(1+s)(1+2s)}.$$

Determine 1. GM and PM 2. GCF and PCF 3. Stability

Solution:

The open-loop t.f. is given as $G(s) = \dfrac{1+4s}{s^2(1+s)(s+2s)}$.

To get the sinusoidal t.f., substitute $s = j\omega$ in $G(s)$.

i.e.,
$$G(s) = \frac{1+j4\omega}{j\omega^2(1+j\omega)(1+j2\omega)}$$

The phase angle is given as

$$\phi = -180 - \tan^{-1}\omega - \tan^{-1}2\omega + \tan^{-1}4\omega.$$

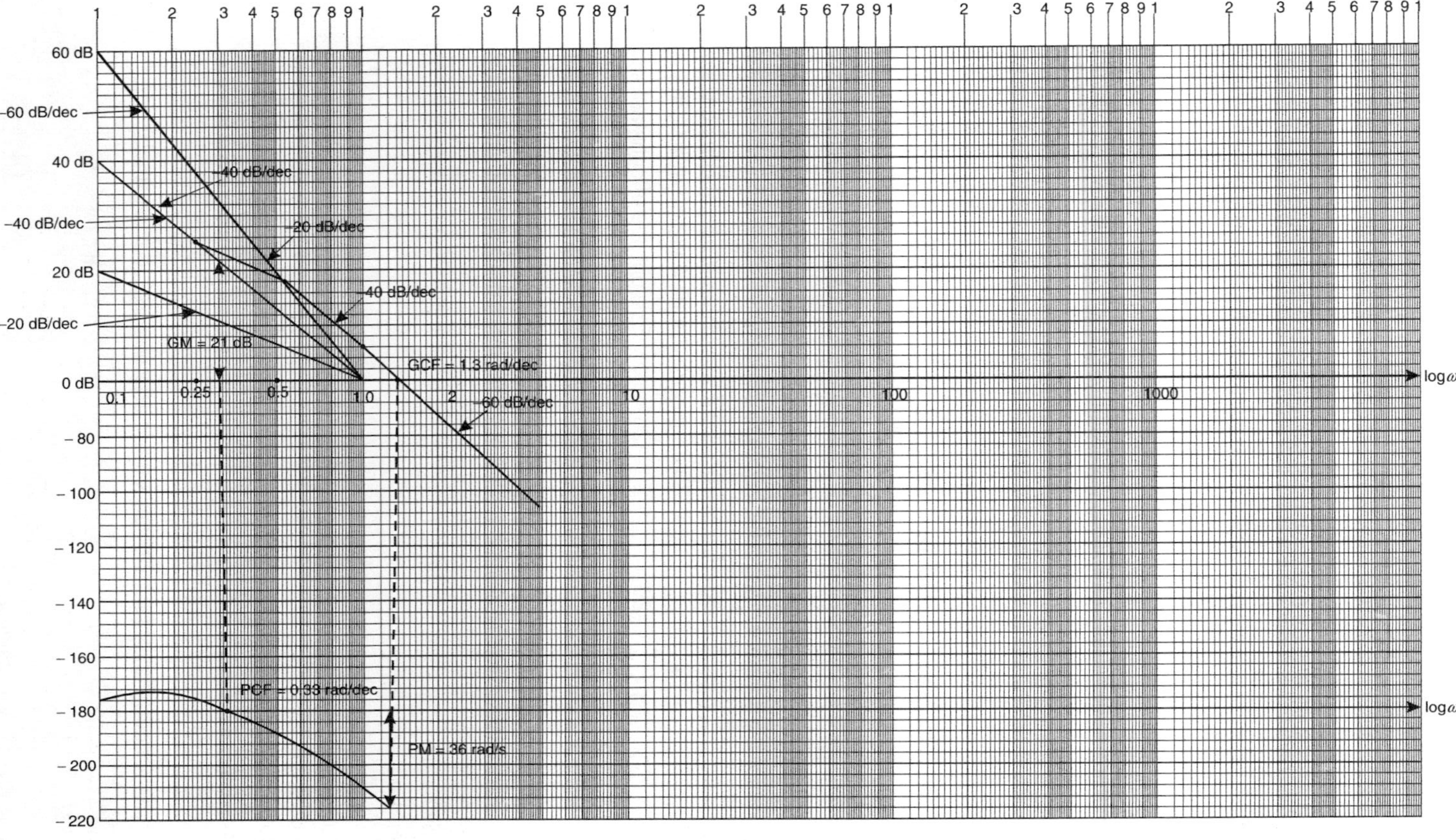

Figure (g) Bode plot for ASP-3

Phase plot

Frequency (ω)	Phase angle (ϕ)
0.1	−175.2
0.5	−188.2
1	−212.4
10	−262.8
15	−265.2
25	−267.2

Magnitude plot

The type of the given system is 2, i.e., $m = 0$, $n = 2$ and hence

1. Initial slope = −40 dB/dec
2. Gain value in dB or intersection point on dB axis

$$40 + 20\log k = 40 + 20\log 1 = 40\text{dB}$$

The corner frequencies are 0.25, 0.5 and 1 rad/s.

Factor	Corner frequency in rad/s	Slope in dB/dec	Change in slope dB/dec	Asymptotic log-magnitude characteristic
$\dfrac{1}{(j\omega^2)}$	None	−40	-	Straight line of constant slope of −40 dB/s passing through $\omega = 1$
$1 + j4\omega$	$\omega_1 = 0.25$	+20	−40 + 20 = −20	Straight line of constant slope +20 dB/dec originating at $\omega_1 = 0.25$
$\dfrac{1}{1 + j2\omega}$	$\omega_2 = 0.5$	−20	−20 − 20 = −40	Straight line of constant slope −20 dB/dec originating at $\omega_2 = 0.5$
$\dfrac{1}{1 + j\omega}$	$\omega_3 = 1$	−20	−40 − 20 = −60	Straight line of constant slope −60 dB/dec originating at $\omega_3 = 1$.

The Bode plot is drawn as shown in Figure (g).
From the Bode plot, we have

1. The gain crossover frequency (GCF) = 0.3 rad/s
   The phase crossover frequency (PCF) = 0.33 rad/s
2. The gain margin (GM) = 21 dB
   The phase margin (PM) = −36°

Stability Analysis

Since GCF > PCF, the system is unstable.

ASP-4: Sketch the Bode plot for the following t.f. and determine the system gain k for the GCF to be 5 rad/s,

$$G(s) = \frac{ke^{-0.1s}}{s(1+s)(1+0.1s)}.$$

Solution:

Let us assume that the gain value $k = 1$

The open-loop t.f. is given as

$$G(s) = \frac{e^{-0.1s}}{s(1+s)(1+0.1s)}.$$

To get the sinusoidal t.f., substitute $s = j\omega$ in $G(s)$.

i.e.,
$$G(s) = \frac{e^{-0.1j\omega}}{j\omega(1+j\omega)(1+0.1\omega)}.$$

Magnitude of $e^{-0.1j\omega} \Rightarrow \left|e^{-0.1j\omega}\right| = 1$

The phase angle of $e^{-0.1j\omega}$ is given as

$$\angle e^{-j0.1\omega} = -0.1\omega \text{ rad } [\therefore \text{ if } x = e^{-j\theta} \text{ the angle of } x, \text{ i.e., } \angle x = -\theta].$$

$$= -0.1\omega \times \frac{180}{\pi}$$

$$= -5.73\ \omega \text{ degrees.}$$

The phase angle of the open-loop t.f. is given as

$$\phi = -90 - 5.73\omega - \tan^{-1}\omega - \tan^{-1}0.1\omega.$$

Phase Plot

Frequency (ω)	Phase angle (ϕ)
0.1	-97
0.5	−122
1	−146
2	−176
2.5	−186.5
5	−223
10	−276

Magnitude plot

The type of the given t.f. is 1, i.e., $m = 0$, $n = 1$ and hence, we have

1. Initial slope = −20 dB/dec
2. Gain value in dB or intersection point on dB axis

$$20 + 20 \log k = 20 + 20 \log 1 = 20 \text{ dB}$$

The corner frequencies are 1 and 10 rad/s.

Factor	Corner frequency in rad/s	Slope in dB/dec	Change in slope dB/dec	Asymptotic log-magnitude characteristic
$\frac{1}{j\omega}$	None	–20	-	Straight line of constant slope of –20 dB/s passing through $\omega = 1$
$e^{-j0.1\omega}$	None	-	-	Coincide with 0 dB line
$\frac{1}{1+j\omega}$	$\omega_1 = 1$	–20	–20 – 20 = –40	Straight line of constant slope –20 dB/dec originating at $\omega_1 = 1$
$\frac{1}{1+j0.1\omega}$	$\omega_2 = 10$	–20	–40 – 20 = –60	Straight line of constant slope –20 dB/dec originating at $\omega_2 = 10$

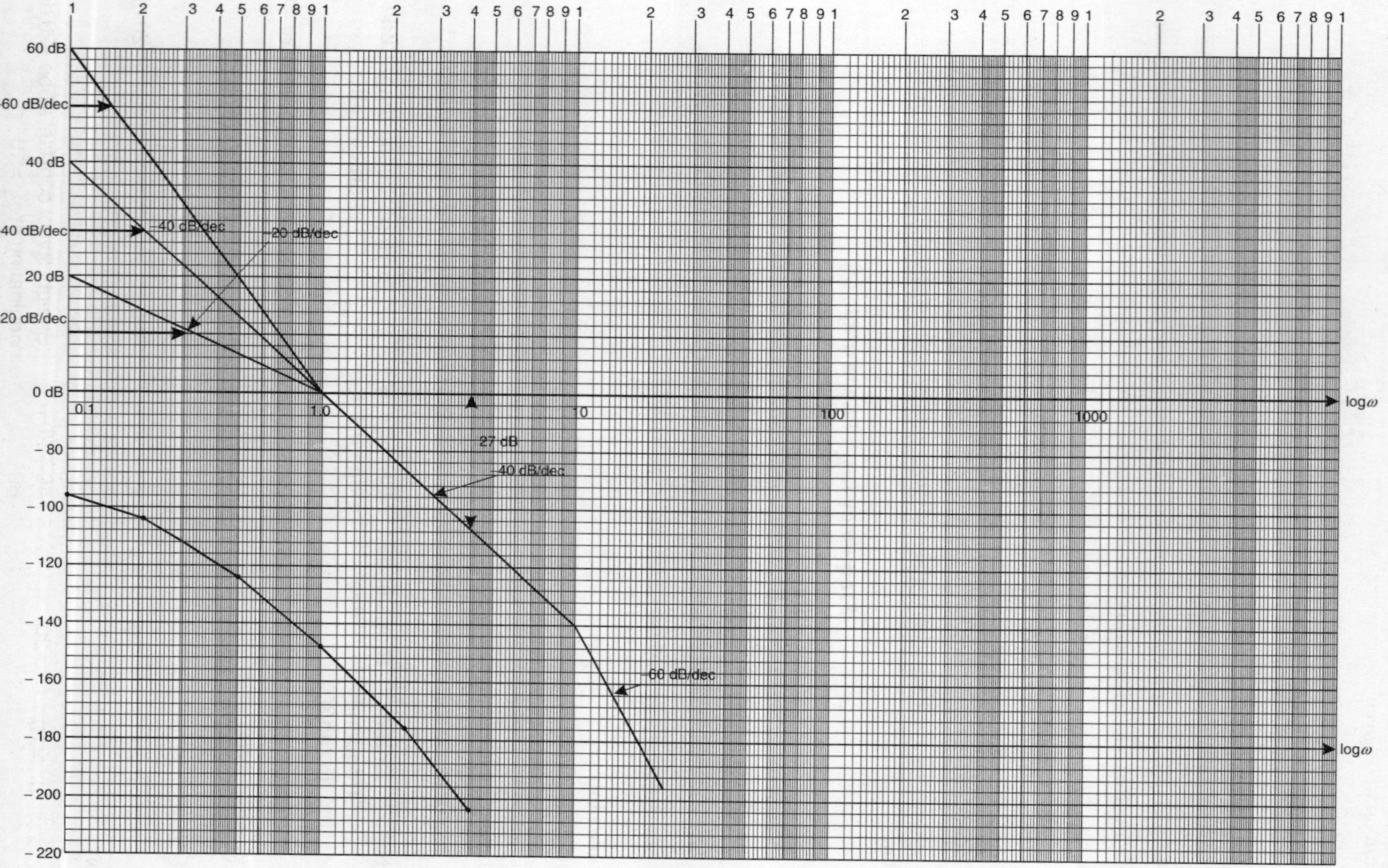

Figure (h) Bode plot for ASP-4

If the GCF is required to be 5 rad/s, then the corresponding magnitude plot is lifted by 27 dB, as shown in Figure (h).

∴
$$20\log k = 27$$
$$\log k = \frac{27}{20} = 1.35$$
$$\Rightarrow k = 10^{1.35} = 22.4.$$

ASP-5: Sketch the Bode plot for the following t.f. and determine phase margin and gain margin $G(s) = \dfrac{75(1+0.2s)}{s(s^2+16s+100)}$.

Solution:

The t.f. of the system is given as

$$G(s) = \frac{75(1+0.2s)}{s(s^2+16s+100)}$$
$$G(s) = \frac{75(1+0.2s)}{s\times 100\left[\dfrac{s^2}{100}+\dfrac{16s}{100}+1\right]}$$
$$G(s) = \frac{0.75\,(1+0.2s)}{s(1+0.01s^2+0.16s)}.$$

To get sinusoidal t.f., substitute $s = j\omega$ in $G(s)$; then, we have

$$G(j\omega) = \frac{0.75(1+0.2j\omega)}{j\omega(1-0.01\omega^2+j0.16\omega)}.$$

The phase angle is given as

$$\phi = \angle G(j\omega) = \tan^{-1} 0.2\omega - 90 - \tan^{-1}\left(\frac{0.16\omega}{1-0.001\omega^2}\right) \text{for } \omega \le \omega_n$$
$$= \phi_1 - 90 - \phi_2$$
$$= \phi_1 - 90 - (\phi_2 + 180) \text{ for } \omega > \omega_n.$$

Phase plot

Frequency (rad/s)	**–90**	$-\tan^{-1} 0.2\omega$ (ϕ_1)	$-\tan^{-1}\dfrac{0.16\omega}{1-0.01\omega^2}$ (ϕ_2)	ϕ
0.1	–90	+1.145	–0.916	–89.771
1	–90	+11.31	–9.18	–87.87
5	–90	+45	–46.84	–91.84
10	–90	+63.43	–90	–116.57
15	–90	75.56	+62.487 – 180	–131.97
20	–90	75.9	+46.847 – 180	–147.253
25	–90	78.69	+37.30 – 180	–154.01
40	–90	82.87	+23.106 – 180	–164.02
50	–90	84.29	+18.43 – 180	–167.20
100	–90	87.13	+9.18 – 180	–173.69
500	–90	90	+1.833 – 180	–178.167

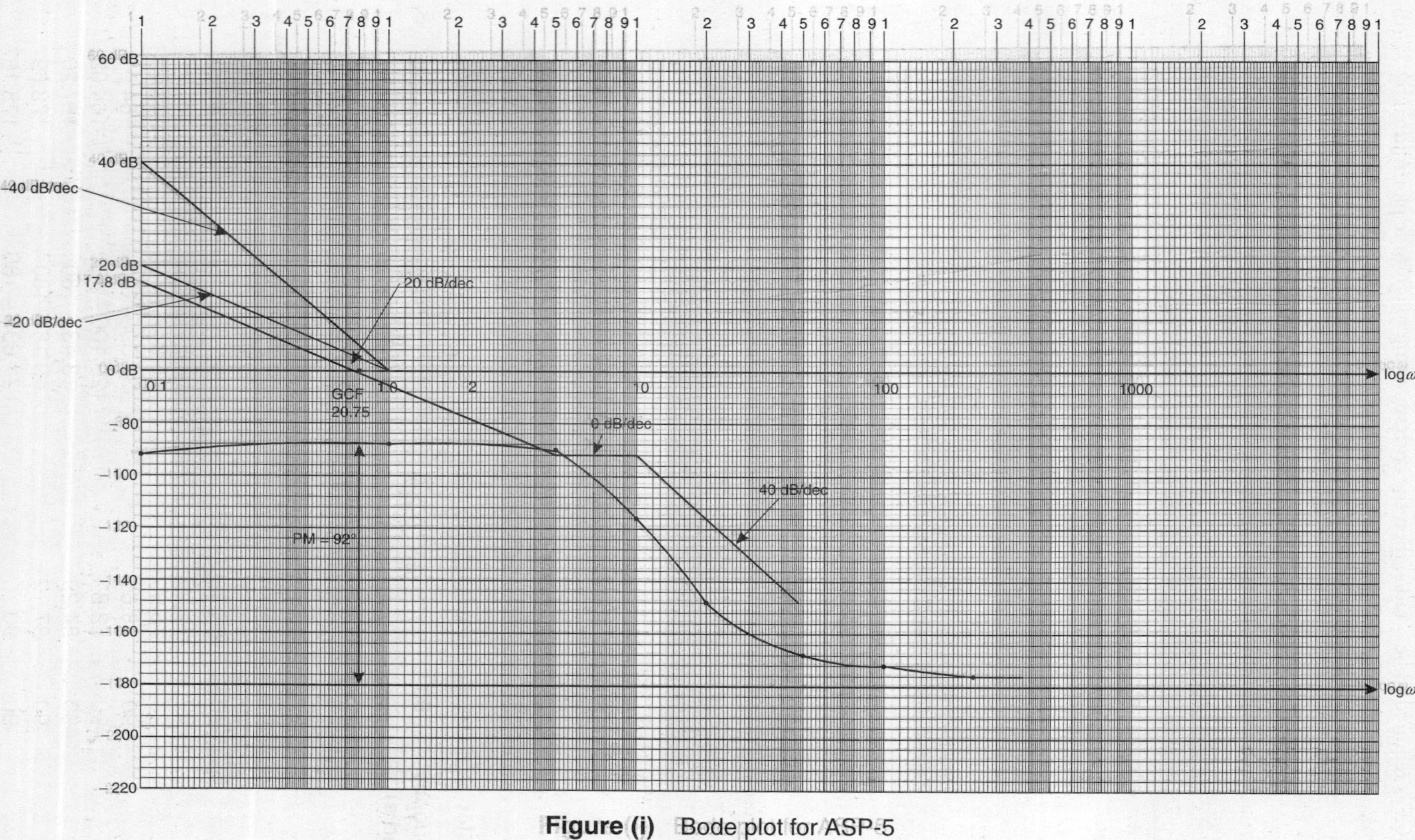

Figure (i) Bode plot for ASP-5

Magnitude plot

Since the system is of Type 1, we have

1. Initial slope = $(m - n)20 = -20$ dB/dec
2. Intersection point on the dB axis $= 20 \log k - (m-n)20$

$$= 20 + 20\log k$$
$$= 20 + 20\log 0.75$$
$$= 17.5\text{dB}.$$

The corner frequencies are 5 and 10 rad/s.

Factor	Corner frequency in rad/s	Slope in dB/dec	Change in slope in dB/ dec	Asymptotic log-magnitude characteristic
$\frac{1}{j\omega}$	–	–20	–20	Straight line of constant slope of –20 dB/dec passing through = 1 rad/s
$1+ j0.2\omega$	$\omega_1 = 0.25$	+20	0	Straight line of constant slope + 20 dB/dec originating at $\omega_1 = 5$ rad/s
$\frac{1}{1-0.01\omega^2 + j0.16\omega}$	$\omega_2 = 0.5$	–40	40	Straight line of constant slope 0 dB/dec for $\omega \le 10$ rad/s and –40 dB/ dec for $\omega > 10$ rad/s

From the Bode plot shown in Figure (i), we have

1. Phase margin = 92°
2. Gain margin = ∞

Stability

Both GM and PM are positive, and hence, the given system is stable.

Note: For the quadratic factor, the corner frequency is ω_n. In quadratic factors, the angle varies from 0° to 180°. However, the calculator calculates $\tan^{-1}$ only between 0° and 90°. Hence, a correction factor of 180° should be added to the phase angle after corner frequency ω_n.

ASP-6: Sketch the Bode plot for a unity feedback system characterised by the open-loop t.f. $G(s) = \dfrac{k(1+0.2s)(1+0.025s)}{s^3(1+0.001s)(1+0.005s)}$.

Show that the system is conditionally stable. Find the range of values of k for which the system is stable.

Solution:

Let us assume that $k = 1$

The open-loop t.f. is given as

$$G(s) = \frac{k(1+0.2s)(1+0.025s)}{s^3(1+0.001s)(1+0.005s)}.$$

To get the sinusoidal t.f., substitute $s = j\omega$ in $G(s)$.

$$G(j\omega) = \frac{(1+0.2j\omega)(1+0.025\omega)}{(j\omega)^3(1+j0.001\omega)(1+j0.005\omega)}.$$

The phase angle is given as

$$\phi = -270 + \tan^{-1} 0.2\omega + \tan^{-1} 0.025\omega - \tan^{-1} 0.001\omega - \tan^{-1} 0.005\omega.$$

Phase plot

Frequency in (rad/s)	Phase angle (ϕ)
0.1	–268°
0.8	–260°
1	–257°
3	–236°
10	–198°
15	–183°
30	–163°
60	–148°
100	–147°
300	–172
400	–182

Magnitude plot

Since the type of the system is 3, i.e., $m = 0$, $n = 3$ we have

1. Initial slope = $(m - n)20 = -60$ dB/dec.
2. Intersection point on dB axis = $20 \log k - (m - n)20 = 60 + 20 \log k = 60$ dB.

The corner frequencies are 5, 40, 200 and 1000 rad/s

Factor	Corner frequency in rad/s	Slope in dB/dec	Change in slope dB/dec	Asymptotic log-magnitude characteristic
$\dfrac{1}{(j\omega)^3}$	–	–60	–60	A straight line of constant slope –60 dB/dec passing through $\omega = 1$.
$1 + j0.2\omega$	5	+20	–40	A straight line of constant slope +20 dB/dec originating from $\omega = 5$
$1 + j0.025\omega$	40	+20	–20	A straight line of constant slope +20 dB/dec originating from $\omega = 40$
$\dfrac{1}{1+j0.005\omega}$	200	–20	–40	A straight line of constant slope –20 dB/dec originating from $\omega = 200$
$\dfrac{1}{1+j0.005\omega}$	1000	–20	–60	A straight line of constant slope –20 dB/dec originating from $\omega = 1000$

From the graph shown in Figure (j), we know that the phase plot crosses the –180° line twice indicating that the system is conditionally stable.

$$20 \log k = 60$$

$$\log k = \frac{60}{20} = 3 \Rightarrow k = 10^3 = 1000$$

Further,

$$20 \log k = 98$$

$$\log k = \frac{98}{20} = 4.9$$

$$k = 10^{4.9} = 79432.8.$$

Hence, the condition for stability is $1000 < k < 79432.8$.

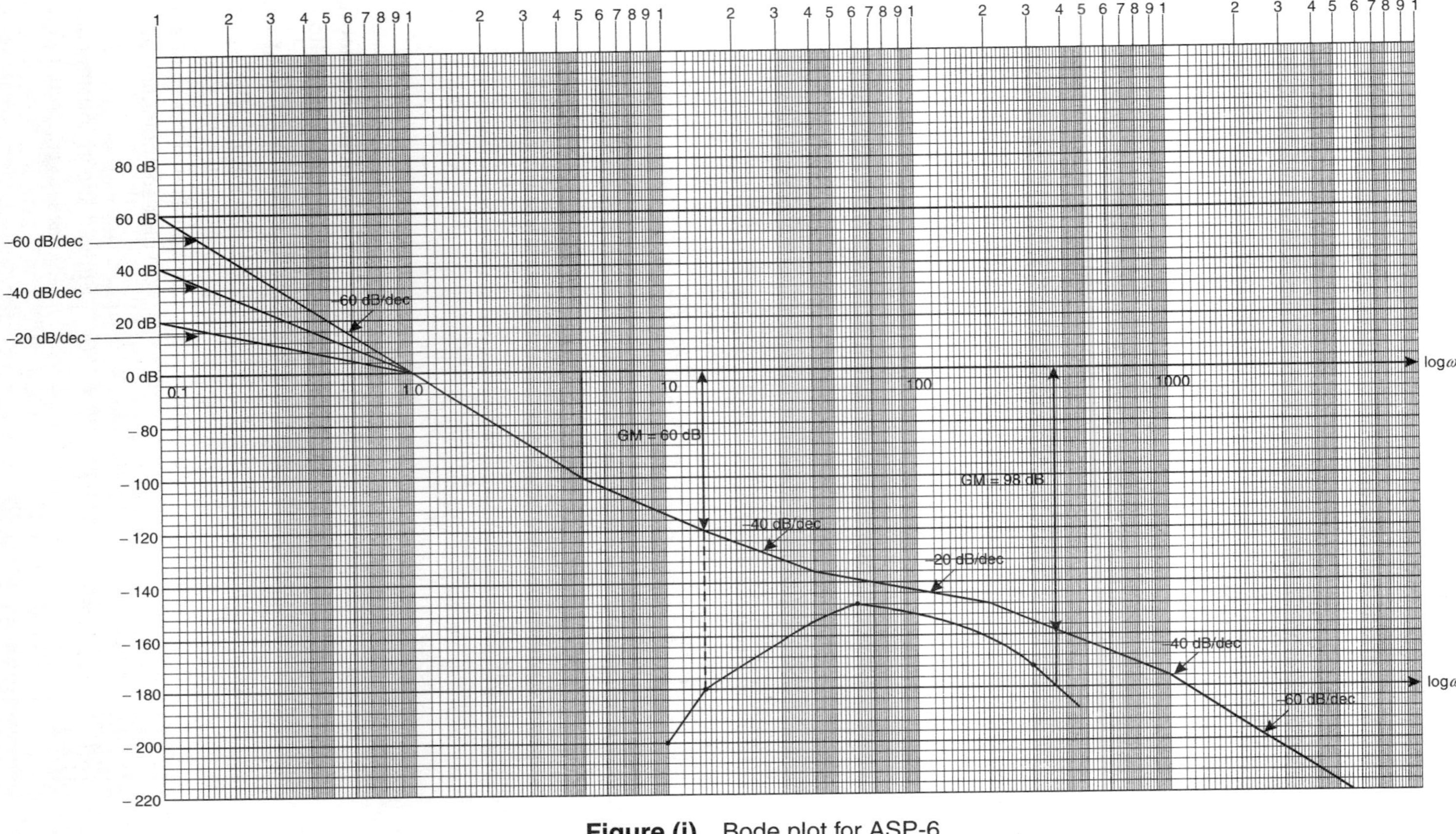

Figure (j) Bode plot for ASP-6

ASP-7: Determine the values of gain k for the following open-loop t.f. so that

1. The gain margin is 15 dB
2. Phase margin is 60°

$$G(s)H(s) = \frac{k}{s(1+0.1s)(1+s)}.$$

Solution:

The open-loop t.f. of the given system by assuming $k = 1$ is

$$G(s)H(s) = \frac{k}{s(1+0.1s)(1+s)} \text{ [substitute } k=1\text{]}.$$

To get sinusoidal t.f., substitute $s = j\omega$ in $G(s)H(s)$.

$$G(j\omega)H(j\omega) = \frac{1}{j\omega(1+j0.1\omega)(1+j\omega)}$$

The phase angle is given as

$$\phi = -90 - \tan^{-1} 0.1\omega - \tan^{-1} \omega.$$

Phase plot

Frequency in (rad/s)	Phase angle (ϕ)
0.1	–96.2835
1	–140.7105
5	–195.25
10	–219.286
15	–232.4
20	–240.54
25	–245.90
30	–249.65
50	–257.54

Magnitude plot

Since the type of the system is 1, we have

1. The initial slope = –20 dB/dec
2. Intersection point on dB axis

$$= 20 + 20 \log k$$
$$= 20 + 20 \log 1 = 20 \text{ dB}$$

The corner frequencies are 1 and 10 rad/s as shown in Figure (k).

Factor	Corner frequency	Slope in dB/dec	Change in slope dB/dec	Asymptotic log-magnitude characteristic
$\frac{1}{j\omega}$	-	–20	–20	Straight line of constant slope of –20 dB/s passing through $\omega = 1$
$\frac{1}{1+j\omega}$	1	–20	–20 – 20 = –40	Straight line of constant slope of –20 dB/dec originating at $\omega_1 = 1$
$\frac{1}{1+0.1j\omega}$	10	–20	–40 – 20 = –60	Straight line of constant slope of –20 dB/s originating at $\omega_2 = 10$

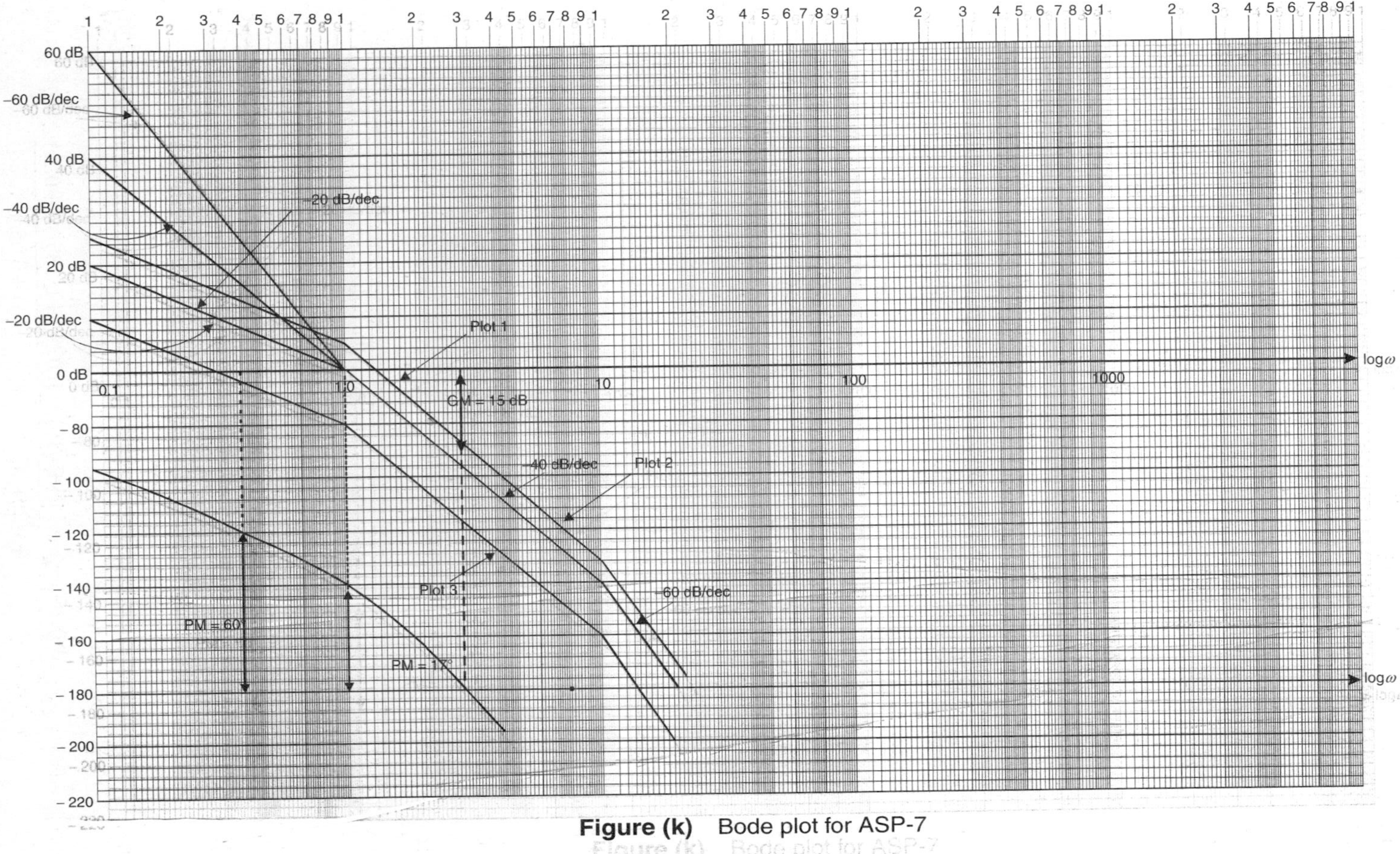

Figure (k) Bode plot for ASP-7

1. For the given gain margin 15dB, we have to lift the magnitude plot upwards as shown in the figure (the upwards lifted plot is given by Plot 2)
From Plot 2, we have

$$20\log k = 25 - 20$$
$$\log k = 15/20 = 0.25$$
$$k = 10^{0.25} = 1.7778$$

2. For given phase margin 60°, we have to lift the magnitude plot downwards as shown in the figure (the downwards lifted plot is given by Plot 3)
From Plot 3, we have

$$20 + 20\log k = 12$$
$$\log k = \frac{-8}{20} = -0.4.$$
$$k = 10^{-0.4} = 0.4$$

ASP-8: Find the t.f. of a system having Bode plot, as shown in the following figure.

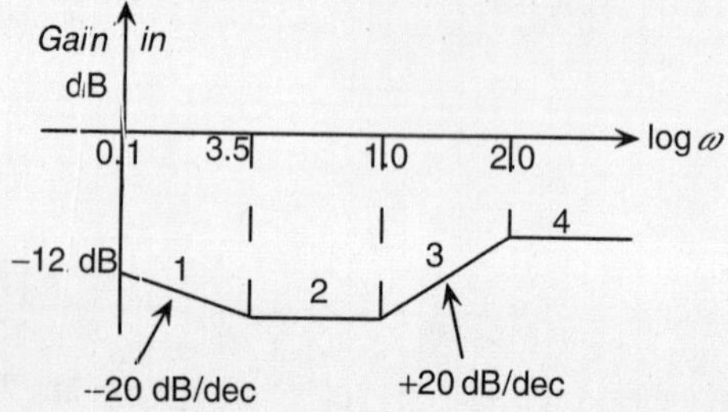

Solution:

For Part 1, at $\omega = 0.1$, dB $= -12$ and the slope of the curve is -20 dB/dec.

$\therefore$

$$\text{t.f.}_1 = \frac{k}{s}$$

$$\text{dB} = 20\log k - 20\log\omega$$

$$-12 = 20\log k - 20\log 0.1 = 20\log k + 20$$

$\Rightarrow$

$$20\log k = -32 \Rightarrow k = 0.025$$

$$\Rightarrow \text{t.f.}_1 = \frac{0.025}{s}.$$

At $\omega = 3.5$, the slope of the curve changes from -20 dB/s to 0 dB/dec, there is a zero at $\omega = 3.5$.
For Parts 1 and 2, we can write

$$\text{t.f.}_{12} = \frac{0.025(1+s/3.5)}{s}.$$

At $\omega = 10$, the slope of the curve changes from 0 dB/dec to +20 dB/dec; there is a zero at $\omega = 10$.
For Parts 1, 2 and 3, we can write

$$\text{t.f.}_{123} = \frac{0.025(1+s/3.5)(1+s/10)}{s}.$$

At $\omega = 20$, the slope of the curve changes from +20 dB/dec to 0 dB/dec, there is a pole at $\omega = 20$.
For Parts 1, 2, 3 and 4, we can write

$$\text{t.f.} = \frac{0.0.25(1+s/3.5)(1+s/10)}{s(1+s/20)} = \frac{0.025(1+0.285s)(1+0.1s)}{s(1+0.05s)}$$

ASP-9: Find the t.f. of a system having Bode plot as shown in the figure.

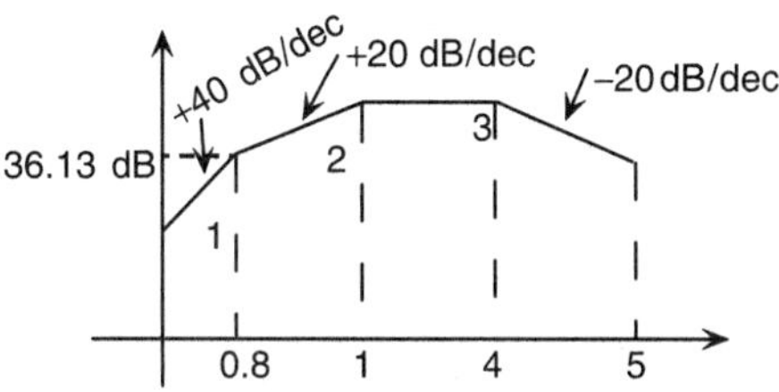

Solution:

For Part 1

$$\text{t.f.}_{\cdot 1} = ks^2$$

$$\text{dB} = 20\log k + 40\log\omega$$

At

$$\omega = 0.8,\ \text{dB} = 36.13$$

$$36.13 = 20 \log k + 40 \log 0.5 = 20 \log k - 3.87$$

$$20 \log k \Rightarrow k = 10^2 = 100$$

$$\therefore \text{t.f.}_{\cdot 1} = 100\, s^2$$

At $\omega = 0.8$, the slope of the curve changes from 40 dB/dec to +20 dB/dec and there is a pole at $\omega = 0.8$. For Parts 1 and 2, we can write

$$\text{t.f.}_{\cdot 12} = \frac{100s^2}{(1+s/0.8)}.$$

At $\omega=1$, the slope of the curve changes from +20 dB/dec to 0 dB/dec and there is a pole at $\omega = 1$. For Parts 1, 2 and 3, we can write

$$\text{t.f.}_{\cdot 123} = \frac{100s^2}{(1+s/0.8)(+s/1)}.$$

At $\omega = 4$, the slope of the curve changes from 0 dB/dec to −20 dB/dec and there is a pole at $\omega = 4$. For Parts 1, 2, 3 and 4, we can write

$$\text{t.f.} = \frac{100s^2}{(1+s/0.8)(1+s/1)(1+s/4)} = \frac{100s^2}{(1+1.25s)(1+s)(1+0.25s)}.$$

ASP-10: Determine the t.f. whose approximate Bode plot is as shown in the figure.

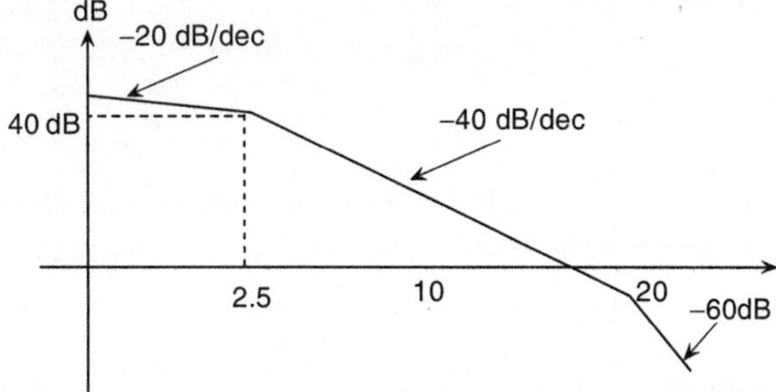

Solution:

For Part 1

$$\text{t.f.}_{\cdot 1} = \frac{k}{s}.$$

The corner frequencies are 2.5 and 40.

At $$\omega = 2.5,\ \text{dB} = 40$$

$$40 = 20 \log k + 20 \log 2.5 = 20 \log k - 7.95$$

$\Rightarrow$ $$20 \log k = 47.95 \Rightarrow k = 10^{2.398} = 250.$$

At $\omega = 2.5$ rad/s, the slope changes from –20 dB/dec to –40 dB/dec due to factor $\dfrac{1}{(1+s/2.5)}$.

At $\omega = 40$ rad/s, the slope changes from –40 dB/dec to –60 dB/dec due to factor $\dfrac{1}{(1+s/40)}$.

The t.f. is

$$G(s) = \frac{250}{s(1+s/2.5)(1+s/40)} = \frac{250}{s(1+0.46)(1+0.025s)}.$$

ASP-11: Find the t.f. of the system where asymptotic approximation is given in the figure.

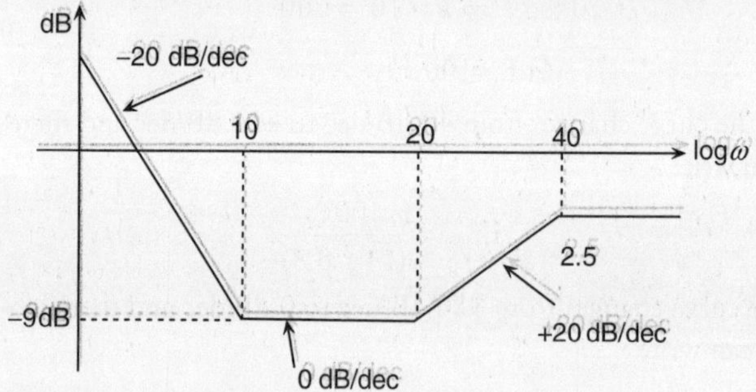

Solution:

For Part 1

$$\text{t.f.}_1 = \frac{k}{s}.$$

The corner frequencies are 10, 20 and 40.

At $$\omega = 10,\ \text{dB} = -9$$

$$\text{dB} = 20 \log k - 20 \log 10$$

$$-9 = 20 \log k - 20 \Rightarrow 20 \log k = 11 \Rightarrow \omega = 10^{11/20} = 3.55$$

$$\text{t.f.}_1 = 3.55/s$$

At $\omega = 10$ rad/s, the slope changes from –20 dB/dec to 0 dB/dec due to factor $(1 + s/\omega)$.
At $\omega = 20$ rad/s, the slope changes from 0 dB/s to +20 dB/dec due to factor $(1+ s/20)$. At $\omega = 40$ rad/s, the slope changes from +20 dB/dec to 0 dB/dec due to factor $\dfrac{1}{(1+s/40)}$.

The t.f. is given as

$$G(s) = \frac{3.55(1+s/10)(1+s/20)}{s(1+s/40)} = \frac{3.55(1+0.1s)(1+0.05s)}{s(1+0.025s)}.$$

REVIEW QUESTIONS

1. What is Bode plot? State the advantages of Bode plot.
2. State the steps for Bode plot.
3. Define gain margin and phase margin. Explain the significance of these terms to find the closed-loop system stability.
4. Define gain crossover frequency and phase crossover frequency. Explain the significance of these terms to find the system stability.
5. State the advantages and disadvantages of frequency response analysis.
6. Define the following terms:
   (i) Resonant frequency (ii) Cut-off rate
   (iii) Resonant peak (iv) Bandwidth
7. What do you mean by resonant peak? How is it related to the relative stability of a closed-loop system?
8. Explain the nature of Bode plot for
   (i) poles at origin (ii) simple pole (iii) simple zero
9. What is meant by corner frequency? What is its significance?
10. Obtain the correlation between the time response and the frequency response.

EXERCISE PROBLEMS

1. Sketch the Bode plot and determine GCF, PCF, GM and PM. Determine the stability of the system for the t.f. $G(s)H(s)=\dfrac{2(s+0.25)}{s^2(s+1)(s+0.5)}$.
2. Determine the value of gain K for the following open-loop t.f. so that
   (a) the gain margin is 20 dB (b) the phase margin is 60°
$$G(s)H(s)=\frac{2(s+0.25)}{s^2(1+0.1s)(1+0.05s)}$$
3. Find the frequency domain specifications for a system having $G(s)=\dfrac{36}{s(s+8)}$ with unity feedback.
4. Sketch the Bode plot for the open-loop t.f. $G(s)H(s)=\dfrac{200(s+2)}{s(s^2+10s+100)}$. Determine the stability of the system.
5. A unit-step input is applied to a unity feedback control system having t.f. $G(s)=\dfrac{K}{s(1+sT)}$. Determine (i) the values of K and T to have $M_p = 20\%$, $\omega_r = 6$ rad/s (ii) the resonant peak M_r.
6. A unity feedback has open-loop t.f. $G(s)=\dfrac{288(s+4)}{s(s+2)(s^2+4.8s+144)}$. Draw the Bode plot for the system.
7. Sketch the Bode plot for the forward path t.f. of a unity feedback system is given as $G(s)=\dfrac{10e^{-0.1s}}{s(1+0.1s)^2}$ and determine the stability of the system.
8. A unity feedback has open-loop t.f. $G(s)=\dfrac{K}{s(s+4)(s+10)}$. Draw the Bode plot for the system and find the value of K for a marginally stable system.
9. Draw the Bode plot for the open-loop t.f. given as $G(s)=\dfrac{4}{(s+2)(s+4)(s+5)}$ and determine GM and PM.

10. Sketch the Bode plot for the open-loop t.f. given as $G(s)H(s) = \dfrac{300(s^2+2s+4)}{s(s+10)(s+20)}$ and determine the stability of the system.

11. Sketch the Bode plot for the open-loop t.f. given as $G(s)H(s) = \dfrac{100K(s+5)(s+40)}{s^3(s+100)(s+200)}$ and determine the range of K for closed-loop stability.

12. Sketch the Bode plot for the open-loop t.f. $G(s)H(s) = \dfrac{2e^{-sT}}{s(s+1)(1+0.5s)}$. Determine the maximum value of T for the system to be stable.

13. Sketch the Bode plot for the open-loop t.f. $G(s)H(s) = \dfrac{75e^{-sT}}{s(s^2+10s+100)}$. Determine the maximum value of T for the system to be stable.

14. For the Bode plot as shown in the figure, find
(i) the open-loop t.f. (ii) ω_{gc1} and ω_{gc2}

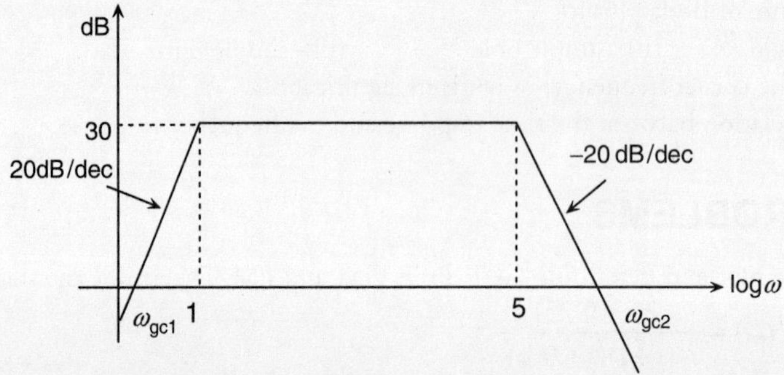

15. Find the open-loop t.f. of a system whose approximate plot is as shown in the figure.

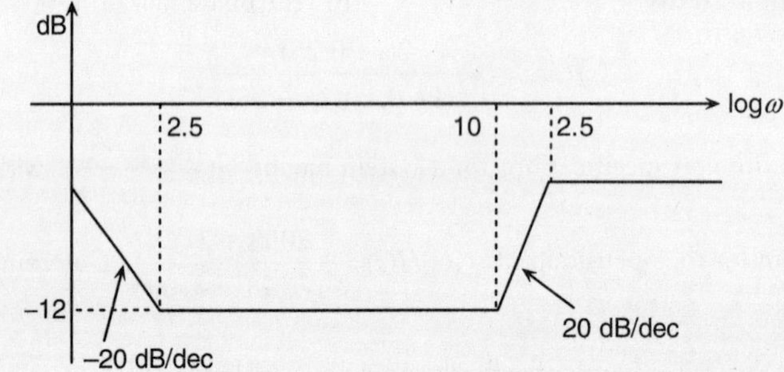

16. Find the open-loop t.f. of a system whose approximate plot is as shown in the figure.

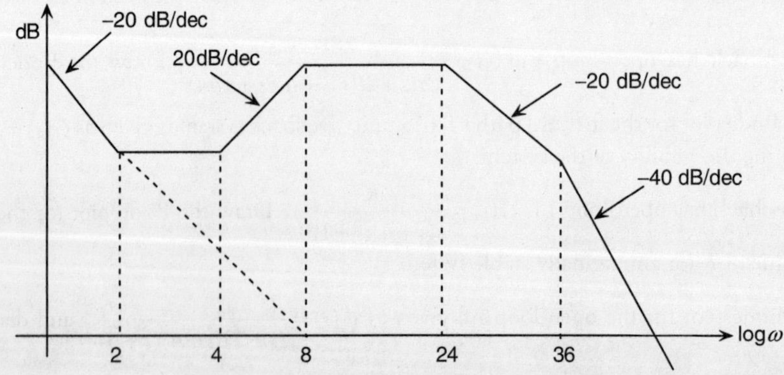

ANSWERS

1. GCF = 1.24 rad/s, PCF = 0.4 rad/s, GM = –20, PM = 18°, Unstable
2. (a) $K = 2.65$ (b) $K = 10$
3. $\omega_r = 2$ rad /s, $M_r = 1.006$, BW = 6.7 rad/s
4. GM = ∞, PM = 47°, and the system is stable
5. (i) $K = 8.6$, $T = 0.14$ (ii) $M_r = 1.234$
7. GM = –5.3 dB, PM = 54°, and the system is unstable
8. For marginally stable system, GCF = PCF, $K = 400$
9. GM = 36 dB, PM = Very high and positive
10. GM = ∞, PM = 96°, the system is stable
11. $2820 < K < 48840$
12. $T = 7$ ms
13. $T = 2$ s
14. $\omega_{gc1} = 0.0316\text{rad/s}, \omega_{gc2} = 158\text{rad/s}$
15. $G(s)H(s) = \dfrac{0.63(1+0.4s)(1+0.1s)}{(1+0.04s)}$
16. $G(s)H(s) = \dfrac{8(1+0.5s)(1+0.0285s)}{s(1+125s)(1+0.042s)(1+0.028s)}$

OBJECTIVE TYPE QUESTIONS

1. The initial slope of the Bode plot for a t.f. having no poles at the origin is
   (a) –10 dB/decade (b) 0 dB/decade
   (c) 10 dB/decade (d) 24 dB/decade
2. The error at corner frequency due to the term $(1 + j\omega T)^{\pm N}$ is
   (a) $\pm 5N$ dB (b) $\pm 2N$ dB
   (c) $\pm 6N$ dB (d) $\pm 3N$ dB
3. The gain for drawing the Bode plot for the t.f. $G(s) = \dfrac{320(s+2)}{s(s+1)(s^2+8s+64)}$.
   (a) 10 (b) 5 (c) 16 (d) 160
4. The initial slope of the Bode plot gives an indication of the
   (a) type of the system (b) nature of the system time response
   (c) system stability (d) gain margin
5. With reference to the Bode plot, the corner frequencies for the t.f.
   $G(s)H(s) = \dfrac{20(1+0.5s)}{s(1+0.2s)(1+s)(s^2+3s+9)}$ are
   (a) 1, 2, 3 and 5 rad/s (b) 1, 2, 3 and 4 rad/s
   (c) 1, 3, 5 and 9 rad/s (d) 1, 2, 5 and 9 rad/s
6. The frequency at which the magnitude of the Bode plot crosses 0-dB axis gives
   (a) natural frequency (b) phase crossover frequency
   (c) gain crossover frequency (d) corner frequency
7. The phase angle at $\omega = \infty$ for the t.f. $G(j\omega) = \dfrac{90(1+j0.5\omega)}{j\omega(1+j\omega)(1+j2\omega)}$ is
   (a) 90° (b) –90° (c) 180° (d) –180°

8. For a stable system,
   (a) both GM and PM positive
   (b) both GM and PM are negative
   (c) GM is positive and PM is negative
   (d) GM is negative and PM is positive
9. For a Type I system, the intersection of the initial slope of the Bode plot with 0-dB axis gives
   (a) steady-state error (b) error constant
   (c) phase margin (d) crossover frequency
10. For a stable system,
   (a) the gain crossover occurs earlier than phase crossover
   (b) the phase crossover occurs earlier than gain crossover
   (c) the gain crossover and phase crossover frequencies are very near to each other
   (d) the gain crossover and phase crossover frequencies are coincident
11. The phase angle for the t.f. $G(j\omega)=\dfrac{1}{(j\omega T)^3}$ at corner frequency is
   (a) 45° (b) −90° (c) −270° (d) −135°
12. The C.E. of a closed-loop system is given as $S^2+4S+16=0$, the resonant peak will be
   (a) 2 (b) $2\sqrt{3}$ (c) 4 (d) $2\sqrt{2}$
13. (I) Nyquist criterion is in frequency domain
   (II) Bode plot is in frequency domain
   (III) Root locus plot is in time domain
   (IV) *R–H* criterion is in time domain
   (a) I, II and III are correct (b) I, III and IV are correct
   (c) I and II are correct (d) All are correct
14. If the gain of the open-loop system is doubled, the gain margin
   (a) is not effected (b) gets doubled
   (c) becomes half (d) becomes one-fourth
15. The formula for resonant frequency ω_r is
   (a) $\omega_r=\omega_n\sqrt{1-\xi^2}$ (b) $\omega_r=\omega_n\sqrt{1+\xi^2}$
   (c) $\omega_r=\omega_n\sqrt{1-2\xi^2}$ (d) $\omega_r=\omega_n\sqrt{1+2\xi^2}$
16. The resonant peak M_r for a second-order system is unity, the damping ratio ξ for the system is
   (a) 1.0 (b) 0.707 (c) 0.78 (d) 1.21
17. The term $G(s)H(s)=s$ on magnitude plot of Bode plot will have a slope
   (a) 6 dB/octave (b) −6 dB/octave
   (c) 12 dB/octave (d) 20 dB/dec
18. Frequency response of a system is defined as
   (a) the transient response to a sinusoidal input
   (b) the steady-state response to a sinusoidal input
   (c) the transient response to a parabolic input
   (d) the steady-state response to a unit step input
19. The Bode amplitude plot of a constant is a
   (a) line with slope −40 dB/dec (b) line with slope −20 dB/dec
   (c) line with slope +20 dB/dec (d) straight line parallel to frequency axis
20. An octave frequency range is given as
   (a) $\dfrac{\omega_1}{\omega_2}=2$ (b) $\dfrac{\omega_1}{\omega_2}=4$ (c) $\dfrac{\omega_1}{\omega_2}=8$ (d) $\dfrac{\omega_1}{\omega_2}=10$

21. The value of 20 dB/dec is equal to
(a) 6 dB/octave
(b) −6 dB/octave
(c) 12 dB/octave
(d) −12 dB/octave
22. If the time constant of a first-order factor of a t.f. is T, the corner frequency is
(a) 10 T
(b) $1/T$
(c) T
(d) T^2
23. At low frequencies, the initial slope of the Bode plot for Type 3 system is
(a) −20 dB/dec
(b) −60 dB/dec
(c) +20 dB/dec
(d) +60 dB/dec
24. A unit change in $\log_{10}\omega$ corresponds to a frequency ratio of
(a) 2π
(b) 3π
(c) 10
(d) 100
25. The initial slope of the Bode plot for a Type 2 system is
(a) −12 dB/octave
(b) −6 dB/octave
(c) −24 dB/octave
(d) +24 dB/octave
26. The t.f. $G(s) = 1/s^2$ has a phase shift of
(a) 90°
(b) −90°
(c) −180°
(d) 180°
27. A system has 14 poles and 2 zeros. The slope of its highest frequency asymptote and its magnitude plot is
(a) −40 dB/dec
(b) −240 dB/dec
(c) −280 dB/dec
(d) −320 dB/dec
28. A decade frequency range is given as
(a) $\frac{\omega_1}{\omega_2} = 2$
(b) $\frac{\omega_1}{\omega_2} = 4$
(c) $\frac{\omega_1}{\omega_2} = 8$
(d) $\frac{\omega_1}{\omega_2} = 10$
29. For the second-order t.f. $T(s) = \frac{4}{s^2 + 2s + 4}$, the resonant peak will be
(a) 4
(b) $\frac{4}{3}$
(c) 2
(d) $\frac{2}{\sqrt{3}}$
30. For the second-order t.f. $T(s) = \frac{4}{s^2 + 2s + 4}$, the resonant frequency will be
(a) 2
(b) $\sqrt{2}$
(c) 3
(d) $\sqrt{3}$
31. The magnitude versus frequency plot is shown in the following figure:

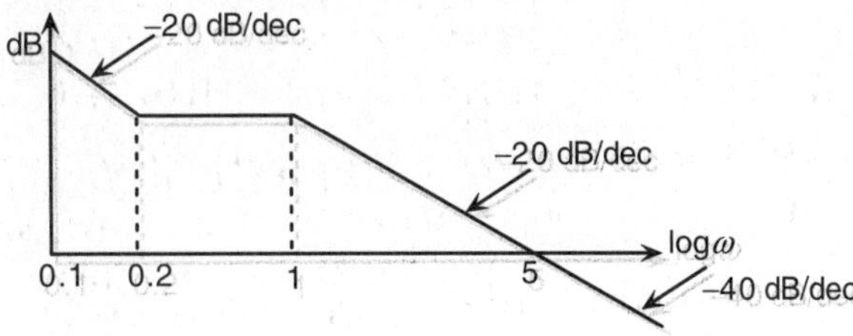

The time constants are
(a) 1, 2 and 5 s
(b) 0.2, 1 and 5 s
(c) 0.1, 0.4 and 0.8 s
(d) None

32. The t.f. is relation to the Bode plot is given in the following figure:

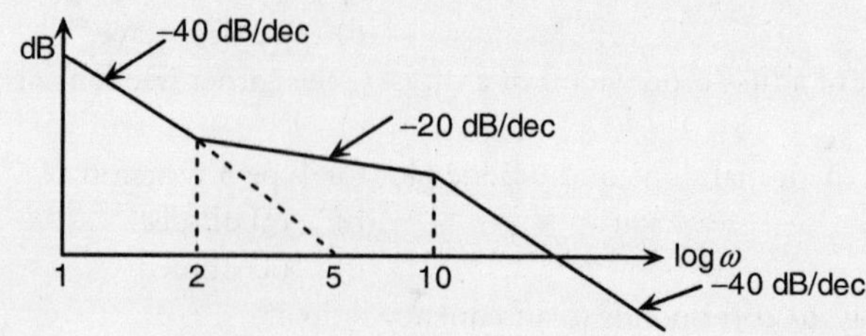

(a) $\dfrac{25(1+0.5s)}{s^2(1+0.1s)}$ (b) $\dfrac{25(1+0.1s)}{s^2(1+0.5s)}$

(c) $\dfrac{25(1+2s)}{s^2(1+10s)}$ (d) None

ANSWERS

1. b	**2.** d	**3.** a	**4.** a	**5.** a	**6.** c	**7.** d	**8.** a	**9.** b	**10.** a
11. d	**12.** d	**13.** d	**14.** c	**15.** c	**16.** b	**17.** a	**18.** b	**19.** d	**20.** a
21. a	**22.** b	**23.** b	**24.** c	**25.** a	**26.** c	**27.** b	**28.** d	**29.** d	**30.** b
31. b	**32.** a								

CHAPTER 12

Stability Analysis of Nyquist Plot Method

Contents

12.1 INTRODUCTION

A major disadvantage of Bode plot is that we have two separate curves showing the variation of gain and phase shift with frequency. The method of combining these two values in a single plot is referred to as the polar plot. In this chapter, in continuation of the discussion on frequency response, the polar plot is discussed, which is the base of Nyquist plot. Using *R–H* stability criterion, only the absolute stability of a system can be found out. On the other hand, the Nyquist criterion uses a different approach for finding the stability of a system. Further, it focuses on the relative stability of the system. It is possible to determine the stability of the closed-loop system from its open-loop frequency response without knowing the roots of the closed-loop system. The aim of this chapter is to introduce the polar plot and Nyquist plot as well as the stability analysis of the system.

12.2 POLAR PLOT

The polar plot of a sinusoidal t.f. $G(j\omega)$ depicts a plot of the magnitude of $G(j\omega)$ versus the phase angle of $G(j\omega)$ on polar coordinates for different values of frequencies ranging from 0 to ∞. Thus, the polar plot is the locus of vectors $|G(j\omega)|\angle G(j\omega)$, as ω is varied from 0 to ∞. The polar plot is very useful in determining the stability of a closed-loop system from its open-loop frequency response. Generally, in polar plots, the positive phase angles are measured in anticlockwise direction, while the negative phase angles are measured in clockwise direction from the reference axis, as shown in Figure 12.1.

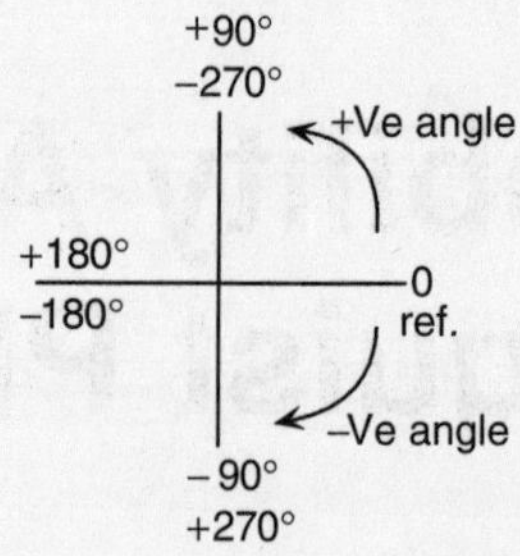

Figure 12.1 Polar graph

Figure 12.2 shows that for minimum phase t.f. with only poles, the type of the system determines the quadrant at which the polar plot starts and the order of the system determines the quadrant at which the plot ends.

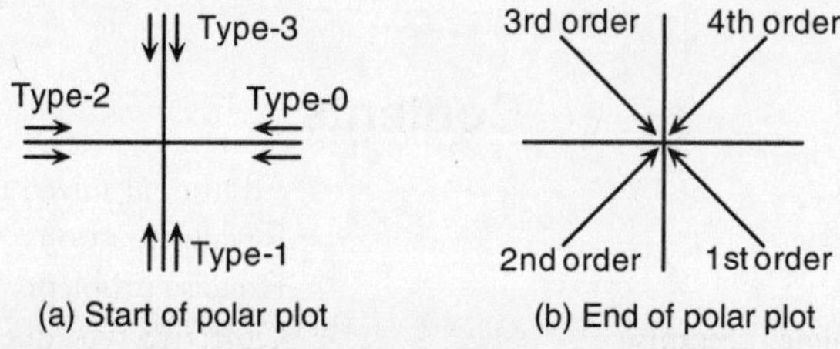

Figure 12.2 Starting and ending points of polar plot

12.2.1 Polar Plot of Some Standard Functions

The polar plots for some standard t.fs. are given as follows.

1. $G(s) = \frac{1}{s}$.

By substituting $s = j\omega$, we have

$$G(j\omega) = \frac{1}{j\omega}.$$

Here,
$$M = |G(j\omega)| = \frac{1}{\omega}$$

$$\phi = \angle G(j\omega) = -90°$$

ω	**M**	ϕ
0	∞	−90
1	1	−90
10	0.1	−90
⋮	⋮	⋮
∞	0	−90

For various values of ω, magnitude M changes but angle ϕ remains constant as 90°. The polar plot of the given t.f. is as shown in Figure 12.3.

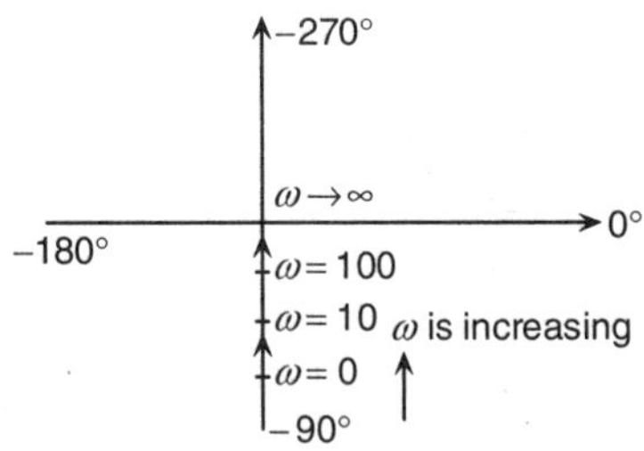

Figure 12.3 Polar plot of $G(s) = \frac{1}{s}$

2. $G(s) = s$

By substituting $s = j\omega$, we have

$$G(j\omega) = j\omega.$$

Here,

$$M = |G(j\omega)| = \omega$$

$$\phi = \angle G(j\omega) = 90°$$

ω	M	ϕ
0	0	90°
10	10	90°
100	100	90°
⋮	⋮	⋮
∞	0	90°

For various values of ω, magnitude M changes but angle ϕ remains constant as 90°. The polar plot of the given t.f. is as shown in Figure 12.4.

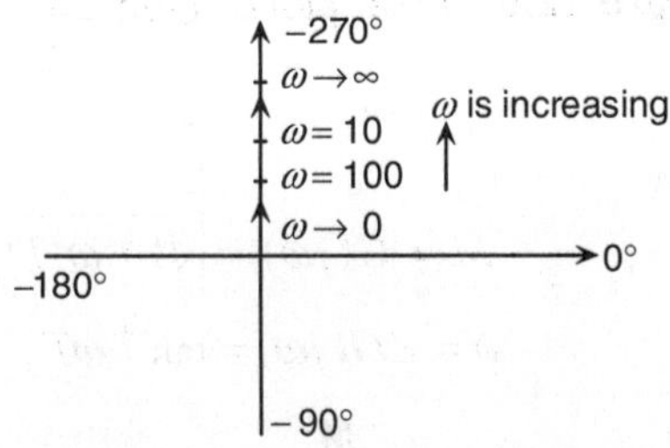

Figure 12.4 Polar plot of $G(s) = s$

3. $G(s) = \frac{1}{1+sT}$

By substituting $s = j\omega$, we have

$$G(j\omega) = \frac{1}{1+j\omega T}.$$

Here,

$$M = |G(j\omega)| = \left|\frac{1}{\sqrt{1+\omega^2T^2}}\right|.$$

$$\phi = \angle G(j\omega) = -\tan^{-1}\omega T$$

ω	M	ϕ
0	1	$-0°$
$\frac{1}{T}$	$\frac{1}{\sqrt{2}}$	$-45°$
$\frac{10}{T}$	$\frac{1}{\sqrt{101}}$	$-84.2°$
$\vdots$	$\vdots$	$\vdots$
∞	0	$-90°$

From the table, we have

At $\omega = 0,\ G(j\omega) = 1\angle 0°$

At $\omega = \infty,\ G(j\omega) = 0\angle -90°$

The polar plot of the given t.f. is as shown in Figure 12.5.

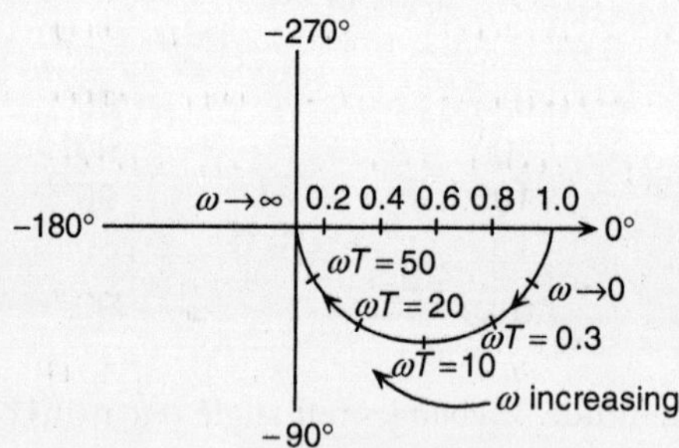

Figure 12.5 Polar plot of $G(s) = \dfrac{1}{1+sT}$

4. $G(s) = 1 + sT$

By substituting $s = j\omega$, we have

Here,
$$M = |G(j\omega)| = \left|\sqrt{1+\omega^2T^2}\right|.$$
$$\phi = \angle G(j\omega) = \tan^{-1}\omega T$$

ω	M	ϕ
0	1	$0°$
$\frac{1}{T}$	$\sqrt{2}$	$45°$
$\frac{10}{T}$	$\sqrt{101}$	$84.2°$
$\vdots$	$\vdots$	$\vdots$
∞	∞	$90°$

From the table, we have

At $\omega = 0,\ G(j\omega) = 1\angle 0°$

At $\omega = \infty,\ G(j\omega) = \infty\angle -90°$

The polar plot of the given t.f. is as shown in Figure 12.6.

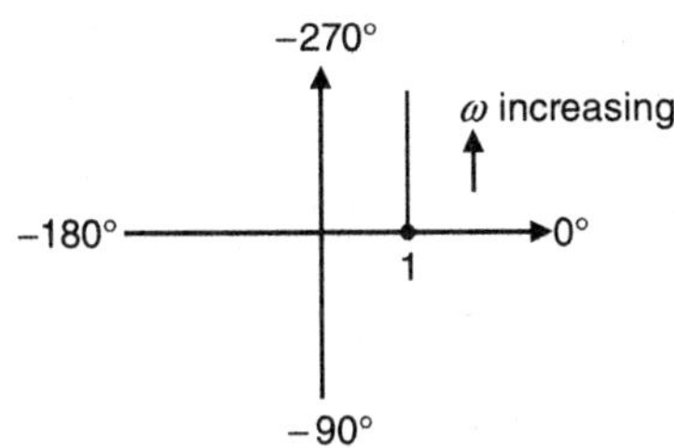

Figure 12.6 Polar plot of $G(s) = 1 + sT$

5. $G(s) = \dfrac{sT}{1+sT}$

By substituting $s = j\omega$, we have

$$G(j\omega) = \frac{j\omega T}{1 + j\omega T}.$$

Here,

$$M = |G(j\omega)| = \left|\frac{\omega T}{\sqrt{1+\omega^2 T^2}}\right|$$

$$\phi = \angle G(j\omega) = 90^{\circ} - \tan^{-1}\omega T$$

ω	M	ϕ
0	0	90°
$\frac{1}{T}$	$\frac{1}{\sqrt{2}}$	45°
$\frac{10}{T}$	$\frac{10}{\sqrt{101}}$	84.3°
⋮	⋮	⋮
∞	1	0°

From the table, we have

At $\omega = 0,\ G(j\omega) = 0\angle 90^0$

At $\omega = \infty,\ G(j\omega) = 1\angle 0^0$

The polar plot of the given t.f. is as shown in Figure 12.7.

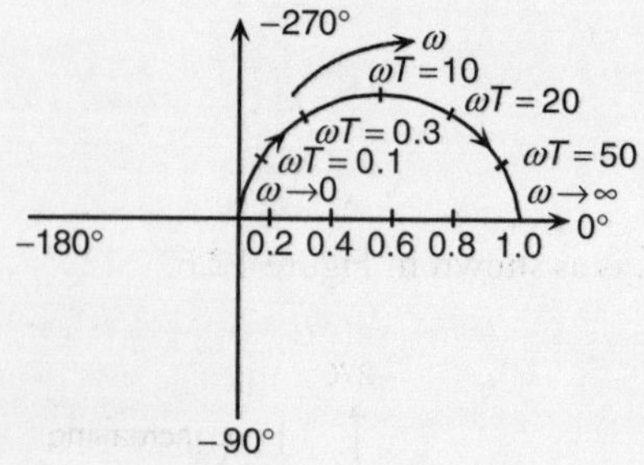

Figure 12.7 Polar plot of $G(s)=\dfrac{sT}{1+sT}$

6. Type-0 system or $G(s)=\dfrac{1}{(1+sT_1)(1+sT_2)}$

By substituting $s = j\omega$, we have

$$G(j\omega)=\frac{1}{(1+j\omega T_1)(1+j\omega T_2)}.$$

Here,

$$M=|G(j\omega)|=\frac{1}{\sqrt{1+\omega^2T_1^2}\sqrt{1+\omega^2T_2^2}}.$$

$$\phi=\angle G(j\omega)=-\tan^{-1}\omega T_1-\tan^{-1}\omega T_2$$

ω	**M**	ϕ
0	1	0°
∞	0	-180°

The polar plot of the given t.f. is as shown in Figure 12.8.

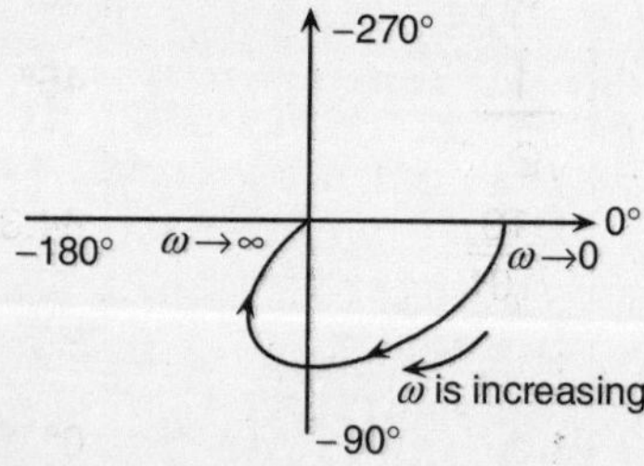

Figure 12.8 Polar plot of $G(s)=\dfrac{1}{(1+sT_1)(1+sT_2)}$

At $\omega = 0$, $|G(j\omega)| = 1$, $\angle G(j\omega) = 0°$ and as the value is changed from 0 to high, the magnitude of $G(j\omega)$ diminishes and the angle of $G(j\omega)$ decreases from 0°. If the value of frequency is increased to infinity, the angle of $G(j\omega)$ becomes $-180°$ and the magnitude falls to 0.

7. Type-1 system or $G(s) = \frac{1}{1+sT}$.

   By substituting $s = j\omega$, we have

$$G(j\omega) = \frac{1}{j\omega(1+j\omega T)}.$$

Here,

$$M = |G(j\omega)| = \frac{1}{\omega\sqrt{1+\omega^2 T^2}}.$$

$$\phi = \angle G(j\omega) = -90^\circ - \tan^{-1}\omega T$$

ω	**M**	ϕ
0	∞	90°
∞	0	–180°

The polar plot of the given t.f. is as shown in Figure12.9.

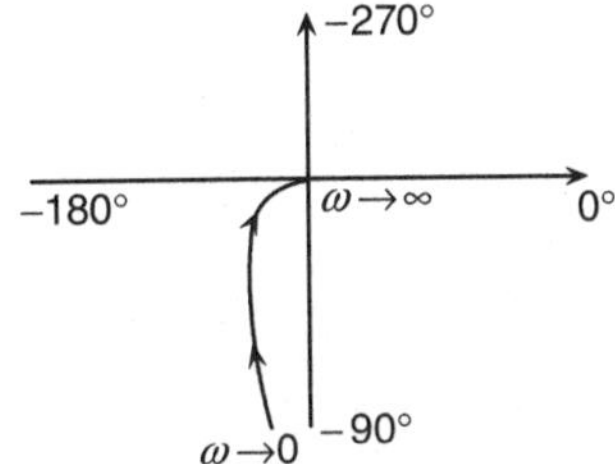

Figure 12.9 Polar plot of $G(s) = \frac{1}{s(1+sT)}$

The polar plot starts at $|G(j\omega)| = \infty$ with an angle $\angle G(j\omega) = -90^\circ$ at $\omega = 0$. As the value of ω increases, the magnitude decreases and approaches 0 as $\omega \to \infty$.

8. Type-2 system or $G(s) = \frac{1}{s^2(1+sT)}$

   By substituting $s = j\omega$, we have

$$G(j\omega) = \frac{1}{(j\omega)^2(1+j\omega T)}.$$

Here,

$$M = |G(j\omega)| = \frac{1}{\omega^2\sqrt{1+\omega^2 T^2}}.$$

$$\phi = \angle G(j\omega) = -180^\circ - \tan^{-1}\omega T$$

ω	**M**	ϕ
0	∞	–180°
∞	0	–270°

The polar plot of the given t.f. is as shown in Figure 12.10.

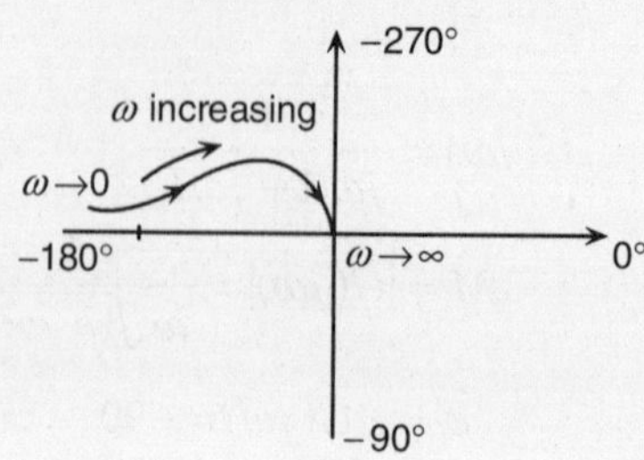

Figure 12.10 Polar plot of $G(s)=\dfrac{1}{s^2(1+sT)}$

As the value of ω is increased, the magnitude decreases from ∞ to 0 and phase angle gradually changes from −180° to −270°.

12.2.2 Procedure for Constructing a Polar Plot

The following procedure is adopted for constructing a polar plot

1. Determine the t.f. $G(s)$ of the system.
2. Substitute $s=j\omega$ in the t.f. $G(s)$ of the system and obtain frequency response function $G(j\omega)$ at $\omega=0$ and $\omega=\infty$ by having

$$\underset{\omega\to 0}{Lt}\left|G(j\omega)\right| \;\&\; \underset{\omega\to\infty}{Lt}\left|G(j\omega)\right|$$
$$\underset{\omega\to 0}{Lt}\angle G(j\omega) \;\&\; \underset{\omega\to\infty}{Lt}\angle G(j\omega).$$

3. Rationalise the complex frequency function $G(j\omega)$ and separate real and imaginary parts.
4. Determine the frequency at which the plot intersects the real axis by equating the imaginary parts of $G(j\omega)$ to zero. Hence, calculate the value of $G(j\omega)$ at the intersection point by substituting the determined value of frequency in the rationalised expression of $G(j\omega)$.
5. Determine the frequency at which the plot intersects the imaginary axis by equating the real part of $G(\omega)$ to zero. Hence, calculate the value of $G(j\omega)$ at the intersection point by substituting the determined value of frequency in the rationalised expression of $G(j\omega)$.
6. Sketch the complete plot of the system that satisfies the above-mentioned specifications.

Problem-12.1: Sketch the polar plot of the t.f. $G(s)=\dfrac{10}{s(s+1)}$.

Solution:

The t.f. of the given system is $G(s)=\dfrac{10}{s(s+1)}$.

To get sinusoidal t.f., substitute $s=j\omega$ in $G(s)$.

$$G(j\omega)=\frac{10}{j\omega(j\omega+1)}$$

$$M=\left|G(j\omega)\right|=\frac{10}{\omega\sqrt{1+\omega^2}}$$

$$\phi=\angle G(j\omega)=-90-\tan^{-1}\omega$$

ω	M	ϕ
0	∞	–90°
∞	0	–180°

Next, let us determine whether $G(j\omega)$ intersects any of the axis of $G(j\omega)$-plane.

$$G(j\omega) = \frac{10}{j\omega(j\omega+1)} = \frac{10}{j\omega-\omega^2} = \frac{10(-j\omega-\omega^2)}{(-j\omega-\omega^2)(j\omega-\omega^2)} = \frac{-10\omega^2}{\omega^4+\omega^2} - j\frac{10\omega}{\omega^4+\omega^2}$$

Let us assign the imaginary part (I.P.) of $G(j\omega) = 0$

$$\Rightarrow \qquad \frac{10\omega}{\omega^4+\omega^2} = 0$$

$$\Rightarrow \qquad \omega = \infty.$$

$\Rightarrow$ The plot intersects the imaginary axis at the origin.

The required polar plot is as shown in the figure.

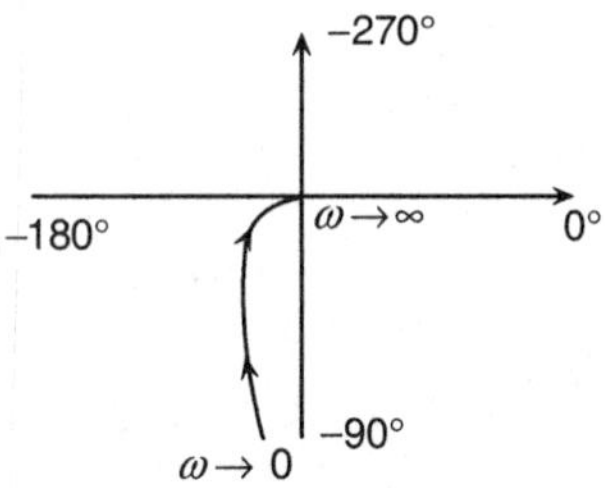

Problem-12.2: Sketch the polar plot of the t.f. $G(s) = \dfrac{6}{(s+1)(s+2)}$.

Solution:

The t.f. of the given system is $G(s) = \dfrac{6}{(s+1)(s+2)}$.

To get sinusoidal t.f., substitute $s = j\omega$ in $G(s)$.

$$G(j\omega) = \frac{6}{(1+j\omega)(j\omega+2)}$$

$$M = |G(j\omega)| = \frac{6}{\sqrt{1+\omega^2}\,\sqrt{4+\omega^2}}$$

$$\phi = \angle G(j\omega) = -\tan^{-1}\omega - \tan^{-1}\omega/2.$$

ω	M	ϕ
0	3	0°
∞	0	–180°

Next, let us determine whether $G(j\omega)$ intersects any of the axis of $G(j\omega)$-plane.

$$G(j\omega) = \frac{6}{(1+j\omega)(j\omega+2)}$$

$$= \frac{6(1-j\omega)(2-j\omega)}{(1+j\omega)(j\omega+2)(1-j\omega)(2-j\omega)}$$

$$= \frac{6(2-\omega^2)}{(1+\omega^2)(4+\omega^2)} - \frac{j18\omega}{(1+\omega^2)(4+\omega^2)}.$$

Let us assign I.P. of $G(j\omega) = 0$.

$$\frac{18\omega}{(1+\omega^2)(4+\omega^2)} \Rightarrow \omega = 0$$

$$\therefore \qquad G(j0) = \frac{6\times 1}{2\times 1} = 3$$

Let us assign real part (R.P.) of $G(j\omega) = 0$.

$$\frac{6(2-\omega^2)}{(1-\omega^2)(4+\omega^2)} = 0 \Rightarrow \omega = \pm\sqrt{2}$$

$$G(j\sqrt{2}) = \frac{6}{(1+j\sqrt{2})(2+j\sqrt{2})} = \frac{6}{j3\sqrt{2}} = -j\sqrt{2}$$

$$G(-j\sqrt{2}) = \frac{6}{(1-j\sqrt{2})(2-j\sqrt{2})} = \frac{-6}{j3\sqrt{2}} = j\sqrt{2}$$

Let us consider only positive value of ω that is, $\omega = 2$.

The required polar plot is as shown in the following figure.

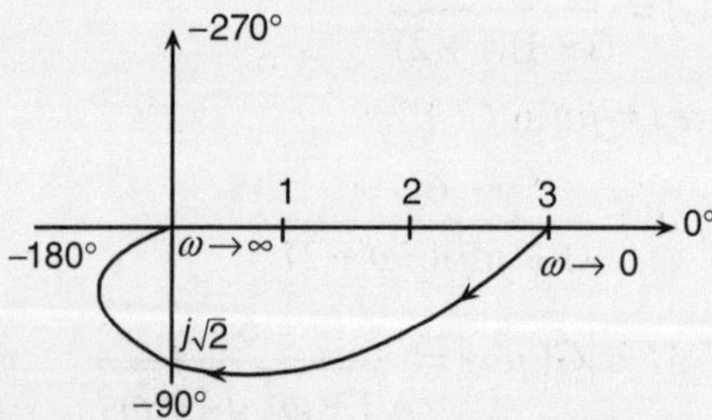

12.2.3 Gain and Phase Crossover Frequencies in Polar Plots

The polar plot intersects the negative real axis at a distance x and the frequency where it crosses the negative real axis is called phase crossover frequency, indicated as ω_{pc} as shown in Figure 12.11. A circle with radius equal to unity is drawn and it intersects the polar plot at a point P, where the frequency is ω_{gc} called gain crossover frequency. Phase crossover frequency ω_{pc} is the frequency at which

$\angle G(j\omega)H(j\omega) = -180°$. In polar plot, we have to find such a point on the plot whose phase angle is $-180°$, that is, a point on the negative real axis; such a point Q is shown in Figure 12.11.

Consider a polar plot as shown in Figure 12.11. To get a point with $M = 1$, draw a circle with radius 1 and centre as origin. The point where this circle intersects polar plot is the point where $|G(j\omega)H(j\omega)| = 1$ and the corresponding frequency is $\omega = \omega_{gc}$. Now, ω_{pc} is the frequency at which $\angle G(j\omega)H(j\omega) = -180°$. In polar plot, we have to find such a point on the plot whose angle is $-180°$, that is, a point on the negative real axis. Therefore, at $\omega = \omega_{pc}$, the polar plot intersects with the negative real axis. Such a point Q is shown in Figure 12.11.

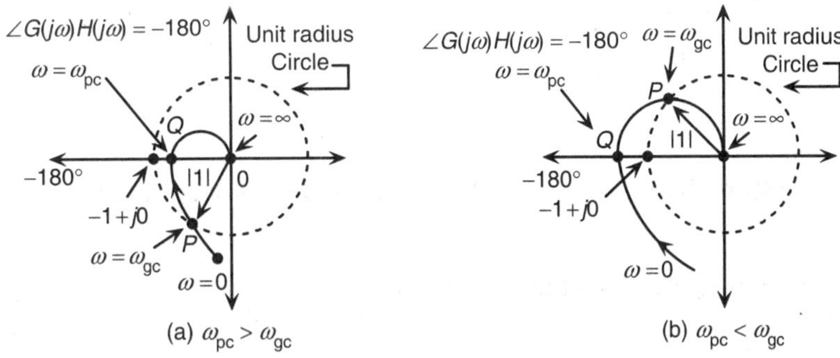

Figure 12.11 ω_{pc} and ω_{gc} in polar plot

In determining the stability, $|G(j\omega)H(j\omega)| = 1$ and $\angle G(j\omega)H(j\omega) = -180$ play a very important role. This point $1\angle -180°$ is nothing but a point $(-1 + j0)$ on the negative real axis and called critical point in polar and Nyquist plot analysis.

12.2.4 Determination of GM and PM from Polar Plots

According to the definition of gain margin (GM), it is the margin in gain that can be introduced in the system until the system reaches the verge of instability.

$$\text{GM} = \frac{1}{|G(j\omega)H(j\omega)|\Big|_{\omega=\omega_{pc}}}.$$

In polar plot, $|G(j\omega)H(j\omega)|_{\omega=\omega_{pc}}$ is nothing but OQ where Q is the intersection point of the polar plot with negative real axis. Q is the point that corresponds to $\omega = \omega_{pc}$.

$$\therefore \qquad \text{GM} = \frac{1}{|OQ|}.$$

According to definition of phase margin (PM), we have

$$\text{PM} = 180° + \angle G(j\omega)H(j\omega)\big|_{\omega=\omega_g}.$$

Now, point P is the point on polar plot corresponding to $\omega = \omega_{gc}$ obtained by drawing unit radius circle. Therefore, if ϕ is the angle of point P corresponding to $\omega = \omega_{gc}$, the PM can be calculated as shown in Figure 12.12.

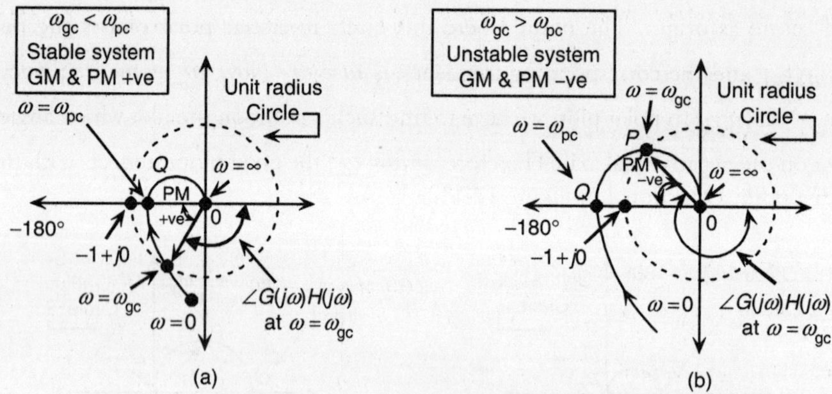

Figure 12.12 Determination of GM and PM from polar plot

Problem-12.3: Sketch the polar plot of the t.f. $G(s) = \dfrac{10}{s(s+1)(s+2)}$.

Solution:

The t.f. of the given system is $G(s) = \dfrac{10}{s(s+1)(s+2)}$.

To get sinusoidal t.f., substitute $s = j\omega$ in $G(s)$.

$$G(j\omega) = \frac{10}{j\omega(1+j\omega)(j\omega+2)}$$

$$M = |G(j\omega)| = \frac{10}{\omega\sqrt{1+\omega^2}\,\sqrt{4+\omega^2}}$$

$$\phi = \angle G(j\omega) = -90° - \tan^{-1}\omega - \tan^{-1}\omega/2$$

ω	M	ϕ
0	∞	−90°
∞	0	−270°

Next, let us determine whether $G(j\omega)$ intersects any of the axes of $G(j\omega)$-plane.

$$G(j\omega) = \frac{10}{j\omega(1+j\omega)(j\omega+2)} = \frac{10}{(-\omega^2+j\omega)(2+j\omega)}$$

$$= \frac{10(-\omega^2-j\omega)(2-j\omega)}{(-\omega^2+j\omega)(2+j\omega)(-\omega^2-j\omega)(2-j\omega)}$$

$$= \frac{10(-2\omega^2 + j\omega^3 - 2j\omega - \omega^2)}{(\omega^4 + \omega^2)(4 + \omega^2)}$$

$$= \frac{-30\omega^2}{(\omega^4 + \omega^2)(4 + \omega^2)} + \frac{j10\omega(\omega^2 - 2)}{(\omega^4 + \omega^2)(4 + \omega^2)}.$$

Let us assign I.P. of $G(j\omega) = 0$

$$\frac{10\omega(\omega^2 - 2)}{(\omega^4 + \omega^2)(4 + \omega^2)} = 0$$

$$10\omega(\omega^2 - 2) = 0$$

$$\omega = 0, \omega = \pm\sqrt{2}$$

Let us consider only the positive value, that is, $\omega = \sqrt{2}$.

$$\therefore \qquad G(j\sqrt{2}) = \frac{-30 \times 2}{(4+2)(4+2)} = \frac{-10}{6} = -1.66.$$

Therefore, point $Q = -1.66$ is the intersection point of polar plot with negative real axis and the required polar plot is as shown in the following figure.

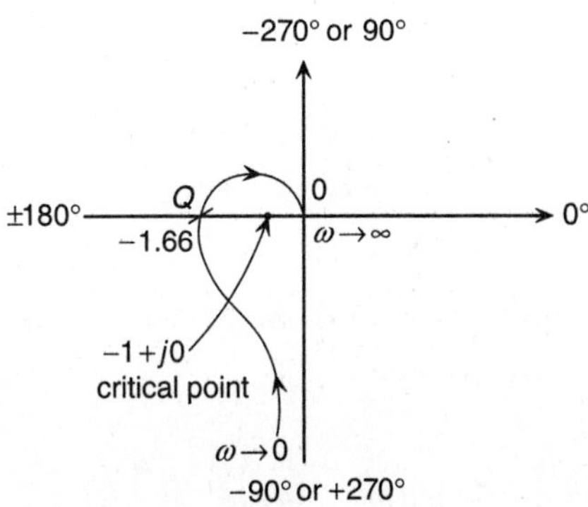

$$\therefore \qquad GM = \frac{1}{|OQ|} = \frac{1}{1.66} = 0.6$$

GM in dB = 20 log (0.6) = −4.43.

As GM is negative and critical point is enclosed, the given system is unstable.

Problem-12.4: Sketch the polar of the t.f. $G(s) = \dfrac{500}{s(s+6)(s+9)}$.

Solution:

The t.f. of the given system is $G(s) = \dfrac{500}{s(s+6)(s+9)}$.

To get sinusoidal t.f., substitute $s = j\omega$ in $G(s)$.

$$G(j\omega) = \frac{500}{j\omega(6+j\omega)(j\omega+9)}$$

$$M = |G(j\omega)| = \frac{500}{\omega\sqrt{36+\omega^2}\sqrt{81+\omega^2}}$$

$$\phi = \angle G(j\omega) = -90° - \tan^{-1}(\omega/6) - \tan^{-1}(\omega/9)$$

ω	**M**	ϕ
0	∞	−90°
∞	0	−270°

Next, let us determine whether $G(j\omega)$ intersects any of the axes of $G(j\omega)$-plane.

$$G(j\omega) = \frac{500}{j\omega(6+j\omega)(j\omega+9)} = \frac{500}{(-\omega^2 + j6\omega)(9+j\omega)}$$

$$= \frac{500(-\omega^2 - j6\omega)(9-j\omega)}{(-\omega^2 + j6\omega)(9+j\omega)(-\omega^2 - j6\omega)(9-j\omega)}$$

$$= \frac{500(-9\omega^2 + j\omega^3 - j54\omega - 6\omega^2)}{(\omega^4 + 36\omega^2)(81+\omega^2)}$$

$$= \frac{-500\times 15\omega^2}{(\omega^4 + 36\omega^2)(81+\omega^2)} + \frac{j500\,\omega(\omega^2 - 54)}{(\omega^4 + 36\omega^2)(81+\omega^2)}.$$

Let us assume I.P. of $G(j\omega) = 0$.

$$\frac{500\omega(\omega^2 - 54)}{(\omega^4 + 36\omega^2)(81+\omega^2)} = 0$$

$$500\,\omega(\omega^2 - 54) = 0$$

$$\omega = 0,\ \omega = \pm 7.35.$$

Let us consider only positive value, that is, $\omega = 7.35$.

$$G(j7.35) = \frac{-500\times 15\times 7.35^2}{4860.8\times 135} = -0.62.$$

Therefore, point $Q = -0.62$ is the intersection point of polar plot with negative real axis. The required polar plot is as shown in the following figure.

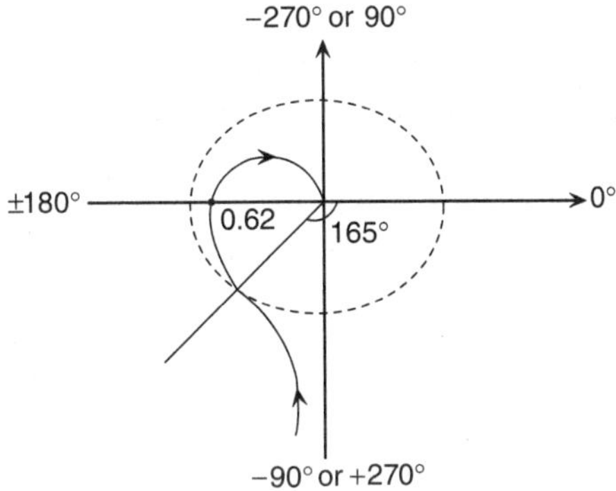

$$\therefore \qquad \text{GM} = \frac{1}{|OQ|} = \frac{1}{0.62} = 1.6121$$

$$\text{GM in dB} = 20 \log (1.6121) = 4.152$$

$$\text{PM} = 180° + \phi = 180° - 165° = 15°$$

Since both GM and PM are positive, the given system is stable.

12.3 NYQUIST PLOT

The concept of Nyquist plot is based on the polar plot, which can be conveniently applied to the stability analysis of any kind of system. The concept is slightly difficult to understand, and hence, the discussion is divided into the following sections to unfold the Nyquist plot in a step-by-step manner.

1. Concept of encirclement
2. Analytic function and its singularities
3. Principle of argument or mapping theorem
4. Nyquist stability criterion.

12.3.1 Concept of Encirclement, Analytic Function and Principle of Argument

Consider a function $q(s)$ that can be expressed as a quotient of two polynomials, which has its own poles and zeros, as follows:

$$q(s) = \frac{(s-\alpha_1)(s-\alpha_2)\ldots(s-\alpha_n)}{(s-\beta_1)(s-\beta_2)\ldots(s-\beta_n)}. \qquad (12.1)$$

Let s be a complex variable, which is represented by $s = (\sigma + j\omega)$ on the complex s-plane. Then, the function $q(s)$ is also complex and may be defined as $q(s) = u + jv$ and represented on the complex $q(s)$-plane with coordinates u and v. Equation (12.1) indicates that for every point s in the s-plane at which $q(s)$ is analytic, we can find a corresponding point $q(s)$ in the $q(s)$-plane. Alternatively, it can be stated that the function $q(s)$ maps the points in the s-plane into the $q(s)$-plane. Since any number of points of the s-plane can be mapped into the $q(s)$-plane, it follows that for a contour in the s-plane which does not go through any singular point, there exists a corresponding contour in the $q(s)$-plane as shown in Figure 12.13. While developing the Nyquist criterion, we are not interested in the exact shape of the $q(s)$-plane contour. An important fact that concerns us is the 'encirclement' of the origin by the $q(s)$-plane contour.

To investigate this, consider an s-plane contour that encloses only one of the zeros of $q(s)$, say $s = \alpha_1$, while all the poles and remaining zeros are distributed in the s-plane outside the contour. From Equation (12.1), point $q(s)$ is given as

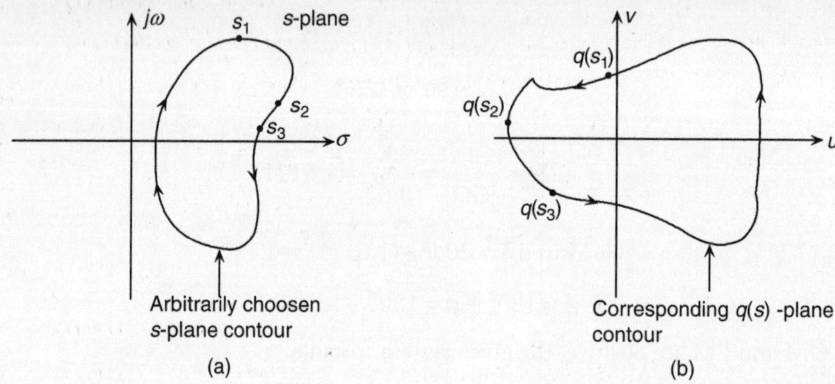

Figure 12.13 Arbitrarily chosen s-plane contour which does not go through singular points and the corresponding $q(s)$-plane contour

$$|q(s)| = \frac{|(s-\alpha_1)||(s-\alpha_2)|\ldots}{|(s-\beta_1)||(s-\beta_2)|\ldots} \tag{12.2}$$

$$\angle q(s) = \angle(s-\alpha_1) + \angle(s-\alpha_2) + \cdots \angle(s-\beta_1) - \angle(s-\beta_2) \cdots \tag{12.3}$$

Note: A function $q(s)$ is said to be 'analytic' in the s-plane, provided the function and all its derivatives exist. The points in the s-plane where the functions (or its derivatives) do not exist are called singular points. The poles of a function $q(s)$ are called singular points.

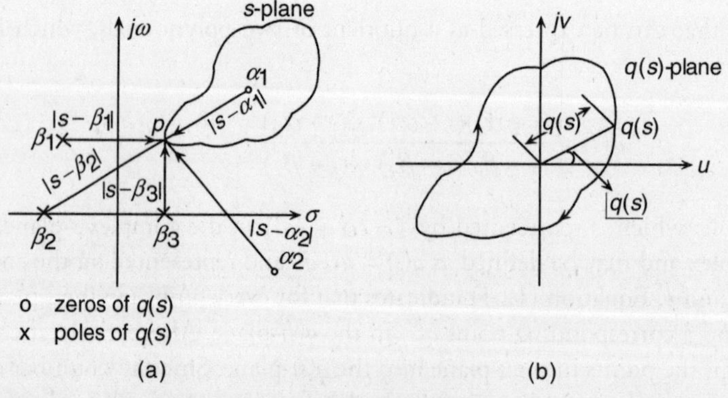

Figure 12.14 Mapping of s-plane and $q(s)$-plane contours

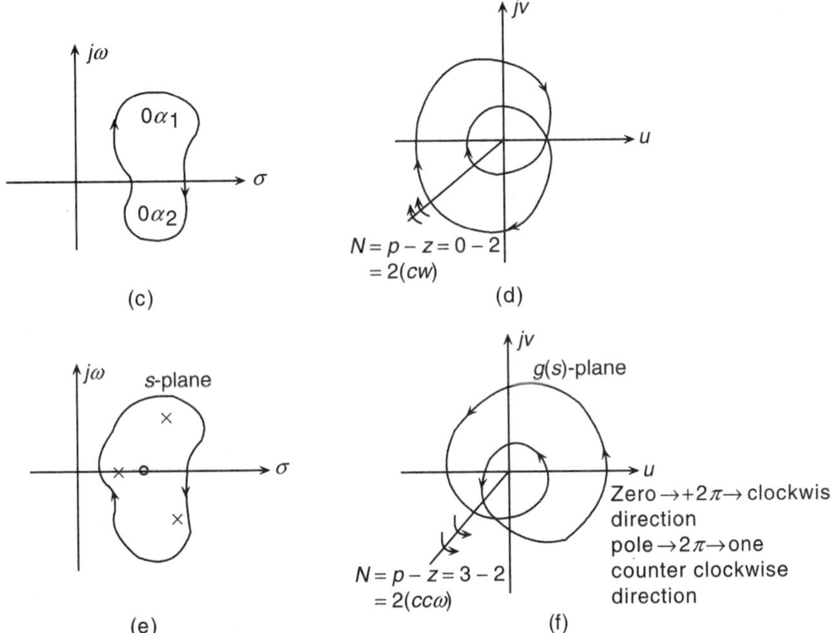

Figure 12.14 *(Continued)*

From Figure 12.14(a) and (b), it is found that as point s follows the prescribed path (i.e., clockwise direction) on the s-plane contour, eventually returning to the starting point, the phasor $(s - \alpha_1)$ generates a net angle of -2π, while all other phasors generate zero net angles. Therefore, $q(s)$-phasor undergoes a net phase change of (-2π). This implies that the tip of the $q(s)$-phasor must describe a closed contour about the origin of the $q(s)$-plane in the 'clockwise direction'.

If the s-plane contour encloses two zeros, say at $s = \alpha_1$ and $s = \alpha_2$, the $q(s)$-plane contour encircles the origin twice in the clockwise direction, as shown in Figure 12.14(c) and (d). Hence, we can say that for each zero of $q(s)$ enclosed by the s-plane contour, the corresponding $q(s)$-plane contour encircles the origin once in the anticlockwise direction.

Thus, if there are P poles and Z zeros of $q(s)$ enclosed by the s-plane contour, then the corresponding $q(s)$-plane contour must encircle the origin Z times in the clockwise direction and P times in the counter clockwise direction, resulting in a net encirclement of the origin, $(P - Z)$ times in the counter clockwise direction.

For example, in the case of 1 zero and 3 poles enclosed by the s-plane contour, the net encirclement of the origin by the $q(s)$-plane contour is $2\pi(3 - 1) = 4\pi$, that is, two counter clockwise revolutions, as shown in Figure 12.14 (e) and (f). This relationship between the enclosure of poles and zeros of $q(s)$ by the s-plane contour and the encirclements of the origin by the $q(s)$-plane contour is commonly known as 'principle of argument'.

Note: Hence, the exact shape of the closed contour in the $q(s)$-plane is not of interest to us, but it is sufficient for us to observe that this contour encircles the origin once. If the contour in the s-plane is so chosen that it does not enclose any zero or pole, the corresponding contour in the $q(s)$-plane will not encircle the origin.

12.3.2 Nyquist Stability Criterion

Consider a single loop-feedback system as shown in Figure 12.15.

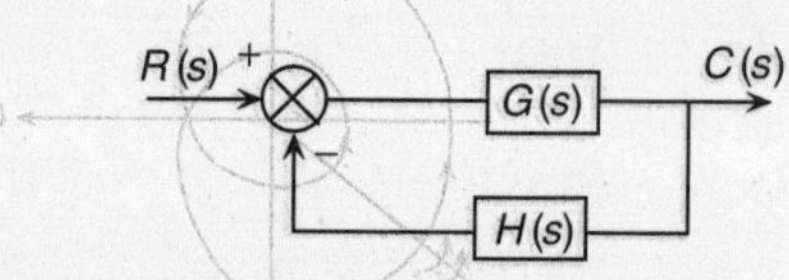

Figure 12.15 Closed-loop control system

The t.f. of the above system is

$$T(s) = \frac{C(s)}{R(s)} = \frac{G(s)}{1+G(s)H(s)}. \tag{12.4}$$

The C.E. of the closed-loop system is

$$1 + G(s)H(s) = 0 \tag{12.5}$$

Let the open-loop t.f. of the system is

$$G(s)H(s) = K\frac{(s+z_1)(s+z_2)\ldots(s+z_m)}{(s+p_1)(s+p_2)\ldots(s+p_n)}; n \geq m$$

$$= K\frac{P(s)}{Q(s)} \tag{12.6}$$

and $$f(s) = 1 + G(s)H(s) \tag{12.7}$$

$$\therefore \quad f(s) = 1 + K\frac{P(s)}{Q(s)} = \frac{Q(s)+KP(s)}{Q(s)}. \tag{12.8}$$

From (12.4) and (12.6), we have

$$T(s) = \frac{G(s)}{1+G(s)H(s)} = \frac{G(s)}{1+K\dfrac{P(s)}{Q(s)}} = \frac{G(s)Q(s)}{Q(s)+KP(s)}. \tag{12.9}$$

From (12.6), (12.8) and (12.9),

1. The poles (P) of $f(s)$ are identical to those of $G(s)H(s)$, that is, the open-loop poles of the system.
2. The zeros (Z) of $f(s)$ are identical to the roots of the C.E. $[1 + G(s)H(s)]$, that is, the closed-loop poles of the system.

For the closed-loop system to be stable, the zeros of $f(s) = 1 + G(s)H(s)$ must lie in the left half of the s-plane. It is important to note that even if some of the open-loop poles lie in the right half of the s-plane, all the zeros of $1+G(s)H(s)$, that is, the closed-loop poles may lie in the left half of the s-plane; in other words, an open-loop unstable system may lead to closed-loop stable operation.

The relationship between the enclosure of poles and zeros of $f(s)$ by the s-plane contour and the encirclement of the origin by the $f(s)$-plane contour is commonly known as the principle of argument that may be stated as follows:

Let $f(s)$ be a ratio of polynomials in s. Let P be the number of poles and Z be the number of zeros of $f(s)$, which are enclosed by a simple closed contour in the s-plane. Let the closed contour be such that it does not pass through any poles or zeros of $f(s)$. The s-plane contour then maps into $f(s)$-plane as a closed curve'.

The number N of clockwise encirclements of the origin of $f(s)$-plane, as a representative point s traces out the entire contour in the s-plane in the clockwise direction, is equal to $(Z - P)$. Therefore,

$$\left.\begin{aligned} N &= Z - P \text{ (in clockwise direction)} \\ &= P - Z \text{ (in anticlockwise direction)} \end{aligned}\right\}. \quad (12.10)$$

Since $f(s) = 1 + G(s)H(s)$, we can obtain $f(s)$-plane contour from the $G(s)H(s)$-plane contour simply by adding +1 to each point of $G(s)H(s)$-plane contour. Figure 12.16 shows a typical $G(s)H(s)$-plane contour Γ_{GH}. The effect of adding +1 to each point of Γ_{GH} to obtain $f(s)$-plane contour Γ_f is accomplished simply by adding +1 to the scale of the real axis, which is given by the number in the parentheses (in Figure 12.16). It is seen that $(-1 + j0)$ point of $G(s)H(s)$ map Γ_{GH} corresponds to the origin of $f(s)$ map Γ_f . We can take $(-1 + j0)$ point of the $G(s)H(s)$-plane as the critical point. Thus, the encirclements of the origin by contour Γ_f is equivalent to the encirclements of the critical point $(-1 + j0)$ by contour Γ_{GH} as shown in Figure 12.17.

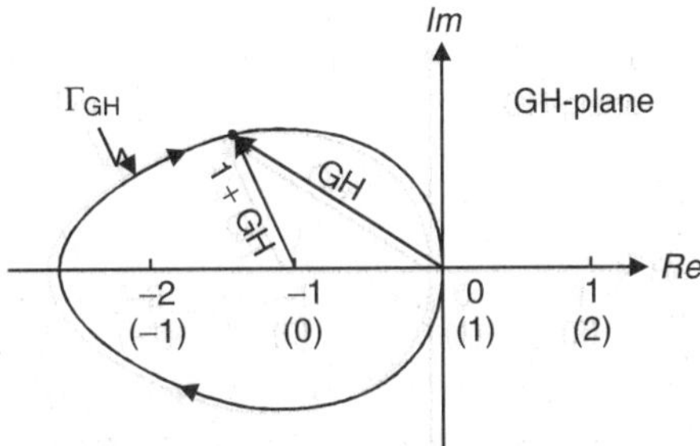

Figure 12.16 Evaluating 1 + $G(s)H(s)$ from the $G(s)H(s)$ map

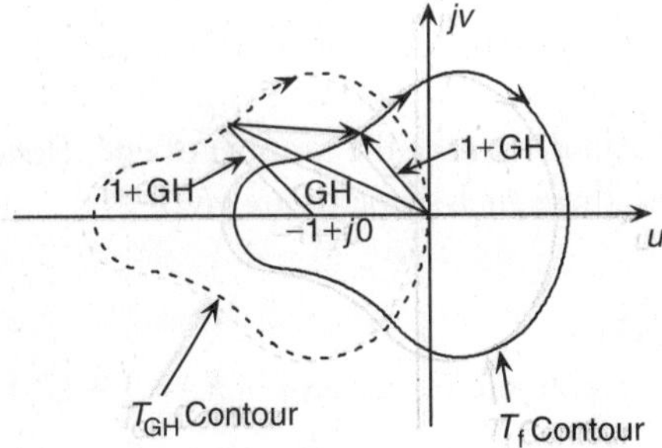

Figure 12.17 Evaluating T_f from the T_{GH} map

A convenient way of determining N with respect to the critical point $(-1 + j0)$ of the $G(s)H(s)$-plane is to draw a radial line from this point. The number of net intersections of the radial line with the Γ_{GH} contour gives the magnitude of N.

Nyquist has suggested that rather than analysing whether all the zeros are located in the left half of the s-plane, it is better to examine the presence of any one zero of $1 + G(s)H(s)$ contour in the right half of the s-plane making the system unstable. Hence, the active region for stability point of view is the right half of the s-plane. In order to investigate the presence of any zeros of $f(s) = 1 + G(s)H(s)$ contour in the right half of the s-plane, let us choose a contour that completely encloses the right half of the s-plane. Such a contour c is called the Nyquist contour, as shown in Figure 12.18(a). It is directed clockwise and comprises an infinite line segment c_1, along the $j\omega$-axis and an arc c_2 of infinite radius. Along c_1, $s = j\omega$ with s varying from $-j\infty$ to $+j\infty$ and along c_2, $s = \lim_{R\to\infty} Re^{j\theta}$ with θ varying from +90° to −90°. The Nyquist contour encloses all the zeros and poles of $f(s) = 1 + G(s)H(s)$ in the right half of the s-plane.

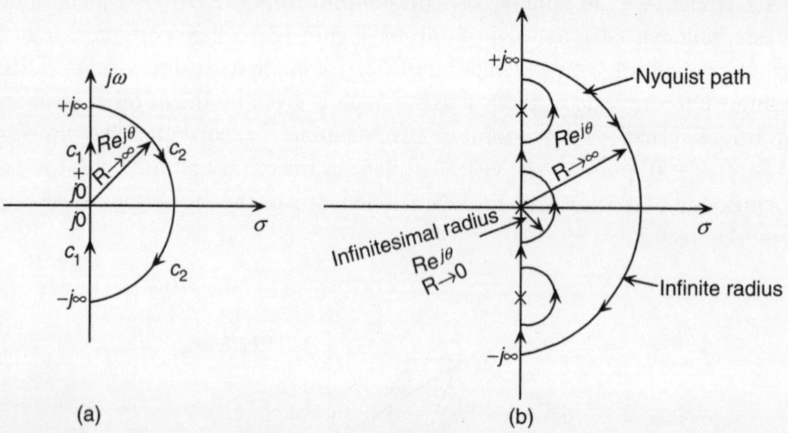

Figure 12.18 Nyquist contour

If the function has poles at origin or poles on the imaginary axis, Nyquist path cannot be selected along the imaginary axis passing through origin. This is because mapping theorem states that at every point on Nyquist path, the function must be analytic. In such a case, Nyquist path is modified in such a way as to bypass these poles by selecting infinitesimal semicircles of radius tending to zero around them but still encircling the entire right half of the s-plane, as shown in Figure 12.18(b). As s moves along the Nyquist contour in the s-plane, a closed contour Γ_{GH} is traversed in the $f(s)$-plane, which encloses the origin.

As this locus is obtained, we can determine the number of encirclements of origin by Nyquist plot, say N. As per mapping theorem, these encirclements are given as

$$N = Z - P. \tag{12.11}$$

As N and P are known, we can get Z.

For the system to be stable, there should be no zeros of $f(s) = 1 + G(s)H(s)$ in the right half of s-plane, that is, $Z = 0$ for stability.

Therefore, Nyquist stability criterion is obtained by substituting $Z = 0$ in $N = Z - P$.

i.e.,

$$N = -P. \tag{12.12}$$

Thus, the Nyquist stability criterion can be stated as follows: 'If the $G(s)H(s)$ contour in the $G(s)H(s)$-plane corresponding to Nyquist contour in the s-plane encircles the point $(-1+j0)$ in the anticlockwise direction as many times as the number of right half of the s-plane poles of $G(s)H(s)$, then the closed-loop system is stable'.

In examining the stability of continuous time control system by using the Nyquist stability criterion, we come across the following three situations:

1. There is no encirclement of $(-1 + j0)$ point. This implies that the system is stable if there are no poles of $G(s)H(s)$ in the right half of the s-plane and the system is unstable if there are poles on the right half of a plane.
2. There is an anticlockwise encirclement (or encirclements) of the point $(-1 + j0)$. This implies that the system is stable if the number of anticlockwise encirclements are same as the number of poles of $G(s)H(s)$ in the right half of the s-plane and the system is unstable if the number of encirclements are not equal to the number of poles on the right half of the s-plane.
3. There is a clockwise encirclement (or encirclements) of the point $(-1 + j0)$. In this case, the system is always unstable.

12.3.3 Procedure for Investigating Stability Using Nyquist Stability Criterion

The following procedure can be followed to investigate the stability of closed-loop system. From the knowledge of open-loop system using Nyquist stability criterion.

1. Count the number of poles P of $G(s)H(s)$ in the right half of the s-plane, that is, with positive real parts.
2. Decide the stability criterion as $N = -P$, that is, the number of times the Nyquist plot should encircle $(1 + j0)$ point for absolute stability.
3. Choose the Nyquist contour (or Nyquist path), as shown in Figure 12.19, which encloses the entire right half of the s-plane except the singular points.

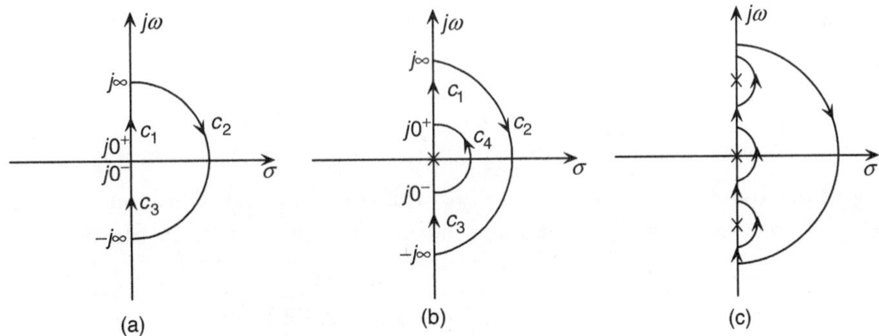

Figure 12.19 Nyquist contour

4. The Nyquist contour should be mapped section-wise in the $G(s)H(s)$-plane using the function $G(s)H(s)$ to determine the encirclement of $(-1 + j0)$ point in the $G(s)H(s)$-plane.
5. Mathematically, ω_{pc} and intersection of Nyquist plot with negative real axis by rationalising $G(j\omega)H(j\omega)$ should be determined.
6. Using the information from Steps 4 to 5, sketch the complete Nyquist plot in the $G(s)H(s)$-plane.
7. Count the number of encirclements (N) of $(-1 + j0)$ point by Nyquist plot. If this matches with the criterion decided in Step (2), the system is stable; otherwise, it is unstable.

8. The gain margin can be calculated by using the following equation:

$\therefore$
$$\text{GM} = \frac{1}{|OQ|}$$

$$\text{GM in dB} = 20\log_{10}\frac{1}{|OQ|}$$

where Q is the intersection point of Nyquist plot with negative real axis obtained in Step (5)

Problem-12.5: A unity feedback control system has $G(s) = \dfrac{10}{s(s+1)(s+2)}$. Draw the Nyquist plot and determine the closed-loop stability.

Solution:

Given that $G(s)H(s) = \dfrac{10}{s(s+1)(s+2)}$ as $H(s) = 1$.

1. Number of poles in the right half of the s-plane, $P = 0$
2. For stability $N = -P = 0$
   The Nyquist plot should not encircle $(-1 + j0)$ point for absolute stability of this system.
3. As there is one pole at origin, the Nyquist contour is as shown in figure, which contains Sections c_1, c_2, c_3 and c_4.

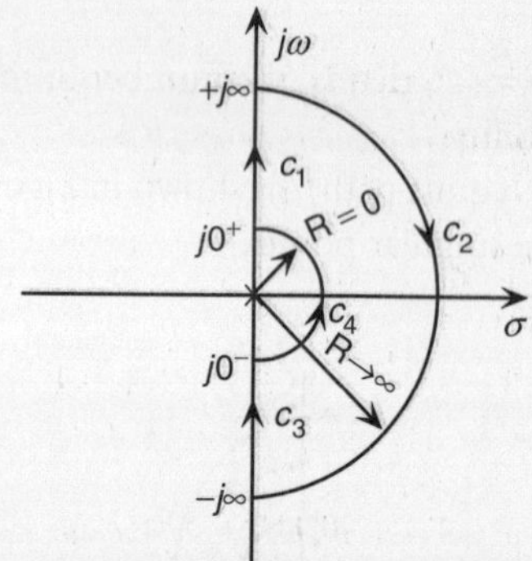

4. **Mapping of Section c_1:** In Section c_1, $\omega \to 0$ to ∞, that is, mapping of Section c_1 gives polar plot of $G(j\omega)H(j\omega)$ in $(u\text{–}v)$-plane. To get the sinusoidal t.f., substitute $s = j\omega$ in $G(s)H(s)$.

$\therefore$
$$G(j\omega)H(j\omega) = \frac{10}{j\omega(1+j\omega)(2+j\omega)}$$

$$M = |G(j\omega)H(j\omega)| = \frac{10}{\omega\sqrt{1+\omega^2}\,\sqrt{4+\omega^2}}$$

$$\phi = \underline{|G(j\omega)H(j\omega)} = -90° - \tan^{-1}\omega - \tan^{-1}\omega/2$$

$$G(j\omega)H(j\omega) = \frac{10}{j\omega(1+j\omega)(2+j\omega)}.$$

ω	M	ϕ
0	∞	–90°
∞	0	–270°

$$= \frac{10}{(-\omega^2 + j\omega)(2 + j\omega)}$$

$$= \frac{-10}{(\omega^2 - j\omega)(2 + j\omega)} \times \frac{(\omega^2 + j\omega)(2 - j\omega)}{(\omega^2 + j\omega)(2 - j\omega)}$$

$$= \frac{-10[3\omega^2 + j\,(2\omega - \omega^3)]}{(\omega^4 + \omega^2)(4 + \omega^2)}.$$

Let us assign I.P. of $G(j\omega)H(j\omega) = 0$.

i.e.,
$$\frac{-10(2\omega - \omega^3)}{(\omega^4 + \omega^2)(4 + \omega^2)} = 0$$

$$\omega^2 - 2 = 0 \Rightarrow \omega = \sqrt{2}$$

$$G(j\omega)H(j\omega) = \frac{-30 \times 2}{(4+2)(4+2)} = \frac{-60}{36} = -1.667$$

Let us assign R.P. of $G(j\omega)H(j\omega) = 0$.

i.e.,
$$\frac{-30\omega^2}{(\omega^4 + \omega^2)(4 + \omega^2)} = 0 \Rightarrow \omega = \infty$$

$$\therefore \qquad G(j\omega)H(j\omega) = 0.$$

Thus, the mapping of Section c_1 in the $(u–v)$-plane gives the following figure.

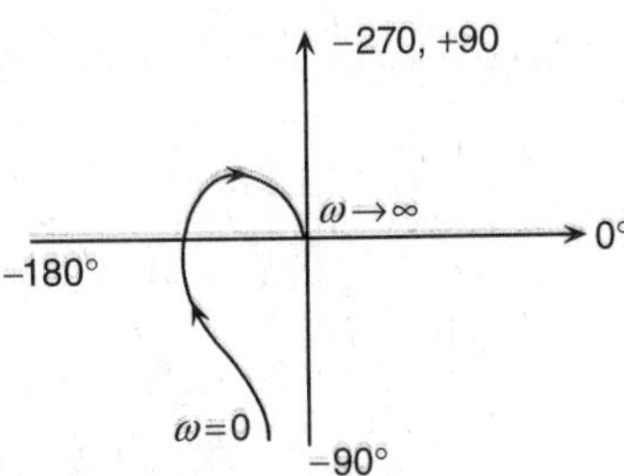

5. **Mapping of Section c_3:** In Section c_3, ω is varying from $-\infty$ to 0. The mapping of Section c_3 is given by the locus of $G(j\omega)H(j\omega)$ where ω is varying from $-\infty$ to 0. The inverse polar plot is given by the mirror image of polar plot with respect to real axis as shown in the following figure.

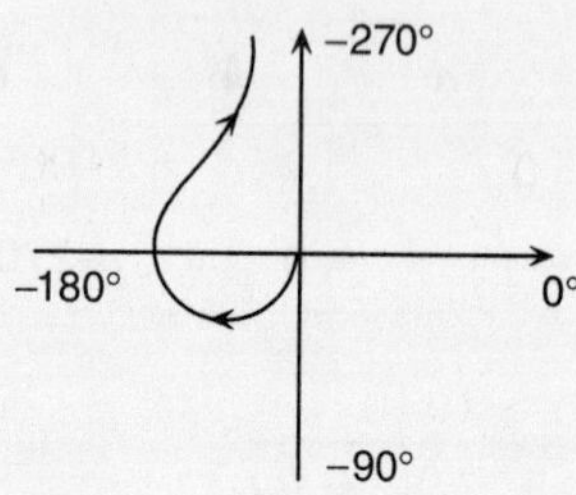

6. **Mapping of Section c_2:** The mapping of Section c_2 from *s*-plane to (*u*–*v*)-plane is obtained by assigning $s = \lim_{R\to\infty} Re^{j\theta}$ in $G(s)H(s)$ and varying θ from $+\frac{\pi}{2}$ to $-\frac{\pi}{2}$. Since $s \to Re^{j\theta}$ and $R \to \infty$, $G(s)H(s)$ can be approximated as (i.e., $1 + sT \approx sT$).

$$G(s)H(s) = \frac{10}{s(s+1)(s+2)}$$

$$= \frac{10}{2s(1+s/2)(1+s)} = \frac{10}{2\times s\times 0.5s\times s} = \frac{10}{s^3}$$

$$G(s)H(s)\big|_{s=\lim_{R\to\infty} Re^{j\theta}} = \frac{10}{\lim_{R\to\infty} R^3 e^{j3\theta}} = 0e^{-j3\theta}$$

$$= 0\angle{-270^\circ} \text{ to } 0^\circ \text{ to } +270^\circ$$

From the afore-mentioned discussion, we can say that Section c_2 in the *s*-plane is mapped as a circular arc of zero radius, that is, an origin in (*u*–*v*)-plane (or $G(s)H(s)$-plane) and not required to be analysed.

7. **Mapping of Section c_4:** The mapping of Section c_4 from *s*-plane to (*u*–*v*)-plane (i.e., $G(s)H(s)$-plane) is obtained by assigning $s = \lim_{R\to 0} Re^{j\theta}$ in $G(s)H(s)$ and varying θ from $-\frac{\pi}{2}$ to $+\frac{\pi}{2}$. Since $s \to Re^{j\theta}$ and $R\to 0$, $G(s)H(s)$ can be approximated as [i.e., $(1+sT) \approx 1$]

$$G(s)H(s) = \frac{5}{s(1+0.5)(1+s)} = \frac{5}{s\times 1\times 1} = \frac{5}{s}$$

$$\therefore \qquad |G(s)H(s)|_{s=\lim_{R\to 0} Re^{j\theta}} = \frac{5}{\lim_{R\to 0} Re^{j\theta}} = \infty e^{-j\theta}$$

$$= \infty\angle{+90^\circ} \text{ to } \infty\angle{-90^\circ}$$

$$= \infty\angle{+90^\circ} \text{ to } 0^\circ \text{ to} \angle{-90^\circ}$$

From the afore-mentioned discussion, we can say that Section c_4 in the *s*-plane is mapped as a circle of infinite radius with argument varying from +90° to −90° as

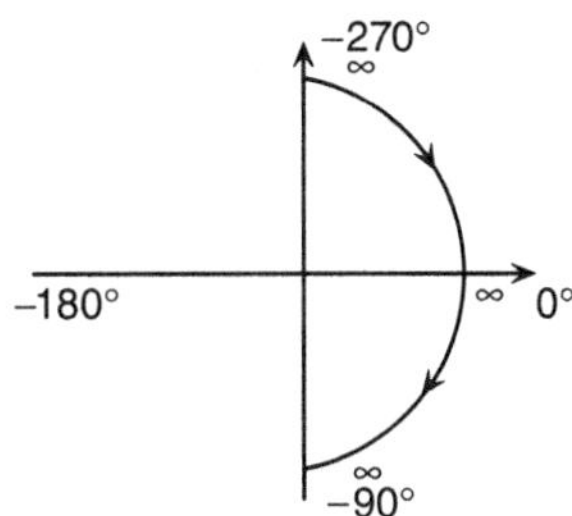

8. **Complete Nyquist Plot:** The complete Nyquist plot in $G(s)H(s)$-plane can be obtained by combining the mappings of individual sections as

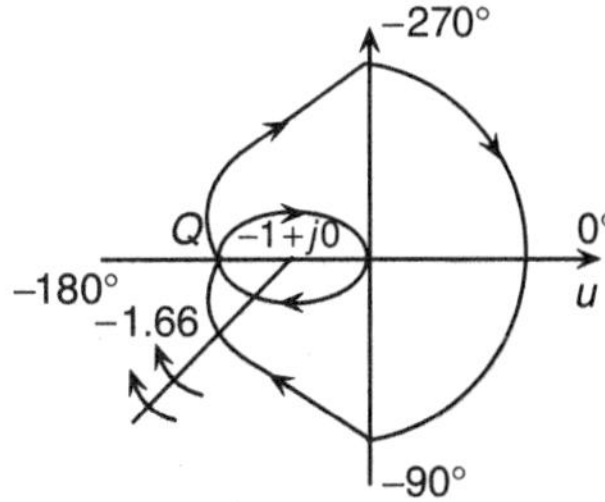

9. The number of encirclements of $(-1+j0)$ are

$$N=+2.$$

However, for stability, $N=0$, the given system is unstable.
According to mapping theorem,

$$N=Z-P$$

$$2=Z-0 \Rightarrow Z=2.$$

Actually, there are 2 zeros of $1 + G(s)H(s)$ encircled by Nyquist path, that is, two closed-loop poles are there in the right half of the s-plane due to which the closed-loop system is unstable.

Note:
It should be noted that for all the systems with

$$G(s)H(s)=\frac{K}{s(1+sT_1)(1+sT_2)}.$$

The shape of the Nyquist plot will remain same as shown in this example. Only the coordinates of point Q will change depending on the values of K, T_1 and T_2, which will decide the encirclement of $(-1+j0)$ point and also the stability.

Problem-12.6: By Nyquist stability criterion, determine the stability of closed-loop system, whose open-loop t.f. is given as $G(s)H(s)=\dfrac{K}{s(s+1)(s-1)}$. Further, determine the stability of open-loop and closed-loop systems.

Solution:

Given that $G(s)H(s) = \dfrac{s+2}{(s+1)(s-1)}$.

1. Number of poles in the right half of the s-plane, $P = 1$.
2. For stability, $N = -P = -1$.
3. As there is no pole at origin, the Nyquist contour is as shown in the figure, which contains Sections c_1, c_2 and c_3.

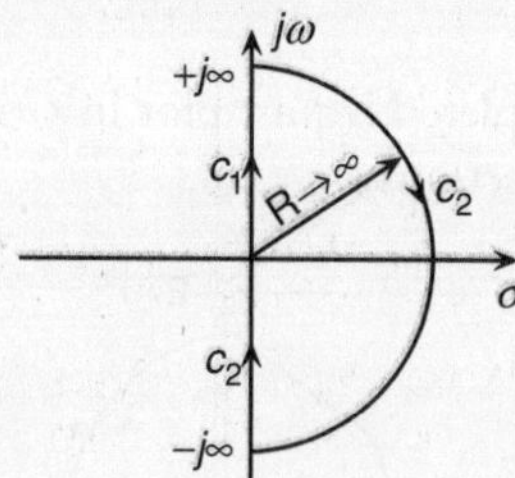

4. **Mapping of Section c_1:** In Section c_1, $\omega \to 0$ to ∞, that is, the mapping of Section c_1 gives the polar plot of $G(j\omega)H(j\omega)$ in (u–v)-plane.
   To get sinusoidal t.f., assign $s = j\omega$ in $G(s)H(s)$.

$$G(j\omega)H(j\omega) = \frac{2+j\omega}{(1+j\omega)(-1+j\omega)}$$

$$M = |G(j\omega)H(j\omega)| = \frac{\sqrt{4+\omega^2}}{\sqrt{1+\omega^2}\,\sqrt{1+\omega^2}}$$

$$\begin{aligned}\phi = \angle G(j\omega)H(j\omega) &= \tan^{-1}(\omega/2) - \tan^{-1}\omega - \tan^{-1}(-\omega)\\ &= \tan^{-1}(0.5\omega) - \tan^{-1}\omega - 180^\circ + \tan^{-1}\omega\\ &= -180^\circ + \tan^{-1}0.5\omega\end{aligned}$$

ω	M	ϕ
0	–2	–180°
∞	0	–90°

$$G(j\omega)H(j\omega) = \frac{-2(1+j0.5\omega)}{(1+\omega^2)}$$

$$= \frac{-2}{1+\omega^2} - \frac{j\omega}{1+\omega^2}$$

Let us assign I.P. of $G(j\omega)H(j\omega) = 0$.

i.e., $$\frac{\omega}{1+\omega^2} = 0 \Rightarrow \omega = 0.$$

$$\therefore \qquad G(j\omega)H(j\omega) = \frac{-2}{1+\omega^2} = -2$$

Let us assign R.P. of $G(j\omega)H(j\omega) = 0$.

i.e.,
$$\frac{-2}{1+\omega^2} = 0 \Rightarrow \omega = 0$$

$$\therefore \qquad G(j\omega)H(j\omega) = 0.$$

Thus, the mapping of Section c_1 in $(u\text{–}v)$-plane is given in the following figure.

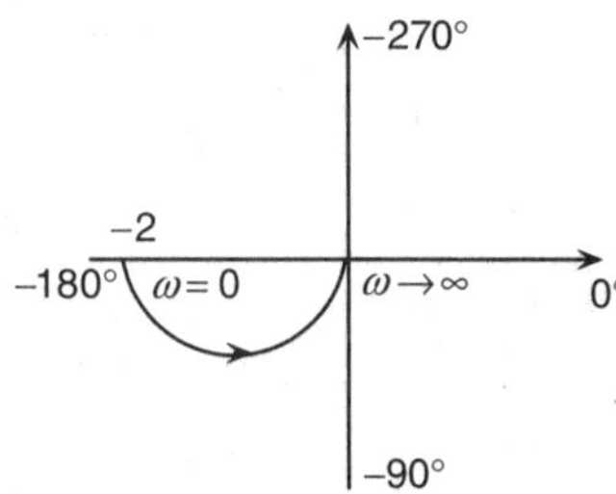

5. **Mapping of Section c_2:** The mapping of Section c_2 from s-plane to $G(s)H(s)$-plane is obtained by assigning $s = \lim_{R\to\infty} Re^{j\theta}$ in $G(s)H(s)$ and varying θ from $+\frac{\pi}{2}$ to $-\frac{\pi}{2}$. Since $s \to Re^{j\theta}$ and $R\to$, $1+sT \approx sT$

$$G(s)H(s) = \frac{2(1+0.5s)}{(1+s)(-1+s)}$$

$$= \frac{2\times 0.5s}{s\times s} = \frac{1}{s}$$

$$G(s)H(s)\Big|_{s=\lim_{R\to\infty} Re^{j\theta}} = \frac{s}{\lim_{R\to\infty} Re^{j\theta}} = 0e^{-j\theta}$$

$$= 0\angle{-90°} \text{ to } 0\angle{+90°}$$

or

$$= 0\angle{-90°} \text{ to } 0° \text{ to } +90°$$

From the afore-mentioned discussion, we can say that section c_2 in the s-plane is mapped into a circular arc of zero radius (i.e., a point or origin) in the $(u\text{–}v)$-plane and are not considered.

6. **Mapping of Section c_3:** In Section c_3, $\omega\to\infty$ to 0, that is, mapping of Section c_3 gives the inverse polar plot of $G(j\omega)H(j\omega)$, as shown in the following figure.

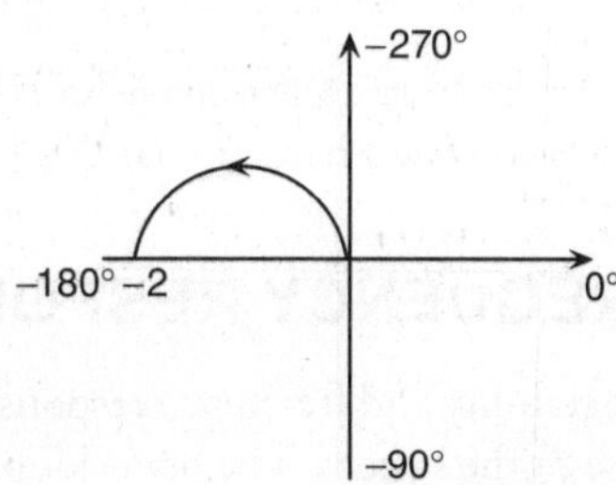

7. **Complete Nyquist plot:** The complete Nyquist plot in $G(s)H(s)$ or $(u\text{–}v)$-plane can be obtained by combining the mappings of individual sections as follows:

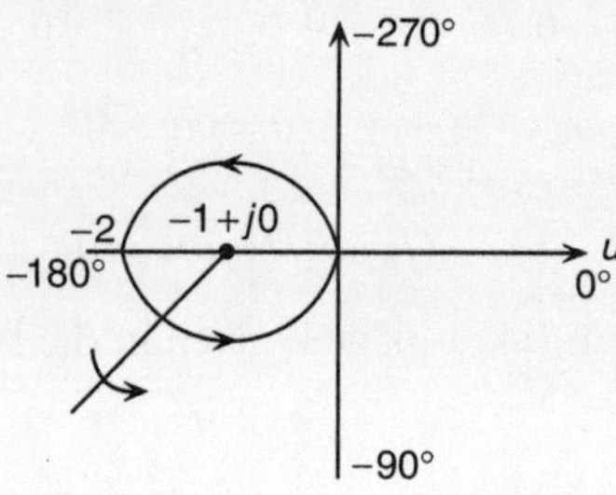

8. The number of encirclements of $(-1+j0)$ are $N=-1$ (counter clockwise encirclement). Since the number of anticlockwise encirclements is equal to the number of open-loop poles on the right half of the s-plane, the closed-loop system is stable.
   In this problem, it is clear that even though the open-loop system is unstable, the closed-loop system is stable. Such a case is a rare example.

12.3.4 Advantages of Nyquist Plot

The following are the advantages of the Nyquist plot

1. It is useful for determining the stability of the closed-loop system from open-loop t.f. without knowing the roots of the C.E.
2. It gives the same information about absolute stability as provided by Routh's criterion.
3. It also indicates the relative stability and gives the values of GM and PM.
4. It is very useful for analysing conditionally stable systems.

12.3.5 Effect of Addition of Poles and Zeros to Nyquist Plot

1. **Addition of poles:** The addition of pole at $s=0$ to a loop t.f. will affect the stability of the closed system adversely. A system that has a loop t.f. with more than one pole at $s=0$ (type 2 or higher) is likely to be unstable or difficult to stabilise.
   Adding a pole at $s=0$ is equivalent to dividing $G(s)H(s)$ by $(j\omega)$. The phase of $G(j\omega)H(j\omega)$ is reduced by 90° at both zero and infinite frequencies. In addition, the magnitude of $G(j\omega)H(j\omega)$ at $\omega=0$ becomes infinite. The effect of adding a finite pole at $s=\frac{-1}{T}$ to the t.f. shifts the phase of the Nyquist plot −90° at $\omega=\infty$. This means it is rotated clockwise by another 90° from that of the original one.
2. **Additions of Zeros:** Adding zero to the loop t.f. will have an effect of reducing the overshoot and general effect of stabilisation. It increases the phase of $G(s)H(s)$ by 90° at $\omega=\infty$.

12.4 CLOSED-LOOP FREQUENCY RESPONSE

We have already seen that closed-loop stability and frequency response specifications can be obtained by plotting open-loop frequency response of the system. The open-loop frequency response is the graph of the magnitude of $G(j\omega)H(j\omega)$ and the phase angle of $G(j\omega)H(j\omega)$ against frequency ω. Methods such as Bode plot, polar plot and Nyquist plot have been already discussed to obtain open-loop frequency response.

For a unity feedback closed-loop stable system, the closed-loop frequency can be obtained from the open-loop frequency response. The closed-loop t.f. of a unity feedback [$H(s) = 1$] system is given as

$$T(s) = \frac{C(s)}{R(s)} = \frac{G(s)}{1+G(s)}$$

$$\therefore \qquad T(j\omega) = \frac{C(j\omega)}{R(j\omega)} = \frac{G(j\omega)}{1+G(j\omega)} = M\angle\alpha$$

where

M is the magnitude of closed-loop t.f. and α is the phase angle of closed-loop t.f.

The magnitude and phase of closed-loop system are functions of frequency ω. The sketch of magnitude and phase of closed-loop system with respect to ω yields closed-loop frequency response plot. The magnitude and phase of closed-loop system for various values of frequency can be evaluated analytically or graphically. The analytical method of determining the frequency response involves tedious calculations. Two graphical methods are available to determine the closed-loop frequency response from open-loop frequency response. They are M and N circles and Nichols chart.

12.4.1 *M* and *N* Circles

The magnitude of closed-loop t.f. with unity feedback system can be shown to be in the form of circle for every value of M. These circles are called M circles.

If the phase of closed-loop t.f. with unity feedback is α, then it can be shown that tan α will be in the form of circle for every value of α. These circles are called N circles.

The M and N circles are used to find the closed-loop frequency graphically from the open-loop frequency response $G(j\omega)H(j\omega)$ without calculating the magnitude and phase of the closed-loop t.f. at each frequency.

1. ***M* circles (Constant Magnitude Loci)**

   The closed-loop t.f. of a unity feedback system is

$$T(s) = \frac{C(s)}{R(s)} = \frac{G(s)}{1+G(s)}.$$

   Let $\qquad G(j\omega) = X + jY.$

$$T(j\omega) = \frac{G(j\omega)}{1+G(j\omega)} = \frac{X+jY}{1+X+jY}.$$

   Let M be magnitude of $T(j\omega)$, so we can write

$$M = \frac{|X+jY|}{|1+X+jY|} = \frac{\sqrt{X^2+Y^2}}{\sqrt{(1+X)^2+Y^2}}$$

$$M^2 = \frac{X^2+Y^2}{(1+X)^2+Y^2}.$$

$$M^2[(1+X)^2+Y^2]=X^2+Y^2$$

$$M^2[(X^2+2X+1+Y^2]=X^2+Y^2$$

$$X^2(1-M^2)+2XM^2+1+Y^2(M^2-1)+M^2=0$$

$$X^2+\frac{2XM^2}{M^2-1}+Y^2+\frac{M^2}{M^2-1}=0$$

$$X^2-\frac{2M^2X}{1-M^2}+Y^2=\frac{M^2}{1-M^2}$$

$$X^2-\frac{2M^2}{1-M}X+\left[\frac{M^2}{1-M^2}\right]+Y^2=\frac{M^2}{1-M^2}+\left[\frac{M^2}{1-M^2}\right]^2$$

$$\left(X-\frac{M^2}{1-M^2}\right)+Y^2=\frac{M^2}{1-M^2}\left(1+\frac{M^2}{1-M^2}\right)$$

$$\left(X-\frac{M^2}{1-M^2}\right)^2+Y^2=\left[\frac{M}{(1-M^2)}\right]^2 \tag{12.13}$$

Equation (12.13) represents the equation of a circle with the centre at $X=\frac{M^2}{1-M^2}$, $Y=0$ having radius $r=\frac{M}{1-M^2}$.

For a particular value of M, a circle will be obtained. The magnitude of a closed-loop t.f. is constant for that particular value of M, and for various values of M, a family of such circles are formed. Hence, these are called constant magnitude loci or M circles. The following table gives the centre and radius of such circles for various values of M.

M	Centre $X=\frac{M^2}{1-M^2}, Y=0$	Radius $r=\left(\frac{M}{1-M^2}\right)$
0.2	(0.041,0)	0.208
0.5	(0.33,0)	0.667
1	∞	∞
2	(–1.33,0)	0.667
5	(–1.041,0)	0.208

From the above table, we can conclude that

1. If M becomes larger and larger when compared to 1, the radius becomes smaller and smaller, ultimately circles converge to the critical point $(-1+j0)$.
2. If $M>1$, the centres are to the left of point $(-1+j0)$.

3. If M becomes smaller and smaller when compared to 1, the radius becomes smaller and smaller, ultimately circles converge to the origin.
4. If $M < 1$, the centres are to the right of the origin.
5. If $M = 1$, the locus is of the points that are equidistant from the origin as well as from point $(-1 + j0)$.

Figure 12.20 shows the family of constant M circles.

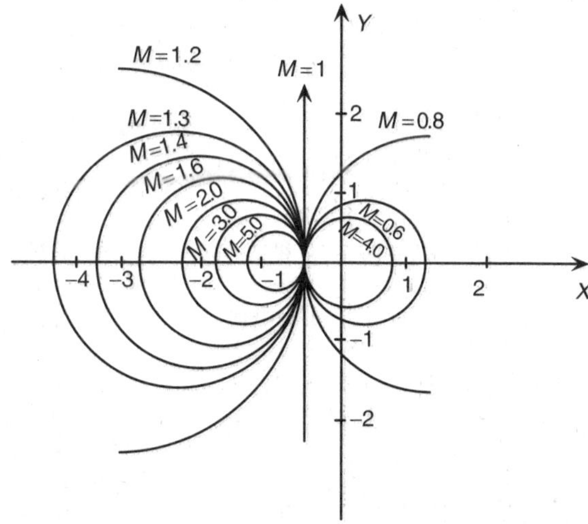

Figure 12.20 A family of constant M circles

1. ***N* circles (constant phase loci)**
   Similar to the magnitude M of $T(j\omega)$, let us obtain the phase angle α in terms of X and Y.

$$\angle T(j\omega) = \angle \frac{G(j\omega)}{1 + G(j\omega)} = \angle \frac{X + jY}{1 + X + jY}$$

$$\alpha = \tan^{-1}\frac{Y}{X} - \tan^{-1}\frac{Y}{1+X}$$

$$\tan\alpha = N = \tan\left[\tan^{-1}\left(\frac{Y}{X}\right) - \tan^{-1}\left(\frac{Y}{1+X}\right)\right]$$

$$= \tan\tan^{-1}\left[\frac{\dfrac{Y}{X} - \dfrac{Y}{1+X}}{1 + \dfrac{Y^2}{X(1+X)}}\right]$$

$$= \frac{Y(1+X) - YX}{X(1+X) + Y^2}$$

$$N = \frac{Y}{X^2 + Y^2 + X}$$

$$X^2+Y^2+X-\frac{Y}{N}=0$$

$$X^2+X+\frac{1}{4}+Y^2-\frac{Y}{N}+\frac{1}{(2N)^2}=\frac{1}{4}+\frac{1}{(2N)^2}$$

$$\left(X+\frac{1}{2}\right)^2+\left(Y-\frac{1}{2N}\right)^2=\frac{1}{4}+\frac{1}{4N^2}. \tag{12.14}$$

Equation (12.14) represents the equation of a circle with the centre at $X=\frac{-1}{2}$ and $Y=\frac{1}{2N}$ with radius $r=\sqrt{\frac{1}{4}+\frac{1}{(2N)^2}}$.

For different values of N, a family of circles is formed. On a particular circle, the value of N, that is, phase angle α, remains constant; hence, these circles are called constant phase loci or N circles. The following table gives the centre and radius of different circles for different values of α, that is, N.

α	$N=\tan\alpha$	**Centre** $X=-\frac{1}{2}, Y=\frac{1}{2N}$	**Radius** $r=\sqrt{\frac{1}{4}+\frac{1}{4N^2}}$
–90°	∞	$\left(\frac{1}{2},0\right)$	0.5
–60°	–1.732	$\left(\frac{-1}{2},0.288\right)$	0.577
–30°	–0.5773	$\left(\frac{-1}{2},-0.866\right)$	1.0
0°	0	$\left(\frac{-1}{2},\infty\right)$	∞
30°	0.5773	$\left(\frac{-1}{2},0.866\right)$	1.0
60°	1.732	$\left(\frac{-1}{2},0.88\right)$	0.577
90°	∞	$\left(\frac{-1}{2},0\right)$	0.5

Equation (12.14) is satisfied for $X=0$, $Y=0$ as well as $X=1$, $Y=0$, irrespective of the value of N. We conclude that each circle represented by (12.14) passes through points (0,0) and $(-1+j0)$. Figure 12.21 shows the family of constant N circles.

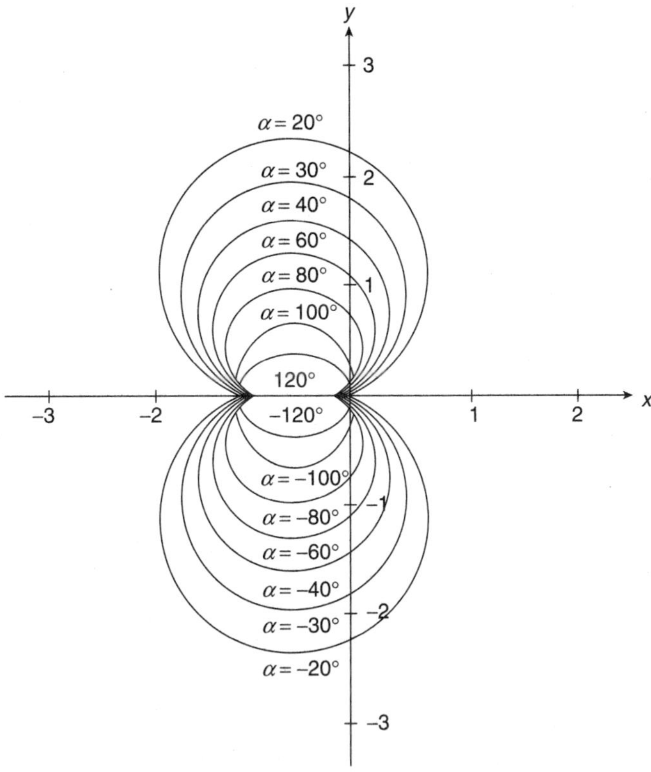

Figure 12.21 A family of constant *N* circles

12.4.2 Nichol's Chart

N-.B. Nichols transformed the constant *M* and *N* circles to log-magnitude and phase angle coordinates and the resulting chart is known as Nichol's chart.

The frequency response can also be obtained by sketching the magnitude in decibels against phase angle for various frequencies. The plot obtained is called gain-phase plot or log-magnitude versus phase plot. Figure 12.22 shows such a gain-phase plot for $G(j\omega)=\dfrac{1}{1+j\omega T}$, that is, $G(s)=\dfrac{1}{1+sT}$.

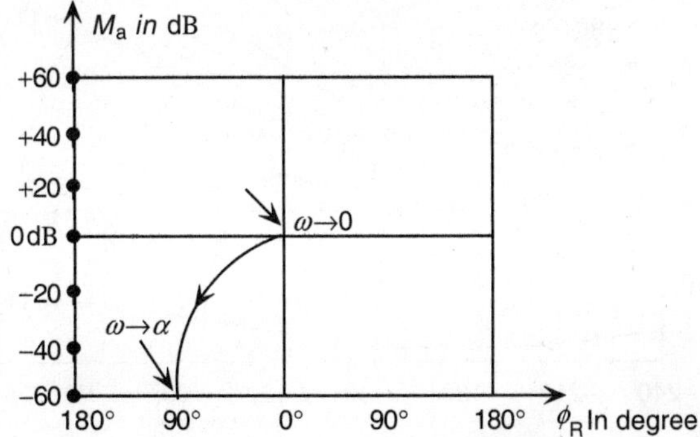

Figure 12.22 Gain-phase plot

If, now, the constant magnitude loci (*M* circles) and constant phase angle loci (*N* circles) are transferred to the gain-phase plot, the resultant chart is called Nichol's chart.

If we superimpose the gain-phase plot of an open-loop t.f. on Nichol's chart, we get very easily the closed-loop frequency response. The magnitude is expressed in decibels, while the phase angle is in degrees. Nichol's chart gives the points of intersection of the gain-phase plot of an open-loop t.f., which are symmetrical about the −180° axis. The *M* and *N* loci are repeated for every 360°, and there is symmetry at every 180° interval. The *M* loci are centred about the critical point (0 dB, −180°)

Using Nichol's chart, the closed-loop frequency response can be determined graphically from the locus of the open-loop system. When the Nichol's plot of $G(j\omega)$ is sketched on Nichol's chart, the locus of $G(j\omega)$ will cut the *M* and *N* contours at various points. The cutting point of the locus of $G(j\omega)$ with the *M* contour gives the magnitude of closed-loop system corresponding to a frequency same as that of $G(j\omega)$ at that point. The cutting point of locus of $G(j\omega)$ and *N* contour gives the phase of closed loop corresponding to a frequency same as that of $G(j\omega)$ at that point. The magnitude *M* and phase angle $\alpha(N = \tan \alpha)$ of closed-loop system are tabulated. The closed-loop frequency response can be plotted on a semi-log graph sheet using the tabulated values. The closed-loop frequency response consists of two plots, namely magnitude versus ω and phase angle versus ω. Figure 12.23 shows Nichol's chart.

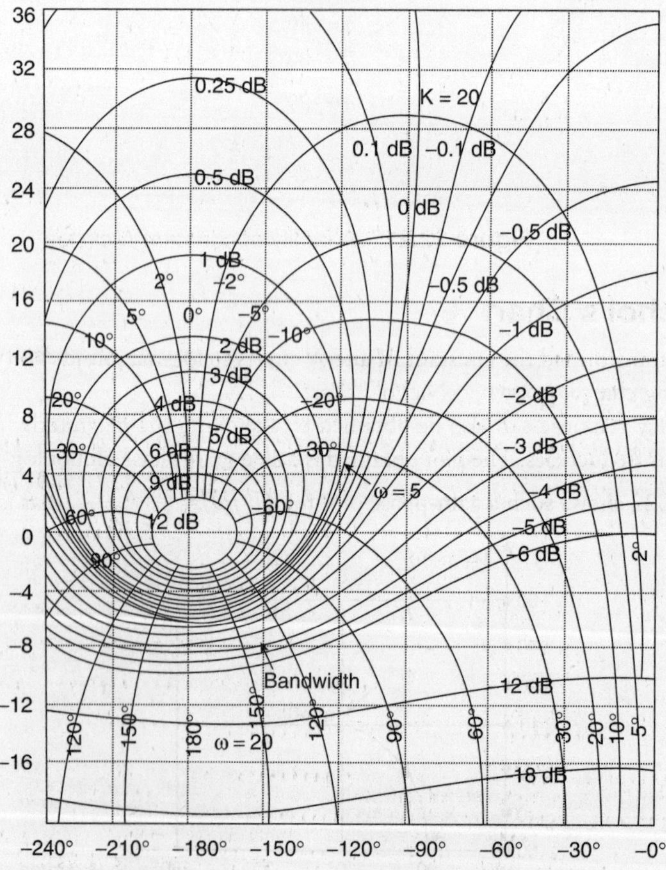

Figure 12.23 Nichol's chart

Nichol's chart is used for determining:

1. The complete closed-loop frequency response.
2. The value of resonant peak M_r of closed-loop system with given value of $G(j\omega)$.
3. The frequency ω_r corresponding to M_r for the closed-loop system.
4. Other frequency and time domain specifications if M_r and ω_r are known.
5. The 3-dB bandwidth of the closed-loop system.
6. The value of K for the given value of M_r.

ADDITIONAL SOLVED PROBLEMS

ASP-1: Sketch the polar plot for the t.f.

$$G(s) = \frac{12}{s^2(s+1)(s+2)}.$$

Solution:

Let the t.f. of the system is given as

$$G(s) = \frac{12}{s^2(1+s)(2+s)}.$$

To get sinusoidal t.f., substitute $s = j\omega$ in $G(s)$

$$G(j\omega) = \frac{12}{(j\omega)^2(1+j\omega)(2+j\omega)}$$

$$M = |G(j\omega)| = \frac{12}{\omega^2\sqrt{1+\omega^2}\sqrt{4+\omega^2}}$$

$$\phi = \angle G(j\omega) = 180 - \tan^{-1}\omega - \tan^{-1}\omega/2$$

ω	**M**	ϕ
0	∞	–180°
∞	0	–360°

Next, let us determine whether $G(j\omega)$ intersects any of the axis of $G(j\omega)$-plane.

$$G(j\omega) = \frac{-12}{\omega^2(1+j\omega)(2+j\omega)}$$

$$= \frac{-12(1-j\omega)(2-j\omega)}{\omega^2(1+j\omega)(2+j\omega)(1-j\omega)(2-j\omega)}$$

$$= \frac{-12(2-j\omega-j2\omega-\omega^2)}{\omega^2(1+\omega^2)(4+\omega^2)}$$

$$= \frac{-12(2-\omega^2)}{\omega^2(1+\omega^2)(4+\omega^2)} + \frac{12\times j3\omega}{\omega^2(1+\omega^2)(4+\omega^2)}$$

Let us assign I.P. of $G(j\omega) = 0$.

$$\Rightarrow \qquad \frac{36\omega}{\omega^2(1+\omega^2)(4+\omega^2)} = 0 \Rightarrow \omega = \infty$$

Let us assign R.P. of $G(j\omega) = 0$

$$\Rightarrow \qquad \frac{12(2-\omega^2)}{\omega^2(1+\omega^2)(4+\omega^2)} = 0 \Rightarrow 2-\omega^2 = 0$$

$$\Rightarrow \qquad \omega = \sqrt{2}$$

$$G(j\sqrt{2}) = \frac{12}{(j\sqrt{2})^2(1+j\sqrt{2})(2+j\sqrt{2})}$$

$$= \frac{-12}{2(-2+j2\sqrt{2}+j\sqrt{2}+2)}$$

$$= \frac{-6}{j3\sqrt{2}} = j\sqrt{2}$$

The polar plot is as shown in the following figure.

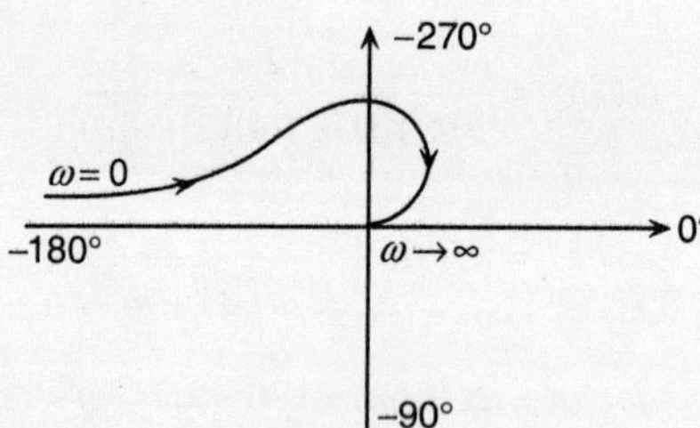

ASP-2: Draw the polar plot for the t.f. $G(s) = \frac{1}{(1+s)^3}$

Solution:

The t.f. of the system is given as

$$G(s) = \frac{1}{(1+s)^3} = \frac{1}{(1+s)(1+s)(1+s)}.$$

To get sinusoidal t.f., substitute $s = j\omega$ in $G(s)$.

$$G(j\omega) = \frac{1}{(1+j\omega)(1+j\omega)(1+j\omega)}$$

$$M = |G(j\omega)| = \frac{1}{\sqrt{1+\omega^2}\sqrt{1+\omega^2}\sqrt{1+\omega^2}}$$

$$\phi = \angle G(j\omega) = -\tan^{-1}\omega - \tan^{-1}\omega - \tan^{-1}\omega = -3\tan^{-1}\omega$$

ω	M	ϕ
0	1	0
∞	0	–270°

Next, let us determine whether $G(j\omega)$ intersects any of the axis of $G(j\omega)$-plane.

$$G(j\omega) = \frac{1}{(1+j\omega)^3}$$

$$= \frac{(1-j\omega)^3}{(1+j\omega)^3(1-j\omega)^3}$$

$$= \frac{1+j\omega^3 - j3\omega - 3\omega^2}{(1+\omega^2)^3}$$

$$= \frac{1-3\omega^2}{(1+\omega^2)^3} + \frac{j\omega(\omega^2-3)}{(1+\omega^2)^3}$$

Let us assign I.P. of $G(j\omega) = 0$

$$\Rightarrow \qquad \frac{\omega\left(\omega^2-3\right)}{\left(1+\omega^2\right)^3} = 0 \Rightarrow \omega\,(\omega^2 - 3) = 0$$

$$\Rightarrow \qquad \omega = 0,\ \omega = \sqrt{3}$$

$$G(j0) = \frac{1}{1\times1\times1} = 1$$

$$G(j\sqrt{3}) = \frac{1}{\left(1+j\sqrt{3}\right)^3} = \frac{1}{1-j3\sqrt{3}+j3\sqrt{3}-9} = \frac{-1}{8} = -0.125.$$

Let us assign R.P. of $G(j\omega) = 0$

$$\Rightarrow \qquad \frac{1-3\omega^2}{\left(1+\omega^2\right)3} = 0 \Rightarrow 1 - 3\omega^2 = 0$$

$$\Rightarrow \qquad \omega = \frac{1}{\sqrt{3}}$$

$$G(j/\sqrt{3}) = \left(\frac{1}{1+j/\sqrt{3}}\right)^3 = \frac{3\sqrt{3}}{(j+\sqrt{3})^3}$$

$$= \frac{3\sqrt{3}}{-j+3\sqrt{3}-3\sqrt{3}+j9} = \frac{3\sqrt{3}}{j8} = \frac{-j3\sqrt{3}}{8} = -j0.649.$$

The polar plot is as shown in the following figure.

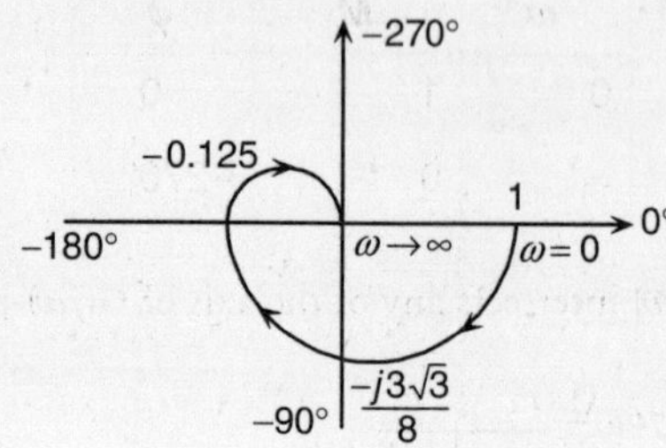

ASP-3: By Nyquist stability criterion, determine the stability of closed-loop system, when open-loop t.f. is given as $G(s)H(s) = \dfrac{3}{(s+1)(s+2)}$.

Solution:

Given that $G(s)H(s) = \dfrac{3}{(s+1)(s+2)}$

1. Number of poles in the right half of the s-plane, $P = 0$.
2. For stability, $N = -P = 0$
3. As there is no pole at the origin, the Nyquist contour is as shown in the figure, which contains Sections c_1, c_2 and c_3.

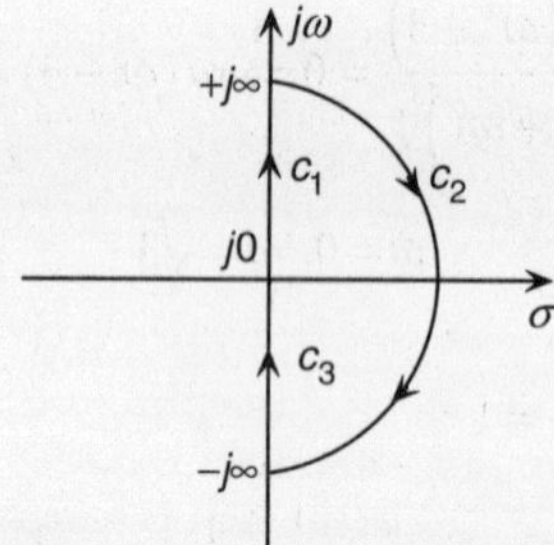

4. Mapping of Section c_1: In Section c_1, $\omega \to 0$ to ∞, that is, the mapping of Section c_1 gives polar plot of $G(j\omega)H(j\omega)$ in $(u\text{–}v)$-plane.

To get the sinusoidal t.f., substitute $s = j\omega$ in $G(s)H(s)$.

$$G(j\omega)H(j\omega) = \frac{3}{(1+j\omega)(2+j\omega)}$$

$$M = |G(j\omega)H(j\omega)| = \frac{3}{\sqrt{1+\omega^2}\sqrt{4+\omega^2}}$$

$$\phi = \angle G(j\omega)H(j\omega) = -\tan^{-1}\omega - \tan^{-1}\omega/2$$

ω	M	ϕ
0	1.5	0
∞	0	−180°

$$G(j\omega)H(j\omega) = \frac{3}{(1+j\omega)(2+j\omega)} \frac{(1-j\omega)(2-j\omega)}{(1-j\omega)(2-j\omega)}$$

$$= \frac{3[2-\omega^2 + j(-2\omega-\omega)]}{(1+\omega^2)(4+\omega^2)}$$

$$= \frac{3(2-\omega^2)}{(1+\omega^2)(4+\omega^2)} - j\frac{3\omega\times 3}{(1+\omega^2)(4+\omega^2)}.$$

Let us assign I.P. of $G(j\omega)H(j\omega) = 0$

i.e.,
$$\frac{9\omega}{(1+\omega^2)(4+\omega^2)} = 0 \Rightarrow \omega = 0$$

$\therefore$
$$G(j\omega)H(j\omega) = \frac{3\times 2}{1\times 4} = 1.5$$

Let us assign R.P. of $G(j\omega)H(j\omega) = 0$

i.e.,
$$\frac{3(2-\omega^2)}{(1+\omega^2)(4+\omega^2)} = 0 \Rightarrow \omega^2 = 2$$

$\Rightarrow$
$$\omega = \sqrt{2}$$

$\therefore$
$$G(j\omega)H(j\omega) = -j\times\frac{3\sqrt{2}\times 3}{3\times 6} = -\frac{j}{\sqrt{2}} = -j0.707.$$

Thus, the mapping of Section c_1 in $(u\text{–}v)$-plane gives the following figure.

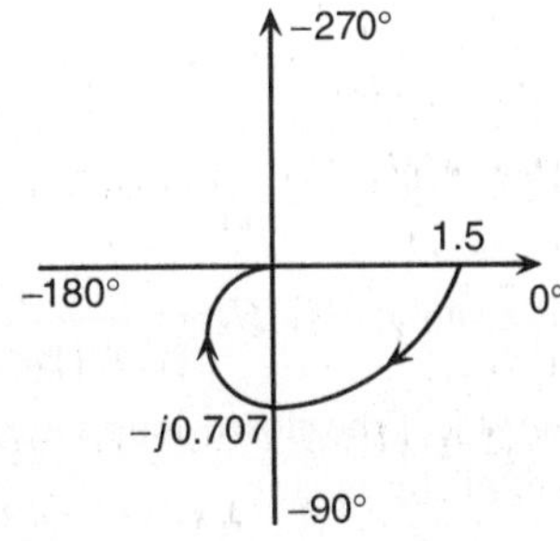

5. **Mapping of Section c_2:** The mapping of Section c_2 from s-plane to $(u\text{–}v)$-plane is obtained by substituting $s = \lim_{R\to\infty} Re^{j\theta}$ in $G(s)H(s)$ and varying θ from +90° to –90°. Since $s \to Re^{j\theta}$ and $R\to\infty$, $1 + sT \approx sT$

$$G(s)G(s) = \frac{3}{(s+1)(s+2)} = \frac{1.5}{(1+s)(1+0.55)} = \frac{1.5}{s\times 0.55} = \frac{3}{s^2}$$

$$G(s)H(s)\big|_{s=\lim_{R\to\infty} Re^{j\theta}} = \frac{3}{\lim_{R\to\infty} R^2 e^{j2\theta}} = 0e^{-j2\theta}$$

$$= 0\angle -180° \text{ to } 0° \text{ to } +180°.$$

From the afore-mentioned discussion, we can say that Section c_2 in s-plane is mapped into a circular arc of zero radius (i.e., a point) in the $(u–v)$-plane and not required to be analysed.

6. **Mapping of Section c_3:** In Section c_3, $\omega \to -\infty$ to 0, that is, mapping of Section c_3 gives the inverse polar plot of $G(j\omega)H(j\omega)$, as shown in the following figure.

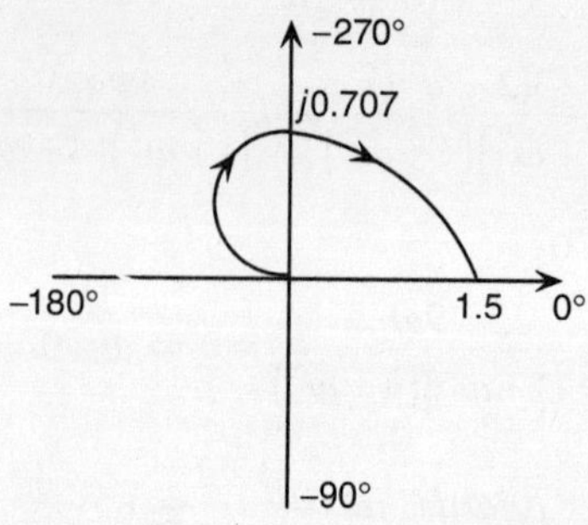

7. **Complete Nyquist plot:** The complete Nyquist plot in $G(s)H(s)$ or $(u–v)$-plane can be obtained by combining the mapping of individual sections as shown in the following figure.

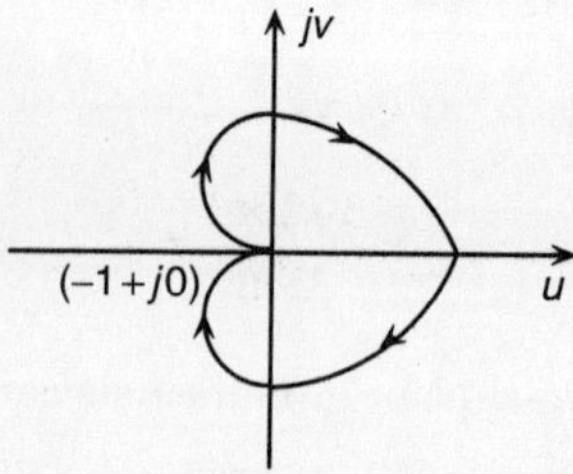

8. The Nyquist plot does not encircle the critical point $(-1 + j0)$ (i.e., $N = 0$). Since the given t.f. does not contain any poles in the right half of the s-plane, the closed-loop system is stable.

ASP-4: The open-loop t.f. of a system is given as $G(s)H(s) = \dfrac{1+4s}{s^2(1+s)(1+2s)}$. Determine the stability of closed-loop system by using Nyquist criterion. If the closed-loop system is not stable, then find the number of closed-loop poles lying on the right half of the s-plane.

Solution:

Given that $G(s)H(s) = \dfrac{1+4s}{s^2(1+s)(1+2s)}$.

1. Number of poles in the right half of the s-plane, $P = 0$.
2. For stability, $N = -P = 0$.
3. As there are two poles at origin, the Nyquist contour is as shown in the figure, which contains Sections c_1, c_2, c_3 and c_4.

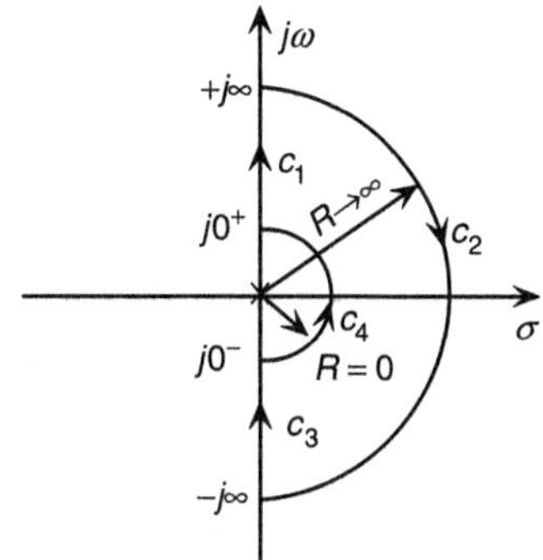

4. **Mapping of Section c_1:** In Section c_1, $\omega \to 0$ to ∞, that is, the mapping of Section c_1 gives the polar plot of $G(j\omega)H(j\omega)$ in $(u\text{–}v)$-plane.
To get sinusoidal t.f., substitute $s = j\omega$ in $G(s)H(s)$.

$$G(j\omega)H(j\omega) = \frac{1 + j4\omega}{(j\omega)^2(1 + j\omega)(1 + j2\omega)}.$$

$$M = |G(j\omega)H(j\omega)| = \frac{\sqrt{1+16\omega^2}}{\omega^2\sqrt{(1+\omega^2)(1+4\omega^2)}}$$

$$\phi = \underline{|G(j\omega)H(j\omega)} = -180° - \tan^{-1}\omega - \tan^{-1}2\omega + \tan^{-1}4\omega$$

ω	**M**	ϕ
0	1.5	–180°
∞	0	–270°

$$G(j\omega)H(j\omega) = \frac{1 + j4\omega}{-\omega^2(1 - 2\omega^2 + 3j\omega)}$$

$$= \frac{-(1 + j4\omega)}{\omega^2(1 - 2\omega^2 + j3\omega)} \times \frac{(1 - 2\omega^2 - j3\omega)}{(1 - 2\omega^2 - j3\omega)}$$

$$= \frac{-[1 + 10\omega^2 + j(\omega - 8\omega^3)]}{\omega^2[(1 - 2\omega^2)^2 + 9\omega^2]}$$

$$= \frac{-(1 + 10\omega^2)}{\omega^2[(1 - 2\omega^2) + 9\omega^2]} - \frac{j(\omega - 8\omega^3)}{\omega^2[(1 - 2\omega^2) + 9\omega^2]}$$

Let us assign I.P. of $G(j\omega)H(j\omega) = 0$

i.e.,
$$\frac{\omega - 8\omega^3}{\omega^2[(1 - 2\omega^2) + 9\omega^2]} = 0$$

$\Rightarrow$
$$\omega(1 - 8\omega^2) = 0$$

$$\omega = 0,\ \omega = \frac{1}{2\sqrt{2}}.$$

$$\therefore \qquad G(j\omega)H(j\omega) = -\frac{\left(1+10\times\frac{1}{8}\right)}{\frac{1}{8}\left[\left(1-2\times\frac{1}{8}\right)^2+\frac{9}{8}\right]} = -10.64.$$

Thus, the mapping of Section c_1 in $(u\text{–}v)$-plane is as given in the following figure.

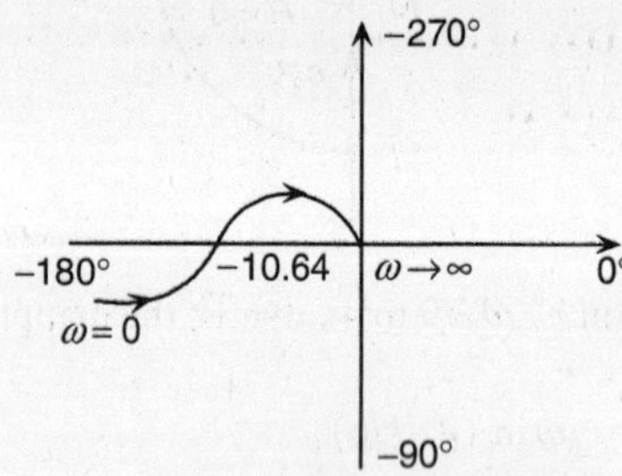

5. **Mapping of Section c_2:** Mapping of Section c_2 from s-plane to $(u\text{–}v)$-plane is obtained by assigning $s = \lim_{R\to\infty} Re^{j\theta}$ in $G(s)H(s)$ and varying θ from $+\pi/2$ to $-\pi/2$. Since $s\to Re^{j\theta}$ and $R\to\infty$, $1 + T \approx sT$.

$$G(s)H(s) = \frac{1+4s}{s^2(1+s)(1+2s)} = \frac{4s}{s^2\times s\times 2s} = \frac{2}{s^3}$$

$$\therefore \qquad G(s)H(s)\big|_{s=\lim_{R\to\infty} Re^{j\theta}} = \frac{2}{\lim_{R\to\infty} R^3 e^{j3\theta}} = 0e^{-j3\theta}$$

$$= 0\angle -270^\circ \text{ to } 0^\circ \text{ to } +270^\circ.$$

From the afore-mentioned discussion, we can say that Section c_2 in s-plane is mapped into a circular area of zero radius (i.e., a point) in the $(u\text{–}v)$-plane and not required to be analysed.

6. **Mapping of Section c_3:** In Section c_3, $\omega\to-\infty$ to 0, that is, the mapping of Section c_3 gives the inverse polar plot of $G(j\omega)H(j\omega)$, as shown in the following figure.

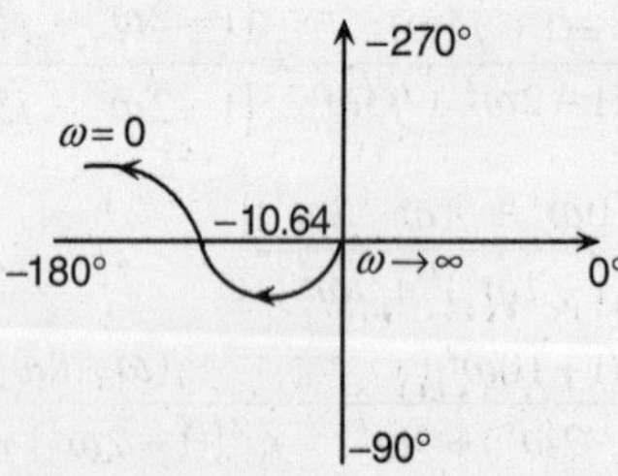

7. **Mapping of Section c_4:** Mapping of Section c_4 from s-plane to $(u\text{–}v)$-plane can be obtained by substituting $s = \lim_{R\to 0} Re^{j\theta}$ in $G(s)H(s)$ and varying θ from $-\pi/2$ to $+\pi/2$. Since $s = Re^{j\theta}$ and $R = 0$, $1 + sT = 1$.

$$G(s)H(s) = \frac{1+4s}{s^2(1+s)(1+2s)} = \frac{1}{s^2\times 1\times 1} = \frac{1}{s^2}.$$

$$(s)H(s)\bigg|_{s-\lim_{R\to 0} Re^{j\theta}} = \frac{1}{\lim_{R\to 0} Re^{j2\theta}} = \infty e^{-j2\theta}$$

$$= \infty\angle +180^\circ \text{ to } 0^\circ \text{ to } -180^\circ.$$

From the afore-mentioned discussion, we can say that Section c_4 in the s-plane is mapped into a circle of infinite radius with argument varying from +180° to −180°, as shown in the following figure.

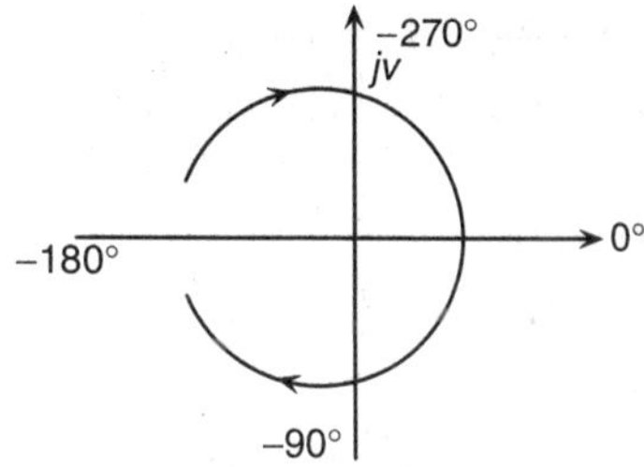

8. **Complete Nyquist plot:** The complete Nyquist plot in $G(s)H(s)$ or $(u–v)$-plane can be obtained by combining the mapping of individual sections as shown in the following figure.

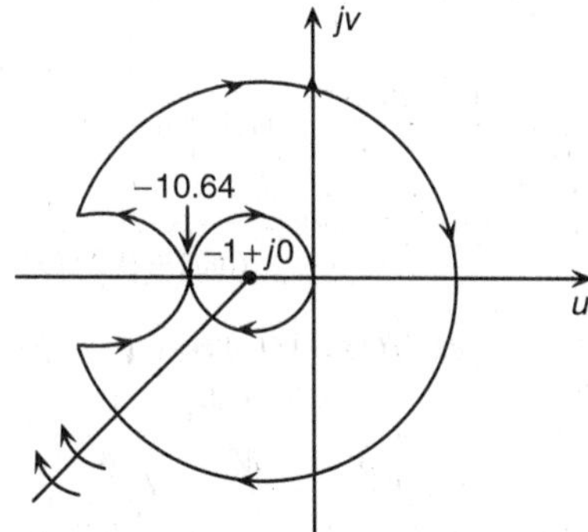

9. The number of encirclements of $(-1 + j0)$ are
   $N = +2$ (clockwise encirclements)
   However, for stability, $N = 0$.
   ∴ The closed-loop system is unstable.
   According to the mapping theorem, we have

$$N = Z - P$$

$$2 = Z - 0 \Rightarrow Z = 2$$

   Actually, there are 2 zeros of $1 + G(s)H(s)$ encircled by Nyquist path, that is, 2 closed-loop poles are there in the right half of the s-plane, due to which the closed-loop system is unstable.

ASP-5: Sketch the Nyquist plot of the system whose open-loop t.f. is given as $G(s)H(s) = \dfrac{5}{s(1+0.5s)}$ and hence determine the gain margin.

Solution:

Given that $G(s)H(s) = \dfrac{5}{s(1+0.5s)}$.

1. Number of poles in the right half of the *s*-plane, $P = 0$.
2. For stability, $N = -P = 0$.
3. As there is one pole at origin, the Nyquist contour is chosen as shown in the figure, which contains Sections c_1, c_2, c_3 and c_4.

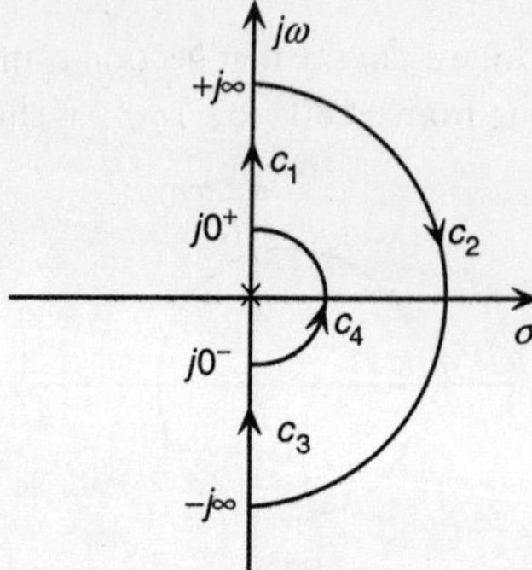

4. **Mapping of Section c_1:** In Section c_1, $\omega \rightarrow 0$ to ∞, that is, the mapping of Section c_1 gives the polar plot of $G(j\omega)H(j\omega)$ in $(u–v)$-plane.
   To get the sinusoidal t.f., substitute $s = j\omega$ in $G(s)H(s)$.

$$G(j\omega)H(j\omega) = \frac{5}{j\omega(1+j0.5\omega)}.$$

$$M = |G(j\omega)H(j\omega)| = \frac{5}{\omega\sqrt{1+0.25\omega^2}}.$$

$$\phi = \angle G(j\omega)H(j\omega) = -90 - \tan^{-1} 0.5\omega \cdot$$

ω	M	ϕ
0	∞	−90°
∞	0	−180°

$$G(j\omega)H(j\omega) = \frac{5}{j\omega(1+j0.5\omega)}$$

$$= \frac{5}{-0.5\omega^2 + j\omega} = -\frac{5}{(0.5\omega^2 - j\omega)}\frac{(0.5\omega^2 + j\omega)}{(0.5\omega^2 + j\omega)}$$

$$= \frac{-5(0.5\omega^2 + j\omega)}{0.25\omega^2 + \omega^2}$$

$$= \frac{-5(0.5\omega^2 + j\omega)}{1.25\omega^2} = \frac{-2.5}{1.25} - j\frac{5}{1.25\omega}.$$

There is no intersection point of the plot with the real or imaginary axis.
Thus, the mapping of Section c_1 in $(u–v)$-plane gives the following figure.

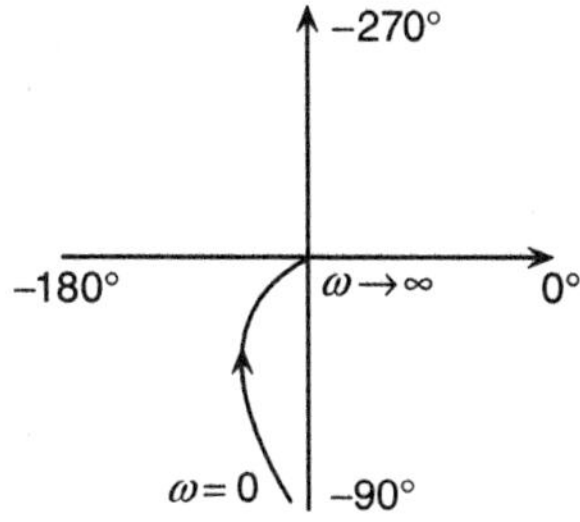

5. **Mapping of Section c_2:** The mapping of Section c_2 from the s-plane to $(u–v)$-plane is obtained by substituting $s = \lim_{R\to\infty} Re^{j\theta}$ in $G(s)H(s)$ and varying θ from $+\pi/2$ to $-\pi/2$. Since $s \to Re^{j\theta}$ and $R\to\infty$, $1 + sT \approx sT$

$$G(s)H(s) = \frac{5}{s(1+0.5s)} = \frac{5}{s\times 0.5s} = \frac{10}{s^2}$$

$$G(s)H(s)\Big|_{s=\lim_{R\to\infty} Re^{j\theta}} = \frac{10}{\lim_{R\to\infty} R^2 e^{j2\theta}} = 0e^{-j2\theta}$$

$$= 0\angle -180° \text{ to } 0° \text{ to } +180°$$

From the afore-mentioned discussion, we can say that Section c_2 in the s-plane is mapped into a circular of zero radius in the $(u–v)$-plane and not analysed.

6. **Mapping of Section c_3:** In Section c_3, $\omega \to -\infty$ to 0, that is, mapping of Section c_3 gives the inverse polar plot of $G(j\omega)H(j\omega)$ as in the following figure.

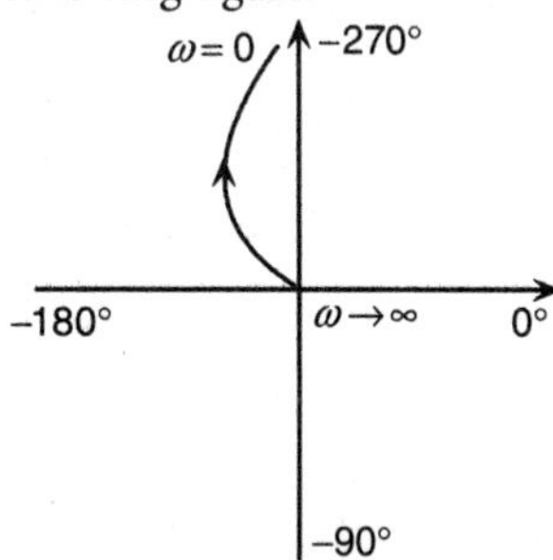

7. **Mapping of Section c_4:** Mapping of Section c_4 from s-plane to $(u–v)$-plane can be obtained by substituting $s = \lim_{R\to 0} Re^{j\theta}$ in $G(s)H(s)$ and varying θ from $-\pi/2$ to $+\pi/2$. Since $s = Re^{j\theta}$ and $R = 0$, $1 + sT = 1$.

$$G(s)H(s) = \frac{5}{s(1+0.5s)} = \frac{5}{s\times 1} = \frac{5}{s}$$

$$G(s)H(s)\Big|_{s=\lim_{R\to 0} Re^{j\theta}} = \frac{5}{\lim_{R\to\infty} Re^{j2\theta}} = \infty e^{-j\theta}$$

$$\infty\angle 90° \text{ to } 0° \text{ to } -90°$$

From the afore-mentioned discussion, we can say that Section c_4 in the s-plane is mapped into a circle of infinite radius with argument varying from +90° to −90° as shown in the following figure.

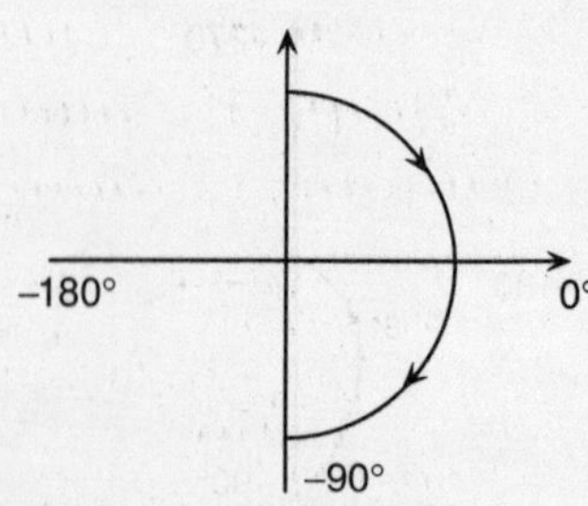

8. **Complete Nyquist plot:** The complete Nyquist plot in (u–v)-plane can be obtained by combining the mapping of individual section as in the following figure.

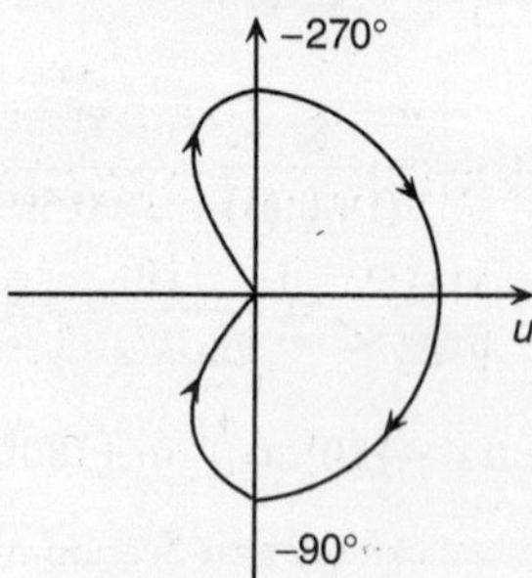

9. The Nyquist plot does not encircle the critical point $(-1+j0)$, that is, $N = 0$. Since the given t.f. does not contain any poles in the right half of the s-plane, the closed-loop system is stable.
As $\omega_{pc} = 0$ for any increase in gain, ω_{pc} cannot approach ω_{gc}, and hence $G.M = +\infty$dB.

ASP-6: For a feedback control system, sketch the Nyquist plot whose open-loop t.f. is given as $G(s)H(s) = \dfrac{40}{(s+4)(s^2+2s+2)}$. Find the gain margin and stability from Nyquist plot.

Solution:

Given that $G(s)H(s) = \dfrac{40}{(s+4)(s^2+2s+2)}$.

1. Number of poles in the right half of the s-plane, $P = 0$.
2. For stability $N = -P = 0$.
3. As there is no pole at the origin, the Nyquist contour is chosen as shown in the figure, which contains Sections c_1, c_2 and c_3.

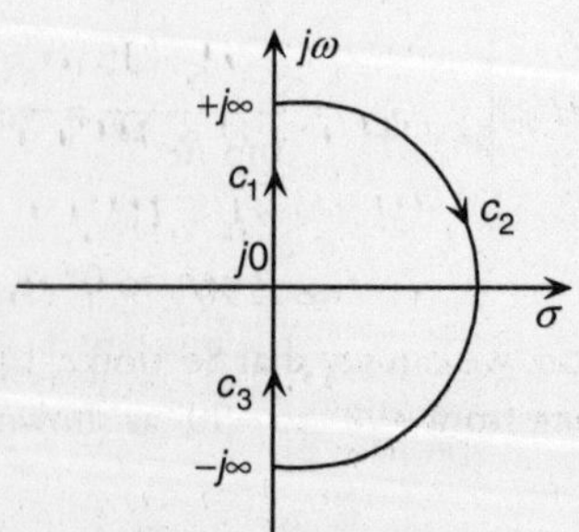

4. **Mapping of Section c_1:** In Section c_1, $\omega \to 0$ to ∞, that is, the mapping of Section c_1 gives the polar plot of $G(j\omega)H(j\omega)$ in $(u\text{–}v)$-plane.

   To get the sinusoidal t.f., substitute $s = j\omega$ in $G(s)H(s)$.

$$G(j\omega)H(j\omega) = \frac{40}{(j\omega+4)(2-\omega^2+j2\omega)}$$

$$M = |G(j\omega)H(j\omega)| = \frac{40}{\sqrt{\omega^2+16}\sqrt{(2-\omega^2)^2+4\omega^2}}$$

$$\phi = \angle G(j\omega)H(j\omega) = -\tan^{-1}\omega/4 - \tan^{-1}\left[\frac{2\omega}{2-\omega^2}\right]$$

Note: A quadratic factor contributes an angle of $\pm180°$ as $\omega \to \infty$ and contributes an angle of $0°$ as $\omega \to 0$.

ω	M	ϕ
0	5	0°
∞	0	–270°

$$G(j\omega)H(j\omega) = \frac{40}{(4+j\omega)(2-\omega^2+j2\omega)} \times \frac{(4-j\omega)(2-\omega^2-2j\omega)}{(4-j\omega)(2-\omega^2-2j\omega)}.$$

$$= \frac{40(4-j\omega)(2-\omega^2-j2\omega)}{(16+\omega^2)[(2-\omega^2)^2+4\omega^2)]}$$

$$= \frac{40[8-4\omega^2-j8\omega-j2\omega+j\omega^3-2\omega^2]}{(16+\omega^2)(4+\omega^4)}$$

$$= \frac{40(8-6\omega^2)}{(16+\omega^2)(4+\omega^4)} + \frac{j(\omega^3-10\omega)}{(16+\omega^2)(4+\omega^4)}.$$

Let us assign I.P. of $G(j\omega)H(j\omega) = 0$.

$$\text{i.e.,} \quad \frac{\omega^3-10\omega}{(16+\omega^2)(4+\omega^4-3\omega^2)} = 0$$

$$\Rightarrow \quad \omega(\omega^2-10) = 0$$

$$\omega = 0, \omega = \sqrt{10}\text{, i.e., } \omega_{pc} = \sqrt{10}.$$

$$\therefore \quad G(s)H(s) = \frac{40(8-60)}{(16+10)(4+100)} = -0.769\text{, i.e., } Q = -0.769.$$

Let us assign R.P. of $G(j\omega)H(j\omega) = 0$.

$$\text{i.e.,} \quad \frac{8-6\omega^2}{(16+\omega^2)(4+\omega^4)} = 0$$

$$\Rightarrow \quad 8-6\omega^2 = 0 \Rightarrow \omega = \sqrt{4/3}.$$

$$G(j\omega)H(j\omega) = \frac{j\sqrt{\frac{4}{3}}\left[\frac{4}{3}-10\right]}{\left[16+\frac{4}{3}\right]\left[4+\frac{16}{9}\right]}$$

$$= \frac{-j\,3\sqrt{3}}{52} = -j0.1$$

Thus, the mapping of Section c_1 in $(u–v)$-plane is shown in the following figure.

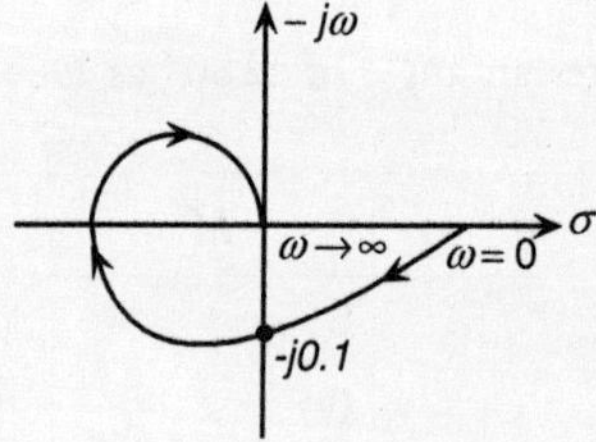

5. **Mapping of Section c_2:** The mapping of Section c_2 from s-plane to $(u–v)$-plane gives a point and need not be considered.
6. **Mapping of Section c_3:** In Section c_3, $\omega \to \infty$ to 0, that is, mapping of Section c_3 gives the inverse polar plot of $G(s)H(s)$, as given in the following figure.

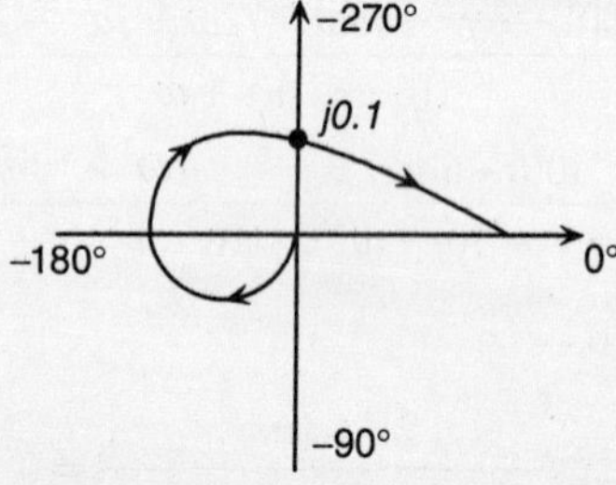

7. **Complete Nyquist plot:** The complete Nyquist plot in $(u–v)$-plane can be obtained by combining the mapping of individual section as in the following figure.

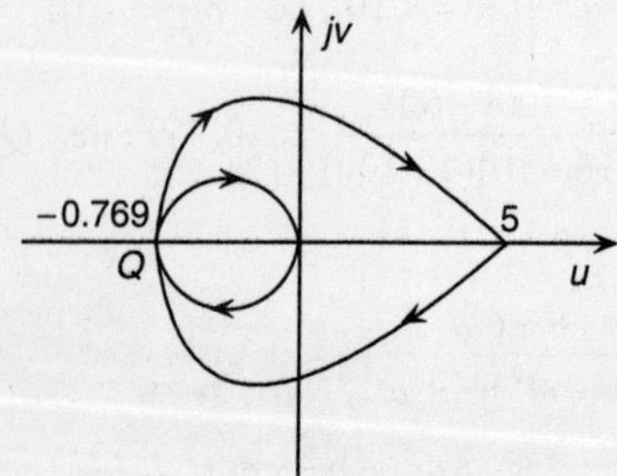

8. The Nyquist plot does not encircle the critical point $(-1 + j0)$, that is, $N = 0$. Since the given t.f. does not contain any poles in the right half of the s-plane, the closed-loop system is stable.

$$\text{GM} = \frac{1}{|OQ|} = \frac{1}{0.769} = 1.3.$$

$$\text{GM in dB} = 20\log\frac{1}{|OQ|}$$

$$= 20\log\left(\frac{1}{0.769}\right) = 2.27\,\text{dB}$$

ASP-7: Draw the Nyquist plot for the system whose open-loop t.f. is $G(s)H(s) = \dfrac{K}{s(s+2)(s+10)}$. Determine the range of K for which the closed-loop system is stable.

Solution:

Given that $G(s)H(s) = \dfrac{K}{s(s+2)(s+10)}$.

1. Number of poles in the right half of the s-plane, $P = 0$.
2. For stability, $N = -P = 0$.
3. As there is one pole at origin, the Nyquist contour is chosen as shown in the figure, which contains Sections c_1, c_2, c_3 and c_4.

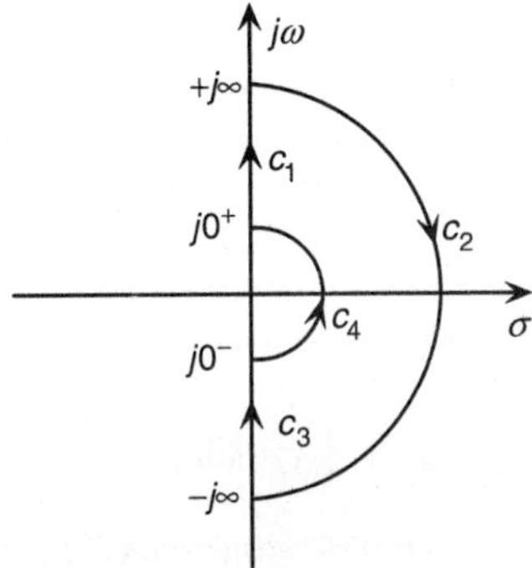

4. **Mapping of Section c_1:** In Section c_1, $\omega \to 0$ to ∞, that is, the mapping of Section c_1 gives the polar plot of $G(j\omega)H(j\omega)$ in $(u–v)$-plane.
   To get the sinusoidal t.f., substitute $s = j\omega$ in $G(s)H(s)$

$$\therefore \quad G(j\omega)H(j\omega) = \frac{K}{j\omega(j\omega+2)(j\omega+10)}$$

$$M = |G(j\omega)H(j\omega)| = \frac{K}{\omega\sqrt{\omega^2+4}\sqrt{\omega^2+100}}$$

$$\theta = \angle G(j\omega)H(j\omega) = -90 - \tan^{-1}\omega/2 - \tan^{-1}\omega/10$$

ω	M	ϕ
0	∞	–90°
∞	0	–270°

$$G(s)H(s)=\frac{K}{s\times 2(1+s/2)10(1+s/10)}$$

$$=\frac{0.05K}{s(1+0.5s)(1+0.1s)}$$

$$G(j\omega)H(j\omega)=\frac{0.05K}{j\omega(1+j0.5\omega)(1+j0.1\omega)}$$

$$=\frac{0.05K}{j\omega(1+j0.6\omega-0.05\omega^2)}$$

$$=\frac{0.05K}{-0.6\omega^2+j\omega(1-0.05\omega^2)}$$

$$=\frac{0.05K[-0.6\omega^2-j\omega(1-0.05\omega^2)]}{[-0.6\omega^2+j\omega(1-0.05\omega^2)][-0.6\omega^2-j\omega(1-0.05\omega^2)]}$$

$$=\frac{-0.05K\times 0.6\omega^2}{0.36\omega^4+\omega^2(1-0.05\omega^2)^2}-\frac{j0.05K\omega(1-0.05\omega^2)}{0.36\omega^4+\omega^2(1-0.05\omega^2)^2}$$

Let us assign I.P. of $G(j\omega)H(j\omega)=0$.

i.e., $$1-0.05\omega^2=0\Rightarrow\omega\frac{1}{\sqrt{0.05}}=4.47$$

i.e., $$\omega_{pc}=4.47\text{ rad/s.}$$

$$\therefore\quad G(j\omega)H(j\omega)=\frac{-0.05K\times 0.6\times 4.47^2}{0.36\times 4.47^2+4.47^2(1-0.05\times 4.47^2)^2}=-0.00417K.$$

Thus, the mapping of Section c_1 in the $(u–v)$-plane gives the following figure.

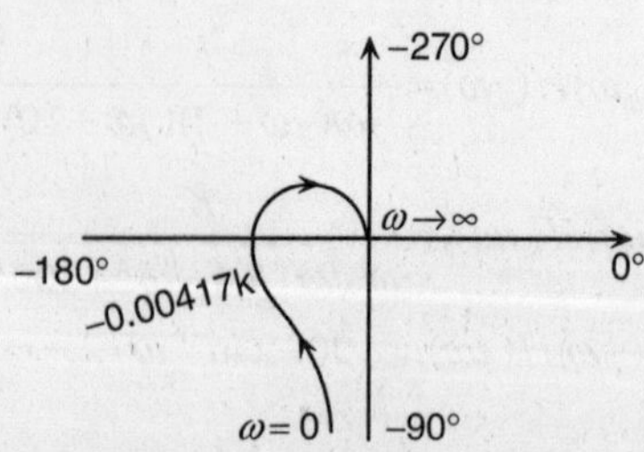

5. **Mapping of Section c_2:** Thus, mapping of Section c_2 from the s-plane to (u–v)-plane can be obtained by substituting $s = \lim_{R\to\infty} Re^{i\theta}$ in $G(s)H(s)$, which gives a circular area of zero radius (i.e., a point). Therefore, it need not be considered.

6. **Mapping of Section c_3:** In Section c_3, $\omega \to -\infty$ to 0, that is, the mapping of Section c_3 gives the inverse polar of $G(j\omega)H(j\omega)$ as given in the following figure.

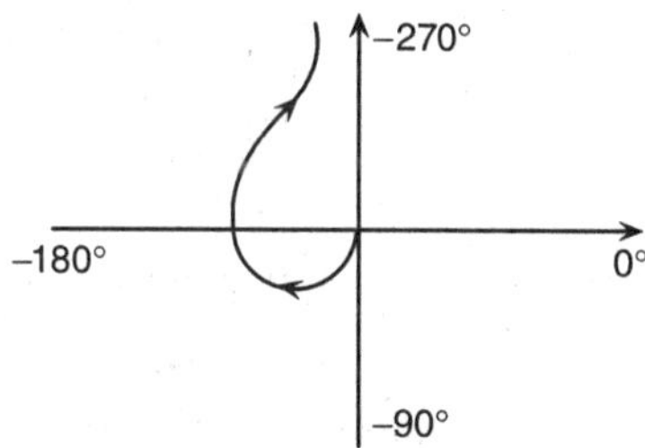

7. **Mapping of Section c_4:** The mapping of Section c_4 from s-plane to (u–v)-plane can be obtained by substituting $s = \lim_{R\to 0} Re^{j\theta}$ in $G(s)H(s)$ and varying θ from $-\pi/2$ to $+\pi/2$. Since $s = Re^{j\theta}$ and $R \to 0$, $1 + sT \approx 1$.

$$\therefore \quad G(s)H(s) = \frac{K}{s(s+2)(s+10)}$$

$$= \frac{0.05K}{s(1+0.05s)(1+0.15)} = \frac{0.05K}{s\times 1\times 1} = \frac{0.05K}{s}$$

$$G(s)H(s)\big|_{s=\lim_{R\to 0} Re^{j\theta}} = \frac{0.05K}{\lim_{R\to 0} Re^{j\theta}} = \infty e^{-j\theta}$$

$$= \infty\angle{+90^\circ} \text{ to } 0^\circ \text{ to } -90^\circ.$$

From the afore-mentioned discussion, we can say that Section c_4 in the s-plane is mapped into a circle of infinite radius with argument varying from +90° to –90°, as shown in the following figure.

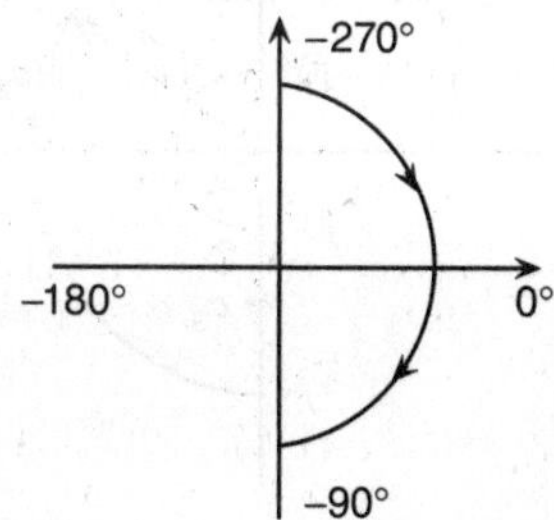

8. **Complete Nyquist plot:** The complete Nyquist plot in (u–v)-plane can be obtained by combining the mapping of individual section as shown in the following figure.

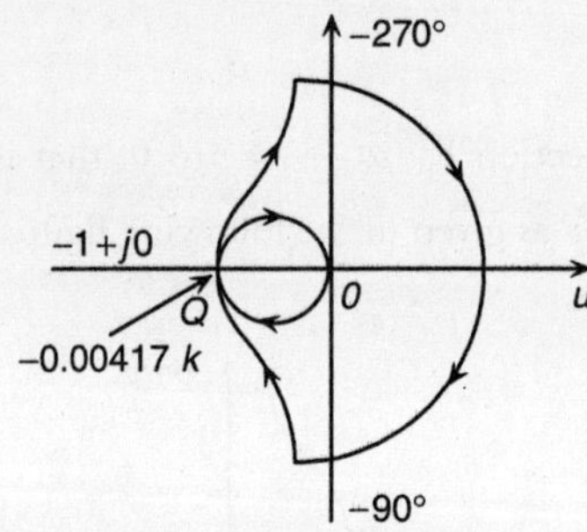

9. Now, for absolute stability, $N = 0$, that is, $(-1 + j0)$ point should be located on the left side of point Q. i.e., $|OQ| < 1$

$$\Rightarrow |0.0041K| < 1$$

$$K < \frac{1}{0.00417} < 240$$

$\therefore$ The range of value of K for stability is

$$0 < K < 240$$

ASP-8: Sketch the Nyquist plot for a system with $G(s)H(s) = \dfrac{10(s+3)}{s(s-1)}$ and also determine the closed-loop stability.

Solution:

Given that $G(s)H(s) = \dfrac{10(s+3)}{s(s-1)}$.

1. Number of poles in the right half of the s-plane, $P = 1$.
2. For stability, $N = -P = -1$
3. As there is one pole at origin, the Nyquist contour is chosen as shown in the figure, which contains Sections c_1, c_2, c_3 and c_4.

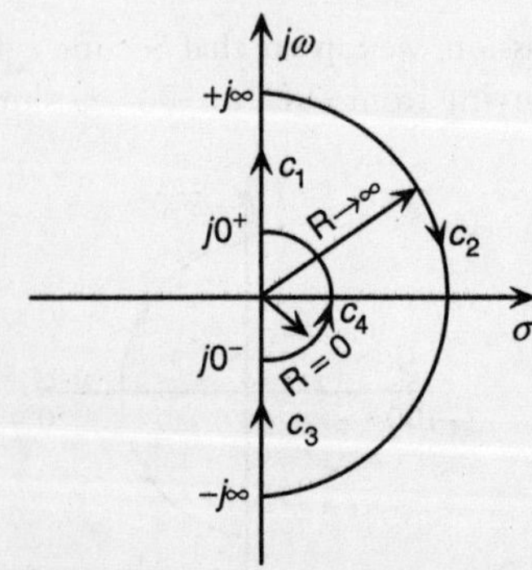

4. **Mapping of Section c_1 :** In Section c_1, $\omega \to 0$ to ∞, that is, the mapping of Section c_1 gives the polar plot of $G(j\omega)H(j\omega)$ in (u–v)-plane.
To get sinusoidal t.f., substitute $s = j\omega$ in $G(s)H(s)$.

$$\therefore \qquad G(j\omega)H(j\omega) = \frac{10(3+j\omega)}{j\omega(j\omega-1)}.$$

$$M = |G(j\omega)H(j\omega)| = \frac{10\sqrt{9+\omega^2}}{\omega\sqrt{1+\omega^2}}$$

$$\phi = \angle G(j\omega)H(j\omega) = -90 + \tan^{-1}\left(\frac{\omega}{3}\right) - \tan\left(\frac{-\omega}{1}\right)$$

$$= -90 + \tan^{-1}\left(\frac{\omega}{3}\right) - \left(180^{\circ} - \tan^{-1}\omega\right)$$

$$= -270^{\circ} + \tan^{-1}\left(\frac{\omega}{3}\right) + \tan^{-1}\omega$$

ω	M	ϕ
0	∞	–270°
∞	0	–90°

$$G(j\omega)H(j\omega) = \frac{-10(3+j\omega)}{j\omega(1-j\omega)} = \frac{-\omega(3+j\omega)}{j\omega+\omega^2}$$

$$= \frac{-10(3+j\omega)(\omega^2-j\omega)}{(\omega^2+j\omega)(\omega^2-j\omega)}$$

$$= \frac{-10(3\omega^2 - j3\omega + j\omega^3 + \omega^2)}{(\omega^4+\omega^2)}$$

$$= \frac{-40\omega^2}{\omega^4+\omega^2} - \frac{10j(\omega^3-3\omega)}{\omega^4+\omega^2}$$

Let us assign I.P. of $G(j\omega)H(j\omega) = 0$.

i.e.,
$$\frac{10j(\omega^3-3\omega)}{\omega^4+\omega^2} = 0$$

$$\Rightarrow \qquad \omega^3 - 3\omega = 0$$

$$\Rightarrow \qquad \omega = 0,\ \omega = \sqrt{3} \ \therefore \text{ i.e } \omega_{pc} = \sqrt{3}$$

$$\therefore \ G(j\omega)H(j\omega) = \frac{-40\times 3}{9+3} = -10 \ \therefore \text{ i.e., } Q = -10.$$

Thus, the mapping of Section c_1 in $(u–v)$-plane gives the following figure.

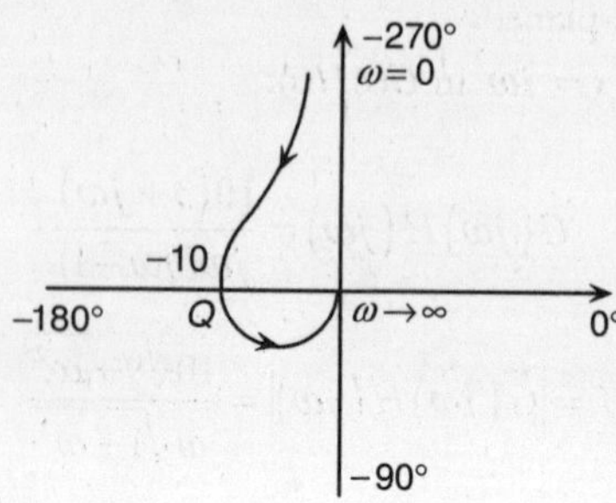

5. **Mapping of Section c_2:** The mapping of section from s-plane to $(u–v)$-plane can be obtained by substituting $s = \lim_{R\to\infty} Re^{j\theta}$ in $G(s)H(s)$, which gives a circular area of zero radius (i.e., a point). Therefore, it is not considered.

6. **Mapping of Section c_3:** In Sectionc_3, $\omega \to -\infty$ to 0, that is, mapping of Section c_3 gives the inverse polar plot of $G(j\omega)H(j\omega)$ as in the following figure.

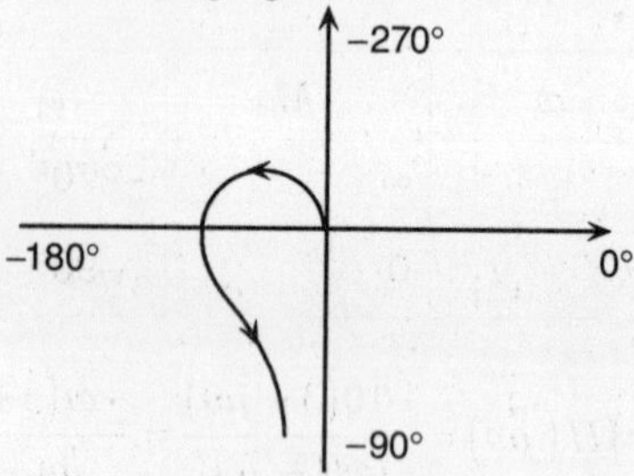

7. **Mapping of Section c_4:** Mapping of Section c_4 from s-plane to $(u–v)$-plane can be obtained by substituting $s = \lim_{R\to 0} Re^{j\theta}$ in $G(s)H(s)$ and varying θ from $-\pi/2$ to $+\pi/2$. Since $s = Re^{j\theta}$ and $R\to 0$, $1+sT \approx 1$.

$$\therefore \quad G(s)H(s) = \frac{10(s+3)}{s(s-1)} = \frac{10\times 3\left(1+\frac{s}{3}\right)}{s\times -1(1-s)} = \frac{10\times 3\times 1}{s\times -1} = \frac{30}{s\times(-1)}$$

$$G(s)H(s)\Big|_{s=\lim_{R\to 0} Re^{j\theta}} = \frac{30}{\lim_{R\to 0} Re^{j\theta}\times(-1)}$$

$$= \frac{30}{\lim_{R\to 0} Re^{j(\theta-180^\circ)}} = \infty e^{-j(\theta-180^\circ)}$$

$$= \infty\angle{+270} \text{ to } 0^\circ \text{ to } +90^\circ.$$

Note:

(–1) introduces –180° phase angle in the phase angle expression.

From the afore-mentioned discussion, we can say that Section c_4 in the s-plane is mapped into a circle of infinite radius with argument varying from +270° to +90°, as shown in the following figure.

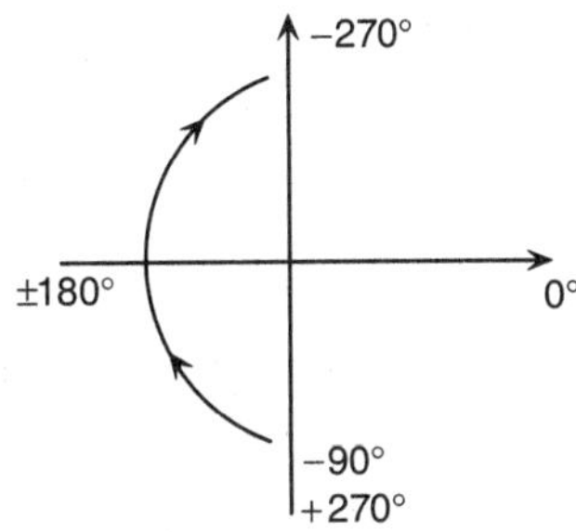

8. **Complete Nyquist plot:** The complete Nyquist plot in $(u\text{–}v)$-plane can be obtained by combining the mapping of individual section as shown in the following figure.

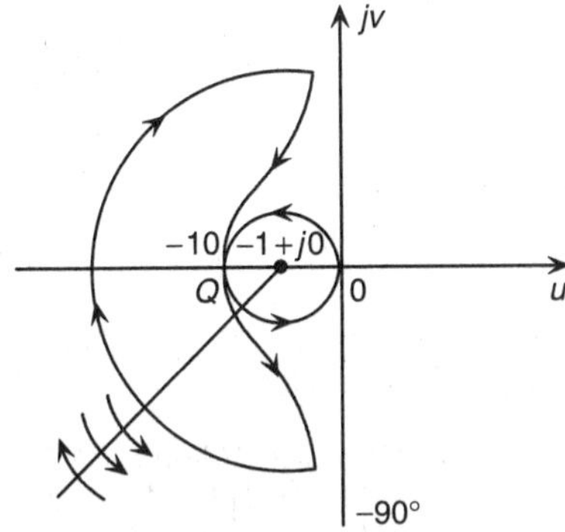

9. The number of encirclements of $(-1 + j0)$ are
   $N = 1 - 2 = -1$ (anticlockwise).
   Since the number of anticlockwise encirclements is equal to the number of open-loop poles on the right half of the s-plane, the closed-loop system is stable.

REVIEW QUESTIONS

1. Explain Nyquist stability criterion clearly.
2. Discuss the effect of adding poles and zeros to Nyquist plot.
3. Define gain margin and phase margin. Explain the significance of these terms to find the closed-loop system stability.
4. Discuss the significance of constant M and N circles to find the closed-loop system stability.
5. Explain the polar plot with an example.
6. Explain how Nyquist contour is selected for stability analysis.
7. What are the advantages of Nyquist method?
8. How can closed-loop stability be predicted from open-loop t.f.?
9. Why Nyquist path does not contain L.H.S of s-plane? Explain.
10. Draw and explain polar plots for Type-0, -1 and -2 systems.
11. How are gain margin and phase margin calculated from Nyquist plot?
12. What are constant M circles? Explain.
13. What are constant N circles? Explain.
14. Show that the loci of constant phase angles are circles.

EXERCISE PROBLEMS

1. For $G(s)H(s)=\dfrac{1}{(s+p_1)(s+p_2)}, p_1, p_2>0$, sketch Nyquist plot and determine the stability of the system.
2. For $G(s)H(s)=\dfrac{12}{s(s+1)(s+2)}$, sketch Nyquist plot and determine the stability of the system. Calculate GM of the system.
3. Using Nyquist stability criterion, determine the closed-loop system stability for a system whose open-loop t.f. is given as $G(s)H(s)=\dfrac{100}{(1+0.1s)(1+0.2s)(1+0.3s)}$.
4. For $G(s)H(s)=\dfrac{40}{(s+4)(s^2+2s+2)}$, sketch Nyquist plot and determine the stability of the system. Calculate GM of the system.
5. Sketch the Nyquist plot for the given open-loop t.f. and discuss the closed-loop system stability $G(s)H(s)=\dfrac{10}{s(1+0.2s)(s-1)}$.
6. The open-loop t.f. of a unity feedback control system is given as $G(s)=\dfrac{2(s+0.25)}{s^2(s+1)(s+0.5)}$. Determine the closed-loop stability by applying Nyquist criterion.
7. Sketch the Nyquist plot for the system whose open-loop t.f. is $G(s)H(s)=\dfrac{Ks^3}{(s+1)(s+2)}$, and determine the stability condition.
8. Construct the Nyquist plot for a feedback control system whose open-loop t.f. is given as $G(s)H(s)=\dfrac{5}{s(1-s)}$ and also determine the stability of open-loop and closed-loop system.

ANSWERS

1. The given system is stable and the Nyquist plot is as shown in figure.

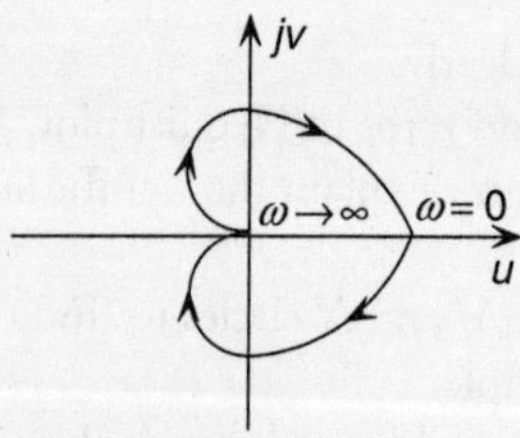

2. The given system is unstable and the GM = 6.2 dB. The Nyquist plot is as shown in the following figure.

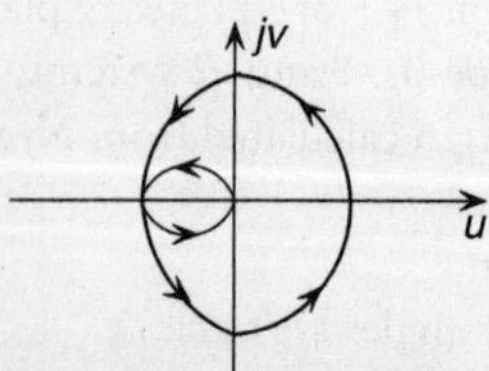

3. The given system is stable.
4. The system is stable is stable and GM = 2.27 dB.

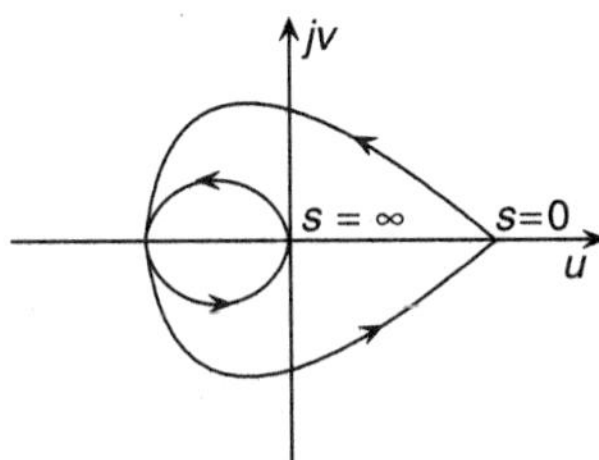

5. The system is unstable. The Nyquist plot is as shown in the following figure.

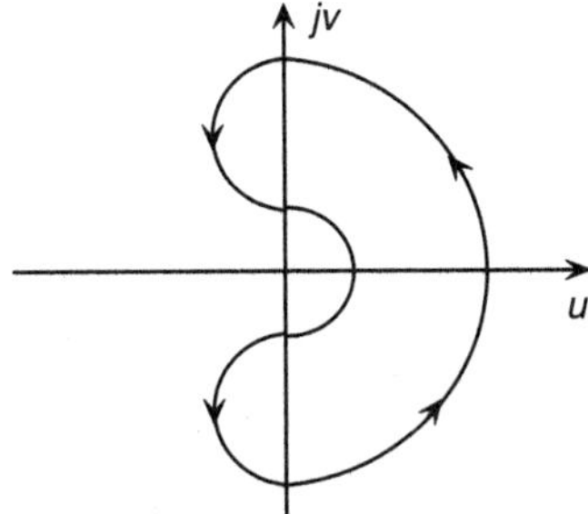

6. The closed-loop system is unstable and the Nyquist plot is as shown in the following figure.

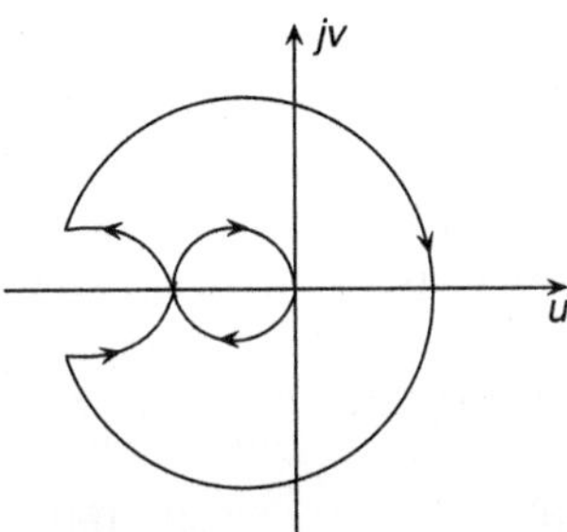

7. For stability, $K<3/2$ and the Nyquist plot is as shown in the following figure.

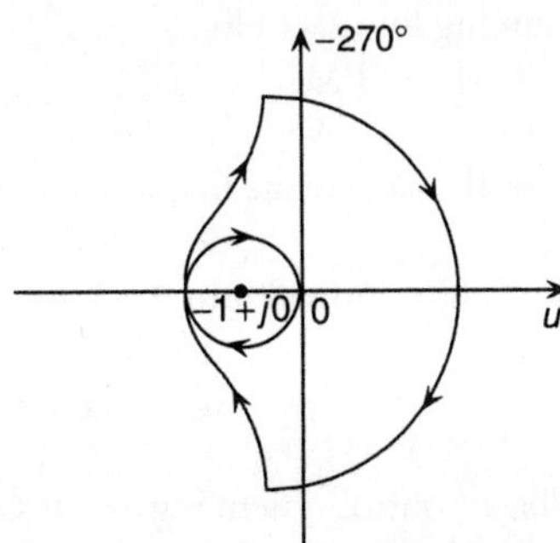

8. Both open-loop and closed-loop systems are unstable and the Nyquist plot is as shown in the following figure.

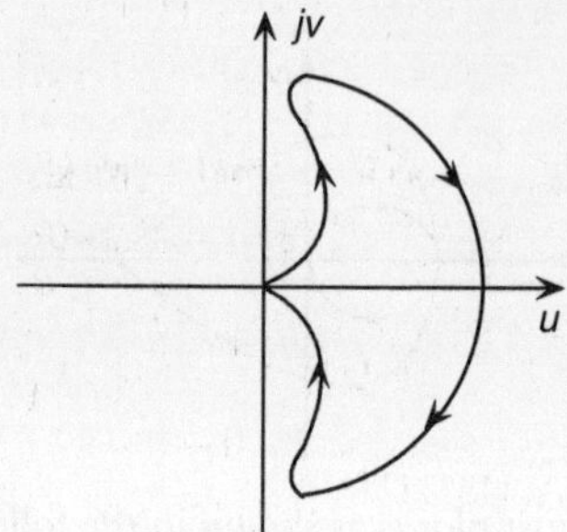

OBJECTIVE TYPE QUESTIONS

1. The origin for the investigation of closed-loop stability in relation to Nyquist stability criterion is
(a) $1+j0$ (b) $-1+j0$ (c) $0+j1$ (d) $n\pi/2$
2. In $G(s)\ H(s)$-plane, the Nyquist plot from $\omega = 0^-$ to $\omega = 0^+$ is closed through an angle
(a) $n\pi$ (b) $2n\pi$ (c) $-n\pi$ (d) $n\pi/2$
3. Adding a pole to the open-loop t.f. $G(s)\ H(s)$ ——— the stability of the system
(a) decreases (b) increases (c) does not affect (d) can't say
4. Adding a zero to the open-loop t.f., $G(s)\ H(s)$ ——— the stability of the system
(a) decreases (b) increases (c) does not affect (d) can't say
5. For Nyquist plot, we use
(a) open-loop t.f. (b) closed-loop t.f. (c) C.E. (d) All
6. If the Nyquist plot crosses at $-p$, the absolute gain margin is
(a) $1/p$ (b) $-1/p$ (c) p (d) $-p$
7. The negative real axis in $G(s)\ H(s)$-plane for Nyquist plot corresponds to
(a) $-90°$ phase (b) $+90°$ phase (c) $-180°$ phase (d) $+180°$phase
8. The phase angle at $\omega = \infty$ for the t.f. $G(j\omega) = \dfrac{90(1+j5\omega)}{j\omega(1+j4\omega)(1+j6\omega)}$ is
(a) 90° (b) −90° (c) 180° (d) −180°
9. The disadvantage of polar plot is that
(a) it is cramped at high frequencies.
(b) its calculations are time consuming for exact plot.
(c) it is very difficult to calculate GM and PM.
(d) all
10. The gain margin is the reciprocal of the gain at the frequency at which the phase angle is
(a) 90° (b) −90° (c) 180° (d) −180°
11. The unit circle in the Nyquist plot is transformed into 0 dB line of Bode plot at
(a) low frequencies (b) high frequencies
(c) all frequencies (d) medium frequencies
12. If the open-loop t.f. of unity feedback control system is given as $G(s) = \dfrac{1}{(s+1)(s+2)(s-1)}$, the encirclement of the point $(-1+j0)$ by the Nyquist plot is
(a) 1 (b) 2 (c) 3 (d) 4

13. The gain margin in dB of a unity feedback system, whose open-loop t.f. is $G(s)=\dfrac{1}{s(s+1)}$, is given as

(a) 0 (b) 1 (c) −1 (d) Infinite

14. I. Nyquist criterion is in frequency domain.
II. Bode plot is in frequency domain.
III. Root locus plot is in time domain.
IV. *R–H* criterion is in time domain.

(a) I, II, III are correct (b) I, III, IV are correct
(c) I, II are correct (d) All are correct

15. If the open-loop t.f. of a unity feedback control system is $G(s)=\dfrac{10}{(s+5)^3}$, the GM of the system will be

(a) 20 dB (b) 40 dB (c) 60 dB (d) 80 dB

16. The $G(s)=\dfrac{10}{(s+5)^2}$ intersects the real axis at $\omega=\omega_0$, then the real part and ω_0 are respectively given as

(a) −5, 1 (b) −2.5, 1 (c) −5, 0.5 (d) −5, 2

17. The line corresponding to negative real axis of Nyquist plot transforms into

(a) 0 dB line of Bode plot (b) 10 dB line of Bode plot
(c) −180° phase shift line of Bode plot (d) −90° phase shift line of Bode plot

18. The term $G(s)H(s)=s$ on magnitude plot of Bode plot will have a slope

(a) 6 dB per octave (b) −6 dB per decade
(c) 12 dB per octave (d) 20 dB decade

19. A system has a gain margin of −12 dB, and the gain corresponding to 180° phase shift is

(a) 12 dB (b) −12 dB (c) 1/12 dB (d) −1/12 dB

20. The advantage of Nyquist stability criterion is that

(a) it guides in stabilising an unstable system.
(b) it enables to predict closed-loop stability from open-loop results.
(c) it is applicable to experimental results of frequency response of open-loop system.
(d) All

21. The polar plot of the open-loop t.f. of a feedback control system intersects the real axis at −2. The GM of the system is

(a) −5 dB (b) 0 dB (c) −6 dB (d) −2 dB

22. If the polar plot of at t.f. passes through a critical point (−1, 0), then the GM is

(a) 0 dB (b) 1 dB (c) −1 dB (d) infinite

23. The open-loop t.f. of a unity feedback system $G(s)=\dfrac{K}{(s+1)(1+2s)(1+3s)}$ is

(a) 1.414 (b) 1 (c) 0 (d) 1.732

24. For a constant *M* circle, the centre will lie at $\left(\dfrac{M^2}{1-M^2},0\right)$ in *G*-plane and has a radius of

(a) M (b) $1-M^2$ (c) $\dfrac{M}{1-M^2}$ (d) $\dfrac{M}{1+M^2}$

25. The constant M circle for $M = 1$ is the
(a) straight line $x = -1/2$
(b) critical point $(-1, j0)$
(c) circle with $r = 0.33$
(d) All

26. The constant M loci plot is symmetrical w.r.t
(a) real axis and imaginary axis
(b) $M = 1$ straight line and real axis
(c) $M = 1$ straight line and imaginary axis
(d) $M = 1$ straight line

27. If the phase angle of an open-loop t.f. at gain crossover point is β, the phase margin is
(a) $180° - \beta$
(b) $180° + \beta$
(c) $\beta - 180°$
(d) $90° - \beta$

28. The centre of M circle is located on x-axis at
(a) M
(b) $1 - M^2$
(c) $\dfrac{M^2}{1-M^2}$
(d) $\dfrac{M^2}{1+M^2}$

29. The radius of constant N circle at centre $\left(-\dfrac{1}{2}, \dfrac{1}{2N}\right)$ for given phase angle is
(a) $\dfrac{1}{N}$
(b) $\dfrac{\sqrt{N^2+1}}{2N}$
(c) $\dfrac{2N}{\sqrt{N+1}}$
(d) N

30. The radius of constant N circle for $N=1$ is
(a) 2
(b) $\sqrt{2}$
(c) 1
(d) $\dfrac{1}{\sqrt{2}}$

31 A radius of constant N circle having centre at $(-1/2, j0)$ in the G-plane represents the phase angle, which is equal to
(a) 90°
(b) −90°
(c) 180°
(d) −180°

32. The type of the system used for the following polar plot is
(a) 0
(b) 1
(c) 2
(d) 3

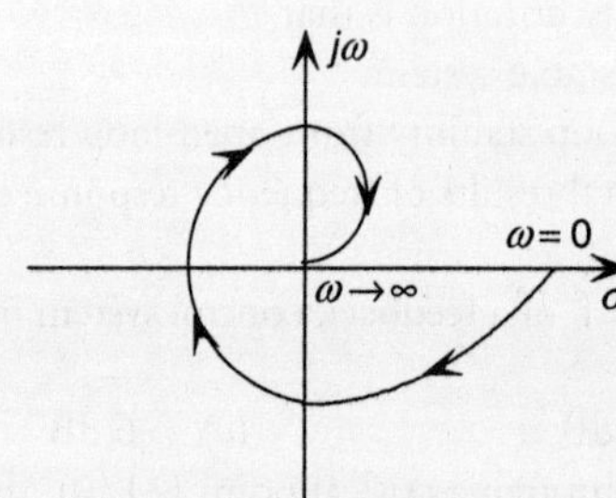

33. The Nyquist plot of Type-1 control system is shown in the following figure. The system is
(a) unstable
(b) marginally stable
(c) critical stable
(d) likely to be stable

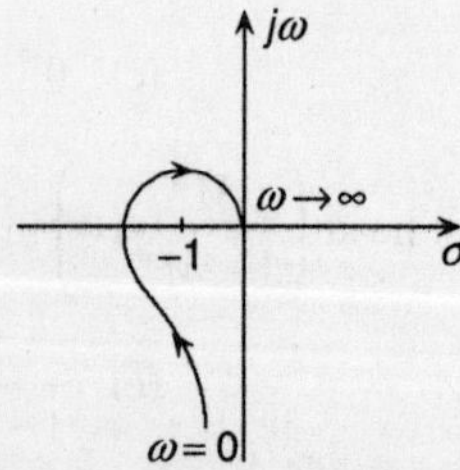

34. The GM of the following system having its polar of open-loop t.f. is given as

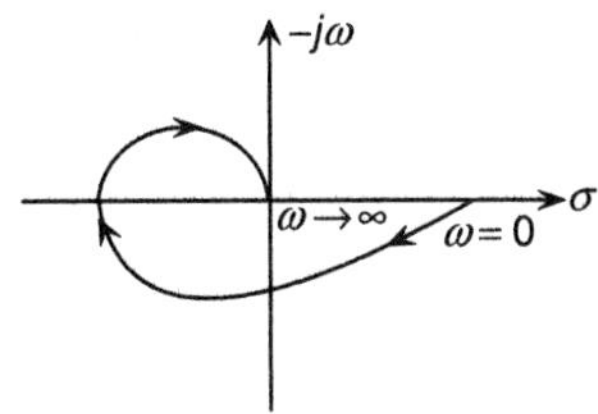

(a) −20 dB (b) 10 dB (c) 20 dB (d) −10 dB

35. When the input frequency is varied from zero to infinity causing variations in magnitude and phase angle, the traced locus by phasor is called
(a) polar plot (b) root locus (c) Bode plot (d) none

36. Nyquist stability criterion is based upon principle of
(a) arguments (b) relative stability (c) *s*-plane (d) all

37. Nyquist stability is employed for the determination of
(a) absolute stability (b) relative stability (c) both (a) and (b) (d) None

38. Which of the following technique provides quick transient and stability response?
(a) Nyquist plot (b) Bode plot (c) Root locus (d) All

ANSWERS

1. b	**2.** c	**3.** a	**4.** b	**5.** a	**6.** a	**7.** c	**8.** d	**9.** d	**10.** c
11. b	**12.** a	**13.** b	**14.** d	**15.** b	**16.** a	**17.** c	**18.** a	**19.** a	**20.** d
21. c	**22.** a	**23.** b	**24.** c	**25.** a	**26.** b	**27.** c	**28.** c	**29.** b	**30.** d
31. a	**32.** a	**33.** a	**34.** c	**35.** a	**36.** a	**37.** c	**38.** a		

34. The GH of the following system having its point of intersection [illegible] is given as [illegible]

35. When an input frequency is varied from zero to infinity the locus traced out by magnitude and phase angle of $G(j\omega)H(j\omega)$ is called

(a) Cauchy's (b) relative stability (c) [illegible] (d) all

38. Which of the following techniques is used to find both the absolute and relative stability?

1. c 2. c 3. c 4. c 5. c 6. a 7. [illegible] 8. d 9. d 10. c
11. a 12. a 13. b 14. a 15. b 16. c 17. c 18. a 19. c 20. d

CHAPTER 13

State-Space Analysis

Contents

13.1 INTRODUCTION

The analysis and design of a control system are carried out using transfer functions along with a variety of graphical techniques such as root locus plot, Bode plot and Nyquist plot based on the input–output relations of the system. They are applicable only to linear time-invariant (LTI) single-input single-output (SISO) systems. Hence, a new approach to control system analysis and design is evolved and it can be applied to the design of optimal and adaptive control systems, which are mostly time varying, with or without non-linear multiple inputs multiple outputs (MIMO). This new approach is based on the concept of state, which includes the initial conditions in the design.

13.1.1 Advantages of State-space Technique

1. It is possible to analyse linear (or non-linear) time-variant (or time-invariant) SISO (or MIMO) systems.
2. State equations are highly compatible for simulation on analogue or digital computer.

3. It is possible to optimise the system useful for optimal design.
4. State-space analysis gives us the information about the internal behaviour of the system, as well as the input and output behaviour.
5. State-space techniques can be used to find the stability of a negative feedback system when the feedback signal is other than the output signal.
6. It is possible to include initial conditions in state-space technique.

13.2 CONCEPT OF STATE, STATE VARIABLES AND STATE VECTOR

State: The state of a dynamic system is the smallest set of variables called state variables; the knowledge of these variables at $t = t_0$ together with the input for $t \geq t_0$ will completely determine the behaviour of the system for any time $t \geq t_0$.

Note that in dealing with LTI systems, we usually choose the reference time to be zero.

State variables: The state variables of a dynamic system are the smallest set of variables that determine the state of the dynamic system. If at least n variables $x_1(t)$, $x_2(t)$,... $x_n(t)$ are needed to completely describe the behaviour of a dynamic system, then such n variables $x_1(t)$, $x_2(t)$,... $x_n(t)$ are called a set of state variables.

State vector: If n state variables are needed to completely describe the behaviour of a given system, then these n state variables can be considered to be the n components of a vector $x(t)$. Such a vector is called a state vector.

State space: The n-dimensional space whose coordinate axes consists of the x_1-axis, x_2-axis ... x_n-axis is called a state space. Any state can be represented by a point in the state space.

13.3 STATE MODEL OF A GENERAL CONTROL SYSTEM

In the state variable formulation of a system, we are concerned with three types of variables such as input variables, output variables and state variables. The state-space representation of a system that has m-inputs, p-outputs and n-state variables may be visualised as shown in Figure 13.1.

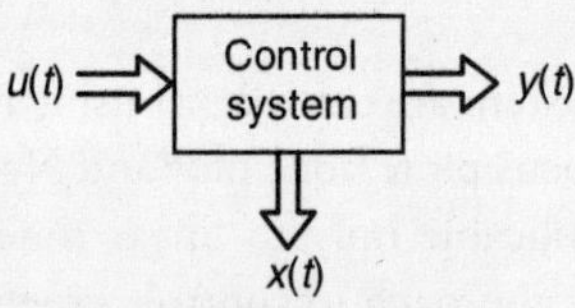

Figure 13.1 Block diagram representation of a general control system

The different variables of the system may be represented by the vectors as

$$u(t) = \text{input vector} = \begin{bmatrix} u_1(t) \\ u_2(t) \\ \vdots \\ u_m(t) \end{bmatrix}.$$

$$x(t) = \text{state variable vector} = \begin{bmatrix} x_1(t) \\ x_2(t) \\ \vdots \\ x_n(t) \end{bmatrix}$$

$$y(t) = \text{output vector} = \begin{bmatrix} y_1(t) \\ y_2(t) \\ \vdots \\ y_p(t) \end{bmatrix}.$$

For a general system of Figure 13.1, the state-space representation can be arranged in the form of *n*-first order differential equations. They are

$$\left.\begin{aligned} \frac{dx_1}{dt} &= \dot{x}_1 = f_1(x_1, x_2, \ldots x_n, u_1, u_2, \ldots u_m) \\ \frac{dx_2}{dt} &= \dot{x}_2 = f_2(x_1, x_2, \ldots x_n, u_1, u_2, \ldots u_m) \\ &\vdots \qquad \vdots \\ \frac{dx_n}{dt} &= \dot{x}_n = f_n(x_1, x_2, \ldots x_n, u_1, u_2, \ldots u_m) \end{aligned}\right\}. \tag{13.1}$$

The *n*-differential equations of (13.1) may be written in the vector form as

$$\dot{x}(t) = f[x(t), u(t)] \tag{13.2}$$

where

$$x(t) = \begin{bmatrix} x_1(t) \\ x_2(t) \\ \vdots \\ x_n(t) \end{bmatrix} = \text{State Vector}$$

$$u(t) = \begin{bmatrix} u_1(t) \\ u_2(t) \\ \vdots \\ u_m(t) \end{bmatrix} = \text{Input Vector}$$

$$f(.) = \begin{bmatrix} f_1(.) \\ f_2(.) \\ \vdots \\ f_n(.) \end{bmatrix} = \text{Function Vector}$$

Equation (13.2) is the state equation for time-invariant system. However, for time-variant systems, the function vector $f(.)$ is dependent on time as well, and the vector equation may be given as

$$\dot{x}(t) = f[x(t), u(t), t]. \tag{13.3}$$

The output, in general, can be expressed in terms of the state vector $x(t)$ and input vector $u(t)$ as follows:

For time-invariant systems:

$$y(t) = g[x(t), u(t)]. \tag{13.4}$$

For time-variant systems,

$$y(t) = g[x(t), u(t), t]. \tag{13.5}$$

where $y(t) = \begin{bmatrix} y_1(t) \\ y_2(t) \\ \vdots \\ y_p(t) \end{bmatrix}$ = Output vector

$$u(t) = \begin{bmatrix} u_1(t) \\ u_2(t) \\ \vdots \\ u_m(t) \end{bmatrix} = \text{Input vector}$$

$$g(.) = \begin{bmatrix} g_1(.) \\ g_2(.) \\ \vdots \\ g_p(.) \end{bmatrix} = \text{Function vector}$$

Equations (13.4) and (13.5) are the output equations for time-invariant and time-variant systems, respectively. The state equations and output equations together constitute the state model of the system. Thus, the state model of a general control system, as shown in Figure (13.1), is given by the following equations.

For time-invariant systems:

$$\dot{x}(t) = f[x(t), u(t)]; \text{ State equation}$$
$$y(t) = g[x(t), u(t)]; \text{ Output equation}$$

For time-variant systems:

$$\dot{x}(t) = f[x(t), u(t), t]; \text{ State equation}$$
$$y(t) = g[x(t), u(t), t]; \text{ Output equation}$$

13.4 STATE MODEL OF A LINEAR SYSTEM

The state model of a system consists of the state equation and output equation. The state equation of a system is a function of state variables and input variables and the output equation is also a function of state variables and input variables. In state-space representation, the derivative of each state variable can be written as a linear combination of system states and inputs, that is,

$$\left.\begin{aligned}
\dot{x}_1 &= a_{11}x_1 + a_{12}x_2 + \ldots + a_{1n}x_n + b_{11}u_1 + b_{12}u_2 + \ldots + b_{1m}u_m \\
\dot{x}_2 &= a_{21}x_1 + a_{22}x_2 + \ldots + a_{2n}x_n + b_{21}u_1 + b_{22}u_2 + \ldots + b_{2m}u_m \\
&\vdots \\
\dot{x}_n &= a_{n1}x_1 + a_{n2}x_2 + \ldots + a_{nn}x_n + b_{n1}u_1 + b_{n2}u_2 + \ldots + b_{nm}u_m
\end{aligned}\right\} \tag{13.6}$$

where the coefficients a_{ij} and b_{ij} are constants.

Equation (13.6) can be expressed in the matrix form as

$$\begin{bmatrix} \dot{x}_1 \\ \dot{x}_2 \\ \vdots \\ \dot{x}_n \end{bmatrix} = \begin{bmatrix} a_{11} & a_{12} & \cdots & a_{1n} \\ a_{21} & a_{22} & \cdots & a_{2n} \\ \vdots & & & \vdots \\ a_{n1} & a_{n2} & \cdots & a_{nn} \end{bmatrix} \begin{bmatrix} x_1 \\ x_2 \\ \vdots \\ x_n \end{bmatrix} + \begin{bmatrix} b_{11} & b_{12} & \cdots & b_{1m} \\ b_{21} & b_{22} & \cdots & b_{2m} \\ \vdots & & & \vdots \\ b_{n1} & b_{n2} & \cdots & b_{nm} \end{bmatrix} \begin{bmatrix} u_1 \\ u_2 \\ \vdots \\ u_m \end{bmatrix}. \tag{13.7}$$

Equation (13.7) may be written in vector matrix form as

$$\dot{x}(t) = A\,x(t) + B\,u(t);\ \text{state equation}$$

where

$$\begin{aligned}
x(t) &= \text{state vector of order } (n \times 1) \\
u(t) &= \text{input vector of order } (m \times 1) \\
A &= \text{system matrix of order } (n \times n) \\
B &= \text{input matrix of order } (n \times m)
\end{aligned}$$

Similarly, the output variables can be written as a linear combination of system states and inputs, that is,

$$\left.\begin{aligned}
y_1 &= c_{11}x_1 + c_{12}x_2 + \ldots + c_{1n}x_n + d_{11}u_1 + d_{12}u_2 + \ldots + d_{1m}u_m \\
y_2 &= c_{21}x_1 + c_{22}x_2 + \ldots + c_{2n}x_n + d_{21}u_1 + d_{22}u_2 + \ldots + d_{2m}u_m \\
&\vdots \\
y_p &= c_{p1}x_1 + c_{p2}x_2 + \ldots + c_{pn}x_n + c_{p1}u_1 + c_{p2}u_2 + \ldots + d_{pm}u_m
\end{aligned}\right\} \tag{13.8}$$

where the coefficients c_{ij} and d_{ij} are constants.

In the matrix form, the above-mentioned equations can be expressed as

$$\begin{bmatrix} y_1 \\ y_2 \\ \vdots \\ y_p \end{bmatrix} = \begin{bmatrix} c_{11} & c_{12} & \cdots & c_{1n} \\ c_{21} & c_{22} & \cdots & c_{2n} \\ \vdots & & & \vdots \\ c_{p1} & c_{p2} & \cdots & c_{pn} \end{bmatrix} \begin{bmatrix} x_1 \\ x_2 \\ \vdots \\ x_n \end{bmatrix} + \begin{bmatrix} d_{11} & d_{12} & \cdots & d_{1m} \\ d_{21} & d_{22} & \cdots & d_{2m} \\ \vdots & & & \vdots \\ d_{p1} & d_{p2} & \cdots & d_{pm} \end{bmatrix} \begin{bmatrix} u_1 \\ u_2 \\ \vdots \\ u_m \end{bmatrix}. \tag{13.9}$$

Equations (13.9) can be expressed in vector matrix form as

$$y(t) = C\,x(t) + D\,u(t);\ \text{output equation}$$

where

$$\begin{aligned}
y(t) &= \text{output vector of order } (p \times 1) \\
C &= \text{output matrix of order } (p \times n) \\
D &= \text{transmission matrix of order } (p \times m)
\end{aligned}$$

The state model of a system consists of state equation and output equation. Hence, the state model of an LTI system is given by the following equations.

$$\dot{x}(t) = A\,x(t) + B\,u(t) \text{..... state equation}$$

$$y(t) = C\,x(t) + D\,u(t) \text{..... output equation.}$$

The block diagram representation of the state model of an LTI system is shown in Figure 13.2.

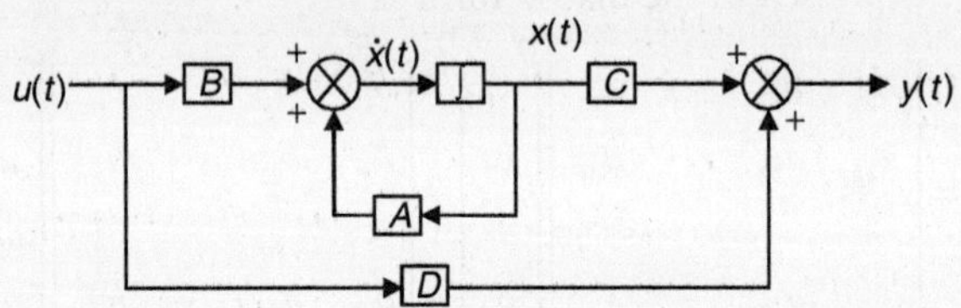

Figure 13.2 Block diagram of a state model

13.5 SELECTION OF STATE VARIABLES

The state variables of a given system are not unique. There are infinitely many choices for any given system. Following are some of the guidelines to choose the state variables of a given system.

1. For a physical system like an electrical system, the number of state variables needed to represent the system must be equal to the number of energy storing elements present in the system.
2. If a system is represented by a linear differential equation, then the number of state variables needed to represent the system must be equal to the order of the differential equation.
3. If a system is represented by a t.f., then the number of state variables needed to represent the system must be equal to the highest power of s in the denominator of the t.f.

13.6 STATE-SPACE REPRESENTATION OF CONTROL SYSTEMS

13.6.1 State-space Representation Using Physical Variables

A minimal number of state variables are chosen for obtaining the state model of the system. For a physical system such as an electrical system or mechanical system, the number of state variables needed to represent the system must be equal to the number of the energy storing elements presented in the system. The best choice of state variables in an electrical system is current and voltage in energy storage elements. The choice of state variables in mechanical system is displacement, velocity or acceleration.

Problem-13.1: Obtain the state model for an *RLC* series circuit.

Solution:

Consider an *RLC* series network, as shown in the figure. The dynamic behaviour of the system is completely defined for $t \geq t_0$.

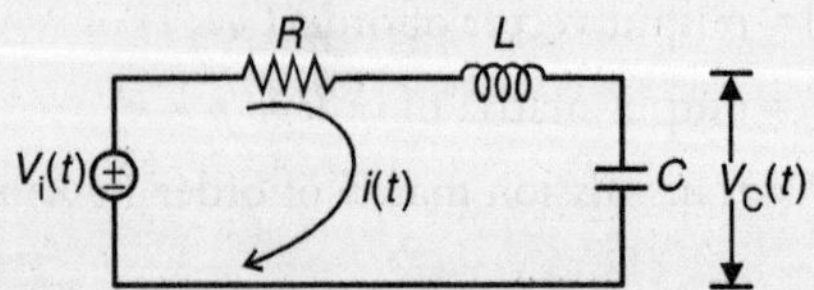

By applying KVL to the circuit, we have

$$V_i = Ri + L\frac{di}{dt} + \frac{1}{C}\int i dt. \quad (1)$$

$$V_c = \frac{1}{C}\int i dt$$

$$\Rightarrow \quad i = C\frac{dV_c}{dt}. \quad (2)$$

Let us choose the state variables as i and V_c

i.e.,
$$x_1 = i$$
$$x_2 = V_c$$

In vector matrix notation, (1) and (2) can be written as

$$V_i = Rx_1 + L\dot{x}_1 + x_2$$

$$\Rightarrow \quad \dot{x}_1 = \frac{-R}{L}x_1 - \frac{1}{L}x_2 + \frac{1}{L}V_i$$

$$x_1 = C\dot{x}_2$$

$$\Rightarrow \quad \dot{x}_2 = \frac{1}{C}x_1.$$

The state-space representation of the above-mentioned system is

$$\begin{bmatrix}\dot{x}_1\\ \dot{x}_2\end{bmatrix} = \begin{bmatrix}-\frac{R}{L} & -\frac{1}{L}\\ \frac{1}{C} & 0\end{bmatrix}\begin{bmatrix}x_1\\ x_2\end{bmatrix} + \begin{bmatrix}\frac{1}{L}\\ 0\end{bmatrix}V_i.$$

Problem-13.2: Obtain the state model for the physical system as shown in the figure.

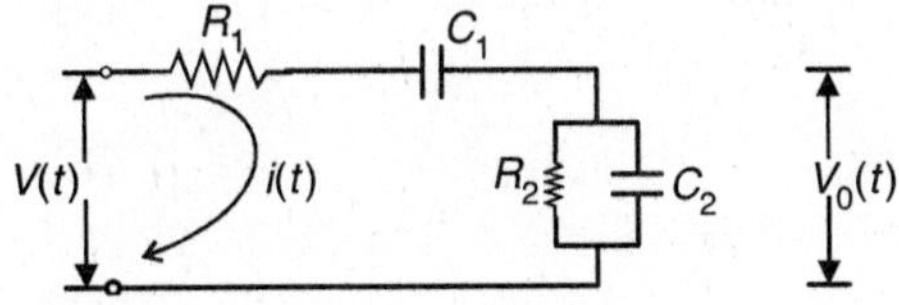

Solution:

Let us choose the state variables as voltage across C_1 and C_2.

i.e.,
$$x_1 = \frac{1}{C_1}\int i dt \quad (1)$$

$$x_2 = V_0(t) \quad (2)$$

$$\Rightarrow \quad C_1\dot{x}_1 = i \quad (3)$$

By applying KVL to the circuit, we have

$$V(t) = R_1 i + \frac{1}{C_1}\int i\,dt + V_0(t)$$

$$V(t) - V_0(t) = R_1 i + x_1. \tag{4}$$

From (3) and (4), we get

$$V(t) - x_2 = R_1 C_1 \dot{x}_1 + x_1$$

$$\Rightarrow \qquad \dot{x}_1 = -\frac{1}{R_1 C_1}x_1 - \frac{1}{R_1 C_1}x_2 + \frac{1}{R_1 C_1}V(t) \tag{5}$$

By applying KCL to the circuit, we have

$$i = \text{current through } R_2 + \text{current through } C_2$$

$$= \frac{V_0(t)}{R_2} + C_2\frac{dV_0(t)}{dt}.$$

$$i = \frac{x_2}{R_2} + C_2\dot{x}_2. \tag{6}$$

From (3) and (6), we get

$$C_1\dot{x}_1 = \frac{x_2}{R_2} + C_2\dot{x}_2$$

$$C_1\left[-\frac{1}{R_1 C_1}x_1 - \frac{1}{R_1 C_1}x_2 + \frac{1}{R_1 C_1}V(t)\right] = \frac{x_2}{R_2} + C_2\dot{x}_2 \; [\because \text{from (5)}]$$

$$-\frac{1}{R_1 C_2}x_1 - \frac{1}{R_1 C_2}x_2 + \frac{1}{R_1 C_2}V(t) = \frac{x_2}{R_2 C_2} + \dot{x}_2$$

$$\Rightarrow \qquad \dot{x}_2 = -\frac{1}{R_1 C_2}x_1 - \left[\frac{1}{R_1 C_2} + \frac{1}{R_2 C_2}\right]x_2 + \frac{1}{R_1 C_2}V(t). \tag{7}$$

Let the output be $V_0(t) = y$.

$$\Rightarrow \qquad y = x_2$$

$$\therefore \qquad \dot{x}_1 = -\frac{1}{R_1 C_1}x_1 - \frac{1}{R_1 C_1}x_2 + \frac{1}{R_1 C_1}V(t)$$

$$\dot{x}_2 = -\frac{1}{R_1 C_2}x_1 - \left[\frac{1}{R_1 C_2} + \frac{1}{R_2 C_2}\right]x_2 + \frac{1}{R_1 C_2}V(t)$$

$$y(t) = x_2.$$

The state-space representation of the system is

$$\begin{bmatrix} \dot{x}_1 \\ \dot{x}_2 \end{bmatrix} = \begin{bmatrix} -\dfrac{1}{R_1C_1} & -\dfrac{1}{R_1C_1} \\ -\dfrac{1}{R_1C_2} & \left(\dfrac{-1}{R_1C_2} + \dfrac{-1}{R_2C_2}\right) \end{bmatrix} \begin{bmatrix} x_1 \\ x_2 \end{bmatrix} + \begin{bmatrix} \dfrac{1}{R_1C_1} \\ \dfrac{1}{R_1C_2} \end{bmatrix} V(t)$$

$$y = [0 \quad 1]\begin{bmatrix} x_1 \\ x_2 \end{bmatrix}.$$

Problem-13.3: Obtain the state model for the given mechanical system as shown in the figure.

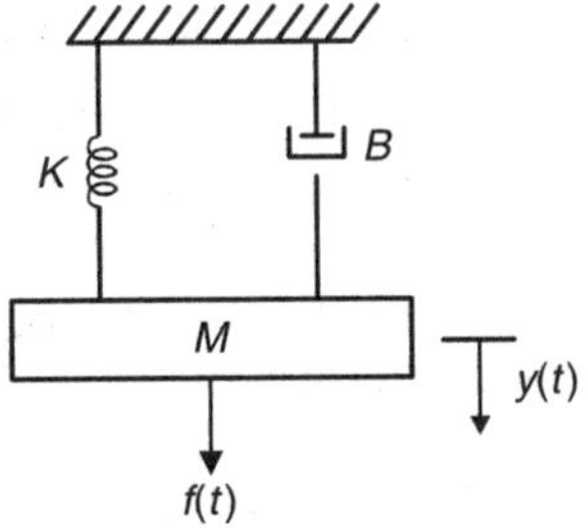

Solution:

By applying Newton's Second Law, the equation of motion is given as

$$M\frac{d^2y}{dt^2} + B\frac{dy}{dt} + Ky = f(t). \tag{1}$$

Let us choose the state variables as x_1 and x_2.

$\therefore$
$$x_1 = y$$

$$x_2 = \frac{dy}{dt} = \dot{x}_1$$

$\Rightarrow$
$$\dot{x}_1 = x_2 \tag{2}$$

From (1), we have

$$M\dot{x}_2 + Bx_2 + Kx_1 = f(t)$$

$$\dot{x}_2 = \frac{-K}{M}x_1 - \frac{B}{M}x_2 + \frac{1}{M}f(t). \tag{3}$$

Let the displacement y be the output

i.e.,
$$y = x_1. \tag{4}$$

The state-space representation of the given system is

$$\dot{x}_1 = x_2$$

$$\dot{x}_2 = \frac{-K}{M}x_1 - \frac{B}{M}x_2 + \frac{1}{M}f(t)$$

$$y = x_1$$

$$\begin{bmatrix} \dot{x}_1 \\ \dot{x}_2 \end{bmatrix} = \begin{bmatrix} 0 & 1 \\ \frac{-K}{M} & \frac{-B}{M} \end{bmatrix} \begin{bmatrix} x_1 \\ x_2 \end{bmatrix} + \begin{bmatrix} 0 \\ \frac{1}{M} \end{bmatrix} f(t)$$

$$y = [1 \quad 0] \begin{bmatrix} x_1 \\ x_2 \end{bmatrix}.$$

13.6.2 State-space Representation Using Phase Variables

The phase variables are defined as those particular state variables that are obtained from one of the system variables and its derivatives. Often, the variable used is the system output and remaining state variables are then the derivatives of the output. The phase-variable state model is easily determined if the system model is already known in the form of differential equation or t.f.

Consider the following n^{th} order differential equation governing the output $y(t)$ to the input $u(t)$ of a system.

$$\overset{n\bullet}{y} + a_1 \overset{(n-1)\bullet}{y} + \ldots + a_{n-1}\dot{y} + a_n y = b_0 \overset{n\bullet}{u} + b_1 \overset{(n-1)\bullet}{u} + \ldots + b_{n-1}\dot{u} + b_n \dot{u} \tag{13.10}$$

The t.f. of the above-mentioned system may be expressed as

$$G(s) = \frac{Y(s)}{U(s)} = \frac{b_0 s^n + b_1 s^{n-1} + \ldots. + b_{n-1}s + b_n}{s^n + a_1 s^{n-1} + \ldots. + a_{n-1}s + a_n}. \tag{13.11}$$

Using phase variables, we can represent the system defined by (13.11) in either of the phase variable canonical forms, as discussed in the following methods.

Method 1: Controllable canonical form

The t.f. of the given system can be written as

$$G(s) = \frac{Y(s)}{U(s)} = \frac{Y(s)}{Q(s)} \cdot \frac{Q(s)}{U(s)} \tag{13.12}$$

where

$$\frac{Q(s)}{U(s)} = \frac{1}{s^n + a_1 s^{n-1} + \ldots + a_{n-1}s + a_n} \tag{13.13}$$

and

$$\frac{Y(s)}{Q(s)} = b_0 s^n + b_1 s^{n-1} + \ldots + b_{n-1}s + b_n. \tag{13.14}$$

Equation (13.13) can be rewritten as

$$(s^n + a_1 s^{n-1} + \ldots + a_{n-1}s + a_n)Q(s) = U(s)$$

$$s^n Q(s) = -a_1 s^{n-1}Q(s) - \ldots - a_{n-1}\, sQ(s) - a_n Q(s) + U(s). \tag{13.15}$$

Let us define the state variables as

$$\left.\begin{aligned} X_1(s) &= Q(s) \\ X_2(s) &= sQ(s) &&= sX_1(s) \\ X_3(s) &= s^2Q &&= sX_2(s) \\ &\vdots && \vdots \\ X_{n-1}(s) &= s^{n-2}Q(s) &&= sX_{n-2}(s) \\ X_n(s) &= s^{n-1}Q(s) &&= sX_{n-1}(s) \end{aligned}\right\}. \tag{13.16}$$

Then, clearly

$$\begin{aligned} sX_1(s) &= X_2(s) \\ sX_2(s) &= X_3(s) \\ &\vdots \\ sX_{n-2}(s) &= X_{n-1}(s) \\ sX_{n-1}(s) &= X_n(s) \\ sX_n(s) &= -a_1X_n(s) - \ldots - a_{n-1}X_2(s) - a_nX_1(s) + U(s). \end{aligned}$$

By taking inverse Laplace transforms to the afore-mentioned set of equations, we have

$$\begin{aligned} \dot{x}_1 &= x_2 \\ \dot{x}_2 &= x_3 \\ &\vdots \\ \dot{x}_{n-1} &= x_n \\ \dot{x}_n &= -a_1x_n - a_2x_{n-1} \cdots - a_{n-1}x_2 - a_nx_1 + u. \end{aligned}$$

The above-mentioned equations can be represented in matrix form as

$$\begin{bmatrix} \dot{x}_1 \\ \dot{x}_2 \\ \vdots \\ \dot{x}_n \end{bmatrix} = \begin{bmatrix} 0 & 1 & 0 & \cdots & 0 \\ 0 & 0 & 1 & \cdots & 0 \\ \vdots & \vdots & \vdots & & \\ -a_n & -a_{n-1} & -a_{n-2} & \cdots & -a_1 \end{bmatrix} \begin{bmatrix} x_1 \\ x_2 \\ \vdots \\ x_n \end{bmatrix} + \begin{bmatrix} 0 \\ 0 \\ \vdots \\ 1 \end{bmatrix} u(t). \tag{13.17}$$

Now, (13.14) can be written as

$$\begin{aligned} Y(s) &= b_0s^nQ(s) + b_1s^{n-1}Q(s) + \ldots + b_{n-1}sQ(s) + b_nQ(s) \\ &= b_0[-a_1s^{n-1}Q(s) - \ldots - a_{n-1}sQ(s) - a_nQ(s) + U(s)] \\ &\quad + b_1s^{n-1}Q(s) + \ldots + b_{n-1}sQ(s) + b_nQ(s) \\ Y(s) &= (b_1 - a_1b_0)s^{n-1}Q(s) + \ldots + (b_{n-1} - a_{n-1}b_0)sQ(s) \\ &\quad + (b_n - a_nb_0)Q(s) + b_0U(s). \end{aligned} \tag{13.18}$$

From (13.16) and (13.18), we have

$$Y(s) = (b_n - a_nb_0)X_1(s) + (b_{n-1} - a_{n-1}b_0)X_2(s) + \ldots.. + (b_1 - a_1b_0)X_n(s) + b_0U(s).$$

In matrix form, after taking inverse Laplace transforms, we get

$$y = [(b_n - a_n b_0)(b_{n-1} - a_{n-1} b_0).....(b_1 - a_1 b_0)] \begin{bmatrix} x_1 \\ x_2 \\ \vdots \\ x_n \end{bmatrix} + b_0 u. \tag{13.19}$$

Problem-13.4: Convert the given t.f. into state–space form:

$$\frac{Y(s)}{U(s)} = \frac{s^2 + 2s + 4}{s^3 + 3s^2 + 6s + 10}.$$

Solution:

Let
$$\frac{Y(s)}{U(s)} = \frac{Y(s)}{X(s)} \cdot \frac{X(s)}{U(s)}$$

where
$$\frac{X(s)}{U(s)} = \frac{1}{s^3 + 3s^2 + 6s + 10}$$

$\Rightarrow$
$$s^3 X(s) + 3s^2 X(s) + 6sX(s) + 10X(s) = U(s) \tag{1}$$

and
$$\frac{Y(s)}{X(s)} = s^2 + 2s + 4$$

$\Rightarrow$
$$Y(s) = s^2 X(s) + 2sX(s) + 4X(s) \tag{2}$$

By taking inverse Laplace transforms to the above-mentioned equations, we have

$$\frac{d^3 x}{dt^3} + 3\frac{d^2 x}{dt^2} + 6\frac{dx}{dt} + 10x = u \tag{3}$$

$$y = \frac{d^2 x}{dt^2} + 2\frac{dx}{dt} + 4x. \tag{4}$$

Let us choose the state variables as x_1, x_2 and x_3.

i.e.,
$$x_1 = x$$

$$x_2 = \frac{dx}{dt} = \dot{x}_1 \Rightarrow \dot{x}_1 = x_2$$

$$x_3 = \frac{d^2 x}{dt^2} = \dot{x}_2 \Rightarrow \dot{x}_2 = x_3.$$

From (3) and (4), we have

$$\dot{x}_3 + 3x_3 + 6x_2 + 10x_1 = u$$

$$\dot{x}_3 = -10x_1 - 6x_2 - 3x_3 + u.$$

$\Rightarrow$
$$y = x_3 + 2x_2 + 4x_1.$$

∴ The state and output equations of the above-mentioned system are

$$\dot{x}_1 = x_2$$
$$\dot{x}_2 = x_3$$
$$\dot{x}_3 = -10x_1 - 6x_2 - 3x_3 + u$$
$$y = 4x_1 + 2x_2 + x_3.$$

The state–space representation of the above system is

$$\begin{bmatrix} \dot{x}_1 \\ \dot{x}_2 \\ \dot{x}_3 \end{bmatrix} = \begin{bmatrix} 0 & 1 & 0 \\ 0 & 0 & 1 \\ -10 & -6 & -3 \end{bmatrix} \begin{bmatrix} x_1 \\ x_2 \\ x_3 \end{bmatrix} + \begin{bmatrix} 0 \\ 0 \\ 1 \end{bmatrix} u.$$

$$y = [4 \quad 2 \quad 1] \begin{bmatrix} x_1 \\ x_2 \\ x_3 \end{bmatrix}.$$

Problem-13.5: Obtain the state–space representation of an nth order differential equation.

Solution:

Consider the nth order differential equation as

$$\overset{n\bullet}{y} + a_1 \overset{(n-1)\bullet}{y} + \ldots + a_{n-1}\dot{y} + a_n y = bu. \tag{1}$$

Let us assume the state variables as $x_1, x_2, x_3 \ldots x_n$, that is, output and its derivatives

∴ $$x_1 = y$$

$$x_2 = \dot{y} = \dot{x}_1 \quad \Rightarrow \dot{x}_1 = x_2$$
$$x_3 = \ddot{y} = \dot{x}_2 \quad \Rightarrow \dot{x}_2 = x_3$$
$$\vdots \quad \vdots \qquad \vdots$$
$$x_n = \overset{(n-1)\bullet}{y} = \dot{x}_{n-1} \Rightarrow \dot{x}_{n-1} = x_n$$

From (1), we have

$$\dot{x}_n + a_1 x_n + a_2 x_{n-1} + \ldots + a_{n-1}x_2 + a_n x_1 = bu$$
$$\dot{x}_n = -a_n x_1 - a_{n-1}x_2 - \ldots - a_2 x_{n-1} - a_1 x_n + bu.$$

The state equations of the above–mentioned system are

$$\dot{x}_1 = x_2$$
$$\dot{x}_2 = x_3$$
$$\vdots$$
$$\dot{x}_{n-1} = x_n$$
$$\dot{x}_n = -a_n x_1 - a_{n-1}x_2 - \ldots - a_2 x_{n-1} - a_n x_n + bu.$$

On arranging the above–mentioned equations in the matrix form, we get

$$\begin{bmatrix}\dot{x}_1\\ \dot{x}_2\\ \vdots\\ \dot{x}_{n-1}\\ \dot{x}_n\end{bmatrix}=\begin{bmatrix}0 & 1 & 0 & \cdots & 0\\ 0 & 0 & 1 & \cdots & 0\\ \vdots & & & & \vdots\\ 0 & 0 & 0 & \cdots & 1\\ -a_n & -a_{n-1} & -a_{n-2} & \cdots & a_1\end{bmatrix}\begin{bmatrix}x_1\\ x_2\\ \vdots\\ x_{n-1}\\ x_n\end{bmatrix}+\begin{bmatrix}0\\ 0\\ \vdots\\ 0\\ b\end{bmatrix}u.$$

The output equation becomes

$$y = x_1 = [1\ 0\ 0\ \cdots\ 0]\begin{bmatrix}x_1\\ x_2\\ \vdots\\ x_n\end{bmatrix}.$$

***Problem-13.6*:** Consider the system defined by $\dddot{y}+6\ddot{y}+11\dot{y}+6y=6u$, where y is the output and u is the input of the system. Obtain a state–space representation, and the block diagram representation of the system.

***Solution*:**

Given that

$$\frac{d^3y}{dt^3}+6\frac{d^2y}{dt^2}+11\frac{dy}{dt}+6y=6u. \tag{1}$$

Let us define the state variables as x_1, x_2 and x_3.

$\therefore$

$$x_1 = y$$

$$x_2 = \frac{dy}{dt} = \dot{x}_1$$

$$x_3 = \frac{d^2y}{dt^2} = \dot{x}_2$$

From (1), we have

$$\dot{x}_3 + 6x_3 + 11x_2 + 6x_1 = 6u$$

$$\dot{x}_3 = -6x_1 - 11x_2 - 6x_3 + 6u$$

$\therefore$ The state equations of the above-mentioned system are

$$\dot{x}_1 = x_2$$

$$\dot{x}_2 = x_3$$

$$\dot{x}_3 = -6x_1 - 11x_2 - 6x_3 + 6u.$$

On arranging the above-mentioned equations in the matrix form, we get

$$\begin{bmatrix}\dot{x}_1\\ \dot{x}_2\\ \dot{x}_3\end{bmatrix}=\begin{bmatrix}0 & 1 & 0\\ 0 & 0 & 1\\ -6 & -11 & -6\end{bmatrix}\begin{bmatrix}x_1\\ x_2\\ x_3\end{bmatrix}+\begin{bmatrix}0\\ 0\\ 6\end{bmatrix}u.$$

The output equation is given as

$$y = x_1 = [1\ 0\ 0]\begin{bmatrix} x_1 \\ x_2 \\ x_3 \end{bmatrix}.$$

The following figure shows the block diagram representation of the present state equation and output equation. Notice that the t.f. of the feedback blocks is identical with negatives of the coefficients of the original differential equation.

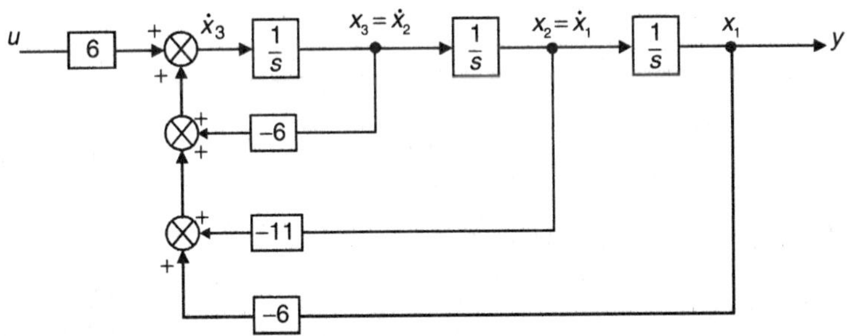

Method 2: Observable canonical form

The t.f. of the general control system is

$$G(s) = \frac{Y(s)}{U(s)} = \frac{b_0 s^n + b_1 s^{n-1} + \ldots + b_{n-1}s + b_n}{s^n + a_1 s^{n-1} + \ldots + a_{n-1}s + a_n}$$

$$s^n Y(s) + a_1 s^{n-1} Y(s) + \ldots + a_{n-1} s Y(s) + a_n Y(s) = b_0 s^n U(s) + b_1 s^{n-1} U(s) + \ldots + b_{n-1} s U(s) + b_n U(s).$$

$$\begin{aligned} Y(s) = b_0 U(s) + \frac{1}{s}\left[b_1 U(s) - a_1 Y(s)\right] + \ldots + \frac{1}{s^{n-1}} \\ \left[b_{n-1} U(s) - a_{n-1} Y(s)\right] + \frac{1}{s^n}\left[b_n U(s) - a_n Y(s)\right]. \end{aligned} \tag{13.20}$$

Let us define the state variables as

$$\left.\begin{aligned} X_n(s) &= \frac{1}{s}[b_1 U(s) - a_1 Y(s) + X_{n-1}(s)] \\ X_{n-1}(s) &= \frac{1}{s}[b_2 U(s) - a_2 Y(s) + X_{n-2}(s)] \\ &\vdots \\ &\vdots \\ X_2(s) &= \frac{1}{s}[b_{n-1} U(s) - a_{n-1} Y(s) + X_1(s)] \\ X_1(s) &= \frac{1}{s}[b_n U(s) - a_n Y(s)] \end{aligned}\right\}. \tag{13.21}$$

From (13.20) and (13.21), we have

$$Y(s) = b_0 U(s) + X_n(s). \tag{13.22}$$

By taking inverse Laplace transforms on both sides, we have

$$y = x_n + b_0 u. \tag{13.23}$$

When we substitute $Y(s)$ from (13.22) in (13.23) and multiply both sides of the equations by s, we get

$$\left.\begin{aligned} sX_n(s) &= b_1 U(s) - a_1[b_0 U(s) + X_n(s)] + X_{n-1}(s) \\ sX_n(s) &= X_{n-1}(s) - a_1 X_n(s) + [b_1 - a_1 b_0]U(s) \\ sX_{n-1}(s) &= X_{n-2}(s) - a_2 X_n(s) + [b_2 - a_2 b_0]U(s) \\ &\vdots \\ sX_2(s) &= X_1(s) - a_{n-1} X_n(s) + [b_{n-1} - a_{n-1} b_0]U(s) \\ sX_1(s) &= -a_n X_n(s) + [b_n - a_n b_0]U(s) \end{aligned}\right\}. \tag{13.24}$$

By taking inverse Laplace transforms on both sides of (13.24) and writing them in the reverse order, we obtain

$$\left.\begin{aligned} \dot{x}_1 &= -a_n x_n + [(b_n - a_n b_0)]\, u \\ \dot{x}_2 &= x_1 - a_{n-1} x_n + (b_{n-1} - a_{n-1} b_0)\, u \\ &\vdots \\ \dot{x}_{n-1} &= x_{n-2} - a_2 x_n + [b_2 - a_2 b_0]\, u \\ \dot{x}_n &= x_{n-1} - a_1 x_n + [b_1 - a_1 b_0]\, u \\ y &= x_n + b_0 u \end{aligned}\right\}. \tag{13.25}$$

The above-mentioned equations can be written in matrix form as

$$\begin{bmatrix} \dot{x}_1 \\ \dot{x}_2 \\ \vdots \\ \dot{x}_{n-1} \\ \dot{x}_n \end{bmatrix} = \begin{bmatrix} 0 & 0 & \cdots & 0 & 0 & -a_n \\ 1 & 0 & & 0 & 0 & -a_{n-1} \\ \vdots & & & & \vdots & \\ 0 & 0 & \cdots & 1 & 0 & -a_2 \\ 0 & 0 & \cdots & 0 & 1 & -a_1 \end{bmatrix} \begin{bmatrix} x_1 \\ x_2 \\ \vdots \\ x_{n-1} \\ x_n \end{bmatrix} + \begin{bmatrix} b_n - a_n b_0 \\ b_{n-1} - a_{n-1} b_0 \\ \vdots \\ b_2 - a_2 b_0 \\ b_1 - a_1 b_0 \end{bmatrix} u. \tag{13.26}$$

$$y = [0 \quad 0 \quad \cdots \quad 0 \quad 1] \begin{bmatrix} x_1 \\ x_2 \\ \vdots \\ x_n \end{bmatrix} + b_0 u. \tag{13.27}$$

Equations (13.26) and (13.27) give the phase-variable state model of the system in observable canonical form.

Method 3: Diagonal canonical form

Consider the t.f. of the system, which is defined as

$$G(s)=\frac{Y(s)}{U(s)}=\frac{b_0s^n+b_1s^{n-1}+\ldots+b_{n-1}s+b_n}{s^n+a_1s^{n-1}+\ldots+a_{n-1}s+a_n} \tag{13.28}$$

Here, we consider the case where the denominator polynomial involves only distinct roots. For the distinct roots case, (13.28) can be written as

$$G(s)=\frac{Y(s)}{U(s)}=\frac{b_0s^n+b_1s^{n-1}+\ldots+b_{n-1}s+b_n}{(s-\alpha_1)(s-\alpha_2)\ldots(s-\alpha_n)}$$

$$=b_0+\frac{k_1}{s-\alpha_1}+\frac{k_2}{s-\alpha_2}+\ldots+\frac{k_n}{s-\alpha_n}. \tag{13.29}$$

where k_i $(i=1, 2\ldots n)$ are the residues of the poles at

$$s=\alpha_i\,(i=1,2,\ldots n).$$

Equation (13.29) can be written as

$$Y(s)=b_0U(s)+\frac{k_1}{s-\alpha_1}U(s)+\frac{k_2}{s-\alpha_2}U(s)+\ldots+\frac{k_n}{s-\alpha_n}U(s). \tag{13.30}$$

Let us define the state variables as

$$X_1(s)=\frac{1}{s-\alpha_1}U(s)$$

$$X_2(s)=\frac{1}{s-\alpha_2}U(s)$$

$$\vdots \qquad \vdots$$

$$X_n(s)=\frac{1}{s-\alpha_n}U(s).$$

Then, clearly

$$sX_1(s)=\alpha_1X_1(s)+U(s)$$

$$sX_2(s)=\alpha_2X_2(s)+U(s)$$

$$\vdots \qquad \vdots \qquad \vdots$$

$$sX_n(s)=\alpha_nX_n(s)+U(s).$$

By taking inverse Laplace transforms to the above-mentioned equations, we have

$$\dot{x}_1=\alpha_1x_1+u$$

$$\dot{x}_2=\alpha_2x_2+u$$

$$\vdots$$

$$\dot{x}_n=\alpha_nx_n+u.$$

The above-mentioned equations can be written in matrix form as

$$\begin{bmatrix} \dot{x}_1 \\ \dot{x}_2 \\ \vdots \\ \dot{x}_n \end{bmatrix} = \begin{bmatrix} \alpha_1 & 0 & 0 & \cdots & 0 \\ 0 & \alpha_2 & 0 & \cdots & 0 \\ \vdots & & & & \vdots \\ 0 & 0 & 0 & \cdots & \alpha_n \end{bmatrix} \begin{bmatrix} x_1 \\ x_2 \\ \vdots \\ x_n \end{bmatrix} + \begin{bmatrix} 1 \\ 1 \\ \vdots \\ 1 \end{bmatrix} u. \tag{13.31}$$

Now, in terms of state variables, (13.30) can be written as

$$Y(s) = b_0 U(s) + k_1 X_1(s) + k_2 X_2(s) + \ldots k_n X_n(s).$$

By taking inverse Laplace transform, we have

$$y = k_1 x_1 + k_2 x_2 + \ldots + k_n x_n + b_0 u.$$

In matrix form, we have

$$y = [k_1 \; k_2 \; \ldots \; k_n] \begin{bmatrix} x_1 \\ x_2 \\ \vdots \\ x_n \end{bmatrix} + b_0 u. \tag{13.32}$$

Equations (13.31) and (13.32) give the state model of the system in diagonal canonical form.

Method 4: Jordan canonical form

Consider the t.f. of the system, which is defined as

$$G(s) = \frac{Y(s)}{U(s)} = \frac{b_0 s^n + b_1 s^{n-1} + \ldots + b_{n-1} s + b_n}{s^n + a_1 s^{n-1} + \ldots + a_{n-1} s + a_n}. \tag{13.33}$$

Here, we consider the case where the denominator polynomial involves multiple roots. Let us assume that the denominator polynomial involves a triple pole at $s = a_1$ and a double pole at $s = a_4$. All other roots are distinct from each other. For this case, (13.33) can be written as

$$G(s) = \frac{Y(s)}{U(s)} = \frac{b_0 s^n + b_1 s^{n-1} + \ldots + b_{n-1} s + b_n}{(s-\alpha_1)^3 (s-\alpha_4)^2 (s-\alpha_6)(s-\alpha_7)\ldots(s-\alpha_n)}. \tag{13.34}$$

The partial fraction expansion becomes

$$\frac{Y(s)}{U(s)} = b_0 + \frac{k_1}{(s-\alpha_1)^3} + \frac{k_2}{(s-\alpha_1)^2} + \frac{k_3}{(s-\alpha_1)} + \frac{k_4}{(s-\alpha_4)^2} + \frac{k_5}{(s-\alpha_4)} + \frac{k_6}{(s-\alpha_6)} + \frac{k_7}{(s-\alpha_7)} + \ldots + \frac{k_n}{(s-\alpha_n)}$$

or

$$Y(s) = b_0 U(s) + \frac{k_1}{(s-\alpha_1)^3} U(s) + \frac{k_2}{(s-\alpha_1)^2} U(s) + \frac{k_3}{(s-\alpha_1)} U(s) + \frac{k_4}{(s-\alpha_4)^2} U(s) + \frac{k_5}{(s-\alpha_4)} U(s) + \frac{k_6}{(s-\alpha_6)} U(s) + \frac{k_7}{(s-\alpha_7)} U(s) + \ldots + \frac{k_n}{(s-\alpha_n)} U(s). \tag{13.35}$$

Let us define the state variables as

$$X_1(s) = \frac{1}{(s-\alpha_1)^3}U(s)$$

$$X_2(s) = \frac{1}{(s-\alpha_1)^2}U(s)$$

$$X_3(s) = \frac{1}{(s-\alpha_1)}U(s)$$

$$X_4(s) = \frac{1}{(s-\alpha_4)^2}U(s)$$

$$X_5(s) = \frac{1}{(s-\alpha_4)}U(s)$$

$$X_6(s) = \frac{1}{(s-\alpha_6)}U(s)$$

$$\vdots \qquad \vdots$$

$$X_n(s) = \frac{1}{(s-\alpha_n)}U(s).$$

Further, note that the following relationship exists among $X_1(s)$, $X_2(s)$ and $X_3(s)$ and in between $X_4(s)$ and $X_5(s)$.

$$\frac{X_1(s)}{X_2(s)} = \frac{1}{s-\alpha_1}$$

$$\frac{X_2(s)}{X_3(s)} = \frac{1}{s-\alpha_1}$$

and

$$\frac{X_4(s)}{X_5(s)} = \frac{1}{s-\alpha_4}.$$

Then, from the preceding definition of state variables and the preceding relationships, we have

$$sX_1(s) = \alpha_1 X_1(s) + X_2(s)$$

$$sX_2(s) = \alpha_1 X_2(s) + X_3(s)$$

$$sX_3(s) = \alpha_1 X_3(s) + U(s)$$

$$sX_4(s) = \alpha_4 X_4(s) + X_5(s)$$

$$sX_5(s) = \alpha_4 X_5(s) + U(s)$$

$$sX_6(s) = \alpha_6 X_6(s) + U(s)$$

$$sX_7(s) = \alpha_7 X_7(s) + U(s)$$

$$\vdots \qquad \vdots$$

$$sX_n(s) = \alpha_n X_n(s) + U(s).$$

By taking inverse Laplace transforms to the above-mentioned equation, we have

$$\dot{x}_1 = \alpha_1 x_1 + x_2$$
$$\dot{x}_2 = \alpha_1 x_2 + x_3$$
$$\dot{x}_3 = \alpha_1 x_3 + u$$
$$\dot{x}_4 = \alpha_4 x_4 + x_5$$
$$\dot{x}_5 = \alpha_4 x_5 + u$$
$$\dot{x}_6 = \alpha_6 x_6 + u$$
$$\dot{x}_7 = \alpha_7 x_7 + u$$
$$\vdots \quad \vdots$$
$$\dot{x}_n = \alpha_n x_n + u.$$

The above-mentioned equation can be expressed in a matrix form as

$$\begin{bmatrix} \dot{x}_1 \\ \dot{x}_2 \\ \dot{x}_3 \\ \dot{x}_4 \\ \dot{x}_5 \\ \dot{x}_6 \\ \dot{x}_7 \\ \vdots \\ \dot{x}_n \end{bmatrix} = \begin{bmatrix} \alpha_1 & 1 & 0 & 0 & 0 & 0 & 0 & \cdots & 0 \\ 0 & \alpha_1 & 1 & 0 & 0 & 0 & 0 & \cdots & 0 \\ 0 & 0 & \alpha_1 & 0 & 0 & 0 & 0 & \cdots & 0 \\ 0 & 0 & 0 & \alpha_4 & 1 & 0 & 0 & \cdots & 0 \\ 0 & 0 & 0 & 0 & \alpha_4 & 0 & 0 & \cdots & 0 \\ 0 & 0 & 0 & 0 & 0 & \alpha_6 & 0 & \cdots & 0 \\ 0 & 0 & 0 & 0 & 0 & 0 & \alpha_7 & \cdots & 0 \\ \vdots & \vdots & \vdots & \vdots & \vdots & \vdots & & & \\ 0 & 0 & 0 & 0 & 0 & 0 & 0 & \cdots & \alpha_n \end{bmatrix} \begin{bmatrix} x_1 \\ x_2 \\ x_3 \\ x_4 \\ x_5 \\ x_6 \\ x_7 \\ \vdots \\ x_n \end{bmatrix} + \begin{bmatrix} 0 \\ 0 \\ 1 \\ 0 \\ 1 \\ 1 \\ 1 \\ \vdots \\ 1 \end{bmatrix} u. \qquad (13.36)$$

Now, in terms of state variables, (13.35) can be written as

$$Y(s) = b_0 U(s) + k_1 X_1(s) + k_2 X_2(s) + k_3 X_3(s) + \ldots + k_n X_n(s).$$

By taking Laplace transforms, we get

$$y = b_0 u + k_1 x_1 + k_2 x_2 + k_3 x_3 + \ldots + k_n x_n.$$

The equation can be expressed in a matrix form as

$$y = [k_1 \quad k_2 \quad k_3 \cdots k_n] \begin{bmatrix} x_1 \\ x_2 \\ x_3 \\ \vdots \\ x_n \end{bmatrix} + b_0 u. \qquad (13.37)$$

The state-space representation in the form is given as (13.36) and (13.37) and it is said to be in the Jordan canonical form.

***Problem-13.7*:** A feedback system has a closed-loop t.f. $\dfrac{Y(s)}{U(s)} = \dfrac{10(s+4)}{s(s+1)(s+3)}$. Construct different state models for this system.

***Solution*:**

The t.f. of the given system is

$$\frac{Y(s)}{U(s)} = \frac{10(s+4)}{s(s+1)(s+3)}.$$

Method 1: Controllable phase-variable canonical form

Let
$$\frac{Y(s)}{U(s)} = \frac{Y(s)}{X(s)}\frac{X(s)}{U(s)}$$

Where
$$\frac{X(s)}{U(s)} = \frac{1}{s(s+1)(s+3)}$$

$\Rightarrow$
$$\frac{X(s)}{U(s)} = \frac{1}{s(s^2+4s+3)}$$

$\Rightarrow$
$$s^3X(s) + 4s^2X(s) + 3sX(s) = U(s) \qquad (1)$$

and
$$\frac{Y(s)}{X(s)} = 10(s+4)$$

$\Rightarrow$
$$Y(s) = 10s\,X(s) + 40X(s) \qquad (2)$$

By taking inverse Laplace transforms to the above-mentioned equations, we have

$$\frac{d^3x}{dt^3} + 4\frac{d^2x}{dt^2} + 3\frac{dx}{dt} = u. \qquad (3)$$

$$y = 10\frac{dx}{dt} + 40x. \qquad (4)$$

Let us choose the state variables as x_1, x_2 and x_3.

i.e.,
$$x_1 = x$$
$$x_2 = \frac{dx}{dt} = \dot{x}_1 \Rightarrow \dot{x}_1 = x_2$$
$$x_3 = \frac{d^2x}{dt^2} = \dot{x}_2 \Rightarrow \dot{x}_2 = x_3.$$

Equations (3) and (4) in terms of state variables can be written as

$$\dot{x}_3 + 4x_3 + 3x_2 = u$$
$$\dot{x}_3 = -3x_2 - 4x_3 + u$$
$$y = 10x_2 + 40x_1.$$

The state and output equations of the above-mentioned system are

$$\dot{x}_1 = x_2$$
$$\dot{x}_2 = x_3$$
$$\dot{x}_3 = -3x_2 - 4x_3 + u$$
$$y = 40x_1 + 10x_2.$$

The state–space representation of the above-mentioned system is

$$\begin{bmatrix}\dot{x}_1\\ \dot{x}_2\\ \dot{x}_3\end{bmatrix}=\begin{bmatrix}0 & 1 & 0\\ 0 & 0 & 1\\ 0 & -3 & -4\end{bmatrix}\begin{bmatrix}x_1\\ x_2\\ x_3\end{bmatrix}+\begin{bmatrix}0\\ 0\\ 1\end{bmatrix}u$$

$$y=\begin{bmatrix}40 & 10 & 0\end{bmatrix}\begin{bmatrix}x_1\\ x_2\\ x_3\end{bmatrix}.$$

Method 2: Diagonal canonical form

$$\frac{Y(s)}{U(s)}=\frac{10(s+4)}{s(s+1)(s+3)}$$

$$=\frac{40}{3}\frac{1}{s}-\frac{15}{s+1}+\frac{5}{3(s+3)}$$

$$Y(s)=\frac{40}{3}\frac{1}{s}U(s)-\frac{15}{s+1}U(s)+\frac{5}{3}\frac{U(s)}{(s+3)}$$

$$Y(s)=\frac{40}{3}X_1(s)-15X_2(s)+\frac{5}{3}X_3(s) \tag{1}$$

where $X_1(s)=\dfrac{U(s)}{s}$

$$X_2(s)=\frac{U(s)}{s+1}$$

$$X_3(s)=\frac{U(s)}{s+3}$$

From the above-mentioned equations, we have

$$\begin{aligned}sX_1(s)&=U(s)\\ sX_2(s)&=-X_2(s)+U(s)\\ sX_3(s)&=-3X_3(s)+U(s).\end{aligned} \tag{2}$$

By taking inverse Laplace transforms to (1) and (2), we have

$$\begin{aligned}\dot{x}_1&=u\\ \dot{x}_2&=-x_2+u\\ \dot{x}_3&=-3x_3+u\\ y&=\frac{40}{3}x_1-15x_2+\frac{5}{3}x_3.\end{aligned}$$

The equations can be expressed in a matrix form as

$$\begin{bmatrix} \dot{x}_1 \\ \dot{x}_2 \\ \dot{x}_3 \end{bmatrix} = \begin{bmatrix} 0 & 0 & 0 \\ 0 & -1 & 0 \\ 0 & 0 & -3 \end{bmatrix} \begin{bmatrix} x_1 \\ x_2 \\ x_3 \end{bmatrix} + \begin{bmatrix} 1 \\ 1 \\ 1 \end{bmatrix} u$$

$$y = \begin{bmatrix} \frac{40}{3} & -15 & \frac{5}{3} \end{bmatrix} \begin{bmatrix} x_1 \\ x_2 \\ x_3 \end{bmatrix}.$$

Method 3: Observable phase-variable canonical form

Given that
$$\frac{Y(s)}{U(s)} = \frac{10s+40}{s^3+4s^2+3s}.$$

$$s^3Y(s) + 4s^2Y(s) + 3sY(s) = 10sU(s) + 40U(s)$$

$$s^3Y(s) = -4s^2Y(s) - 3sY(s) + 10sU(s) + 40U(s)$$

$$Y(s) = \frac{1}{s}[-4Y(s)] + \frac{1}{s^2}[10U(s) - 3Y(s)] + \frac{1}{s^3}40U(s) \tag{1}$$

Let us define the state variables.

$$\left.\begin{aligned} X_3(s) &= \frac{1}{s}[-4Y(s) + X_2(s)] \\ X_2(s) &= \frac{1}{s}[10U(s) - 3Y(s) + X_1(s)] \\ X_1(s) &= \frac{1}{s}40U(s) \end{aligned}\right\}. \tag{2}$$

From (1) and (2), we have

$$Y(s) = X_3(s). \tag{3}$$

By taking inverse Laplace transform to the above-mentioned equation, we have

$$y = x_3. \tag{4}$$

From (2) and (3), we have

$$sX_1(s) = 40U(s)$$
$$sX_2(s) = X_1(s) - 3X_3(s) + 10U(s)$$
$$sX_3(s) = X_2(s) - 4X_3(s).$$

By taking inverse Laplace transforms to the above-mentioned equations, we have

$$\dot{x}_1 = 40u$$
$$\dot{x}_2 = x_1 - 3x_3 + 10u$$
$$\dot{x}_3 = x_2 - 4x_3.$$

The equation can be expressed in matrix form as

$$\begin{bmatrix} \dot{x}_1 \\ \dot{x}_2 \\ \dot{x}_3 \end{bmatrix} = \begin{bmatrix} 0 & 0 & 0 \\ 1 & -3 & 0 \\ 0 & 1 & -4 \end{bmatrix} \begin{bmatrix} x_1 \\ x_2 \\ x_3 \end{bmatrix} + \begin{bmatrix} 40 \\ 10 \\ 0 \end{bmatrix} u$$

$$y = [0 \quad 0 \quad 1] \begin{bmatrix} x_1 \\ x_2 \\ x_3 \end{bmatrix}.$$

Problem-13.8: Obtain the state–space representation in canonical form for the following differential equation

$$\dddot{y} + 6\ddot{y} + 11\dot{y} + 6y = \dddot{u} + 8\ddot{u} + 17\dot{u} + 8u.$$

Solution:

By taking Laplace transforms to the above-mentioned equation under zero initial condition, we have

$$s^3Y(s) + 6s^2Y(s) + 11sY(s) + 6Y(s) = s^3U(s) + 8s^2U(s) + 17sU(s) + 8U(s)$$

$$\frac{Y(s)}{U(s)} = \frac{s^3 + 8s^2 + 17s + 8}{s^3 + 6s^2 + 11s + 6}$$

$$s^3 + 6s^2 + 11s + 6 \;\Big|\; s^3 + 8s^2 + 17s + 8 \;\Big|\; 1$$
$$\underline{s^3 + 6s^2 + 11s + 6}$$
$$2s^2 + 6s + 2$$

$$Y(s) = 1 + \frac{2s^2 + 6s + 2}{s^3 + 6s^2 + 11 + 6}$$

$$= 1 + \frac{2s^2 + 6s + 2}{(s+1)(s+2)(s+3)}$$

$$= 1 - \frac{1}{s+1} + \frac{2}{s+2} + \frac{1}{s+3}$$

$$Y(s) = U(s) - \frac{U(s)}{s+1} + 2\frac{U(s)}{s+2} + \frac{U(s)}{s+3}$$

$$Y(s) = U(s) - X_1(s) + 2X_2(s) + X_3(s) \qquad (1)$$

where $$X_1(s) = \frac{U(s)}{s+1}$$

$$X_2(s) = \frac{U(s)}{s+2}$$

$$X_3(s) = \frac{U(s)}{s+3}.$$

From the above-mentioned equations, we have

$$\left.\begin{aligned} sX_1(s) &= -X_1(s) + U(s) \\ sX_2(s) &= -2X_2(s) + U(s) \\ sX_3(s) &= -3X_3(s) + U(s) \end{aligned}\right\}. \tag{2}$$

By taking Laplace transforms to (1) and (2), we have

$$\begin{aligned} \dot{x}_1 &= -x_1 + u \\ \dot{x}_2 &= -2x_2 + u \\ \dot{x}_3 &= -3x_3 + u \\ y &= -x_1 + 2x_2 + x_3 + u. \end{aligned}$$

The equation can be expressed in a matrix form as

$$\begin{bmatrix} \dot{x}_1 \\ \dot{x}_2 \\ \dot{x}_3 \end{bmatrix} = \begin{bmatrix} -1 & 0 & 0 \\ 0 & -2 & 0 \\ 0 & 0 & -3 \end{bmatrix} \begin{bmatrix} x_1 \\ x_2 \\ x_3 \end{bmatrix} + \begin{bmatrix} 1 \\ 1 \\ 1 \end{bmatrix} u$$

$$y = [-1 \quad 2 \quad 1] \begin{bmatrix} x_1 \\ x_2 \\ x_3 \end{bmatrix} + u.$$

Problem-13.9: Obtain the state-space representation in Jordan canonical form for the given t.f. $\dfrac{Y(s)}{U(s)} = \dfrac{2s^2 + 6s + 7}{(s+1)^2(s+2)}$.

Solution:

Notice that the system has repeated roots.

$$\frac{Y(s)}{U(s)} = \frac{2s^2 + 6s + 7}{(s+1)^2(s+2)} = \frac{A}{(s+1)^2} + \frac{B}{s+1} + \frac{C}{s+2}$$

$$A = \underset{s\to -1}{\text{Lt}} \left[\frac{(2s^2 + 6s + 7)(s+1)^2}{(s+1)^2(s+2)}\right]$$

$$= \underset{s\to -1}{\text{Lt}} \left[\frac{(2s^2 + 6s + 7)}{s+2}\right] = \frac{2 - 6 + 7}{-1 + 2} = 3$$

$$B = \underset{s\to -1}{\text{Lt}} \frac{d}{ds}\left[(s+1)^2 \times \frac{(2s^2 + 6s + 7)}{(s+1)^2(s+2)}\right]$$

$$= \underset{s\to -1}{\text{Lt}} \frac{d}{ds}\left[\frac{(2s^2 + 6s + 7)}{s+2}\right]$$

$$= \underset{s\to-1}{\text{Lt}}\left[\frac{(s+2)(4s+6)-(2s^2+6s+7)}{(s+2)^2}\right]$$

$$= \frac{(-4+6)-(2-6+7)}{(-1+2)^2} = \frac{2-3}{1} = -1$$

$$C = \underset{s\to-2}{\text{Lt}}\left[(s+2)\frac{(2s^2+6s+7)}{(s+1)^2(s+2)}\right]$$

$$= \underset{s\to-2}{\text{Lt}}\left[\frac{(2s^2+6s+7)}{(s+1)^2}\right]$$

$$= \frac{8-12+7}{(-2+1)^2} = 3$$

$$\frac{Y(s)}{U(s)} = \frac{3}{(s+1)^2} + \frac{-1}{s+1} + \frac{3}{s+2}$$

$$Y(s) = \frac{3U(s)}{(s+1)^2} - \frac{U(s)}{s+1} + \frac{3U(s)}{s+2}$$

Let us define the state variables as

$$X_1(s) = \frac{U(s)}{(s+1)^2}$$

$$X_2(s) = \frac{U(s)}{s+1}$$

$$X_3(s) = \frac{U(s)}{s+2}.$$

However, $X_1(s)$ and $X_2(s)$ are related as

$$\frac{X_1(s)}{X_2(s)} = \frac{1}{s+1}.$$

Then, clearly,

$$sX_1(s) = -X_1(s) + X_2(s)$$

$$sX_2(s) = -X_2(s) + U(s)$$

$$sX_3(s) = -2X_3(s) + U(s)$$

and

$$Y(s) = 3X_1(s) - X_2(s) + 3X_3(s).$$

By taking inverse Laplace transforms to the above-mentioned equations, we get

$$\dot{x}_1 = -x_1 + x_2$$

$$\dot{x}_2 = -x_2 + u$$

$$\dot{x}_3 = -2x_3 + u$$

$$y = 3x_1 - x_2 + 3x_3.$$

The equation can be expressed in a matrix form as

Jordan block

$$\begin{bmatrix}\dot{x}_1\\ \dot{x}_2\\ \dot{x}_3\end{bmatrix}=\left[\begin{array}{cc:c}-1 & 1 & 0\\ 0 & -1 & 0\\ \hdashline 0 & 0 & -2\end{array}\right]\begin{bmatrix}x_1\\ x_2\\ x_3\end{bmatrix}+\begin{bmatrix}0\\ 1\\ 1\end{bmatrix}u$$

$$y=\begin{bmatrix}3 & -1 & 3\end{bmatrix}\begin{bmatrix}x_1\\ x_2\\ x_3\end{bmatrix}.$$

13.7 NON-UNIQUENESS OF STATE-SPACE REPRESENTATION

For a given t.f., the state–space representation is not unique. However, the state equations are related to each other by similarity transformation.

Similarity transformation: The square matrices A and B are said to be similar if a non-singular matrix P exist such that $P^{-1}AP = B$. The process of transformation is called similarity transformation and it is a linear transformation. The matrix P is called transformation matrix. Further, matrix A can be obtained from B by a similarity transformation with transfer matrix P^{-1}, that is, $A = PBP^{-1}$. The similarity transformation can be used for diagonalisation of a square matrix.

Consider the system described as

$$\dot{x}(t) = Ax(t) + B\,u(t). \tag{13.38}$$

$$y(t) = Cx(t) + D\,u(t). \tag{13.39}$$

Let us define a new state vector $\bar{x}(t)$ as

$$x(t) = P\bar{x}(t) \tag{13.40}$$

where P is a non-singular matrix

$$\dot{x}(t) = P\dot{\bar{x}}(t) \tag{13.41}$$

$$\therefore \qquad P\dot{\bar{x}}(t) = AP\bar{x}(t) + Bu(t).$$

By multiplying both sides with P^{-1}, we get

$$P^{-1}P\dot{\bar{x}}(t) = P^{-1}AP\bar{x}(t) + P^{-1}Bu(t)$$

$$\dot{\bar{x}}(t) = P^{-1}AP\bar{x}(t) + P^{-1}Bu(t).$$

Let

$$P^{-1}AP = \bar{A},\ P^{-1}B = \bar{B}$$

$$\therefore \qquad \dot{\bar{x}}(t) = \bar{A}\bar{x}(t) + \bar{B}u(t)$$

$$\begin{aligned} y(t) &= Cx(t) + Du(t)\\ &= CP\bar{x}(t) + Du(t)\\ &= \bar{C}\bar{x}(t) + \bar{D}u(t)\end{aligned}$$

where $\bar{C} = CP$, $\bar{D} = D$

$\therefore$ The new form of state–space representation is

$$\dot{\bar{x}}(t) = \bar{A}\bar{x}(t) + \bar{B}u(t)$$
$$y(t) = \bar{C}\bar{x}(t) + \bar{D}u(t).$$

From the above-mentioned state-space representation, it is clear that state-space representation is not unique.

Let the C.E. of the system be

$$|\lambda I - A| = 0$$
$$|\lambda I - \bar{A}| = |\lambda I - P^{-1}AP|$$
$$= |\lambda P^{-1}IP - P^{-1}AP|$$
$$= |P^{-1}||\lambda I - A||P|$$
$$= |\lambda I - A|$$
$$|\lambda I - \bar{A}| = |\lambda I - A|$$

$\Rightarrow$ $$|\lambda I - \bar{A}| = |\lambda I - A|.$$

Similarity transformation has the following properties:

1. The characteristic values of the matrix are unchanged under the transformation.
2. $|\bar{A}| = |A|$.
3. Trace $\bar{A}$ = Trace A

13.8 SOLUTION OF STATE EQUATION

The state equation of a dynamic system is given as

$$\dot{x}(t) = Ax(t) + Bu(t). \tag{13.42}$$

Case (i): If A is a constant matrix and $u(t)$ is zero vector, that is, no control forces are applied to the system, then (13.42) represents a homogenous LTI system.
Case (ii): If A is a constant and $u(t)$ is non-zero vector, that is, control forces are applied to the system, then (13.42) represents a non-homogenous LTI system.

13.8.1 Solution of Non-homogenous State Equation

The time response of a system represented by (13.42) on application of input function u along with given initial conditions can be considered in two parts.

1. Zero input response, where only initial conditions are considered and input function u is zero.
2. Zero state response, where only input function is considered and initial conditions are zero.

Let the state equation of the system be

$$\dot{x}(t) = Ax(t) + Bu(t). \tag{13.43}$$

The above-mentioned equation can be rearranged as

$$\dot{x}(t) - Ax(t) = Bu(t).$$

Pre-multiplying both sides with e^{-At}, we get

$$e^{-At}[\dot{x}(t) - Ax(t)] = e^{-At}\, Bu(t)$$

$$e^{-At}\,\dot{x}(t) + e^{-At}(-A)x(t) = e^{-At}\, Bu(t)\,. \tag{13.44}$$

Equation (13.44) can be written as

$$\frac{d}{dt}[e^{-At}\, x(t)] = e^{-At}\, Bu(t)$$

$$d[e^{-At}\, x(t)] = e^{-At}\, Bu(t)\, dt. \tag{13.45}$$

On integrating Equation (13.45) between limits 0 to t, we get

$$e^{-Ax}x(t) = x(0) + \int_0^t e^{-A\tau}\, Bu(\tau)\, d\tau \tag{13.46}$$

where $x(0)$ is the initial condition vector or integral constant and τ is the dummy variable substituted for t.

Pre-multiplying both sides of (13.46) by e^{At}, we get

$$e^{-At}e^{At}x(t) = e^{At}x(0) + e^{At}\int_0^t e^{-A\tau}\, Bu(\tau)\, d\tau. \tag{13.47}$$

$$x(t) = e^{At}x(0) + \int_0^t e^{A(t-\tau)}\, Bu(\tau)\, d\tau. \tag{13.48}$$

(zero input response) (zero state response)

Equation (13.48) is the solution of state equation, when the initial conditions are known at $t = 0$. If initial conditions are known at $t = t_0$, then the solution of state equation is given as

$$x(t) = e^{A(t-t_0)}x(t_0) + \int_{t_0}^t e^{A(t-\tau)}\, Bu(\tau)\, d\tau. \tag{13.49}$$

Here, $e^{At} = \phi(t)$ is called state transition matrix.

13.8.2 Solution of Homogenous State Equation

If input function is not taken into account, then (13.42) is a homogenous state equation, which gives the zero input response.

$$\dot{x}(t) = Ax(t). \tag{13.50}$$

The solution of the above-mentioned equation is

$$x(t) = e^{At}x(0). \tag{13.51}$$

However, $$e^{At} = I + At + \frac{A^2t^2}{2!} + \ldots \tag{13.52}$$

where A is the constant square matrix and I is the unit matrix.

From (13.51) and (13.52), we have

$$x(t) = \left[I + At + \frac{A^2t^2}{2!} + \ldots \right] x(0)$$

$$\frac{dx(t)}{dt} = \left[A + \frac{A^2\,2t}{2!} + \ldots \right] x(0)$$

$$\dot{x}(t) = A\left[I + At + \frac{A^2t^2}{2!} + \ldots \right] x(0)$$

$$= A\,e^{At}\,x(0).$$

$\therefore$ $$\dot{x}(t) = Ax(t) [\because x(t) = e^{At}x(0)] \tag{13.53}$$

The equation shows that $x(t) = e^{At}\,x(0)$ is the solution of homogeneous state equation (13.50).

13.9 STATE TRANSITION MATRIX (STM)

The state equation of an LTI system is given as

$$\dot{x}(t) = Ax(t) + Bu(t). \tag{13.54}$$

For homogenous system, $u(t) = 0$.

We have

$$\dot{x}(t) = Ax(t). \tag{13.55}$$

By taking Laplace transforms on both sides, we have

$$sX(s) - x(0) = AX(s)$$

$$(sI - A)X(s) = x(0).$$

$$X(s) = [sI - A]^{-1}x(0) \tag{13.56}$$

On taking inverse Laplace transform, we get

$$x(t) = L^{-1}[(sI - A)]^{-1}x(0)$$

$$\Rightarrow x(t) = e^{At}x(0). \tag{13.57}$$

Equation (13.57) gives the solution of the LTI homogenous state equation (13.55). From (13.57), it is observed that the initial state $x(0)$ at $t = 0$ is driven to a state $x(t)$ at time t. This transition in state is carried out by the matrix exponential e^{At}. Because of this property, e^{At} is termed as state transition matrix and denoted by $\phi(t)$.

Thus, $$\phi(t) = e^{At} = L^{-1}[(sI - A)]^{-1} = STM \tag{13.58}$$

13.9.1 Properties of State Transition Matrix

1. $\phi(0) = I$
   Substitute $t = 0$
   $\phi(0) = e^{A\times 0} = e^0 = I$
   $\Rightarrow \quad \phi(0) = I$

2. $\phi(-t) = \phi^{-1}(t)$
   $\phi(t) = e^{At}$
   $\phi(-t) = e^{A(-t)} = e^{-At} = [e^{At}]^{-1} = \phi^{-1}(t)$

3. $\phi(t_2 - t_1)\phi(t_1 - t_0) = \phi(t_2 - t_0)$

$$\begin{aligned}\phi(t_2 - t_1)\phi(t_1 - t_0) &= e^{A(t_2 - t_1)}e^{A(t_1 - t_0)} \\ &= e^{A(t_2 - t_1 + t_1 - t_0)} \\ &= e^{A(t_2 - t_0)}\end{aligned}$$

4. $\phi(t_1 + t_2) = \phi(t_1).\phi(t_2)$
   $\phi(t_1 + t_2) = e^{A(t_2 + t_1)} = e^{At_1} \cdot e^{At_2} = \phi(t_1)\phi(t_2)$

5. $[\phi(t)]^k = \phi(kt)$, where k = integer
   $[\phi(t)]^k = [e^{At}]^k = e^{Akt} = \phi(kt)$

6. $\dot{\phi}(t) = A\phi(t)$
   $\dot{\phi}(t) = \dfrac{d}{dt}[e^{At}] = Ae^{At} = A\phi(t)$

13.10 TRANSFER MATRIX OR TRANSFER FUNCTION FROM STATE MODEL

The matrix relating Laplace transform of the output to the Laplace transform of the input of state–space representation of a control system is known as transfer matrix.

The state–space representation of a control system is given as

$$\dot{x}(t) = Ax(t) + Bu(t)$$
$$y(t) = Cx(t) + Du(t).$$

By taking Laplace transforms to the above-mentioned equations, we get

$$sX(s) - x(0) = AX(s) + Bu(s)$$
$$Y(s) = Cx(s) + DU(s).$$

Under zero initial conditions, we have

$$sX(s) = AX(s) + BU(s)$$

$$\Rightarrow \qquad (sI - A)X(s) = BU(s)$$

$$\Rightarrow \qquad X(s) = (sI - A)^{-1} BU(s)$$

$$Y(s) = CX(s) + DU(s)$$

$$= C(sI - A)^{-1} BU(s) + DU(s)$$

$$= [C(sI - A)^{-1} B + D]\, U(s)$$

$$\Rightarrow \qquad \frac{Y(s)}{U(s)} = G(s) = C(sI - A)^{-1} B + D.$$

$$\Rightarrow \qquad \text{Transfer matrix } G(s) = C(sI - A)^{-1} B + D \tag{13.59}$$

Problem-13.10: Compute state transition matrix, when $A = \begin{bmatrix} 0 & 1 \\ -2 & -3 \end{bmatrix}$.

Solution:

Given that

$$A = \begin{bmatrix} 0 & 1 \\ -2 & -3 \end{bmatrix}$$

$$sI - A = \begin{bmatrix} s & 0 \\ 0 & s \end{bmatrix} - \begin{bmatrix} 0 & 1 \\ -2 & -3 \end{bmatrix} = \begin{bmatrix} s & -1 \\ 2 & s+3 \end{bmatrix}$$

$$(sI - A)^{-1} = \frac{\text{Adj}(sI - A)}{|sI - A|}$$

$$= \frac{1}{s^2 + 3s + 2} \begin{bmatrix} s+3 & 1 \\ -2 & s \end{bmatrix}$$

$$= \begin{bmatrix} \dfrac{s+3}{(s+1)(s+2)} & \dfrac{1}{(s+1)(s+2)} \\ \dfrac{-2}{(s+1)(s+2)} & \dfrac{s}{(s+1)(s+2)} \end{bmatrix}$$

$$= \begin{bmatrix} \dfrac{2}{s+1} - \dfrac{1}{s+2} & \dfrac{1}{s+1} - \dfrac{1}{s+2} \\ \dfrac{-2}{s+1} + \dfrac{2}{s+2} & -\dfrac{1}{s+1} + \dfrac{2}{s+2} \end{bmatrix}.$$

$\therefore$ State transition matrix (STM) is given as

$$\phi(t) = L^{-1}[(sI - A)^{-1}]$$

$$= \begin{bmatrix} L^{-1}\left[\dfrac{2}{s+1} - \dfrac{1}{s+2}\right] & L^{-1}\left[\dfrac{1}{s+1} - \dfrac{1}{s+2}\right] \\ L^{-1}\left[\dfrac{-2}{s+1} + \dfrac{2}{s+2}\right] & L^{-1}\left[-\dfrac{1}{s+1} + \dfrac{2}{s+2}\right] \end{bmatrix}$$

$$= \begin{bmatrix} 2e^{-t} - e^{-2t} & e^{-t} - e^{-2t} \\ -2e^{-t} + 2e^{-2t} & -e^{-t} + 2e^{-2t} \end{bmatrix}.$$

Problem-13.11: An LTI system is characterised by the homogeneous state equation.

$$[\dot{x}] = \begin{bmatrix} 1 & 0 \\ 1 & 1 \end{bmatrix} [x].$$

Find the time response of the system. Let us assume the initial state vector as

$$x(0) = \begin{bmatrix} 1 \\ 0 \end{bmatrix}.$$

Solution:

From the given state-space representation, we have

$$A = \begin{bmatrix} 1 & 0 \\ 1 & 1 \end{bmatrix}$$

$$sI - A = \begin{bmatrix} s & 0 \\ 0 & s \end{bmatrix} - \begin{bmatrix} 1 & 0 \\ 1 & 1 \end{bmatrix} = \begin{bmatrix} s-1 & 0 \\ -1 & s-1 \end{bmatrix}$$

$$[sI - A]^{-1} = \frac{\text{Adj}(sI - A)}{|sI - A|}$$

$$= \frac{1}{(s-1)^2} \begin{bmatrix} s-1 & 0 \\ 1 & s-1 \end{bmatrix} = \begin{bmatrix} \dfrac{1}{s-1} & 0 \\ \dfrac{1}{(s-1)^2} & \dfrac{1}{s-1} \end{bmatrix}.$$

∴ STM is given as

$$\phi(t) = L^{-1}[(sI - A)^{-1}]$$

$$= \begin{bmatrix} L^{-1}\left[\dfrac{1}{s-1}\right] & 0 \\ L^{-1}\left[\dfrac{1}{(s-1)^2}\right] & L^{-1}\left[\dfrac{1}{s-1}\right] \end{bmatrix}.$$

$$= \begin{bmatrix} e^{t} & 0 \\ te^{t} & e^{t} \end{bmatrix}$$

The time response of an LTI system is given as

$$x(t) = \phi(t)\, x(0)$$

$$= \begin{bmatrix} e^{t} & 0 \\ te^{t} & e^{t} \end{bmatrix} \begin{bmatrix} 1 \\ 0 \end{bmatrix} = \begin{bmatrix} e^{t} \\ te^{t} \end{bmatrix}.$$

***Problem-13.12*:** Consider the control system with state model

$$\begin{bmatrix}\dot{x}_1\\ \dot{x}_2\end{bmatrix}=\begin{bmatrix}0 & 1\\ -2 & -3\end{bmatrix}\begin{bmatrix}x_1\\ x_2\end{bmatrix}+\begin{bmatrix}0\\ 2\end{bmatrix}u.$$

Given $x_1(0) = 0$, $x_2(0) = 1$ and u = unit step input. Obtain the time response of the system at $t > 0$.

***Solution*:**

From the given state–space representation, we have

$$A=\begin{bmatrix}0 & 1\\ -2 & -3\end{bmatrix}, B=\begin{bmatrix}0\\ 2\end{bmatrix}, x(0)=\begin{bmatrix}0\\ 1\end{bmatrix}$$

$$sI-A=\begin{bmatrix}s & 0\\ 0 & s\end{bmatrix}-\begin{bmatrix}0 & 1\\ -2 & -3\end{bmatrix}=\begin{bmatrix}s & -1\\ 2 & s+3\end{bmatrix}$$

$$[sI-A]^{-1}=\frac{1}{s^2+3s+2}\begin{bmatrix}s+3 & 1\\ -2 & s\end{bmatrix}$$

$$=\begin{bmatrix}\dfrac{s+3}{(s+1)(s+2)} & \dfrac{1}{(s+1)(s+2)}\\ \dfrac{-2}{(s+1)(s+2)} & \dfrac{s}{(s+1)(s+2)}\end{bmatrix}$$

$$=\begin{bmatrix}\dfrac{2}{s+1}-\dfrac{1}{s+2} & \dfrac{1}{s+1}-\dfrac{1}{s+2}\\ \dfrac{-2}{s+1}+\dfrac{2}{s+2} & -\dfrac{1}{s+1}+\dfrac{2}{s+2}\end{bmatrix}.$$

$$\therefore \text{STM} = \phi(t) = L^{-1}(sI-A)^{-1}$$

$$=\begin{bmatrix}2e^{-t}-e^{-2t} & e^{-t}-e^{-2t}\\ -2e^{-t}+2e^{-2t} & -e^{-t}+2e^{-2t}\end{bmatrix}$$

The time response of the LTI, non-homogenous system is given as

$$x(t)=\phi(t)x(0)+\int_0^t \phi(t-\tau)Bu(\tau)\,d\tau.$$

Now, with $u(t) = 1$, $t > 0$, we may write

$$x(t)=\phi(t)x(0)+\phi(t)\int_0^t \phi(-\tau)\,Bu(\tau)\,d\tau$$

$$\phi(-\tau)Bu(\tau)=\begin{bmatrix}(2e^{\tau}-e^{2\tau}) & (e^{\tau}-e^{2\tau})\\ (-2e^{\tau}+2e^{2\tau}) & (-e^{\tau}+2e^{2\tau})\end{bmatrix}\begin{bmatrix}0\\ 2\end{bmatrix}$$

$$=\begin{bmatrix}2e^{\tau}-2e^{2\tau}\\ -2e^{\tau}+4e^{2\tau}\end{bmatrix}$$

$$\therefore \quad \int_0^t \phi(-\tau)Bu(\tau)d\tau = \begin{bmatrix} (2e^{\tau} - e^{2\tau}) \\ (-2e^{\tau} + 2e^{2\tau}) \end{bmatrix}_0^t$$

$$= \begin{bmatrix} (2e^{t} - e^{2t}) \\ (-2e^{t} + 2e^{2t}) \end{bmatrix} - \begin{bmatrix} 2-1 \\ -2+2 \end{bmatrix}$$

$$= \begin{bmatrix} 2e^{t} - e^{2t} - 1 \\ -2e^{t} + 2e^{2t} \end{bmatrix}$$

$$x(t) = \begin{bmatrix} 2e^{-t} - e^{-2t} & e^{t} - e^{-2t} \\ -2e^{-t} + 2e^{-2t} & -e^{-t} + 2e^{-2t} \end{bmatrix}\begin{bmatrix} 0 \\ 1 \end{bmatrix} + \begin{bmatrix} (2e^{-t} - e^{-2t}) & (e^{-t} - e^{-2t}) \\ (-2e^{-t} + 2e^{-2t}) & (-e^{-t} + 2e^{-2t}) \end{bmatrix}$$

$$\begin{bmatrix} 2e^{t} - e^{2t} - 1 \\ -2e^{t} + 2e^{2t} \end{bmatrix}$$

$$= \begin{bmatrix} e^{-t} - e^{-2t} \\ -e^{-t} + 2e^{-2t} \end{bmatrix} + \begin{bmatrix} 1 - 2e^{-t} + e^{-2t} \\ 2e^{-t} - 2e^{-2t} \end{bmatrix}$$

$$= \begin{bmatrix} 1 - e^{-t} \\ e^{-t} \end{bmatrix}.$$

13.11 CHARACTERISTIC EQUATION, EIGENVALUES AND EIGENVECTORS

The state equation of the system is given as

$$\dot{x}(t) = Ax(t) + Bu(t). \tag{13.60}$$

The equation $|\lambda I - A| = 0$ is called C.E. of the system.

13.11.1 Eigenvalues

The roots of the C.E. are called eigenvalues of the system.

Properties of eigenvalues

1. The trace of matrix A is the sum of eigenvalues, that is, Trace $(A) = \lambda_1 + \lambda_2 + ... + \lambda_n$, where λ_i, $i = 1, 2, ... n$ are the eigenvalues of matrix A.
2. The determinant of matrix A is the product of eigenvalues, that is, $\det(A) = \lambda_1 \lambda_2 \lambda_3 ......\lambda_n$.
3. If A is non-singular with eigenvalues λ_i, then $\frac{1}{\lambda_i}$, $i = 1, 2, 3, ... n$ are the eigenvalues of A^{-1}.
4. If λ_i are the eigenvalues of A, then they are the eigenvalues of A^T.
5. If A is a real symmetric matrix, then its eigenvalues are all real.

13.11.2 Eigenvectors

Vector x_i is termed to be an eigenvector of matrix A associated with eigenvalue $\lambda_i (i = 1, 2....n)$ and satisfies the following equation

$$[\lambda_i I - A]x_i = 0. \tag{13.61}$$

The eigenvectors of the matrix are determined by solving (13.61).

Properties of eigenvectors

1. The eigenvector x_i cannot be a null vector.
2. The eigenvector x_i is given by any non-zero column of the matrix adj $(\lambda_i I - A)$, where λ_i denotes the ith distinct eigenvalue of A.
3. If x_i is an eigenvector of A, then kx_i is also an eigenvector of A, where k is a scalar quantity.

Eigenvectors of multiple order eigenvalues

When one or more eigenvalues of A is of multiple order, a full set of n linearly independent eigenvectors may or may not exist. The number of linearly independent eigenvectors associated with eigenvalue λ_i of multiplicity m_i is equal to the degeneracy d_i of $(\lambda_i I - A)$.

The degeneracy d_i of $(\lambda_i I - A)$ is defined as

$$d_i = n - r_i \tag{13.62}$$

where n is the dimension of A and r_i is the rank of $\lambda_i I - A$.

The eigenvectors of a matrix A with multiple-order eigenvalues are determined according to the following methods.

Full degeneracy ($d_i = m_i$)

For the eigenvalue λ_i, which has a multiplicity m_i, the fully degenerated case has a full set of m_i eigenvectors associated with λ_i. These eigenvectors are found from non-zero columns of

$$\frac{1}{(m_i - 1)!}\left[\frac{d^{m_i - 1}}{d\lambda^{m_i - 1}} \quad \text{Adj}(\lambda I - A)\right]_{\text{at } \lambda = \lambda_i}. \tag{13.63}$$

Simple degeneracy

When the degeneracy of $(\lambda_i I - A)$ is equal to one (simple degeneracy), only one eigenvector is associated with λ_i regardless of multiplicity of λ_i. The eigenvector associated with λ_i can be determined using the same method as for the case of the distinct eigenvalues. However, for m_i^{th}-order eigenvalue, there are $(m_i - 1)$ additional vectors called generalised vectors. The $(m_i - 1)$ generalised eigenvectors $p_{i2}, p_{i3}, \ldots, p_{im_i - 1}$ are found from the following $(m_i - 1)$ equations.

$$\begin{aligned}
(\lambda_i I - A)p_{i2} &= -p_{i1} \\
(\lambda_i I - A)p_{i3} &= -p_{i2} \\
&\vdots \\
(\lambda_i I - A)p_{im_i} &= -p_{im_i - 1}
\end{aligned}$$

where pi_1 is the eigenvector of λ_i determined by solving the following set of matrix equations.

$$(\lambda_i I - A)p_{i1} = 0$$

13.12 CAYLEY–HAMILTON THEOREM

This theorem states that every square matrix must satisfy its own C.E.

Let us consider the C.E. as

$$s^n + a_1 s^{n-1} + ... + a_{n-1}s + a_n = 0.$$

If A is a square matrix, then

$$A^n + a_1 A^{n-1} + ... + a_{n-1}A + a_n I = 0$$

$$A^n = -[a_1 A^{n-1} + a_2 A^{n-2} + ... + a_{n-1}A + a_n I].$$

Problem-13.13: For the system matrix, $A = \begin{bmatrix} 3 & 0 & 0 \\ 2 & 4 & 1 \\ 2 & 1 & 4 \end{bmatrix}$, find the eigenvalues and eigenvectors.

Solution:

The C.E. of the system is

$$|\lambda I - A| = 0$$

$$\left| \begin{bmatrix} \lambda & 0 & 0 \\ 0 & \lambda & 0 \\ 0 & 0 & \lambda \end{bmatrix} - \begin{bmatrix} 3 & 0 & 0 \\ 2 & 4 & 1 \\ 2 & 1 & 4 \end{bmatrix} \right| = 0$$

$$\begin{vmatrix} \lambda - 3 & 0 & 0 \\ -2 & \lambda - 4 & -1 \\ -2 & -1 & \lambda - 4 \end{vmatrix} = 0$$

$$(\lambda - 3)[(\lambda - 4)^2 - 1] = 0$$

$$(\lambda - 3)(\lambda^2 - 8\lambda + 15) = 0$$

$$(\lambda - 3)(\lambda - 3)(\lambda - 5) = 0.$$

The eigenvalues are $\lambda_1 = \lambda_2 = 3,\ \lambda_3 = 5$. Thus, the eigenvalue $\lambda_1 = 3$ has a multiplicity of 2 and $\lambda_3 = 5$ is of simple order.

To check the degeneracy of $(\lambda I - A)$ for $\lambda_1 = 3$, we form the matrix $(\lambda_1 I - A)$ as

$$\lambda_1 I - A = \begin{bmatrix} \lambda - 3 & 0 & 0 \\ -2 & \lambda - 4 & -1 \\ -2 & -1 & \lambda - 4 \end{bmatrix}_{\text{at } \lambda = 3} = \begin{bmatrix} 0 & 0 & 0 \\ -2 & -1 & -1 \\ -2 & -1 & -1 \end{bmatrix},$$

which has a rank of 1. Thus, the degeneracy of $(\lambda_1 I - A)$ is

$$d_1 = n - r_1 = 3 - 1 = 2.$$

Since λ_1 is of multiplicity 2, we say that $(\lambda_1 I - A)$ is of full degeneracy.

$$\frac{d}{d\lambda}[\text{Adj}\,(\lambda I - A)]_{\lambda=3} = \frac{d}{d\lambda}\begin{bmatrix} (\lambda-3)(\lambda-5) & 0 & 0 \\ 2(\lambda-3) & (\lambda-3)(\lambda-4) & \lambda-3 \\ 2(\lambda-3) & \lambda-3 & (\lambda-3)(\lambda-4) \end{bmatrix}_{\text{at } \lambda=3}$$

$$= \begin{bmatrix} 2\lambda-8 & 0 & 0 \\ 2 & 2\lambda-7 & 1 \\ 2 & 1 & 2\lambda-7 \end{bmatrix}_{\lambda=3} = \begin{bmatrix} -2 & 0 & 0 \\ 2 & -1 & 1 \\ 2 & 1 & -1 \end{bmatrix}.$$

Thus, the two independent columns of the last matrix are selected as the eigenvectors.

$$\therefore \qquad p_1 = \begin{bmatrix} -1 \\ 1 \\ 1 \end{bmatrix} \qquad p_2 = \begin{bmatrix} 0 \\ -1 \\ 1 \end{bmatrix}$$

For the eigenvalue $\lambda_3 = 5$, the eigenvector is found by setting $(\lambda_3 - A)p_3 = 0$

$$\begin{bmatrix} \lambda_3-3 & 0 & 0 \\ -2 & \lambda_3-4 & -1 \\ -2 & -1 & \lambda_3-4 \end{bmatrix}\begin{bmatrix} x_1 \\ x_2 \\ x_3 \end{bmatrix} = 0$$

$$\begin{bmatrix} 2 & 0 & 0 \\ -2 & 1 & -1 \\ -2 & -1 & 1 \end{bmatrix}\begin{bmatrix} x_1 \\ x_2 \\ x_3 \end{bmatrix} = 0$$

$$x_1 = 0$$

$$-2x_1 + x_2 - x_3 = 0 \Rightarrow x_2 - x_3 = 0$$

$$-2x_1 - x_2 + x_3 = 0 \Rightarrow x_2 - x_3 = 0$$

$$\Rightarrow x_2 = x_3 = 1.$$

$$\therefore \qquad p_3 = \begin{bmatrix} x_1 \\ x_2 \\ x_3 \end{bmatrix} = \begin{bmatrix} 0 \\ 1 \\ 1 \end{bmatrix}$$

***Problem-13.14*:** Find the eigenvalues and eigenvectors for the given matrix $A = \begin{bmatrix} 1 & 2 \\ -2 & -3 \end{bmatrix}$.

***Solution*:**

The C.E. of the system is
$$|\lambda I - A| = 0$$

$$\begin{bmatrix} \lambda-1 & -2 \\ 2 & \lambda+3 \end{bmatrix} = 0$$

$$(\lambda-1)(\lambda+3) + 4 = 0.$$

$$\lambda^2 + 2\lambda + 1 = 0$$

$$(\lambda+1)^2 = 0$$

$$\Rightarrow \qquad \lambda = -1$$

Thus, the eigenvalue $\lambda = -1$, has a multiplicity of 2. To check the degeneracy of the matrix $\lambda I - A$,we form

$$\lambda I - A = \begin{bmatrix} \lambda - 1 & -2 \\ 2 & \lambda + 3 \end{bmatrix}_{\lambda=-1} = \begin{bmatrix} -2 & -2 \\ 2 & 2 \end{bmatrix},$$

which has a rank of 1.

The degeneracy $d = n - r = 2 - 1 = 1$

This means that we can find only one independent eigenvector for the eigenvalue λ_1 from

$$\text{Adj}(\lambda_1 I - A) = \text{Adj}\begin{bmatrix} -2 & -2 \\ 2 & 2 \end{bmatrix} = \begin{bmatrix} 2 & 2 \\ -2 & -2 \end{bmatrix}.$$

Thus, the eigenvector corresponding to eigenvalue $\lambda_1 = -1$ is

$$p_1 = \begin{bmatrix} 1 \\ -1 \end{bmatrix}.$$

To find the generalised eigenvector, we set

$$(\lambda_1 I - A)\, p_2 = -p_1$$

$$\begin{bmatrix} -2 & -2 \\ 2 & 2 \end{bmatrix}\begin{bmatrix} x_1 \\ x_2 \end{bmatrix} = \begin{bmatrix} -1 \\ 1 \end{bmatrix}$$

$$-2x_1 - 2x_2 = -1 \Rightarrow x_1 = 0,\; x_2 = 0.5$$

$$2x_1 + 2x_2 = 1.$$

$$\therefore \qquad p_2 = \begin{bmatrix} x_1 \\ x_2 \end{bmatrix} = \begin{bmatrix} 0 \\ 0.5 \end{bmatrix}$$

Problem-13.15: If $A = \begin{bmatrix} 3 & 2 \\ 2 & 3 \end{bmatrix}$, then find A^4 by using Cayley–Hamilton theorem.

Solution:

The C.E. of the system is

$$|\lambda I - A| = 0$$

$$\left|\begin{bmatrix} \lambda & 0 \\ 0 & \lambda \end{bmatrix} - \begin{bmatrix} 3 & 2 \\ 2 & 3 \end{bmatrix}\right| = 0$$

$$\begin{vmatrix} \lambda - 3 & -2 \\ -2 & \lambda - 3 \end{vmatrix} = 0$$

$$(\lambda - 3)^2 - 4 = 0$$

$$\lambda^2 - 6\lambda + 5 = 0.$$

By applying Cayley–Hamilton theorem, we have

$$A^2-6A+5I=0$$
$$A^2=6A-5I$$
$$A^3=A^2.A=(6A-5I)A=6A^2-5A$$
$$=6(6A-5I)-5A$$
$$=31A-30I$$
$$A^4=A^2.A^2$$
$$=(6A-5I)^2$$
$$=36A^2-60A+25I$$
$$=36(6A-5I)-60A+25I$$
$$=156A-155I$$
$$=156\begin{bmatrix}3 & 2\\ 2 & 3\end{bmatrix}-155\begin{bmatrix}1 & 0\\ 0 & 1\end{bmatrix}$$
$$=\begin{bmatrix}313 & 312\\ 312 & 313\end{bmatrix}.$$

13.13 DIAGONALISATION

The physical variable-state model, in general, is not convenient for the investigation of system properties and evaluation of time response. The diagonal canonical form or normal form of state model, wherein matrix A turns out to be a diagonal matrix, is most suitable for this purpose. Thus, it is useful to study the techniques by means of which a general state model can be transformed into a diagonal canonical form. These techniques are often referred to as diagonalisation techniques.

Consider an MIMO state model

$$\dot{x}(t)=Ax(t)+Bu(t) \tag{13.64 (a)}$$
$$y(t)=Cx(t)+Du(t) \tag{13.64 (b)}$$

In the abovementioned state model, the matrix A is non-diagonal. Let us define a new state vector $z(t)$ such that

$$x(t)=M\,z(t)$$

where M is a non-singular matrix.

Under this similarity transformation, the original state model given by (13.63) is modified to

$$\dot{z}(t)=M^{-1}AMz(t)+M^{-1}Bu(t) \tag{13.65 (a)}$$
$$y(t)=CMz(t)+Du(t), \tag{13.65 (b)}$$

If we select the matrix M such that $M^{-1}AM$ is a diagonalised matrix of matrix A, then the model given by (13.64) is a state model in diagonal canonical form. Under this condition, matrix M is called the diagonalising matrix or the model matrix, which is constructed by placing the eigenvectors of matrix A together.

If m_1, m_2 ... m_n be the eigenvectors of matrix A, corresponding to the eigenvalues $\lambda_1, \lambda_2, \ldots . \lambda_n$, respectively, then the model matrix is given as

$$M = \left[m_1 \vdots m_2 \vdots \ldots\ldots\ldots \vdots m_n\right]_{nxn}.$$

Thus, the general state model given by (13.64) modifies to a new state model (in diagonal canonical), which is given as

$$\dot{z}(t) = \Lambda z(t) + \tilde{B}u(t) \qquad (13.66)\ (a)$$

$$y(t) = \tilde{C}z(t) + Du(t) \qquad (13.66)\ (b)$$

where $\Lambda = M^{-1}AM$; a diagonal matrix

$$\tilde{B} = M^{-1}B$$

$$\tilde{C} = CM$$

In fact, matrix Λ could be obtained directly without the need to compute $M^{-1}AM$, since the diagonal elements of matrix Λ are given by distinct eigenvalues $\lambda_1, \lambda_2, \ldots \lambda_n$ of matrix A.

i.e., $$\Lambda = \begin{bmatrix} \lambda_1 & 0 & 0 & \ldots & 0 \\ 0 & \lambda_2 & 0 & \ldots & \ldots \\ \vdots & & & & \\ 0 & 0 & 0 & \ldots & \lambda_n \end{bmatrix} = M^{-1}AM$$

Note that A and Λ matrices have the same C.E., and therefore, the eigenvalues are invariant under the transformation.

If matrix A is given in bush or phase-variable or companion form, that is,

$$A = \begin{bmatrix} 0 & 1 & 0 & \ldots & 0 \\ 0 & 0 & 1 & \ldots & 0 \\ \vdots & \vdots & \vdots & & \\ 0 & 0 & 0 & \ldots & 1 \\ -a_n & -a_{n-1} & -a_{n-2} & \ldots & -a_1 \end{bmatrix}$$

then the model matrix with reference to bush or phase-variable form of matrix A can be shown to be a special matrix called Vandermonde Matrix and is given as

$$V = \begin{bmatrix} 1 & 1 & \ldots & 1 \\ \lambda_1 & \lambda_2 & \ldots & \lambda_n \\ \lambda_1^2 & \lambda_2^2 & \ldots & \lambda_n^2 \\ \vdots & \vdots & \vdots & \vdots \\ \lambda_1^{n-1} & \lambda_2^{n-1} & \ldots & \lambda_n^{n-1} \end{bmatrix}$$

where $\lambda_1, \lambda_2, \ldots \lambda_n$ are distinct eigenvalues of matrix A. An advantage of diagonal matrix is that the inverse of such a matrix can be obtained merely by inspection.

For example $\begin{bmatrix} \alpha & 0 & 0 \\ 0 & \beta & 0 \\ 0 & 0 & \gamma \end{bmatrix}^{-1} = \begin{bmatrix} 1/\alpha & 0 & 0 \\ 0 & 1/\beta & 0 \\ 0 & 0 & 1/\gamma \end{bmatrix}$

Now, consider the case in which matrix A involves multiple eigenvalues. It is impossible to diagonalise the matrix. However, there exists a similarity transformation $x(t) = Sz(t)$ such that the matrix $J = S^{-1}$, which is almost a diagonal matrix. The matrix J is said to be in Jordan Canonical form. Under this similarity transformation, the state model in Jordan Canonical form can thus be expressed as

$$\dot{z}(t) = Jz(t) + \tilde{B}u(t) \tag{13.67 (a)}$$

$$y(t) = \tilde{C}z(t) + Du(t) \tag{13.67 (b)}$$

where $J = S^{-1}AS$; a Jordan Matrix (almost diagonal)

$\tilde{B} = S^{-1}B$

$\tilde{C} = CB$

Note that for each λ_i of multiplicity q, the Jordan matrix $J = S^{-1}AS$ will have $q \times q$ Jordan Block corresponding to eigenvalue λ_i.

Problem-13.16: Obtain the eigenvalues, eigenvectors and state model in canonical form for the following system

$$\dot{x}(t) = \begin{bmatrix} 0 & 1 & 0 \\ 3 & 0 & 2 \\ -12 & -7 & -6 \end{bmatrix} x(t) + \begin{bmatrix} 1 \\ 0 \\ 2 \end{bmatrix} u(t)$$

$$y(t) = \begin{bmatrix} 1 & 0 & 0 \end{bmatrix} x(t)$$

Solution:

From the above-mentioned state–space representation, we have

$$A = \begin{bmatrix} 0 & 1 & 0 \\ 3 & 0 & 2 \\ -12 & -7 & -6 \end{bmatrix}, \quad B = \begin{bmatrix} 1 \\ 0 \\ 2 \end{bmatrix}.$$

Eigenvalues:

$$\lambda I - A = \begin{bmatrix} \lambda & 0 & 0 \\ 0 & \lambda & 0 \\ 0 & 0 & \lambda \end{bmatrix} - \begin{bmatrix} 0 & 1 & 0 \\ 3 & 0 & 2 \\ -12 & -7 & -6 \end{bmatrix} = \begin{bmatrix} \lambda & -1 & 0 \\ -3 & \lambda & -2 \\ 12 & 7 & \lambda+6 \end{bmatrix}$$

$$|\lambda I - A| = 0 \quad \Rightarrow \begin{vmatrix} \lambda & -1 & 0 \\ -3 & \lambda & -2 \\ 12 & 7 & \lambda+6 \end{vmatrix} = 0$$

$$\lambda[\lambda(\lambda+6)+14]+1[-3(\lambda+6)+24]=0$$
$$\lambda(\lambda^2+6\lambda+14)-3\lambda-18+24=0$$
$$\lambda^3+6\lambda^2+11\lambda+6=0$$
$$(\lambda+1)(\lambda^2+5\lambda+6)=0$$
$$(\lambda+1)(\lambda+2)(\lambda+3)=0$$
$$\lambda_1=-1,\ \lambda_2=-2,\ \lambda_3=-3$$

∴ The eigenvalues are -1, -2 and -3.

Eigenvectors:

When $\lambda_1=-1$, the corresponding eigenvector is

$$\begin{bmatrix}-1 & -1 & 0\\ -3 & -1 & -2\\ 12 & 7 & 5\end{bmatrix}\begin{bmatrix}x_{11}\\ x_{12}\\ x_{13}\end{bmatrix}=0$$

$$x_{11}+x_{12}=0 \Rightarrow x_{12}=-x_{11}$$
$$3x_{11}+x_{12}+2x_{13}=0 \Rightarrow x_{13}=-x_{11}$$
$$12x_{11}+7x_{12}+5x_{13}=0.$$

If $x_{11}=1$, then $x_{12}=-1,\ 2x_{13}=-1$

The eigenvector is $x_1=\begin{bmatrix}1\\ -1\\ -1\end{bmatrix}$

When $\lambda=-2$, the corresponding eigenvector is

$$\begin{bmatrix}\lambda & -1 & 0\\ -3 & \lambda & -2\\ 12 & 7 & \lambda+6\end{bmatrix}\begin{bmatrix}x_{21}\\ x_{22}\\ x_{23}\end{bmatrix}=0$$

$$\begin{bmatrix}-2 & -1 & 0\\ -3 & -2 & -2\\ 12 & 7 & 4\end{bmatrix}\begin{bmatrix}x_{21}\\ x_{22}\\ x_{23}\end{bmatrix}=0$$

$$2x_{21}+x_{22}=0 \Rightarrow x_{22}=-2x_{21}$$
$$3x_{21}+2x_{22}+2x_{23}=0 \Rightarrow x_{23}=\tfrac{1}{2}x_{21}.$$

If $x_{21}=2$, then $x_{22}=-4$ and $x_{23}=1$

The eigenvector is $x_2 = \begin{bmatrix} 2 \\ -4 \\ 1 \end{bmatrix}$.

When $\lambda = -3$, the corresponding eigenvector is

$$\begin{bmatrix} \lambda & -1 & 0 \\ -3 & \lambda & -2 \\ 12 & 7 & \lambda+6 \end{bmatrix}\begin{bmatrix} x_{31} \\ x_{32} \\ x_{33} \end{bmatrix} = 0$$

$$\begin{bmatrix} -3 & -1 & 0 \\ -3 & -3 & -2 \\ 12 & 7 & 3 \end{bmatrix}\begin{bmatrix} x_{31} \\ x_{32} \\ x_{33} \end{bmatrix} = 0$$

$$3x_{31} + x_{32} = 0 \Rightarrow x_{32} = -3x_{31}$$

$$3x_{31} + 3x_{32} + 2x_{33} = 0 \Rightarrow x_{33} = 3x_{31}.$$

If $x_{31} = 1$, then $x_{32} = -3$ and $x_{33} = 3$

The eigenvector is $x_3 = \begin{bmatrix} 1 \\ -3 \\ 3 \end{bmatrix}$

$\therefore$ The model matrix is $M = \begin{bmatrix} 1 & 2 & 1 \\ -1 & -4 & -3 \\ -1 & 1 & 3 \end{bmatrix}$.

$$\text{Adj}\, M = \begin{bmatrix} -9 & 6 & -5 \\ -5 & 4 & -3 \\ -2 & 2 & -2 \end{bmatrix}^T = \begin{bmatrix} -9 & -5 & -2 \\ 6 & 4 & 2 \\ -5 & -3 & -2 \end{bmatrix}$$

$$\text{Det}\, M = \begin{vmatrix} 1 & 2 & 1 \\ -1 & -4 & -3 \\ -1 & 1 & 3 \end{vmatrix} = -9 + 12 - 5 = -2$$

$$M^{-1} = \frac{\text{Adj}\, M}{\text{Det}\, M} = \frac{-1}{2}\begin{bmatrix} -9 & -5 & -2 \\ 6 & 4 & 2 \\ -5 & -3 & -2 \end{bmatrix}$$

$$M^{-1}AM = -\frac{1}{2}\begin{bmatrix} -9 & -5 & -2 \\ 6 & 4 & 2 \\ -5 & -3 & -2 \end{bmatrix}\begin{bmatrix} 0 & 1 & 0 \\ 3 & 0 & 2 \\ -12 & -7 & -6 \end{bmatrix}\begin{bmatrix} 1 & 2 & 1 \\ -1 & -4 & -3 \\ -1 & 1 & 3 \end{bmatrix}$$

$$= \begin{bmatrix} -1 & 0 & 0 \\ 0 & -2 & 0 \\ 0 & 0 & -3 \end{bmatrix}$$

$$M^{-1}B = \frac{-1}{2}\begin{bmatrix} -9 & -5 & -2 \\ 6 & 4 & 2 \\ -5 & -3 & -2 \end{bmatrix}\begin{bmatrix} 1 \\ 0 \\ 2 \end{bmatrix} = \begin{bmatrix} 13/2 \\ -5 \\ 9/2 \end{bmatrix}.$$

$$\text{CM} = \begin{bmatrix} 1 & 0 & 0 \end{bmatrix}\begin{bmatrix} 1 & 2 & 1 \\ -1 & -4 & -3 \\ -1 & 1 & 3 \end{bmatrix} = \begin{bmatrix} 1 & 2 & 1 \end{bmatrix}$$

The state representation in canonical form is

$$\begin{bmatrix} \dot{z}_1 \\ \dot{z}_2 \\ \dot{z}_3 \end{bmatrix} = \begin{bmatrix} -1 & 0 & 0 \\ 0 & -2 & 0 \\ 0 & 0 & -3 \end{bmatrix}\begin{bmatrix} z_1 \\ z_2 \\ z_3 \end{bmatrix} + \begin{bmatrix} 13/2 \\ -5 \\ 9/2 \end{bmatrix} u$$

$$y = \begin{bmatrix} 1 & 2 & 1 \end{bmatrix}\begin{bmatrix} z_1 \\ z_2 \\ z_3 \end{bmatrix}.$$

13.14 CONTROLLABILITY AND OBSERVABILITY

Controllability is concerned with the problem of whether it is possible to steer a system from a given initial state to an arbitrary state. A system is said to be controllable if it is possible by means of an unbounded control vector to transfer the system from any initial state to any other state in a finite time period. Observability is concerned with the problem of determining the state of a dynamic system from the observation of the output and control vectors in a finite time period. A system is said to be observable if, with the system in state $x(0)$, it is possible to determine this state from the observation of the output and control vectors over a finite time period.

The concepts of controllability and observability were introduced by R. E. Kalman. They play an important role in the optimal control of multivariable systems. The conditions of controllability and observability may govern the existence of a complete solution to the design problems of the control system.

13.14.1 Controllability

A control system is said to be completely state controllable if it is possible to transfer the system from any arbitrary initial state to any desired state in a finite time period. In other words, a control system is controllable if every state variable can be controlled in a finite time period by some unconstrained control signal. If any state variable is independent of the control signal, then it is impossible to control this state variable, and therefore, the system is uncontrollable.

1. **Complete State Controllability:** Consider a multiple-input LTI control system, which is defined as

$$\left.\begin{aligned}\dot{x}(t) &= Ax(t) + Bu(t)\\ y(t) &= Cx(t) + Du(t)\end{aligned}\right\}. \tag{13.68}$$

The system described by (13.68) is said to be completely state controllable or simply state controllable, if it is possible to transfer the system state from initial state $x(t_0)$ to the final state $x(t_f)$ in a specified time period by the application of the control signal $u(t)$.

The system described by (13.68) is completely state controllable if and only if the matrix $Q_c = [A^0 B : AB \ : A^2 B : \ldots : A^{n-1} B]$ is of rank n. Here, the matrix Q_c is called state controllability matrix.

2. **Complete Output Controllability:** The system described by (13.68) is said to be completely output controllable, if for any initial time $t = 0$, there exists a set of unconstrained control signal $u(t)$ such that any final output $y(t)$ can be reached from arbitrary initial states in a finite time, t.

The system described by (13.68) is completely output controllable if and only if the following matrix is of rank p, where p is the dimension of the output vector $y(t)$.

$$Q_{co} = [D : CB : CAB : \ldots : CA^{n-1} B]$$

Here, matrix Q_{co} is referred to as the output controllability matrix.

***Problem-13.17*:** Check whether the system represented by

$$\dot{x}(t) = \begin{bmatrix} -0.5 & 0 \\ 0 & -2 \end{bmatrix}\begin{bmatrix} x_1(t) \\ x_2(t) \end{bmatrix} + \begin{bmatrix} 0 \\ 1 \end{bmatrix} u(t) \text{ is controllable or not?}$$

Solution:

From the given state–space representation, we have

$$A = \begin{bmatrix} -0.5 & 0 \\ 0 & -2 \end{bmatrix}, B = \begin{bmatrix} 0 \\ 1 \end{bmatrix}$$

$$AB = \begin{bmatrix} -0.5 & 0 \\ 0 & -2 \end{bmatrix}\begin{bmatrix} 0 \\ 1 \end{bmatrix} = \begin{bmatrix} 0 \\ -2 \end{bmatrix}$$

$$Q_c = \begin{bmatrix} B & AB \end{bmatrix}\begin{bmatrix} 0 & 0 \\ 1 & -2 \end{bmatrix}.$$

Rank of $Q_c = r = 1$

Here, we have $n = 2$.

Hence, $r \neq n$ and the system is uncontrollable.

***Problem-13.18*:** A linear dynamic time-invariant system is represented as

$$\dot{x}(t) = Ax(t) + Bx(t)$$

where

$$A = \begin{bmatrix} 0 & 1 & 0 \\ 0 & 0 & 1 \\ 0 & -2 & -3 \end{bmatrix}, B = \begin{bmatrix} 0 & 1 \\ 0 & 0 \\ 1 & 0 \end{bmatrix}.$$

Find if the system is completely controllable or not.

***Solution*:**

For the system, $n = 3$, we have

$$AB = \begin{bmatrix} 0 & 1 & 0 \\ 0 & 0 & 1 \\ 0 & -2 & -3 \end{bmatrix}\begin{bmatrix} 0 & 1 \\ 0 & 0 \\ 1 & 0 \end{bmatrix} = \begin{bmatrix} 0 & 0 \\ 1 & 0 \\ -3 & 0 \end{bmatrix}$$

$$A^2B = A(AB)\begin{bmatrix} 0 & 1 & 0 \\ 0 & 0 & 1 \\ 0 & -2 & -3 \end{bmatrix}\begin{bmatrix} 0 & 0 \\ 1 & 0 \\ -3 & 0 \end{bmatrix} = \begin{bmatrix} 1 & 0 \\ -3 & 0 \\ 7 & 0 \end{bmatrix}$$

$$Q_c = [B\ AB\ A^2B] = \begin{bmatrix} 0 & 1 & 0 & 0 & 1 & 0 \\ 0 & 0 & 1 & 0 & -3 & 0 \\ 1 & 0 & -3 & 0 & 7 & 0 \end{bmatrix}.$$

Consider the determinant, $\begin{vmatrix} 0 & 1 & 0 \\ 0 & 0 & 1 \\ 1 & 0 & 3 \end{vmatrix} = 1 \neq 0$

Here, the rank of the matrix, $Q_c = 3 = n$

Thus, the given system is completely controllable.

13.14.2 Observability

The system is said to be completely observable if every initial state $x(t_0)$ can be completely determined with the knowledge of output $y(t)$ over a finite time interval $t_0 \leq t \leq t_f$. Essentially, a system is completely observable, if every state variable of the system affects the output. If any of the states cannot be observed from the measurement of the output, the state is said to be unobservable and the system is not completely observable or is simply unobservable.

1. **Complete Observability:**

   Consider an SISO LTI control system defined by

$$\left.\begin{aligned} \dot{x}(t) &= Ax(t) + Bu(t) \\ y(t) &= Cx(t) \end{aligned}\right\}. \tag{13.69}$$

   It is completely observable if and only if the following matrix is of rank n.

$$Q_o = \left[C^T : A^T C^T : \ldots\ldots : (A^T)^{n-1} C^T\right]$$

***Problem-13.19*:** Check whether the system represented by

$$\dot{x}(t) = \begin{bmatrix} -3 & 1 & 1 \\ -1 & 0 & 1 \\ 0 & 0 & 1 \end{bmatrix} x(t) + \begin{bmatrix} 0 & 1 \\ 0 & 0 \\ 2 & 1 \end{bmatrix} u(t)$$

$$y(t) = \begin{bmatrix} 0 & 0 & 1 \\ 1 & 1 & 0 \end{bmatrix} x(t)$$

is observable or not.

Solution:

From the given state-space representation, we have

$$A=\begin{bmatrix}-3 & 1 & 1\\ -1 & 0 & 1\\ 0 & 0 & 1\end{bmatrix}, B=\begin{bmatrix}0 & 1\\ 0 & 0\\ 2 & 1\end{bmatrix} C=\begin{bmatrix}0 & 0 & 1\\ 1 & 1 & 0\end{bmatrix}$$

$$A^{\mathrm{T}}=\begin{bmatrix}-3 & -1 & 0\\ 1 & 0 & 0\\ 1 & 1 & 1\end{bmatrix}, C^{\mathrm{T}}=\begin{bmatrix}0 & 1\\ 0 & 1\\ 1 & 0\end{bmatrix}$$

$$A^{\mathrm{T}}C^{\mathrm{T}}=\begin{bmatrix}-3 & -1 & 0\\ 1 & 0 & 0\\ 1 & 1 & 1\end{bmatrix}\begin{bmatrix}0 & 1\\ 0 & 1\\ 1 & 0\end{bmatrix}=\begin{bmatrix}0 & -4\\ 0 & 1\\ 1 & 2\end{bmatrix}$$

$$(A^{\mathrm{T}})^2C^{\mathrm{T}}=A^{\mathrm{T}}(A^{\mathrm{T}}C^{\mathrm{T}})=\begin{bmatrix}-3 & -1 & 0\\ 1 & 0 & 0\\ 1 & 1 & 1\end{bmatrix}\begin{bmatrix}0 & -4\\ 0 & 1\\ 1 & 2\end{bmatrix}=\begin{bmatrix}0 & 11\\ 0 & -4\\ 1 & -1\end{bmatrix}$$

$$Q_o=[C^{\mathrm{T}}:A^{\mathrm{T}}C^{\mathrm{T}}:(A^{\mathrm{T}})^2C^{\mathrm{T}}]=\begin{bmatrix}0 & 1 & 0 & -4 & 0 & 11\\ 0 & 1 & 0 & 1 & 0 & -4\\ 1 & 0 & 1 & 2 & 1 & -1\end{bmatrix}$$

We are able to choose a determinant from Q_0 where the value is not zero, which implies that the rank of Q_0 is 3. Here, the system is completely observable.

Problem-13.20: Evaluate the controllability and observability of the following state model

$$A=\begin{bmatrix}-1 & 0 & 0\\ 0 & -2 & 0\\ 0 & 0 & -3\end{bmatrix}, B=\begin{bmatrix}1 & 0\\ 1 & 2\\ 2 & 1\end{bmatrix}, C=\begin{bmatrix}1 & 1 & 2\\ 3 & 1 & 5\end{bmatrix}.$$

Solution:

For controllability:

$$Q_c=[B{:}\ AB{:}\ A^2B] \text{ as } n=3$$

$$AB=\begin{bmatrix}-1 & 0 & 0\\ 0 & -2 & 0\\ 0 & 0 & -3\end{bmatrix}\begin{bmatrix}1 & 0\\ 1 & 2\\ 2 & 1\end{bmatrix}=\begin{bmatrix}-1 & 0\\ -2 & -4\\ -6 & -3\end{bmatrix}$$

$$A^2B=A(AB)=\begin{bmatrix}-1 & 0 & 0\\ 0 & -2 & 0\\ 0 & 0 & -3\end{bmatrix}\begin{bmatrix}-1 & 0\\ -2 & -4\\ -6 & -3\end{bmatrix}=\begin{bmatrix}-1 & 0\\ 4 & 8\\ 18 & 9\end{bmatrix}$$

$$Q_c = [B\colon AB\colon A^2B] = \begin{bmatrix} 1 & 0 & -1 & 0 & -1 & 0 \\ 1 & 2 & -2 & -4 & 4 & 8 \\ 2 & 1 & -6 & -3 & 18 & 9 \end{bmatrix}$$

$$\begin{vmatrix} 1 & 0 & -1 \\ 1 & 2 & -2 \\ 2 & 1 & -6 \end{vmatrix} = -7 \neq 0; \text{ hence, the rank of } Q_c = 3 = n$$

Thus, the system is completely controllable.

For observability:

$$Q_o = [C^T : A^TC^T : (A^T)^2C^T] \text{ as } n = 3$$

$$A^TC^T = \begin{bmatrix} -1 & 0 & 0 \\ 0 & -2 & 0 \\ 0 & 0 & 3 \end{bmatrix}\begin{bmatrix} 1 & 3 \\ 1 & 1 \\ 2 & 5 \end{bmatrix} = \begin{bmatrix} -1 & -3 \\ -2 & -2 \\ -6 & -15 \end{bmatrix}$$

$$(A^T)^2C^T = A^T(A^TC^T) = \begin{bmatrix} -1 & 0 & 0 \\ 0 & -2 & 0 \\ 0 & 0 & -3 \end{bmatrix}\begin{bmatrix} -1 & -3 \\ -2 & -2 \\ -6 & -15 \end{bmatrix} = \begin{bmatrix} 1 & 3 \\ 4 & 4 \\ 18 & 45 \end{bmatrix}$$

$$Q_o = \begin{bmatrix} 1 & 3 & -1 & -3 & 1 & 3 \\ 1 & 1 & -2 & -2 & 4 & 4 \\ 2 & 5 & -6 & -15 & 18 & 45 \end{bmatrix}$$

$$|Q_o| = \begin{bmatrix} 1 & 3 & -1 \\ 1 & 1 & -2 \\ 2 & 5 & -6 \end{bmatrix} = -7 \neq 0; \text{ hence, the rank of } Q_o \text{ is } 3, \text{which is equal to } n.$$

Thus, the system is completely observable.

***Problem-13.21*:** Examine whether the discrete data system

$$x(t) = \begin{bmatrix} 0 & 1 \\ -2 & -2 \end{bmatrix} x(t) + \begin{bmatrix} 1 \\ -1 \end{bmatrix} u(t)$$

$$y(t) = \begin{bmatrix} 1 & 0 \end{bmatrix} x(t)$$

is (i) state controllable

(ii) output controllable and

(iii) observable.

***Solution*:**

From the given state-space representation, we have

$$A = \begin{bmatrix} 0 & 1 \\ -2 & -2 \end{bmatrix},\ B = \begin{bmatrix} 1 \\ -1 \end{bmatrix}, C = \begin{bmatrix} 1 & 0 \end{bmatrix}.$$

(i) **For state controllability**

$$Q_c = \left[A^0 B : AB : A^2 B : \ldots A^{n-1} B\right]$$

$$AB = \begin{bmatrix} 0 & 1 \\ -2 & -2 \end{bmatrix}\begin{bmatrix} 1 \\ -1 \end{bmatrix} = \begin{bmatrix} -1 \\ 0 \end{bmatrix}$$

$$Q_c = [B : AB] = \begin{bmatrix} 1 & -1 \\ -1 & 0 \end{bmatrix}$$

Rank of $\quad Q_c = r = 2$

Here, we have $n = 2$.

Therefore, the given system is completely state controllable.

(ii) **For output controllability:**

$$Q_{co} = \left[D : CA^0 B : CAB \ldots . CA^{n-1} B\right]$$

$$CB = [1 \quad 0]\begin{bmatrix} 1 \\ -1 \end{bmatrix} = [1]$$

$$CA = [1 \quad 0]\begin{bmatrix} 0 & 1 \\ -2 & -2 \end{bmatrix} = [0 \quad 1]$$

$$CAB = [0 \quad 1]\begin{bmatrix} 1 \\ -1 \end{bmatrix} = [-1]$$

$$Q_{co} = [CB : CAB] = [1 \quad -1]$$

$$|Q_c| \neq 0, \text{ rank} = 1$$

Therefore, the system is completely output controllable.

(iii) **For observability:**

$$Q_o = \left[C^T : A^T C^T : \ldots (A^T)^{n-1} C^T\right]$$

$$A^T = \begin{bmatrix} 0 & -2 \\ 1 & -2 \end{bmatrix}, C^T = \begin{bmatrix} 1 \\ 0 \end{bmatrix}$$

$$A^T C^T = \begin{bmatrix} 0 & -2 \\ 1 & -2 \end{bmatrix}\begin{bmatrix} 1 \\ 0 \end{bmatrix} = \begin{bmatrix} 0 \\ 1 \end{bmatrix}$$

$$Q_o = \left[C^T : A^T C^T\right] = \begin{bmatrix} 1 & 0 \\ 0 & 1 \end{bmatrix}$$

$$Q_o \neq 0, \text{ and hence, rank } r = 2 = n$$

Therefore, the given system is completely observable.

Problem-13.22: Consider the system defined by

$$\dot{x}(t) = \begin{bmatrix} a & b \\ c & d \end{bmatrix} x(t) + \begin{bmatrix} 1 \\ 1 \end{bmatrix} u(t)$$

$$y(t) = \begin{bmatrix} 1 & 0 \end{bmatrix} x(t).$$

Determine the conditions *a*, *b*, *c* and *d* for complete state controllability and complete observability.

Solution:

From the given state-space representation, we have

$$A = \begin{bmatrix} a & b \\ c & d \end{bmatrix},\ B = \begin{bmatrix} 1 \\ 1 \end{bmatrix}, C = \begin{bmatrix} 1 & 0 \end{bmatrix}$$

$$Q_c = \begin{bmatrix} A^0 B : AB : A^2 B : \ldots A^{n-1} B \end{bmatrix}$$

$$AB = \begin{bmatrix} a & b \\ c & d \end{bmatrix} \begin{bmatrix} 1 \\ 1 \end{bmatrix} = \begin{bmatrix} a+b \\ c+d \end{bmatrix}$$

$$Q_c = [B : AB] = \begin{bmatrix} 1 & a+b \\ 1 & c+d \end{bmatrix}.$$

For complete state controllability, $\begin{vmatrix} 1 & a+b \\ 1 & c+d \end{vmatrix} \neq 0 \Rightarrow a+b \neq c+d$

$$Q_o = \begin{bmatrix} C^T : A^T C^T : \ldots (A^T)^{n-1} C^T \end{bmatrix}$$

$$A^T = \begin{bmatrix} a & c \\ b & d \end{bmatrix},\ C^T = \begin{bmatrix} 1 \\ 0 \end{bmatrix}$$

$$A^T C^T = \begin{bmatrix} a & c \\ b & d \end{bmatrix} \begin{bmatrix} 1 \\ 0 \end{bmatrix} = \begin{bmatrix} a \\ b \end{bmatrix}$$

$$Q_o = \begin{bmatrix} C^T : A^T C^T \end{bmatrix} = \begin{bmatrix} 1 & a \\ 0 & b \end{bmatrix}.$$

For complete observability,

$$|Q_o| = \begin{vmatrix} 1 & a \\ 0 & b \end{vmatrix} = b \neq 0.$$

ADDITIONAL SOLVED PROBLEMS

ASP-1: Construct a state model for a system characterised by the differential equation

$$\frac{d^3 y}{dt^3} + 4\frac{d^2 y}{dt^2} + 9\frac{dy}{dt} + 4y + u = 0.$$

Solution:

Let us choose the state variables as x_1, x_2 and x_3

$\therefore$ $$x_1 = y.$$

$$x_2 = \frac{dy}{dt} = \dot{x}_1 \Rightarrow \dot{x}_1 = x_2$$

$$x_3 = \frac{d^2y}{dt^2} = \dot{x}_2 \Rightarrow \dot{x}_2 = x_3$$

$$\frac{d^3y}{dt^3} + 4\frac{d^2y}{dt^2} + 9\frac{dy}{dt} + 4y = -u$$

$$\dot{x}_3 + 4x_3 + 9x_2 + 4x_1 = -u$$

$$\dot{x}_3 = -4x_1 - 9x_2 - 4x_3 - u.$$

The state equations are

$$\dot{x}_1 = x_2$$

$$\dot{x}_2 = x_3$$

$$\dot{x}_3 = -4x_1 - 9x_2 - 4x_3 - u$$

$$\begin{bmatrix} \dot{x}_1 \\ \dot{x}_2 \\ \dot{x}_3 \end{bmatrix} = \begin{bmatrix} 0 & 1 & 0 \\ 0 & 0 & 1 \\ -4 & -9 & -4 \end{bmatrix} \begin{bmatrix} x_1 \\ x_2 \\ x_3 \end{bmatrix} + \begin{bmatrix} 0 \\ 0 \\ -1 \end{bmatrix} u.$$

Here, the output is y

$\therefore$ $$y = x_1$$

$$y = \begin{bmatrix} 1 & 0 & 0 \end{bmatrix} \begin{bmatrix} x_1 \\ x_2 \\ x_3 \end{bmatrix}.$$

ASP-2: Construct the state model for the mechanical system as shown in the figure.

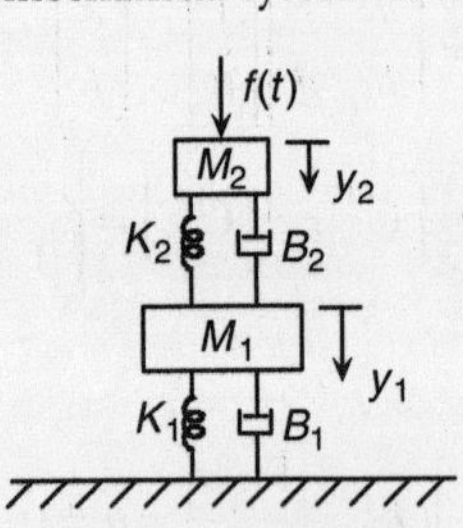

f(t)
M2
y2
K2
B2
M1
y1
K1
B1

Solution:

By Newton's Second Law, the force balance equation at mass M_1 is given as

$$M_1 \frac{d^2y_1}{dt^2} + B_1 \frac{dy_1}{dt} + B_2\left(\frac{dy_1}{dt} - \frac{dy_2}{dt}\right) + K_1 y_1 + K_2\left(y_1 - y_2\right) = 0. \quad (1)$$

By Newton's Second Law, the force balance equation at mass M_2 is given as

$$f(t) = M_2 \frac{d^2y_2}{dt^2} + B_2\left(\frac{dy_2}{dt} - \frac{dy_1}{dt}\right) + K_2(y_1 - y_2). \quad (2)$$

Let us choose four state variables as x_1, x_2, x_3 and x_4

i.e.,

$$x_1 = y_1$$

$$x_2 = y_2$$

$$x_3 = \frac{dy_1}{dt} = \dot{x}_1 \Rightarrow \dot{x}_1 = x_3 \quad (3)$$

$$x_4 = \frac{dy_2}{dt} = \dot{x}_2 \Rightarrow \dot{x}_2 = x_4 \quad (4)$$

From (1), we have

$$M_1\dot{x}_3 + B_1x_3 + B_2x_3 - B_2x_4 + K_1x_1 + K_2x_1 - K_2x_2 = 0$$

$$M_1\dot{x}_3 = -(K_1 + K_2)x_1 + K_2x_2 - (B_1 + B_2)x_3 + B_2x_4.$$

$$\dot{x}_3 = \frac{-(K_1 + K_2)}{M_1}x_1 + \frac{K_2}{M_1}x_2 - \frac{(B_1 + B_2)}{M_1}x_3 + \frac{B_2}{M_1}x_4. \quad (5)$$

From (2), we have

$$f(t) = M_2\dot{x}_4 + B_2x_4 - B_2x_3 + K_2x_2 - K_2x_1$$

$$M_2\dot{x}_4 = K_2x_1 - K_2x_2 + B_2x_3 - B_2x_4 + f(t)$$

$$\dot{x}_4 = \frac{K_2}{M_2}x_1 - \frac{K_2}{M_2}x_2 + \frac{B_2}{M_2}x_3 - \frac{B_2}{M_2}x_4 + \frac{1}{M_2}f(t). \quad (6)$$

The state equations of the mechanical system are

$$\dot{x}_1 = x_3$$

$$\dot{x}_2 = x_4$$

$$\dot{x}_3 = -\left(\frac{K_1 + K_2}{M_1}\right)x_1 + \frac{K_2}{M_1}x_2 - \left(\frac{B_1 + B_2}{M_1}\right)x_3 + \frac{B_2}{M_1}x_4$$

$$\dot{x}_4 = \frac{K_2}{M_2}x_1 - \frac{K_2}{M_2}x_2 + \frac{B_2}{M_2}x_3 - \frac{B_2}{M_2}x_4 + \frac{1}{M_2}f(t)$$

$$\begin{bmatrix} \dot{x}_1 \\ \dot{x}_2 \\ \dot{x}_3 \\ \dot{x}_4 \end{bmatrix} = \begin{bmatrix} 0 & 0 & 1 & 0 \\ 0 & 0 & 0 & 1 \\ -\frac{(K_1 + K_2)}{M_1} & \frac{K_2}{M_1} & -\frac{(B_1 + B_2)}{M_1} & \frac{B_2}{M_1} \\ \frac{K_2}{M_2} & \frac{-K_2}{M_2} & \frac{B_2}{M_2} & \frac{B_2}{M_2} \end{bmatrix} \begin{bmatrix} x_1 \\ x_2 \\ x_3 \\ x_4 \end{bmatrix} + \begin{bmatrix} 0 \\ 0 \\ 0 \\ \frac{1}{M} \end{bmatrix} f(t).$$

Let the displacements y_1 and y_2 be the outputs of the system.

$\therefore$ $$y_1 = x_1,\ y_2 = x_2$$

$$\begin{bmatrix} y_1 \\ y_2 \end{bmatrix} = \begin{bmatrix} 1 & 0 & 0 & 0 \\ 0 & 1 & 0 & 0 \end{bmatrix} \begin{bmatrix} x_1 \\ x_2 \\ x_3 \\ x_4 \end{bmatrix}.$$

ASP-3: Obtain the state model for the mechanical system as shown in the figure.

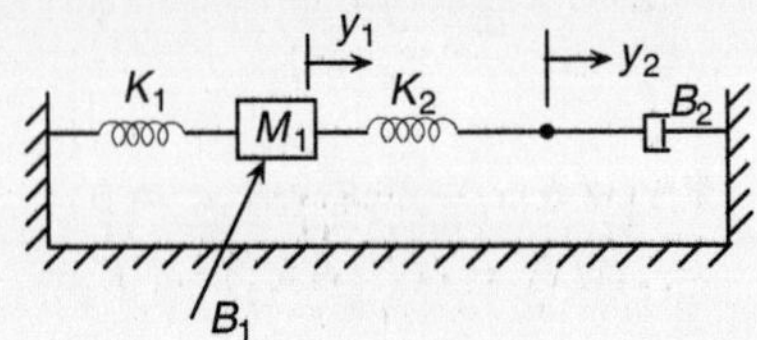

Solution:

By Newton's Second Law, the force balance equation at mass M_1 (i.e., at displacement y_1) is given as

$$M_1 \frac{d^2 y_1}{dt^2} + K_1 y_1 + K_2(y_1 - y_2) + B_1 \frac{dy_1}{dt} = 0. \tag{1}$$

By Newton's Second Law, the force balance equation at point mass (i.e., at displacement y_2) is given as

$$B_2 \frac{dy_2}{dt} + K_2(y_2 - y_1) = 0. \tag{2}$$

Let us choose three state variables as x_1, x_2 and x_3

$$x_1 = y_1$$

$$x_2 = y_2$$

$$x_3 = \frac{dy_1}{dt} = \dot{x}_1 \Rightarrow \dot{x}_1 = x_3. \tag{3}$$

From (1), we have

$$M_1 \dot{x}_3 + (K_1 + K_2)x_1 - K_2 x_2 + B_1 x_3 = 0$$

$$\dot{x}_3 = -\frac{(K_1 + K_2)}{M_1} x_1 + \frac{K_2}{M_2} x_2 - \frac{B_1}{M_1} x_3. \tag{4}$$

From (2), we have

$$B_2 \dot{x}_2 + K_2 x_2 - K_2 x_1 = 0$$

$$\dot{x}_2 = \frac{K_2}{B_2} x_1 - \frac{K_2}{B_2} x_2. \tag{5}$$

The state equations are

$$\dot{x}_1 = x_3$$
$$\dot{x}_2 = \frac{K_2}{B_2}x_1 - \frac{K_2}{B_2}x_2$$
$$\dot{x}_3 = -\frac{(K_1 + K_2)}{M_1}x_1 + \frac{K_2}{M_2}x_2 - \frac{B_1}{M_1}x_3$$

$$\begin{bmatrix} \dot{x}_1 \\ \dot{x}_2 \\ \dot{x}_3 \end{bmatrix} = \begin{bmatrix} 0 & 0 & 1 \\ \frac{K_2}{B_2} & -\frac{K_2}{B_2} & 0 \\ \frac{-(K_1 + K_2)}{M_1} & \frac{K_2}{M_2} & -\frac{B_1}{M_1} \end{bmatrix} \begin{bmatrix} x_1 \\ x_2 \\ x_3 \end{bmatrix}.$$

If the desired outputs are y_1 and y_2, then

$$y_1 = x_1$$
$$y_2 = x_2$$
$$\begin{bmatrix} y_1 \\ y_2 \end{bmatrix} = \begin{bmatrix} 1 & 0 & 0 \\ 0 & 1 & 0 \end{bmatrix} \begin{bmatrix} x_1 \\ x_2 \\ x_3 \end{bmatrix}.$$

ASP-4: The dynamic equation of the system is represented by

$$\begin{bmatrix} \dot{x}_1 \\ \dot{x}_2 \end{bmatrix} = \begin{bmatrix} -2 & 1 \\ 0 & -3 \end{bmatrix} \begin{bmatrix} x_1 \\ x_2 \end{bmatrix} + \begin{bmatrix} 4 \\ 1 \end{bmatrix}$$
$$y = \begin{bmatrix} 1 & 0 \end{bmatrix} \begin{bmatrix} x_1 \\ x_2 \end{bmatrix}.$$

Is the system is controllable or observable?

Solution:

From the above-mentioned state-space representation, we have

$$A = \begin{bmatrix} -2 & 1 \\ 0 & -3 \end{bmatrix} \quad B = \begin{bmatrix} 4 \\ 1 \end{bmatrix} \quad C = \begin{bmatrix} 1 & 0 \end{bmatrix}$$
$$AB = \begin{bmatrix} -2 & 1 \\ 0 & -3 \end{bmatrix} \begin{bmatrix} 4 \\ 1 \end{bmatrix} = \begin{bmatrix} -7 \\ -3 \end{bmatrix}$$
$$Q_c = [B : AB] = \begin{bmatrix} 4 & -7 \\ 1 & -3 \end{bmatrix}.$$

$|Q_c| \neq 0 \Rightarrow Q_c$ is singular and the system is controllable.

$$A^T = \begin{bmatrix} -2 & 0 \\ 1 & -3 \end{bmatrix}, C^T = \begin{bmatrix} 1 \\ 0 \end{bmatrix}$$

$$A^T C^T = \begin{bmatrix} -2 & 0 \\ 1 & -3 \end{bmatrix}\begin{bmatrix} 1 \\ 0 \end{bmatrix} = \begin{bmatrix} -2 \\ 1 \end{bmatrix}$$

$$[Q_0] = \left[C^T : A^T C^T\right] = \begin{bmatrix} 1 & -2 \\ 0 & 1 \end{bmatrix}$$

$|Q_0| \neq 0 \Rightarrow Q_0$ is singular and the system is observable.

ASP-5: Obtain the state model of the system whose t.f. is given as

$$\frac{Y(s)}{U(s)} = \frac{10}{s^3 + 2s^2 + 4s + 10}.$$

Solution:

$$\frac{Y(s)}{U(s)} = \frac{10}{s^3 + 2s^2 + 4s + 10}$$

$$s^3 Y(s) + 2s^2 Y(s) + 4sY(s) + 10Y(s) = 10U(s)$$

By taking inverse Laplace transforms on both sides, we get

$$\frac{d^3 y}{dt^3} + 2\frac{d^2 y}{dt^2} + 4\frac{dy}{dt} + 10y = 10u. \tag{1}$$

Let us choose the state variables as x_1, x_2 and x_3.

$$x_1 = y$$

$$x_2 = \frac{dy}{dt} = \dot{x}_1 \Rightarrow \dot{x}_1 = x_2. \tag{2}$$

$$x_3 = \frac{d^2 y}{dt^2} = \dot{x}_2 \Rightarrow \dot{x}_2 = x_3. \tag{3}$$

From the differential equation, we have

$$\dot{x}_3 + 2x_3 + 4x_2 + 10x_1 = 10u$$

$$\dot{x}_3 = -10x_1 - 4x_2 - 2x_3 + 10u. \tag{4}$$

The state equations are

$$\dot{x}_1 = x_2$$
$$\dot{x}_2 = x_3$$
$$\dot{x}_3 = -10x_1 - 4x_2 - 2x_3 + 10u.$$

The output equation is

$$y = x_1.$$

The state model in the matrix form is

$$\begin{bmatrix} \dot{x}_1 \\ \dot{x}_2 \\ \dot{x}_3 \end{bmatrix} = \begin{bmatrix} 0 & 1 & 0 \\ 0 & 0 & 1 \\ -10 & -4 & -2 \end{bmatrix} \begin{bmatrix} x_1 \\ x_2 \\ x_3 \end{bmatrix} + \begin{bmatrix} 0 \\ 0 \\ 10 \end{bmatrix} u$$

$$y = \begin{bmatrix} 1 & 0 & 0 \end{bmatrix} \begin{bmatrix} x_1 \\ x_2 \\ x_3 \end{bmatrix}.$$

ASP-6: A feedback system has a closed-loop t.f. $\frac{Y(s)}{U(s)} = \frac{10(s+2)}{s(s+1)(s+4)}$. Construct the state model.

Solution:

Given that

$$\frac{Y(s)}{U(s)} = \frac{10(s+2)}{s(s+1)(s+4)}.$$

Let

$$\frac{Y(s)}{U(s)} = \frac{Y(s)X(s)}{X(s)U(s)} = \frac{10(s+2)}{s(s+1)(s+4)}$$

$$\frac{X(s)}{U(s)} = \frac{1}{s(s+1)(s+4)}$$

$$[s(s+1)(s+4)]X(s) = U(s)$$

$$s^3X(s) + 5s^2X(s) + 4sX(s) = U(s).$$

By taking inverse Laplace transforms on both sides, we get

$$\dddot{x} + 5\ddot{x} + 4\dot{x} = u.$$

Let us choose the state variables as x_1, x_2 and x_3.

$\therefore$

$$x_1 = x$$

$$x_2 = \frac{dx}{dt} = \dot{x}_1 \Rightarrow \dot{x}_1 = x_2$$

$$x_3 = \frac{d^2x}{dt^2} = \dot{x}_2 \Rightarrow \dot{x}_2 = x_3$$

$$\dot{x}_3 + 5x_3 + 4x_2 = u$$

$$\dot{x}_3 = -4x_2 - 5x_3 + u.$$

The state equations are

$$\dot{x}_1 = x_2$$
$$\dot{x}_2 = x_3$$
$$\dot{x}_3 = -4x_2 - 5x_3 + u$$
$$\frac{Y(s)}{X(s)} = 10(s+2)$$
$$Y(s) = 10sX(s) + 20X(s).$$

By taking inverse Laplace transforms on both sides, we have

$$y = 10\dot{x} + 20x$$
$$= 10x_2 + 20x_1.$$

The output equation is

$$y = 20x_1 + 10x_2.$$

The state-space representation is

$$\begin{bmatrix} \dot{x}_1 \\ \dot{x}_2 \\ \dot{x}_3 \end{bmatrix} = \begin{bmatrix} 0 & 1 & 0 \\ 0 & 0 & 1 \\ 0 & -4 & -5 \end{bmatrix} \begin{bmatrix} x_1 \\ x_2 \\ x_3 \end{bmatrix} + \begin{bmatrix} 0 \\ 0 \\ 1 \end{bmatrix} u$$

$$y = \begin{bmatrix} 20 & 10 & 0 \end{bmatrix} \begin{bmatrix} x_1 \\ x_2 \\ x_3 \end{bmatrix}.$$

ASP-7: Let the state model of a system be given as

$$\begin{bmatrix} \dot{x}_1 \\ \dot{x}_2 \end{bmatrix} = \begin{bmatrix} 0 & 1 \\ -12 & -7 \end{bmatrix} \begin{bmatrix} x_1 \\ x_2 \end{bmatrix} + \begin{bmatrix} 0 \\ 1 \end{bmatrix} u$$

$$y = \begin{bmatrix} -10 & -4 \end{bmatrix} \begin{bmatrix} x_1 \\ x_2 \end{bmatrix} + u.$$

Under zero initial conditions, determine the

(i) Transfer function of the system, $G(s)$

(ii) State transition matrix, $\phi(t)$.

Solution:

From the state model, the following matrices can be obtained as

$$A = \begin{bmatrix} 0 & 1 \\ -12 & -7 \end{bmatrix}, B = \begin{bmatrix} 0 \\ 1 \end{bmatrix}, C = \begin{bmatrix} -10 & -4 \end{bmatrix}, D = [1].$$

(i) **State transition matrix:**

$$sI - A = s\begin{bmatrix} 1 & 0 \\ 0 & 1 \end{bmatrix} - \begin{bmatrix} 0 & 1 \\ -12 & -7 \end{bmatrix} = \begin{bmatrix} s & -1 \\ 12 & s+7 \end{bmatrix}$$

$$(sI - A)^{-1} = \begin{bmatrix} s & -1 \\ 12 & s+7 \end{bmatrix}^{-1}$$

$$= \frac{1}{s^2 + 7s + 12}\begin{bmatrix} s+7 & 1 \\ -12 & 5 \end{bmatrix}$$

$$= \frac{1}{(s+3)(s+4)}\begin{bmatrix} s+7 & 1 \\ -12 & 5 \end{bmatrix}$$

$$= \begin{bmatrix} \dfrac{s+7}{(s+3)(s+4)} & \dfrac{1}{(s+3)(s+4)} \\ \dfrac{-12}{(s+3)(s+4)} & \dfrac{s}{(s+3)(s+4)} \end{bmatrix}$$

$$= \begin{bmatrix} \dfrac{4}{(s+3)} - \dfrac{3}{(s+4)} & \dfrac{1}{(s+3)} - \dfrac{1}{(s+4)} \\ \dfrac{-12}{(s+3)} - \dfrac{-12}{(s+4)} & \dfrac{-3}{(s+3)} + \dfrac{4}{(s+4)} \end{bmatrix}$$

State transition matrix is

$$\phi(t) = L^{-1}(sI - A)^{-1}$$

$$= \begin{bmatrix} L^{-4}\left(\dfrac{4}{(s+3)} - \dfrac{3}{(s+4)}\right) & L^{-1}\left(\dfrac{1}{(s+3)} - \dfrac{1}{(s+4)}\right) \\ L^{-1}\left(\dfrac{-12}{(s+3)} - \dfrac{-12}{(s+4)}\right) & L^{-1}\left(\dfrac{-3}{(s+3)} + \dfrac{4}{(s+4)}\right) \end{bmatrix}$$

$$= \begin{bmatrix} 4e^{-3t} - 3e^{-4t} & e^{-3t} - e^{-4t} \\ -12e^{-3t} + 12e^{-4t} & -3e^{-3t} + 4e^{-4t} \end{bmatrix}.$$

(ii) **Transfer function of the system $G(s)$:**
The t.f. of the system is given as

$$G(s) = C(sI - A)^{-1} B + D$$

$$C(sI - A)^{-1} = [-10 - 4]\begin{bmatrix} \dfrac{s+7}{\Delta} & \dfrac{1}{\Delta} \\ \dfrac{-12}{\Delta} & \dfrac{s}{\Delta} \end{bmatrix}$$

$$= \begin{bmatrix} \dfrac{-10s - 70 + 48}{\Delta} & \dfrac{-10 - 4s}{\Delta} \end{bmatrix}$$

$$= \begin{bmatrix} \dfrac{-10s - 22}{\Delta} & \dfrac{-10 - 4s}{\Delta} \end{bmatrix}$$

$$C(sI-A)^{-1}B=\begin{bmatrix}\dfrac{-10s-22}{\Delta} & \dfrac{-10-4s}{\Delta}\end{bmatrix}\begin{bmatrix}0\\1\end{bmatrix}$$

$$=\frac{-10-4s}{\Delta}=\frac{-10-4s}{(s+3)(s+4)}.$$

$\therefore$

$$G(s)=C(sI-A)^{-1}B+D$$

$$=\frac{-10-4s}{(s+3)(s+4)}+1$$

$$=\frac{-10-4s+s^2+7s+12}{(s+3)(s+4)}$$

$$=\frac{s^2+3s+2}{(s+3)(s+4)}=\frac{(s+1)(s+2)}{(s+3)(s+4)}$$

ASP-8: Obtain the time response of the following system:

$$\dot{x}=\begin{bmatrix}0 & 1\\-2 & 0\end{bmatrix}\begin{bmatrix}x_1\\x_2\end{bmatrix}, x(0)=[1-1]^T$$

$$y=\begin{bmatrix}1 & -1\end{bmatrix}\begin{bmatrix}x_1\\x_2\end{bmatrix}.$$

Solution:

The time response of the system is given as

$$x(t)=\phi(t)x(0)$$

$$\phi(t)=L^{-1}(sI-A)^{-1}$$

$$sI-A=\begin{bmatrix}s & 0\\0 & s\end{bmatrix}-\begin{bmatrix}0 & 1\\-2 & 0\end{bmatrix}=\begin{bmatrix}s & -1\\2 & s\end{bmatrix}$$

$$(sI-A)^{-1}=\begin{bmatrix}s & -1\\2 & s\end{bmatrix}^{-1}=\frac{1}{s^2+2}\begin{bmatrix}s & 1\\-2 & s\end{bmatrix}=\begin{bmatrix}\dfrac{s}{s^2+2} & \dfrac{1}{s^2+2}\\ \dfrac{-2}{s^2+2} & \dfrac{s}{s^2+2}\end{bmatrix}$$

$$\phi(t)=L^{-1}(sI-A)^{-1}=L^{-1}\begin{bmatrix}\dfrac{s}{s^2+2} & \dfrac{1}{s^2+2}\\ \dfrac{-2}{s^2+2} & \dfrac{s}{s^2+2}\end{bmatrix}$$

$$=\begin{bmatrix}\cos\sqrt{2}t & \dfrac{1}{\sqrt{2}}\sin\sqrt{2}t\\ -\sqrt{2}\sin\sqrt{2}t & \cos\sqrt{2}t\end{bmatrix}.$$

$$\therefore \quad x(t) = \phi(t)x(0) = \begin{bmatrix} \cos\sqrt{2}t & \frac{1}{\sqrt{2}}\sin\sqrt{2}t \\ -\sqrt{2}\sin\sqrt{2}t & \cos\sqrt{2}t \end{bmatrix}\begin{bmatrix} 1 \\ -1 \end{bmatrix}$$

$$= \begin{bmatrix} \cos\sqrt{2}t - \frac{1}{\sqrt{2}}\sin\sqrt{2}t \\ -\sqrt{2}\sin\sqrt{2}t - \cos\sqrt{2}t \end{bmatrix}$$

ASP-9: A control system is described by the dynamic equation

$$\dot{x}(t) = \begin{bmatrix} 1 & 4 \\ -2 & -5 \end{bmatrix} x(t) + \begin{bmatrix} 0 \\ 1 \end{bmatrix} u(t)$$

$$y = \begin{bmatrix} 1 & 0 \end{bmatrix} x(t).$$

Find the eigenvalues and eigenvectors of the system.

Solution:

From the given state–space representation, we have

$$A = \begin{bmatrix} 1 & 4 \\ -2 & -5 \end{bmatrix} B = \begin{bmatrix} 0 \\ 1 \end{bmatrix} C = \begin{bmatrix} 1 & 0 \end{bmatrix}$$

$$\lambda I - A = \begin{bmatrix} \lambda & 0 \\ 0 & \lambda \end{bmatrix} - \begin{bmatrix} 1 & 4 \\ -2 & -5 \end{bmatrix} = \begin{bmatrix} \lambda - 1 & -4 \\ 2 & \lambda + 5 \end{bmatrix}.$$

The C.E. is

$$|\lambda I - A| = 0 \Rightarrow \begin{vmatrix} \lambda - 1 & -4 \\ 2 & \lambda + 5 \end{vmatrix} = 0$$

$$(\lambda - 1)(\lambda + 5) + 8 = 0$$

$$\lambda^2 + 4\lambda + 3 = 0$$

$$(\lambda + 1)(\lambda + 3) = 0$$

$$\lambda_1 = -1, \lambda_2 = -3.$$

When $\lambda_1 = -1$, the eigenvector is

$$\begin{bmatrix} \lambda - 1 & -4 \\ 2 & \lambda + 5 \end{bmatrix} [x_1] = 0$$

$$\begin{bmatrix} -2 & -4 \\ 2 & 4 \end{bmatrix} \begin{bmatrix} x_{11} \\ x_{12} \end{bmatrix} = 0$$

$$-2x_{11} + 4x_{12} = 0$$

$$x_{11} + 2x_{12} = 0.$$

When $x_{12} = 1$, then $x_{11} = -2$.

The eigenvector is $x_1 = \begin{bmatrix} x_{11} \\ x_{12} \end{bmatrix} = \begin{bmatrix} -2 \\ 1 \end{bmatrix}$.

When $\lambda_1 = -3$, the eigenvector is

$$\begin{bmatrix} \lambda_2 - 1 & -4 \\ 2 & \lambda_2 + 5 \end{bmatrix} [x_2] = 0$$

$$\begin{bmatrix} -4 & -4 \\ 2 & 2 \end{bmatrix} \begin{bmatrix} x_{21} \\ x_{22} \end{bmatrix} = 0$$

$$4x_{21} + 4x_{22} = 0$$

$$x_{21} + x_{22} = 0.$$

When $x_{21} = 1$, then $x_{22} = -1$.

The eigenvector is $x_2 = \begin{bmatrix} x_{21} \\ x_{22} \end{bmatrix} = \begin{bmatrix} 1 \\ -1 \end{bmatrix}$.

ASP-10: Obtain the state variable representation of (i) armature-controlled DC motor and (ii) field-controlled DC motor

(i) **State model of an armature-controlled DC motor**

The speed of a DC motor is directly proportional to armature voltage and inversely proportional to flux. In armature-controlled DC motor, the desired speed is obtained by varying the armature voltage. This speed control system is an electro-mechanical control system.

The electrical system consists of the armature and the field circuit; however, for analysis purpose, only the armature circuit is considered because the field is excited by a constant voltage. The mechanical system consists of the rotating part of the motor and load connected to the shaft of the motor.

The armature-controlled DC motor speed control system is shown in the Figure 13.3.

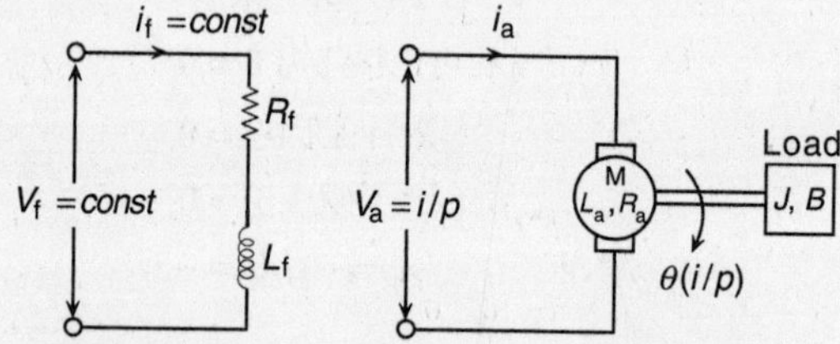

Figure 13.3 Armature-controlled DC motor

Let R_a = Armature resistance, Ω
L_a = Armature inductance, H
i_a = Armature current, A
V_a = Armature voltage, V
e_n = back emf, v
K_t = torque constant, Nm/A
T = torque developed by motor, N-m
θ = angular displacement of shaft, rad

$$\omega = \frac{d\theta}{dt} = \text{angular displacement of shaft, rad/s}$$
$$J = \text{moment of inertia of motor and load kgm}^2\text{/rad}$$
$$B = \text{Frictional coefficient of motor and load Nm/(rad/s)}$$
$$K_b = \text{Back emf constant } V\text{/(rad/s)}$$

The equivalent circuit of armature is shown in the following figure

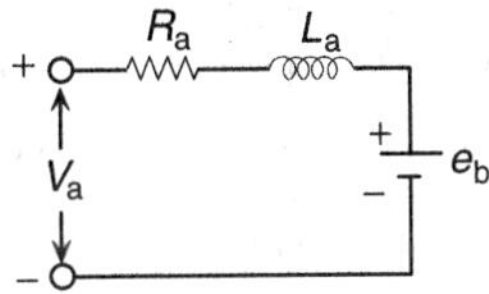

By applying KVL, we get

$$i_a R_a + L_a \frac{dia}{dt} + e_b = V_a. \tag{1}$$

Torque of DC motor is proportional to the product of flux and current. Since flux is constant in this system, the torque is proportional to i_a alone.

$$T_a \propto i_a$$

$$T_a = K_t i_a. \tag{2}$$

The mechanical system of the motor is shown in the figure. The differential equation governing the mechanical system of motor is given as

$$J\frac{d^2\theta}{dt^2} + B\frac{d\theta}{dt} = T_a. \tag{3}$$

The back emf of DC machine is proportional to the speed (angular velocity) of shaft

$$\therefore \quad e_b \propto \frac{d\theta}{dt} \Rightarrow e_b = K_b \frac{d\theta}{dt}. \tag{4}$$

From (1) and (4), we have

$$i_a R_a + L_a \frac{di_a}{dt} + K_b \frac{d\theta}{dt} = V_a. \tag{5}$$

From (2) and (3), we have

$$J\frac{d^2\theta}{dt^2} + B\frac{d\theta}{dt} = K_t i_a \tag{6}$$

Equations (5) and (6) are the differential equations governing the armature-controlled DC motor.
Let us choose i_a, ω and θ as state variables to model the armature-controlled DC motor.

$$\therefore \quad x_1 = i_a$$

$$x_2 = \omega = \frac{d\theta}{dt} = \dot{x}_3 \Rightarrow \dot{x}_3 = x_2 \tag{7}$$

$$x_3 = \theta$$

From (5), we have

$$R_a x_1 + L_a \dot{x}_1 + K_b x_2 = v_a$$

$$L_a \dot{x}_1 = -R_a x_1 - K_b x_2 + V_a$$

$$\dot{x}_1 = -\frac{R_a}{L_a} x_1 - \frac{K_b}{L_a} x_2 + \frac{1}{L_a} V_a. \tag{8}$$

From (6), we have

$$J\dot{x}_2 + Bx_2 = K_t x_1$$

$$J\dot{x}_2 = K_t x_1 - Bx_2$$

$$\dot{x}_2 = \frac{K_t}{J} x_1 - \frac{B}{J} x_2. \tag{9}$$

The state equations are

$$\dot{x}_1 = -\frac{R_a}{L_a} x_1 - \frac{K_b}{L_a} x_2 + \frac{1}{L_a} V_a$$

$$\dot{x}_2 = \frac{K_t}{J} x_1 - \frac{B}{J} x_2$$

$$\dot{x}_3 = x_2$$

$$\begin{bmatrix} \dot{x}_1 \\ \dot{x}_2 \\ \dot{x}_3 \end{bmatrix} = \begin{bmatrix} -\frac{R_a}{L_a} & \frac{-K_b}{L_a} & 0 \\ \frac{K_t}{J} & -\frac{B}{J} & 0 \\ 0 & 1 & 0 \end{bmatrix} \begin{bmatrix} x_1 \\ x_2 \\ x_3 \end{bmatrix} + \begin{bmatrix} \frac{1}{L_a} \\ 0 \\ 0 \end{bmatrix} V_a.$$

Let the desired outputs be i_a, ω and θ. Let us equate the desired output quantities to standard notation y_1, y_2 and y_3 as

$$y_1 = i_a, y_2 = \omega = \frac{d\theta}{dt}, y_3 = \theta$$

$$y_1 = x_1, y_2 = x_2, y_3 = x_3$$

$$\begin{bmatrix} y_1 \\ y_2 \\ y_3 \end{bmatrix} = \begin{bmatrix} 1 & 0 & 0 \\ 0 & 1 & 0 \\ 0 & 0 & 1 \end{bmatrix} \begin{bmatrix} x_1 \\ x_2 \\ x_3 \end{bmatrix}.$$

(ii) State model of field-controlled DC motor

The speed of a DC motor is directly proportional to armature voltage and inversely proportional to flux.

In the field-controlled DC motor, the armature voltage is kept constant and the speed is varied by varying the flux of the machine. Since the flux is directly proportional to the field current, it is varied by varying the field current.

The speed control system is an electromechanical control system. The electrical system consists of armature and field circuit; however, for analysis purpose, only field circuit is considered because the armature is excited by constant voltage. The mechanical system consists of the rotating part of the motor and the load connected to the shaft of the motor.

The field-controlled DC motor speed control system is shown in Figure 13.4.

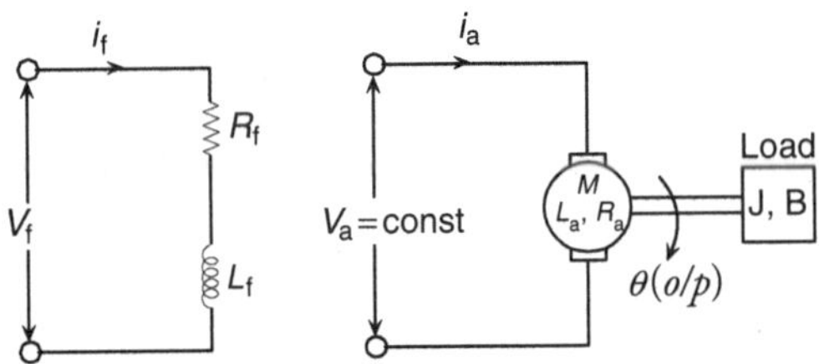

Figure 13.4 Field-controlled DC motor

Let R_f = Field resistance, Ω

L_f = Field inductance, H

i_f = Field current, A

V_f = Field voltage, V

θ = Angular displacement of the motor shaft, rad

$\omega = \dfrac{d\theta}{dt}$ = Angular velocity of the motor shaft, rad/s

T = Torque developed by motor, Nm

k_t = Torque constant, Nm/A

J = Moment of inertia of rotor and load, kgm^2/rad

B = Frictional coefficient of rotor and load, Nm/(rad/s)

By applying KVL to the field circuit, we have

$$R_f i_f + L_f \frac{di_f}{dt} = V_f. \tag{1}$$

The torque of a DC motor is proportional to the product of flux and armature current. Since armature current is constant in this system, the torque is proportional to the flux alone; however, the flux is proportional to the field current.

$\therefore$
$$T \propto i_f$$

$\Rightarrow$
$$T = K_t i_f. \tag{2}$$

The mechanical system of the motor is shown in the figure. The differential equation governing the mechanical system of the motor is given as

$$J\frac{d^2\theta}{dt^2}+B\frac{d\theta}{dt}=T. \quad (3)$$

From (2) and (3), we have

$$J\frac{d^2\theta}{dt^2}+B\frac{d\theta}{dt}=K_t i_f. \quad (4)$$

Equations (1) and (4) are the differential equations governing the field-controlled DC motor.

Let us choose ω and θ as state variables to model the field-controlled DC motor.

i.e.,

$$x_1 = i_f$$

$$x_2 = \omega = \frac{d\theta}{dt} = \dot{x}_3 \Rightarrow \dot{x}_3 = x_2. \quad (5)$$

$$x_3 = \theta$$

From (1), we have

$$R_f i_f + L_f\frac{di}{dt} = V_f$$

$$R_f x_1 + L_f \dot{x}_1 = V_f$$

$$L_f \dot{x}_1 = -R_f x_1 + V_f.$$

$$\dot{x}_1 = -\frac{R_f}{L_f}x_1 + \frac{1}{L_f}V_f \quad (6)$$

From (4), we have

$$J\frac{d^2\theta}{dt^2}+B\frac{d\theta}{dt}=K_t i_f$$

$$J\dot{x}_2 + Bx_2 = K_t x_i$$

$$\dot{x}_2 = \frac{K_t}{J}x_1 - \frac{B}{J}x_2. \quad (7)$$

The state equations are

$$\dot{x}_1 = -\frac{R_f}{L_f}x_1 + \frac{1}{L_f}V_f$$

$$\dot{x}_2 = \frac{K_t}{J}x_1 - \frac{B}{J}x_2$$

$$\dot{x}_3 = x_2$$

$$\begin{bmatrix}\dot{x}_1\\ \dot{x}_2\\ \dot{x}_3\end{bmatrix} = \begin{bmatrix} -\frac{R_f}{L_f} & 0 & 0\\ \frac{K_t}{J} & -\frac{B}{J} & 0\\ 0 & 1 & 0\end{bmatrix}\begin{bmatrix}x_1\\ x_2\\ x_3\end{bmatrix} + \begin{bmatrix}\frac{1}{L_f}\\ 0\\ 0\end{bmatrix}[V_f]$$

Let the desired output be ω and θ. Let us equate the desired output quantities to standard notation y_1 and y_2 as

$$y_1 = \omega, y_2 = \theta$$

$$y_1 = x_2, y_2 = x_3$$

$$\begin{bmatrix} y_1 \\ y_2 \end{bmatrix} = \begin{bmatrix} 0 & 1 & 0 \\ 0 & 0 & 1 \end{bmatrix} \begin{bmatrix} x_1 \\ x_2 \\ x_3 \end{bmatrix}.$$

ASP-11: Obtain the eigenvalues, eigenvectors and state model in canonical form for the system described as

$$\dot{x}(t) = \begin{bmatrix} 0 & 1 & 0 \\ 3 & 0 & 2 \\ -12 & -7 & -6 \end{bmatrix} x(t) + \begin{bmatrix} 1 \\ 0 \\ 2 \end{bmatrix} u(t).$$

$$y(t) = \begin{bmatrix} 1 & 0 & 0 \end{bmatrix} x(t)$$

Solution:

From the above-mentioned state-space representation, we have

$$A = \begin{bmatrix} 0 & 1 & 0 \\ 3 & 0 & 2 \\ -12 & -7 & -6 \end{bmatrix}, \; B = \begin{bmatrix} 1 \\ 0 \\ 2 \end{bmatrix}.$$

Eigenvalues:

$$\lambda I - A = \begin{bmatrix} \lambda & 0 & 0 \\ 0 & \lambda & 0 \\ 0 & 0 & \lambda \end{bmatrix} - \begin{bmatrix} 0 & 1 & 0 \\ 3 & 0 & 2 \\ -12 & -7 & -6 \end{bmatrix} = \begin{bmatrix} \lambda & -1 & 0 \\ -3 & \lambda & -2 \\ 12 & 7 & \lambda+6 \end{bmatrix}$$

$$|\lambda I - A| = 0 \Rightarrow \begin{bmatrix} \lambda & -1 & 0 \\ -3 & \lambda & -2 \\ 12 & 7 & \lambda+6 \end{bmatrix} = 0$$

$$\lambda[\lambda(\lambda+6)+4] + 1[-3(\lambda+6)+24] = 0$$

$$\lambda(\lambda^2+6\lambda+14) - 3\lambda - 18 + 24 = 0$$

$$\lambda^3 + 6\lambda^2 + 11\lambda + 6 = 0$$

Substitute $\lambda = -1 \Rightarrow -1 + 6 - 11 + 6 = 0$.

$$\lambda = -1 \left| \begin{array}{cccc} 1 & 6 & 11 & 6 \\ 0 & -1 & -5 & -6 \\ \hline 1 & 5 & 6 & 0 \end{array} \right.$$

$$(\lambda+1)(\lambda^2+5\lambda+6) = 0$$

$$(\lambda+1)(\lambda+2)(\lambda+3) = 0$$

$$\lambda_1 = -1, \lambda_2 = -2, \lambda_3 = -3$$

$\therefore$ The eigenvalues are $-1, -2$ and -3

Eigenvectors:

When $\lambda_1 = -1$, the eigenvector is

$$\begin{bmatrix} -1 & -1 & 0 \\ -3 & -1 & -2 \\ 12 & 7 & 5 \end{bmatrix}\begin{bmatrix} x_{11} \\ x_{12} \\ x_{13} \end{bmatrix} = 0$$

$$x_{11} + x_{12} = 0 \Rightarrow x_{12} = -x_{11}.$$

$$3x_{11} + x_{12} + 2x_{13} = 0 \Rightarrow x_{13} = -x_{11}$$

$$12x_{11} + 7x_{12} + 5x_{13} = 0$$

If $x_{11} = 1$, then $x_{12} = -1$ and $x_{13} = -1$.

The eigenvector is $x_1 = \begin{bmatrix} 1 \\ -1 \\ -1 \end{bmatrix}$.

When $\lambda = -2$, the eigenvector is

$$\begin{bmatrix} \lambda & -1 & 0 \\ -3 & \lambda & -2 \\ 12 & 7 & \lambda+6 \end{bmatrix}\begin{bmatrix} x_{21} \\ x_{22} \\ x_{23} \end{bmatrix} = 0$$

$$\begin{bmatrix} -2 & -1 & 0 \\ -3 & -2 & -2 \\ 12 & 7 & 4 \end{bmatrix}\begin{bmatrix} x_{21} \\ x_{22} \\ x_{23} \end{bmatrix} = 0.$$

$$2x_{21} + x_{22} = 0 \Rightarrow x_{22} = -2x_{21}$$

$$3x_{21} + 2x_{22} + 2x_{22} = 0 \Rightarrow x_{23} = \frac{1}{2}x_{21}$$

If $x_{12} = 2$, then $x_{22} = -4$ and $x_{23} = 1$

The eigenvector is $x_2 = \begin{bmatrix} 2 \\ -4 \\ 1 \end{bmatrix}$.

When $\lambda = -3$, the eigenvector is

$$\begin{bmatrix} \lambda & -1 & 0 \\ -3 & \lambda & -2 \\ 12 & 7 & \lambda+6 \end{bmatrix}\begin{bmatrix} x_{31} \\ x_{32} \\ x_{33} \end{bmatrix} = 0$$

$$\begin{bmatrix} -3 & -1 & 0 \\ -3 & -3 & -2 \\ 12 & 7 & 3 \end{bmatrix}\begin{bmatrix} x_{31} \\ x_{32} \\ x_{33} \end{bmatrix} = 0$$

$$3x_{31} + x_{32} = 0 \Rightarrow x_{32} = -3x_{31}.$$

$$3x_{31} + 3x_{32} + 2x_{33} = 0 \Rightarrow x_{33} = 3x_{31}$$

$$\text{if } x_{31} = 1, \text{ then } x_{32} = -3 \text{ and } x_{33} = 3$$

The eigenvector is $x_3 = \begin{bmatrix} 1 \\ -3 \\ 3 \end{bmatrix}$.

$\therefore$ The model matrix is $M = \begin{bmatrix} 1 & 2 & 1 \\ -1 & -4 & -3 \\ -1 & 1 & 3 \end{bmatrix}$.

$$\text{Adj}\, M = \begin{bmatrix} -9 & 6 & -5 \\ -5 & 4 & -3 \\ -2 & 2 & -2 \end{bmatrix}^T = \begin{bmatrix} -9 & -5 & -2 \\ 6 & 4 & 2 \\ -5 & -3 & -2 \end{bmatrix}$$

$$\text{Det}\, M = \begin{vmatrix} 1 & 2 & 1 \\ -1 & -4 & -3 \\ -1 & 1 & 3 \end{vmatrix} = -9 + 12 - 5 = -2$$

$$M^{-1} = \frac{\text{Adj}\, M}{\text{Det}\, M} = -\frac{1}{2}\begin{bmatrix} -9 & -5 & -2 \\ 6 & 4 & 2 \\ -5 & -3 & -2 \end{bmatrix}$$

$$M^{-1}AM = -\frac{1}{2}\begin{bmatrix} -9 & -5 & -2 \\ 6 & 4 & 2 \\ -5 & -3 & -2 \end{bmatrix}\begin{bmatrix} 0 & 1 & 0 \\ 3 & 0 & 2 \\ -12 & -7 & -6 \end{bmatrix}\begin{bmatrix} 1 & 2 & 1 \\ -1 & -4 & -3 \\ -1 & 1 & 3 \end{bmatrix}$$

$$= \begin{bmatrix} -1 & 0 & 0 \\ 0 & -2 & 0 \\ 0 & 0 & -3 \end{bmatrix}$$

$$M^{-1}B = -\frac{1}{2}\begin{bmatrix} -9 & -5 & -2 \\ 6 & 4 & 2 \\ -5 & -3 & -2 \end{bmatrix}\begin{bmatrix} 1 \\ 0 \\ 2 \end{bmatrix} = \begin{bmatrix} 13/2 \\ -5 \\ 9/2 \end{bmatrix}$$

$$\text{CM} = \begin{bmatrix} 1 & 0 & 0 \end{bmatrix}\begin{bmatrix} 1 & 2 & 1 \\ -1 & -4 & -3 \\ -1 & 1 & 3 \end{bmatrix} = \begin{bmatrix} 1 & 2 & 1 \end{bmatrix}$$

The state representation in canonical form is

$$\begin{bmatrix} \dot{z}_1 \\ \dot{z}_2 \\ \dot{z}_3 \end{bmatrix} = \begin{bmatrix} -1 & 0 & 0 \\ 0 & -2 & 0 \\ 0 & 0 & -3 \end{bmatrix}\begin{bmatrix} z_1 \\ z_2 \\ z_3 \end{bmatrix} + \begin{bmatrix} 13/2 \\ -5 \\ 9/2 \end{bmatrix} u.$$

$$y = \begin{bmatrix} 1 & 2 & 1 \end{bmatrix}\begin{bmatrix} z_1 \\ z_2 \\ z_3 \end{bmatrix}$$

ASP-12: Obtain the state model for a system represented by an electric circuit as shown in the figure.

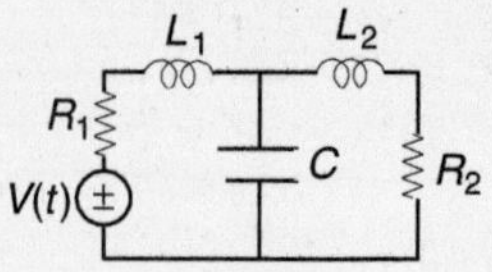

Solution:

Let us choose the current in inductors i_1 and i_2 and voltage across the capacitor V_C as state variables.

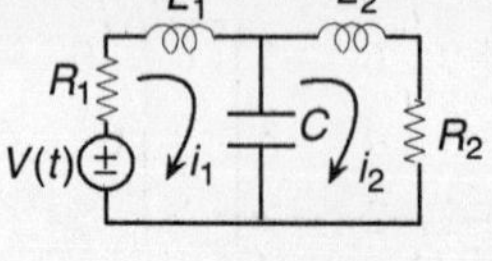

$$x_1 = i_1$$

$$x_2 = i_2$$

$$x_3 = V_C = \frac{1}{C}\int (i_1 - i_2)\,dt$$

By applying KVL to Loop 1, we have

$$V(t) = R_1 i_i + L_1 \frac{di_1}{dt} + \frac{1}{C}\int (i_1 - i_2)\,dt$$

$$V(t) = R_1 x_i + L_1 \dot{x}_1 + x_3$$

$$\Rightarrow \qquad \dot{x}_1 = \frac{-R_1}{L_1} x_1 - \frac{1}{L_1} x_3 + \frac{1}{L_1} V(t). \qquad (1)$$

By applying KVL to Loop 2, we have

$$L_2 \frac{di_2}{dt} + R_2 i_2 + \frac{1}{C}\int (i_2 - i_1)\,dt = 0$$

$$L_2 \dot{x}_2 + R_2 x_2 - x_3 = 0$$

$$\Rightarrow \qquad \dot{x}_2 = -\frac{R_2}{L_2} x_2 + \frac{1}{L_2} x_3 \qquad (2)$$

$$\dot{x}_3 = \frac{1}{C}(i_1 - i_2)$$

$$\dot{x}_3 = \frac{1}{C} x_1 - \frac{1}{C} x_2 \qquad (3)$$

$$\dot{x}_1 = -\frac{R_1}{L_1} x_1 - \frac{1}{L_1} x_3 + \frac{1}{L_1} V(t)$$

$$\dot{x}_2 = -\frac{R_2}{L_2} x_2 + \frac{1}{L_2} x_3$$

$$\dot{x}_3 = \frac{1}{C} x_1 - \frac{1}{C} x_2$$

The state equation is

$$\begin{bmatrix} \dot{x}_1 \\ \dot{x}_2 \\ \dot{x}_3 \end{bmatrix} = \begin{bmatrix} -R_1/L_1 & 0 & -1/L_1 \\ 0 & -R_2/L_2 & 1/L_2 \\ 1/C & -1/C & 0 \end{bmatrix} \begin{bmatrix} x_1 \\ x_2 \\ x_3 \end{bmatrix} + \begin{bmatrix} 1/L_1 \\ 0 \\ 0 \end{bmatrix} V(t).$$

Let us choose the voltage across the resistances as output variables, which are denoted as y_1 and y_2.

$$y_1 = i_1 R_1 = R_1 x_1$$

$$y_2 = i_2 R_2 = R_2 x_2$$

The output equation is

$$\begin{bmatrix} y_1 \\ y_2 \end{bmatrix} = \begin{bmatrix} R_1 & 0 \\ 0 & R_2 \end{bmatrix} \begin{bmatrix} x_1 \\ x_2 \end{bmatrix}.$$

ASP-13: Obtain the state model of the electrical network as shown in the figure by choosing $V_1(t)$ and $V_2(t)$ as the state variables.

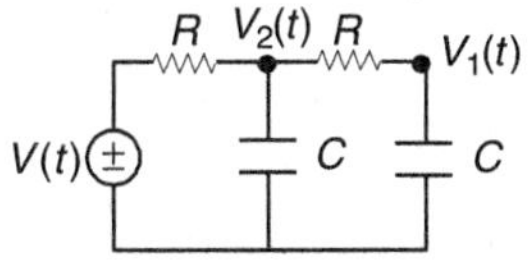

Solution:

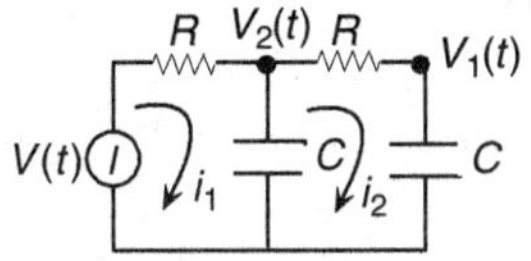

As per the given data, by choosing $V_1(t)$ and $V_2(t)$ as the state variables, we get

$$x_1 = V_1(t) = \frac{1}{C}\int i_2 dt$$

$$x_2 = V_2(t) = \frac{1}{C}\int (i_1 - i_2) dt$$

$$\Rightarrow \qquad \dot{x}_1 = \frac{1}{C} i_2 \Rightarrow i_2 = C\dot{x}_1 \tag{1}$$

$$\dot{x}_2 = \frac{1}{C}(i_1 - i_2) \Rightarrow i_1 - i_2 = C\dot{x}_2 \tag{2}$$

By applying KVL to Loop 1, we have

$$V(t) = Ri_1 + \frac{1}{C}\int (i_1 - i_2) dt.$$

$$V(t) = Ri_1 + x_2. \tag{3}$$

By applying KVL to Loop 2, we have

$$Ri_2 + \frac{1}{C}\int i_2 dt + \frac{1}{C}\int (i_2 - i_1)dt = 0$$

$$Ri_2 + x_1 - x_2 = 0.$$

$$Ri_2 = -x_1 + x_2 \tag{4}$$

$$RC\dot{x}_1 = -x_1 + x_2 \qquad [\therefore \text{ From (1)}]$$

$$\dot{x}_1 = -\frac{1}{RC}x_1 + \frac{1}{RC}x_2 \tag{5}$$

From (2), we have

$$i_1 = i_2 + C\dot{x}_2$$

$$i_1 = C\dot{x}_1 + C\dot{x}_2 \qquad [\therefore \text{ From (1)}]$$

From (3) and (6), we have

$$V(t) = R(C\dot{x}_1 + C\dot{x}_2) + x_2$$

$$V(t) = RC\left(-\frac{1}{RC}x_1 + \frac{1}{RC}x_2\right) + RC\dot{x}_2 + x_2$$

$$= -x_1 + x_2 + RC\dot{x}_2 + x_2$$

$$\Rightarrow \qquad \dot{x}_2 = \frac{1}{RC}x_1 - \frac{2}{RC}x_2 + \frac{1}{RC}V(t)$$

$$\therefore \qquad \dot{x}_1 = -\frac{1}{RC}x_1 - \frac{1}{RC}x_2$$

$$\dot{x}_2 = \frac{1}{RC}x_1 - \frac{2}{RC}x_2 + \frac{1}{RC}V(t).$$

The state equation is

$$\begin{bmatrix} \dot{x}_1 \\ \dot{x}_2 \end{bmatrix} = \begin{bmatrix} -1/RC & 1/RC \\ 1/RC & -2/RC \end{bmatrix}\begin{bmatrix} x_1 \\ x_2 \end{bmatrix} + \begin{bmatrix} 0 \\ 1/RC \end{bmatrix}V(t).$$

The output is $y = V_1(t) = x_1$

$$y = \begin{bmatrix} 1 & 0 \end{bmatrix}\begin{bmatrix} x_1 \\ x_2 \end{bmatrix}.$$

ASP-14: Obtain the state model of the system by drawing the signal flow graph whose t.f. is given as

$$\frac{Y(s)}{U(s)} = \frac{10}{s^3 + 4s^2 + 2s + 1}.$$

Solution:

$$\frac{Y(s)}{U(s)} = \frac{10}{s^3 + 4s^2 + 2s + 1} = \frac{10}{s^3\left(1 + \frac{4}{s} + \frac{2}{s^2} + \frac{1}{s^3}\right)}$$

$$= \frac{10/s^3}{1 - \left(-\frac{4}{s} - \frac{2}{s^2} - \frac{1}{s^3}\right)}$$

The signal flow graph for the above-mentioned t.f. can be constructed as shown in Figure 13.5 with a single forward path consisting of three integrators and with path gain $10/s^3$. The graph will have three individual loops with gains $-\frac{4}{s}, -\frac{2}{s^2}$ and $-\frac{1}{s^3}$.

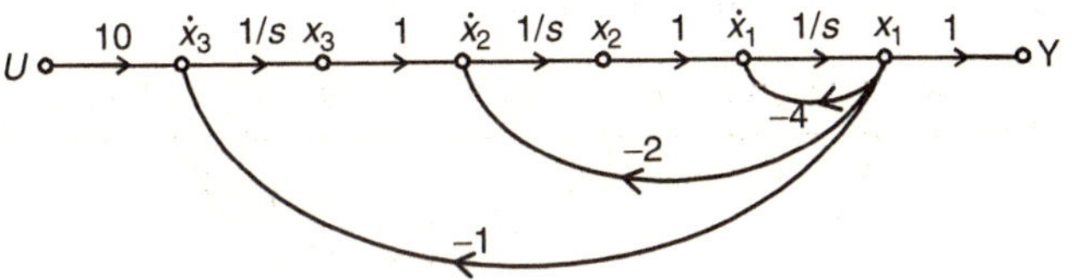

Figuwre 13.5 Signal flow graph

Let us assign a state variable at the output of the integrator $\left(\frac{1}{s}\right)$. The state equations are obtained by summing the signals to the input of the integrators and equating them to the corresponding first derivative of the state variable.

x_2 → $\dot{x}_1$ ← $-4x_1$ x_3 → $\dot{x}_2$ ← $-2x_1$ $10u$ → $\dot{x}_3$ ← $-x_1$

The state equations are

$$\dot{x}_1 = -4x_1 + x_2$$

$$\dot{x}_2 = -2x_1 + x_3 \cdot$$

$$\dot{x}_3 = -x_1 + 10u$$

The output equation is $y = x_1$.
The state model in the matrix form is

$$\begin{bmatrix} \dot{x}_1 \\ \dot{x}_2 \\ \dot{x}_3 \end{bmatrix} = \begin{bmatrix} -4 & 1 & 0 \\ -2 & 0 & 1 \\ -1 & 0 & 0 \end{bmatrix} \begin{bmatrix} x_1 \\ x_2 \\ x_3 \end{bmatrix} + \begin{bmatrix} 0 \\ 0 \\ 10 \end{bmatrix} u.$$

$$y = \begin{bmatrix} 1 & 0 & 0 \end{bmatrix} \begin{bmatrix} x_1 \\ x_2 \\ x_3 \end{bmatrix}$$

ASP-15: For the given t.f. $T(s) = \frac{b_0}{s^3 + a_2 s^2 + a_1 s + a_0}$, obtain the state model by drawing signal flow graph.

Solution:

$$T(s) = \frac{Y(s)}{U(s)} = \frac{b_0}{s^3 + a_2 s^2 + a_1 s + a_0}$$

$$= \frac{b_0}{s^3\left(1 + \frac{a_2}{s} + \frac{a_1}{s^2} + \frac{a_0}{s^3}\right)}$$

$$= \frac{b_0 / s^3}{1 - \left(-\frac{a_2}{s} - \frac{a_1}{s^2} - \frac{a_0}{s^3}\right)}$$

The signal flow consists of (i) three feedback loops (touching each other) with gains $\frac{a_2}{s}, -\frac{a_1}{s^2}$ and $-\frac{a_0}{s^3}$ and (ii) one forward path that touches the loops and has gain $\frac{b_0}{s^3}$. A signal flow graph that satisfies the above-mentioned conditions is drawn as shown in Figure 13.6.

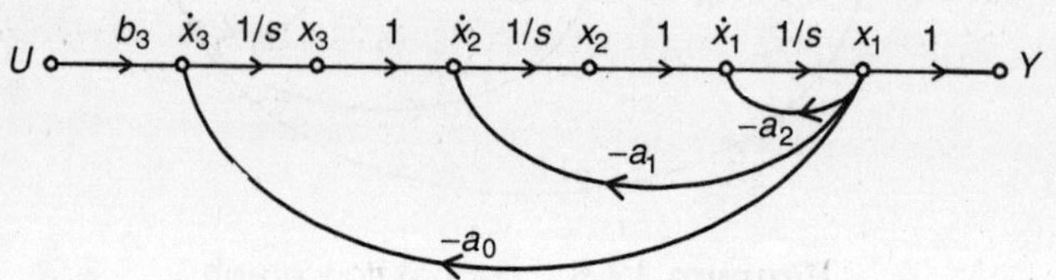

Figure 13.6 Signal flow graph

Assign a state variable at the input of the integrator $\left(\frac{1}{s}\right)$. The state equations are obtained by summing the signals to the input of the integrators and equating them to the corresponding first derivative of the state variable.

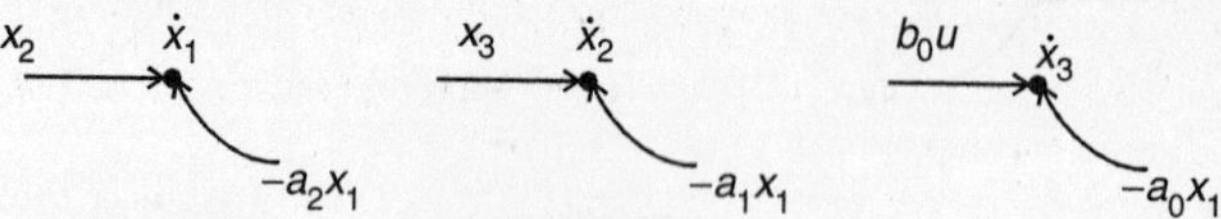

The state equations are

$$\dot{x}_1 = -a_2x_1 + x_2$$

$$\dot{x}_2 = -a_1x_1 + x_3$$

$$\dot{x}_3 = -a_0x_1 + b_0u.$$

The output equation is $y = x_1$.

The state model in the matrix form is

$$\begin{bmatrix} \dot{x}_1 \\ \dot{x}_2 \\ \dot{x}_3 \end{bmatrix} = \begin{bmatrix} -a_2 & 1 & 0 \\ -a_1 & 0 & 1 \\ -a_0 & 0 & 0 \end{bmatrix} \begin{bmatrix} x_1 \\ x_2 \\ x_3 \end{bmatrix} + \begin{bmatrix} 0 \\ 0 \\ b_0 \end{bmatrix} u.$$

$$y = \begin{bmatrix} 1 & 0 & 0 \end{bmatrix} \begin{bmatrix} x_1 \\ x_2 \\ x_3 \end{bmatrix}$$

REVIEW QUESTIONS

1. What are the advantages of state model when compared to t.f. model?
2. List the properties of state transition matrix.
3. Define the following terms:
   (i) State variable (ii) State vector
   (iii) State space (iv) State transition matrix

4. Does the solution of state equation yield the output response of the system? Explain.
5. Show that state transition matrix is a non-singular matrix.
6. Prove that the state-space representation is not unique.
7. State Cayley–Hamilton theorem.
8. Obtain the state-space representation of
   (i) armature-controlled DC motor (ii) field-controlled DC motor
9. Obtain the solution of non-homogeneous state equation by using
   (i) Laplace transform method (ii) Infinite series method
10. Derive the t.f. from the state-space representation.

EXERCISE PROBLEMS

1. Determine the state model for the given electrical circuit as shown in the figure.

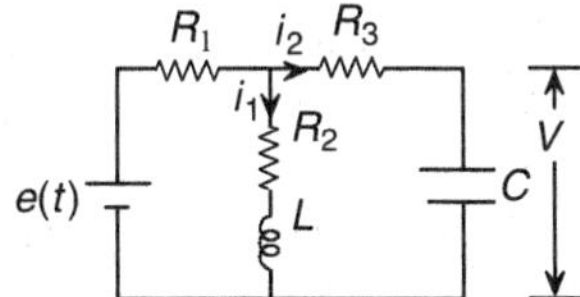

2. Determine the state model for the given electrical circuit as shown in the figure.

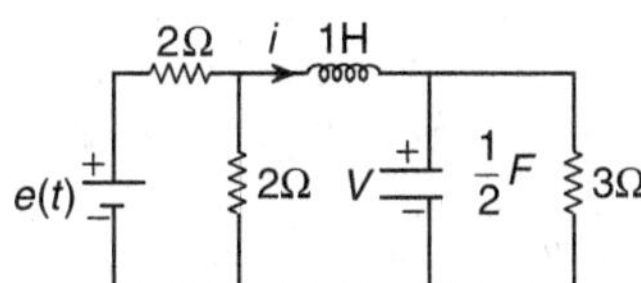

3. Obtain the state transition matrix for the following system

$$\begin{bmatrix} \dot{x}_1 \\ \dot{x}_2 \end{bmatrix} = \begin{bmatrix} 0 & 1 \\ -2 & -3 \end{bmatrix} \begin{bmatrix} x_1 \\ x_2 \end{bmatrix}$$

4. Represent the given differential equation in the state–space model

$$\frac{d^3y}{dt^3} + 2\frac{d^2y}{dt^2} + 5\frac{dy}{dt} + 7y = 4u$$

5. Consider the system having state-space representation as

$$\begin{bmatrix} \dot{x}_1 \\ \dot{x}_2 \end{bmatrix} = \begin{bmatrix} -2 & -3 \\ 4 & 2 \end{bmatrix} \begin{bmatrix} x_1 \\ x_2 \end{bmatrix} + \begin{bmatrix} 3 \\ 5 \end{bmatrix} u(t),$$

$$y = \begin{bmatrix} 1 & 1 \end{bmatrix} \begin{bmatrix} x_1 \\ x_2 \end{bmatrix}$$

Obtain the t.f. of the system.

6. Determine the transfer matrix for the following MIMO system

$$\begin{bmatrix} \dot{x}_1 \\ \dot{x}_2 \end{bmatrix} = \begin{bmatrix} 0 & 3 \\ -2 & -5 \end{bmatrix} \begin{bmatrix} x_1 \\ x_2 \end{bmatrix} + \begin{bmatrix} 1 & 1 \\ 1 & 1 \end{bmatrix} \begin{bmatrix} u_1 \\ u_2 \end{bmatrix}, y = \begin{bmatrix} 2 & 1 \\ 1 & 0 \end{bmatrix} \begin{bmatrix} x_1 \\ x_2 \end{bmatrix}$$

7. Find the canonical form of representation (or diagonalise) of the following system:

$$\begin{bmatrix} \dot{x}_1 \\ \dot{x}_2 \\ \dot{x}_3 \end{bmatrix} = \begin{bmatrix} 0 & 1 & 0 \\ 3 & 0 & 2 \\ -12 & -7 & -6 \end{bmatrix} \begin{bmatrix} x_1 \\ x_2 \\ x_3 \end{bmatrix} + \begin{bmatrix} 1 \\ 0 \\ 2 \end{bmatrix} u(t),$$

$$y = \begin{bmatrix} 1 & 0 & 0 \end{bmatrix} \begin{bmatrix} x_1 \\ x_2 \\ x_3 \end{bmatrix}$$

8. For the following matrix, find the diagonalisation matrix $\begin{bmatrix} 3 & -2 \\ -1 & 2 \end{bmatrix}$.

9. An LTI system is characterised by the state equation $\begin{bmatrix} \dot{x}_1 \\ \dot{x}_2 \end{bmatrix} = \begin{bmatrix} 1 & 0 \\ 0 & 1 \end{bmatrix} \begin{bmatrix} x_1 \\ x_2 \end{bmatrix} + \begin{bmatrix} 0 \\ 1 \end{bmatrix} u(t)$, where u is the unit step function.

   Compute the solution by assuming initial conditions as $x(0) = \begin{bmatrix} 1 & 0 \end{bmatrix}^T$.

10. Obtain the state-space representation for the given t.f.

$$\frac{Y(s)}{U(s)} = \frac{24}{s^3 + 9s^2 + 26s + 24}.$$

11. Obtain the state–space representation for the given t.f.

$$\frac{Y(s)}{U(s)} = \frac{s^2 + 7s + 2}{s^3 + 9s^2 + 26s + 24}.$$

12. A system is represented by the state and output equation is given as

$$\begin{bmatrix} \dot{x}_1 \\ \dot{x}_2 \end{bmatrix} = \begin{bmatrix} -3 & -2 \\ -1 & -2 \end{bmatrix} \begin{bmatrix} x_1 \\ x_2 \end{bmatrix} + \begin{bmatrix} 1 \\ 1 \end{bmatrix} u(t), y = \begin{bmatrix} 1 & 2 \end{bmatrix} \begin{bmatrix} x_1 \\ x_2 \end{bmatrix}.$$

   Find (i) the C.E. (ii) poles of the t.f.

13. Consider the system having state-space representation as

$$\begin{bmatrix} \dot{x}_1 \\ \dot{x}_2 \\ \dot{x}_3 \end{bmatrix} = \begin{bmatrix} -2 & 1 & 0 \\ 0 & -3 & 1 \\ -3 & -4 & -5 \end{bmatrix} \begin{bmatrix} x_1 \\ x_2 \\ x_3 \end{bmatrix} + \begin{bmatrix} 0 \\ 0 \\ 1 \end{bmatrix} u(t), \; y = \begin{bmatrix} 0 & 1 & 0 \end{bmatrix} \begin{bmatrix} x_1 \\ x_2 \\ x_3 \end{bmatrix}.$$

   Obtain the t.f of the system.

14. Find the free solution for the following state equation:

$$\begin{bmatrix} \dot{x}_1 \\ \dot{x}_2 \end{bmatrix} = \begin{bmatrix} 0 & 1 \\ -3 & -4 \end{bmatrix} \begin{bmatrix} x_1 \\ x_2 \end{bmatrix} + \begin{bmatrix} 0 \\ 2 \end{bmatrix} u(t), \text{ where the initial condition is } x(0) = \begin{bmatrix} 1 & 0 \end{bmatrix}^T.$$

15. Check for the controllability and observability for the following system

$$\begin{bmatrix} \dot{x}_1 \\ \dot{x}_2 \end{bmatrix} = \begin{bmatrix} -0.5 & 0 \\ 0 & -0.2 \end{bmatrix} \begin{bmatrix} x_1 \\ x_2 \end{bmatrix} + \begin{bmatrix} 1 \\ 1 \end{bmatrix} u(t),\ y = \begin{bmatrix} 1 & 2 \end{bmatrix} \begin{bmatrix} x_1 \\ x_2 \end{bmatrix}.$$

16. Check for the controllability and observability for the following system

$$\begin{bmatrix} \dot{x}_1 \\ \dot{x}_2 \\ \dot{x}_3 \end{bmatrix} = \begin{bmatrix} 1 & 0 & 0 \\ 0 & 0 & 1 \\ -1 & 0 & -2 \end{bmatrix} \begin{bmatrix} x_1 \\ x_2 \\ x_3 \end{bmatrix} + \begin{bmatrix} 0 \\ 0 \\ 1 \end{bmatrix} u(t),\ y = \begin{bmatrix} 1 & 1 & 0 \end{bmatrix} \begin{bmatrix} x_1 \\ x_2 \\ x_3 \end{bmatrix}.$$

17. Find the state transition matrix for the system

$$\begin{bmatrix} \dot{x}_1 \\ \dot{x}_2 \\ \dot{x}_3 \end{bmatrix} = \begin{bmatrix} -2 & 1 & 0 \\ 0 & -2 & 1 \\ 0 & 0 & -2 \end{bmatrix} \begin{bmatrix} x_1 \\ x_2 \\ x_3 \end{bmatrix}.$$

18. Obtain eigenvalues and eigenvectors for the following system

$$\begin{bmatrix} \dot{x}_1 \\ \dot{x}_2 \\ \dot{x}_3 \end{bmatrix} = \begin{bmatrix} 0 & 1 & 0 \\ 3 & 0 & 2 \\ -12 & -7 & -6 \end{bmatrix} \begin{bmatrix} x_1 \\ x_2 \\ x_3 \end{bmatrix} + \begin{bmatrix} 1 \\ 0 \\ 2 \end{bmatrix} u(t),\ y = \begin{bmatrix} 1 & 0 & 0 \end{bmatrix} \begin{bmatrix} x_1 \\ x_2 \\ x_3 \end{bmatrix}.$$

ANSWERS

1. By choosing i_2 and V as state variables, we have

$$\begin{bmatrix} \dot{x}_1 \\ \dot{x}_2 \end{bmatrix} = \begin{bmatrix} -\dfrac{R_1R_2 + R_2R_3 + R_3R_1}{L(R_1 + R_3)} & \dfrac{R_1}{L(R_1 + R_3)} \\ \dfrac{-R_1}{C(R_1 + R_3)} & \dfrac{+1}{C(R_1 + R_3)} \end{bmatrix} \begin{bmatrix} x_1 \\ x_2 \end{bmatrix} + \begin{bmatrix} \dfrac{R_3}{L(R_1 + R_3)} \\ \dfrac{1}{C(R_1 + R_3)} \end{bmatrix} \begin{bmatrix} e \\ e \end{bmatrix}.$$

$$y = \begin{bmatrix} 0 & 1 \end{bmatrix} \begin{bmatrix} x_1 \\ x_2 \end{bmatrix}$$

2. By choosing inductor current and capacitor voltage as state variables, we have

$$\begin{bmatrix} \dot{x}_1 \\ \dot{x}_2 \end{bmatrix} = \begin{bmatrix} -1 & -1 \\ 2 & -2/3 \end{bmatrix} \begin{bmatrix} x_1 \\ x_2 \end{bmatrix} + \begin{bmatrix} 0.5 \\ 0 \end{bmatrix} e(t).$$

3. $$\phi(t) = e^{At} = \begin{bmatrix} 2e^{-t} - e^{-2t} & e^{-t} - e^{-2t} \\ -2e^{-t} + 2e^{-2t} & -e^{-t} + 2e^{-2t} \end{bmatrix}$$

4. $$\begin{bmatrix} \dot{x}_1 \\ \dot{x}_2 \\ \dot{x}_3 \end{bmatrix} = \begin{bmatrix} 0 & 1 & 0 \\ 0 & 0 & 1 \\ -7 & -5 & -2 \end{bmatrix} \begin{bmatrix} x_1 \\ x_2 \\ x_3 \end{bmatrix} + \begin{bmatrix} 1 \\ 0 \\ 4 \end{bmatrix} u(t)$$

5. Transfer function $T(s) = \dfrac{8s+1}{s^2+8}$

6. For MIMO system, we cannot find t.f. However, we can calculate transfer matrix that gives the relationship between each input and each output mathematically.

$$\text{Transfer matrix} = \begin{bmatrix} \dfrac{3s+14}{(s+2)(s+3)} & \dfrac{3s+14}{(s+3)(s+2)} \\ \dfrac{s+8}{(s+2)(s+3)} & \dfrac{s+8}{(s+2)(s+3)} \end{bmatrix}.$$

7. $\begin{bmatrix} \dot{z}_1 \\ \dot{z}_2 \\ \dot{z}_3 \end{bmatrix} = \begin{bmatrix} -1 & 0 & 0 \\ 0 & -2 & 0 \\ 0 & 0 & -3 \end{bmatrix} \begin{bmatrix} z_1 \\ z_2 \\ z_3 \end{bmatrix} + \begin{bmatrix} 6.5 \\ -5 \\ 4.5 \end{bmatrix} u(t),\ y = [1 \quad 2 \quad 1] \begin{bmatrix} z_1 \\ z_2 \\ z_3 \end{bmatrix}$

8. Diagonalisation matrix = $\begin{bmatrix} 1 & 0 \\ 0 & 4 \end{bmatrix}$

9. $x(t) = \begin{bmatrix} e^t \\ 1 - e^t \end{bmatrix}$

10. $\begin{bmatrix} \dot{x}_1 \\ \dot{x}_2 \\ \dot{x}_3 \end{bmatrix} = \begin{bmatrix} 0 & 1 & 0 \\ 0 & 0 & 1 \\ -24 & -26 & -9 \end{bmatrix} \begin{bmatrix} x_1 \\ x_2 \\ x_3 \end{bmatrix} + \begin{bmatrix} 0 \\ 0 \\ 24 \end{bmatrix} u(t), \quad y = [1 \quad 0 \quad 0] \begin{bmatrix} x_1 \\ x_2 \\ x_3 \end{bmatrix}$

11. $\begin{bmatrix} \dot{x}_1 \\ \dot{x}_2 \\ \dot{x}_3 \end{bmatrix} = \begin{bmatrix} 0 & 1 & 0 \\ 0 & 0 & 1 \\ -24 & -26 & -9 \end{bmatrix} \begin{bmatrix} x_1 \\ x_2 \\ x_3 \end{bmatrix} + \begin{bmatrix} 0 \\ 0 \\ 1 \end{bmatrix} u(t), \quad y = [2 \quad 7 \quad 1] \begin{bmatrix} x_1 \\ x_2 \\ x_3 \end{bmatrix}$

12. (i) The C.E. is $s^2 + 5s + 4 = 0$ (ii) Poles are $-4, -1$

13. Transfer function $T(s) = \dfrac{s+2}{s^3 + 10s^2 + 35s + 41}$

14. $x(t) = \begin{bmatrix} 1.5e^{-t} - 0.5e^{-3t} \\ -1.5e^{-t} + 1.5e^{-3t} \end{bmatrix}$

15. State controllable and observable

16. The system is state observable but not controllable

17. $\phi(t) = \begin{bmatrix} e^{-2t} & te^{-2t} & 0.5e^{-2t} \\ 0 & e^{-2t} & te^{-2t} \\ 0 & 0 & e^{-2t} \end{bmatrix}$

18. Eigenvalues: $-1, -2, -3$

Eigenvectors: $\begin{bmatrix} 1 \\ -1 \\ 1 \end{bmatrix}, \begin{bmatrix} 2 \\ -4 \\ 1 \end{bmatrix}, \begin{bmatrix} 1 \\ -3 \\ 3 \end{bmatrix}$

OBJECTIVE TYPE QUESTIONS

1. In general, state equations are represented by
   (a) trigonometric equations (b) algebraic equations
   (c) differential equations (d) integral equations
2. The solution of linear homogeneous state equation is
   (a) output equation (b) state equation
   (c) initial conditions need not be zero (d) state transition equation
3. The advantage of state-space representation
   (a) is the application to MIMO systems (b) can be applied to time variant system
   (c) is that the initial conditions need not be zero (d) All
4. A system is represented by three state variables x_1, x_2 and x_3 in phase-variable form, then $\dot{x}_1$ is equal to ______
   (a) x_1 (b) x_2 (c) x_3 (d) None
5. If $\phi(t)$ is the state transition matrix, then $\phi^{-1}(t)$ is _____
   (a) $\phi(t)$ (b) $\phi(-t)$ (c) $-\phi(t)$ (d) 0
6. $\phi(0) =$ _____
   (a) 0 (b) I (c) A (d) 0
7. The eigenvalues of the matrix $A = \begin{bmatrix} 3 & 4 \\ 2 & 1 \end{bmatrix}$ are
   (a) -1 and -5 (b) -1 and 5 (c) 1 and 5 (d) None
8. For an SISO system, B matrix contains ______ number of columns
   (a) 1 (b) 2 (c) 3 (d) 4
9. The state transition matrix $\phi(t)$ is given as
   (a) $sI - A$ (b) $(sI - (A))^{-1}$ (c) $L^{-1}[(sI - A)^{-1}]$ (d) None
10. In exponential series form, the state transition matrix is ______
    (a) e^{-At} (b) e^{At} (c) e^{A} (d) None
11. For the state model
    $\dot{x} = Ax + Bu;\ y = C.x + Du,$ the transfer matrix is ______
    (a) $C(sI - (A)^{-1})\,B + D$ (b) $C\,(sI - (A))\,B + D$
    (c) $CB[(sI - (A))]^{-1} + D$ (d) None
12. If a system is represented by a set of state variables X, then which of the following is true.
    (a) This representation is unique.
    (b) This representation is not unique.
    (c) This representation is unique if it is in phase-variable form.
    (d) None
13. As compared to t.f. approach, state-space approach will have the following limitation.
    (a) Stability (b) Performance
    (c) Difficult to judge state variables (d) None
14. In general, state or system matrix in state-space theory is represented by
    (a) A (b) B (c) C (d) D
15. If the state variable formulation of a system is given by equations ______
    $\begin{bmatrix} \dot{x}_1 \\ \dot{x}_2 \end{bmatrix} = \begin{bmatrix} -2 & 1 \\ 0 & -3 \end{bmatrix}\begin{bmatrix} x_1 \\ x_2 \end{bmatrix} + \begin{bmatrix} 0 \\ 1 \end{bmatrix} u \quad y = \begin{bmatrix} 1 & 1 \end{bmatrix}\begin{bmatrix} x_1 \\ x_2 \end{bmatrix}$, then the t.f. of the system is_____
    (a) $1/(s + 1)(s + 3)$ (b) $1/(s + 1)$ (c) $1/(s + 3)$ (d) $1/(s + 2)$

16. The analysis of MIMO system is conveniently studied by

(a) state-space approach (b) root locus approach
(c) characteristic equation approach (d) Nichols chart

17. Order of a system is equal to

(a) number of output variables (b) number of input variables
(c) number of state variables (d) number of zeros

18. In general, number of poles for a t.f. is equal to ______ in state-space analysis

(a) number of state variables (b) number of input variables
(c) number of output variables (d) number of zeros

19. State transition matrix can be found out by

(a) Laplace transform method (b) using power series method
(c) both (a) and (b) (d) None

20. The t.f. of the following state-space representation is

$$A=\begin{bmatrix}-3 & 1\\ 0 & -1\end{bmatrix} B=\begin{bmatrix}1\\1\end{bmatrix} C=\begin{bmatrix}1 & 1\end{bmatrix} \text{ and } D=0$$

(a) $\dfrac{2s+5}{(s+1)(s+3)}$ (b) $\dfrac{2s+5}{(s+2)(s+3)}$

(c) $\dfrac{2s+5}{(s+1)(s+2)}$ (d) $\dfrac{2s+5}{(s+1)^2}$

21. The three types of variables used in state-space representation are

(a) the controllable phase variables, state-space variables and output variables.
(b) the state-space variables, output variables and state variables
(c) the controllable phase variables, state variables and output variables
(d) the input variables, output variables and state variables

22. Phase variables are defined as those state variables that are obtained from

(a) one of the system variables and its limits
(b) one of the system variables and its derivatives
(c) one of the system variables and its integrations
(d) None

23. The C.E. of an armature-controlled DC motor is of ______order equation

(a) zero (b) first (c) second (d) third

24. The C.E. of a field-controlled DC motor is of ______ order equation

(a) zero (b) first (c) second (d) third

25 The value of matrix A in the form of state-space representation $\dot{x} = Ax + Bu$ for the system described by the differential equation $\dfrac{d^2y}{dt^2}+2\dfrac{dy}{dt}+3y=0$ is

(a) $\begin{bmatrix}1 & 0\\ -2 & -1\end{bmatrix}$ (b) $\begin{bmatrix}1 & 0\\ -1 & -2\end{bmatrix}$

(c) $\begin{bmatrix}1 & 1\\ -2 & 1\end{bmatrix}$ (d) $\begin{bmatrix}0 & 1\\ -3 & -2\end{bmatrix}$

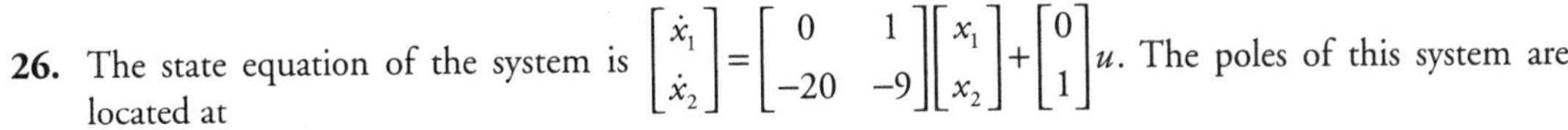

26. The state equation of the system is $\begin{bmatrix} \dot{x}_1 \\ \dot{x}_2 \end{bmatrix} = \begin{bmatrix} 0 & 1 \\ -20 & -9 \end{bmatrix} \begin{bmatrix} x_1 \\ x_2 \end{bmatrix} + \begin{bmatrix} 0 \\ 1 \end{bmatrix} u$. The poles of this system are located at
   (a) −1, −9 (b) −20, 9
   (c) 20, 9 (d) −4, −5
27. In general, output matrix in state-space theory is represented by
   (a) A (b) B
   (c) C (d) D
28. In general, the number of state variables in an electrical system is
   (a) the number of resistors (b) the number of inductors
   (c) the number of capacitors (d) both *b* and *c*
29. The eigenvalues of the matrix $\begin{bmatrix} 1 & 0 & 0 \\ 0 & 2 & 0 \\ 0 & 0 & 3 \end{bmatrix}$ are
   (a) 1, −2, 3 (b) 1, 2, 3 (c) −1, −2, −3 (d) −1, 2, −3

ANSWERS

1. c	**2.** d	**3.** d	**4.** b	**5.** b	**6.** b	**7.** b	**8.** a	**9.** c	**10.** b
11. a	**12.** b	**13.** a	**14.** a	**15.** d	**16.** a	**17.** c	**18.** a	**19.** c	**20.** a
21. d	**22.** b	**23.** d	**24.** d	**25.** d	**26.** d	**27.** c	**28.** d	**29.** b	

CHAPTER 14

Basic Compensation Techniques

Contents

14.1 INTRODUCTION

Generally, control systems are designed to perform specific jobs. If the performance of a control system is not up to expectations as per the desired specifications, then it is required that some change in the system is needed to obtain the desired performance. The change can be accomplished by adjusting forward path gain or inserting a compensating device in the control systems. It is necessary sometimes to compensate an unstable system to make it stable or it may be necessary to improve an existing system to satisfy or to meet specifications on the steady-state and transient performance. It is necessary to introduce some kind of corrective subsystems to force the chosen plant to meet the given specifications. These subsystems are known as compensators and their job is to compensate for the deficiency in the performance of the plant. The compensator may be an electrical, mechanical, hydraulic, pneumatic or other type of device or network. Usually, an electrical network serves as a compensator in many control systems.

The process of redesigning a system by adding an external device is called compensation of a control system. The external device is known as a compensator.

14.2 TYPES OF COMPENSATION

An external device termed as compensator can be added in a system anywhere depending upon the requirement. Depending upon the introduction of a compensator, the compensation techniques are classified into following types.

1. **Series compensation:** When the compensator is placed in series with the forward path t.f. of the plant, the scheme is known as series compensation or cascade compensation. Figure 14.1 shows the arrangement of series compensation.

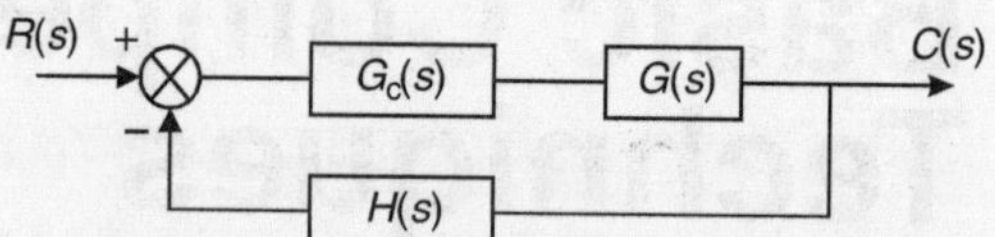

Figure 14.1 Series compensation

In this scheme, the signal flows from lower energy level to higher energy level. Therefore, additional amplifiers are required to increase gain as well as to provide necessary isolation. A series scheme has more number of components when compared to a parallel scheme.

2. **Parallel compensation:** If the feedback is taken from some internal element and a compensator introduced in such a feedback path to provide an additional internal feedback loop, this scheme is known as feedback compensation or parallel compensation. This arrangement is shown in Figure 14.2.

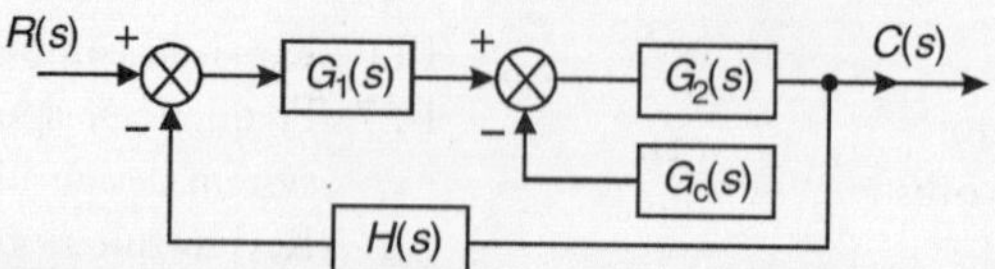

Figure 14.2 Parallel compensation

In this scheme, the flow of signal is from a higher energy level to a lower energy level. Therefore, additional amplifiers are not required.

3. **Series–parallel compensation:** In some cases, it is necessary to provide both types of compensations, series as well as parallel (feedback). Such a scheme is called series–parallel compensation. The arrangement is shown in Figure 14.3. To select the proper compensation scheme, the nature of signals available in the system, the power levels at various points, the available components, the economic considerations and the designer's experience are important.

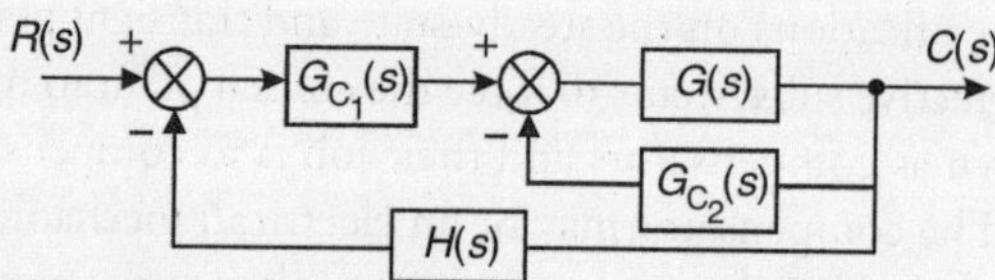

Figure 14.3 Series–parallel compensation

Once the configuration of the overall system has been selected, the next step is to select the structure of the compensator, and then, to obtain the parameters of the compensator that satisfy the selected specifications in the best possible manner. The compensation network introduces additional poles or zeros in the original system, thereby altering the original t.f.

14.3 COMPENSATING NETWORKS

The compensator is a physical device that may be an electrical network, mechanical unit, pneumatic, hydraulic or a combination of various types of devices. Here, we discuss only the electrical networks that are used for series compensation.

The following electrical compensating networks are generally used:

1. Lead network or lead compensator
2. Lag network or lag compensator
3. Lag–lead network or lag–lead compensator.

When a sinusoidal input is applied to a network, a sinusoidal steady-state output having a phase shift is obtained. If the steady-state output leads the input, the network is called lead network. If the steady-state output lags behind the input, the network is called lag network. Both phase lag and lead occur in the lag–lead network in different frequency regions. The phase lag occurs in the low frequency region, whereas the phase lead occurs in the high frequency region.

14.4 LEAD COMPENSATOR

Lead compensator improves the transient response of a system by adding a zero nearer to origin to the system's t.f. This is known as derivative effect. This compensator network is basically high-pass filter (low frequency signals are attenuated) network. Figure 14.4 shows a phase-lead network where in the phase of the output voltage leads the phase of the input voltage for sinusoidal inputs.

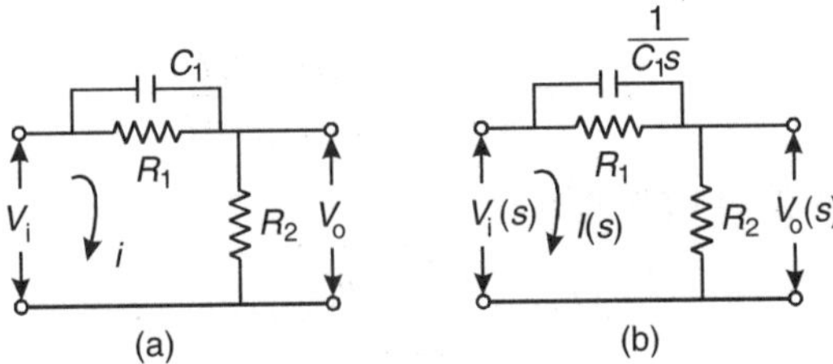

Figure 14.4 Phase-lead network and its transformed network

The t.f. of the phase-lead network is derived as follows.
From Figure 14.4 (b), we have

$$R_{eq} = R_2 + \frac{R_1 \times 1/C_1 s}{R_1 + 1/C_1 s} = R_2 + \frac{R_1}{1 + sR_1C_1} \tag{14.1}$$

$$V_i(s) = R_{eq} I(s) = \left(R_2 + \frac{R_1}{1 + sR_1C_1} \right) I(s) \tag{14.2}$$

$$V_0(s) = R_2 I(s)$$

$$T(s) = \frac{V_0(s)}{V_i(s)} = \frac{R_2 I(s)}{\left(R_2 + \dfrac{R_1}{1 + sR_1C_1} \right) I(s)} = \frac{R_2(1 + sR_1C_1)}{R_1 + R_2 + sR_1C_1R_2}$$

$$\Rightarrow \quad T(s) = \frac{R_2(1 + sR_1C_1)}{(R_1 + R_2)\left(1 + \dfrac{R_1R_2}{R_1 + R_2} C_1 s\right)} = \frac{R_2}{R_1 + R_2} \cdot \frac{(1 + sR_1C_1)}{\left(1 + \dfrac{R_1R_2}{R_1 + R_2} C_1 s\right)} = \alpha \frac{(1 + sT)}{1 + \alpha sT}$$

$$\Rightarrow \quad T(s) = \frac{V_0(s)}{V_i(s)} = \alpha \frac{(1 + sT)}{1 + \alpha sT} = \frac{s + 1/T}{s + 1/\alpha T} \tag{14.3}$$

where
$$\alpha = \frac{R_2}{R_1 + R_2} < 1,\ T = R_1 C_1$$

Equation (14.3) gives the t.f. of the lead network.

The lead compensator has a zero at $s = -1/T$ and a pole at $s = -1/aT$. Since $0 < a < 1$, the zero is always located to the right of the pole. Figure 14.5 shows the pole–zero pair of a lead compensator.

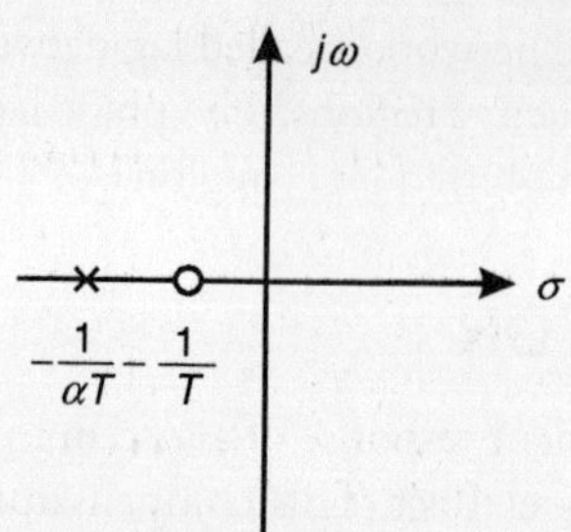

Figure 14.5 Pole–zero plot of the lead compensator

The sinusoidal t.f. is given by substituting $s = j\omega$ in $T(s)$

$$T(s) = \frac{V_0(s)}{V_1(s)} = \alpha \frac{(1 + j\omega T)}{1 + j\alpha\omega T} \tag{14.4}$$

The magnitude and phase angle are given as

$$M = \alpha \frac{\sqrt{1 + \omega^2 T^2}}{\sqrt{1 + \alpha^2 \omega^2 T^2}} \tag{14.5}$$

$$\phi = \tan^{-1}\omega T - \tan^{-1}a\omega T \tag{14.6}$$

If $a < 1$, ϕ is positive; hence, the name phase-lead network.

The two corner frequencies are given as

$\omega = 1/T$ = lower corner frequency

$\omega = 1/aT$ = upper corner frequency

1. **Maximum phase-lead angle (ϕ_m) and a**

   $\phi = \tan^{-1}\omega T - \tan^{-1}a\omega T$

   To get ω_m for which the phase angle is maximum, let us substitute $\frac{d\phi}{d\omega} = 0$.

$$\Rightarrow \quad \frac{T}{1 + \omega^2 T^2} - \frac{\alpha T}{1 + \alpha^2 \omega^2 T^2} = 0$$

$$\Rightarrow \quad \frac{T}{1 + \omega^2 T^2} = \frac{\alpha T}{1 + \alpha^2 \omega^2 T^2} = 0$$

$$\Rightarrow \quad 1 + a^2\omega^2 T^2 = a(1 + \omega^2 T^2)$$

$$\Rightarrow \quad \omega^2 T^2 a = 1$$

$$\Rightarrow \quad \omega^2 = \frac{1}{T^2 \alpha}$$

$$\Rightarrow \qquad \omega_m = \frac{1}{T\sqrt{\alpha}} = \sqrt{\left(\frac{1}{T}\right)\left(\frac{1}{\alpha T}\right)} \tag{14.7}$$

Equation (14.7) gives the frequency at which the phase lead is maximum and ω_m is the geometric mean of the two corner frequencies of the compensator.

From (14.6), we can get

$$\tan\phi = \tan(\tan^{-1}\omega T - \tan^{-1}\alpha\omega T)$$

$$= \tan\left(\tan^{-1}\frac{\omega T - \alpha\omega T}{1+\alpha\omega^2T^2}\right) = \frac{\omega T(1-\alpha)}{1+\alpha\omega^2T^2}$$

When $\omega = \omega_m = \dfrac{1}{T\sqrt{\alpha}}$, then $\phi = \phi_m$.

$$\therefore \qquad \tan\phi_m = \frac{\frac{1}{\sqrt{\alpha}}(1-\alpha)}{1+\alpha\times\frac{1}{\alpha}} = \frac{1-\alpha}{2\sqrt{\alpha}}$$

$$\sin\phi_m = \frac{1-\alpha}{1+\alpha}.$$

2. Polar plot of a lead compensator

We know that

$$M = \alpha\frac{\sqrt{1+\omega^2T^2}}{\sqrt{1+\alpha^2\omega^2T^2}}.$$

$$\phi = \tan^{-1}\omega T - \tan^{-1}\alpha\omega T$$

At $\omega = 0$, $M = a$ and $\phi = 0$.

At $\omega = \infty$, $M = 1$ and $\phi = 0$.

Here, the points are starting as well as terminating on the positive real axis. The magnitude is always positive for $0 < \omega < \infty$ and ϕ will be always positive for real $a < 1$, $\tan^{-1}\omega T > \tan^{-1}a\omega T$. Therefore, all the points of a polar plot are always in the first quadrant of the complex plane. Figure 14.6 shows the polar plot of a lead compensator.

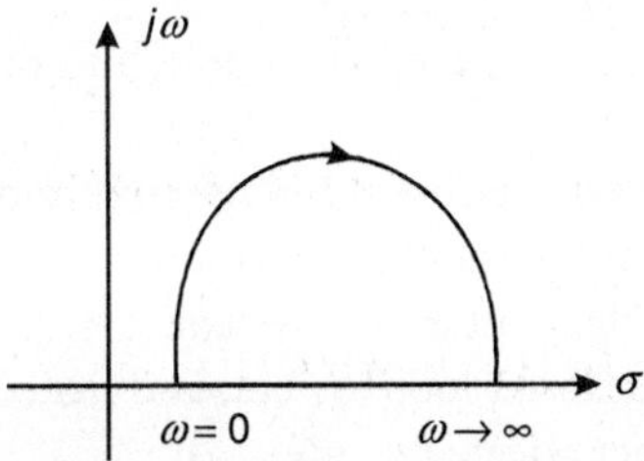

Figure 14.6 Polar plot of lead compensator

3. **Bode plot of a lead compensator:** The lead compensator has the following corner frequencies

$\omega_{c1} = \frac{1}{T}$ for a simple zero at $s = \frac{-1}{T}$ = Lower corner frequency

$\omega_{c2} = \frac{1}{\alpha T}$ for a simple pole at $s = \frac{-1}{\alpha T}$ =Upper corner frequency

Figure 14.7 shows the Bode plot of a lead compensator.

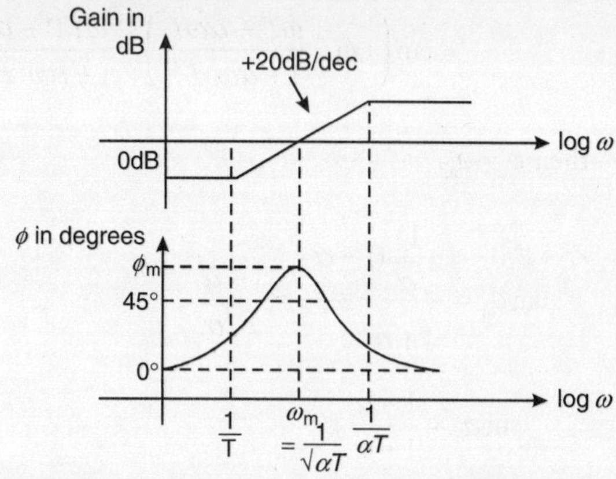

Figure 14.7 Bode plot of lead compensator

It can be noted that at $\omega = \omega_m$

$$M = \alpha\frac{\sqrt{1+\omega^2T^2}}{\sqrt{1+\alpha^2\omega^2T^2}} = \alpha\frac{\sqrt{1+\frac{1}{\alpha}}}{\sqrt{1+\alpha}} = \sqrt{\alpha}$$

Thus, at $\omega = \omega_m$, the magnitude in decibels is

$$\text{dB} = 20\log\sqrt{\alpha} = 10\log a = -10\log\frac{1}{\alpha}. \tag{14.8}$$

14.4.1 Effects of Lead Compensator

The various effects of lead compensator are as follows

1. The lead compensator adds a dominant zero and a pole, and therefore, the damping of the closed-loop system is increased.
2. Due to the increase in damping, less overshoot, less rise time and less settling time are obtained, and hence, there is an improvement in the transient response.
3. It improves the phase margin of the closed-loop system.
4. It increases the bandwidth of the closed-loop system, and hence, the response is fast.
5. The steady-state error does not get affected.

14.4.2 Limitations of Lead Compensator

The various limitations of lead compensator are

1. Lead compensator requires an additional increase in gain to offset the attenuation inherent in the lead network. In most of the cases, larger gain requirement means larger space, more elements, greater weight and higher cost.
2. More bandwidth is sometimes not desirable. This is because the noise entering the system at the input may become objectionable due to large bandwidth. It makes the system more susceptible to noise signals because of the increase in high frequency gain.
3. The compensated system may have a larger undershoot than overshoot. Therefore, the tendency to overcompensate the system may lead to a conditionally stable system.

14.5 LAG COMPENSATOR

Lag compensator improves the steady-state stability (by reducing the steady-state error) of the system by adding a pole near the origin. This is known as integral control effect. It is basically a low-pass filter (high frequency signals are attenuated). Figure 14.8 shows a phase-lag network wherein the phase of the output voltage lags the phase of the input voltage for sinusoidal input.

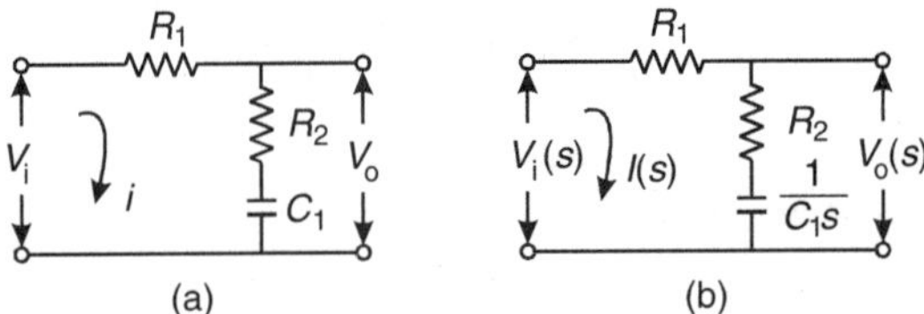

Figure 14.8 Phase-lag network and its transformed network

The t.f. of the phase-lag network is derived as follows

From Figure 14.8(b), we have

$$V_0(s) = \left(R_2 + \frac{1}{C_1 s}\right) I/(s) \tag{14.9}$$

$$V_i(s) = \left(R_1 + R_2 + \frac{1}{sC_1}\right) I(s) \tag{14.10}$$

$$T(s) = \frac{V_0(s)}{V_i(s)} = \frac{\left(R_2 + \frac{1}{C_1 s}\right) I(s)}{\left(R_1 + R_2 + \frac{1}{sC_1}\right) I(s)} = \frac{(1 + sR_2C_1)}{1 + (R_1 + R_2)sC_1}$$

$$= \frac{(1 + sR_2C_1)}{1 + \frac{(R_1 + R_2)}{R_2} sR_2C_1} = \frac{1 + sT}{1 + s\beta T}$$

$$\Rightarrow \qquad T(s) = \frac{V_0 s}{V_i(s)} = \frac{1 + sT}{1 + s\beta T} \tag{14.11}$$

where $$\beta = \frac{R_1 + R_2}{R_2} > 1, T = R_2 C_1.$$

Equation (14.11) gives the t.f. of the lag network.
The lag compensator has a zero at $s = -1/T$ and a pole at $S = 1/\beta T$. Since $0 < \beta < 1$, the zero is always located to the right of the pole. Figure 14.9 shows the pole–zero of a lag compensator.

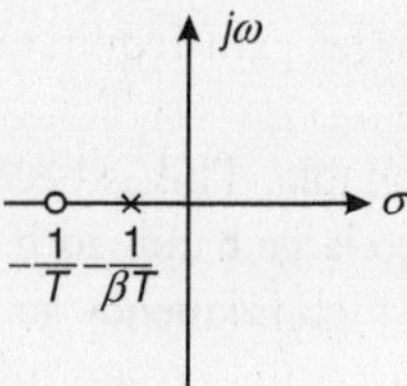

Figure 14.9 Pole–zero plot of the lag compensator

From the pole–zero configuration of phase-lag network, it is observed that the pole is near to origin when compared to zero; hence, the effect of pole is dominant; therefore, when the phase-lag network is introduced in cascade with forward path t.f., the phase shift is reduced. The introduction of lag compensator in the system sometimes increases the steady–state error coefficient and the system may become unstable in the closed-loop operation. Hence, a zero should be added very near to the origin.

The sinusoidal t.f. is given by substituting $s = j\omega$ in $T(s)$.

$$T(j\omega) = \frac{V_0(j\omega)}{V_i(j\omega)} = \frac{(1 + j\omega T)}{1 + j\beta\omega T} \tag{14.12}$$

The magnitude and phase angle are given as

$$M = \frac{\sqrt{1 + \omega^2 T^2}}{\sqrt{1 + \beta^2\omega^2 T^2}}. \tag{14.13}$$

$$\phi = \tan^{-1}\omega T - \tan^{-1}\beta\omega T \tag{14.14}$$

If $\beta > 1$, ϕ is negative, the output lags the input, and hence the name phase-lag network. The two corner frequencies are given as

$\omega = 1/T$ = upper corner frequency
$\omega = 1/\beta T$ = lower corner frequency

1. **Maximum phase-lag angle ϕ_m and β**
   $\phi = \tan^{-1}\omega T - \tan^{-1}\beta\omega T$
   To get ω_m for which the phase angle is maximum, let us substitute $\frac{d\phi}{d\omega} = 0$

$$\Rightarrow \quad \frac{T}{1 + \omega^2 T^2} - \frac{\beta T}{1 + \beta^2\omega^2 T^2} = 0$$

$$\Rightarrow \quad \frac{T}{1 + \omega^2 T^2} = \frac{\beta T}{1 + \beta^2\omega^2 T^2}$$

$$\Rightarrow \quad 1 + \beta^2\omega^2 T^2 = \beta(1 + \omega^2 T^2)$$

$$\Rightarrow \quad \omega^2 T^2 \beta = 1$$

$$\Rightarrow \qquad \omega^2 = \frac{1}{T^2\beta}$$

$$\Rightarrow \qquad \omega_m = \frac{1}{T\sqrt{\beta}} = \sqrt{\left(\frac{1}{T}\right)\left(\frac{1}{\beta T}\right)} \qquad (14.15)$$

Equation (14.15) gives the frequency at which the phase lag is maximum and ω_m is the geometric mean of the two corner frequencies of the compensator.

From (14.14), we can get

$$\tan\phi = \tan(\tan^{-1}\omega T - \tan^{-1}\beta\omega T)$$

$$= \tan\left(\tan^{-1}\frac{\omega T - \beta\omega T}{1+\beta\omega^2T^2}\right) = \frac{\omega T(1-\beta)}{1+\beta\omega^2T^2}$$

When $\omega = \omega_m = \dfrac{1}{T\sqrt{\beta}}$, then $\phi = \phi_m$

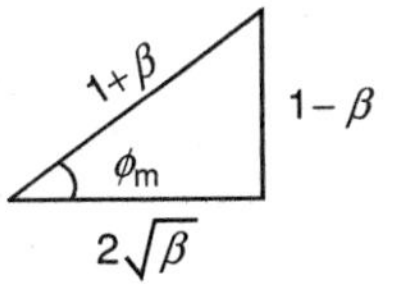

$$\therefore \qquad \tan\phi_m = \frac{\frac{1}{\sqrt{\beta}}(1-\beta)}{1+\beta\times\frac{1}{\beta}} = \frac{1-\beta}{2\sqrt{\beta}}$$

$$\sin\phi_m = \frac{1-\beta}{1+\beta}$$

2. **Polar plot of a lag compensator:**

We know that

$$M = \frac{\sqrt{1+\omega^2T^2}}{\sqrt{1+\beta^2\omega^2T^2}}$$

$$\phi = \tan^{-1}\omega T - \tan^{-1}\beta\omega T$$

$$\text{At } \omega = 0,\ M = 1 \text{ and } \phi = 0$$

$$\text{At } \omega = \infty,\ M = \frac{1}{\beta} \text{ and } \phi = 0$$

Here, both the points are starting as well as terminating on the positive real axis. The magnitude is always positive for $0 < \omega < \infty$ and the phase angle ϕ is negative for $\beta > 1$, $\tan^{-1}\omega T < \tan^{-1}\beta\omega T$. Thus, the resultant ϕ will be always negative for $0 < \omega < \infty$. Therefore, all the points of a polar plot are always in the fourth quadrant of the complex plane. Figure 14.10 shows the polar plot of a lag compensator.

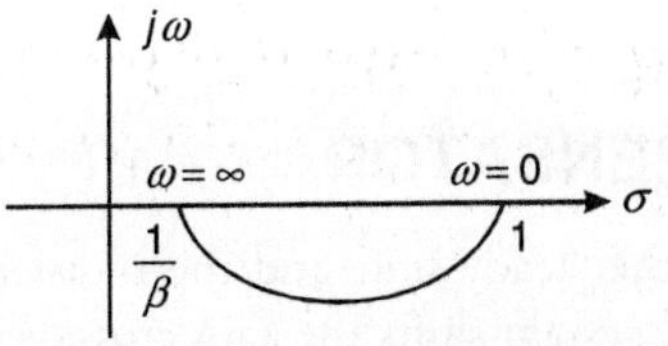

Figure 14.10 Polar plot of lag compensator

3. **Bode plot of a lag compensator**

The lag compensator has following corner frequencies

$\omega_{c1} = \dfrac{1}{\beta T}$ for a simple pole at $s = \dfrac{-1}{\beta T}$ = lower corner frequency

$\omega_{C_2} = \dfrac{1}{T}$ for a simple zero at $s = \dfrac{-1}{T}$ = upper corner frequency

Figure 14.11 shows the Bode plot of a lag compensator.

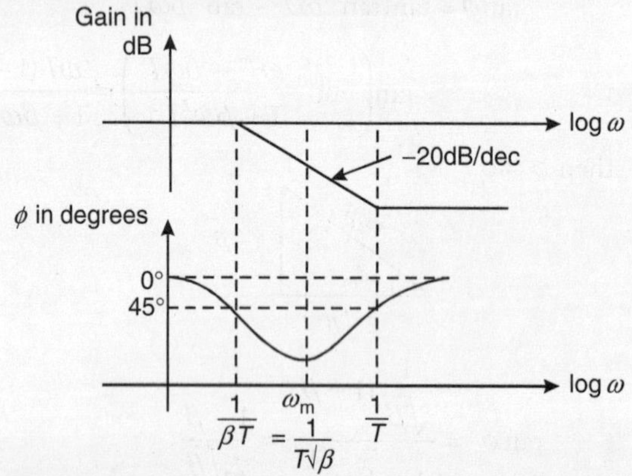

Figure 14.11 Bode plot of lag compensator

14.5.1 Effects of Lag Compensator

The various effects of lag compensator are as follows:

1. The lag compensator allows high gain at low frequencies, and thus, it is basically a low-pass filter. Hence, it improves the steady-state response.
2. The attenuation provided by the lag compensator shifts the gain crossover frequency to a low frequency point. Thus, the bandwidth of the system gets reduced and slow response occurs. Therefore, rise time and settling time are usually long, and hence, the transient response lasts for long time.

14.5.2 Limitations of Lag Compensator

The various limitations of lag compensator are as follows:

1. The system becomes more sensitive to the parameter variations.
2. Lag compensator approximately acts as proportional plus integral controller, and thus, it tends to make the system less stable.

14.6 LAG–LEAD COMPENSATOR

This compensator improves both the steady-state and the transient response. The use of phase-lead compensation network in a control system shifts the gain crossover frequency point to a higher value and therefore, the bandwidth is increased; thus, this factor improves the speed of the response and reduces the overshoot. However, it does not significantly improve the steady-state error. The use of phase-lag compensation network in a control system shifts the gain crossover frequency point to a lower value, thereby decreasing the band-width and improving the steady-state error; however, the speed of the response is reduced.

The speed of the response and steady-state error can be simultaneously improved if both phase-lag and phase-lead compensation networks are used. If, instead of using two separate lag and lead networks, a single network known as phase lag–lead network shown in Figure 14.12 is used, it combines the effects of both lag and lead networks.

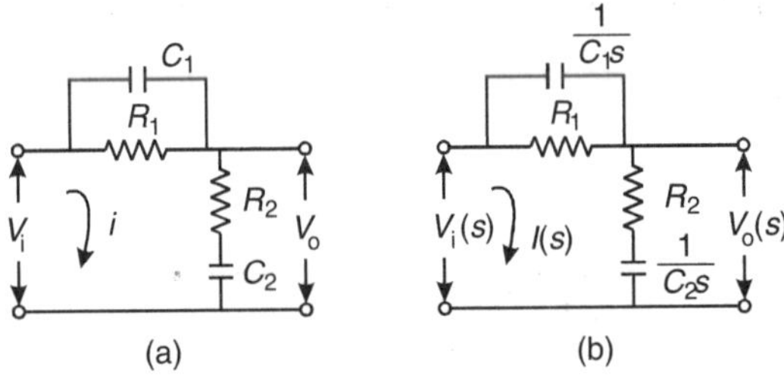

Figure 14.12 Lag–lead network and its transformed network

The t.f. of the lag–lead network is derived as follows:
From Figure 14.12 (a and b), we have

$$V_i(s) = \frac{R_1 \times 1/sC_1}{R_1 + 1/sC_1} I(s) + (R_2 + I/sC_2)I(s)$$

$$= \frac{R_1C_2s + (1 + R_1C_1s)(1 + R_2C_2s)}{(1 + R_1C_1s)sC_2} I(s) \tag{14.16}$$

$$V_0(s) = \left(R_2 + \frac{1}{C_2 s}\right) I(s) = \frac{(1 + R_2C_2s)}{sC_2} I(s) \tag{14.17}$$

$$T(s) = \frac{V_0(s)}{V_i(s)} = \frac{\dfrac{(1 + R_2C_2s)}{sC_2} I(s)}{\dfrac{R_1C_2s + (1 + R_1C_1s)(1 + R_2C_2s)}{(1 + R_1C_1s)sC_2} I(s)}$$

$$= \frac{(1 + R_1C_1s)(1 + R_2C_2s)}{R_1C_2s + (1 + R_1C_1s)(1 + R_2C_2s)}$$

$$\therefore \quad T(s) = \frac{(s + \frac{1}{T_1})(s + \frac{1}{T_2})}{s^2 + s\left(\dfrac{1}{\alpha T_1} + \dfrac{1}{\beta T_2}\right) + \dfrac{1}{\alpha\beta T_1 T_2}} \tag{14.18}$$

where $T_1 = R_1C_1$, $T_2 = R_2C_2$

$$\frac{1}{\alpha T_1} + \frac{1}{\beta T_2} = \frac{1}{R_1C_1} + \frac{1}{R_2C_2} + \frac{1}{R_2C_1}$$

$$\alpha\beta T_2 T_1 = R_1 R_2 C_1 C_2$$

$\Rightarrow$
$$\alpha\beta = 1 \Rightarrow \beta = 1/\alpha$$

$\therefore$
$$T(s) = \frac{(s+\frac{1}{T_1})(s+\frac{1}{T_2})}{(s+\frac{1}{\alpha T_1})(s+\frac{1}{\beta T_2})}, \ \beta > 1 \text{ and } \alpha < 1 \tag{14.19}$$

Equation (14.19) gives the t.f. of the lag–lead network.

The phase-lead portion adds the phase-lead angle and phase-lag portion provides attenuation near and above the gain crossover frequency. Figure 14.13 shows the pole–zero plot of the lag–lead compensator.

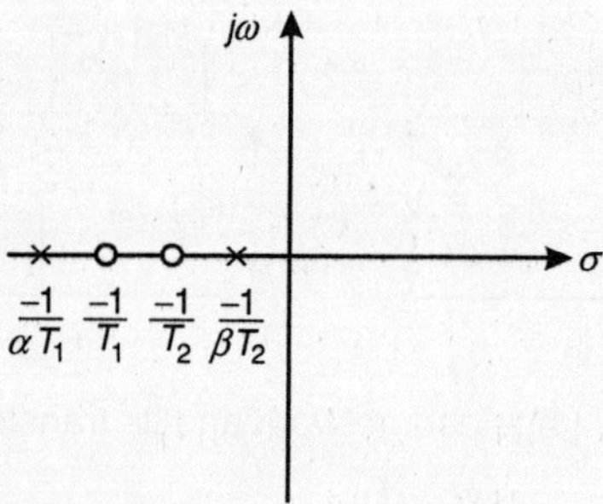

Figure 14.13 Pole–zero plot of lag–lead compensator

The sinusoidal t.f. is given by substituting $s = j\omega$ in $T(s)$

$\therefore$
$$T(j\omega) = \frac{\left(j\omega+\frac{1}{T_1}\right)\left(j\omega+\frac{1}{T_2}\right)}{\left(j\omega+\frac{1}{\alpha T_1}\right)\left(j\omega+\frac{1}{\beta T_2}\right)}. \tag{14.20}$$

The magnitude and phase angle are given as

$$M = \frac{\sqrt{\omega^2+1/T_1^2}\,\sqrt{\omega^2+1/T_2^2}}{\sqrt{\omega^2+1/\alpha^2T_1^2}\,\sqrt{\omega^2+1/\beta^2T_2^2}} \tag{14.21}$$

$$\phi = \tan^{-1}(\omega T_1) + \tan^{-1}(\omega T_2) - \tan^{-1}(\alpha\omega T_1) - \tan^{-1}(\beta\omega T_2) \tag{14.22}$$

1. **Polar plot of a lag–lead compensator**

We know that

$$M = \frac{\sqrt{\omega^2+1/T_1^2}\,\sqrt{\omega^2+1/T_2^2}}{\sqrt{\omega^2+1/\alpha^2T_1^2}\,\sqrt{\omega^2+1/\beta^2T_2^2}}.$$

$$\phi = \tan^{-1}(\omega T_1) + \tan^{-1}(\omega T_2) - \tan^{-1}(\alpha\omega T_1) - \tan^{-1}(\beta\omega T_2)$$

$$\text{At } \omega = 0,\ M = \frac{1}{\alpha\beta} = 1 \text{ and } \phi = 0$$

$$\text{At } \omega = \infty,\ M = 1 \text{ and } \phi = 0$$

Figure 14.14 shows the polar plot of lag–lead compensator. Thus, the starting and terminating points of polar plot is same as $1\angle 0°$. However, there is certain frequency ω_1 up to which it acts as a lag network giving polar plot in the fourth quadrant. While for $\omega > \omega_1$ to ∞, it acts as a lead network giving polar plot in the first quadrant. Hence, the entire polar plot is circular in nature.

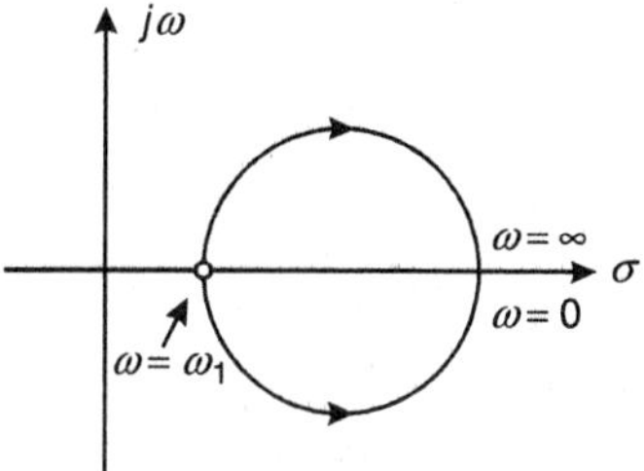

Figure 14.14 Polar plot of lag–lead compensator

In the lag–lead network, the frequency ω_1 is given as

$$\omega_1 = \frac{1}{\sqrt{T_1 T_2}}.$$

2. Bode plot of a lag–lead compensator:

The lag–lead compensator has following corner frequencies

$$\omega_{c_1} = \frac{1}{\beta T_2} \text{ for a simple pole}$$

$$\omega_{c_2} = \frac{1}{T_2} \text{ for a simple zero}$$

$$\omega_{c_3} = \frac{1}{T_1} \text{ for a simple zero}$$

$$\omega_{c_4} = \frac{1}{\alpha T_1} \text{ for a simple pole}$$

Therefore, the line of zero slope continues until ω_{c_1}. At ω_{c_1} the slope becomes -20 dB/decade. At ω_{c_2}, the slope again becomes zero and continues till ω_{c_3}. At ω_{c_3}, the slope becomes +20 dB/decade and continues till ω_{c_4}, and at ω_{c_4}, it again becomes zero. The Bode plot of lag–lead compensator is as shown in Figure 14.15.

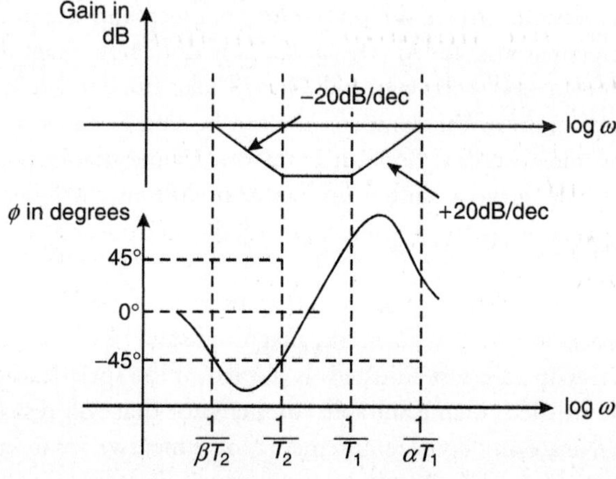

Figure 14.15 Bode plot of lag–lead compensator

14.6.1 Effects of Lag–Lead Compensator

The various effects of lag–lead compensator are

1. Lag–lead compensator is used when both fast response and good static accuracy are desired.
2. The use of lag–lead compensator increases the low frequency gain, which in turn improves the steady state. At the same time, it also increases the bandwidth of the system, thus making the system response very fast.
3. In general, the phase-lead portion of this compensator is used to achieve large bandwidth, and hence short rise time and settling time. However, the phase-lag portion provides the major damping of the system.

14.7 FREQUENCY-RESPONSE APPROACH TO CONTROL SYSTEM DESIGN

Generally, in control system design, transient-response performance is usually very important. In the frequency-response approach, we specify the transient-response performance in an indirect manner. In other words, the transient-response performance is specified in terms of the phase margin, gain margin, resonant peak (these give a rough estimate of the system damping), gain crossover frequency, resonant frequency, bandwidth (these give a rough estimate of the speed of transient response) and static error constants (for steady-state accuracy). Although the correlation between the transient response and the frequency response is indirect, the frequency domain specifications can be conveniently met in the Bode diagram approach.

After the open-loop system has been designed by the frequency-response method, the closed-loop poles and zeros can be determined. The transient-response characteristics must be checked to see whether the designed system satisfies the requirements in the time domain. If it does not satisfy the requirements, then the compensator must be modified and the analysis repeated until a satisfactory result is obtained.

Design in the frequency domain is simple and straightforward. The frequency-response plot indicates clearly the manner in which the system should be modified, although the exact quantitative prediction of the transient-response characteristics cannot be made. There are basically two approaches in the frequency-domain design. One is the polar-plot approach and the other is the Bode diagram approach. When a compensator is added, the polar plot does not retain the original shape, and therefore, we need to draw a new polar plot, which will take time and it is also inconvenient. On the other hand, a Bode diagram of the compensator can be simply added to the original Bode diagram, and thus, plotting the complete Bode diagram is a simple matter. Further, if the open-loop gain is varied, the magnitude curve is shifted up or down without changing the slope of the curve, and the phase curve remains the same. Therefore, for design purposes, it is better to work with the Bode diagram.

A common approach to the Bode diagram is to first adjust the open-loop gain so that the requirement on the steady-state accuracy is met. Then, the magnitude and phase curves of the uncompensated open loop (with the open-loop gain just adjusted) is plotted. If the specifications on the phase margin and gain margin are not satisfied, then a suitable compensator that will reshape the open-loop t.f. is determined. Finally, if there are any other requirements to be met, we try to satisfy them, unless some of them are mutually contradictory.

14.7.1 Design of a Lead Compensator

The main objective of designing a lead compensator is to reshape the frequency curve to provide sufficient phase-lead angle. Consider the system as shown in Figure 14.16. Assume that the performance specifications are given in terms of phase margin, gain margin, static velocity error constants, etc. Let $G_c(s)$ be the lead compensator for the given uncompensated system $G(s)$. The parameters of $G_c(s)$ have to be found from the given performance specifications. The following procedure is adopted to design a lead compensator by frequency-response approach.

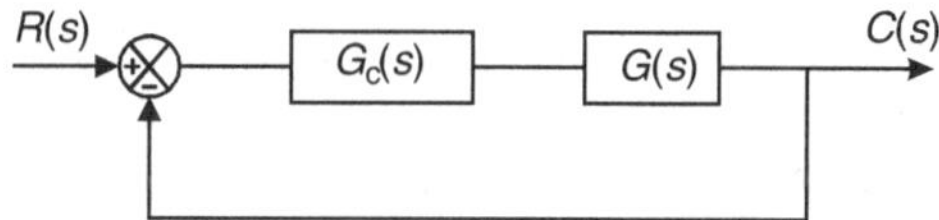

Figure 14.16 Closed-loop compensated system

1. At zero frequency, the lead compensator has gain α. However, as $\alpha < 1$, it provides an attenuation. To cancel this attenuation, the practical lead compensator is realised with an amplifier having gain K_c in series with basic lead network. Hence, from the designer's point of view, the general t.f. of the lead compensator is assumed to be

$$G_c(s) = K_c\alpha\frac{(1+sT)}{(1+\alpha sT)}. \tag{14.23}$$

Let $K_c\alpha = K = \text{DC gain}$

$$\therefore \quad G_c(s) = K\frac{(1+sT)}{(1+\alpha sT)}. \tag{14.24}$$

Thus, the open-loop t.f. of the compensated system becomes

$$G_c(s)G(s) = K\frac{(1+sT)}{(1+\alpha sT)}G(s) = \frac{(1+sT)}{(1+\alpha sT)}KG(s) = \frac{(1+sT)}{(1+\alpha sT)}G_1(s) \tag{14.25}$$

where $G_1(s) = KG(s)$.

The value of gain K is found from the given error constant.

2. Using the gain K thus determined, draw a Bode diagram of $G_1(j\omega)$ for the gain adjusted but uncompensated system. From this plot, obtain the phase margin.
3. If the phase margin satisfies the prescribed phase margin, then the system does not need the compensation. Otherwise, determine the necessary phase-lead angle (ϕ_m) to be added to the system by lead compensator to make-up for the deficiency in the prescribed phase margin (ϕ_d) (i.e., due to the addition of the lead compensator, gain crossover frequency is shifted to the right and the phase margin is decreased).

i.e.,
$$\phi = \phi_d - \phi_{uc} + \varepsilon \tag{14.26}$$

where ϕ_{uc} is the phase margin obtained in Step 2; ε is the margin of safety (5° to 12°) as cross-over frequency may shift due to compensation

4. Determine the attenuation factor α by using the following equation:

$$\sin\phi_m = \frac{1-\alpha}{1+\alpha} \Rightarrow \alpha = \frac{1-\sin\phi_m}{1+\sin\phi_m}. \tag{14.27}$$

5. Determine the frequency ω_m where the magnitude of the uncompensated system $G(j\omega)$ is equal to $-20\log(1/\sqrt{\alpha})$. Select this frequency as the new gain crossover frequency. This frequency is

$$\omega_m = \frac{1}{T\sqrt{\alpha}}.$$

   As α is known, determine $1/T$.
6. Determine the corner frequencies of the lead compensator as follows:
   Zero of lead compensator: $\omega = 1/T$.
   Pole of lead compensator: $\omega = 1/\alpha T$.
7. As $K = K_C\,\alpha$, determine the value of K_C.
8. Draw the Bode plot of the compensated system (i.e., $G_c(s)\ G(s)$) and again find the phase margin of the compensated system. If the phase margin is still low, increase the value of ε and repeat from Step 3 until a satisfactory result is obtained.

14.7.2 Design of a Lag Compensator

The main objective of designing a lag compensator is to provide attenuation in the high frequency range components for determining the sufficient phase margin. The procedure for designing lag compensators for the system, as shown in Figure 14.16, by the frequency-response approach may be stated as follows.

1. Assume the following lag compensator:

$$G_c(s) = K\frac{(1+sT)}{(1+s\beta T)}. \tag{14.28}$$

   Thus, the open-loop t.f. of the compensated system becomes

$$G_c(s)G(s) = K\frac{(1+sT)}{(1+\beta sT)}G(s) = \frac{(1+sT)}{(1+\beta sT)}KG(s) = \frac{(1+sT)}{(1+\beta sT)}G_1(s) \tag{14.29}$$

   where $G_1(s) = KG(s)$.
   Find the value of gain K from the given error constant.
2. Using the gain K thus determined, draw a Bode diagram of $G_1(j\omega)$ for the gain adjusted but uncompensated system. From this plot, obtain the gain margin and phase margin.
3 If the gain-adjusted but uncompensated system does not satisfy the specifications on the phase and gain margins, then find the phase margin ω as

$$\phi = \phi_d + \varepsilon \tag{14.30}$$

   where ε is the margin of safety (5° to 12°)
   The ε compensates for the phase lag of the lag compensator.
4. Find the new gain crossover frequency ω'_{gc} corresponding to phase margin ϕ degrees, that is, the frequency at which phase angle of open-loop t.f. is $-180 + \phi$.
5. To prevent detrimental effects of phase lag due to the lag compensator, the pole and zero of the lag compensator must be located substantially lower than the new gain crossover frequency. Therefore, choose the corner frequency $\omega = 1/T$ (corresponding to the zero of the lag compensator), that is, 1 decade below the new gain crossover frequency.

$$\therefore \qquad \omega = 1/T = \omega'_{gc}/10 \Rightarrow T = 10/\omega'_{gc}. \qquad (14.31)$$

6. Determine the attenuation necessary to bring the magnitude curve down to 0 dB at the new gain crossover frequency. Noting that this attenuation is –20 log β, determine the value of β.
   Then, the other corner frequency (corresponding to a pole of the lag compensator) is determined from $\omega = 1/(\beta T)$.
7. Thus, once t.f. of lag compensator is known, draw the Bode plot of compensated system [i.e., $G_c(s) \times G(s)$] and check the performance specifications. If specifications are not satisfied, repeat the design by modifying ε value.

14.7.3 Design of a Lag–Lead Compensator

The design of a lag–lead compensator by the frequency-response approach is based on the combination of the design techniques discussed under lead compensation and lag compensation.

Let us assume that the lag–lead compensator is of the following form

$$G_c(s) = K\frac{(1+sT_1)}{(1+s\alpha T_1)}\frac{(1+sT_2)}{(1+s\beta T_2)} \qquad (14.32)$$

where $\beta > 1$, $\alpha < 1$, $\beta\alpha = 1$. The phase-lead portion of the lag-lead compensator (the portion involving T_1) alters the frequency-response curve by adding phase-lead angle and increasing the phase margin at the gain crossover frequency. The phase-lag portion (the portion involving T_2) provides attenuation near and above the gain crossover frequency, thereby allowing an increase in gain at the low-frequency range to improve the steady-state performance.

REVIEW QUESTIONS

1. What is a compensator? Explain the need of a compensator in control systems.
2. What is compensation? List the various compensation schemes used in practice.
3. Discuss lead compensator. Sketch the Bode plot of a lead compensator. Give the design steps of a lead compensator.
4. Discuss lag compensator. Sketch the Bode plot of a lead compensator. Give the design steps of a lead compensator.
5. Derive the t.f. of
   (i) Lag network
   (ii) Lead network
   (iii) Lag–lead network
6. List the important electrical networks used for the compensation of control systems.
7. Draw and explain Bode plot of the following:
   (i) Lag network
   (ii) Lead network
   (iii) Lag–lead network
8. Draw and explain polar plot of the following:
   (i) Lag network
   (ii) Lead network
   (iii) Lag–lead network
9. Compare the characteristics of the three types of compensators.
10. Explain the effects and limitations of the three types of electrical compensators.

OBJECTIVE TYPE QUESTIONS

1. Lead compensator network is
   (a) Low-pass filter (b) High-pass filter
   (c) Equaliser (d) None
2. Lag compensator network is
   (a) Low-pass filter (b) High-pass filter
   (c) Equaliser (d) None
3. The following compensator increases the bandwidth
   (a) lag (b) lead (c) lag–lead (d) All
4. The lead compensator in a feedback system
   (a) increases the margin of stability
   (b) speeds up the transient response
   (c) increases the system error contract to some extent
   (d) All
5. The maximum phase shift that can be provided by lead compensator with t.f. $G_c(s)=\dfrac{1+6s}{1+2s}$ is
   (a) **15°** (b) **30°** (c) **45°** (d) **60°**
6. The t.f. of a compensating network is of the form $G_c=\dfrac{1+\alpha Ts}{1+Ts}$. If this is a phase-lag network, the value of α will be
   (a) $\alpha = \mathbf{0}$ (b) $\alpha < 1$
   (c) $\alpha > \mathbf{1}$ (c) $\alpha = 1$
7. Indicate which of the following t.f. represents the phase-lead compensator.
   (a) $G_c(s)=\dfrac{s+1}{s+2}$ (b) $G_c(s)=\dfrac{3+6s}{2+6s}$
   (c) $G_c(s)=\dfrac{s+5}{3s+2}$ (d) None
8. The t.f. $G_c(S)=\dfrac{1+0.5s}{s+1}$ represents
   (a) a lead network (b) a lag network
   (c) a lag–lead network (d) None
9. Lag–lead compensation is required for
   (a) steady-state response improvement (b) transient-response improvement
   (c) both (a) and (b) (d) None
10. A lag network for compensation generally consists of
    (a) R only (b) R and C elements
    (c) R and L elements (d) None
11. A lead network for compensation generally consists of
    (a) R only (b) R and C elements
    (c) R and L elements (d) None
12. A phase lag–lead network introduced in the output at low and high frequencies, respectively, are
    (a) lead, lag (b) lag, lead (c) lag, lag (d) lead, lead

13. If poles are added in a t.f., it will cause
(a) lag compensation
(b) lead compensation
(c) lag–lead compensation
(d) None

14. If zeros are added in a t.f., it will cause
(a) lag compensation
(b) lead compensation
(c) lag–lead compensation
(d) None

15. The phase-lag network
(a) maintains constant voltage gain
(b) decreases bandwidth
(c) increases system stability
(d) All

16. Lag compensator improves appreciably
(a) steady-state accuracy
(b) transient response
(c) Both (a) and (b)
(d) None

17. Lead compensator improves appreciably
(a) steady-state accuracy
(b) transient response
(c) Both (a) and (b)
(d) None

18. A phase-lag compensator will
(a) improve relative stability
(b) increase the speed of the response
(c) increase bandwidth
(d) increase overshoot

ANSWERS

1. b	**2.** a	**3.** b	**4.** d	**5.** b	**6.** b	**7.** a	**8.** b	**9.** c	**10.** b
11. b	**12.** b	**13.** a	**14.** b	**15.** d	**16.** a	**17.** b	**18.** a		

CHAPTER

15 Digital Control Systems

Contents

15.1 INTRODUCTION

The analysis of linear control system is based on the fact that the signals at various points in the system are continuous with respect to time. However, in some applications, it is convenient to use any one or more control signals at discrete time intervals of time; for example, in some industrial process control applications, the signal is available only in sampled data form as a sequence of pulses. Control systems using one or more signals at discrete time intervals are known as sampled data or digital or discrete-time control systems. Digital controllers are used for achieving optimal performance; for example, in the form of maximum productivity, maximum profit, minimum cost or minimum energy use.

Generally, controllers are used in control system to modify the error signal for better control action. They are classified into two types:

1. **Analog Controllers**
   (a) These are constructed using analogue elements and their input and output are analogue signals, which are continuous function of time.

(b) They are complex and expensive; once fabricated, it is difficult to alter the controllers.

2. **Digital Controllers**
   (a) These are constructed using non-programmable devices, microprocessor-based systems or computer-based systems.
   (b) These are used in complex and time shared control functions.
   (c) They are simple, versatile, programmable, fast acting and less expensive.
   (d) It is easy to alter the control functions by modifying the program instructions.

15.2 CONTINUOUS-TIME VERSUS DISCRETE-TIME CONTROL SYSTEMS

If all the system variables of a control system are functions of time, it is termed as a continuous-time control system.

Example: The speed control of a DC motor with tacho-generator feedback.

If one or more system variables of a control system are known at a certain discrete time, it is termed as a discrete-time control system.

Example: The microprocessor or computer-based system.

The input and output signals of discrete-time systems are digital or discrete, but the input and output signals of continuous-time systems are analogue or continuous-time signals.

Continuous-time systems, whose signals are continuous in time, may be described by differential equations; while in discrete-time systems, the signals are digital signals or sampled data signals that are described by difference equations.

15.3 BLOCK DIAGRAM OF A DIGITAL CONTROL SYSTEM

A control system that uses a digital computer as a controller or compensator is known as digital control system. The advantages of using a digital computer for compensation include accuracy, reliability, economy and most importantly, flexibility.

A block diagram of a digital control system is shown in Figure 15.1. The basic elements of the system are shown by the blocks. The controller operation is controlled by a clock. The input and

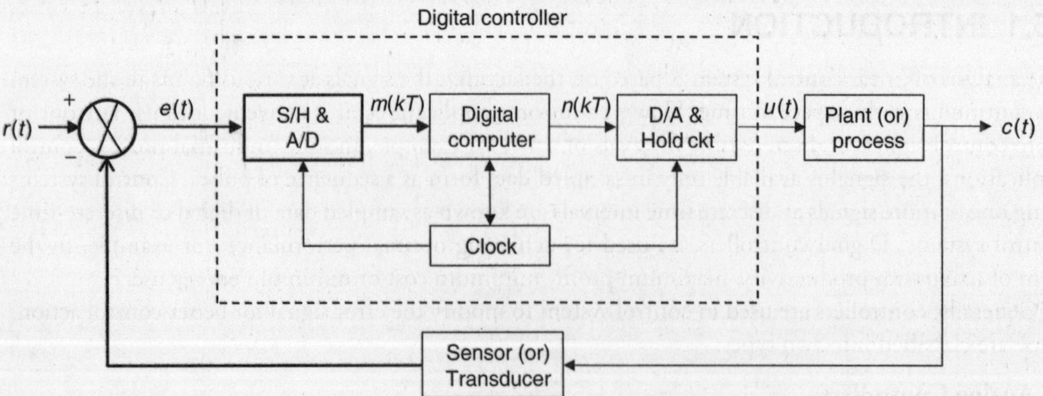

Figure 15.1 Block diagram of a digital control system

output signal in a digital computer will be digital signal, but the error signal (input to the controller) to be modified by the controller and the control signal (output of the controller) to drive the plant are analogue in nature. Hence, a sample and hold circuit and an analogue to digital converter (ADC) are provided at the digital computer input. A digital to analogue converter (DAC) and a hold circuit are provided at the digital computer output.

The sampler (S/H circuit) converts the continuous-time error signal into a sequence of pulses and ADC produces a binary code (binary number) of each sample. These codes are input data to the digital computer that processes the binary code by means of an algorithm and produces another stream of binary codes as output. The DAC and hold circuit converts the output binary code to continuous-time signal (analogue signal) called control signal and feeds it to the plant either directly or through an actuator to drive the plant (or to control its dynamics).

The operation that transforms continuous-time signals into discrete-time data is called sampling or discretisation or encoding. The inverse operation is the operation that transforms discrete-time data into a continuous-time signal and it is called data hold or decoding; it amounts to a reconstruction of a continuous-time signal from the sequence of discrete-time data.

The function of each block in the block diagram is explained below:

Sample and Hold circuit: Sample-and-hold circuit is a general term used for sample-and-hold amplifier. It describes a circuit that receives an analogue input signal and holds this signal at a constant value for a specified period of time. Usually, the signal is electrical, but other forms are possible such as optical and mechanical.

Analogue-to-Digital converter (ADC): An analogue-to-digital converter, also called an encoder, is a device that converts an analogue signal into a digital signal, usually a numerically coded signal. Such a converter is needed as an interface between an analogue component and a digital component. A sample-and-hold circuit is often an integral part of a commercially available A/D converter.

The operation of A/D conversion can be explained as in the block diagrams shown in Figure.15.2. The A/D conversion operation is carried out in three stages: sample and hold, quantisation and encoding.

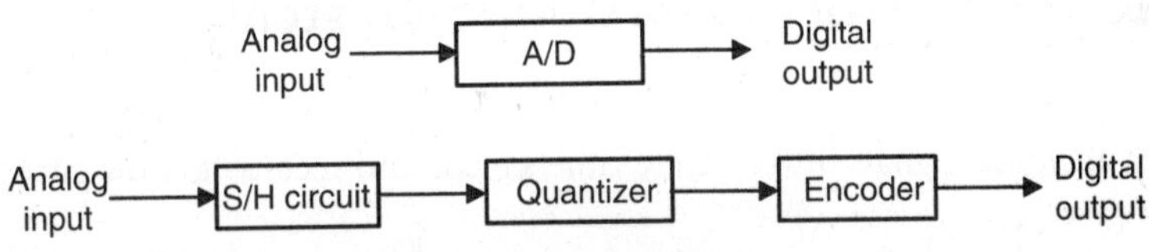

Figure 15.2 A/D conversion

Digital-to-Analogue converter (DAC): A digital-to-analogue converter, also called as a decoder, is a device that converts a digital signal into an analogue signal. Such a converter is needed as an interface between a digital component and an analogue component.

The operation of D/A conversion can be explained as in the following block diagrams (see Figure 15.3). Two stages are involved in the D/A conversion process: decoding and holding.

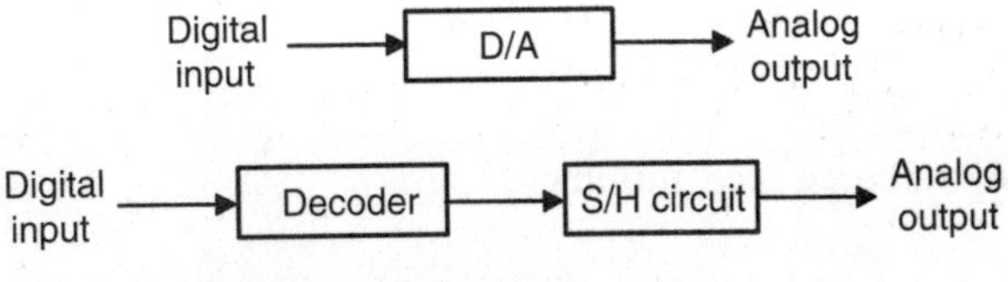

Figure 15.3 D/A conversion

Plant or process: A plant is any physical object to be controlled, like a furnace, a chemical reactor or a set of mechanical parts functioning together to perform a particular operation, such as a servosystem or a spacecraft.

A process is generally defined as a progressive operation or development marked by a series of gradual changes that succeed one another in a relatively fixed way and lead towards a particular result or end.

Transducer: A transducer is a device that converts an input signal into an output signal of another form, such as a device that converts a pressure signal into a voltage output. The output signal, in general, depends on past history of the input. Transducers may be classified as analogue transducers and sampled-data transducers or digital transducers.

15.4 ADVANTAGES, DISADVANTAGES AND APPLICATIONS OF DIGITAL CONTROL SYSTEM

15.4.1 Advantages of Digital Control System

The following are the advantages of digital control system:

1. Digital components are less susceptible to ageing and environmental variations.
2. They are less sensitive to noise and disturbance.
3. Digital processors are more compact and light weight.
4. They are highly accurate, fast and flexible and more reliable.
5. They are becoming increasingly inexpensive.
6. They provide high sensitivity to parameter variations.
7. They allow more flexibility in programming without an alternation in hardware.
8. Digital coded signals can be stored, transmitted, retransmitted, detected, analysed or processed as desired.

15.4.2 Disadvantages of Digital Control System

Some of the disadvantages of digital control systems are as follows:

1. Conversion of analogue signals into discrete-time signals and reconstruction introduces noise and errors in the signal.
2. There are limitations on computing speed and signal resolution.
3. Time delays are caused in the control loops due to the limitation on computing speed.
4. Limit cycles in the closed loop due the finite word length of the processor may cause system instability.

15.4.3 Applications of Digital Control System

The following are the applications of digital control system

1. Telecommunications
   Multiplexing, Compression, Echo control etc.
2. Audio processing
   Music, Speed generation, Speed recognition etc.
3. Echo location
   Radar, Sonar, Reflection seismology etc.

4. Image processing
   Medical, Space, Commercial image products etc.

15.5 SAMPLING PROCESS

Sampling is the conversion of continuous-time signal into a discrete-time signal obtained by taking samples of continuous-time signal at discrete time instants. A sampling process is used whenever a control system involves a digital controller, since a sampling operation and quantisation are necessary for entering data in such a controller.

Figure 15.4 shows a switch being used as a sampler. The input to the switch is a continuous-time signal denoted by $f(t)$, as shown in Figure 15.4(b). The switch is closed for a short duration of time, say p, and then it remains open for some duration of time. This operation is repeated at intervals of T called sampling period or sampling interval. The reciprocal of T, that is, $F_s = 1/T$ is called sampling rate or sampling frequency. The switch output appears only for the closing duration (p) of switch. The signal $f(t)$ is thus sampled at regular intervals of time, as shown in Figure 15.4(c). The sampled signal is denoted as $f^*(t)$ or $f(kT)$ meaning that $f^*(t)$ is obtained after sampling the input signal $f(t)$ at regular intervals of time.

A practical sampler acts like a switch closing every T seconds for a short duration of p seconds. Therefore, sampled signal can represented as follows:

$$f_p^*(t) = f(t)\sum_{k=0}^{\infty}\left[u(t-kT) - u(t-kT-p)\right] \tag{15.1}$$

where $u(t)$ is a unit step function.

The output of an ideal sampler is given as

$$f^*(t) = \sum_{k=0}^{\infty}\left[f(kT)\delta(t-kT)\right]. \tag{15.2}$$

By applying Laplace transform to the sampled signal (above-mentioned equation), we have

$$F^*(s) = \sum_{k=0}^{\infty}\left[f(kT)e^{-k.T.s}\right]. \tag{15.3}$$

We know from signal and system classes that the relationship between z and Laplace transforms is given as

$$z = e^{s \cdot T}$$

$$\Rightarrow \qquad s = \frac{1}{T}\ln z$$

$$F^*\left(s = \frac{1}{T}lnz\right) = F(z) = \sum_{k=0}^{\infty}\left[f(kT)z^{-k}\right] \tag{15.4}$$

where $F(z) = Z$-transform of $f^*(t) = Z\{r^*(t)\}$

$$= \text{Laplace transform of } f^*(t)\Big|_{s=\frac{1}{T}\ln z}$$

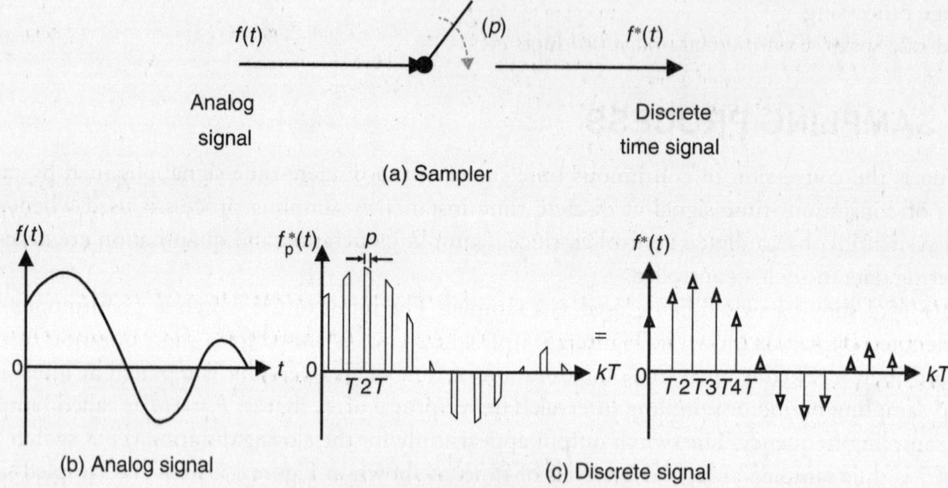

Figure 15.4 Sampling of an analogue signal

15.6 TYPES OF SAMPLING OPERATIONS

A sampling operation is the process of transforming a continuous-time signal into discrete-time signal. The different types of sampling operations are as follows:

1. **Periodic sampling:** In this case, the sampling instants are equally spaced or $t_s = kT$ ($k = 0, 1, 2\ldots$). It is the most conventional type of sampling operation.
2. **Multiple-order sampling:** The difference between two consecutive sampling instants is repeated periodically; (or) a particular sampling pattern (t_k) is repeated periodically, that is, ($t_{k+1} - t_k$) is constant for all k.
3. **Multiple-rate sampling:** In this sampling, two simultaneous sampling operations with different time periods are carried out on the signal to produce the sampled output. Thus, a digital control system may have different sampling periods in different feedback paths or may have multiple sampling rates.
4. **Random sampling:** In this case, the sampling instants are random or t_k is a random variable.

15.7 HOLD DEVICE

The function of hold device is to convert sampled signal into continuous signal. The values of continuous-time signal in between the sampling instants are calculated by extrapolation. The sampled signal provided by the sampling process is a weighted impulse train of $x^*(t)$, which is being the input signal to the system t.f. The system output is also an impulse train, the envelope of which gives the output, $x^*(t)$ or $x(kT)$ at the sampling instants only, as shown in Figure 15.5. The original signal is reconstructed from the sampled signal by using the hold circuit.

The hold circuit brings smoothness in the sampled output. At the sampling instants, the hold signal and the original signal have the same value. The use of the hold circuit enables the signal to be held

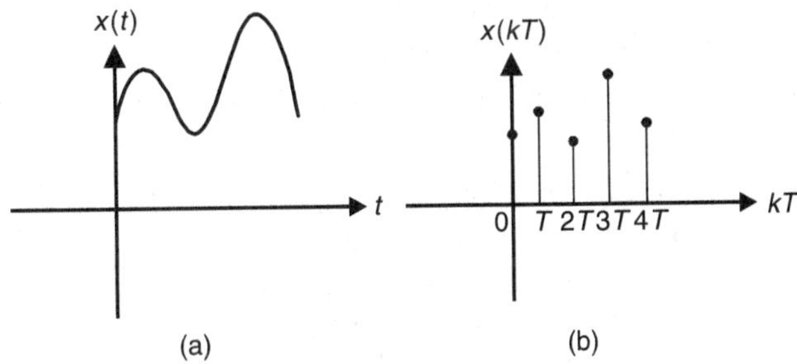

Figure 15.5 Sampled data control system input and output

between two consecutive sampling instants at the preceded value until the next sampling instant is reached.

15.8 SAMPLING AND HOLDING PROCESS

The basic concept of the sample and hold circuit is shown in Figure 15.6.

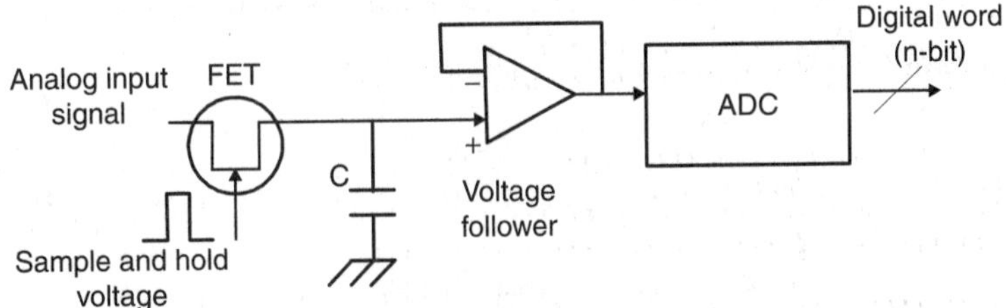

Figure 15.6 Sample and hold circuit

The sample and hold is connected to the input of ADC. When the electronic switch (field effect transistor) is closed, the capacitor voltage will track the input voltage. After sometime, when a conversion of the input signal is desired, the electronic switch is opened, isolating the capacitor from the input signal. Thus, the capacitor will hold (be charged) to the voltage when the switch was closed. The voltage follower allows this voltage to be impressed upon the ADC input, but the capacitor does not discharge because of very high input impedance of the follower. The start convert is then issued, and the conversion proceeds with the input voltage, which is kept constant in the capacitor. When the conversion is completed, the electronic switch is re-closed to capture a new sample and the above-mentioned sequence is repeated.

15.9 Z-TRANSFORM

A mathematical tool commonly used for the analysis and synthesis of the discrete-time systems is the *z*-transform. The role of the *z*-transform in discrete-time systems is similar to that of the Laplace transform in continuous-time systems. The *z*-transform provides a method for the analysis of discrete-time systems in the frequency domain since it is more useful than time domain analysis. In a linear discrete-time control system, a linear difference equation characterises the dynamics of the system. To determine the system's response to a given input, such a difference equation must be solved. With the *z*-transform method, the solution to the linear difference equation becomes algebraic in nature.

Definition:

The *z*-transform method is an operational method that is very powerful when working with discrete-time systems. For a given sequence values of $x(kT)$, its *z*-transform is defined by $X(z)$ and is given as

$$X(z) = Z[x(kT)] = \sum_{k=-\infty}^{k=+\infty} x(kT)z^{-k} \tag{15.5}$$

where *z* is a complex variable.

The sequence of (15.5) is considered to be two-sided and the transform is called two-sided *z*-transform, since the time index *k* is defined for both positive and negative values. If the sequence $x(kT)$ is one-sided sequence [i.e., $x(kT)$ is defined only for the positive values of *k*], then the *z*-transform is called one-sided *z*-transform.

$$X(z) = Z[x(kT)] = \sum_{k=0}^{\infty} x(kT)z^{-k} \tag{15.6}$$

15.10 Z-TRANSFORMS OF SOME STANDARD FUNCTIONS

1. **Unit-step function:** Consider the unit-step function

$$x(t) = \begin{cases} 1, & t \geq 0 \\ 0, & t < 0 \end{cases}$$

$$X(z) = Z[1] = \sum_{k=0}^{\infty} 1 \times z^{-k}$$

$$= 1 + z^{-1} + z^{-2} + \ldots$$

$$= \frac{1}{1-z^{-1}} = \frac{z}{z-1}$$

i.e., if $x(t) = 1$, then $X(z) = \dfrac{z}{z-1}$

2. **Unit-ramp function:** Consider the unit-ramp function

$$x(t) = \begin{cases} t, & t \geq 0 \\ 0, & t < 0 \end{cases}$$

Substitute $t = kT, k = 0, 1, 2, \cdots$

$$X[z] = Z[kT] = \sum_{k=0}^{\infty} (kT)z^{-k}$$

$$= T\sum_{k=0}^{\infty} kz^{-k}$$

$$= T\left[z^{-1} + 2z^{-2} + 3z^{-3} + \ldots\right]$$

We know that

$$X(z) = Z[1] = [1 + z^{-1} + z^{-2} + \ldots] = \frac{z}{z-1}. \tag{15.7}$$

By differentiating both sides with respect to z, we have

$$-z^{-2} - 2z^{-3} + 3z^{-4} - \ldots = \frac{(z+1) - z}{(z-1)^2} = -\frac{1}{(z-1)^2}$$

$$\Rightarrow \qquad z^{-1} + 2z^{-2} + 3z^{-3} + \ldots. = \frac{z}{(z-1)^2}. \tag{15.8}$$

From (15.7) and (15.8), we have

$$X(z) = T \times \left[\frac{z}{(z-1)^2}\right] = \frac{zT}{(z-1)^2}$$

i.e. if, $x(kT) = kT$, then $X(z) = \dfrac{zT}{(z-1)^2}$

3. **Exponential function:** Consider the exponential function

$$\cdot(t) = \begin{cases} e^{-at}, & t \geq 0 \\ 0, & t < 0 \end{cases}$$

Substitute $t = kT$, $k = 0, 1, 2, \ldots$

$$x(kT) = e^{-akT}, k = 0, 1, 2\ldots$$

$$X(z) = Z\left[e^{-at}\right] = \sum_{k=0}^{\infty} x(kT)z^{-k}$$

$$= \sum_{k=0}^{\infty} e^{-akT} z^{-k}$$

$$= 1 + e^{-aT}z^{-1} + e^{-2aT}z^{-2} + e^{-3aT}z^{-3} + \ldots$$

$$= 1 + [e^{-aT}z^{-1}]^1 + [e^{-aT}z^{-1}]^2 + \ldots$$

$$= \frac{1}{1 - e^{-aT}z^{-1}}$$

$$= \frac{z}{z - e^{-aT}}$$

i.e., if $x(t) = e^{-at}$, then $X(z) = \dfrac{z}{z - e^{-aT}}$

Note: Sum of the infinite series, $1 + r^2 + r^3 + r^4 + \ldots = \dfrac{1}{1-r}$

4. **Polynomial function, a^t:** Let us obtain the z-transform of $x(t)$, which is defined as

$$x(t) = \begin{cases} a^t, & t \geq 0 \\ 0, & t < 0. \end{cases}$$

Substitute $t = kT$, $k = 0, 1, 2,\ldots$

$$x(kT) = a^{kT}$$

$$X(z) = Z[a^{kT}] = \sum_{k=0}^{\infty} a^{kT} z^{-k}$$

$$= 1 + a^T z^{-1} + a^{2T} z^{-2} + a^{3T} z^{-3} + \ldots$$

$$= 1 + (a^T z^{-1})^1 + (a^T z^{-1})^2 + \ldots$$

$$= \frac{1}{1 - a^T z^{-1}} = \frac{z}{z - a^T}$$

i.e.,
$$\text{if } x(t) = a^t, \text{ then } X(z) = \frac{z}{z - a^T}$$

5. **Sinusoidal function:** Consider the sinusoidal function

$$x(t) = \begin{cases} \sin \omega t, & t \geq 0 \\ 0, & t < 0 \end{cases}$$

Noting that

$$e^{j\omega t} = \cos \omega t + j \sin \omega t$$

$$e^{-j\omega t} = \cos \omega t - j \sin \omega t$$

we have

$$\sin \omega t = \frac{1}{2j}\left[e^{j\omega t} - e^{-j\omega t}\right]$$

$$\cos \omega t = \frac{1}{2}\left[e^{j\omega t} + e^{-j\omega t}\right]$$

Substitute $t = kT, k = 0, 1, 2\ldots$

$$x(kT) = \sin \omega kT \text{ for } k = 0, 1, 2\ldots$$

$$X(z) = Z[\sin \omega t] = Z\left[\frac{1}{2j}(e^{j\omega t} - e^{-j\omega t})\right]$$

$$= \frac{1}{2j}\left[\frac{z}{z - e^{j\omega T}} - \frac{z}{z - e^{-j\omega T}}\right]$$

$$= \frac{z}{2j}\left[\frac{z - e^{-j\omega T} - z + e^{j\omega T}}{z^2 - z\left(e^{j\omega T} + e^{-j\omega T}\right) + 1}\right]$$

$$= \frac{z}{2j} \times \frac{2j \sin \omega T}{z^2 - 2z\cos \omega T + 1}$$

$$= \frac{z \sin \omega T}{z^2 - 2z\cos \omega T + 1}$$

If $$x(t) = \sin \omega t, \text{then } X(z) = \frac{z \sin \omega T}{z^2 - 2z\cos \omega T + 1}.$$

6. **Cosine function:** Consider the cosine function

$$x(t) = \begin{cases} \cos \omega t, & t \ge 0 \\ 0, & t < 0 \end{cases}$$

Substitute $t = kT,\ k = 0, 1, 2\ldots$

$$x(kT) = \cos \omega kT,\ k = 0, 1, 2, 3\ldots$$

$$X(z) = Z[\cos \omega t] = Z\left[\frac{e^{j\omega t} + e^{-j\omega t}}{2}\right]$$

$$= \frac{1}{2}\left[\frac{z}{z - e^{j\omega T}} + \frac{z}{z + e^{-j\omega T}}\right]$$

$$= \frac{z}{2}\left[\frac{z - e^{-j\omega T} + z - e^{j\omega T}}{z^2 - z(e^{j\omega T} + e^{-j\omega T}) + 1}\right]$$

$$= \frac{z}{2}\left[\frac{2z - e^{j\omega t} - e^{-j\omega t}}{z^2 - 2z\cos \omega T + 1}\right]$$

$$= \frac{z(z - \cos \omega T)}{z^2 - 2z\cos \omega T + 1}$$

i.e., $$\text{if } x(t) = \cos \omega t, \text{ then } X(z) = \frac{z(z - \cos \omega T)}{z^2 - 2z\cos \omega T + 1}$$

7. **$x(k) = k^P$, P being a positive integer:**

$$X(z) = Z\left[k^P\right]$$

$$= \sum_{k=0}^{\infty} k^P z^{-k}$$

$$= z\sum_{k=0}^{\infty} k^{P-1} z^{-(k+1)} \tag{15.9}$$

Further, by definition,

$$Z\left[k^{P-1}\right] = \sum_{k=0}^{\infty} k^{P-1} z^{-k}. \tag{15.10}$$

Differentiating (15.10) with respect to z, we have

$$\frac{d}{dz}\left\{Z\left[k^{P-1}\right]\right\}=\sum_{k=0}^{\infty}k^{P-1}(-k)z^{-(k+1)} \tag{15.11}$$

From (15.9) and (15.11), we have

$$Z(k^{P}]=-z\frac{d}{dz}\left\{Z\left[k^{P-1}\right]\right\} \tag{15.12}$$

This gives a recurrence formula.

Example:

$$Z[k]=\frac{z}{(z-1)^2}$$

We know that

$$Z[k^{P}]=-z\frac{d}{dz}\left\{Z\left[k^{P-1}\right]\right\}$$

Substitute $P=1$

$$Z[k]=-z\frac{d}{dz}\{Z[1]\}$$

$$=-z\frac{d}{dz}\left(\frac{z}{z-1}\right)$$

$$=-z\left[\frac{(z-1)\times 1-z\times 1}{(z-1)^2}\right]=\frac{z}{(z-1)^2}$$

$$\Rightarrow \qquad Z[k]=\frac{z}{(z-1)^2}$$

A list of standard z-transforms is given in Table 15.1.

Table 15.1 Some standard Z-Transforms

S. No	*x*(*t*)	*x*(*kT*) or *x*(*k*)	*z*-Transform, *X*(*z*)
1	–	$\delta(k)$	1
2	–	$u(k)$ or 1	$\frac{z}{z-1}$
3	–	a^k	$\frac{z}{z-a}$
4	–	ka^k	$\frac{az}{(z-a)^2}$
5	–	$(k+1)a^k$	$\frac{z^2}{(z-a)^2}$

(continued)

Table 15.1 *(continued)*

S. No	$x(t)$	$x(kT)$ or $x(k)$	z-Transform, $X(z)$
6	–	$\frac{a^k}{k!}$	$e^{az^{-1}}$
7	t	kT	$\frac{zT}{(z-1)^2}$
8	t^2	k^2T^2	$\frac{T^2z(z+1)}{(z-1)^3}$
9	e^{-at}	e^{-akT}	$\frac{z}{z-e^{-aT}}$
10	te^{-at}	kTe^{-akT}	$\frac{Tze^{-aT}}{(z-e^{-aT})^2}$
11	$\sin\omega t$	$\sin\omega kT$	$\frac{z\sin\omega T}{z^2-2z\cos\omega T+1}$
12	$\cos\omega t$	$\cos\omega kT$	$\frac{z(z-\cos\omega T)}{z^2-2z\cos\omega T+1}$
13	$e^{-at}\sin\omega t$	$e^{-atT}\sin\omega kT$	$\frac{ze^{aT}\sin\omega T}{z^2e^{2aT}-2ze^{aT}\cos\omega T+1}$
14	$e^{-at}\cos\omega t$	$e^{-atT}\cos\omega kT$	$\frac{ze^{aT}(ze^{aT}-\cos\omega T)}{z^2e^{2aT}-2ze^{aT}\cos\omega T+1}$
15	–	k^P	$-z\frac{d}{dz}\left\{Z\left[k^{P-1}\right]\right\}$
16	–	$\frac{1}{k!}$	$e^{1/z}$

***Problem-15.1*:** Obtain the z-transform of $X(s)=\dfrac{1}{s(s+1)}$.

***Solution*:**

$$x(t)=L^{-1}\left[X(s)\right]=L^{-1}\left[\frac{1}{s(s+1)}\right]=L^{-1}\left[\frac{1}{s}-\frac{1}{s(s+1)}\right]=1-e^{-t}$$

Substitute $t = kT$,

$$x(kT) = 1 - e^{-kT}$$
$$X(z) = Z[1 - e^{-kT}]$$
$$= Z[1] - Z[e^{-kT}]$$
$$= \frac{z}{z-1} - \frac{z}{z-e^{-T}}$$
$$= \frac{z(z - e^{-T} - z + 1)}{(z-1)(z-e^{-T})}$$
$$= \frac{z(1-e^{-T})}{(z-1)(z-e^{-T})}$$

Problem-15.2: Obtain the z-transform of k^2.

Solution:

$$Z[k^2] = \sum_{k=0}^{\infty} k^2 \; z^{-k}$$

From the recurrence relation, we have

$$Z\left[k^P\right] = -z\frac{d}{dz}\left\{Z\left[k^{P-1}\right]\right\}$$

Substitute $P = 2$,

$$Z[k^2] = -z\frac{d}{dz}\{Z[k]\} = -z\frac{d}{dz}\left[\frac{z}{(z-1)^2}\right]$$
$$= -z\left[\frac{(z-1)^2 \times 1 - z \times 2 \times (z-1)}{(z-1)^4}\right]$$
$$= -z\frac{(1-z^2)}{(z-1)^4} = \frac{-z(1-z)(1+z)}{(z-1)^4} = \frac{z(z+1)}{(z-1)^3}$$

Problem-15.3: Obtain the z-transform of $\frac{1}{(k+1)!}$

Solution:

$$Z\left[\frac{1}{(k+1)!}\right] = \sum_{k=0}^{\infty}\frac{1}{(k+1)!}\,z^{-k}$$
$$= \sum_{k=0}^{\infty}\frac{1}{(k+1)!\,z^k}$$
$$= 1 + \frac{1}{2!z} + \frac{1}{3!z^2} + \frac{1}{4!z^3} + \ldots$$
$$= z\left[\frac{1}{z} + \frac{1}{z^2 2!} + \frac{1}{z^3 3!} + \ldots\right]$$

$$= z\left[1 + \frac{\frac{1}{z}}{1!} + \frac{\frac{1}{z^2}}{2!} + \frac{\frac{1}{z^3}}{3!} + \ldots\right] - z$$

$$= ze^{\frac{1}{z}} - z = z\left(e^{\frac{1}{z}} - 1\right)$$

Note: $e^x = 1 + \frac{x}{1!} + \frac{x^2}{2!} + \ldots$

15.11 PROPERTIES AND THEOREMS OF Z-TRANSFORM

The use of the *z*-transform method in the analysis of discrete-time control systems is very helpful if the theorems of the *z*-transforms are discussed. In this section, we present important properties and useful theorems of the *z*-transforms. We assume that the time function $x(t)$ is *z*-transformable and $x(t)$ is zero for $t < 0$.

1. **Multiplication by a constant:** If $X(z)$ is the *z*-transform of $x(t)$, then

$$Z[ax(t)] = aZ[x(t)] = aX(z), \text{ where } a \text{ is a constant.}$$

Proof:

$$Z[ax(t)] = \sum_{k=0}^{\infty} ax(kT)z^{-k} = a\sum_{k=0}^{\infty} x(kT)z^{-k} = aX(z)$$

$$\therefore \quad Z[ax(t)] = aX(z) \tag{15.13}$$

2. **Linearity of the *z*-transform:** If $f(k)$ and $g(k)$ are *z*-transformable and a and b are constants, then $x(k)$ is formed by a linear combination, $x(k) = af(k) + bg(k)$ has the *z*-transform $X(z) = aF(z) + bG(z)$, where $F(z)$ and $G(z)$ are the *z*-transforms of $f(k)$ and $g(k)$, respectively.

Proof:

$$X(z) = Z[x(k)] = Z[af(k)] + bg(k)] = \sum_{k=0}^{\infty}[af(k) + bg(k)]z^{-k}$$

$$= a\sum_{k=0}^{\infty} f(k)z^{-k} + b\sum_{k=0}^{\infty} g(k)z^{-k}$$

$$= aZ[f(k)] + bZ[g(k)] = aF(z) + bG(z)$$

$$\therefore \quad Z[af(k) + bg(k)] = aF(z) + bG(z) \tag{15.14}$$

3. **Multiplication by a^k:** If $X(z)$ is the *z*-transform of $x(k)$, then the *z*-transform of $a^k\,x(k)$ can be given as $X(a^{-1}z)$

i.e., $$Z[a^k x(k)] = X(a^{-1}z) = X(z/a)$$

Proof:

$$Z[a^k x(k)] = \sum_{k=0}^{\infty} a^k x(k) z^{-k}$$

$$= \sum_{K=0}^{\infty} x(k)(a^{-1}z)^{-k} = X(a^{-1}z) = X(z/a) \tag{15.15}$$

4. **Shifting theorem or real translation theorem:** If $x(t) = 0$ for $t < 0$ and $x(t)$ has the z-transform $X(z)$, then

(a) $Z[x(t - nT)] = z^{-n} X(z)$

(b) $Z[x(t + nT)] = z^{n}\left[X(z) - \sum_{k=0}^{n-1} x(kT) z^{-K}\right]$

Proof:

(a)
$$Z[x(t - nT)] = \sum_{K=0}^{\infty} x(kT - nT) z^{-k}$$

$$= \sum_{k=0}^{\infty} x(k - n)T z^{-k+n-n}$$

$$= z^{-n} \sum_{k=0}^{\infty} x[(k - n)T) z^{-(k-n)}$$

Substitute $k - n = m$

If $k = 0, m = -n$

If $k = \infty, m = \infty$

$\therefore$
$$Z[x(t - nT)] = z^{-n} \sum_{m=-n}^{\infty} x(mT) z^{-m}$$

Since $x(mT) = 0$ for $m < 0$, we may change the lower limit on the summation from $m = -n$ to $m = 0$. Hence,

$$Z[x(t - nT)] = z^{-n} \sum_{m=0}^{\infty} x(mT) z^{-m} = z^{-n} X(z)$$

$\therefore$
$$Z[x(t - nT)] = z^{-n} X(z) \tag{15.16}$$

(b)
$$Z[x(t + nT)] = \sum_{k=0}^{\infty} x(kT + nT) z^{-k}$$

$$= \sum_{k=0}^{\infty} x(k + n)T z^{-k-n+n}$$

$$= z^{n} \sum_{k=0}^{\infty} x[(k + n)T) z^{-(k+n)}$$

Substitute $m = k + n$

If $k = 0, m = n$

If $k = \infty, m = \infty$

$$\therefore \quad Z[x(t + nT)] = z^n \sum_{m=n}^{\infty} x(mT)z^{-m}$$

$$= z^n \left[\sum_{m=n}^{\infty} x(mT)z^{-m} + \sum_{m=0}^{n-1} x(mT)z^{-m} - \sum_{m=0}^{n-1} x(mT)z^{-m} \right]$$

$$= z^n \left[\sum_{m=0}^{\infty} x(mT)z^{-m} - \sum_{m=0}^{n-1} x(mT)z^{-m} \right]$$

$$= z^n \left[X(z) - \sum_{k=0}^{n-1} x(kT)z^{-k} \right]$$

$$\therefore \quad Z[x(t + nT)] = z^n \left[X(z) - \sum_{k=0}^{n-1} x(kT)z^{-k} \right] \tag{15.17}$$

From the above-mentioned equation, we obtain

$$Z[x(k+1)] = zX(z) - zx(0)$$

$$z[x(k+2)] = z.Z[x(k+1)] - zx(1)$$

$$= z^2 X(z) - z^2 x(0) - zx(1)$$

Similarly,

$$Z[x(k+n)] = z^n X(z) - z^n x(0) - z^{n-1} x(1) \ldots zx(n-1) \tag{15.18}$$

The z-transform $x(k + m)$ and $x(k - m)$ are given in Table 15.2.

Table 15.2 Z-transform of $x(k + m)$ and $(k - m)$

Discrete function	z-Transform
$x(k+4)$	$z^4X(z) - z^4x(0) - z^3x(1) - z^2x(2) - zx(3)$
$x(k+3)$	$z^3X(z) - z^3x(0) - z^2x(1) - zx(2)$
$x(k+2)$	$z^2X(z) - z^2x(0) - zx(1)$
$x(k+1)$	$zX(z) - zx(0)$
$x(k)$	$X(z)$
$x(k-1)$	$z^{-1}X(z)$
$x(k-2)$	$z^{-2}X(z)$
$x(k-3)$	$z^{-3}X(z)$
$x(k-4)$	$z^{-4}X(z)$

5. **Complex translation theorem:** If $x(t)$ has the z-transform $X(z)$, then the z-transform of $e^{\mp at}x(t)$ can be given as $X(ze^{\pm aT})$.

Proof:

Substitute
$$t = kT$$

$$Z[e^{-at}x(t)] = Z[e^{-akT}x(kT)]$$
$$= \sum_{k=0}^{\infty} e^{-akT}x(kT)z^{-k}$$
$$= \sum_{k=0}^{\infty} x(kT)(ze^{aT})^{-k} = X(ze^{aT})$$

$$\therefore \quad Z[e^{\mp at}x(t)] = X(ze^{\pm aT}) \tag{15.19}$$

Thus, we see that replacing z in $X(z)$ by ze^{aT} gives the z-transform of $e^{-at}x(t)$.

Problem-15.4: Obtain the z-transform of te^{-at}

Solution:

We know that

$$Z[t] = \frac{zT}{(z-1)^2} = X(z).$$

According to complex translation theorem, we have

$$Z[te^{-aT}] = X(ze^{aT}) = \frac{ze^{aT}T}{(ze^{aT}-1)^2} = T\frac{ze^{aT}}{(ze^{aT}-1)^2}.$$

Problem-15.5: Given the z-transforms of $\sin \omega t$ and $\cos \omega t$, obtain the z-transform of $e^{-at}\sin \omega t$ and $e^{-at}\cos \omega t$, respectively, by using the complex translation theorem.

Solution:

We know that

$$Z[\sin \omega T] = \frac{z\sin \omega T}{z^2 - 2z\cos \omega T + 1} = X(z)$$

$$\therefore \quad Z[e^{-at}\sin \omega t] = X(ze^{aT}) = \frac{ze^{aT}\sin \omega T}{z^2 - 2ze^{aT}\cos \omega T + 1}$$

$$Z[\cos \omega T] = \frac{z(z-\cos \omega T)}{z^2 - 2z\cos \omega T + 1} = X(z)$$

$$\therefore \quad z[e^{-at}\cos \omega t] = X(ze^{aT}) = \frac{ze^{aT}(ze^{aT} - \cos \omega T)}{z^2e^{aT} - 2ze^{aT}\cos \omega T + 1}$$

Problem-15.6: Find the z-transform of the sequence

$$x(k) = \left(\frac{1}{2}\right)^k, \text{ for } k = 0, 1, 2, 3.....$$

Solution:

$$X(z) = Z[x(k)] = \sum_{k=0}^{\infty}\left(\frac{1}{2}\right)^k z^{-k}$$

$$= 1 + \frac{1}{2}z^{-1} + \frac{1}{4}z^{-2} + \frac{1}{8}z^{-3} + \ldots$$

$$= 1 + \frac{1}{2z} + \left(\frac{1}{2z}\right)^2 + \left(\frac{1}{2z}\right)^3 + \ldots$$

$$= \left(\frac{1}{1-\left(\frac{1}{2z}\right)}\right) = \frac{z}{z-0.5}$$

Problem-15.7: Obtain the z-transform of $f(a) = \begin{cases} a^{k-1}, & k = 1,2,3\ldots \\ 0, & k \le 0 \end{cases}$.

Solution:

Let

$$x(k) = a^k$$

$$Z[x(k)] = Z[a^k] = \frac{z}{z-a}$$

We know that

$$Z[x(k-1)] = Z[a^{k-1}] = z^{-1}X(z)$$

$\therefore$

$$Z[f(a)] = Z[a^{k-1}] = z^{-1} \times \frac{z}{z-a} = \frac{1}{z-a} = \frac{z^{-1}}{1-az^{-1}}$$

Problem-15.8: Consider the function $y(k)$, which is a sum of functions $x(h)$, where $h = 0, 1, 2, \ldots k$, such that $y(k) = \sum_{h=0}^{k} x(h), k = 0, 1, 2\ldots$

where $y(k) = 0$ for $k < 0$. Obtain the z-transform of $y(k)$.

Solution:

$$y(k) = \sum_{h=0}^{k} x(h) = x(0) + x(1) + \ldots + x(k-1) + x(k)$$

$$y(k-1) = \sum_{h=0}^{k-1} x(h) = x(0) + x(1) + \ldots + x(k-1)$$

$\therefore$

$$y(k) - y(k-1) = x(k)$$

$$Z[y(k) - y(k-1)] = Z[x(k)]$$

$$Y(z) - z^{-1}Y(z) = X(z)$$

$$Y(z) = \frac{1}{1-z^{-1}}X(z)$$

where

$$X(z) = Z[x(k)]$$

6. **Partial differentiation theorem:** Let the z-transform of the function $x(t,a)$ be represented by $X(z, a)$, where a is an independent variable or constant. The z-transform of the partial derivative of $x(t, a)$ with respect to a is given as

$$Z\left\{\frac{\partial}{\partial a}[x(t,a)]\right\} = \frac{\partial}{\partial a}[X(z,a)]$$

Proof:
From the z-transform definition, we have

$$Z\left\{\frac{\partial}{\partial a}[x(t,a)]\right\} = \sum_{k=0}^{\infty}\frac{\partial}{\partial a}x(kT,a)z^{-k}$$

$$= \frac{\partial}{\partial a}\sum_{k=0}^{\infty}x(kT,a)z^{-k}$$

$$= \frac{\partial}{\partial a}X(z\,,a)$$

$$\therefore \quad Z\left\{\frac{\partial}{\partial a}[x(t,a)]\right\} = \frac{\partial}{\partial a}X(z,a) \tag{15.20}$$

***Problem-15.9*:** Find the z-transform of $f(t) = te^{-at}$ by using partial differentiation theorem.

***Solution*:**
We know that

$$\frac{\partial}{\partial a}[-e^{-at}] = te^{-at}$$

$$Z[te^{-at}] = Z\left[\frac{\partial}{\partial a}(-e^{-at})\right]$$

$$= Z\left[-\frac{\partial}{\partial a}[e^{-at}]\right]$$

$$= -\frac{\partial}{\partial a}X(z,a) \text{ (Using partial differential theorem)}$$

where $$X(z,a) = Z[e^{-at}] = \frac{z}{z-e^{-at}}$$

$$\therefore \quad Z[te^{-at}] = -\frac{\partial}{\partial a}\left[\frac{z}{z-e^{-aT}}\right] = -\left[\frac{(z-e^{-aT})\times 0 - ze^{-aT}(-T)}{(z-e^{-aT})^2}\right] = \frac{zTe^{-aT}}{(z-e^{-aT})^2}$$

***Problem-15.10*:** Find the z-transform of $f(t) = t^2e^{-at}$ by using partial differentiation theorem.

***Solution*:**
We know that

$$\frac{\partial}{\partial a}[-te^{-at}] = t^2e^{-at}$$

$$Z[t^2e^{-at}] = Z\left[\frac{\partial}{\partial a}(-te^{-at})\right] = -Z\left[\frac{\partial}{\partial a}te^{-at}\right] = -\frac{\partial}{\partial a}X(z,a)$$

where $$X(z,a) = Z[te^{-at}] = \frac{zTe^{-at}}{(z-e^{-aT})^2}$$

$\therefore$ $$Z[t^2e^{-at}] = -\frac{\partial}{\partial a}\left[\frac{Tze^{-aT}}{(z-e^{-aT})^2}\right]$$

$$= -Tz\left[\frac{\left(z-e^{at}\right)e^{-aT}(-T) - e^{-aT}2\left(z-e^{-aT}\right)\left(-e^{-aT}\right)(-T)}{\left[z-e^{-aT}\right]^4}\right]$$

$$= -zT(z-e^{-aT})\frac{\left[-Te^{-aT} - 2Te^{-2aT}\right]}{\left[z-e^{-aT}\right]^4}$$

$$= T^2z\frac{[ze^{-aT} + e^{-2aT}]}{[z-e^{-aT}]^3}$$

7. **Initial value theorem:** If $x(t)$has the z-transform $X(z)$and $\underset{z\to\infty}{\text{Lt}}\,X(z)$ exists, then the initial value $x(0)$of $x(t)$ or $x(k)$ is given as

$$\underset{k\to 0}{\text{Lt}}\,x(k) = x(0) = \underset{z\to\infty}{\text{Lt}}\,X(z)$$

Proof:

From the z-transform definition, we have

$$x(k)] = X(z) = \sum_{k=0}^{\infty} x(k)z^{-k}$$

$$= x(0) + x(1)z^{-1} + x(2)z^{-2} + \ldots$$

By applying $\underset{z\to\infty}{\text{Lt}}$ on both sides, we have

$$\underset{z\to\infty}{\text{Lt}}\,X(z) = \underset{z\to\infty}{\text{Lt}}\left[x(0) + \frac{x(1)}{z} + \frac{x(2)}{z^2}\ldots\right]$$

$$= x(0) + 0 + 0 + \ldots$$

$$= x(0)$$

$\therefore$ $$\underset{k\to 0}{\text{Lt}}\,x(k) = x(0) = \underset{z\to\infty}{\text{Lt}}\,X(z) \qquad (15.21)$$

8. **Final value theorem:** If the function $x(k)$ has the z-transform $X(z)$ and if the function $(1-z^{-1})X(z)$ does not have poles on or outside the unit circle $|z| = 1$ in the z-plane, then $\underset{k\to\infty}{\text{Lt}}\,x(k) = \underset{z\to 1}{\text{Lt}}(1-z^{-1})X(z)$

Proof:

$$Z[x(k)] = X(z) = \sum_{k=0}^{\infty} x(k)z^{-k}$$

$$= x(0) + x(1)z^{-1} + x(2)z^{-2} + \ldots \qquad (15.22)$$

$$Z[x(k-1)] = z^{-1}X(z)$$

$$= \sum_{k=0}^{\infty} x(k-1)z^{-k} \tag{15.23}$$

On subtracting (15.23) from (15.22), we have

$$X(z) - z^{-1}X(z) = \sum_{k=0}^{\infty} x(k)z^{-k} - \sum_{k=0}^{\infty} x(k-1)z^{-k}.$$

By applying $\underset{z\to 1}{\text{Lt}}$ on both sides, we have

$$\underset{z\to l}{\text{Lt}}[X(z) - z^{-1}X(z)] = \underset{z\to l}{\text{Lt}}[\sum_{k=0}^{\infty} x(k)z^{-k} - \sum_{k=0}^{\infty} x(k-1)z^{-k}]$$

$$= \sum_{k=0}^{\infty}[x(k) - x(k-1)]$$

$$= [x(0) - x(-1)] + [x(1) - x(0)] + [x(2) - x(1)] \ldots + x(\infty)$$

$$= x(\infty)$$

i.e.,

$$\underset{z\to 1}{\text{Lt}}[1 - z^{-1}]X(z) = x(\infty) = \underset{k\to\infty}{\text{Lt}}\ x(k)$$

$\therefore$

$$x(\infty) = \underset{k\to\infty}{\text{Lt}}\ x(k) = \underset{z\to 1}{\text{Lt}}(1 - z^{-1})X(z) \tag{15.24}$$

Problem-15.11: Determine the initial value $x(0)$ if the z-transforms of $x(t)$ is given as

$$X(z)\frac{(1-e^{-T})z^{-1}}{(1-z^{-1})(1-e^{-T}z^{-1})}.$$

Solution:

According to the initial value theorem, we have

$$x(0) = \underset{z\to\infty}{\text{Lt}}\ X(z) = \underset{z\to\infty}{\text{Lt}}\left[\frac{(1-e^{-T})z^{-1}}{(1-z^{-1})(1-e^{-T}z^{-1})}\right] = 0$$

Problem-15.12: Determine the final value $x(\infty)$ of $X(z) = \frac{1}{1-z^{-1}} - \frac{1}{1-e^{-aT}z^{-1}}, a > 0$ by using final value theorem.

Solution:

According to the final value theorem, we have

$$x(\infty) = \underset{z\to 1}{\text{Lt}}(1 - z^{-1})X(z)$$

$$= \underset{z\to 1}{\text{Lt}}(1 - z^{-1})\left[\frac{1}{1-z^{-1}} - \frac{1}{1-e^{-aT}z^{-1}}\right]$$

$$= \underset{z\to 1}{\text{Lt}}\left[1 - \frac{1-z^{-1}}{1-e^{-aT}z^{-1}}\right] = 1$$

Problem-15.13: Consider the difference equation $x(k) - ax(k-1) = 1, -1 < a < 1$. Determine the initial and final values of $x(k)$ by using initial and final value theorems.

Solution:

$$x(k) - ax(k-1) = 1$$

On taking z-transforms both sides, we have

$$Z[x(k) - ax(k-1)] = Z[1]$$

$$X(z) - az^{-1}X(z) = \frac{z}{z-1} = \frac{1}{1-z^{-1}}$$

$$\Rightarrow \qquad X(z)[1 - az^{-1}] = \frac{1}{1-z^{-1}}$$

$$\Rightarrow \qquad X(z) = \frac{1}{(1-z^{-1})(1-az^{-1})}$$

Initial value,

$$x(0) = \underset{z\to\infty}{\text{Lt}}\, X(z)$$

$$= \underset{z\to\infty}{\text{Lt}} \frac{1}{(1-z^{-1})(1-az^{-1})} = 1.$$

Final value,

$$x(\infty) = \underset{z\to 1}{\text{Lt}}(1-z^{-1})X(z)$$

$$= \underset{z\to 1}{\text{Lt}}\left[(1-z^{-1}) \times \frac{1}{(1-z^{-1})(1-az^{-1})}\right] = \frac{1}{1-a}.$$

Problem-15.14: Determine the initial value and final value of the sequence $f(k)$ to which the z-transform is $F(z) = \dfrac{4z^3 - 5z^2 + 8z}{(z-1)(z-0.5)^2}$.

Solution:
Initial value is given as

$$f(0) = \underset{z\to\infty}{\text{Lt}}\, F(z)$$

$$= \underset{z\to\infty}{\text{Lt}} \frac{4z^3 - 5z^2 + 8z}{(z-1)(z-0.5)^2}$$

$$= \underset{z\to\infty}{\text{Lt}}\left[\frac{z^3\left(4 - 5/z + 8/z^2\right)}{z^3\left(1 - 1/z\right)\left(1 - 0.5/z\right)^2}\right] = \frac{4}{1} = 4$$

Final value is given as

$$f(\infty) = \underset{z\to 1}{\text{Lt}}(1-z^{-1})F(z)$$

$$= \underset{z\to 1}{\text{Lt}}\left\{\frac{(z-1)}{z}\left[\frac{4z^3 - 5z^2 + 8z}{(z-1)(z-0.5)^2}\right]\right\}$$

$$= \underset{z\to 1}{\text{Lt}}\left[\frac{4z^2 - 5z + 8}{(z-0.5)^2}\right] = \frac{4-5+8}{(1-0.5)^2} = 28$$

9. **Complex differentiation theorem**

The region of convergence of a power series in z may be differentiated with respect to z any number of times to get a convergent series. The derivative of $X(z)$ converges in the same region as $X(z)$.

i.e.,
$$Z[k^m x(k)] = \left(-z\frac{d}{dz}\right)^m X(z)$$

Proof:

Let us consider

$$X(z) = \sum_{k=0}^{\infty} x(k)z^{-k}.$$

Differentiating $X(z)$ with respect to z, we have

$$\frac{d}{dz}X(z) = \sum_{k=0}^{\infty}(-k)x(k)z^{-k-1}.$$

By multiplying both sides with respect to $-z$, we have

$$-z\frac{d}{dz}X(z) = \sum_{k=0}^{\infty}(k)x(k)z^{-k} = Z[kx(k)] \tag{15.25}$$

Differentiating (15.25) with respect to z on both sides, we have

$$\frac{d}{dz}\left[-z\frac{d}{dz}X(z)\right] = \sum_{k=0}^{\infty}(-k^2)x(k)z^{-k-1}$$

$$-z\frac{d}{dz}\left[-z\frac{d}{dz}X(z)\right] = \sum_{k=0}^{\infty}k^2x(k)z^{-k}.$$

$\therefore$
$$\left(-z\frac{d}{dz}\right)^2 X(z) = Z[k^2x(k)]$$

The operation $\left(-z\frac{d}{dz}\right)^2$ implies that we apply the operator $-z\frac{d}{dz}$ twice.

Similarly, in general, we can write

$$Z[k^m x(k)] = \left(-z\frac{d}{dz}\right)^m X(z). \tag{15.26}$$

Such a complex differentiation theorem enables us to obtain the new z-transform pairs from the known z-transform pairs.

Problem-15.15: Obtain the z-transform of unit ramp of $x(k)$ with the knowledge of z transform of unit-step signal.

Solution:

For the unit-step signal, $x(k) = 1$

$\therefore$
$$X(z) = Z[1] = \frac{z}{z-1}$$

For the ramp signal,
$$x_1(k) = tx(k) = kT.1$$

$$X_1(z) = Z[kT.1]$$

$$= -Tz\frac{d}{dz}X(z) = -Tz\frac{d}{dz}\left(\frac{z}{z-1}\right) = -Tz\left[\frac{(z-1)-z}{(z-1)^2}\right] = \frac{Tz}{(z-1)^2}.$$

Problem-15.16: Find the *z*-transform of $x(t) = t\cos t$.

Solution:

Substitute $t = kT$

$$x(kT) = kT\cos kT$$

$$Z[x(kT)] = Z[kT\cos kT]$$

$$= T\times -z\frac{d}{dz}\{z(\cos kT)\}$$

$$= -Tz - \frac{d}{dz}\left[\frac{z(z-\cos T)}{z^2 - 2z\cos T + 1}\right]$$

$$= -Tz\left[\frac{(z^2 - 2z\cos T + 1)(2z - \cos T) - (z^2 - z\cos T)(2z - 2\cos T)}{(z^2 - 2z\cos T + 1)^2}\right]$$

$$= -Tz\left[\frac{z^2\sin T - 2z^2\cos T + 2z - \sin T}{(z^2 - 2z\cos T + 1)^2}\right]$$

Problem-15.17: Find the *z*-transform of $x(t) = te^{-at}$

Solution:

Substitute $t = kT$

$$x(kT) = kTe^{-akt}$$

$$Z[x(kT)] = Z\left[kTe^{-akt}\right] = T(-z)\frac{d}{dz}\{Z[e^{-akt}]\}$$

$$= -Tz\frac{d}{dz}\left[\frac{z}{z-e^{-aT}}\right] = -Tz\left[\frac{(z-e^{-aT})-z}{(z-e^{-aT})^2}\right] = zT\left[\frac{e^{-aT}}{(z-e^{-aT})^2}\right]$$

Table 15.3 lists the theorems on *z*-transforms discussed so far.

Table 15.3 Theorems on *Z*-Transforms

S. No	Theorem	Name
1.	$Z\{ax(t)\} = aX(z)$	Multiplication by a constant
2.	$Z\{x_1(t) + x_2(t)\} = X_1(z) + X_2(z)$	Linearity theorem
3.	$Z\{e^{-at}x(t)\} = X(e^{aT}z)$	Complex translation theorem
4.	$Z\{x(t-nT)\} = z^{-n}X(z)$	Real translation theorem

(continued)

Table 15.3 *(continued)*

S. No	Theorem	Name
5.	$Z\{tx(t)\} = -Tz\dfrac{dX(z)}{dz}$	Complex differentiation theorem
6.	$x(0) = \lim_{z\to\infty} X(z)$	Initial value theorem
7.	$x(\infty) = \lim_{z\to 1}(1-z^{-1})X(z)$	Final value theorem

Note: kT may be substituted for t in the table.

15.12 INVERSE Z-TRANSFORMS

z-Transforms serve the same role for discrete-time control systems as that the Laplace transforms serve for continuous-time control systems. For the z-transform to be useful, we must be familiar with the methods of finding inverse z-transform. The notation for the inverse z-transform is Z^{-1}. The inverse z-transform of $X(z)$ gives the corresponding time sequence $x(k)$ at sampling instants. Thus, the inverse z-transform of $X(z)$ yields a unique $x(k)$, but does not yield a unique $x(t)$. This means that the inverse z-transform yields a time sequence that specifies the values of $x(t)$ only at discrete instants of time $t = 0, T, 2T, \ldots\ldots$, and says nothing about the values of $x(t)$ at all other times. In other words, many different time functions $x(t)$ can have same $x(kT)$.

Given the z-transform function $X(z)$, the inverse z-transform of $X(z)$ is denoted by $x(k) = z^{-1}[X(z)]$. In general, the inverse z-transform can be carried out with one of the following four methods.

1. ***Z*-Transform table**

   For simple functions, the inverse z-transform can be looked up from the z-transform table.

Problem-15.18: Evaluate the inverse z-transform of $e^{1/z} + \dfrac{z(z-\cos\theta)}{z^2 - 2z\cos\theta + 1}$.

Solution:

$$z^{-1}\left[e^{1/z} + \frac{z(z-\cos\theta)}{z^2-2z\cos\theta+1}\right] = z^{-1}\left[e^{1/z}\right] + z^{-1}\left[\frac{z(z-\cos\theta)}{z^2-2z\cos\theta+1}\right] = \frac{1}{k!} - \cos k\theta$$

2. **Partial fraction expansion method**

   The partial fraction expansion method used here is similar to the partial fraction expansion method used in the Laplace transform with a minor modification. The z-transform $X(z)$ can be expanded into partial fractions and using z-transform table, the corresponding inverse z-transform can be obtained. As indicated in z-transform table, z appears in the numerator of the z-transform and as such it is convenient to expand $\dfrac{X(z)}{z}$ into partial fractions instead of $X(z)$.

Problem-15.19: Evaluate $Z^{-1}\left[\dfrac{3z^2+z}{(5z-1)(5z+2)}\right]$

Solution:

Let
$$X(z) = \frac{3z^2+z}{(5z-1)(5z+2)}$$

$$\frac{X(z)}{z} = \frac{3z+1}{(5z-1)(5z+2)} = \frac{A}{5z-1} + \frac{B}{5z+2}$$

$$= \frac{8}{15}\cdot\frac{1}{5z-1} + \frac{1}{15}\cdot\frac{1}{5z+2}$$

$$= \frac{8}{75}\left(\frac{1}{z-\frac{1}{5}}\right) + \frac{1}{75}\frac{1}{\left(z+\frac{2}{5}\right)}$$

$$\Rightarrow \qquad X(z) = \frac{8z}{75(z-0.2)} + \frac{z}{75(z+0.4)}$$

$$\therefore \qquad Z^{-1}[X(z)] = Z^{-1}\left[\frac{8z}{75(z-0.2)} + \frac{z}{75(z+0.4)}\right] = \frac{8}{75}(0.2)^k + \frac{1}{75}(-0.4)^k$$

Problem-15.20: Find the inverse *z*-Transform of $\frac{4z^2-2z}{z^3-5z^2+8z-4}$.

Solution:

Let $$X(z) = \frac{4z^2-2z}{z^3-5z^2+8z-4} = \frac{z(4z-2)}{(z-1)(z-2)^2}$$

$$\frac{X(z)}{z} = \frac{4z-2}{(z-1)(z-2)^2} = \frac{A}{z-1} + \frac{B}{z-2} + \frac{C}{(z-2)^2}$$

$$= \frac{2}{z-1} - \frac{2}{z-2} + \frac{6}{(z-2)^2}$$

$$\Rightarrow \qquad X(z) = \frac{2}{z-1} - \frac{2}{z-2} + \frac{6}{(z-2)^2}$$

$$\therefore \qquad x(k) = Z^{-1}[X(z)] = Z^{-1}\left[\frac{2}{z-1} - \frac{2}{z-2} + \frac{6}{(z-2)^2}\right] = 2(1)^k - 2(2)^k + 3k(2)^k$$

3. **Power series or direct division method:**
   In the direct division method, we obtain the inverse *z*-transform by expanding $X(z)$ into an infinite power series in $(z - 1)$. This method is useful when it is difficult to obtain the closed form expression for the inverse *z*-transform or it is desired to find only the first several terms of $x(k)$.
   In general, the *z*-transform equation can be given as

$$X(z) = \sum_{k=0}^{\infty} x(kT)z^{-k} = x(0) + x(T)z^{-1} + \ldots + x(kT)z^{-k}\,. \qquad (15.27)$$

The given function $X(z)$ can be expressed as a ratio of two polynomials in z as follows:

$$X(z) = \frac{a_0z^n + a_1z^{n-1} + a_2z^{n-2} + \ldots + a_n}{b_0z^m + b_1z^{m-1} + b_2z^{m-2} + \ldots + b_m}\,. \qquad (15.28)$$

If $n \leq m$, then dividing the numerator by denominator in (15.28), we get

$$X(z) = x_0z^0 + x_1z^{-1} + x_2z^{-2} + \ldots \qquad (15.29)$$

By comparing the coefficients of power series of (15.29) and the z-transforms expression of (15.27), it is concluded that the value of $x(kT)$ at sampling instants, that is, $x(0)$, $x(T)$, $x(kT)$.... is determined. However, the partial fraction expansion method gives an expression for $x(k)$ in closed form.

Problem-15.21: Find $x(k)$ for $k = 0, 1, 2, 3, 4$ when $X(z)$ is given as $X(z)=\dfrac{10z+5}{(z-1)(z-0.2)}$ by using direct division method.

Solution:

$$X(z)=\frac{10z+5}{(z-1)(z-0.2)}=\frac{z\left(10+\frac{5}{z}\right)}{z^2\left(1-\frac{1}{z}\right)\left(1-\frac{0.2}{z}\right)}=\frac{10z^{-1}+5z^{-2}}{1-1.2z^{-2}+0.2^{-2}}$$

$$\begin{array}{r|l}
1-1.2z^{-1}+0.2z^{-2} & 10z^{-1}+5z^{-2}+10z^{-1}-12z^{-2}+2z^{-3} \quad 10z^{-1}+17z^{-2}+18.4z^{-3}+18.68z^{-4} \\
& 17z^{-2}-2z^{-3} \\
& 17z^{-2}-20.4z+3.4z^{-4} \\ \hline
& 18.42z^{-3}-3.4z^{-4} \\
& 18.42z^{-3}-22.08z^{-4}+3.666z^{-5} \\ \hline
& 18.68z^{-4}-3.66z^{-5} \\
& 18.68z^{-4}-22.25z^{-5}+3.736z^{-6} \\ \hline
& 18.73z^{-5}-3.736z^{-6}
\end{array}$$

$$\therefore \qquad X(z)=10z^{-1}+17z^{-2}+18.4z^{-3}+18.68z^{-4}+\frac{18.73z^{-5}-3.736z^{-6}}{1-1.2z^{-1}+0.2z^{-2}}$$

$$x(0) = 0,\ x(1) = 10,\ x(2) = 17,\ x(3) = 18.4,\ x(4) = 18.68$$

4. **Inversion integral method**

This method is a very useful technique for obtaining the inverse z-transform. The inversion integral for the z-transform $X(z)$ is given as

$$Z^{-1}[X(z)]=X(kT)=x(k)=\frac{1}{2\pi j}\oint_c X(z)z^{k-1}dz$$

where c is a circle with its centre at the origin of the z-plane such that all poles of $X(z)z^{k-1}$ are inside it.

The equation for giving the inverse z-transform in terms of residues can be derived by using the theory of complex variables. It can be obtained as follows

$$x(k)=k_1+k_2+k_3+\ldots+k_n$$

$$=\sum_{i=1}^{n}\left[\text{Residue of } X(z)z^{k-1} \text{at pole } z=z_i \text{ of } X(z)z^{k-1}\right]$$

where k_1, k_2, k_3,..., k_n denote the residues of $x(z)z^{k-1}$ at poles z_1, z_2, z_3,..., z_n, respectively. In evaluating residues, we come across the following two cases:

Case (i): If $X(z)z^{k-1}$ contains a simple pole $z = z_i$ then the corresponding residue is given as

$$K_i = \underset{z\to z_i}{\text{Lt}}\left[(z-z_i)X(z)z^{k-1}\right].$$

Case (ii): If $X(z)z^{k-1}$ contains multiple pole z_j of order q, then the residue is given as

$$K_j = \frac{1}{(q-1)!} \underset{z \to z_j}{\text{Lt}} \frac{d^{q-1}}{dz^{q-1}} \left[(z - z_j)^q X(z) z^{k-1} \right].$$

Problem-15.22: Find the inverse z-transform of the following by using inverse integral method.

1. $X(z) = \dfrac{z}{(z-1)^2}$

Solution:
Here, there is a multiple pole, $z_1 = 1$ of order 2
Residue corresponding to multiple pole is given as

$$K_j = \frac{1}{(q-1)!} \underset{z \to z_j}{\text{Lt}} \frac{d^{q-1}}{dz^{q-1}} \left[(z - z_j)^q X(z) z^{k-1} \right]$$

$$= \frac{1}{(2-1)!} \underset{z \to 1}{\text{Lt}} \frac{d}{dz} \left[(z-1)^2 \frac{z}{(z-1)^2} z^{k-1} \right]$$

$$= \underset{z \to 1}{\text{Lt}} \frac{d}{dz} \left[z^k \right] = \underset{z \to 1}{\text{Lt}}\, k z^{k-1} = k$$

$\therefore$ $$Z^{-1} \left[\frac{z}{(z-1)^2} \right] = k.$$

2. $X(z) = \dfrac{\left(1 - e^{-aT}\right) Z}{(z-1)\left(z - e^{-aT}\right)}$

Solution:
Here, the simple poles are given as

$$z_1 = 1, z_2 = e^{-aT}.$$

Residue corresponding to simple pole is given as

$$K_i = \underset{z \to z_i}{\text{Lt}} \left[(z - z_i) X(z) z^{k-1} \right].$$

Residue, $$K_1 = \underset{z \to z_1}{\text{Lt}} \left[(z - z_1) X(z) z^{k-1} \right]$$

$$= \underset{z \to z_1}{\text{Lt}} \left[(z - z_1) \frac{(1 - e^{-aT}) z \times z^{k-1}}{(z-1)(z - e^{-aT})} \right] = \frac{(1 - e^{-aT})}{(1 - e^{-aT})} \times 1 = 1$$

Residue, $$K_2 = \underset{z \to z_2}{\text{Lt}} \left[(z - z_2) X(z) z^{k-1} \right]$$

$$= \underset{z \to e^{-aT}}{\text{Lt}} \left[(z - e^{-aT}) \frac{(1 - e^{-aT}) z \times z^{k-1}}{(z-1)(z - e^{-aT})} \right] = \frac{(1 - e^{-aT})}{(1 - e^{-aT})} \times e^{-akT} = -e^{-akT}$$

$\therefore$ $$Z^{-1} \left[\frac{(1 - e^{-aT}) z}{(z-1)(z - e^{-aT})} \right] = 1 - e^{-akT}$$

3. $X(z) = \dfrac{z^2}{(z-1)(z-0.02)}$

Solution:

Here, 14the simple poles are given as

$$z_1 = 1, z_2 = 0.02$$

Residue, $$K_1 = \underset{z\to 1}{\text{Lt}}\left[(z-1)\frac{z^2}{(z-1)(z-0.02)} \times z^{k-1}\right]$$

$$= \underset{z\to 1}{\text{Lt}}\left[\frac{z^{k+1}}{(z-0.02)}\right] = \frac{1}{1-0.02} = \frac{1}{0.98} = 1.02$$

Residue, $$K_2 = \underset{z\to 0.02}{\text{Lt}}\left[(z-0.02)\frac{z^2}{(z-1)(z-0.02)} \times z^{k-1}\right]$$

$$= \underset{z\to 0.02}{\text{Lt}} \frac{z^{k+1}}{(z-1)} = \frac{(0.02)^{k+1}}{-0.98} = \frac{0.02}{-0.98} \times (0.02)^k = -0.0204 \times (0.02)^k$$

$\therefore$ $$Z^{-1}\left[\frac{z^2}{(z-1)(z-0.02)}\right] = 1.02 - 0.0204 \times (0.02)^k$$

15.13 Z-TRANSFORM METHOD FOR SOLVING DIFFERENCE EQUATIONS

Difference equations can be solved easily by the use of a digital computer, provided the numerical values of all coefficients and parameters are given. However, closed form expressions for $x(k)$ cannot be obtained from the computer solution, except for special cases. The usefulness of the z-transform method is that it enables us to obtain the closed form expression $x(k)$.

Consider the linear time-invariant discrete-time system characterised by the following linear difference equation.

$$x(k) + a_1 x(k-1) + \ldots + a_n x(k-n) = b_0 u(k) + b_1 u(k-1) + \ldots + b_n u(k-n)$$

where $u(k)$ and $x(k)$ are the system's input and output, respectively, at the kth iteration.

In describing such a difference equation in the z-plane, we take the z-transform of each term in the equation.

***Problem-15.23*:** Solve the following difference equation by use of the z-transform method.

$$x(k+2) + 3x(k+1) + 2x(k) = 0,\ x(0) = 0,\ x(1) = 1$$

Solution:

By taking the z-transform on both sides of the given difference equation, we obtain

$$z^2 X(z) - z^2 x(0) - zx(1) + 3zx(z) - 3zx(0) + 2X(z) = 0$$

$$z^2 X(z) - z + 3zX(z) + 2X(z) = 0$$

$$X(z)\left[z^2 + 3z + 2\right] = z$$

$\Rightarrow$ $$X(z) = \frac{z}{z^2 + 3z + 2} = \frac{z}{(z+1)(z+2)}$$

$$= z\left[\frac{1}{z+1} - \frac{1}{z+2}\right] = \frac{z}{z+1} - \frac{z}{z+2}$$

On taking inverse z-transform on both sides, we have

$$x(k) = (-1)^k - (-2)^k.$$

15.14 LIMITATIONS OF THE Z-TRANSFORM METHOD

The following considerations should be kept in mind while applying the z-transform.

1. **Ideal sampler assumption:** The derivation of the z-transform of a continuous data function $x(t)$ is based on the sampling of the function by an ideal sampler. The result of this is that the z-transform $X(z)$ represents the function $x(t)$ only at the sampling instants.
2. **Non-uniqueness of the inverse z-transform:** Given $X(z)$, the inverse z-transform of $X(z)$ gives only a unique solution to $x(kT)$. Strictly, the solution to $x(t)$ is unknown.
3. **Accuracy of the z-transform method** The accuracy of the z-transform method depends on the magnitude of the sampling frequency ω_s or the sampling period T, relative to the highest frequency component contained in the function $x(t)$. If the sampling period is too large or sampling frequency is too low, relative to the variation of $x(t)$, the z-transform solution may be erroneous, since $x^*(t)$ would not be a good representative of $x(t)$ at the sampling instants.

ADDITIONAL SOLVED PROBLEMS

ASP-1: Find out $Z^{-1}\left[\frac{10}{(z+1)(z+2)}\right]$

Solution:

Let
$$X(z) = \left[\frac{10}{(z+1)(z+2)}\right]$$

The simple poles are given as

$$z_1 = -1, z_2 = -2$$

Residue,
$$K_1 = \underset{z\to -1}{\text{Lt}}\left[(z+1)\frac{10}{(z+1)(z+2)} \times z^{k-1}\right]$$

$$= \underset{z\to -1}{\text{Lt}}\left[\frac{10z^{k-1}}{(z+2)}\right] = \frac{10(-1)^{k-1}}{1} = -10(-1)^k.$$

Residue,
$$K_2 = \underset{z\to -2}{\text{Lt}}\left[(z+2)\frac{10}{(z+1)(z+2)} \times z^{k-1}\right]$$

$$= \underset{z\to -2}{\text{Lt}}\left[\frac{10z^{k-1}}{(z+1)}\right] = \frac{10(-2)^{k-1}}{1} = 5(-2)^k$$

$\therefore$
$$Z^{-1}\left[\frac{10}{(z+1)(z+2)}\right] = -10(-1)^k + 5(-2)^k$$

ASP-2: Given $X(s)=\dfrac{1}{s^2(s+1)}$, obtain the $X(z)$ by using residues.

Solution:

Residue Method:

- If $X(s)$ has poles $s_1, s_2, \ldots s_n$ and $s = s_i$ is a simple pole, then the corresponding residue is given as $K_i = \underset{s \to s_i}{\text{Lt}}\left[(s-s_i)X(s)\dfrac{z}{z-e^{Ts}}\right]$.
- If $X(s)$ has multiple pole s_j of order q, then the corresponding residue is given as

$$K_j = \frac{1}{(q-1)!}\underset{s \to s_j}{\text{Lt}}\frac{d^{q-1}}{ds^{q-1}}\left[(s-s_j)^q X(s)\frac{z}{z-e^{Ts}}\right].$$

Given that $$X(s)=\frac{1}{s^2(s+1)}.$$

The given system has simple pole at $s_1 = -1$ and multiple poles at $s_2 = 0$.

Residue, $$K_1 = \underset{s \to -1}{\text{Lt}}\left[\frac{1}{(s+1)(s^2)}\times\frac{z}{z-e^{Ts}}\right]$$

$$= \underset{s \to -1}{\text{Lt}}\left[\frac{1}{s^2}\times\frac{z}{z-e^{Ts}}\right] = \frac{z}{z-e^{-T}}$$

Residue, $$K_2 = \frac{1}{(2-1)!}\underset{s \to 0}{\text{Lt}}\frac{d}{ds}\left[(s-0)^2\frac{1}{s^2(s+1)}\frac{z}{z-e^{Ts}}\right]$$

$$= \underset{s \to 0}{\text{Lt}}\frac{d}{ds}\left[\frac{1}{s+1}\left(\frac{z}{z-e^{Ts}}\right)\right]$$

$$= \underset{s \to 0}{\text{Lt}}\, z\frac{d}{ds}\left[(s+1)^{-1}(z-e^{Ts})^{-1}\right]$$

$$= \underset{s \to 0}{\text{Lt}}\left[-(s+1)^{-1}(z-e^{Ts})^{-2}(-e^{Ts})+(z-e^{Ts})^{-1}(-1)(s+1)^{-2}\right]$$

$$= z\left[\frac{T}{z-1)^2}-\frac{1}{z-1}\right]$$

$$\therefore \quad X(z)=Z\left[\frac{1}{s^2(s+1)}\right] = \text{sum of residues} = \frac{z}{z-e^{-1}}+\frac{Tz}{(z-1)^2}-\frac{z}{z-1}$$

ASP-3: Find out $Z\left[\dfrac{10}{s(s+2)}\right]$.

Solution:

Let $$X(s)=\left[\frac{10}{s(s+2)}\right]$$

The simple poles are given as

$$s_1 = 0, s_2 = -2.$$

Residue, $$K_1 = \underset{s \to 0}{\text{Lt}}\left[s\frac{10}{(s+2)(s)}\times\frac{z}{z-e^{Ts}}\right]$$

$$= \underset{s\to 0}{\text{Lt}}\left[\frac{10}{s+2}\times\frac{z}{z-e^{Ts}}\right] = \frac{10z}{2(z-1)} = \frac{5z}{(z-1)}$$

Residue, $$K_2 = \underset{s\to -2}{\text{Lt}}\left[(s+2)\frac{10}{(s+2)(s)}\times\frac{z}{z-e^{Ts}}\right]$$

$$= \underset{s\to -2}{\text{Lt}}\left[\frac{10}{s}\times\frac{z}{z-e^{Ts}}\right] = \frac{10z}{-2(z-e^{-2T})} = \frac{-5z}{(z-e^{-2T})}$$

∴ $$X(z) = Z\left[\frac{10}{s(s+2)}\right] = \text{Sum of residues} = \frac{5z}{(z-1)} - \frac{-5z}{(z-e^{-2T})}$$

ASP-4: Find the *Z*-transform of $X(s) = \dfrac{\omega}{(s^2-\omega^2)}$.

Solution:

Given that

$$X(s) = \frac{\omega}{s^2-\omega^2} = \frac{\omega}{(s+\omega)(s-\omega)}$$

The simple poles are given as

$$s_1 = \omega, s_2 = -\omega.$$

Residue, $$K_1 = \underset{s\to \omega}{\text{Lt}}\left[(s-\omega)\frac{\omega}{(s+\omega)(s-\omega)}\times\frac{z}{z-e^{Ts}}\right]$$

$$= \underset{s\to \omega}{\text{Lt}}\left[\frac{\omega}{2\omega}\times\frac{z}{z-e^{Ts}}\right] = \frac{z}{2(z-e^{T\omega})}$$

Residue, $$K_2 = \underset{s\to -\omega}{\text{Lt}}\left[(s+\omega)\frac{\omega}{(s+\omega)(s-\omega)}\times\frac{z}{z-e^{Ts}}\right]$$

$$= \underset{s\to -\omega}{\text{Lt}}\left[\frac{\omega}{-2\omega}\times\frac{z}{z-e^{Ts}}\right] = \frac{-z}{2(z-e^{-T\omega})}$$

∴ $$X(z) = Z\left[\frac{\omega}{s^2+\omega^2)}\right] = \text{sum of residues} = \frac{z}{2(z-e^{T\omega})} + \frac{-z}{2(z-e^{-T\omega})}$$

$$= \frac{z\sin h\omega T}{z^2 - 2z\cos h\omega T}$$

ASP-5: Given *z*-transform $G(z) = \dfrac{z(z-2)}{3(z-1)(z-4)}$.

Solution:

Let $$G(z) = \frac{z(z-2)}{3(z-1)(z-4)}$$

$$\frac{G(z)}{z} = \frac{z(z-2)}{3(z-1)(z-4)} = \frac{1}{3}\left[\frac{A}{z-1}+\frac{B}{z-4}\right]$$

$$= \frac{1}{9}\cdot\frac{1}{z-1} + \frac{2}{9}\cdot\frac{1}{z-4}$$

⇒ $$G(z) = \frac{1}{9}\cdot\frac{z}{z-1} + \frac{2}{9}\cdot\frac{z}{z-4}$$

∴ $$g(kT) = Z^{-1}[G(z)] = Z^{-1}\left[\frac{1}{9}\cdot\frac{z}{z-1}+\frac{2}{9}\cdot\frac{z}{z-4}\right] = \frac{1}{9}(1)^k + \frac{2}{9}(4)^k$$

ASP-6: Find the z-transform for the following.

(a) $X(s) = \dfrac{s+2}{s^2(s+1)(s+4)}$

(b) $F(s) = \dfrac{s+2}{s^2(s^2+2s+3)}$

Solution:

(a) Given that $X(s) = \dfrac{s+2}{s^2(s+1)(s+4)}$

The given system has simple poles at $s_1 = -1$, $s_2 = -4$ and multiple pole at $s_3 = 0$.

Residue corresponding to simple pole, $s_1 = -1$

Residue, $$K_1 = \underset{s\to -1}{\text{Lt}}\left[(s+1)\frac{s+2}{s^2(s+1)(s+4)} \times \frac{z}{z-e^{Ts}}\right]$$

$$= \underset{s\to -1}{\text{Lt}}\left[\frac{s+2}{s^2(s+4)} \times \frac{z}{z-e^{Ts}}\right] = \frac{1}{3}\frac{z}{(z-e^{-T})}$$

Residue corresponding to simple pole, $s_2 = -4$

Residue, $$K_2 = \underset{s\to -4}{\text{Lt}}\left[(s+4)\frac{s+2}{s^2(s+1)(s+4)} \times \frac{z}{z-e^{Ts}}\right]$$

$$= \underset{s\to -4}{\text{Lt}}\left[\frac{s+2}{s^2(s+4)} \times \frac{z}{z-e^{Ts}}\right] = \frac{1}{24}\frac{z}{(z-e^{-4T})}$$

Residue corresponding to multiple pole, $s_3 = 0$

Residue, $$K_3 = \frac{1}{(2-1)!}\underset{s\to 0}{\text{Lt}}\frac{d}{ds}\left[(s-0)^2\frac{s+2}{(s+1)(s+4)}\frac{z}{z-e^{Ts}}\right]$$

$$= \underset{s\to 0}{\text{Lt}}\frac{d}{ds}\left[\frac{s+2}{(s+1)(s+4)}\left(\frac{z}{z-e^{Ts}}\right)\right]$$

$$= \underset{s\to 0}{\text{Lt}}\frac{d}{ds}\left[\frac{s+2}{(s^2+5s+4)}\left(\frac{z}{z-e^{Ts}}\right)\right]$$

$$= \underset{s\to 0}{\text{Lt}}\frac{d}{ds}\left[\left(\frac{z}{z-e^{Ts}}\right)\frac{s^2+5s+4-(s+2)(2s+5)}{(s^2+5s+4)} + \frac{s+2}{(s^2+5s+4)}\cdot\left(\frac{z}{z-e^{Ts}}\right)(-Te^{Ts})\right]$$

$$= \frac{z}{z-1}\cdot\frac{-6}{4^2} + \frac{2}{(1)(4)}\frac{zT}{(z-1)^2}$$

$$= \frac{-3}{8}\left(\frac{z}{z-1}\right) + \frac{1}{2}\frac{zT}{(z-1)^2}$$

$\therefore$ $$X(z) = Z\left[\frac{s+2}{s^2(s+1)(s+4)}\right] = \text{sum of residues} = K_1 + K_2 + K_3$$

$$= \frac{1}{3}\frac{z}{z-e^{-T}} + \frac{1}{24}\frac{z}{z-e^{-4T}} - \frac{3}{8}\left(\frac{z}{z-1}\right) + \frac{1}{2}\frac{zT}{(z-1)^2}$$

(b) Given that $F(s) = \dfrac{s+2}{s^2(s^2+2s+3)}$

The given system has simple poles at $s_1 = -1 + j1.414$, $s_2 = -1 - j1.414$ and multiple poles at $s_3 = 0$

Residue corresponding to simple pole, $s_1 = -1 + j1.414$

Residue,
$$K_1 = \underset{s \to s_1}{\text{Lt}} \left[(s - s_1) \frac{s+2}{s^2(s-s_1)(s-s_2)} \times \frac{z}{z - e^{Ts}} \right]$$
$$= \underset{s \to s_1}{\text{Lt}} \left[\frac{s+2}{s^2(s-s_2)} \times \frac{z}{z - e^{Ts}} \right]$$
$$= \left[\frac{-1 + j1.414 + 2}{(-1 + j1.414)^2(-1 + j1.414 + 1 + j1.414)} \times \frac{z}{z - e^{T(-1+j1.414)}} \right]$$
$$= \left[\frac{(0.056 + j0.196)z}{z - e^{T(-1+j1.414)}} \right]$$

Residue corresponding to simple pole, $s_2 = -1 - j1.414$

Residue,
$$K_2 = \underset{s \to s_2}{\text{Lt}} \left[(s - s_2) \frac{s+2}{s^2(s-s_1)(s-s_2)} \times \frac{z}{z - e^{Ts}} \right]$$
$$= \underset{s \to s_2}{\text{Lt}} \left[\frac{s+2}{s^2(s-s_2)} \times \frac{z}{z - e^{Ts}} \right]$$
$$= \left[\frac{-1 - j1.414 + 2}{(-1 - j1.414)^2(-1 - j1.414 + 1 - j1.414)} \times \frac{z}{z - e^{T(-1-j1.414)}} \right]$$
$$= \left[\frac{(0.056 - j0.196)z}{z - e^{T(-1-j1.414)}} \right]$$

Residue corresponding to multiple pole, $s_3 = 0$

Residue,
$$K_3 = \frac{1}{(2-1)!} \underset{s \to 0}{\text{Lt}} \frac{d}{ds} \left[(s-0)^2 \frac{s+2}{s^2(s^2+2s+3)} \frac{z}{z - e^{Ts}} \right]$$
$$= \underset{s \to 0}{\text{Lt}} \frac{d}{ds} \left[\frac{s+2}{(s^2+2s+3)} \left(\frac{z}{z - e^{Ts}} \right) \right]$$
$$= \underset{s \to 0}{\text{Lt}} \frac{d}{ds} \left[\left(\frac{z}{z - e^{Ts}} \right) \left[\frac{s^2 + 2s + 3 - (s+2)(2s+2)}{(s^2+2s+3)^2} \right] + \frac{(s+2)}{(s^2+2s+3)} \cdot \left(\frac{-z}{z - e^{Ts}} \right) (-Te^{Ts}) \right]$$
$$= \frac{z}{z-1} \cdot \frac{-1}{3^2} + \frac{2}{3} \frac{zT}{(z-1)^2}$$
$$= \frac{-1}{9} \left(\frac{z}{z-1} \right) + \frac{2}{3} \frac{zT}{(z-1)^2}$$

$$\therefore \quad X(z) = Z \left[\frac{s+2}{S^2(s+2s+3)} \right] = \text{sum of residues} = K_1 + K_2 + K_3$$
$$= \left[\frac{(0.056 + j0.196)z}{z - e^{T(-1+j1.414)}} \right] + \left[\frac{(0.056 - j0.196)z}{z - e^{T(-1-j1.414)}} \right] - \frac{1}{9} \left(\frac{z}{z-1} \right) + \frac{2}{3} \frac{zT}{(z-1)^2}$$

ASP-7: Find the inverse z-transform for the following

(a) $F(z) = \dfrac{z^2}{(z-2)(z-3)}$

(b) $F(z) = \dfrac{z-2}{z^2+2z+1}$

(c) $F(z) = \dfrac{5z^2+z+2}{z(z^2+4z+3)}$

Solution:

(a) $X(z) = \dfrac{z^2}{(z-2)(z-3)}$

Here, the simple poles are given as

$$z_1 = 2,\ z_2 = 3$$

Residue,
$$K_1 = \underset{z\to 2}{\text{Lt}}\left[(z-2)\frac{z^2}{(z-2)(z-3)} \times z^{k-1}\right]$$
$$= \underset{z\to 2}{\text{Lt}}\left[\frac{z^{k+1}}{(z-3)}\right] = \frac{2^{k+1}}{2-3} = -2\times(2)^k$$

Residue,
$$K_2 = \underset{z\to 3}{\text{Lt}}\left[(z-3)\frac{z^2}{(z-2)(z-3)} \times z^{k-1}\right]$$
$$= \underset{z\to 3}{\text{Lt}}\left[\frac{z^{k+1}}{z-2}\right] = \frac{(3)^{k+1}}{3-2} = 3\times(3)^k$$

$$\therefore \qquad Z^{-1}\left[\frac{z^2}{(z-2)(z-3)}\right] = -2\times(2)^k + 3\times(3)^k = -2^{k+1} + 3^{k+1}$$

(b) $F(z) = \dfrac{z-2}{z^2+2z+1}$

Here, there is a multiple pole, $z_1 = -1$ of order 2.
Residue corresponding to multiple pole is given as

$$K_j = \frac{1}{(q-1)!}\underset{z\to z_j}{\text{Lt}}\frac{d^{q-1}}{dz^{q-1}}[(z-z_j)^q X(z) z^{k-1}]$$
$$= \frac{1}{(2-1)!}\underset{z\to -1}{\text{Lt}}\frac{d}{dz}\left[(z+1)^2\frac{z-2}{(z+1)^2}z^{k-1}\right]$$
$$= \underset{z\to -1}{\text{Lt}}\frac{d}{dz}[(z-2)z^{k-1}] = \underset{z\to -1}{\text{Lt}}\frac{d}{dz}[z^k - 2z^{k-1}]$$
$$= \underset{z\to -1}{\text{Lt}}[kz^{k-1} - 2(k-1)z^{k-2}] = -k(-1)^k - 2(k-1)(-1)^k$$
$$= (2-3k)(-1)^k$$

$$\therefore \qquad Z^{-1}\left[\frac{z-2}{(z+1)^2}\right] = (2-3k)(-1)^k.$$

(c) $F(z)=\dfrac{5z^2+z+2}{z(z^2+4z+3)}=\dfrac{5z^2+z+2}{z(z+3)(z+1)}$

Here, the simple poles are given as

$$z_1=0,\ z_2=-1,\ z_3=-3$$

Residue K_1 corresponding to simple pole, $z_1=0$ is

$$K_1=\underset{z\to 0}{\text{Lt}}\left[(z-0)\frac{5z^2+z+2}{z(z+1)(z+3)}\times z^{k-1}\right]$$

$$=\underset{z\to 0}{\text{Lt}}\left[\frac{5z^2+z+2}{(z+1)(z+3)}z^{k+1}\right]=0$$

Residue K_2 corresponding to simple pole, $z_2=-1$ is

$$K_2=\underset{z=-1}{\text{Lt}}\left[(z+1)\frac{5z^2+z+2}{z(z+1)(z+3)}\times z^{k-1}\right]$$

$$=\underset{z\to 0}{\text{Lt}}\left[\frac{5z^2+z+2}{z(z+3)}z^{k+1}\right]=\frac{5-1+2}{-1(-1+3)}(-1)^{k+1}=3(-1)^k$$

Residue K_3 corresponding to simple pole, $z_3=-3$ is

$$K_3=\underset{z=-3}{\text{Lt}}\left[(z+3)\frac{5z^2+z+2}{z(z+1)(z+3)}\times z^{k-1}\right]$$

$$=\underset{z\to 0}{\text{Lt}}\left[\frac{5z^2+z+2}{z(z+1)}z^{k+1}\right]=\frac{45-3+2}{-3(-3+1)}(-3)^{k+1}=\frac{22}{3}(-3)^{k+1}$$

$$\therefore\quad Z^{-1}\left[\frac{5z^2+z+2}{z(z^2+4z+3)}\right]=3(-1)^k+\frac{22}{3}(-3)^{k+1}$$

ASP-8: Find the response $y(k)$ when the input $x(k)$ is a unit-step sequence for the following difference equation:

$$y(k+1)+0.5y(k)=x(k) \text{ and } y(0)=0.$$

Solution:

The given equation is

$$y(k+1)+0.5y(k)=x(k).$$

By applying the z-transform on both the sides, we get

$$zY(z)-zy(0)+0.5Y(z)=\frac{z}{z-1}$$

$$zY(z)-z\times 0+0.5Y(z)=\frac{z}{z-1}$$

$$[z+0.5]Y(z)=\frac{z}{z-1}$$

$$\Rightarrow\quad Y(z)=\frac{z}{(z-1)(z+0.5)}$$

$$\frac{Y(z)}{z} = \frac{1}{(z-1)(z+0.5)} = \frac{3}{2}\frac{1}{(z-1)} - \frac{3}{2}\frac{1}{(z+0.5)}$$

$$\Rightarrow \qquad Y(z) = \frac{3}{2}\frac{1}{(z-1)} - \frac{3}{2}\frac{1}{(z+0.5)}$$

On taking inverse z-transform on both the sides, we get

$$y(k) = \frac{3}{2}(1)^k - \frac{3}{2}(-0.5)^k.$$

ASP-9: Find the solution of the following difference equation $x(k+2) - 1.3\,x(k+1) + 0.4(k) = u(k)$, where $x(0) = x(1) = 0$ and $x(k) = 0$ for $k < 0$. For the input function $u(k)$, consider the following two cases:

$$u(k) = \begin{cases} 1, & k = 0,1,2,\ldots \\ 0, & k < 0 \end{cases} \quad \text{and } u(0) = 1,\, u(k) = 0,\, k \neq 0.$$

Solution:

Taking the z-transform on both sides of the given difference equation, we have

$$z^2X(z) - z^2x(0) - zx(1) - 1.3zx(z) + 1.3zx(0) + 0.4X(z) = \frac{z}{z-1}$$

$$z^2X(z) - 1.3zX(z) + 0.4X(z) = \frac{z}{z-1}$$

$$X(z)[z^2 - 1.3z + 0.4] = \frac{z}{z-1}.$$

$$X(z) = \frac{z}{(z-1)(z^2-1.3z+0.4)} = \frac{z}{(z-1)(z-0.5)(z-0.8)}$$

$$= z\left[\frac{10}{z-1} + \frac{3.33}{z-0.5} - \frac{13}{z-0.8}\right] = \frac{10z}{z-1} + \frac{3.33z}{z-0.5} - \frac{13z}{z-0.8}$$

On taking inverse z-transform on both sides, we have

$$x(k) = 10(1)^k + 3.33(0.5)^k - 13(0.8)^k.$$

ASP-10: Solve the difference equation

$$y(k+2) + 3y(k+1) + 2y(k) = r(k), r(k) = (1)^k, y(0) = 1, y(k) = 0, k < 0.$$

Solution:

Given that

$$y(k+2) + 3y(k+1) + 2y(k) = r(k).$$

Substitute

$$k = -1$$

$$y(1) + 3y(0) + 2y(-0) = (1)^{-1} = 1$$

$$y(1) = 1 - 3 = -2$$

On taking z-transforms to the above-mentioned equation, we have

$$z^2Y(z) - z^2y(0) - zy(1) + 3[zY(z) - zy(0)] + 2Y(z) = \frac{z}{z-1}$$

$$Y(z)[z^2 + 3z + 2] - z^2 - z(-2) - 3z = \frac{z}{z-1}$$

$$Y(z) = \frac{z^3}{(z-1)(z^2+3z+2)}$$

$$\frac{Y(z)}{z} = \frac{z^2}{(z-1)(z+1)(z+2)}$$

$$= \frac{1}{6}\cdot\frac{1}{z-1} - \frac{1}{2}\cdot\frac{1}{z+1} + \frac{4}{3}\cdot\frac{1}{z+2}$$

$$Y(z) = \left[\frac{1}{6}\cdot\frac{z}{z-1} - \frac{1}{2}\cdot\frac{z}{z+1} + \frac{4}{3}\cdot\frac{z}{z+2}\right]$$

By taking inverse z-transforms on both sides, we get

$$y(k) = \frac{1}{6}(1)^k - \frac{1}{2}(-1)^k + \frac{4}{3}(-2)k.$$

REVIEW QUESTIONS

1. What are the advantages and disadvantages of digital control system?
2. Explain clearly the configuration of basic digital control scheme with the help of a neat block diagram.
3. Explain the different types of sampling operations.
4. State and prove the following properties or theorems of z-transform.
   (i) Shifting theorem (ii) Complex translation theorem (iii) Final value theorem
5. Obtain the z-transform of the following functions:
   (i) Unit-step function (ii) Exponential function (iii) Sinusoidal function
6. Explain the operation of ideal sampler.
7. State and prove any four important theorems of z-transform.
8. What are the limitations of z-transform method?
9. Explain the sample-and-hold operations with suitable circuit diagram and waveforms.
10. What are the various methods used to determine inverse z-transform? Explain partial fraction inverse z-transform method.

EXERCISE PROBLEMS

1. Find the z-transforms of
   (a) $f(t) = e^{-at}\sin\omega t$ (b) $F(s) = \dfrac{2}{s^2(s+2)}$ (c) k^2 (d) ka^{k-1}
2. Find the inverse z-transforms of
   (a) $F(z) = \dfrac{z^2}{(z-1)(z-0.2)}$ (b) $F(z) = \dfrac{z(1-e^{-aT})}{(z-1)(z-e^{-aT})}$
   (c) $F(z) = \dfrac{z^2}{(z-1)^2(z-e^{-aT})}$
3. Obtain the inverse z-transform of
   (a) $X(z) = \dfrac{z^{-2}}{(1-z^{-1})^3}$ (b) $X(z) = \dfrac{z+2}{z^2(z-2)}$

   Using (i) direct division method and (ii) partial fraction expansion method.

4. Find the inverse z-transform of $X(z)=\dfrac{z}{1+z^2}$.
5. Obtain the z-transform of unit-step functions that are delayed by 1 sampling period and 4 sampling periods.
6. Obtain the z-transform of

$$x(k)=a^k \text{ for } k= \ ,2,3\ldots$$
$$=0 \text{ for } k\le 0.$$

7. Obtain the z-transform of the curve $x(t)$, as shown in the figure. Assume that the sampling period T is 1 s.

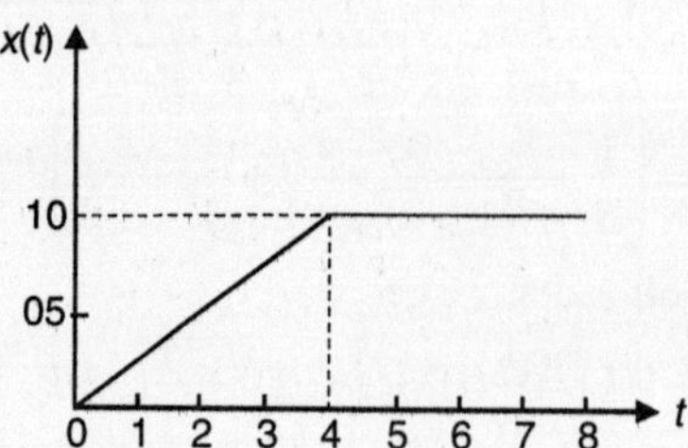

8. Solve the following difference equation:
$$y(k+2)+3y(k+1)+2y(k)=0 \text{ and } y(-1)=-1/2, y(-2)=3/4$$
9. Obtain the solution of the following difference equation in terms of $x(0)$ and $x(1)$ $x(k+2)+(a+b)x(k+1)+abx(k)=0$ where a and b are constants and $k = 0, 1, 2....$
10. Given the Z-transform

$$X(z)=\frac{z(1-e^{-aT})}{(z-1)(z-e^{-aT})}$$

where a is a constant and T is the sampling period, determine the inverse z-transform $x(kT)$ by using the partial fraction expansion method.
11. Find the solution of the following difference equation $x\ (k+2)-1.3x(k+1)+0.4x(k)=u(k)$, where $x(0)=x(1)=0$ and $x(k)=0$ for $k<0$. For the input function $u(k)$, consider the following two cases:

$$u(k)=\begin{cases}1, & k=0,1,2,\cdots\\ 0, & k>0\end{cases} \quad \text{and } u(0)=1,\ u(k)=k\neq 0$$

12. Find the inverse z-transform of

$$F(z)=\frac{8z-19}{(z-2)(z-3)}$$

ANSWERS

1. (a) $\dfrac{ze^{aT}\sin\omega T}{z^2e^{2aT}-2ze^{aT}\cos\omega T+1}$ (b) $\dfrac{z}{2(z-e^{-2T})}-\dfrac{z}{2(z-1)}+\dfrac{Tz}{(z-1)^2}$

(c) $\dfrac{(z+1)z}{(z-1)^3}$ (d) $\dfrac{z}{(z-a)^2}$

2. (a) $f(k)=1.25-0.25(0.2)^k$ (b) $f(kT)=1-e^{-akT}$

(c) $f(kT)=\dfrac{kT}{T(1-e^{-aT})}-\dfrac{e-akT}{(1-e^{-aT})^2}$

3. (a) $x(k)=\dfrac{1}{2}k(k-1)$ (b) $x(k)=1$ for $k=2$

$= 2^{k-1}$ for $k=3,4,5,\cdots$

4. $x(k)=\dfrac{1}{2}(j)^{k-1}+\dfrac{1}{2}(-j)^{k-1}$

5. $\dfrac{1}{z-1},\dfrac{z^{-4}}{1-z^{-1}}$

6. $\dfrac{1}{z-a}$

7. $X(z)=\dfrac{1}{4}z^{-1}\left[\dfrac{1-z^{-4}}{(1-z^{-1})^2}\right]$

8. $y(k)=(-1)^k-(-2)^k$

9. $x(k)=\dfrac{bx(0)+x(1)}{b-a}(-a)^k+\dfrac{ax(0)+x(1)}{a-b}(-b)^k, a\neq b,\text{where } k=0,1,2\cdots$

$x(k)=x(0)(-a)^k+[ax(0)+x(1)]k(-a)^{k-1}, a=b,\text{where } k=0,1,2\cdots$

10. $x(kT)=1-e^{-akT}$

11. $x(k)=10(1)^k+3.33(0.5)^k-13(0.8)^k$

12. $f(k)=-\dfrac{19}{6}(1)^k+\dfrac{3}{2}(2)^k+\dfrac{5}{3}(3)^k$

OBJECTIVE TYPE QUESTIONS

1. The z-transform of unit step function is

(a) $\dfrac{z}{z-1}$ (b) $\dfrac{z}{z+1}$ (c) $\dfrac{z^2}{z-1}$ (d) $\dfrac{z^2}{z+1}$

2. Find the inverse z-transform of $F[z]=\dfrac{z}{z-0.2}$

(a) $(2)^k$ (b) $(-2)^k$ (c) $(0.2)^k$ (d) $(-0.2)^k$

3. One of the limitation of z-transform is

(a) ideal sampler assumption (b) non-uniqueness of the inverse z-transform
(c) accuracy of the z-transform method (d) All

4. The z-transform of te^{-at}

(a) $\dfrac{zTe^{aT}}{(ze^{aT}-1)^2}$ (b) $\dfrac{zTe^{2aT}}{(ze^{aT}-1)}$ (c) $\dfrac{zT}{(ze^{aT}-1)^2}$ (d) None

5. The z-transform of the sequence $x(k)=\left(\dfrac{1}{2}\right)^k$, for $k=0, 1, 2, 3\cdots$ is

(a) $\frac{z}{z+0.5}$ (b) $\frac{z}{z-0.5}$ (c) $\frac{z^2}{z-1}$ (d) $\frac{z}{z^2+1}$

6. The inverse z-transform can be obtained by using following r method
   (a) Partial fraction expression method
   (b) Power series or direct division method
   (c) Counter integration or inversion integral method
   (d) All

7. The z-transform of $\left[\frac{10}{s(s+2)}\right]$
   (a) $\frac{5z}{(z-1)}$ (b) $\frac{-5z}{(z-e^{-2T})}$ (c) $\frac{5z}{(z-1)} - \frac{-5z}{(z-e^{-2T})}$ (d) None

8. The initial theorem states that
   (a) $x(0) = \underset{z\to\infty}{\text{Lt}}\, X(z)$ (b) $x(\infty) = \underset{z\to 1}{\text{Lt}}(1-z^{-1})X(z)$
   (c) $x(0) = \underset{z\to 0}{\text{Lt}}\, X(z)$ (d) $x(0) = \underset{z\to\infty}{\text{Lt}}\, zX(z)$

9. The final theorem states that
   (a) $x(0) = \underset{z\to\infty}{\text{Lt}}\, X(z)$ (b) $x(\infty) = \underset{z\to 1}{\text{Lt}}(1-z^{-1})X(z)$
   (c) $x(0) = \underset{z\to 0}{\text{Lt}}\, X(z)$ (d) $x(0) = \underset{z\to\infty}{\text{Lt}}\, zX(z)$

10. The z-transforms of k^2
    (a) $\frac{5z}{(z-1)}$ (b) $\frac{(z+1)}{(z-1)^2}$ (c) $\frac{(z+1)z}{(z-1)^3}$ (d) None

11. z-Transform for $X(z)$ of the sequence $x(k)$ is given as
    (a) $X(z) = \sum_{k=0}^{\infty} x(k)z^{-k}$ (b) $X(z) = \sum_{k=1}^{\infty} x(k)z^{-k}$ (c) $X(z) = \sum_{k=0}^{\infty} z^{-k}$ (d) none

12. Z-transform of $\delta(k)$ is
    (a) z (b) $1/z$ (c) 1 (d) None

13. In multiple rate sampling,
    (a) the sampling instants are random
    (b) two concurrent sampling operations occur at $t_k = pT_1$ and qT_2 where T_1 and T_2 are constants and p and q are integers
    (c) the sampling instants are equally spaced, or $t_k = kT$ (k = 0, 1, 2...)
    (d) the pattern of the t_k is repeated periodically

14. The initial and final values of the time function corresponding to the z-transform $\frac{4z^3-5z^2+8z}{(z-1)(z-0.5)^2}$ are
    (a) zero and indeterminate (b) 2 and 14
    (c) 4 and 28 (d) 8 and 56

ANSWERS

1. a	**2.** c	**3.** d	**4.** a	**5.** b	**6.** d	**7.** c	**8.** a	**9.** b	**10.** c
11. a	**12.** c	**13.** b	**14.** c						

APPENDIX MATLAB Fundamentals

Contents

A.1 INTRODUCTION

The name MATLAB stands for '**MAT**rix **LAB**oratory'. MATLAB was written originally to provide easy access to matrix software developed by the LINPACK (linear system package) and EISPACK (Eigen system package) projects. It is a high-performance language for technical computing and integrates computation, visualisation and programming environment. Furthermore, it is a modern programming language environment; it has sophisticated data structures, contains inbuilt editing and debugging tools, and supports object-oriented programming. These factors make MATLAB an excellent tool for teaching and research.

MATLAB is used in every facet of computational mathematics. Some commonly used mathematical calculations where MATLAB is used most frequently are matrices and arrays, 2-D and 3-D plotting and graphics, linear algebra, algebraic equations, non-linear functions, statistics, data analysis, calculus and differential equations, numerical calculations, integration, curve fitting and various other special functions.

MATLAB is widely used as a computational tool in science and engineering encompassing the fields of physics, chemistry, mathematics and all engineering streams. It is used in a range of applications including signal processing and communications, image and video processing, control systems, test and measurement, computational finance, computational biology, etc. This chapter is a brief introduction to MATLAB basics and its application to control systems.

A.2 STARTING AND QUITTING MATLAB

After logging into your account, you can enter MATLAB by double-clicking on the MATLAB shortcut icon on your Windows desktop. When you **start** MATLAB, a special window called MATLAB desktop appears. The desktop is a window that contains other windows like command window, command history, current directory, workspace and help browser.

In order to **quit** MATLAB, type 'quit' or 'exit' after the command prompt, followed by pressing the enter or return key.

Let us see some basic commands at first.

For example, if we want to exit from MATLAB, then we have to follow this step.

>> ***exit*** ***or*** >> ***quit***

>> ***help*** – it is the most important function for learning MATLAB on our own. It shows all the help topics, and we have to choose one of them depending on our requirement.

For example, if we know the topic that we want to search for, then we have to follow the command.

>> ***help topic*** ***or*** >> ***help function-name***

If we do not know anything about MATLAB, then we can have a demo by giving the following command.

>> ***demo***

A.3 DISPLAY WINDOWS

MATLAB has the following three display windows:

(i) **Command window:** This is the main area where you enter commands at the command line, indicated by the command prompt (>>). It is used to enter the commands and data to display plots and graphs.

(ii) **Graphics window:** This is used to display plots and graphs.

(iii) **Edit or debugger window:** This is used to create and modify m-files. These m-files contain program or script of MATLAB commands.

A.4 GENERAL PURPOSE COMMANDS IN MATLAB

A user new to MATLAB must know some of the useful commands. As such, there are numerous commands in MATLAB but the list of commands shown in Table A.1 would be quite useful for novice user.

Table A.1 General purpose commands

S.No.	Command	Description
1	demos	Demonstrates about the capabilities of MATLAB and inbuilt demos can be run for better understanding
2	ver	Gives information about MATLAB version
3	clc	Clears the command window
4	clf	Clears the current figure
5	clear all	Removes all functions, variables and global variables from the MATLAB
6	quit	Exits MATLAB

(Continued)

Table A.1 *(Continued)*

S.No.	Command	Description
7	who	Lists the variables currently in the memory
8	inmem	Lists the functions in MATLAB memory
9	pack	Saves all variables on the disk after clearing the memory and then reloads the variables
10	recycle	Moves deleted files to recycle bin
11	symb	Defines a variable as a symbol

A.5 PREDEFINED VARIABLES IN MATLAB

MATLAB includes a number of predefined variables. Some of the predefined variables that are available to use in MATLAB programs are listed in Table A.2.

Table A.2 Predefined variables

S.No.	Variable	Description
1	pi	Represents the value of π
2	i, j	$\sqrt{-1}$, imaginary number
3	inf	Infinity, represents a very large number
4	NaN	Stands for not a number and typically occurs as a result of an expression being undefined, as in the case of division by zero
5	ans	Stores the result of an output by default
6	eps	Represents the smallest difference between two numbers
7	clock	Represents the current time in a six-element row vector containing year, month, day, hour, minute and second
8	date	Represents the current date like 21-March-2015

A.6 PERCENTAGE (%) AND SEMICOLON (;) IN MATLAB

(i) **Percentage (%) symbol:** When per cent symbol (%) is typed in the beginning of a line, then the line is designated as a command. When **enter** key is pressed, the line does not get executed.

(ii) **Semicolon (;) symbol:** If a semicolon (;) is typed at the end of a command, then the output of the command is not displayed.

A.7 VARIABLES

A variable is a symbolic name used to address the storage of data. It is linked with a value. The linked value can be changed during the course of running the program. In MATLAB, assigning values to the variables is simple. If no variable is assigned to the output value, then **ans** is taken as a variable by default. For instance,

```
>> 20-2^4
ans = 12
>> ans*5
ans = 60
```

The result of the first calculation is labelled as 'ans' by MATLAB and is used in the second calculation where its value is changed.

We can use our own names to store numbers.

```
>> x = 20-2^4
   x = 12
>> y = x*5
   y = 60
```

In MATLAB, variables are case sensitive, i.e., variable *x* is not same as *X*, so this should always be kept in mind before using assigned variables. Furthermore, the name of the variable should start from a letter, but it can be followed by numbers, digits or underscores.

The following examples are allowed:

NetCost, Left2Pay, x3, X3, z25c5

The following examples are not allowed:

Net-Cost, 2pay, %x, @sign

A.8 ARRAYS

An array is the fundamental unit of data in a MATLAB program. All data processed in MATLAB as an input data, output data or for computational work is in the form of arrays. Normally, a vector or a one-dimensional matrix is referred to as an array.

A vector is a row or column of elements. In a row vector, the elements are entered with a space or a comma between the elements inside the square brackets.

For example,

```
>> A = [5 10 15]
   A = 5 10 15
>> B = [5, 10, 15]
   B = 5 10 15
```

In a column vector, the elements are entered with a semicolon between the elements inside the square brackets.

For example,

```
>> A = [5; 10; 15]
   A =
   5
   10
   15
```

In order to get the transpose of vector A, we can write

>> V = A′ % Find the transpose of array A

It will display V as

```
V = 5 10 15
```

An array can also be defined by the following syntax:

Initial: increment: terminator

For example, an array `X = [ 1 3 5 7 9 11]` can be defined as

```
>> X = [1:2:11]
   X = 1  3  5  7  9  11
```

Example 1

```
clc
clear all
t =0:10e-6:0.08; % time is taken from 0sec to 0.08 sec in steps of 10 micro
    seconds
    f =50;            % frequency is taken as 50Hz
    w =2*pi*f;        % angular frequency in tad/sec
  Vm =230;            % maximum value of 230V
Vout =Vm*sin(w*t)% output voltage wave form 0 sec to 0.08 sec
plot(t,Vout)
```

The output will be:

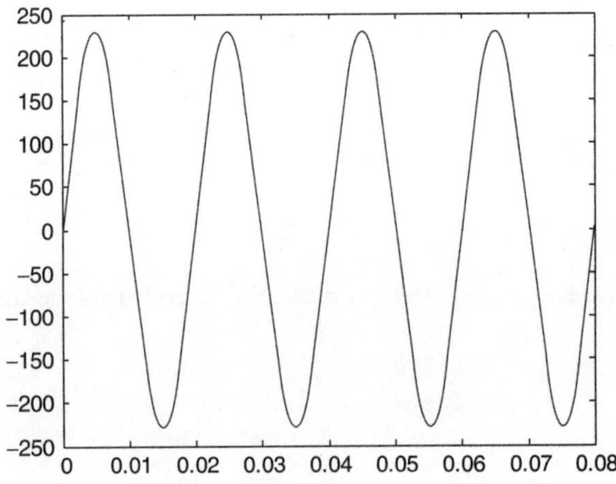

A.9 MATRICES

A matrix is a rectangular array of numbers. By using MATLAB, we can create matrices and can perform different tasks on that. Some of the following commands will help us in creating matrices and doing some operations on them.

For example, if we want to create one row matrix, then we have to follow these commands.

```
>> A = [1 2 3]
   A = 1 2 3
```

Now, we want to create one column matrix.

```
>> B = [1; 2; 3]
B =
    1
    2
    3
```

If we want to design a 2×2 matrix, then the command will be:

```
>> C = [1  2; 3  4]
C =
    1  2
    3  4
```

If we want to design a 3×3 matrix, then the command will be:

```
>> D = [1  2  3; 4  5  6; 7 8 9]
C =
    1  2  3
    4  5  6
    7  8  9
```

To experience the matrix arithmetic in MATLAB, enter the following **A** and **B** matrices.

```
>> A = [1 2 3; 4 5 6; 7 8 9], B = [2; 4; 6]
```

The transpose of a matrix is the result of interchanging rows and columns. MATLAB denotes the transpose by following the matrix with the single-quote [apostrophe].

```
>> C = A'
C =
    1  4  7
    2  5  8
    3  6  9
```

Scalars multiply matrices as expected, and matrices may be added in the usual way; both are done 'element by element'.

```
>>2*A
ans =
        2     4     6
        8    10    12
       14    16    18
```

Table A.3 provides some useful commands that can be utilised for computations involving matrices.

Table A.3 Some useful commands of matrix

S.No.	Command	Description
1	mldivide \	Divides the left matrix
2	mldivide /	Divides the right matrix
3	mpower ^	Power of the matrix

(continued)

Table A.3 *(continued)*

S.No.	Command	Description
4	mtimes *	Multiplies the matrices
5	company	Companion matrix
6	diag	Diagonal matrices and diagonals of a matrix
7	eye	To get identity matrix
8	expm	To get exponential of a matrix
9	inv	Inverse of a matrix
10	norm	Norm of a matrix
11	rank	Rank of a matrix
12	transpose'	To get transpose matrix
13	ctrb	To compute the controllability matrix
14	obsv	To compute the observability matrix
15	diag	Creates a matrix with diagonal elements
16	det	Determinant of a matrix
17	eig	Eigen values and eigenvectors
18	rank	Number of linearly independent
19	zeros	Matrix of zeros
20	ones	Matrix of ones

A.9.1 Colon Operator in a Matrix

The colon operator can also be used to pick out a certain row or column. For example, the statement A(m:n, p:q) specifies rows m to n and column p to q. Table A.4 gives the use of a colon in addressing arrays in a matrix.

Table A.4 Colon used for a matrix

S.No.	Command	Description	Example

1	A(:,n)	Refers to the elements in all the rows of a columns *n* of the matrix A	If $A = \begin{bmatrix} 1 & 2 & 3 \\ 4 & 5 & 6 \\ 7 & 8 & 9 \end{bmatrix}$ then
			`>> A(:, 2)`
			`ans=  2`
			`5`
			`8`

2	A(n,:)	Refers to the elements in all the columns of a row *n* of the matrix A	`>> A(2, :)`
			`ans=`
			`4 5 6`

(continued)

Table A.4 *(continued)*

3	A(:, m:n)	Refers to the elements in all the rows between columns 'm' and 'n' of the matrix A	>> A(:,2:3) ans= 2 3 5 6 8 9
4	A(m:n, :)	Refers to the elements in all the columns between rows 'm' and 'n' of the matrix A	>> A(2:3,:) ans= 4 5 6 7 8 9
5	A(m:n, p:q)	Refers to the elements in rows 'm' through 'n' and columns 'p' through 'q' of the matrix A	>> A(2:3,2:3) ans = 5 6 8 9

Example 2

```
clc
clear all
A = [1 0 2; 1 2 3; 2 1 3]; % defining  a matrix A of order 3x3
B = [2 3 1; 2 1 5; 4 2 3]; % defining  a matrix B of order 3x3
display('Matrix addition is')
C = A + B                     % matrix addition
display('Matrix multiplication is')
D = A * B                       % matrix multiplication
display('Matrix transpose is')
E = D'                        % transpose of matrix D
display('Matrix determinant is')
F = det(D)                    % determinant of matrix D
```

The output of the above program will be:

Matrix addition is

```
C =
     3     3     3
     3     3     8
     6     3     6
```

Matrix multiplication is

```
D =
    10     7     7
    18    11    20
    18    13    16
```

Matrix transpose is

```
E =
    10    18    18
     7    11    13
     7    20    16
```

Matrix determinant is

```
F =
   -84
```

A.10 MATLAB OPERATORS

Different types of MATLAB operators are defined below.

A.10.1 Arithmetic Operators

These operators are used for numerical computational work involving numerals, arrays and matrices. The symbols for arithmetic operations with scalars are summarised in Table A.5.

Table A.5 Arithmetic operators

S.No.	Arithmetic operation	Symbol	Example
1	Addition	+	4 + 2 = 6
2	Subtraction	–	4 – 2 = 2
3	Multiplication	*	4 * 2 = 8
4	Exponentiation	^	4^3(4 * 4 * 4 = 64)
5	Right division	/	4/2 = 2
6	Left division	\	4\2 = 1\2

A.10.2 Relational Operators

Relational operators are used for comparing the operands quantitatively for conditional branching of execution. Table A.6 gives a description of the relational operators available in MATLAB.

Table A.6 Relational operators

S.No.	Command/operator	Description
1	eq == (equal to)	Right-hand side operand is equal to the left-hand side operand of the operator
2	ne ~= (not equal)	Right-hand side operand is not equal to the left-hand side operand of the operator
3	lt < (less than)	Left-hand side operand is less than the right-hand side operand of the operator
4	gt > (greater than)	Left-hand side operand is greater than the right-hand side operand of the operator
5	le <= (less than or equal)	Left-hand side operand is less than or equal to the right-hand side operand of the operator
6	gt >= (greater than or equal)	Left-hand side operand is greater than or equal to the right-hand side operand of the operator

A.10.3 Logical Operators

A relational operator compares two numbers by determining whether a comparison is true or false. Relations may be connected or quantified by the logical operators. Table A.7 gives a description of the logical operators available in MATLAB.

Table A.7 Logical operators

S.No.	Command/operator	Description
1	& (and)	Returns 1 for every element location that is true in both arrays, otherwise returns 0, e.g. A&B or and(A, B)
2	| (or)	Returns 1 for every element location that is true in either one or other array, otherwise returns 0, e.g. A|B or or(A, B)
3	~ (not)	Returns complement of each element of the input array, e.g., ~A or not(A)
4	xor	Returns 1 for every element location that is true in only one array, otherwise returns 0, e.g., xor(A, B)

A.10.4 Operator Precedence

The expressions are normally evaluated from left to right. Precedence rules determine the order in which MATLAB evaluates an expression. These are shown in Table A.8.

Table A.8 Operator Precedence

Precedence	Operator
1	Parentheses ()
2	Transpose (.'), power (.^), matrix power (^)
3	Unary plus (+), unary minus (–), logical negation (~)
4	Multiplication (.*), right division (./), left division (.\), matrix multiplication (*), matrix right division (/), matrix left division (\)
5	Addition (+), subtraction (–)
6	Colon operator (:)
7	Less than (<), less than or equal to (<=), greater (>),greater than or equal to (>=), equal to (==), not equal to (~=)
8	Element-wise AND, (&)
9	Element-wise OR, (|)

Example 3

```
clc
clear all
%Program to demonstrate athematic operators
r = [1 2 3];            %different radii
Area = pi*r.^2          % areas
Volume = (1/3)*pi*r.^3 %volume
```

```
The output of the above program will be:
Area =
    3.1416   12.5664   28.2743
Volume =
    1.0472    8.3776   28.2743
```

Example 4

```
clc
clear all
%Program to demonstrate few relational operators
A = [1 1 2; 1 2 3; 2 1 3]; % defining  a matrix A of order 3x3
B = [2 1 1; 2 1 3; 4 2 3]; % defining  a matrix B of order 3x3
A == B                 % Is A equal to B, then it will give 1, otherwise 0
A ~= B                 % Is A not equal to B, then it will give 0, otherwise
```

The output of the above program will be:

```
ans =
     0     1     0
     0     0     1
     0     0     1
ans =
     1     0     1
     1     1     0
     1     1     0
```

A.11 ELEMENTARY MATHEMATICAL BUILT-IN FUNCTIONS

MATLAB offers many predefined mathematical functions for technical computing. Typing ***help elfun*** and ***help specfun*** calls up full lists of elementary and special functions, respectively.

There is a long list of mathematical functions that are built into MATLAB. These functions are called built-ins. Many standard mathematical functions, such as sin(x), cos(x), tan(x), e^x and ln(x) are evaluated by the functions sin, cos, tan, exp and log, respectively, in MATLAB. List of these commonly used elementary MATLAB mathematical built-in functions are given in Table A.9.

Table A.9 Mathematical functions

S.No.	Function	Description
1	abs(x)	Absolute value of x
2	sqrt(x)	Square root of x
3	sign(x)	Signum function—returns a value of –1 if x is less than zero, a value of zero if x equals zero, and a value of 1 otherwise.
4	round(x)	Rounds x to the nearest integer
5	floor(x)	Rounds x towards $-\infty$
6	ceil(x)	Rounds x towards $+\infty$

(continued)

Table A.9 *(continued)*

S.No.	Function	Description
7	rem(x, y)	Returns the reminder of x/y
8	angle(x)	Phase angle of x
9	log(x)	Computes ln x, the natural logarithm of x to the base e.
10	log10(x)	Computes log x, the common logarithm of x to the base 10.
11	exp(x)	Exponential of e^x
12	cos(x)	Computes the cosine of x
13	sin(x)	Computes the sine of x
14	tan(x)	Computes the tangent of x
15	acos(x)	Computes the arccosine or inverse cosine of x
16	asin(x)	Computes the arcsine or inverse sine of x
17	atan(x)	Computes the arctangent or inverse tangent of x
18	cosh(x)	Computes the hyperbolic cosine of x
19	sinh(x)	Computes the hyperbolic sine of x
20	tanh(x)	Computes the hyperbolic tangent of x
21	acosh(x)	Computes the inverse hyperbolic cosine of x
22	asinh(x)	Computes the inverse hyperbolic sine of x
23	atanh(x)	Computes the inverse hyperbolic tangent of x
24	atan2(y,x)	Computes the arctangent or inverse tangent of value y/x

A.12 DISPLAY FORMATS

MATLAB has several different screen output formats for displaying numbers. These formats can be found by typing the help command; help format in the command window. Table A.10 shows some of these formats.

Table A.10 Display formats

S.No.	Command	Description	Example
1	format short	Fixed point with 4 decimal digits	>>format short, pi ans = 3.1416
2	format long	Fixed point with 14 decimal digits	>>format long, pi ans = 3.14159265358979
3	format short e	Scientific notation with 4 decimal digits	>>format short e, pi ans = 3.1416e + 000
4	format long e	Scientific notation with 14 decimal digits	>>format long e, pi ans = 3.141592653589793e + 000
5	format short g	Best of fixed or floating point format with 5 digits	>> format short g, pi ans = 3.1416

(continued)

Table A.10 (*continued*)

S.No.	Command	Description	Example
6	format long g	Best of fixed or floating point format with 15 digits for double and 7 digits for single	>> format long g, pi ans = 3.14159265358979
7	format bank	Two decimal digits	>> format bank, pi ans = 3.14
8	format compact	Suppress extra line feeds	>> format compact, pi ans = 3.14

A.13 MATLAB GRAPHICS

MATLAB has an excellent set of graphic tools. Plotting a given data set or the results of computation is possible with very few commands. Users are highly encouraged to plot mathematical functions and results of analysis as often as possible. Trying to understand mathematical equations with graphics is an enjoyable and very efficient way of learning mathematics.

A.13.1 Plots

Two-dimensional plot can be obtained using the function **plot**. By typing **help plot**, you can see the various capabilities of this main command for two-dimensional plotting. Some of which are illustrated below.

If x and y are two vectors of the same length, then plot(x, y) plots x versus y.

For example, if we want to draw a plot on y versus x, then the command will be as follows:

The MATLAB command to plot a graph is plot(x, y). The vectors x = [1; 2; 3; 4; 5; 6] and y = [3; 4; 5; 4; 5; 8] produce the picture as follows.

```
>> x = [1 2 3 4 5 6];
>> y = [3 4 5 4 5 8];
>> plot(x, y)
```

Then the plot will be

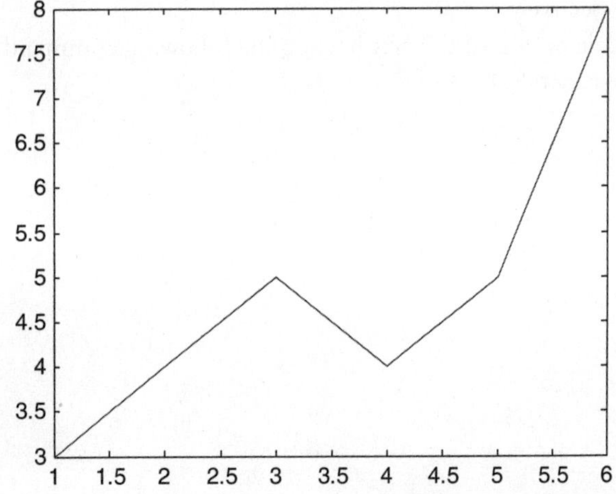

Or else, if we want to draw a plot for sine function, i.e., a pure sinusoidal wave, then the command will be

```
>> x = 0: pi/200:2*pi;
>> y = sin(x);
>> plot(x, y)
```

Then the plot will be as follows:

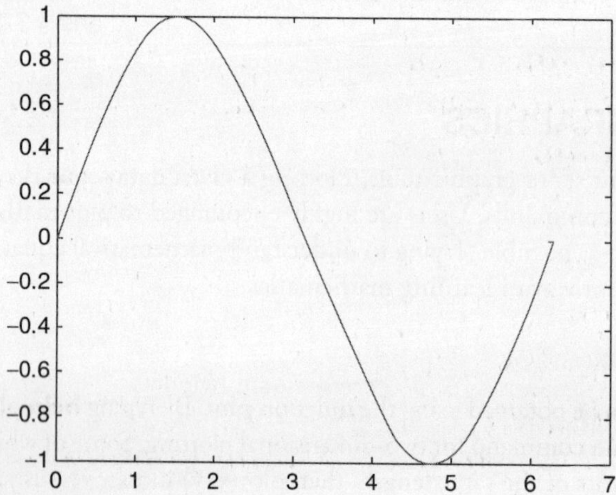

It is good practice to label the axis on a graph and if applicable, indicate what each axis represents. This can be done with the xlabel and ylabel commands.

```
>> xlabel ('x')
>> ylabel ('y = sin(x)')
```

Inside parentheses, and enclosed within single quotes, we can type the text that we wish to be displayed along the *x* and *y* axis, respectively.

We could even put a title on top of the graph using the following command:
>> title ('graph of sine function')

Example 5

```
clc
clear all
x = 0:pi/200:4*pi;
y = sin(x);
plot(x,y,'r')
axis([0 10 -1 1])
xlabel('delta')
ylabel('sinx')
```

The output will be:

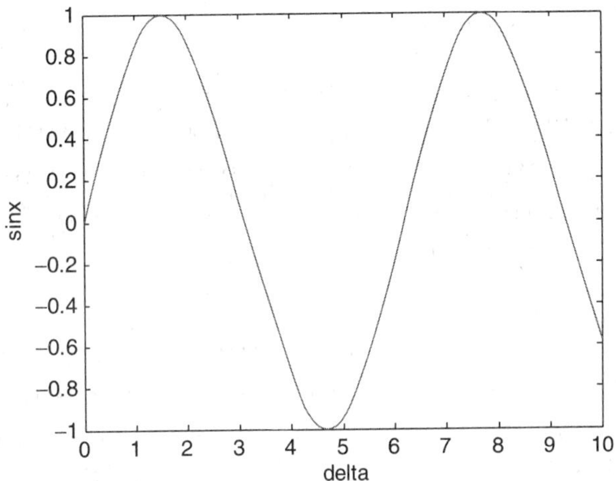

A.13.2 Subplots

If we want to plot more than one graph in a figure window, we can use subplot. The subplot function divides the figure window into a matrix of addressable sub-windows. This command has the following syntax:

(i) ***subplot (m, n, p)*:** It breaks the figure window into small axes of dimension m × n and selects the pth axis for the current path.
(ii) ***subplot (m, n, p, 'replace')*:** It replaces the already existing axis and creates a new one.
(iii) ***subplot (m, n, p, 'align')*:** It aligns the axes such that they do not overlap.

Axis scaling and appearance can also be set by using the following syntax:

axis({[xmin xmax ymin ymax])
axis({[xmin xmax ymin ymax zmin zmax])

Example 6

```
clc
clear all
x = 0:pi/200:4*pi;
y = sin(x);
z = cos(x);
subplot(2,1,1)
plot(x,y,'r')
axis([0 10 -1 1])
xlabel('delta')
ylabel('sinx')
subplot(2,1,2)
plot(x,z,'b')
axis([0 10 -1 1])
xlabel('delta')
ylabel('cosx')
```

The output will be:

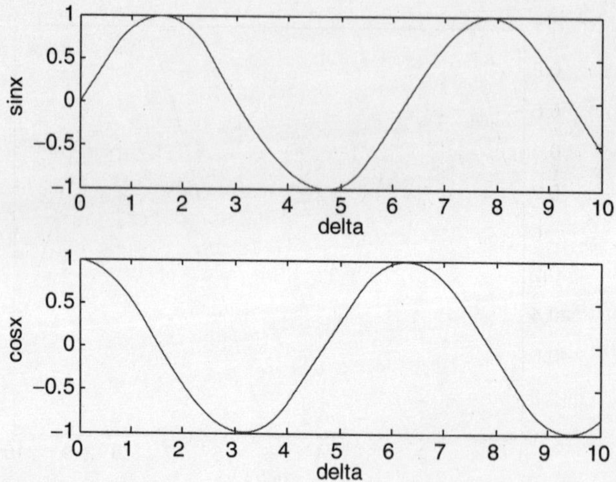

A.13.3 Other types of Plots

(a) **Multiple plots in a graph**

Many plots can be plotted in a graph having common axis. This allows us to create multiple graphs in a single window by using a single plot command.

Example 7

```
clc
clear all
x = 0:pi/200:4*pi;
r = sin(x);
y = sin(x-(2*pi/3));
b = sin(x+(2*pi/3));
plot(x,r,x,y,x,b)
```

The output will be:

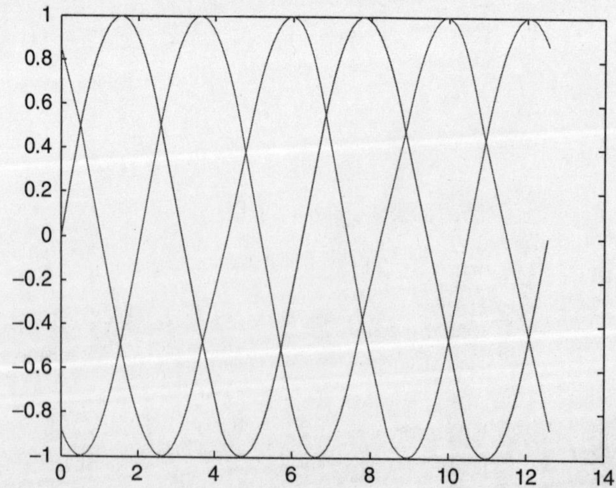

(b) **Logarithmic Plots**

In MATLAB, data can be plotted on logarithmic scales. These scales are normally used to plot a data set that covers a wide range of values. It can also be used to find a certain trend in the dada set. For example, in frequency plots, we normally take log of the frequency, i.e., log ω on the x-axis while the y-axis is linear. When one axis has log scale, the graph is called semi-log graph. In case if both the axes have log scales, then it is called log-log graph. The following commands are available in MATLAB for logarithmic plot:

(i) ***Semilogx(x,y)*:** This command produces a semi-log graph of log$x$ versus y.
(ii) ***Semilogy(x,y)*:** This command produces a semi-log graph of log$y$ versus x.
(iii) ***Loglog(x,y)*:** This command produces a semi-log graph of log$x$ versus log$y$.

Example 8

```
G = 10/(1+0.1*s)
%Program to demonstrate semi-log function
w = 1:10:10000; % angular frequency
G = 10./(1+i*0.1*w);
mag = abs(G); %magnitude of the function
semilogx(w,mag)
xlabel('logw')
ylabel('log magnitude')
```

The output will be:

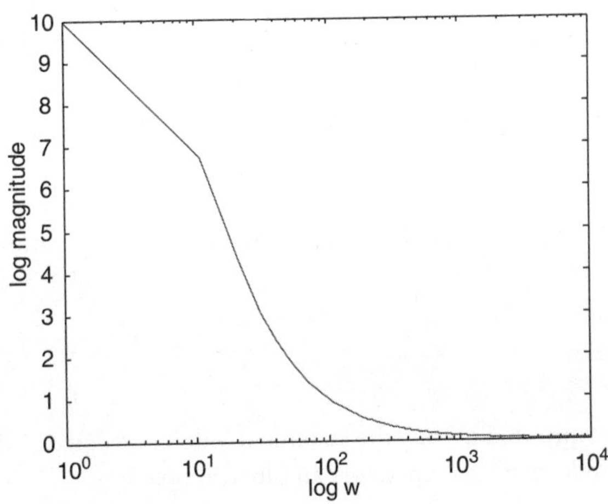

A.14 PROGRAMMING IN MATLAB

One of the most significant features of MATLAB is its extendibility through user-written programs such as m-files. M-files are ordinary ASCII text files written in MATLAB language. The MATLAB commands available for writing loops, conditional loops and conditional statements are **if-else, switch-case, for** and **while** commands. Basically, these MATLAB flow control statements operate like those in most computer languages.

(i) **if-else function:** An **if** statement evaluates an expression and executes a group of statements when the expression is true. The **elseif** and **else** are optional and executes only when previous expressions in the **if** statement are false.

The if-else function has the following syntax:

if expression
statements
elseif expression
statements
else
statements
end

(ii) **switch-case function:** A **switch** block conditionally executes one set of statements from several choices. Each choice is a **case**. If the **case** is true, MATLAB executes the corresponding statements and then exit the **switch** block. Otherwise, it is optional and executes only when no **case** is true. The switch function has the following syntax:

switch expression
case expression
statements
case expression
statements
: :
otherwise
statements
end

(iii) **for loop:** This command executes statements for specified number of times. The for loop function has the following syntax:

for variable = expression
statements
end

(iv) **while loop:** This command repeatedly executes statements while condition is true. The while loop function has the following syntax:

while expression
statements
end

When an evaluated expression is true, the statement will be executed and the result is non-empty. Expressions can include relational operators and logical operators.

A.15 SCRIPT FILES

A script file is an external file that contains a sequence of MATLAB statements. Script files have a filename extension .m and are often called m-files. It consists of a set of MATLAB instructions for performing a particular task. It is run by simply typing its name in the MATLAB command window. The MATLAB editor has numerous menus for tasks such as *saving*, *viewing* and *debugging*. As it performs some simple checks and also uses colour to differentiate between various elements of codes, this text editor is recommended as the tool of choice for writing and editing m-files.

The commands in the command window cannot be saved and executed again. In addition, the command window is not interactive. To overcome these difficulties, we have to create a file with a list of commands, save it, and then run the file. In this way, the commands contained are executed in the order they are listed when the file is run. In addition, as a need arises, one can change or modify the commands in the file, and then the file can be saved and run again. The files that are used in this fashion are known as script files.

A.15.1 Creating and Saving a Script File

Any text editor can be used to create script files. In MATLAB, script files are created and edited in the editor/debugger window. This window can be opened from the command window. From the command window, select file new and then m-file. Once the window is open, the commands of the script file are typed line by line. The second type of m-files is the function file. The function file enables the user to extend the basic library functions by adding one's own computational procedures. Before a script file can be executed, it must be saved.

A.15.2 Running a Script File

A script file can be executed either by typing its name in the command window and then pressing the enter key or directly from the editor window by clicking on the run icon. The file is assumed to be in the current directory or in the search path.

A.15.3 Input to a Script File

When a script file is executed, the variables that are used in the calculations within the file must have assigned values. The assignment of a value to a variable can be done in three ways.

(i) The variable is defined in the script file.
(ii) The variable is defined in the command prompt.
(iii) The variable is entered when the script is executed (using 'input' command).

A.15.4 Output Commands

There are two commands that are commonly used to generate output. They are as follows:

(i) **The 'disp' command:** It displays the elements of a variable without displaying the name of the variable and displays text.
(ii) **The 'fprintf' command:** It displays output (text and data) on the screen or saves it to a file. The output can be formatted using this command.

A.16 USES OF MATLAB IN CONTROL SYSTEMS

MATLAB has an extensive set of functions for the analysis and design of control systems. They involve transfer function representation, matrix operations, root determination, model conversions and plotting of complex functions. These functions are found in MATLAB's control systems toolbox. MATLAB uses models in the form of transfer-functions or state-space equations. Before you begin the next sections, it would be a good idea to run the MATLAB Control System Toolbox demo. This is done by typing 'demo ('toolbox', 'control')' at the MATLAB prompt. In the control systems, we should become familiar with the following commands:

tf: This command is used to enter transfer functions.
series or *: This command is used to combine two transfer functions that are in series.

feedback: This command is used to combine two transfer functions that are in feedback.
step: This command is used to plot the step response of a system.
bode: This command is used to plot the frequency response plot (Bode plot).
rlocus: This command is used to plot the root locus.
acker: This command computes the K matrix to place the poles of A-BK.
c2dm: Continuous system to discrete system
conv: Convolution (useful for multiplying polynomials)
ctrb: The controllability matrix
obsv: The observability matrix
det: This command finds the determinant of a matrix.
dimpulse: Impulse response of discrete-time linear systems
dlqr: Linear-quadratic regulator design for discrete-time sys
lsim: This command simulates the response of the system to sinusoidal inputs.
dstep: Step response of discrete-time linear systems
eig: This command computes the Eigen values of a matrix.
figure: This command creates a new figure or redefine the current figure.
format: Number format (significant digits, exponents)
grid: This command draws the grid lines on the current plot.
hold: This command holds the current graph, see also figure.
impulse: Impulse response of continuous-time linear system
inv: This command finds the inverse of a matrix.
length: Length of a vector, see also size
linspace: This command returns a linearly spaced vector.
log: Natural logarithm
loglog: This command plots using log-log scale and also semilogx/semilogy.
logspace: This command returns a logarithmically spaced vector.
lqr: Linear quadratic regulator design for continuous sys
margin: This command returns the gain margin and phase margin.
norm: Norm of a vector
ones: This command returns a vector or matrix of ones.
zeros: This command returns a vector or matrix of zeros.
place: This command computes the K matrix to place the poles of A-BK.
pzmap: Pole-zero map of linear systems
rank: This command finds the number of linearly independent rows/column.
rlocfind: This command finds the value of k and the poles at the selected point.
roots: This command finds the roots of a polynomial.
ss: This command creates state-space models or convert LTI model.
ss2tf: State-space to transfer function representation
ss2zp: State-space to pole-zero representation
subplot: This command divides the plot window up into pieces.
tf2ss: Transfer function to state-space representation
tf2zp: Transfer function to Pole-zero representation
title: This command adds a title to the current plot.
xlabel/ylabel: This command adds a label to the horizontal/vertical axis.
zgrid: This command generates grid lines of constant damping ratio (zeta).

zp2ss: Pole-zero to state-space representation
zp2tf: Pole-zero to transfer function representation

(i) **Transfer function**

(a) **tf:** This command is used to enter or create transfer functions.

(b) **sys = tf (num, den):** This command creates a continuous-time transfer function 'sys' with numerator(s) 'num' and denominator(s) 'den'. The output SYS is a TF object.

(c) **sys = tf (num, den, ts):** This command creates a discrete-time transfer function with sample time 'ts' (set ts = −1, if the sample time is undetermined).

(d) **[z, p, k] = tf2zp (num, den):** Transfer function to Pole-zero representation.

(e) **pzmap (num, den):** To obtain pole-zero map.

(f) **[r, p, k] = residue (num, den):** To find the partial fractions of the transfer function.

Example 9

```
clc
clear all
%Program to demonstrate transfer function representation
num = [1 5 7];
den = [1 3 6 7 10];
disp('The transfer function of the system is')
sys = tf(num,den)
disp('The transfer function of the discrete time system is')
ts = 0.5;
sys = tf(num,den,ts)
[z,p,k] = tf2zp(num,den)
pzmap(num,den)
```

```
The output of the above program will be:

The transfer function of the system is
Transfer function:
                      s^2 + 5 s + 7
              ------------------------------
              s^4 + 3 s^3 + 6 s^2 + 7 s + 10

The transfer function of the discrete time system is
Transfer function:
                      z^2 + 5 z + 7
              ------------------------------
              z^4 + 3 z^3 + 6 z^2 + 7 z + 10

Sampling time: 0.5

z =
  -2.5000 + 0.8660i
  -2.5000 - 0.8660i
p =
  -1.5860 + 1.2650i
```

```
   -1.5860 - 1.2650i
    0.0860 + 1.5564i
    0.0860 - 1.5564i
k =
     1
```

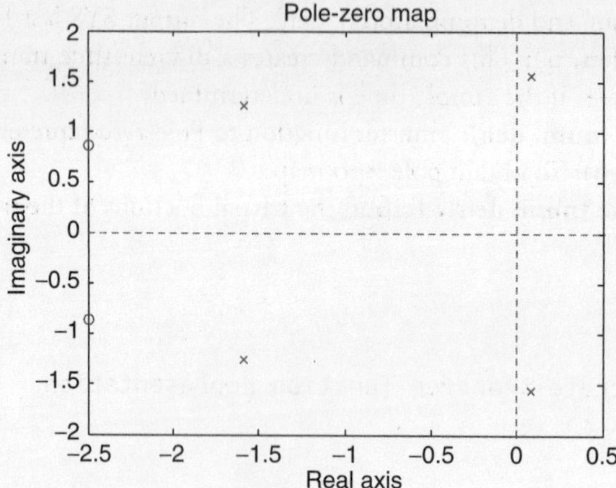

(ii) **Block-diagrams**

(a) **sys = series(sys1,sys2):** It is used to combine two transfer functions that are in series.

(b) **sys = parallel(sys1,sys2):** It is used to combine two transfer functions that are in parallel.

(c) **sys = feedback(sys1,sys2):**It is used to combine two transfer functions that are in feedback.

Example 10

```
clc
clear all
%Program to combine blocks in series, parallel and minor feedback loop
num1 = [4];
den1 = [1 2 1];
sys1 = tf(num1,den1);
num2 = [7];
den2 = [1 5];
sys2 = tf(num2,den2);
disp('Series connection equivalent transfer function is')
sys = series(sys1,sys2)    %series connection
disp('Parallel connection equivalent transfer function is')
sys = parallel(sys1,sys2) %parallel connection
disp('Eliminating minor feedback, equivalent transfer function is')
sys = feedback(sys1,sys2) %eliminate feedback loop
```

The output of the above program will be:

```
Series connection equivalent transfer function is
```

```
Transfer function:
                          28
               ----------------------
               s^3 + 7 s^2 + 11 s + 5

Parallel connection equivalent transfer function is
Transfer function:
                   7 s^2 + 18 s + 27
               ----------------------
               s^3 + 7 s^2 + 11 s + 5

Eliminating minor feedback, equivalent transfer function is
Transfer function:
                       4 s + 20
               -----------------------
               s^3 + 7 s^2 + 11 s + 33
```

(iii) **Time-response analysis**

Example 11

```
clc
clear all
%program to find the step response of a given second order system
%having G(s) = 3/s(s + 4), H(s) = 1
num = [3]
den = [1 4 0]
G = tf(num,den)
H = 1
T = feedback(G,H)
t = 0:0.001:10;
xlabel('time in sec')
ylabel('Amplitude')
step(T)
```

The output of the above program will be:

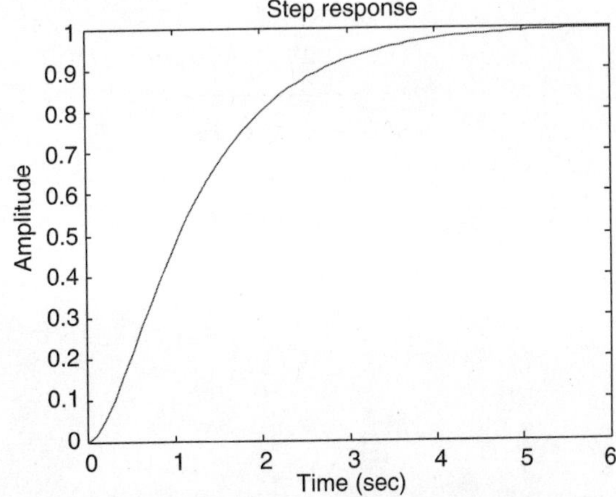

Example 12

```
clc
clear all
%program to find the time domain specifications of any
%second order system, i.e 25/(s² + 6s + 25)
num = input('Enter the num of second order tf  = ')
den = input('Enter den of second order tf = ')
t = tf(num,den)
step(t)
xlabel('t in sec')
ylabel('v in volts')
title('Step response of  a second order system')
wn = sqrt(den(3))
z = den(2)/(2*wn)
wd = wn*sqrt(1-z^2)
disp('The rise time is')
tr = (pi-atan(sqrt(1-z^2)/z))/wd
disp('The peak time is')
tp = pi/wd
disp('the settling time is')
ts = 4/(z*wn)
disp('The peak over shoot is')
pos = exp(-z*pi/sqrt(1-z^2))*100
disp('The steady state error is' )
ess = 2*z/wn
```

The output of the above program will be:

```
Enter the num of second order tf = 25

num = 25

Enter den of second order tf = [1 6 25]

den =    1   6   25

Transfer function:

                                   25
                             --------------
                             s^2 + 6 s + 25

wn = 5
z = 0.6000
wd =  4
The rise time is

tr = 0.5536

The peak time is

tp =  0.7854

The settling time is

ts = 1.3333
```

```
The peak over shoot is
pos =   9.4780
The steady state error is
ess =   0.2400
```

(iv) Root Locus

(a) **rlocus(sys):** This command computes and plots the root locus of the single-input, single-output LTI model **sys**.

(b) **rlocus(sys,k):** This command uses a user-specified vector K of gain values.

(c) **rlocus(sys1,sys2,...):** This command draws the root loci of multiple LTI models **sys1, sys2,...** on a single plot.

(d) **[r,k] = rlocus(sys)** or **r = rlocus(sys,k):** This command returns the matrix R of complex root locations for the gain value K.

Example 13

```
clc
clear all
%program to draw root locus plot having G(s)H(s) = 150/s(s + 4)(s^2 + 4s + 13)
num = [150]
den = [1 8 29 52 0]
sys = tf(num,den)
rlocus(sys)
```

The output of the above program will be:

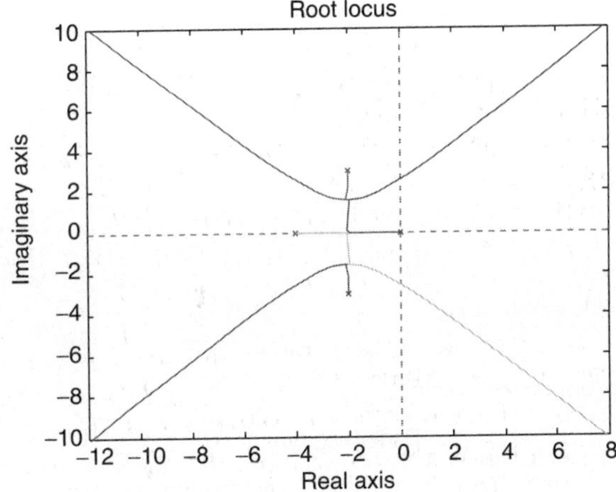

Example 14

```
clc
clear all
%program to draw root locus plot having G(s)H(s) = k(s + 2)(s + 3)/(s + 1)(s - 1)
num = [1 5 6]
den = [1 0 -1]
sys = tf(num,den)
rlocus(sys)
```

The output of the above program will be:

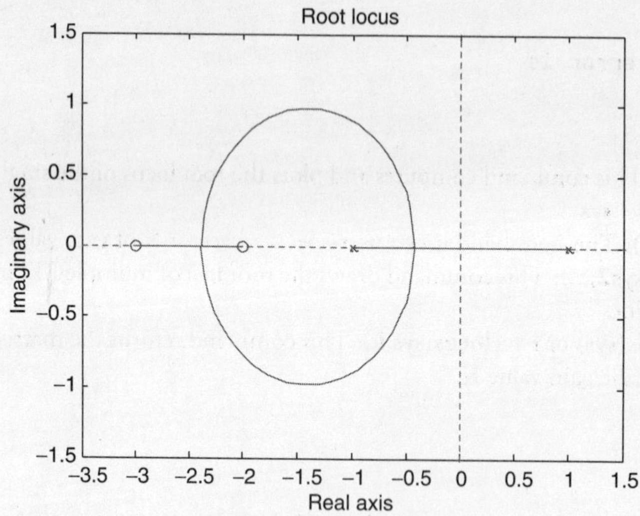

(v) **Bode Plot**

(a) **bode(sys):** This command is used to plot the Bode plot for the given **sys.**

(b) **margin(sys):** This command indicates GM, PM, GCF and PCF on Bode plot.

(c) **[gm,pm,wpc,wgc]=margin(sys):** This command is used to find GM, PM, GCF and PCF of a given system.

Example 15

```
clc
clear all
%program to draw Bode plot having G(s)H(s) = 50/(s^3 + 9s^2 + 30s + 40)
num = [50]
den = [1 9 30 40]
sys = tf(num,den)
bode(sys)
grid
margin(sys)
```

The output of the above program will be:

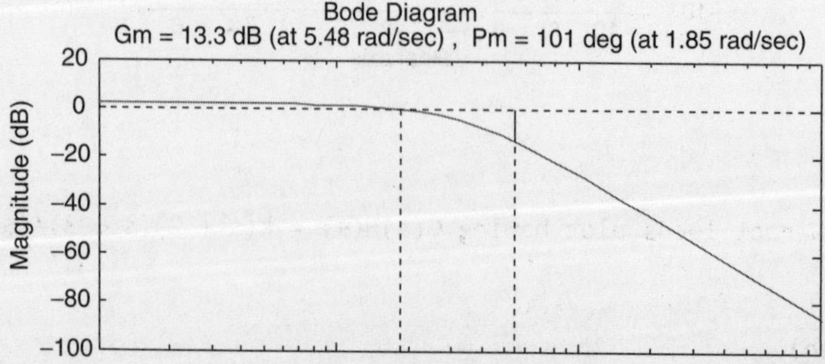

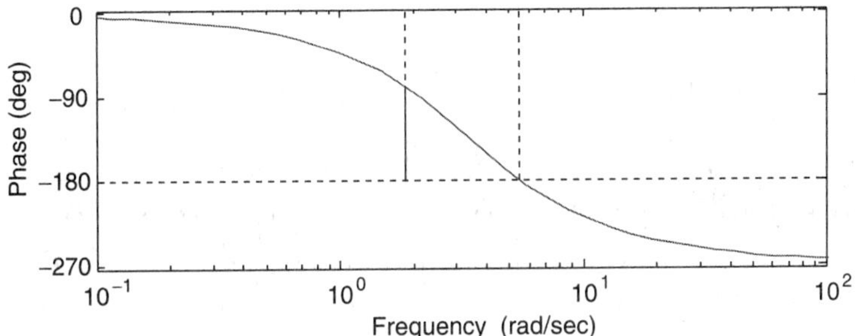

Example 16

```
clc
clear all
% program to draw Bode plot for the given t.f
% here we are taken t.f as G(s)H(s) = (1 + 4s)/(s^2(1 + s)(1 + 2s))
num = input('Enter the numerator')
den = input('Enter the denominator')
sys = tf(num,den)
bode(sys)
grid
[gm,pm,wpc,wgc] = margin(sys)
gmdb=20*log10(gm)
 if(gmdb > 0 & pm > 0)
    disp('system is stable')
    else if(gmdb == 0&pm == 0)
          disp('system is marginally stable')
        else
            disp('system is unstable')
          end
  end
```

The output of the above program will be:

```
Enter the numerator [4 1]
num =
     4     1

Enter the denominator [2 3 1 0 0]
den =
     2     3     1     0     0
Transfer function:
                                  4 s + 1
                           -------------------
                           2 s^4 + 3 s^3 + s^2
```

```
  gm = 0.0937
  pm = -36.6648
 wpc = 0.3535
 wgc = 1.1170
gmdb = -20.5607
system is unstable
```

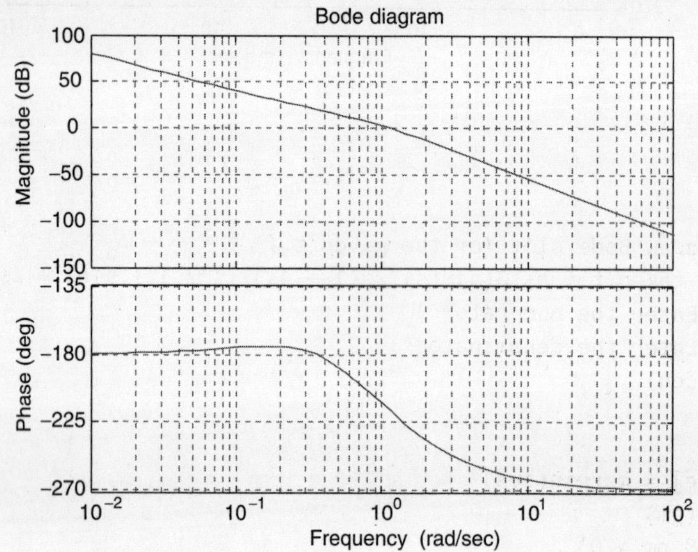

(vi) **Nyquist Plot**

(a) **nyquist(sys):** This command is used to plot the Nyquist plot for the given **sys.**

(b) **nyquist(sys,{wmin,wmax}):** This command draws the Nyquist plot for frequencies between wmin and wmax (in radians/second).

(c) **nyquist(sys,w):** This command uses the user-supplied vector w of frequencies (in radian/second) at which the Nyquist response is to be evaluated.

(d) **nyquist(sys1,sys2,...,w):** This command plots the Nyquist response of multiple LTI models **sys1,sys2,...** on a single plot.

(e) **[gm,pm,wpc,wgc]=margin(sys):**This command is used to find GM, PM, GCF and PCF of a given system.

Example 17

```
clc
clear all
%program to draw Nyquist plot having G(s)H(s) = 3/(s + 1)(s + 2)
num = [3]
den = [1 3 2]
sys = tf(num,den)
nyquist(sys)
```

The output of the above program will be:

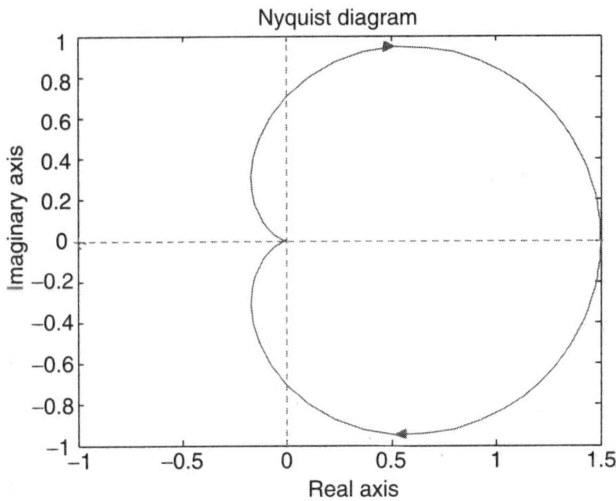

Example 18

```
clc
clear all
%program to Nyquist Bode plot for the given t.f
% here we are taken t.f as G(s)H(s) = (s + 2)/(s + 1)(s - 1)
num = input('Enter the numerator')
den = input('Enter the denominator')
sys = tf(num,den)
nyquist(sys)
[gm,pm,wpc,wgc] = margin(sys)
gmdb = 20*log10(gm)
 if(gmdb > 0 & pm > 0)
    disp('system is stable')
    else if(gmdb == 0&pm == 0)
          disp('system is marginally stable')
        else
            disp('system is unstable')
          end
  end
```

The output of the above program will be:

```
enter the numerator [1 2]

num =
    1     2

enter the denominator [1 0 -1]

den =
    1     0    -1
```

```
Transfer function:
s + 2
-------
s^2 - 1
gm = 0.5000
pm = 29.7131
wpc = 0
wgc = 1.1414
gmdb = -6.0206
system is unstable
```

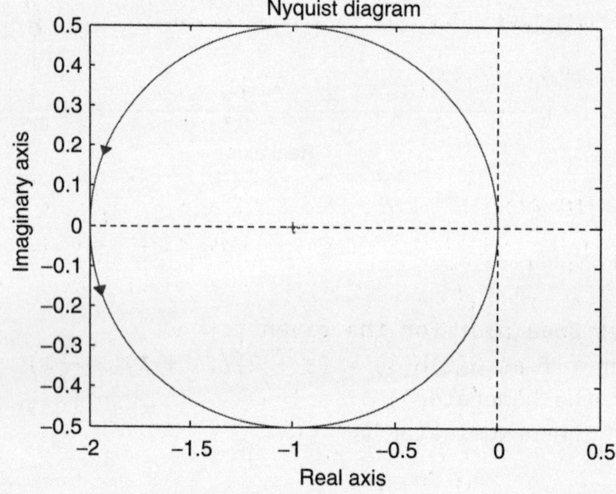

(vii) **State-space analysis**

(a) **[A, B, C, D]= tf2ss(num, den):** This command is used to represent the given system into state space form.

(b) **[num, den] = ss2tf (A, B, C, D):** This command is used to calculate transfer function from the given state-space representation.

(c) **ctrb(sys):** This command returns the controllability matrix of the state-space model **sys** with realisation (A, B, C, D).

(d) **obsv(sys):** This command returns the observability matrix of the state-space model **sys** with realisation (A, B, C, D).

(e) **eig(X):** It gives vector containing the Eigen values of a square matrix X.

(f) **[V, D] = eig(X):** It produces a diagonal matrix D of Eigen values and a full matrix V whose columns are the corresponding Eigen vectors so that X*V = V*D.

Example 19

```
clc
clear all
%program to convert transfer function to state space model
%T(s) = (s^2 + 2s + 4)/(s^3 + 3s^2 + 6s + 10)
num = [1 2 4]
den = [1 3 6 10]
```

```
sys = tf(num,den)
[A, B, C, D] = tf2ss(num,den)
```

The output of the above program will be:

```
num =

     1     2     4

den =

     1     3     6    10

Transfer function:
           s^2 + 2 s + 4
     ----------------------
     s^3 + 3 s^2 + 6 s + 10

A =
    -3    -6   -10
     1     0     0
     0     1     0
B =
     1
     0
     0
C =
     1     2     4
D =
     0
```

Example 20

```
clc
clear all
%program to convert state space model to transfer function
A = [-1 0 0; 0 -2 0; 0 0 -3]
B = [1;1;1]
C = [-1 2 1]
D = [1]
[num,den] = ss2tf(A, B, C, D)
sys = tf(num,den)
```

The output of the above program will be:

```
A =
    -1     0     0
     0    -2     0
     0     0    -3
B =
     1
     1
     1
```

```
C =
    -1     2     1
D =
     1
num =
    1.0000    8.0000   17.0000    8.0000
den =
     1     6    11     6
Transfer function:
                            s^3 + 8 s^2 + 17 s + 8
                            ----------------------
                            s^3 + 6 s^2 + 11 s + 6
```

Example 21

```
clc
clear all
n =input('enter 1 for converting tf to ss &  2 for converting ss to tf')
switch n
case 1
num =input('enter the numerator of tf')
den =input('enter the denominator of tf')
t =tf(num,den)
[A,B,C,D] =tf2ss(num,den)
case 2
A =input('enter the system matrix A')
B =input('enter the independent matrix B')
C =input('enter the output matrix C')
D =input('enter the transmission matrix D')
[num,den] =ss2tf(A,B,C,D)
tf(num,den)
otherwise
disp('the choice is invalid')
end
```

The output of the above program will be:

```
enter 1 for converting tf to ss & 2 for converting ss to tf 1
n =
     1
enter the numerator of tf[10]
num =
    10
enter the denominator of tf[1 2 4 10]
den =
     1     2     4    10
```

```
Transfer function:
                              10
                     ----------------------
                     s^3 + 2 s^2 + 4 s + 10
A =
    -2    -4   -10
     1     0     0
     0     1     0
B =
     1
     0
     0
C =
     0     0    10
D =
     0
enter 1 for converting tf to ss &  2 for converting ss to tf 2
n =
     2
enter the system matrix A [0 1 0; 0 0 1; -10 -4 -2]
A =
     0     1     0
     0     0     1
   -10    -4    -2
enter the independent matrix B [0; 0; 10]
B =
     0
     0
    10
enter the output matrix C [1 0 0]
C =
     1     0     0
enter the transmission matrix D [0]
D =
     0
num =
         0   -0.0000   -0.0000   10.0000
den =
    1.0000    2.0000    4.0000   10.0000
Transfer function:
                          -3.109e-015 s^2 - 6.217e-015 s + 10
                          -----------------------------------
                                s^3 + 2 s^2 + 4 s + 10
```

Example 22

```
clc
clear all
% Program to find controllability
A = [0 1;-2 -2]
B = [1; -1]
C = [1 0]
D = [0]
n = 2
r = rank([B A*B])
m = rank([C*B C*A*B])
if r == n
    disp('The system is completely state controllable')
else
    disp('The system is not completely state controllable')
end
if m == 1
    disp('The system is completely output controllable')
else
    disp('The system is not completely output controllable')
end
```

The output of the above program will be:

```
A =
     0     1
    -2    -2
B =
     1
    -1
C =
     1     0
D =
     0
n = 2
r = 2
m = 1
The system is completely state controllable.
The system is completely output controllable.
```

Example 23

```
clc
clear all
% Program to find observability
A = [-3 1 1;-1 0 1;0 0 1]
```

```
B = [0 1;0 0;2 1]
C = [0 0 1; 1 1 0]
n = 3
r = rank([C' A'*C' A'*A'*C' ])
if r == n
    disp('The system is completely state controllable')
else
    disp('The system is not completely state controllable')
end
```

The output of the above program will be:

```
A =
     -3     1     1
     -1     0     1
      0     0     1
B =
      0     1
      0     0
      2     1
C =
      0     0     1
      1     1     0
n = 3
r = 3
```

The system is completely state controllable.

Example 24

```
clc
clear all
% Program to diagonalize the system matrix
A = [0 1 0; 0 0 1;-6 -11 -6]
[V,D] = eig(A)
Ad = inv(P)*A*P;
disp('The diagonal matrix is:')
disp(Ad)
```

The output of the above program will be:

```
A =
      0     1     0
      0     0     1
     -6   -11    -6
V =
    -0.5774    0.2182   -0.1048
     0.5774   -0.4364    0.3145
    -0.5774    0.8729   -0.9435
```

```
D =
   -1.0000         0         0
         0   -2.0000         0
         0         0   -3.0000
The diagonal matrix is:
   -1.0000   -0.0000    0.0000
   -0.0000   -2.0000   -0.0000
   -0.0000   -0.0000   -3.0000
```

Bibliography

Ackermann J, *Sampled-Data Control Systems*, 2nd edition, Springer-Verlag, Berlin, 1985.

Azzo JJD and Houpis CH, *Linear Control Systems Analysis and Design*, McGraw–Hill, New York, 1998.

Bakshi UA and Bakshi VU, *Control Systems*, Technical Publications, 2003.

Bayliss LE, *Living Control System*, The English Universities Press Ltd, London, 1966.

Bhattacharya SK, *Control Systems Engineering*, Pearson Education, 2005.

Blacklock JH, *Automatic Control of Aircraft and Missiles*, John Wiley & Sons, New York, 1965.

Bogler PL, *Radar Principles with Applications to Tracking Systems*, John Willy & Sons, New York, 1990.

Brockett RW and Willems JL, "Frequency Domain Stability Criteria, Part I", *IEEE Transactions on Automatic Control, AC,* 10 (7), pp 255–261, July 1965.

Canon RH, *Dynamics of Physical Systems*, McGraw–Hill, New York, 1967.

Chen CT, *Analog & Digital Control Design: Transfer Function, State Space & Algebraic Methods*, Saunders College Publishing, Orlando, 1993.

Dorf RC and Bishof RH, *Modern Control Systems*, Pearson Education, 2004.

Dukkapati RV, *Control Systems*, Narosa Publishing House, 2005.

Evans WR, "Graphical Analysis of Control Systems", *AIEE Transactions*, Part II, 1948.

Evans WR, *Control Systems Dynamics*, McGraw–Hill, New York, 1954.

Gajic Z, *Modern Control Systems Engineering*, Prentice Hall, 2000.

Ghosh S, *Control Systems*, Pearson Education, 2004.

Gopal M, *Control Systems Principles and Design*, Tata McGraw–Hill, 2004.

Healey M, *Principles of Automatic Control*, The English Universities Press Ltd, London, 2002.

Jagan NC, *Control Systems*, BS Publications, 2005.

Kuo BC, *Automatic Control systems*, Prentice Hall, 1995.

Kurfess TR and Nagurka ML, "Understanding the Root Locus Using the Gain Plots", *IEEE Control Systems*, 37–40, 1991.

Lathi BP, *Linear Systems and Signals*, Oxford University Press, 2005.

Lynch WA and Truxal JG, *Signals and Systems in Electrical Engineering*, McGraw–Hill, New York, 1962.

Manke BS, *Linear Control Systems*, Khanna Publishers, 2002.

Mayar O, *The Origins of Feedback Control*, MIT Press, Cambridge, 1970.

Mohandas KP, *Modern Control Engineering*, Sanguine Technical Publisher, 2006.

Nagarath IJ and Gopal M, *Control Systems Engineering*, New Age International, 2004.

Nagoor Kani A, *Control Systems*, RBA Publications, 2002.

Nise NS, *Control Systems Engineering*, Wiley, 2000.

Ogata K, *Modern Control Engineering*, Pearson Education, 2002.

Parr EA, *Industrial Control Handbook*, BSP Professional Books, Oxford, Vol. 1–3, 1987.

Raven FH, *Automatic Control Engineering*, McGraw–Hill, New York, 1995.

Rohrs CE, Melsa JL and Schultz D, *Linear Control Systems*, McGraw–Hill, New York, 1993.

Shinners SM, *Modern Control Systems Theory and Design*, John Willy & Sons, New York, 1992.

Stefani RT et al, *Design of Feed-back Control Systems*, Oxford University Press, 2005.

Subbarao A et al, *A Course in Control Engineering*, Dhanpat Rai and Sons, Delhi, 1984.

Thaler GJ, *Automatic Control Systems*, West Publishing Company, St. Paul, 1989.

Varma SN, *Automatic Control Systems*, Khanna Publishers, 2002.

Yeung KS, "A Reformulation of Nyquist's Criterion", *IEEE Transaction on Education, E*, 28(2), pp 58–60, February 1985.

Zak SH, *Systems and Control*, Oxford University Press, 2005.

Index

E

F

G

H

I

J

L

M

N

O

P

R

S

T

U

V

Z